Dynamical Geology of Salt and Related Structures

ACADEMIC PRESS RAPID MANUSCRIPT REPRODUCTION

Dynamical Geology of Salt and Related Structures

Edited by

I. Lerche

Department of Geology
University of South Carolina
Columbia, South Carolina

J. J. O'Brien

Standard Alaska Production Company
Anchorage, Alaska

1987

ACADEMIC PRESS, INC.
Harcourt Brace Jovanovich, Publishers
Orlando San Diego New York Austin
Boston London Sydney Tokyo Toronto

ACADEMIC PRESS, INC.
Orlando, Florida 32887

United Kingdom Edition published by
ACADEMIC PRESS INC. (LONDON) LTD.
24–28 Oval Road, London NW1 7DX

Library of Congress Cataloging in Publication Data

Dynamical geology of salt and related structures.

Includes index.
1. Salt. 2. Salt domes. I. Lerche, I. (Ian)
II. O'Brien, J. J.
QE471.15.S2D96 1987 553.6'3 86-47989
ISBN 0–12–444170–X (alk. paper)

PRINTED IN THE UNITED STATES OF AMERICA

87 88 89 90 9 8 7 6 5 4 3 2 1

We dedicate this book to the memory
of the late Sir Peter Kent,
whose fundamental work on salt behavior
has benefited all. Thank you, Peter.

Contents

Section B
Structural Impact of Salt on Surrounding Formations

Section C
Caprock

Section D
Fluid Flow, Salt Dissolution Heat Flow, and Hydrocarbon Migration

Preface

Salt has three major properties which cause it to play a dominant role in sedimentary basins:

1. On a geological timescale salt flows as a nearly incompressible fluid under applied stress, thereby both distorting sedimentary patterns and influencing further basin evolution.
2. Salt has a density of approximately 2.2 gm/cm^3, which is intermediate between sedimentary densities at deposition (1.6–1.9 gm/cm^3) and the densities of fully compacted sedimentary formations (2.6–2.8 gm/cm^3). The density of salt varies little during burial under an increasing overburden weight. Thus at some point in a basin's evolution the salt will become buoyant and will attempt to rise up through the overlying formations.
3. Salt has a thermal conductivity approximately three times greater than that of "typical" sedimentary formations and so salt bodies act as conduits for heat transport from depth. In the vicinity of the salt local thermal effects, caused or modified by this conductivity contrast, impact on chemical precipitation and dissolution, hydrocarbon maturity, and fluid flow.

As a consequence, the dynamical evolution of subsurface salt bodies has an important impact on basin evolution. This evolution may be reflected in changes in sedimentary patterns induced by the salt, uplifting and thinning of overlying beds, development of faulting in the overburden, fracturing of formations, and rim syncline development; it may also be reflected in the chemical, thermal, and gravitational effects influenced by, or produced by, the salt; and finally, it may be reflected in the trapping of hydrocarbons often found in commercial quantities in association with salt structures.

There has long been a need for a compendium of papers dealing with many aspects of the dynamical evolution of salt bodies in sedimentary basins. There has also been a need to provide at least some quantitative estimates of the parameters relevant to determining when salt may be expected to have an influence in a basin relative to other basinal processes (such as, for instance, tectonic stress, fracturing of formations, and hydrocarbon generation). Excellent books are available on salt physiognomy (Halbouty, "Salt Domes, Gulf Reigon, United States and Mexico," 1979) and on chemistry (Sonnenfeld,

"Brines and Evaporites," 1984). While "Diapirism and Diapirs" (edited by Braunstein and O'Brien, 1968) and "Geodynamics: Application of Continuum Physics to Geologic Problems" (Turcotte and Schubert, 1982) include contributions on individual aspects of salt dome development, there is a need for a text which treats the broad range of dynamical interactions between subsurface mobile salt masses and the surrounding sedimentary formations.

The purpose of this edited treatise is to fill these lacunae. This book consists of four major sections dealing with different aspects of the development of salt and salt-related structures. We provide case histories in each section which illuminate how individual facets of the underlying physics, chemistry, and geology are being exhibited by nature.

Section A deals with salt dynamics and focuses attention on the motion of salt. The emphasis here is on salt, with the sedimentary behavior playing a secondary role. This section includes a discussion of the surface expression of salt plugs; the internal structure of salt diapirs; a case history and a laboratory centrifuge study, both of which describe the impact of differential sediment loading on salt tectonics; and a quantitative discussion of the factors influencing buoyant salt diapirism.

In Section B we address the impact of a mobile salt mass on the structural development of the overlying formations. We are concerned here with processes such as faulting and fracturing in the formations overlying mobile or incompetent salt masses, and with upturning and thinning of sedimentary layers surrounding salt diapirs. This section includes a number of case studies from diverse geographic locations documenting the many ways in which salt can impact the structural development within the overburden.

Section C turns attention to the development of caprock, which is commonly found overlying salt diapirs. The contributions in this section describe the composition of caprock and discuss possible mechanisms for emplacement of various caprock components, the sources of caprock material, and the timing of caprock emplacement.

Section D addresses what may be the most relevant aspects of subsurface salt accumulations from the oil industry viewpoint: the interrelationships between fluid flow, salt dissolution, and heat flow in the vicinity of a salt diapir; and the connections with maturation of source rocks, migration, and trapping of hydrocarbons in salt-related structures. This section includes two contributions on dissolution of salt in the subsurface, one based on well information and the other on seismic data; one article describes the enhanced heat flow occurring in the vicinity of massive salt diapirs and the impact of such thermal anomalies on the maturation of associated organic-rich formations and hydrocarbons; the remaining two articles integrate many of the physical and chemical processes which occur around salt diapirs, emphasizing the interrelationships between these processes and how they each contribute to the development of hydrocarbon reservoirs.

Our goal in selecting contributions for this book is to present a comprehensive treatment of the various aspects of the genesis and development of salt structures, of their impact on sedimentary structural evolution, and of the impact of sedimentary structural development on salt masses, at a level appropriate for a broad audience of geoscientists. In our opinion, at least 80–90% of the material presented should be comprehensible by an able geology graduate student. The professional geoscientist should find a cogent overview of the processes involved in salt dynamics, illustrated with case histories chosen for both geologic content and clarity of presentation.

Given the vast literature on salt, we do an injustice to a large number of researchers by selecting papers other than those that might have been included. We are, however, limited by book length, by topic, and by the theme we wanted to express. In the last analysis the choice of what to include and what to leave out is always a personal one: in this instance the choice reflects to a large extent our own desires and wishes.

We thank Donna Black, who did such an excellent job in keeping the files of correspondence organized, and particularly for her heroic efforts in converting manuscripts to camera-ready copy. Additional typing by Joyce Goodwin was much appreciated. Encouragement to perservere with the project from our editor at Academic Press and from Jim Durig, Doug Williams, John Pantano, Tom McKenna, Song Cao, Jim Iliffe, Kazuo Nakayama, Chieh (Jim) Kuo, and Ken Smith was of great comfort. Lastly, we thank our families and friends for their patience and forbearance during the course of this endeavor.

Partial support for this project came from the Industrial Associates of the Basin Analysis Group at the University of South Carolina: Sun Oil, Litton Resources Group, Statoil, Saga Petroleum, Norsk Hydro, Arco, Phillips Petroleum, Chevron, Texaco, Petrobras, Unocal, Japex, and Marathon.

Ian Lerche
Columbia, South Carolina

J. J. O'Brien
Anchorage, Alaska

SECTION A

SALT STRUCTURES

ISLAND SALT PLUGS IN THE MIDDLE EAST AND THEIR TECTONIC IMPLICATIONS

P. E. Kent+
38 Rodney Road
West Bridgeford
Nottingham NG2 6JH
England

I. INTRODUCTION

The Middle East includes one of the world's largest salt plug provinces. The older exotic material was named after Hormuz Island by early geologists, and this name (Hormuz Complex) will be used here for convenience to cover the polygenetic assemblage. The province is unique in two respects. Firstly it is one of the most favorable areas of the world for seeing exposed rocks, permitting exceptional opportunities for analysis and understanding of structural relationships and development. Secondly the arid climate has inhibited rapid surface solution of the extrusive evaporites which consequently outcrop widely (Fig. 1).

Of the various levels at which evaporites were deposited the most important was at the Cambrian/Precambrian boundary in southern Iran and the adjoining area of the Arabian (or Persian) Gulf. This province was later subject to heavy sedimentation (10-20 km) from the Permian to the Miocene, with only minor phases of faulting to interrupt continuous subsidence. The sedimentary load resulted in tectonic instability with halokinesis from late Mesozoic times onwards.

+Deceased: 1986 July 9

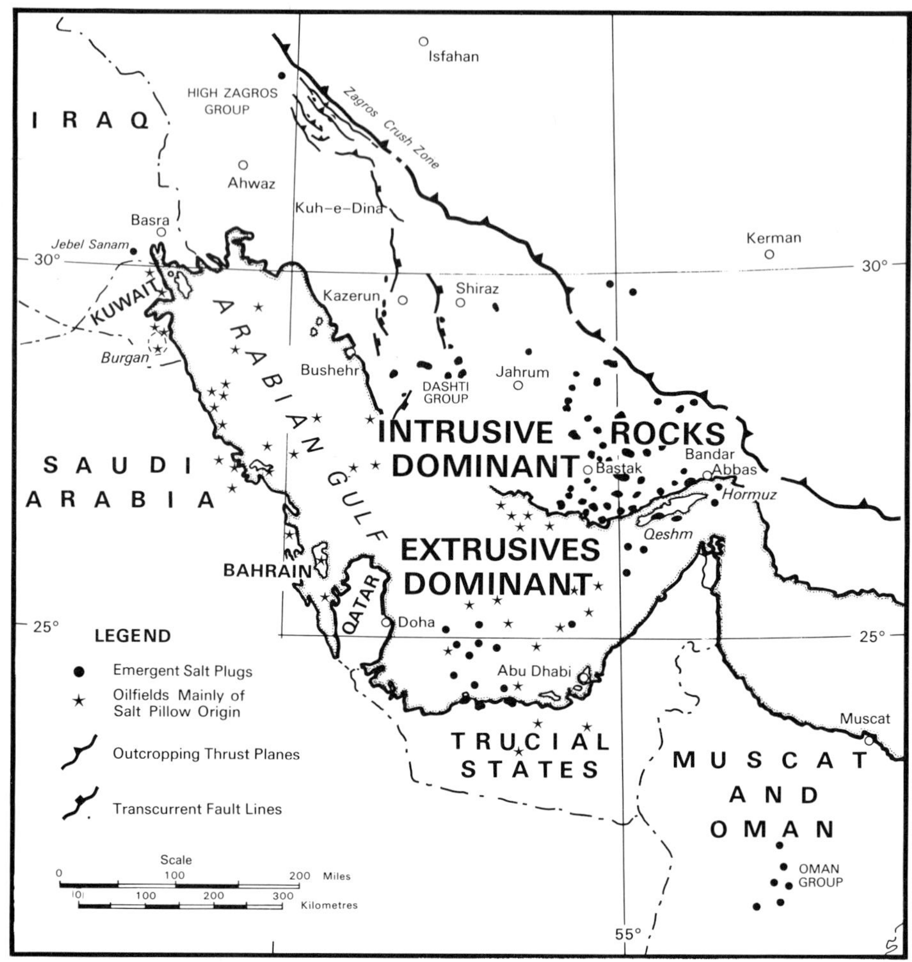

Figure 1. Distribution of Hormuz extrusions - the Arabian Gulf region indicating local dominance of intrusive and extrusive rock clasts.

Overall the Hormuz salt province extended from the outer part of the Arabian Shield, across what was later to be the Zagros orogenic belt, on to the internal plateau as far as the Lut Block of inland Iran, thus providing opportunities for examining the tectonic effect of the same series of interbedded evaporites under a range of different tectonic conditions. Of these contrasting regions the geology of the Zagros belt in Iran is by far the best known and studied. Its various halokinetic features are therefore here outlined before considering the Gulf area.

II. THE IRANIAN MODEL

A. Zagros Tectonics in relation to the Hormuz Complex

A large section of the Hormuz salt basin is occupied by the Zagros range of Iran, fold mountains which arose as a result of late Pliocene orogeny on the site of a NW-SE basin containing a fill of up to 20,000 m of largely conformable sediments. From the Permian through the early Oligo-Miocene marine carbonates predominated. Terrestrial bedded clastic sediments followed, with flysch elements of northeasterly origin appearing locally in the Tertiary, and massive intermontane sandstones of molasse type finally deposited during the Pliocene orogeny (Fig. 2) (Koop and Stoneley, 1982).

When crustal shortening began it was controlled by two major factors - the existence of a potential detachment surface at depth, and the effect of the thousands of feet of nearly continuous massive limestones ranging from the Permian to middle Tertiary. The result was the development at surface of large open folds, with the inner parts of each faulted or crushed in varying amounts. Folds are characteristically 300-1000 meters in amplitude, appearing as enormous limestone whalebacks eroded clear of the later clastic Tertiary formations (Fig. 3).

The fold system of the Zagros would not have developed its characteristic "Jura-type" of folding without an underlying plastic layer. Numerous boreholes in the oilfield belt show this <u>abscherung</u> to be at least deeper than the Jurassic, and deeply eroded anticlines further east and southeast show it to be deeper than the Cambrian. The existence of more than a hundred "Hormuz" salt plugs in the south, and leakages of Hormuz material along fault lines further north, identify the critical incompetent layer to be the Infra-Cambrian evaporite complex.

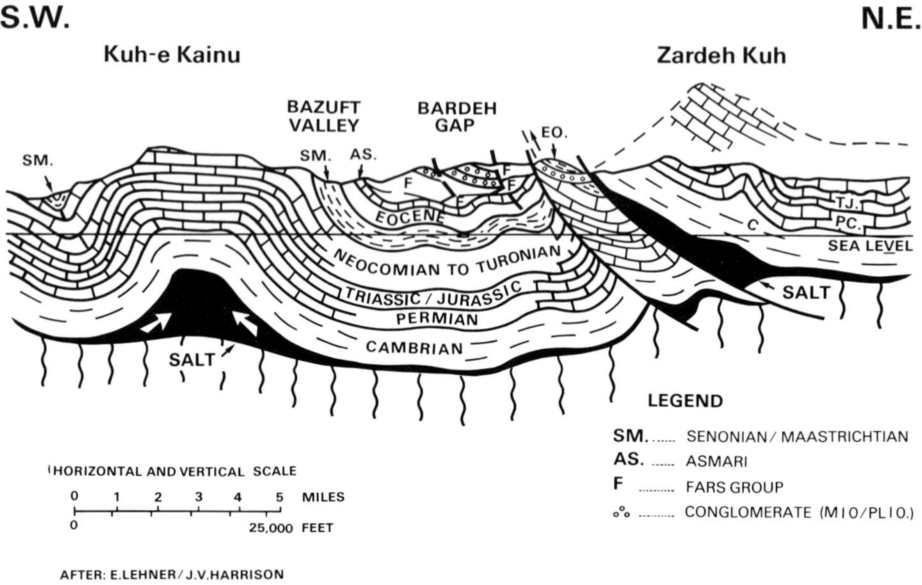

Figure 2. Fold type in the Inner Zagros near Kuh-e Kainu showing overthrust fold and the location of Hormuz "leaks".

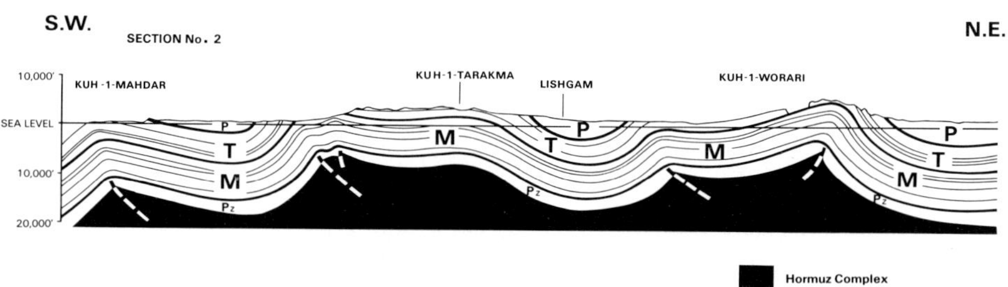

Figure 3. Fold type in the southeastern Zagros near Asalu.

In the northwesterly part of the Zagros the folds are relatively narrow, linear, regular and each is many tens of miles in length (British Petroleum Co. 1964). From Busheyr (Bushire) southwards there is a dominance of short anticlines with numerous extrusive subcircular salt plugs, these being located at random with relation to the Pliocene folds. Additionally the majority of the folds in the south have broad synclines and shattered cores. This contrast between the fold habit of the northern and southern areas may be ascribed to a much greater thickness of the incompetent layer in the south, for the country rocks penetrated in the two areas (Cambrian upwards) are not greatly different in either facies or thickness.

There is nevertheless a strong contrast between the short folds of the southern Zagros (Fig. 2) and the sub-circular diapiric folds of the Gulf itself (Arabian Shield), showing that lateral compression still played an important part in their development.

B. The Iranian Salt Plugs

The exposed salt plugs of southern Iran attracted attention as being topographically distinctive, and those near the coast easily accessible to earlier geologists. Their surface size varies from three to fifteen kilometers across, a diameter which (despite some "mushrooming") is indicated by the size of the larger exotic blocks to be related fairly closely to the size of the necks. Some, mostly those plugs whose activity ended early, are represented by insoluble detritus only (in one or two cases by an empty crater), but many form halite-bearing topographic domes of which the most striking - Kuh-e Namak in Dashti province[1] rises 1200 meters above plains level.

---

[1]Kuh-e Namak means "mountain of salt" and is a common name for the Iranian plugs, necessitating the addition of the province name for identification.

Plate 1. Kuh-e-Namak salt plug, Dashti, with broad "salt glacier" (namakier) flowing from the crest down the south side across flanking limestone ridges.

Several plugs - most notably this one - have overflowed the confines of their flanking limestone country rocks and produced glaciers of solid salt. These show many of the same characteristics as conventional ice glaciers, including ice falls, crevasses, "gendarmes" and perched blocks. The salt glaciers were originally described by Lees (1927) and Harrison (1930) and more recently studied in detail by Talbot (1979) who has proposed the name "namakiers" for these features.

Many salt plugs are located on the axes of the large limestone folds of the Zagros, and it is among these that the active and semi-active intrusions occur. Others are sited on anticlinal noses, on structural flanks or even in synclines;

these are more often moribund or even dead, a contrast which may reflect lack of final activation by the Pliocene orogeny.

A different type of salt intrusion is seen along the oblique fault belts which cross the Zagros (Kent, 1970, Figs. 10, 11). These are mostly in remote areas and have been little studied. They are developed as discontinuous lines of extrusions of varying size from old degraded masses of intrusion could be of Hormuz breccia to more domal later cases. The geometry explained by evaporite fill of varying spaces developed between non-planar fault surfaces on opposite walls of tears, tear faulting being indicated by axial fold displacement as well as by regional concepts (Bostrom, 1985). Slivers of Hormuz material are also associated with thrusts in the High Zagros (Fig. 2) and further evidence of the wide extent of this deep plastic layer is provided by the small sub-circular Hormuz plugs N.E. of the main Zagros Crush Zone e.g. at Dalnashin (Gray, 1950).

C. Age of the Iranian Hormuz Rocks

On the evidence of the less disturbed "Hormuz" rocks inland of the Zagros on the edge of the Lut desert (Stocklin, 1968), of the cyclic development of Cambrian/Precambrian dolomites in Oman on the southern edge of the salt embayment (Kent, 1970, p. 82), and of the interbedding of Hormuz lithologies seen in the very large transported blocks in some of the southern plugs (Kent, 1969) it has been concluded that the Hormuz includes the brecciated product of a series of cyclically interbedded, evaporites and sediments deposited over a considerable period. Since the Hormuz series outcrops in plugs close to typical sections of the undisturbed Zagros from Lower Cambrian upwards (e.g. at Kuh-e Dina), it is deduced that the series is essentially older than the local Cambrian and is mainly Pre-Cambrian.

Trilobite evidence has been held to show that the top of the series may range up into the Cambrian near the Gulf coast. There is no evidence of such late date further inland where entrained blocks with fossils are limited to stromatoporoids of Infracambrian aspect, and the possibility cannot be excluded that the Cambrian blocks are from the vent sides.

D. The Age of Emplacement

In some plugs there is clear evidence of pre-orogenic emergence provided by interbedded Hormuz detritus in rocks dating from Cretaceous to Miocene. Occasionally there is seismic evidence of stratigraphic thinning peripheral to a plug, although this is too infrequent to explain easily the penetration of the limestone overburden package. Nevertheless the virtual absence of the displaced vent rock from any of the plugs (all of which penetrated thousands of feet of the Phanerozoic limestones) is best explained by its erosion after extrusion, together with the original contemporary cover of Miocene clastics. Thus, most of the plugs in fact penetrated a much deeper "layer cake" pre-orogenic and pre-erosional stratigraphy.

The plugs thus did not "pierce at random through the Tertiary folds" (a matter which worried Stocklin, 1968, p. 162) but were older. The pre-existing plugs were located (as various authors had noted) on a rectilinear pattern which must have been much older and which probably related to Basement trends (Fig. 4).

Their location in fact influenced in turn the trend of Zagros axes, which in Laristan frequently originate from or change direction at, individual plugs (Furst, 1979). Being older, the intrusions represented points of weakness in the competent cover rocks.

Some of the plugs are dead, surviving only as chaotic heaps of the characteristic insoluble foetid dolomite, sandstones and igneous detritus. Others, especially those with glaciers, rise domally above the surrounding country rocks and show exposures of

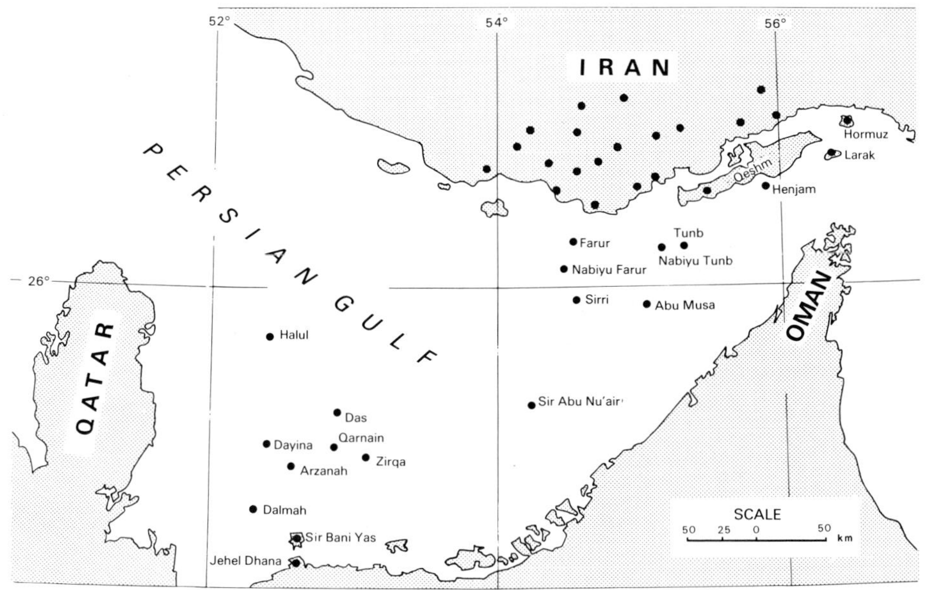

Figure 4. Salt plug alignments in the SE Zagros showing the independence of Pliocene folding from Hormuz trends.

shining halite. These are clearly still being fed from below, balancing the real (if small) steady action of modern atmospheric solution.

E. Igneous rocks and Mineralization in Iran

Igneous rocks are widely (although not universally) distributed in the plugs of the Iranian mainland. The great majority there are of "greenstone" (diabase) which is seen veined into the Hormuz sediments as well as abundant clasts. At the Chah Benu plug, moreover, this rock occurs both as veins and pebble beds interbedded in red Hormuz shale, implying a closely similar date to the associated sediment, and an age of intrusion pre-dating the Zagros Phanerozoic rocks. Extrusive volcanic

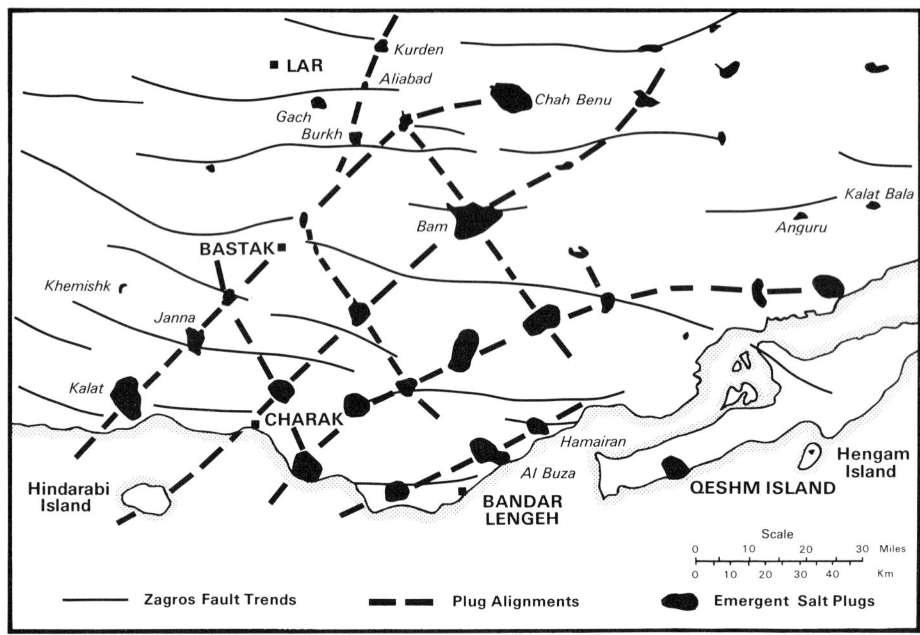

Figure 5. Locations of Hormuz salt plug islands in the Arabian Gulf.

rocks are rare on the mainland, but a notable exception is provided by abundant fragments of ignimbrite at a dead plug near Bandar Abbas, and pillow lava is recorded from the Kalat plug.

In addition to conventional igneous rocks, Walther (1968) has concluded that the widespread crystalline haematite, frequently seen as replacing sedimentary rock, was initially of magmatic origin, having been originally deposited as volcanogenic sediment interbedded with the salt.

Both intrusive and extrusive rocks thus long pre-date the Pliocene Zagros orogeny. Their effect on tectonics is no more significant than that of the clastic, but non-volcanic, foreign rocks in the evaporite complex.

III. THE GULF ISLAND SALT PLUGS

About 20 of the islands in the southern Gulf owe their existence to Hormuz extrusion (Fig. 5). There has been no modern overall account of these, and the following description draws on unpublished reports of visits by E.J. White and M.H. Lowson of BP (1927), in the south reports by T.J. Harris and J.W. Halse of A.D.P.C. (1962), Zirka by R. D. Hawkins and Das by A.J. Martin of BP (1962, 1961) together with visits by the present writer at various times, and in some cases published accounts by A. Gansser (1960).

The island salt plugs are described here in brief outline. Most are evidently small erosional relics of originally much larger extrusions, now below sea level or effectively blanketed by Neogene/Recent sediments. Namakdam (on Qeshm Island), the Hormuz Island plug, Sir Abu Nu'air and Jebel Dhanna (on the Abu Dhabi coast) still demonstrate at surface their full diameters as extruded, with peripheral outcrops of the country rock.

A. General Description

The Hormuz Complex shows a high degree of uniformity in its unusual composition, and to save repetition the characteristic make-up of island plugs is here listed to limit the further description to those less usual characters which merit special mention.

In no case is there any relic of the cubic kilometers of Palaeozoic and later rocks which must have been displaced in the course of vent formation - an anomaly familiar on the Iranian mainland (see above). Instead of the rocks of that sequence, one of the most characteristic rocks (as in Iran) is bedded black dolomite, often coarsely crystallised, often finely laminated and usually foetid. This may well have been a sabkha facies and probably includes the equivalent of the Soltanieh dolomite of northern Iran. Also widespread are sandstones and siltstones, mainly reddish, grey or greenish grey, variably indurated, which may include the equivalents of the Lalun and Barut formations of the north (Kashfi, 1985).

Either of these types may occur as rafts measurable in hundreds of meters across in the evaporite melange, the latter usually being seen in surface exposures as gypsum, having lost most of the halite element by solution.

Igneous rocks occur in quantity in the Gulf islands, but they differ from the igneous accompaniment of the Iranian mainland plugs in being dominantly acid tuffs (trachytes and rhyolites) with acid intrusions. The basic "greenstone", familiar in Iran, is relatively infrequent.

A noticeable feature in all the plugs, here as in Iran, is the amount of red and specular haematite present, varying from ochre to bladed crystals, which may be replacive in any member of the suite.

Each plug shows part, often all, of this assemblage, contrasting totally with the Mesozoic and Tertiary rocks penetrated in the adjacent offshore oilfields.

Another feature common to nearly all the Gulf plugs is the contact with the country rocks. Since the Gulf area has not suffered the post-orogenic uplift and deep erosion seen on the Iranian mainland, island plugs are commonly seen cutting Neogene at surface. Where the latter (usually Middle Miocene) is present it is seen to rest directly on an eroded, channelled and weathered surface of the Hormuz Complex. Often also the Miocene is turned up sharply at the junction, indicative of a late phase of plug movement.

Where the Middle Miocene is absent the discontinuous carapace of the plugs is provided by Pliocene/Recent calcarenite, commonly termed "miliolite", a shallow water limestone-sand assemblage, often coral-bearing. Since this is frequently found up to 200 ft. above present sea level it marks relatively recent uplift, but the beds are not precisely dated and probably cover a considerable period. The cover, of whichever age, commonly contains Hormuz pebbles and boulder beds, indicating contemporary prominence of the old complex.

The islands are here catalogued from the north south-wards, concentrating on those particularly well investigated.

Geographical nomenclature presents a problem. Both Arabic and Iranian (Farsi) names are in use for many of the islands, and each has been variously transliterated into English. As far as possible the most commonly recognized names are used here.

B. Local Details

On Qeshm Island the large Namakdan plug pierces centrally a broad dome on the Miocene outcrops, which are turned up in a well-developed "collar" to the Hormuz plug. Deep test drilling showed that the early Tertiary limestones - a continuous massive group on land - have here passed at depth into an argillaceous facies. In consequence this plug has penetrated in its higher section a mainly incompetent series with a structural effect comparable to that of a Louisiana salt dome.

The Hengam (Henjam) plug forms a small island (6.5 X 3 km) seaward of Qeshm, with the Hormuz rocks only locally exposed in the core of a domal outcrop of uncomformable conglomeratic Neogene. The Hormuz itself consists of banded salt and red haematitic tuff and rhyolite. The description by Gansser (1960, p. 22) illustrated the central part of the dome, with blocks of tonalite in overlying Neogene limestone.

Three kilometers to the southwest of this point a very large isolated block of igneous rock was described by Pilgrim (1908, p. 133) as an 800 sq. ft. mass of altered quartz diorite rising to 40 ft., which he compared with the "serpentine igneous series of Oman". Gansser, on the other hand, mentioned large uniform hornblende gabbro blocks in the Neogene basal breccia, and further northwest a 3000 cu. meter block of gneissose granite, unique in the salt plug islands. This evidence, that large but varied "Basement" masses are involved, led Gansser to suggest that this particular intrusion had penetrated an olistostrome (loc. cit., p. 40) as an alternative explanation to displacement of blocks from a Basement fault scarp.

Larak, 8 km east of Qeshm, is an 8 x 6 km craggy island. Structurally Larak has a core predominantly of salt surrounded by a ring of steeply-dipping clastic Hormuz rocks (including volcanics and the ubiquitous specular iron ore), overlain by Pliocene and Quaternary limestones which are gently domed (Gansser, 1960). This is again evidence that the final movement was relatively recent.

Hormuz Island, located in the re-entrant of the mainland coast east of Qeshm, is the most informative of the eastern group of plugs and deserves full description. The best published description is that of Gansser summarized briefly here.

From the air the island shows a series of annular ridges parallel to the coast, and the coast itself is controlled by a discontinuous ring of upturned Lower and Middle Miocene rocks (Gansser, 1960; fig. 16).

The outermost ring of the plug itself is of haematitic tuffs, locally so rich as to be mined as a source of pigment. Within this the next ring is of trachyte and white rhyolite eruptives, its surface locally littered with blocks of laminated black sulferous dolomite. The third ring occupying the central area of the plug is recorded as being predominantly gypsiferous salt.

Such strong concentric structure is uncommon, but Sir Abu Nu'air in Abu Dhabi waters is another case. On the mainland the Jahrum plug near Firuzabad shows a ring of steeply-dipping red marl concentric with the plug outline, separating chaotic breccias inside and out (Kent, 1958). This concentric habit is comparable to the very strong flow lines seen in some of the

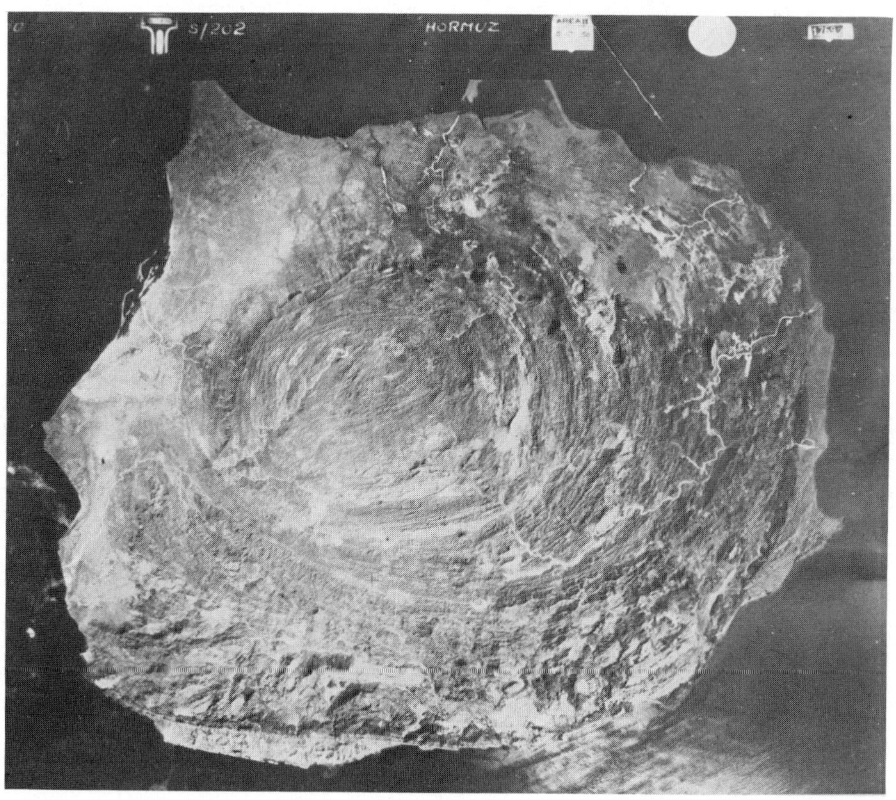

Plate 2. Air view of Hormuz Island showing concentric ridging within salt plug. Miocene rocks flank the plug on the coast opposite the point.

Miocene plugs in the Kavir desert of inner Iran, and is probably analogous to the vertical flow structures observable in the pure salt of some Louisiana salt mines.

Large idiomorphic apatite crystals which occur locally in a ferruginous crust on the Hormuz salt have recently been the object of fission track dating (Hurford et al., 1984), which concludes that the tracks were formed at a depth of 3-5000m at the level of the Upper Jurassic-Triassic at the time when the first salt-pillowing developed, and that the data are further consistent with an origin for the bulk of the Hormuz complex at the later Precambrian level.

Plate 3. Hormuz Island. Mass of acid volcanic tuff within the salt plug.

The writer is indebted to Dr. R. Walls of BP Sunbury for identifying rocks from the tuffaceous series including quartz feldspar porphyry, altered felsite, apatite and pumice tuffs. He noted that some of the specular haematite specimens were muscovite-bearing and that the loose packing of the crystals showed that they had finally crystallized under low load - a phase of crystallization which must have been late in the history of plug movement. A silicified siltstone showed joint faces encrusted with malachite, as in some Iranian mainland plugs. One unusual specimen was a block of granite porphyry, adding to the cases of "Basement" rocks in the salt plugs of the region.

Greater Farur, an island rising from deep water, was well described by Gansser (1960). It is notable among the Gulf islands for a particularly large raft (up to 2 km long) of well-bedded carbonate rocks in the Hormuz breccia, for the occurrence of horneblende diabase with relative rarity of eruptive rocks or of trachyte (compare the Iranian mainland).

Little Farur and the Tunb Islands show the more usual Gulf island assemblage.

Abu (or Bu) Musa is a subcircular salt plug island 5 km broad, on the flank of the Mubarrak oilfield. The northern part includes hills of soda trachyte and rhyolite lavas, the latter cut by quartz-porphyry dykes. The flatter southern part of the island is occupied by richly haematitic bedded red tuffs overlain by rhyolite, an arrangement compatible with Walther's (1968) postulate that the ubiquitous Hormuz haematite was originally of sedimentary-magmatic origin.

A loose block of fine-grained grey rock cut by curving glassy surfaces has been identified as a spilite, with small scale "pillowing" due to minor lava lobes, a discovery in line with pillow lava found in some Iranian mainland plugs (Kent, 1979).

Sirri is a broad (8 km) low island with Hormuz rocks mainly blanketed by sub-Recent coralliferous limestones. Old records seem contradictory, but White and Lowson (1927) record a 350m sequence of bedded sandstones with minor intercalations of gypsum, and bedded limestone. Thick beds of trachyte and trachytic lava were seen.

Halul is a small (2 x 1 km) isolated island near the Qatar coast. According to White and Lowson (1927) it shows the familiar Hormuz mixture, its two peaks being of black platey limestone interbedded with white and reddish-stained gypsum.

Das Island, 2.5 by 1.5 km, has been thoroughly surveyed in development as an industrial base. The northern half consists of exposed Hormuz Complex in an eroded and degraded state, but the exotic rocks (in a gypsiferous ground mass) are so heterogeneous and fractured as to throw little light on the extrusion process (Fig. 6). Only in the northwest does verticality suggest organized flow and proximity to the vent wall.

The assemblage is entirely typical. Rhyolite veins are seen locally intruded into sandstone and shale, but fail to cut the adjoining matrix - hence predating the later plug movement.

Structurally the Das outcrop rises from a gentle dome at Jurassic levels, and the island is probably only the erosional remnant of a larger intrusion.

Qarnain Island is a small degraded island plug, a little less than 3 km long, with three conical hills near the northern end and a shelf of Plio-Pleistocene calcarenite extending southwards. The hills are composed of the familiar Hormuz suite of rocks, but in each case owe their prominence to igneous intrusives. The westerly hill is mainly quartzite shot through with anastomising veins of dolerite, the central peak consists of sediments into which has been intruded a mass of quartz-porphyry, and the third (southeast) peak is of greenish glassy trachyte.

The island gives the appearance of being little more than a relic of a former plug, and seismic survey in fact indicates that it rises from a Cretaceous domal structure some 10 km long.

Zirka (Zirqa, Zirca or Zarakuh) is one of the larger islands, 4.7 x 2.5 km, largely occupied by a hill mass of strongly folded and faulted sediments, essentially a single Hormuz raft. It is flanked by relics of the usual gypsiferous breccia containing dolerite and foetid dolomite blocks (Fig. 7).

On the west flank the present author recorded 200 m of red and grey shales and siltstones with occasional feature-forming sandstones, moderately indurated. On the east Harris and Halse (1962) traversing a different fault block, measured a 160 m sandstone, flagstone and shale succession overlain (?conformably) by 140 m thick amygdaloidal lava. The sharp folding of the sediments is a feature unique to Zirka. Dips vary from 60° to vertical either to NE or SW. It is not clear how such a degree of lateral shortening could have been caused by the process of simple elevation from depth.

Seismic survey shows that Zirka Island marks the summit of a steep-sided dome at depth, indicating that the originating salt plug is likely to be only slightly broader than the island.

Arzanah, 3 km N-S and 1.7 km E-W, is a pear-shaped island in which the Hormuz rocks are represented mainly by the gypsum melange (Fig. 8). Exposures are shallow, but show the usual wide variety of clasts embedded in the gypsum. Ophitic dolerite, trachyte and tuffs are subordinate in quantity to dark dolomite which forms much of the higher ground.

Structure contours on a deep Cretaceous reservoir show that the Arzanah Island is central to a domal uplift, and suggest that the island is almost as large as the underlying salt plug

Dalma, one of the largest salt plug islands, is 8.5 km long and is almost entirely occupied by the 5.5 km wide Hormuz outcrop, flanked by a narrow belt of outward dipping Middle Miocene ("Lower Fars") on east and west. Poorly bedded gypsum predominates in most of the plug, measuring at least 300 m thick. In the gypsum mass are the usual large exotic blocks of platey and banded dolomites, often measuring tens of meters across and occasionally stromatolitic, together with acid extrusives.

A striking feature is the concentric lineations in the steeply-dipping melange, particularly in the southern part. As elsewhere this suggests proximity to the vent wall. There is a suggestion from the structural trends that the plug shape is partly controlled by NNW striking faults.

As at Jebel Dhanna, the plug had been both extruded and deeply eroded before Miocene deposition, and arching of the Miocene indicates still later activity.

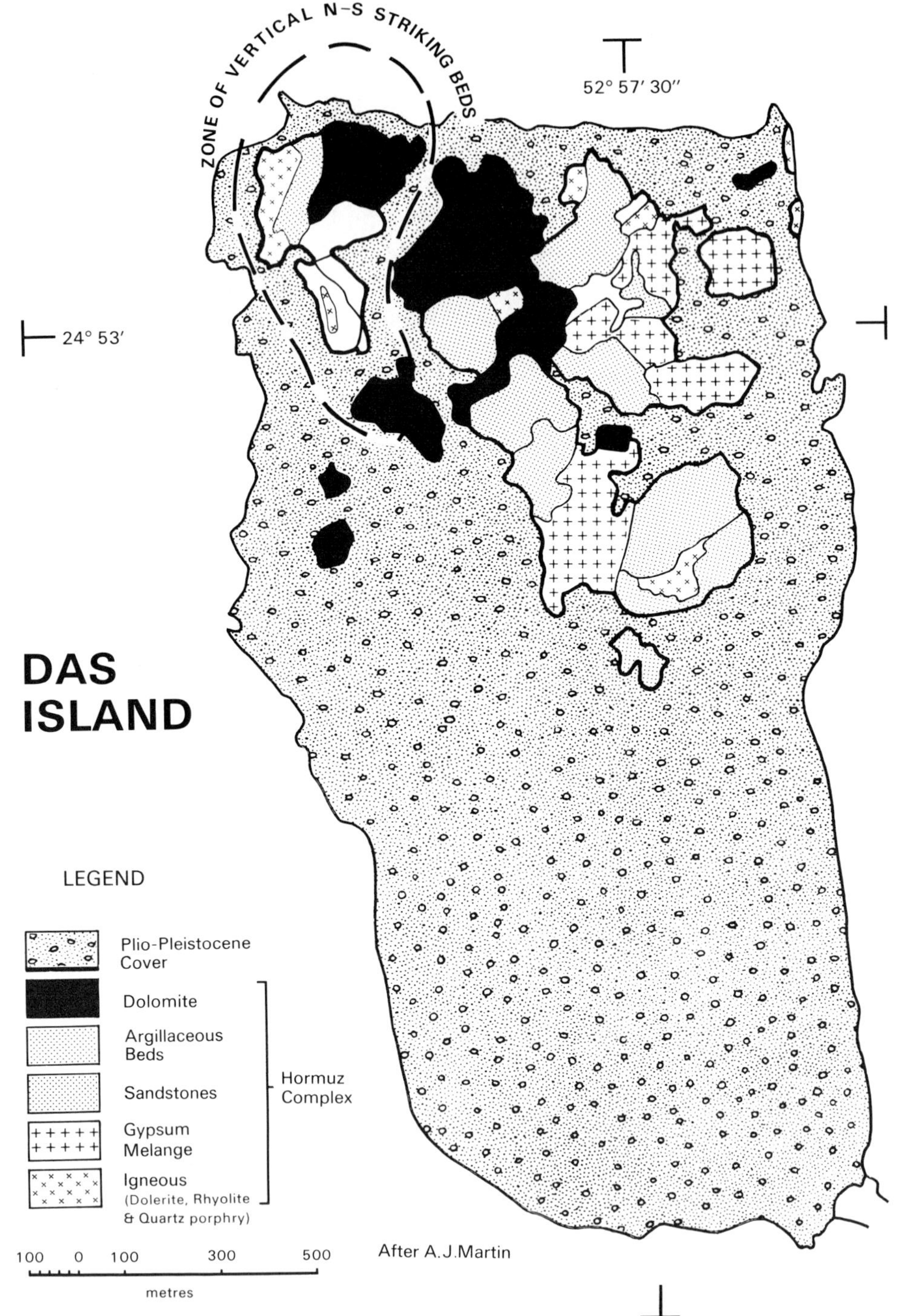
ZONE OF VERTICAL N–S STRIKING BEDS
52° 57′ 30″
24° 53′
DAS
ISLAND
LEGEND
Plio-Pleistocene
Cover
Dolomite
Argillaceous
Beds
Sandstones
Gypsum
Melange
Igneous
(Dolerite, Rhyolite
& Quartz porphry)
Hormuz
Complex
100 0 100 300 500
metres
After A.J.Martin

Plate 4. Qarnain Island, hills of igneous rocks relict from the salt plug mass. Das (another salt plug island) is marked by the distant smoke plume.

Sir Bani Yas, an island 11 km N-S by 8.5 km E-W, has a central circular Hormuz outcrop 3 km across flanked by unconformable Miocene and this in turn by later calcarenites.

The Hormuz assemblage carries a greater proportion of transported rafts than other island plugs: dolomites (partly stromatolitic) and bedded clastic sediments comparable to those of Zirka predominate, with igneous rock masses concentrated in the central section. Halite is sufficiently shallow to have been mined and is still visible in a sink hole some 7 meters below surface. A large block of (biotite) granite was seen near the mine.

Figure 6. Das Island showing chaotic Hormuz outcrop (facing page).

53° 05′
24° 53′
92
70
75
35
65
50
85
75
75
ZIRKA
ISLAND
LEGEND
Sand & Gravel
Neogene
(" Miliolite")
Gypsum Melange
Bedded Ssts,
and siltst.
Igneous
Hormuz
Complex
Strikelines of
Hormuz Sediments
200 0 200 600 1000
metres
After Harris & Halse

Like Hormuz Island, Sir Bani Yas shows a strong development of concentric ridges, with verticality of the rafted blocks near the southern and eastern sides, presumably reflecting flow structure of the plastic intrusion near the vent wall.

Partial seismic coverage of the Cretaceous rocks east of the island fails to show any simple relation to the plug location.

Five kilometers to the south a salt plug uplift on the Abu Dhabi coast, Jebel Dhanna, has been developed as an industrial area and consequently well-surveyed. The Hormuz outcrop, 4 km N-S x 2.5 km E-W is located centrally in the peninsula, flanked by outwardly-dipping Miocene turned up sharply at the contact. Higher beds of this carapace consistently overlap lower strata, symptomatic of intra-Miocene movement.

The Hormuz assemblage covers the range familiar in the island plugs; igneous rock masses are often hundreds of meters across. Dyke rocks are rare and "greenstones" are absent.

Notable features in the northerly part of the plug are bevelled areas of gypsum melange with strong concentric structure, 100-200 meters across, which may be interpreted as independently moving "spines" of evaporite seen in cross section.

The plug is bridged centrally by Miocene, and distribution of steep dips in the melange suggests that the plug may have had twin vents - or more probably twin culminations. Harris and Halse (1962) suggested that the elongation could have been due to extrusion between parallel meridional faults. Alternately the whole Jebel Dhanna - Sir Bani Yas line could be the surface expression of an irregular salt wall, controlled initially by a deep-seated fault. This interpretation would compare with the more common north-south elongation of the salt swells responsible

Figure 7. Zirka Island - a Hormuz outcrop largely occupied by a single Infra-Cambrian sedimentary raft (facing page).

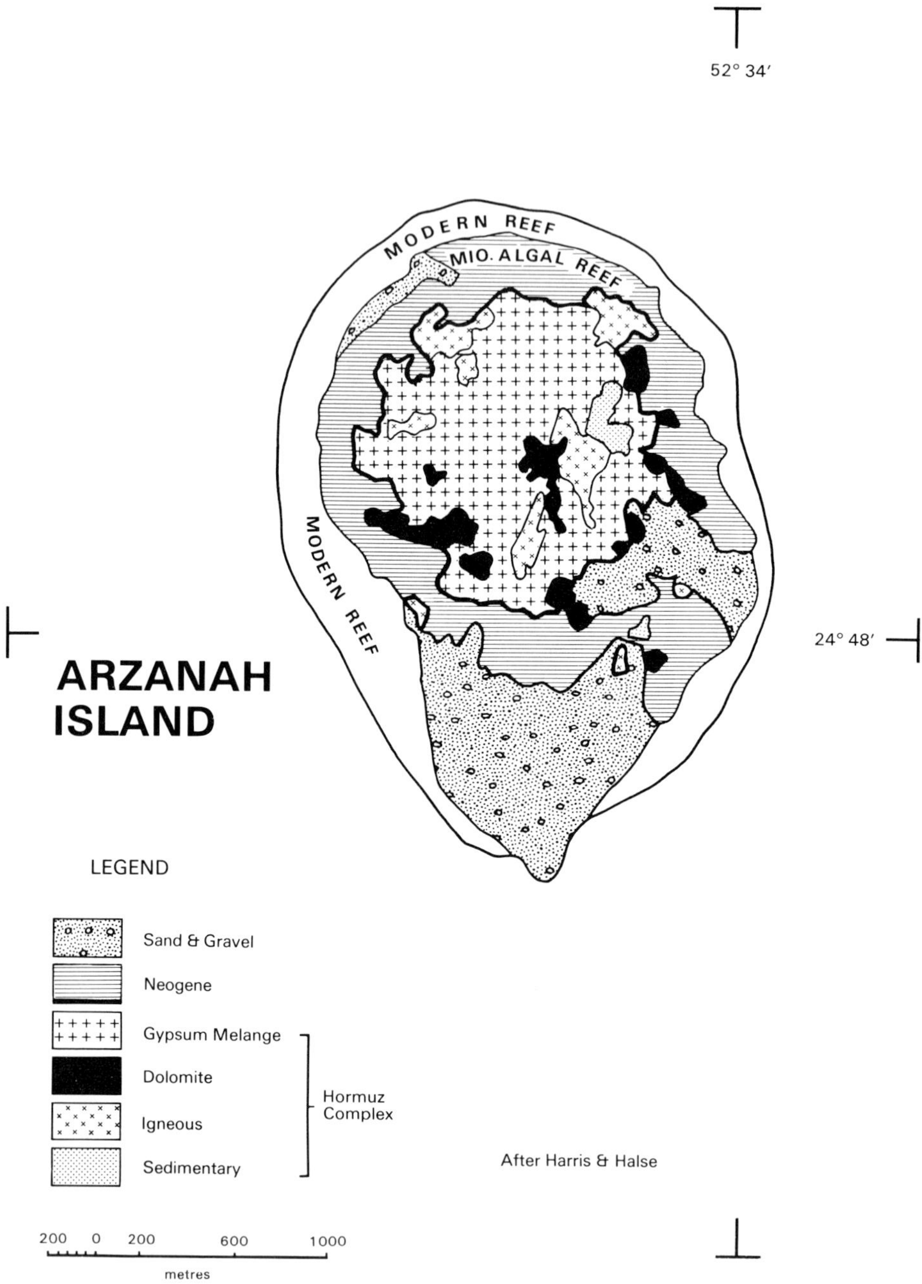

52° 34′

MODERN REEF
MIO. ALGAL REEF
MODERN REEF

24° 48′

ARZANAH ISLAND

LEGEND

Sand & Gravel
Neogene
Gypsum Melange
Dolomite
Igneous
Sedimentary
Hormuz Complex

After Harris & Halse

200 0 200 600 1000
metres

for the genesis of such structures as Damman, west of Qatar, and with the elongate N-S oil-bearing anticlines of the Arabian mainland.

V. SALT PLUGS OF THE ARABIAN MAINLAND

Apart from the coastal Jebel Dhanna salt plug, described above with the salt plug islands, there are extrusive Hormuz outcrops in two areas on the Arabian peninsula - a group of six plugs in Oman and a single salt plug mountain (Jebel Sanam) on the southern border of Iraq. It has been deduced also that the domal structures of landward Abu Dhabi and of the east coastal belt of Saudi Arabia mostly originated by salt pillow development (Dominguez, 1965; Elder and Grieves, 1965). The inference is that the Hormuz depositional basin of evaporites, dolomites, tuffs, etc. extends from Iran across the Gulf to a line within the edge of the Arabian mainland.

The salient features of the Arabian mainland outcrops are described in the following short summary.

A. Iraq: Jebel Sanam

Iraq Jebel Sanam is an isolated oval hill 2 km across on the Kuwait/Iraq border rising 150 m above the Dibdibba gravel plain. It has a domal limestone and limestone-breccia carapace (Neogene?) dissected by radial ravines which expose inliers of the Hormuz complex.

Jebel Sanam was visited by H.G. Busk in 1916; later by H.T. Mayo of Anglo-Persian Oil Company who, in a Company Report, first suggested the Hormuz date by comparison with Qeshm Island; subsequently by de Bockh, Lees and Richardson (1929), by Owen and Nasr (1958) and most recently by al Naquib (1963, 1970). The Jebel lies 130 km north of the great Burgan oilfield dome, which

Figure 8. Arzanah Island. A small salt plug outcrop surrounded by outward dipping Miocene sediments (facing page).

itself is suspected of originating as a very deep salt pillow on the evidence of its symmetry and radial fault pattern, although (in common with many Iranian salt plugs) it fails to produce a gravity "low".

The Hormuz rocks of Jebel Sanam consist of gypsum interbedded with thin limestone and red marl, apparently resting unconformably on a green marl unit, but the beds are heavily faulted and the sequence uncertain. Haematite veining is well developed. Clasts recorded include dolomite, chert, quartzite, and diorite. Owen and Nasr (1958) record an altered dolerite dyke which has since been quarried for roadstone.

None of these rocks matches the local Phanerozoic sequence, which has been explored to below 16,000 feet, and there can be no doubt that it is an extrusion of underlying Hormuz strata.

B. Oman

The desert plains of inner Oman, part of the virtually undisturbed Arabian shield, are the site of six emergent salt plugs characterized by extrusive rocks of Hormuz type comparable with those of Iran. Gravity surveys have indicated the existence of ten further concealed salt intrusions, ascribed to the same series, lying in a belt extending northeast and southwest from the emergent plugs (Tschopp, 1967a; fig. 4).

Brief mention has been made of the geology of the plugs by Morton (1950) and others, and, through the courtesy of Shell and the staff of their associate Petroleum Development (Oman) Ltd., the writer was enabled to visit four of the plugs (1970). Specimens then collected were investigated by Dr. R. Walls and Dr. W. J. Clarke of BP.

1. Lithology

In summary, the Oman plugs show a less varied rock assemblage than those of the Gulf or the Iranian mainland. Dolomite is by far the dominant normal sediment; red marls occur infrequently. Sandstone - of arkosic type (Lalun?) - was observed at Qarn Sahmah only.

The dolomites are frequently very finely laminated, several laminations occurring per millimetre of thickness. In some cases (as at Qarat Kibrit) the laminations are strongly undulant, with an appearance suggesting organic rather than ripple origin. It seems most likely that the extremely fine lamination (in a mainly evaporite series) is due to algal mat layers in a sabkha environment; the close similarity between some salt plug specimens from Qarat al Milh with the Cretaceous of offshore Abu Dhabi and from modern sabkha is striking. In association with the laminated beds are blotchy dolomites and dolomites with a suggestion of enterolithic structure, both possibly due to pene-contemporaneous chemical action.

The best evidence of a bedded sequence is that recorded above at Qarat al Milh, where the alternation of dolomite (partly stromatoporoidal) with evaporites is reminiscent of the cyclic Iranian succession. A cyclic arrangement is also indicated by the dolomite/clastic alternation in the Cambrian/Infra-Cambrian shelf facies of the marginal Haushi/Huqf area in eastern Oman. Haematitization occurs but is infrequent.

The only igneous rocks seen were altered porphyrites and rhyolite tuffs at Qarn Sahmah, indicating the former presence of an outlying isolated volcanic vent. This may belong to the same province as those in the Gulf islands, but the Oman area presumably lay south of the main volcanic group.

2. Age of Oman Diapirs

Deep drilling has shown that the first evaporite group of significance in Oman occurs beneath the Cambrian, as for example at Fahud (Tschopp, 1967(b)). It is therefore fair to assume that the salt plugs originate from Cambrian and/or Precambrian horizons, and it is understood that this is confirmed by seismic evidence of a deep source.

Additionally, the stromatoporoids collected during the present writer's visit include floras (notably *Conophyton*) which characterize the Hormuz of the southern Iran plugs and also occur in the Infra-Cambrian inland there. Although these (as preserved) are not closely diagnostic of horizon, they add to the negative evidence of absence of later forms, giving a broad indication of an Infra-Cambrian date and suggesting that the Hormuz is the basinal equivalent of the dolomite/clastic alternation which conformably underlies the Cambrian in the Haushi/Huqf outcrop area to the east.

The known belt of plugs is 400 km long, but only some 20 km wide. Salt certainly occurs more widely in the early Palaeozoic or late Precambrian as demonstrated by the Fahud record, but (as far as is known to the writer) the positive structures defined in oil exploration are all gentle folds. The limitation of plugs to a narrow belt consequently suggests that thick and therefore mobile salt was developed in a graben, presumably due to a late Precambrian fault phase in the Arabian shield. It was not necessarily part of the main Hormuz depositional area despite the identity of sedimentary facies.

V. REGIONAL IMPLICATIONS: THE AGE OF THE HORMUZ COMPLEX

As noted above, in Iran an Infra-Cambrian age for the greater part of the Hormuz Complex has been established by correlation with outcrops (Stocklin, 1961), by the cross cutting relationships with the Phanerozoic column of the Zagros, and by limited palaeontological evidence (Kent, 1979). Originally a Cambrian date had been assumed (de Bockh, Lees and Richardson, 1929), and it is posible that locally the Complex ranges up to this level, but the rare cases of Cambrian fossils could relate to material entrained during extrusion.

Investigation of the salt plug islands described above, leaves no doubt that the Hormuz of the Gulf area correlates directly with that in southern Iran and can be ascribed to the same broad age range as the Infra-Cambrian sequence described by Kashfi (1985).

Cyclicity at the Zagros outcrops is clearly marked by an alternation of dolomites/evaporites and terrestrial deposits. A thick evaporite unit must have existed at the base of the known Hormuz rocks to provide the halokinetic drive. At least one distinctive evaporite/dolomite unit also overlies terrestrial clastics and conglomerates (as at Chah Benu), with a degree of interbedding, and we know from undisturbed outcrops that a major redbed group (the Lalun) terminated the Infra-Cambrian succession. How far lithological correlation can be taken is still uncertain; the Hamairan facies of siltstones and sandstones (Kent, 1979; p. 123) for example may be unique and equivalent to the inland Barut, or may recur in the succession.

All these rocks, it should be noted, are of shallow water facies or even paralic type; across the whole Hormuz basin there is no indication of oceanic or deep water sediments.

The igneous rocks provide further problems. In the Gulf islands, and some of the southernmost plugs on land, acid volcanics occur in quantity, and most of the associated dyke rocks seen are also acidic, but they are at least older than the Lalun type of sandstones (which lack significant tuffs) and from their position must have post-dated major halite deposits.

In the Zagros area the dominant igneous rock is diabase, usually highly fragmented but seen in various places to be intrusive into the Hormuz sediments. This period of dyke formation could relate to early development of rifting along what was later to become the Zagros Crush Zone (Stocklin, 1961, pp. 878-9; Falcon, 1969), for which the rectilinearity demands an initial tear or tension-fault origin.

The next problem relates to the date of inception of movement. The Zagros rocks show interbedding with Hormuz detritus spreads (indicating surfacing plugs), in rare cases as early as Cretaceous, but more commonly in the Middle Miocene (in the Gach Saran and Mishan formations). Evidence of early thinning is rare, but attenuated strata from Jurassic upwards are seen at Kuh-e Namak and Kuh-e Dirang in Dashti. These latter anomalies are on different Zagros fold lines, but a major gravity low indicates that both anomalous thicknesses relate to a single broad salt pillow oriented approximately NE-SW - a feature only slightly older than the initial salt swells indicated in the offshore oilfields.

A. Infra-Cambrian Palaeogeography

The evidence that the Hormuz Complex includes several dolomite/evaporite alternations and that the dolomites themselves frequently show the fine laminations typical of algal mats (as well as local algal reefs), establishes that the facies marks a broad shallow water and sabkha environment - a shallow shelf at maximum measuring more than 900 km wide. (These figures do not take into account a possible extension to the plugs of middle Oman, which may have been in a detached basin).

This shelf - including much of Iran - was the northeastern margin of Arabia at the time of the evaporite/dolomite deposition. As noted above, the red, purple and greenish sandstones and siltstones at other periods within the Hormuz range were also shallow water or terrestrial deposits.

The area of the evaporite shelf of the Hormuz salt which provides the motive for the Gulf salt pillow oilfields decollement plus the Zagros folding thus amounted to some 1.1 million sq. km, and the dolomite facies to the north in Iran (in deeper water?) covered a comparable area.

Current theory of continental plate movement postulates a former ocean between Africa and Asia, a requirement which has been met in various ways, not always based on firm factual evidence. The earliest reconstructions assumed a break hidden conveniently beneath the modern Arabian Gulf, a thesis demolished by the results of the numerous offshore boreholes with the uninterrupted Mesozoic and Tertiary development now known from Arabia to Iran. More recently the former inter-continental suture is assumed to be marked by the Zagros thrust, better termed the Zagros Crush Zone (Falcon, 1969), Asia having overridden the Afro-Arabian plate along this line, marked as it is by a narrow discontinuous outcrop of ophiolites and radiolarites.

In fieldwork in the Iranian interior however Stocklin (1968a,b) and more recently Kashfi (1985) showed that there was initially stratigraphical continuity across the Zagros Crush Zone, that the Hormuz facies recurred inland north of Kerman, and that its equivalent Infra-Cambrian dolomite facies (Soltanieh) formerly extended across Central Iran to the Alborz mountains without a significant break (Stocklin, 1974 - Fig. 2). Any oceanic suture between Arabia and Asia must have been at least as far north as the southern edge of the Caspian in Infra-Cambrian times.

The data on the Hormuz seen by the writer is entirely consonant with Stocklin's conclusions. There is no stratigraphic break in the Hormuz Complex visible in the extrusive salt plugs from the southern shore of the Gulf through Iran to the Zagros Crush Zone, and northeast of this (after a 30 mile gap) the identical facies assemblage outcrops extensively near Kerman on the Iranian plateau.

In a different setting the thick acid tuffs of the island plugs could indicate an island arc, but in the absence of a contemporary ocean this is difficult to envisage. The evidence of Walther (1968) that the ubiquitous iron oxides of the plugs are likely to have originated in highly ferruginous tuffs might fit better with deposition near the circulation system of an incipient mid- ocean ridge, which would not necessarily imply deep water.

It should be observed however that the spilites which the present writer has recorded in the Hormuz Complex are notable for their rarity. Lava was evidently occasionally extruded below sea level, but no major package of Hormuz pillow lavas is indicated.

B. Hormuz the Relic of a Proto-Ocean?

From the clear-cut evidence that over the area of the Gulf, as in Iran, major diapiric bodies were available to punch vents ahead of, and then float up, very large rafts of sedimentary rock, it is concluded that beds of halite formed a major part of the basal strata of the Hormuz Complex as deposited.

As noted above, associated sedimentary series included very widespread dolomites, most of them foetid, many of them showing sabkha type lithology, often themselves interbedded with evaporites - a coastal plain depositional facies which extended over an area many hundreds of miles across.

Additionally the associated transported masses of well-bedded colored sandstones and siltstones (Lalun and older groups) are suggestive of internal drainage-basin deposits, found at intervals in plugs from inland Iran to the south of the Gulf. Locally also there are inclusions of very thick syngenetic conglomerates, such as might reflect local contemporary faulting, as at Chah Benu in Iran.

All these features would be consistent with the model of a Proto-Ocean as described by Kinsman (1975), pointing to the Hormuz sequence as having originated on the wide rifted shelf edge of a continental margin.

Whether an ocean later developed in the region is still controversial. The main area of the Zagros continued in a shallow water facies through the Permian (coralliferous), Trias (paralic), Jurassic and early Cretaceous (shallow water limestones and evaporites), with a narrow transition to a deep water facies on the edge of a trough which - probably in Cretaceous times - developed radiolarites and ophiolites in what became the Zagros Crush Zone. As Stocklin has emphasized, the trough was very narrow, the bulk of the deep water rocks was not large, and the recurrence of rock of similar facies beyond the trough militates against the northeastern side being on a different continent.

The Zagros sea way was thus something of an accident within the original limits of the Hormuz continental shelf, and it should not be identified with a continental margin. The relationship with the Oman structure, and the identification of Tethys beyond the northern limits of Iran, are beyond the scope of this paper, but have been discussed at length in a paper by Stocklin (1984), to which the interested reader is referred.

VII. AKNOWLEDGEMENTS

The writer is deeply indebted to The British Petroleum Company, p.l.c., Abu Dhabi Petroleum Company, Abu Dhabi Marine Areas Limited, the Iraq Petroleum Company and Associates, and to Shell International Petroleum Co. Ltd. for facilitating field visits to relatively inaccessible locations, and for permitting quotation from reports by M.H. Lowson, the late E.J. White, A.J. Martin and R.D. Hawkins of BP; T.J. Harris, J.W. Halse and D.H.

Morton of IPC. The writer also owes thanks to BP in preparation of the figures, and to Miss M.L. Bransdon for once again typing a difficult manuscript.

Editorial Note: With Peter Kent's untimely demise I have completed revision of Peter's original manuscript. I hope I have done justice to Peter's work. I.L.

REFERENCES

Bockh, H. de and others (1929). Contributions to the stratigraphy and tectonics of the Iranian Ranges. In: The structure of Asia by Gregory, J.W. (ed.), Methuen & Co. Ltd., London.

Bostrom, R.C. (1985). Zonal wrenching in the Tethys orogeny. Geodynamics Res. Symp., Texas A & M Univ., April 25-26 1985, 4. pp.

British Petroleum Co. Ltd. (1964). Geological maps, columns and sections of the High Zagros of SW Iran. Int. Geol. Cong. 22, Delhi, Proc. Scale 1: 250,000.

Dominguez, J.R. (1965). Offshore fields of Qatar. Arab Petrol. Cong. 5, Cairo, Paper 57 (B-1).

Elder, S. and Grieves, K.F.C. (1965). Abu Dhabi Marine Areas geology. Internat. Cong. "Petroleum and the Sea", Monaco, Sect. 1, Paper 127.

Falcon, N.L. (1969). Problems of the relationship between surface structure and deep displacements illustrated by the Zagros Range. In: Time and place in orogeny by Kent, P.E. and others (eds.) Spec. Publ. Geol. Soc. Lond. 3: 9-22.

Furst, M. (1976). Tektonik und Diapirismus der ostlichen Zagrosketten Z. dt. Geol. Ge. 127: 183-225.

Gansser, A. (1960). Uber Schlammvulkane und Salzdome. Vjschr. Naturf. Ges. Zurich 105: 1-46.

Gray, K.W. (1950). A tectonic window in South-western Iran. Q.J. Geol. Soc. Lond. 105: 2.

Harris, T.J. and Halse, J.W. (1962). Geological report on the salt-plug features of Jebel Dhanna and related Islands. Abu Dhabi Petroleum Co. Ltd., unpublished report.

Harrison, J.V. (1930). The geology of some salt-plugs in Laristan (Southern Persia). Q. J. Geol. Soc. Lond. 84: 463-522.

Hurford, A.J., Grunau, H.R. and Stocklin, J. (1984). Fission track dating of an apatite crystal from Hormuz Island, Iran. J. Petrol. Geol 7: 365-80.

Jackson, M.P.A. and Talbot, C.J. (1986). External shapes, strain rates, and dynamics of salt structures. Bull. Geol. Soc. Amer. 97: 305-323.

Kashfi, M.S. (1983). Variations in tectonic styles in the Zagros geosyncline and their relation to the diapirism of salt in southern Iran. J. Petrol. Geol. 6: 195-206.

Kashfi, M.S. (1985). The pre-Zagros integrity of the Iranian platform. J. Petrol. Geol 8: 353-360.

Kent, P.E. (1958). Recent studies of South Persian salt plugs. Bull. Am. Ass. Petrol. Geol. 42: 2951-2972.

Kent, P.E., (1970). The salt plugs of the Persian Gulf region. Trans. Leics. Lit. & Phil. Soc. vol. LXIV: 56-88.

Kent, P.E. (1979). The emergent Hormuz salt plugs of southern Iran. J. Petrol. Geol. 2: 117-144.

Kinsman, D.J.J. (1975). Salt floors to geosynclines. Nature 255: 375-378.
Koop, W.J. and Stoneley, R. (1982). Subsidence history of the Middle East Zagros Basin, Permian to Recent. Phil. Trans. Roy. Soc. Lond. A305: 149-168.
Lees, G.M. (1927). Salzgletscher in Persien. Mitt. Geol. Ges. Wien 22: 29-34.
Morton, D.M. (1959). The geology of Oman. World Petrol. Congr. 5, New York, Proc. Sec. 1, 277-294.
al Naquib, R.M. (1963). Geology of the Arabian Peninsula, Southwestern Iraq. U.S. Geol. Surv. Prof. Paper 560-G.
al Naquib, R.M. (1970). Geology of Jebel Sanam, southern Iraq. J. Geol. Soc. Iraq 3: 9-36.
Owen, R.H.S. and Nasr, S.N. (1958). Stratigraphy of the Kuwait-Basra area. in Habitat of Oil by Weeks, L.G. (ed.) pp. 1252-1278. A.A.P.G. Tulsa.
Pilgrim, G.E. (1908). The geology of the Persian Gulf and the adjoining portions of Persia and Arabia. Mem. Geol. Surv. India 34: 1-177.
Stocklin, J. (1961). Lagoonal formations and salt domes in East Iran. Bull. Petroleum Inst. Teheran, Iran 3: 29-46.
Stocklin, J. (1968a). Structural history and tectonics of Iran: a review. Bull. Am. Ass. Petrol. Geol. 52: 1229-1258.
Stocklin, J. (1968b). Salt deposits of the Middle East. Spec. Pap. Geol. Soc. Am. 88: 157-181.
Stocklin, J. (1974). Possible ancient continental margins in Iran. In: The Geology of Continental Margins by Burk, C.A. (ed.) pp. 873-887, Springer Verlag, New York.
Stocklin, J. (1984). Orogeny and Tethys evolution in the Middle East. An appraisal of current concepts. Int. Geol. Congr. 27, Moscow, Coll. 05-Tectonics of Asia, Reports vol. 5, pp. 65-84.
Talbot, C.J. (1979). Fold trains in a glacier of salt in southern Iran. Structural Geology 1: 5-18.
Tschopp, R.H. (1967a). The general geology of Oman. World Petrol. Congr. 7, Mexico, Proc. vol. 2, pp. 231-241.
Tschopp, R.H. (1967b). Development of the Fahud Field. World Petrol. Congr. 7, Mexico, Proc. vol. 2, pp. 243-250.
Walther, H. (1968). The genesis of the iron ore of the Hormuz Series near Bandar Abbas (Cambrian, SE Iran). Int. Geol. Congr. 23, Prague, Sect. 7.
White, E.J. and Lowson, M.H. (1927). Preliminary report on the Gulf islands reconnaissance 1927. (Anglo-Iranian Oil Co. Ltd. unpublished report).

DEFORMATION WITHIN SALT BODIES

Gerhard Richter-Bernburg
Haarstraße 8
D3000 Hannover 1
Germany

I. INTRODUCTION

What we call "salt formation" usually describes a considerable occurrence of salt consisting mainly of NaCl, mixed or interbedded with chlorides and sulfates of K, Mg, Ca and other, less important, elements, and alternating or interspersed with anhydrite and/or mudstone. These rocks constitute salt bodies, the shapes of which generally differ fundamentally from other sedimentary formations, primarily because they are not usually competent members of a normal stratigraphic sequence. Showing extreme diversification, salt formations must be resolved mostly on the base of drilling or geophysical measurements. Even then the shape of a salt body raises problems comparable to viewing a sculpture of a person, the physique being perfectly reproduced but the interior organs remaining unknown. Even a medical doctor would not know the position of the different organs without previous anatomical studies. A similar problem arises when the interior structure of a salt body comes into question. Nevertheless, there is a fundamental difference: The position of the main organs of a being, their sizes, functions, etc., are basically the same in all individuals of the same species. In comparison, the internal structure of a salt body is individual, most often showing little likeness with the internal structure of a comparable salt body even though both may have very similar external appearances.

This paper will present observations from exposures in mines from many countries, some of which are difficult to access at the present day, in order to illustrate the diversity of internal structural responses which can occur.

II. PHYSICAL REASONS FOR SALT DEFORMATION

The shape of a salt body is caused by the physical properties of the salt rocks. The salt reaction to any mechanical force resembles more that of a liquid than a solid. However, the degree of deformation depends upon the physical factors: the overburden load (p) and the salt temperature (T). Generally, the rocks overlying a salt formation have a higher specific weight than the underlying rock salt (Lachmann, 1911). Experiments have shown that with a load of 400 to 600 kilopounds the halite ceases to be a solid rock and passes into a state of plasticity or viscoplasticity. The higher the temperature, the more the salt becomes mobile (Gussow, 1966). Finally the product (p.T) controls the deformation (Dreyer, 1974; Langer, 1978; Langer and Kern, 1979). Accordingly rock salt will normally be mobile at sub-surface depths of about 1500-2000 m.

However, there are also examples where salt creeps even under more "normal" conditions. In figures 1-2, for which the overburden, composed of Upper-Mesozoic sandstones and limestones, was a maximum of 400 m thick, we see that despite the small load and low temperature, so that the product, pT, is relatively small, the salt nevertheless moved into the holes at the base of the covering anhydrite cap (probably formed during Lower Cretaceous). While the movement of the salt is clear the reason for the motion is not yet unequivocally determined. We bring up

Figure 1. Micro-diapir, produced by creeping activity of the halite member of Zechstein 2. The very gently dipping white halite which is interbedded with anhydrite flakes is cut nearly horizontally by sub-erosion and is covered by solid residual gypsum, the normal cap rock. The even basal boundary of the latter contains conic holes (of unknown origin) of 0.5 - 1.0 m height and about 50-60 cm diameter at the base. All of these holes are filled from below with rock salt. The halite forms a classic salt dome, the substance of which is augmented in the center (the single beds are thicker there), and is squeezed at the flanks. After production of the salt and after deposition of about 1200 m of Mesozoic sediments tectonic movements occurred, and the greatest part of the overlying formations were eroded at the surface. The cap rock had probably been formed in Cretaceous time, by subsurface dissolution. It is remarkable that since that time, i.e. since about 80-100 My, no events fundamentally different from today were felt. The movement of the halite happened under more or less recent conditions at about 200 m depth below surface.

this point here because it demonstrates that our endeavors to understand quantitatively the complete dynamics of geological systems involving salt are far from complete.

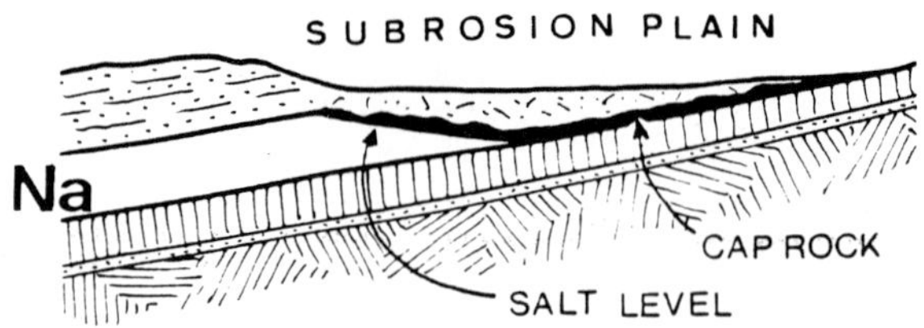

Figure 2. Supplement to fig. 1, showing the geologic position of the photo: subcrop of a salt formation (Na) below the cap rock (black). The left arrow marks the position of the feature shown in fig. 1.

The example shows that the dominant physical property of the salt is its tendency to move like a viscous fluid, a commonality of response which we have to bear in mind in all cases. We must strive to understand not only the irregular varimorphic geological shapes of observed salt masses (Kupfer, 1968) but also the extremely high deformations inside of salt bodies, which are in obvious contrast to the structures of the surrounding solid rock environments.

III. GEOLOGICAL CONDITIONS FOR SALT MOTION

We proceed from the fact that all rock salt bodies below about 1500 m (and in some instances even shallower) tend to move, the deeper parts of them more than the shallower parts because of the greater (p.T) product. The salt will try to follow the path of smallest resistance. Such paths may be openings in the overlying formations such as tension faults and graben structures, or the cores of anticlines. Thus, the salt movement may be initiated by local or regional tectonic impulses, even if these are quite weak. Even slow large-scale bending of the crust produces differential tensions. Consequently salt movement may be initiated even while the strain field is still relatively small. And "the salt goes its own ways" (Stille, 1917; Gignoux, 1930), dependent on the circumstances of the geological environment such as time-dependent evolution of folds or faults where places of lower stress abound. Thus salt will be accumulated at low stress sites by in-flow, whereas on the flanks of the folds, for instance, the normal salt thickness will be reduced by withdrawal.

A. Salt Accumulations

Salt accumulations in axial arches of anticlines are known to occur on a large scale in the Pennsylvanian formation of the Paradox Basin in Colorado and Utah (Shoemaker et al., 1958; Jones, 1959; Hite, 1960; Hite and Cater, 1972), in the folded range of the Zagros Mountains in Iran (Neff, 1960; Trusheim, 1974), and in other areas. From the classic anticline at Stassfurt, central Germany, which has been known for many decades it was, perhaps, first recognized that an accommodation of volume occurs in faults with dilatation effect or in graben systems. Many salt domes in Germany are accumulations sitting upon such faults. Also wall-like salt masses in the subsurface are known which follow the tension faults in geotectonic directions NNE as

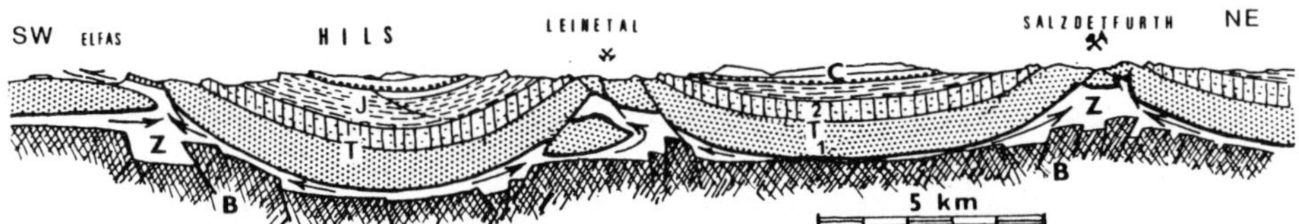

Figure 3. Smooth folds in the Saxonian System of NW-Germany. Weser-Leine-Hills SW of Hannover. The salt formations of Zechstein (Z), following the gravity law, accentuate the gentle anticlines. The salt tendency to rise as high as possible causes formation piercement or even diapirism. Salt in-flows to the cores of the folds where the stress is lowest, and produces folds within the salt formation which are independent of the shape of the salt body. B = pre-Permian basement, Z = Zechstein, mostly salt, T = Triassic, J = Jurassic, C = Cretaceous.

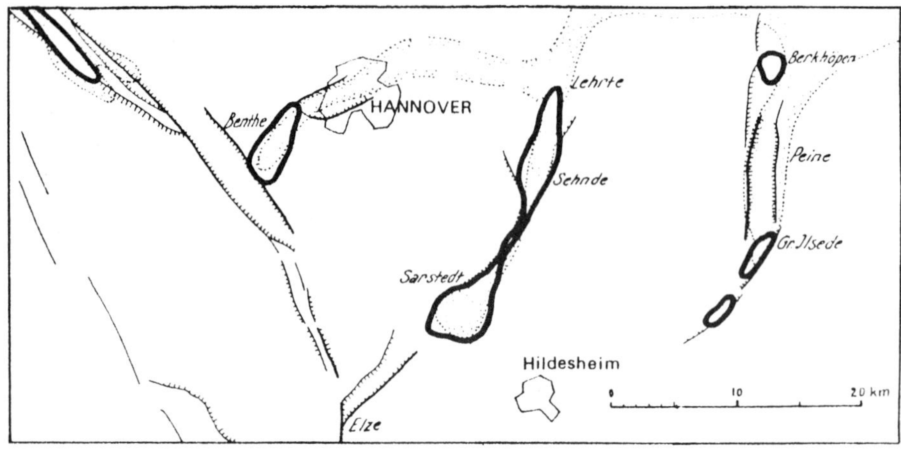

Figure 4. Salt diapirs, initiated by tension faults and graben structures in the strike directions SSW - NNE and NW - SE. Area: Hannover/NW-Germany. Dotted: extent of marine Tertiary.

well as WNW (see figs. 3 and 4 and also Richter-Bernburg (1968, 1972, 1977), Jaritz, (1973)). Comparable occurrences exist along the Jordan cleft near the Dead Sea and elsewhere.

The augmentation of salt in such tectonically "prepared" zones of lower stress can take place in different ways. First, the thickening of salt beds through granular transformation of the NaCl by internal dissolution and recrystallization can occur. Such an effect has been seen in great detail: halite beds of a normal thickness of about 10 cm increase in a short distance to more than 50 cm in the turn of a fold, whereas the salt thins to less than 1 cm on the flanks of the fold as seen in figure 5.

There are also places in tabular salt formations where there is no apparent reason for augmentation but where, nevertheless, an accumulation of many meters thickness happens by many-fold repetition of isoclinal recumbent folding as shown in figure 7.

Figure 5. Particular anticline in white, regular-bedded halite inside of a large salt fold. Halite of Zechstein 3 in mine "Asse" near Braunschweig/NW-Germany. The wall is about 10 m high.

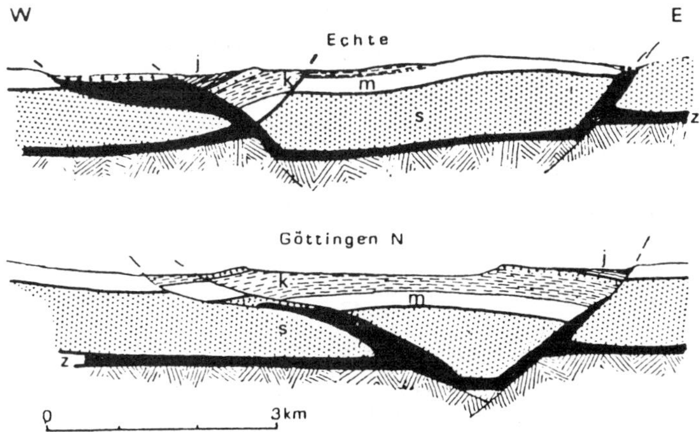

Figure 6. The so-called "Leinetal-Graben" near Gottingen, Germany. The fractures in the overlying formations (s = Buntsandstein, m = Muschelkalk, k = Keuper, j = Jurassic), produced by tectonic tension, enable the Zechstein salt (= z, black) to rise into the "open" gaps and to move upward, because of the salt's lower specific density. In this way, huge salt accumulations form along tension faults.

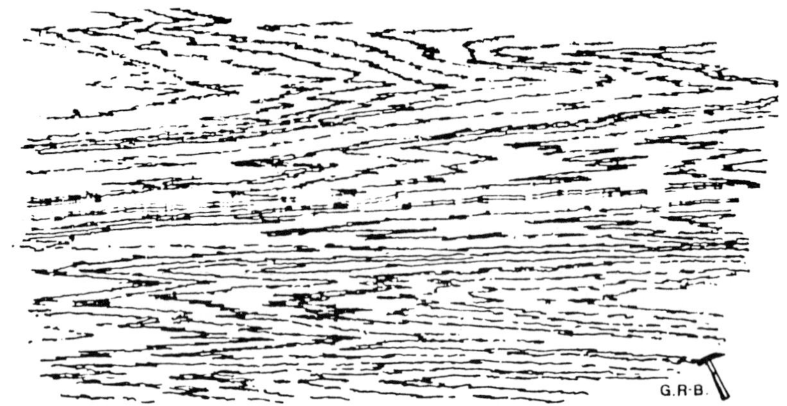

Figure 7. Zigzag folds with recumbent axes in white halite of the Salina formation which is, on average, horizontal. "Whiskey" salt mine near Cleveland/Ohio, USA.

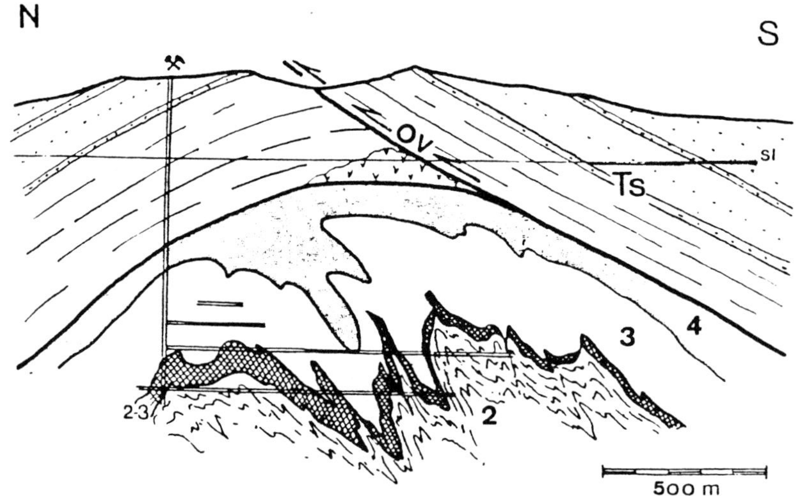

Figure 8. The "Salzdefurth-Anticline" S of Hannover/NW-Germany, showing slight overthrust (OV) in Buntsandstein(Ts) but specific folding of the salt formations in Upper Permian: 2, 3 and 4 = salt of Zechstein 2; Z3 and Z4 respectively; 2-3 = potash bed of Zechstein 2 + anhydrite of Z3. sl = sea level. Shaft and potash mining galleries.

Usually all such accumulations are produced by folding on a regional scale. In anticlinal structures the axes of the salt folds are parallel, or subparallel, with the main structural elements (fig. 8). Salt masses which are initiated by tectonic dilatation effects flow in the direction of tension, i.e. perpendicular to the strike of the main fault or the graben. Large amounts of salt may be accumulated in this way. When they contain minable salts, in the form of clean halite or beds of potash salt, the mine provide an opportunity to study the interior deformations in detail. Profiles like those shown in figs. 8 and 9 are typical patterns.

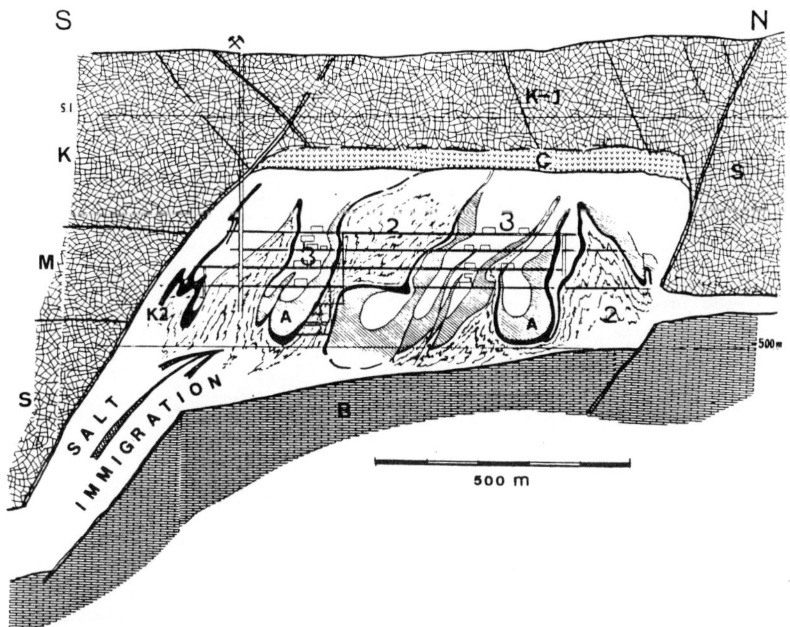

Figure 9. Cross section through a diapiric salt body. Former shaft "Bartensleben" E of Braunschweig, Germany. The salt rose up on a primary tectonic structure, the so-called "Allertal-Graben", a WNW - ESE striking fault system, caused by tension in the upper crust. While Keuper (Upper Triassic) and Jurassic (= K+J) sediments tried to fill the gap from above, the salts of Zechstein in-flow moved quickly into the opening from below. Because of their lower specific weight these salt rocks accumulate by in-flow to produce a system of steep breakers. B = pre-Permian basement, 2 and 3 = salt formations of Zechstein 2 and Zechstein 3, K2 = potash bed "Stass-furt" of Zechstein 2 (mining levels in different depths), A = anhydrite of Z3, C = Cap Rock, S = Buntsandstein, M = Muschelkalk, K = Keuper (all Triassic), sl = sea level.

B. Diapiric Structures

It may happen that during the time of salt movement particular locales are extremely suitable as low stress sites for salt in-flow. Continued flow towards these locations increases the salt concentration, often to the point where the salt will pierce the overlying sediments and a diapir will rise, occasionally reaching the sedimentary surface. This is the origin of the particular shape of salt bodies, usually called "salt domes", with roughly circular horizontal cross-sections. Usually these diapirs occur in groups of many, sometimes hundreds, of individual domes as seen in the Texax Gulf Coast area, in NW-Germany, between the Ural Mts. and the Caspian Sea, in the arctic NW of Canada, in Iran and elsewhere.

When salt flows towards the places of lowest stress we should expect that the strike of the folds is perpendicular to the movement. It might, therefore, be supposed that the folds in a diapir will form a circle, contrary to observation. In fact the salt formations are deformed into a bundle of folds with more or less vertical axes. This type of deformation is very common in all salt diapirs as far as we know. We call this form "drape folds" or "curtain folds" (in German "Kulissen-Falten"). The recognition of these structures is difficult because they can be observed only in mines. There it is possible to see elaborate horizontal sections (Balik, 1949, 1953, Muehlberger, 1962, Kupfer, 1962). In Germany, the stratigraphic sequence has been known since the beginning of the present century, and therefore the tectonic deformation could be interpreted early on with great reliability. Three dimensional constructions are possible due to several mining levels in the same diapir. Thus, the dips of the near vertical fold axes give rather precise information about the whole salt structure (see figs. 10 through 19).

We generally observe a fundamental predominance of the older parts of the salt formation (for instance Zechstein 2 salt) in the centre of the diapir and, at more lateral positions, the younger salt sequences (for instance Zechstein 3-4 salts). The reason seems to be that the older salts were originally several hundreds of meters deeper in position than the younger salts and therefore experienced higher temperature and pressure, producing a greater mobility and, therefore, a faster diapiric movement. Thus, the older salts usually occupy the central parts of diapirs.

A comparison of the diapiric structures of German and Iranian salt bodies is also educational. Iranian salt diapirs are best observed from the air, whereas in Germany only subsurface mapping in mines for potash and rocksalt provides results (figs. 12-14).

In both regions we know precisely the primary sequence and have the opportunity to draw detailed conclusions about the salt movements. In contrast the Texas Gulf Coast area does not offer enough internal salt data for decisions to be made between syncline and anticline in the curtain-folded salt, as shown in figs. 16 and 17.

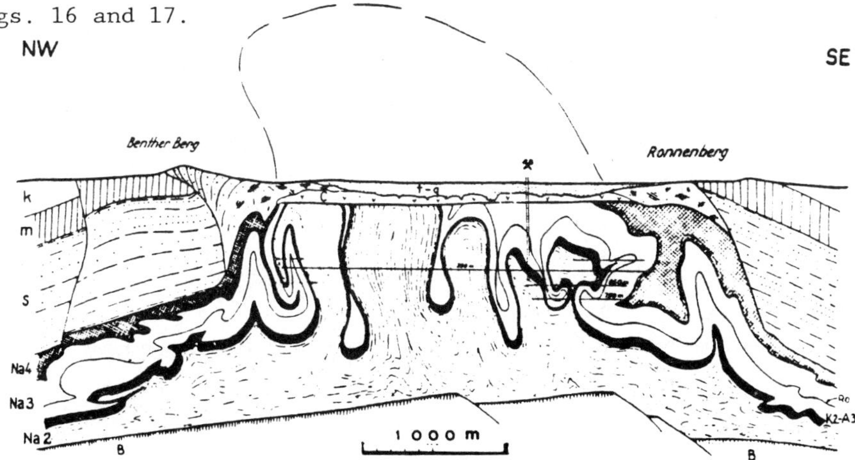

Figure 10. Cross section through the diapir of Benthe/Hannover. Salt in-flow into a primary tension structure. Accumulation of three salt formations Na 2 - Na 3 - Na 4 of three stratigraphic sections of Zechstein. B = pre-Permian basement, Triassic with s = Buntsandsein, m = Muschelkalk, k = Keuper, C = Cap Rock. Shaft "Ronnenberg".

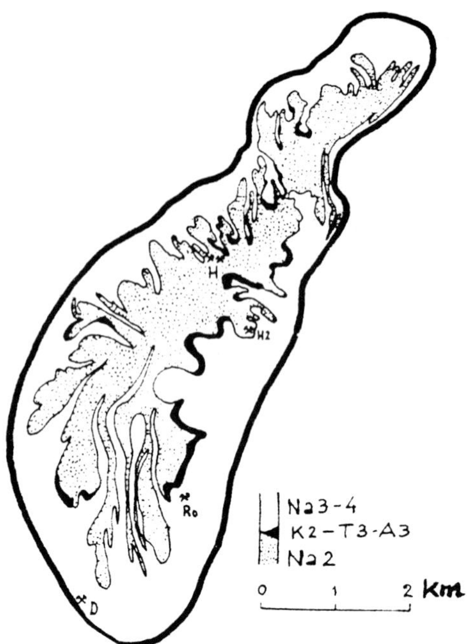

Figure 11. Diapir of Benthe near Hannover/Germany. Horizontal section at about 650 m depth. The central part of the salt body consists of the oldest salt unit, Zechstein 2. In-flow from all sides produced folds with vertical or subvertical axes, so-called curtain folds. See also figs. 12, 13, 15.

Figure 12. Sylvinite bed "Ronnenberg"; In earlier times exploited in the potash mine Ronnenberg near Hannover. The figure shows a plan of a gallery at 750 m depth, mapped in detail by the author for illustrating the intensive curtain-folding.

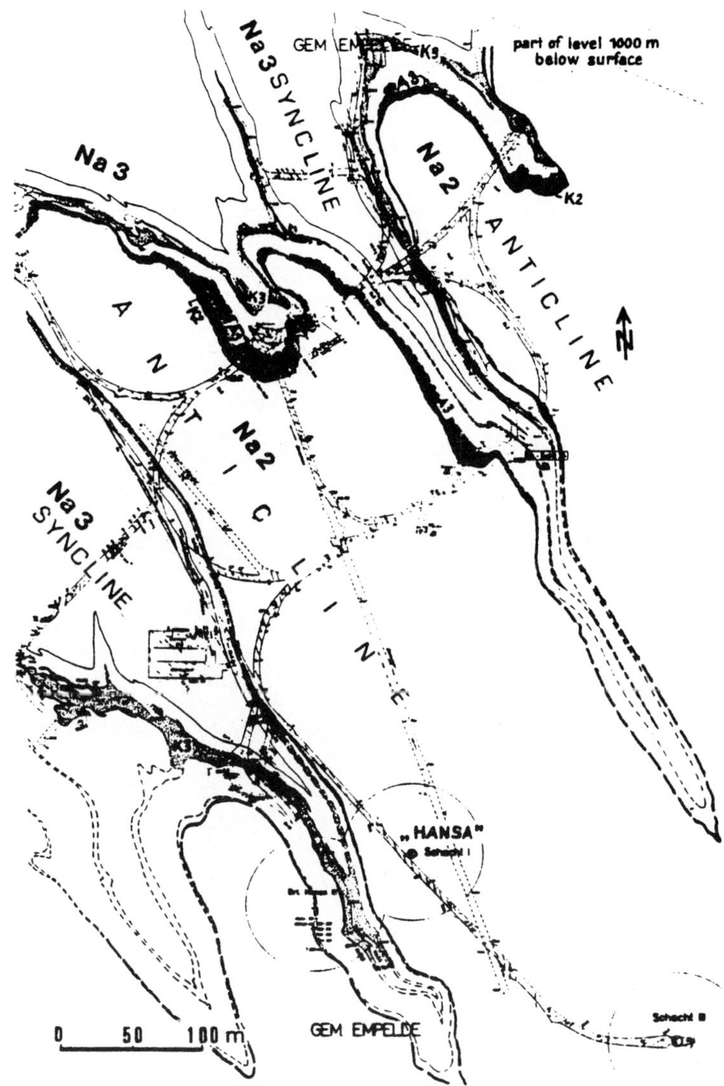

Figure 13. Curtain folds in horizontal section through a salt dome about 1000 m below surface. Former potash mine "Hansa" near Hannover/Germany. The same potash beds could be exploited in vertical position between 400 m and more than 1000 m depth. The dip of the folding axes is also subvertical through this distance. Na2 (Na3) = halite member of Zechstein 2 (Z3), K2 = potash bed "Stassfurt", A3 = main anhydrite of Zechstein 3, K3 = potash bed "Ronnenberg".

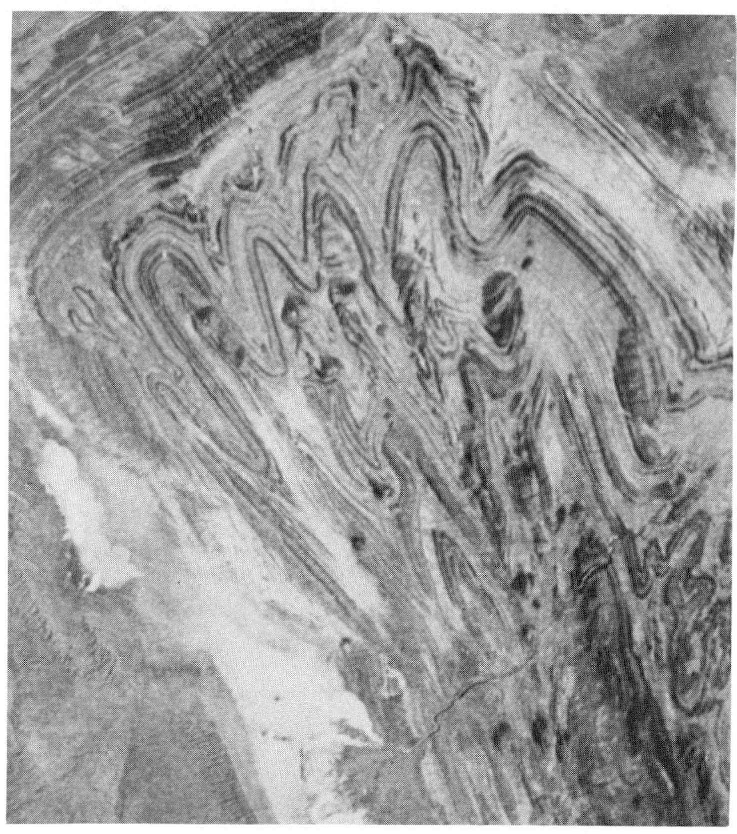

Figure 14. Curtain folds (similar to those in fig. 13), exposed in a diapir of central Iran. Air photo. The salt formation here is of Miocene Age, and is cut by young erosional events.

Investigations of this detailed nature confirm that the salt movement is similar to that of a viscous liquid. As the observed deformations represent a present day "snapshot" of salt flow, we can use them to reconstruct salt movements, and the stresses which induced these movements, occasionally even in detail. For instance, from the observed swirls of the curtain folds on a relatively small scale in fig. 17 we can determine that, as well as planar flow, there was also a rotational flow component. Similar features can be seen on a larger scale in figs. 18-19.

Figure 15. Horizontal section through a part of a salt dome in NW Germany. The distribution of the individual stratigraphic members was mapped by visiting galleries and investigating horizontal test wells in the mining level 750 m below surface. The dip of beds and the curtain fold axes are approximately vertical.

C. Diapirs in Central Iran

From air photos of the central Iranian salt domes we can determine different aspects of salt flow; however the interpretation of air photos alone may sometimes be misleading. The air photo of the Diapir Kuh-i-Namak (= mound of salt) near Qum shows structures which closely resemble the salt glaciers of South Iran, described by Lees (1927), Dunnington (1962), Kent (1970) and Walther (1972). More detailed field investigations reveal that all the salt beds stand vertically or subvertically, and that they belong to the curtain fold type, while the air photo of Kuh-i-Namak at Qum does not capture this more detailed information (fig. 20).

In the Semnan area, about 100 km East of Teheran, a whole family of diapirs occurs in the Mio-Pliocene, which is about 3000-5000 m thick (fig. 21-23). Two saline formations exist there: first, a mass of fairly clean halite, interbedded with some grey anhydrite intercalations; second, a series of thick salt beds in alternation with redbrown mudstone or pelite, most likely an old sabkha sediment. The first formation, of considerable thickness, is probably Eocene; the second, perhaps 1000-2000 m thick, is probably Miocene (Stocklin, 1962; 1968). A group of diapirs, with diameters of around 5 km, and composed of these two salt formations of different ages, pierced the overlying sediments from a depth of several thousand meters - like molasses flowing through a griddle from below as seen in fig. 21.

Figure 16. A Tertiary diapir, horizontally cut by erosion, in central Iran. It consists of two salt units: a) clean halite of Eocene age (right) in the core of the diapir, and b) halite which is interbedded with red claystone of Miocene age. The curtain folds are obvious.

Figure 17. Curtain fold in a USA salt dome. Mine "Grand Saline", Tex. The "ceiling" shows a fold (syncline?, anticline?) with subvertical axis.

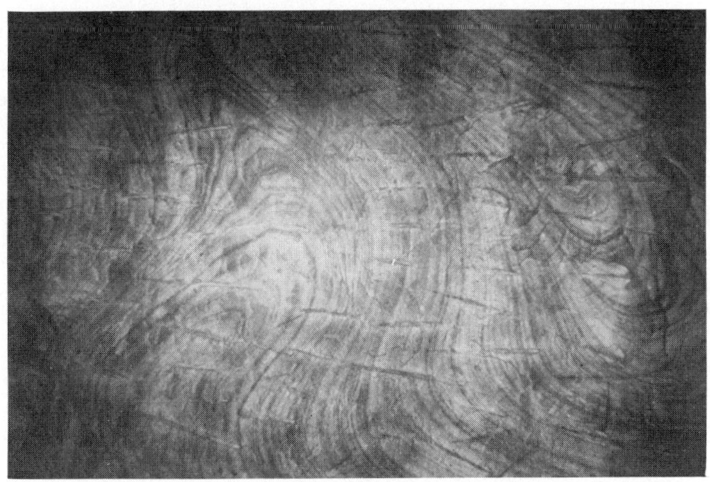

Figure 18. Curtain folds, rotating round a vertical axes. The ceiling of a huge salt mining room (about 15 x 20 m2), seen from below. Weeks Island, Louisiana.

It is remarkable that the general sequence of both units is preserved during a vertical rise over several thousand meters to the surface. The younger salt formation reacted in a more competent fashion, whereas the older salt is assumed to have undergone more turbulent viscous flow. This older salt is also found near the centre of the diapir - similar to the older salts in the German salt domes. The older salt also occupies peripheral positions on the outer side of the salt body as seen from the central axis of the master syncline (fig. 24). Thus we determine that the migration of this deeper and older salt unit had a preferential upward trend as it flowed.

Comparable patterns have been observed in NW-Germany where, around Hannover, there is also a trend for the salt to move from relatively deep positions within the basin to situations where accumulation is preferable (fig. 25).

The detailed structures of the Semnan diapirs are difficult to investigate because of the inaccessibility in the field. Nevertheless, in some places it is clear that the salt body is overturned towards its periphery so that the central older salt formation is moved beyond the younger salt stock (fig. 26).

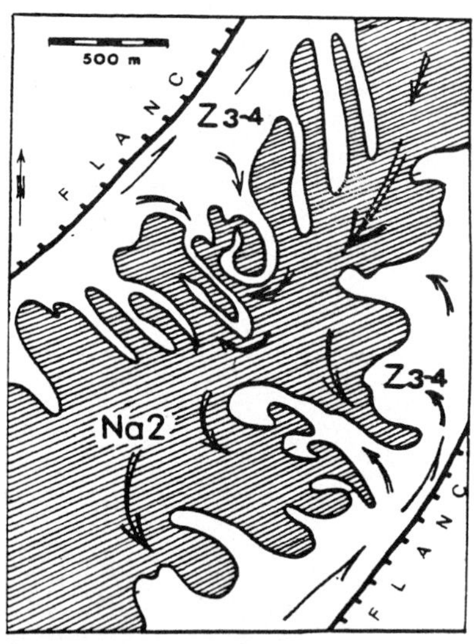

Figure 19. Beginning whirls by differential piercing speed of older and younger parts of the salt formation. Salt dome Benthe/Hannover, NW Germany. Boundary Na2 (Zechstein 2)/Zechstein 3-4. Horizontal section in 600-800 m depth.

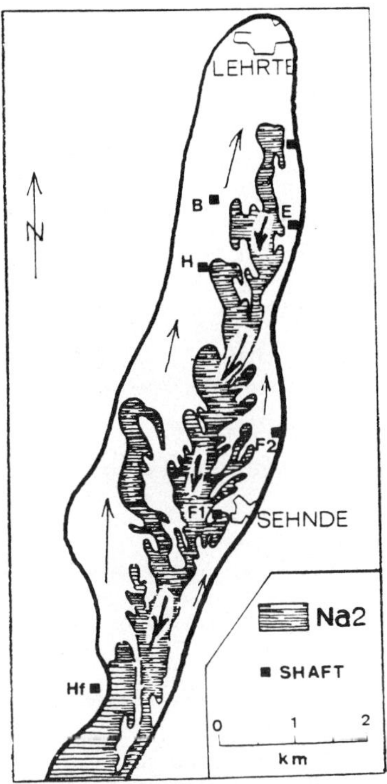

Figure 20. Echelon folds within a diapir, resulting from the different rising speed of the older (dark) and younger (white) parts of the Zechstein salt. Salt dome Sarstedt - Lehrte, NW Germany. Shafts B = Bergmanssegen, H = Hugo, E = Erichssegen, F = Friedrichshall, Hf = Hohenfels. Arrows show the relative movement direction.

Comparable patterns are observed in NW-Germany. For example the salt dome of Haenigsen, near Celle, is distinguished by extrusion of Zechstein-2 salt which overthrusts all the younger salt units of the higher Zechstein. In a pattern of geometrical behavior similar to that of a tectonic nappe, the older salt has a larger lateral extent at shallow depths above the younger rock salt in the "Riedel" mine (fig. 27).

We conclude that, generally, the mobility of the lower (older) salt members, which have been more intensively heated at greater depth than the younger salt, moved faster from their original bedding to low stress accumulation sites than did the

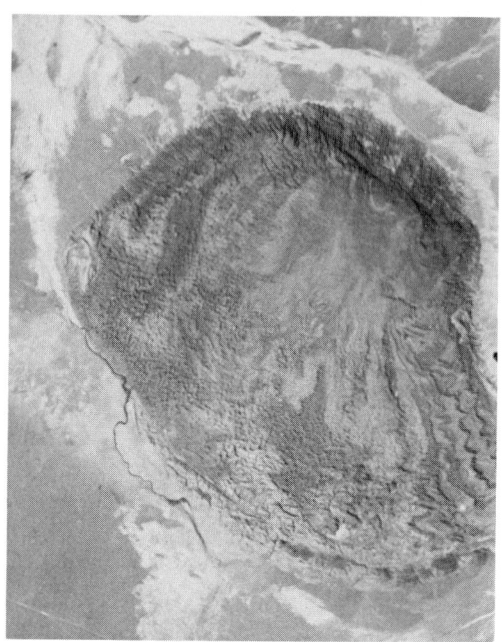

Figure 21. Curtain folds with subvertical axis position (not salt glaciers!). Kuh-i-Namak/Qum, central Iran. Vertical air photo.

younger, shallower, salt. Thus, the deformation of such kinds of salt bodies usually demonstrates an "overflowing" of older salt units on top of the younger members.

D. Tectonic Force Effects

While it is sufficient to have only a local heating at depth in order to promote salt flow, nevertheless the most fundamental deformation takes place if tectonic forces act on salt. We can observe the consequences of such effects in several places.

An extremely intensive destruction of salt is seen in the foothills of the Himalaya Mts., near the city of Mandi. There, a saline formation of Permian or Carboniferous age is situated at the main overthrust fault along the southern rim of the Himalaya range. Consequently, the salt's original structure is completely destroyed. Including marly and anhydrite intercalations, the salt is transformed into a fairly fine-grained mylonite. No original structural information is identifiable in this nearly homogeneous mass of impure salt, which is primitively mined in open air workings and used after solution refining in pans.

Solution refining of salt has also been used for centuries in the Austrian Alps (Salzkammergut). Here, too, the Permo-Triassic salt formation is situated at a tectonic zone of the Kalkalpen, i.e. at the boundary of two tectonic units, and functions as a lubricant for the northward thrusting nappes. The saline formation is accumulated in protected and preserved spaces during alpine orogenesis, and today occurs as more or less rounded forms. In detail the structures are very complicated (Fig. 28).

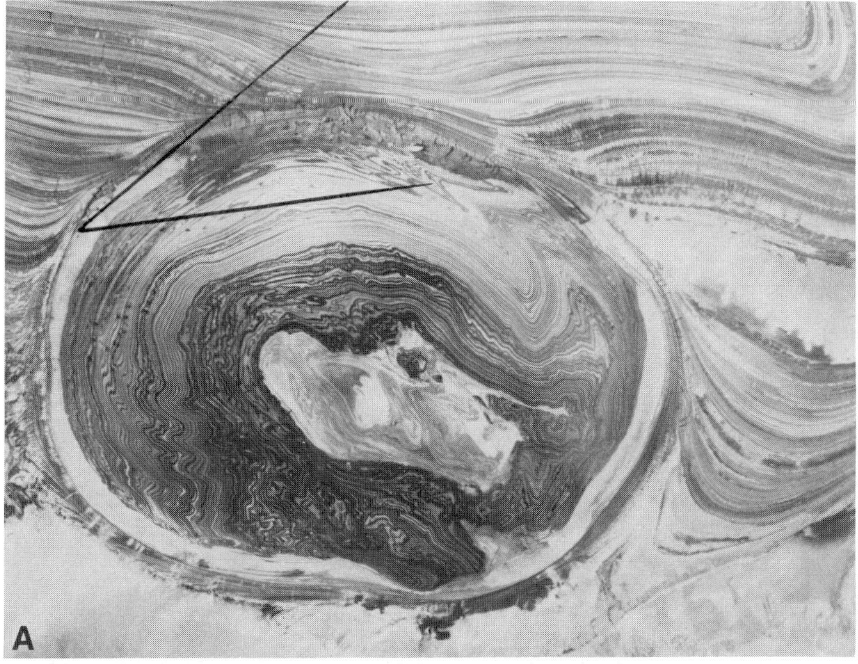

Figure 23 a and b. The "bull's eye" salt dome "S" of the Semnan diapirs (see Fig. 22). The relatively clean halite in the center belongs to the Eocene, the surrounding rocks are alternating salt and brown-red pelite of Miocene age. Intensive folding is mainly

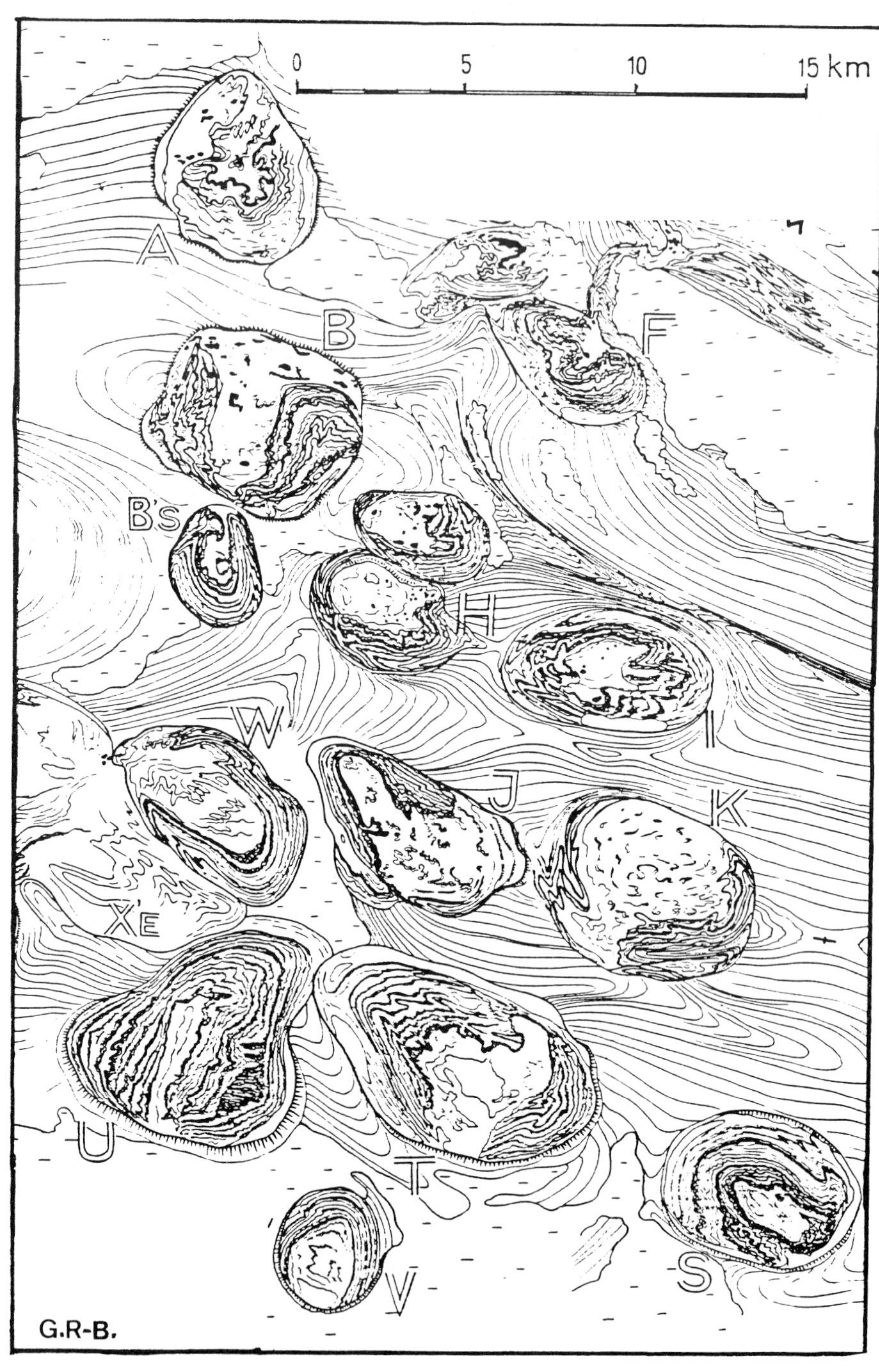
0
5
10
15 km
A
B
F
B's
H
I
W
J
K
X'E
U
T
V
S
G.R-B.

Figure 23b (continued). close to the rim. a) Vertical air photo, b) helicopter photo by the author.

IV. DISHARMONIC STRUCTURES

Another important behavior can be observed in the alpine salt formation, which illustrates the very different reactions of salt to high tectonic stress. Whereas the salt shows flow-like deformation characteristics, the interbedded anhydrite and

Figure 22. Salt diapirs in the area of Semnan, central Iran, about 100 km East of Teheran. Interpretation of vertical air photos by the author. Contours show the competent folds of the several thousand meter thick series of Mio-Pliocene. Inside the diapirs the salt formation of Eocene (white) is surrounded by Miocene salt which is mixed up with pelitic intercalations.

dolomitic mudstones are broken into pieces which "swim" in the halite and give the impression of being passively transported in the moving salt mass.

This effect can also be seen in saline formations at places of low tectonic activity, i.e. in deposits which have not been subjected to any particularly large compression or tension influence. It is often just the deformation of the "hard rock" intercalations which shows what happened to the salt. For instance, a weak dilatation will completely disappear within a normal halite sequence. Any salt flow will not be recorded.

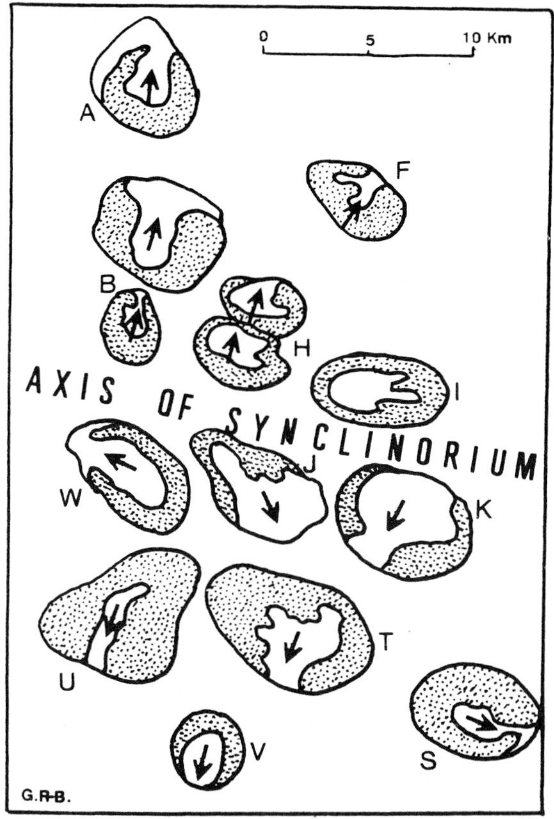

Figure 24. Tertiary diapirs (A - W) in the Semnan area, central Iran. See also fig. 22. Within the diapirs' bodies: white = Eocene salt; dotted = Miocene salt formation. The arrows mark the main movement direction of the more mobile uplifting, older, salt mass which pierces the younger salt.

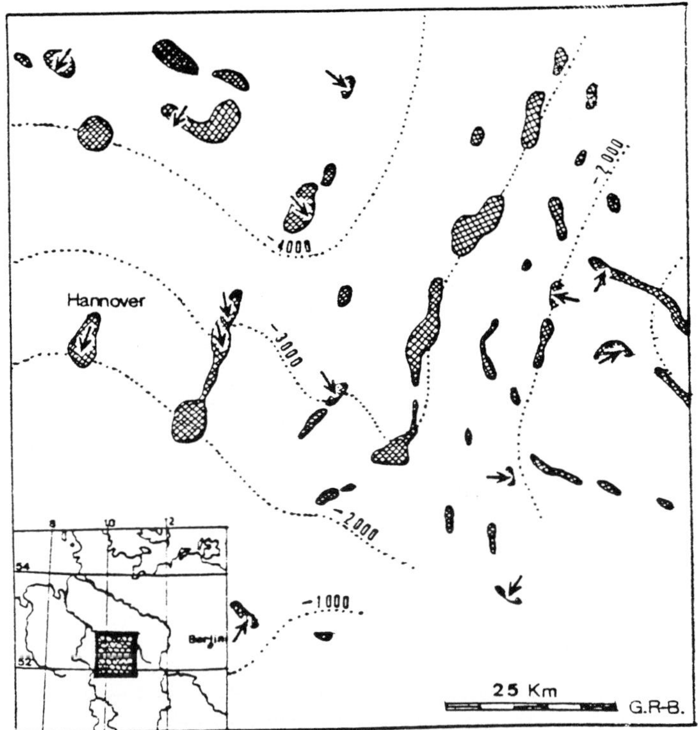

Figure 25. Some salt domes in NW Germany. The arrows show the rising tendency of the diapiric salt, the lower (= older) part of which moves more intensely because of its lower (= warmer) position and its consequent higher mobility. The main direction of motion is from originally deeper position upwards, recognizable from the isobaths of the salt base (dotted curves).

However, when there are intercalations of anhydrite beds, or clay strata, the salt movement becomes evident (fig. 29). We call these phenomena "boudinage", and they are observed on almost every spatial scale.

In the German Zechstein salt formations the interbedded "main anhydrite" (about 50 m thick) of cycle 3 is generally torn asunder by salt movements, and drifts within the flowing salt as giant blocks greater than house size (see e.g. fig. 9). On the other hand, some minerals and salt rocks are more mobile, and

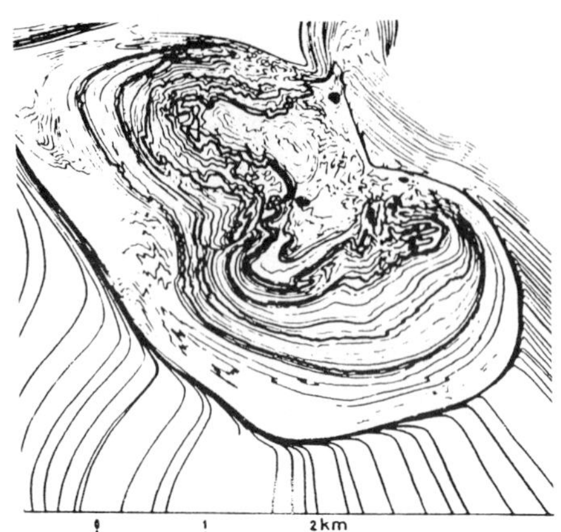

Figure 26. Diapir "F" in the Semnan area, central Iran; see Fig. 22. The outer part of the body is built up by Miocene salt, the bright center consists of cleaner Eocene salt. This partly overlaps the younger member like a volcanic extrusion.

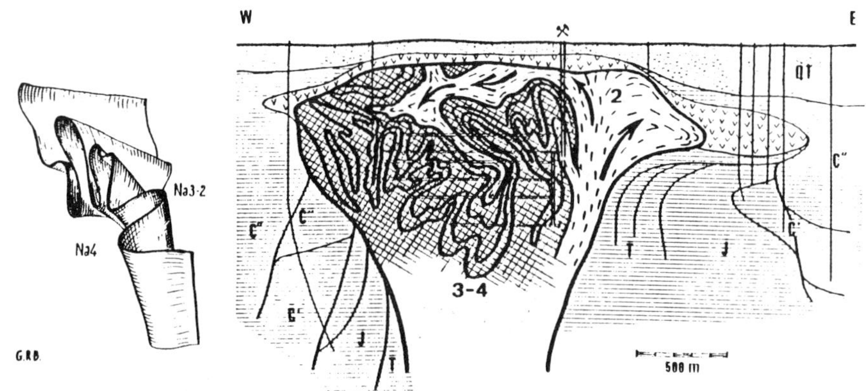

Figure 27. Extrusion of the older part of the Zechstein salt formation (2). This salt, which lies originally deeper and is therefore warmer and more mobile, first overtakes and then overwhelms the younger units (3-4). The axes of the curtain folds are vertical in the stem of the diapir, they are overturned to subhorizontal position in its overhang. (The diagram on the left shows a real - not schematic - construction). Salt dome "Riedel" near Celle/NW Germany.

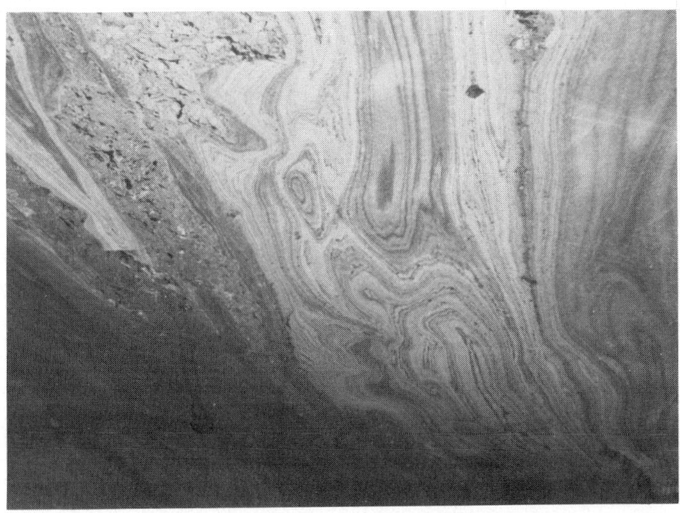

Figure 28. Different type of reaction inside a tectonically disfigured salt formation. The halite has been deformed like a viscous fluid, the intercalated anhydrite is broken into angular pieces. Ceiling of a solution mining room in the Permo-Triassic salt formation of the Alps, Salzkammergut, Hallein/Austria.

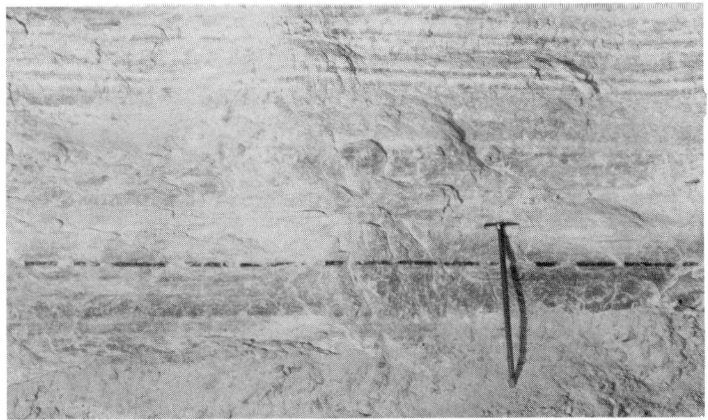

Figure 29. Originally continuous anhydritic pelite within white halite of Zechstein 1. The clay bed is rent to pieces by horizontal creeping movements, not visible in the halite itself. Salt mine "Borth"/Rhein, Germany.

flexible, than halite. Thus when the halite is normally folded, 9magnesium sulfate (kieserite) can be extremely pleated and deformed to produce wrinkles (fig. 30).

A particular type of halite fold appears in some places. When the bedding in the salt is very marked by intercalations of mobile and lubricant clay material, the halite folds easily in an accordian-like manner, sometimes closely spaced, in other places more widely separated (Richter-Bernburg, 1955, pl. 18; Roth, 1955, pl. 21; Ramberti, 1980) (figs. 4, 7, 9, 10).

V. CARNALLITE DEFORMATION

There are many disharmonic movements within a displaced salt formation. A very common phenomenon in potash salt seams is the different kind of motion of carnallitic salt and sylvinitic halite. In the German Zechstein 1 and Zechstein 2 this is seen in all areas where the minable potash bed shows various facies.

Figure 30. Various kind of deformation: An evenly bent threefold bed of halite (dark grey) is overlain by some beds of Kieserite ($MgSO4.H2$), white. These formations are haphazardly mixed although affected by the same stress. The red carnallitic matrix looks grey here. Potash bad "Stassfurt" of Zechstein 2. Mine "Solvayhall", Bernburg/central Germany.

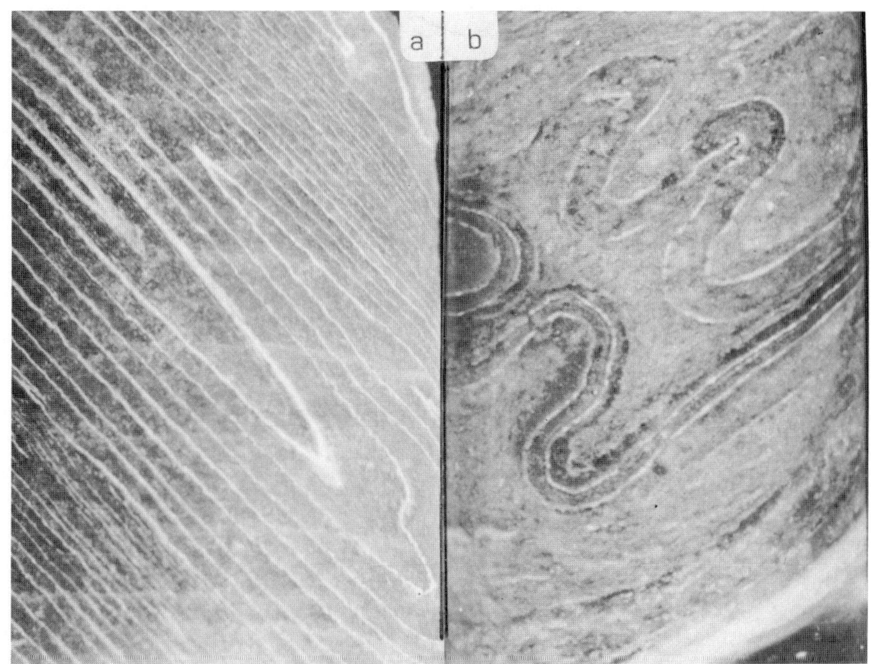

Figure 31. Creep fold within the potash bed "Stassfurt" (Zechstein 2). This member consists of facies of sylvinitic and carnallitic salt in laterally alternating facies. It usually happens that the carnallitite is overwhelmed by an overturned fold of "hartsalz" (sylvite + halite + kieserite). White beds = kieserite. a) The photo shows creep folding in about 12 m high mining room in mine "Rossleben"/ SE-Harz, central Germany. b) Sketch of the same. 1 = "hartsalz", 2 = carnallitite, M = marker bed of halite.

A lateral transition of "hartsalz" (= sylvite + halite ± Mg/Ca-sulfates) to carnallitite (= carnallite + kieserite + halite) is usual. If there is a little salt movement then the hartsalz-facies is, in all observed cases, thrust over the carnallitic facies (fig. 31). All folds show this kind of overlapping, but the carnallite never overthrusts the hartsalz-facies (fig. 32).

In some cases, it seems probable that a secondary metamorphosis of the potash seam may be responsible for the diversification of the facies (Borchert, 1959, 1964). However, the typical shape of deformation, as described above, is incompatible with this presumption because the overthrusting hartsalt above carnallitite is always connected with salt movements.

However, the cause for this more or less horizontal movement may not always be a movement on a larger scale. Arguments can be advanced for a very early event, e.g. for slumping or gliding during the deposition of the potash bed itself. Such arguments would have the hartsalt, a sediment of shallower water, creeping shortly after its precipitation towards the deeper water where the carnallite was more recently deposited. A related behavior is the typical deformation when halite beds are intercalated within a massive mainly carnallitic rock. Then, we have the possibility to discern the type of motion from the observed halite deformation. If the halite is normally folded but the carnallitic rock is unaffected we may then infer the direction of movement (figs. 33, 34, 35, 36). However, in pursuing this idea, in the mine observations we often observe that the halite is

Figure 32. Examples of overthrusts by creeping movements of sylvinitic salt (S) and halite (H) over carnallitic salts (C), all in the seam "Stassfurt" of Zechstein 2. a) mine "Solvayhall"/Bernburg, b) mine "Bleicherode" and 3) "Rossleben", S of Harz Mts., Germany.

suddenly torn and/or that it is missing for many meters. When the halite reappears again it is often curled into a ball-like structure which does not appear to have any connection with the normally folded halite bed. This sequence of events recurrs several times over with spatial gaps often for several hundred meters (fig. 37).

It also happens that there is often no connection at all, the halite occuring in single pieces the size of nuts, fists, or heads which are bent or rolled, and which "swim" within the carnallitic "matrix". The carnallitite is then often a conglomerate or a clastite. For a long time such events have been called "truemmercarnallite" in Germany (Everding, 1970) (fig. 38). As the deformation increases the structures more closely mimic those of a fluid (fig. 39). It should be remarked

Figure 33. Halite, intensively folded inside of the Muribeca formation, the tectonic position of which is quite even. Cores of vertical drill holes, in natural size. Aracaju/Sergipe, Brazil. a) halite of a rock salt member; b) cm-beds of halite within a structureless red carnallite member; scale 5 cm.

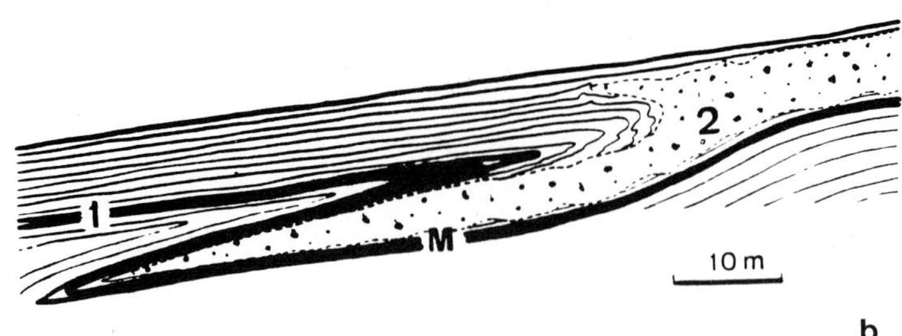

Figure 33b.

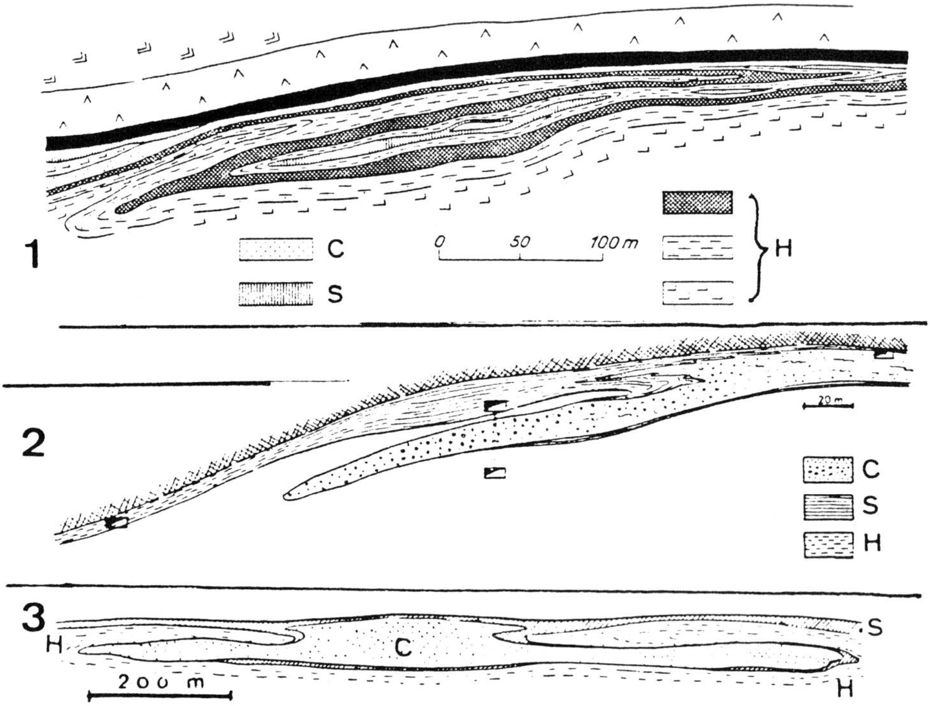

Figure 34. Halite beds (grey) interbedded in carnallitic and sylvinitic salt of a potash bed in Zechstein 1, the orientation of which is subhorizontal. The intensive folds of 2.5 m height are restricted to the potash seam. Mine "Neuhof-Ellers"/Fulda, Germany.

that such effects are also in accord with the high degree of tectonic deformation of the halite in the alpine orogenesis as we have seen previously in fig. 28.

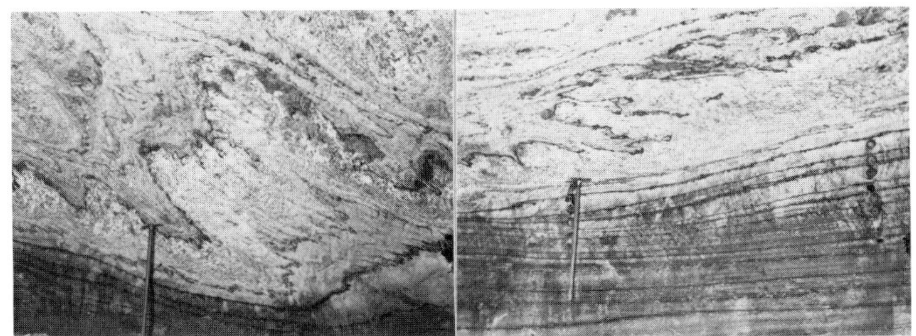

Figure 35. Creeping deformation of the "threefold kieserite bed" in the lower potash seam of Zechstein 1. Mine "Neuhof-Ellers"/Fulda, Germany.

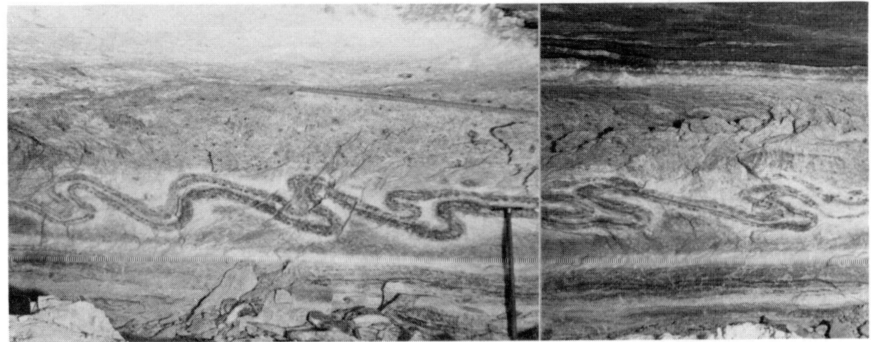

Figure 36. Halite bed, intensely folded by horizontal motion, within the flat lying carnallitic upper potash bed of Zechstein 1. Mine "Heringen", Werra, Germany.

VI. CONCLUSIONS

Based on the observations reported above, we conclude that all salt rocks are mobile and tend, like a viscous fluid, towards locations of lowest-possible stress. The mobility depends on the petrographic properties. The water-containing minerals (for instance carnallite) are the most mobile, halite shows a more normnal viscoplasticity, while anhydrite reacts as a solid rock. The degree of deformation is fundamentally a product of the physical parameters p times T (load times temperature). The greater the depth, the higher will be the load p and the temperature T. Thus even in the same salt formation, the

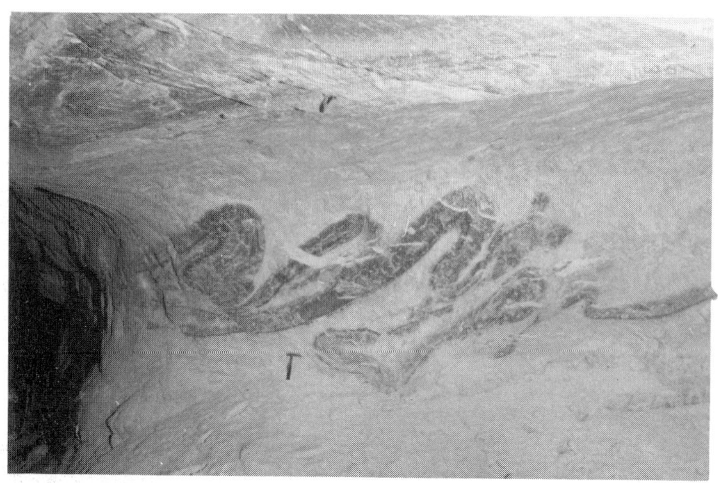

Figure 37. So-called "swimming halite", a rock salt bed of about 0.20 m thickness which is folded or rolled into a ball and then torn apart by horizontal movement. Upper potash seam of Zechstein 1, Mine "Hattorf",Werra, Germany.

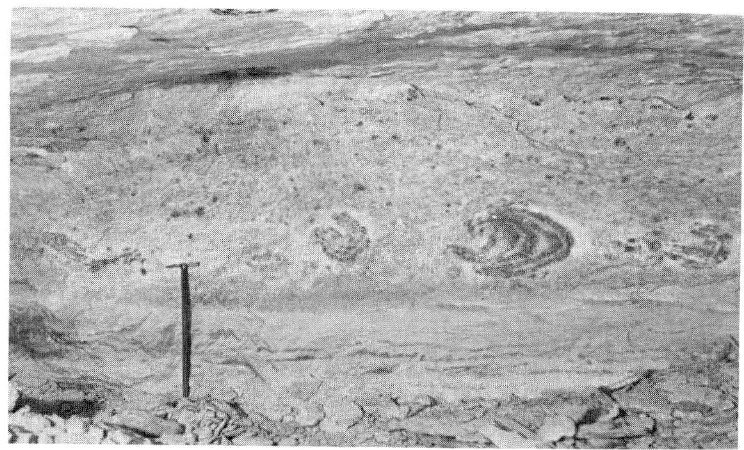

Figure 38. "Truemmer-Carnallite" = rubble carnallite or conglomeratic carnallite. Bent or rolled halite pieces, relicts of originally regular beds, within the "matrix" mass of carnallite (+ kieserite) representing a response to horizontal motions. Potash seam of Zechstein 1. Mine "Hattorf", Werra, Germany.

deeper/older units will be more active than the shallower/younger units. The kinds of deformation produced inside the salt bodies are strongly impacted by the particular "mixes" of young/old

Figure 39. The last state of deformation: carnallitite, with flowing motion. The salt rock consists mainly of red and white carnallite and contains some intergranular white kieserite. It also includes single angular pieces (here dark) of halite which is mechanically added rubble from originally intercalated continuous beds. Lower potash bed of Zechstein 1. Mine "Neuhof-Ellers", Fulda, Germany.

salts and mineral contents. But it would seem that no possibility exists to predict the interior structures, even when we know exactly the external shape of a salt body. The inside structures may only be guessed at in an extremely rough fashion even when based upon subsurface knowledge from mines or field investigation. To determine the intensity, direction and detailed features of the salt movement, extremely detailed studies appear to be unavoidable.

REFERENCES

Balk, R. (1949). Structure of Grand Saline Salt Dome, Texas. AAPG Bull 33, 1791-1829.

de Boer, H.U. (1971). Gefugeregelung in Salzstocken und in ihren Hullgesteinen. Z. Kali & Steins. 5, 413-425.

Borchert, H. (1950). Ozeane Salzlagerstatten, Grundzuge der Entstehung und Metamorphose. Verl. Borntrager, Berlin.

Borchert, H. and Muir, R.O. (1964). Salt deposits. Van Nostrad Co. London.

Dunnington, H.W. (1962). Salt tectonic features of Northern Iraq. G.S.A. Spec. Pap. 88, 183-227.

Everding, H. (1907). Zur Geologie der deutschen Zechsteinsalze. Abh. geol. L. A., N.F. 52.

Fulda, E. (1927). Salztektonik. Z. deutsch. geol. Ges. 79, 178-196.

Fulda, E. (1928). Das Kali II. Stuttgart.

Fulda, E. (1935). Zechstein. In Handb.d.vergleich.Stratigr. Deutschl. Geol. L. A. Berlin.

Gignoux, M. (1930). La tectonique des terrains saliferes. Livre jubil. Soc. geol. France.

Hite, R.J. (1962). Salt Deposits of the Paradox Basin in Southeast Utah and Southwest Colorado. G.S.A. Spec. Pap. 88, 319-330.

Hite, R.J. (1960). Stratigraphy of the saline facies of the Paradox Member of the Hermosa Formation of Southeast Utah and Southwest Colorado. Four Corners Geol. Soc., III Field Conf. 86-89.

Hite, R.J. and Cater, F.W. (1972). Pennsylvanian rocks and Salt anticlines, Paradox basin, Utah and Colorado. Geol. Atlas of the Rocky Mts., Rocky Mts. Geol. Assoc. Geologists.

Jones, R.W. (1950). Origin of the Salt Anticlines of Paradox Basin. AAPG Bull. 43, 1869-1895.

Kent, P.E. (1970). The salt plugs of the Persian Gulf region. Leicester Lit. Phil. Soc. 64, 56-80.

Kupfer, D.H. (1962). Structure of Morton Salt Co. Mine, Weeks Island Salt Dome, Louisiana. AAPG Bull. 46, 1460-1467.

Lachmann, R. (1911). Ekzeme und Tektonik. Centralbl. Min. Geol. Pal. 1917, 414-426.

Lees, G.M. (1927). Salzgletscher in Persien. Mitt. geol. Ges. Wien 20, 29-34.

Lees, G.M. (1931). Salt dome depositional and deformational problems. J. Inst. Petr. Technol. 17, 259-280.

Loffler, J. (1964). Die Carnallitgesteine des Raumes Aschersleben-Schierstedt. Freib. Forsch. Hefte, C 87.

Muehlberger, W.R. (1962). Internal structures and Mode of Uplift of Texas and Louisiana Salt Domes. G.S.A. Spec. Pap. 88, 359-364.

Neff, A.W. (1960). Comparisons between the Salt Anticlines of South Persia and those of the Paradox Basin. Four Corners Geol. Soc., III. Field Conf., 56-68.

Ramberti, L. (1980). I giassimenti salini siciliani. Boll. Assoc. miner. subalpina,. 17,2, 327-353.

Renner, O. (1914). Salzlager und Gebirgsbau im mittleren Leinetal. Arch. f. Lagerst. 13.

Richter (=Richter-Bernburg), G. (1934). Hauptanhydrit und Salzfaltung. Z. Kali,verwandte Salze und Erdol, 1934, 1-7. Halle/Saale.

Richter-Bernburg, G. (1968). Saxonische Tektonik als Indikator erdtiefer Bewegungen. Geol. Jb. 85, 997-1030.

Richter-Bernburg, G. (1972). Saline deposits in Germany, a review and general introduction to the excursions. In Earth Sciences 7. Geology of saline deposits, Proc. Hanover Sympos., May 1968, 275-287. UNESCO Paris.

Richter-Bernburg, G. (1974). The Oberrhein-Graben in its European and global setting. In Approaches to Taphrogenesis. Proc. Internat. Rift Symposium 1972, 13-43.

Richter-Bernburg, G. (1977). "Saxonische Tektonik" HANS STILLE's Begriff in heutiger Sicht. Z.d. deutsch. Geol. Ges. 128, 11-23.

Richter-Bernburg, G. (1980). Salt Tectonics, Interior Structures of Salt Bodies. Bull. Centre Rech., Elf-Aquit. 4, 373-393.

Schauberger, O. (1931). Die Fliesstrukturen im Hallstaetter Salzlager. Jb. Berg und Huettenwesen 79, 57-68.

Schauberger, O. (1953). Zur Genese des alpinen Haselgebirges. Z. deutsch. Geol. Ges. 105, 736-751.

Seidl, E. (1977). Die Salzstoecke des deutschen (germanischen) und des alpinen Perm-Salzes, ein allgemein-wissenschaftliches Problem. Z. Kali 21, 34 ff. Halle.

Shoemaker, E.M., Case, J.E. and Elston, D.P. (1958). Salt Anticlines of the Paradox Basin. Guidebook 9. Field Conf. Internat. Ass. Petr. Geol., 39-59.

Shoemaker, E.M. and Landis, E.R. (1962). Uncompahgre Front and Salt Anticlines of Paradox Basin. AAPG Bull. 44, 24-36.

Stille, H. (1917). Injektivfaltung und damit zusammen-hangende Erscheinungen. Geol. Rundschau 8, 89-142.

Stocklin, J. (1968). Structural history and tectonics of Iran. A review. Bull. A.A.P.G. 52, 1229-1258.

Stocklin, J. (1968). Salt deposits of the Middle East. Geol. Soc. America, Sp. Pap. 88, 157-181.

Stolle, E. (1959). Disharmonische Tektonik im Zechstein. Ber. deutsch. Ges. Geol. Wiss. 4 299-309.

Walther, H.W. (1972). Uber Salzdiapire in Sudost-Iran. Geol. Jb. 90, 359-383.

Zak, I. and Freund, R. (1980). Strain measurements in Eastern Marginal Shear Zone of Mount Sedom Salt Diapir, Israel. AAPG: Bull. 64,4.

INFLUENCE OF DIFFERENTIAL SEDIMENT LOADING ON SALT TECTONICS IN THE EAST TEXAS BASIN

David W. Harris[1]
Mary K. McGowen[2]

[1]Marathon Oil Company
P. O. Box 2690
Cody, Wyoming 82414

[2]ARCO Oil and Gas Company
2300 W. Plano Parkway
Dallas, Texas 75075

I. INTRODUCTION

The Cotton Valley Group (Upper Jurassic) and Hosston Formation (Lower Cretaceous) were studied to investigate the effect of early basin infilling on salt mobilization in the East Texas Basin.

An area in the northwestern part of the East Texas Basin consisting of seven counties--Hunt, Hopkins, Wood, Rains, Kaufman, Van Zandt, and Henderson--was selected for the study of the relationship between salt movement and the influx of Upper Jurassic terrigenous clastic sediment.

Salt movement began at different times in different parts of the basin. The earliest movement occurred around the margins of the basin during Smackover deposition (Jackson and Harris, 1981). At that time, increased subsidence toward the center of the basin caused basinward tilting that, induced by downward creep, mobilized salt.

More extensive salt movement occurred after the influx of Cotton Valley clastic sediment during the Late Jurassic (Fig. 1). Before that time, deposition in the East Texas Basin was dominated by carbonates, evaporites, and marine mudstones and

Era	System	Series	Group	Formation	Member / Unit
MESOZOIC	CRETACEOUS	LOWER CRETACEOUS	TRINITY	PALUXY	
				GLEN ROSE	UPPER GLEN ROSE
					MASSIVE ANHYDRITE
					LOWER GLEN ROSE: Rodesso Member
					James Limestone Mbr
					Pine Island Shale Member
					Pettet (Sligo) Member
				TRAVIS PEAK (HOSSTON)	
	JURASSIC	UPPER JURASSIC	COTTON VALLEY	SCHULER	
				BOSSIER	
			LOUARK	GILMER LIMESTONE (COTTON VALLEY LIMESTONE)	
				BUCKNER	
				SMACKOVER	
		MIDDLE JURASSIC	LOUANN	NORPHLET	
				LOUANN SALT	
		L.Jurassic		WERNER	
	Upper Triassic			EAGLE MILLS	
PALEOZOIC				OUACHITA	

Figure 1. Stratigraphic succession and nomenclature in the East Texas Basin (Wood, 1981).

claystones. Salt movement apparently was controlled by differential loading of Upper Jurassic and Lower Cretaceous fluvial-deltaic systems, as well as by the position of the subjacent Smackover-Gilmer carbonate shelf complex (Jackson and Harris, 1981; McGowen and Harris, 1981, 1984).

Electric logs from 232 wells (Fig. 2), supplemented by Bouguer residual gravity maps and two dip-oriented, six-fold conventional CDP seismic profiles, served as a data base for this study. When possible, well data were integrated with seismic data by using velocity conversion tables. Five seismic reflectors within the Mesozoic were used, including the base of

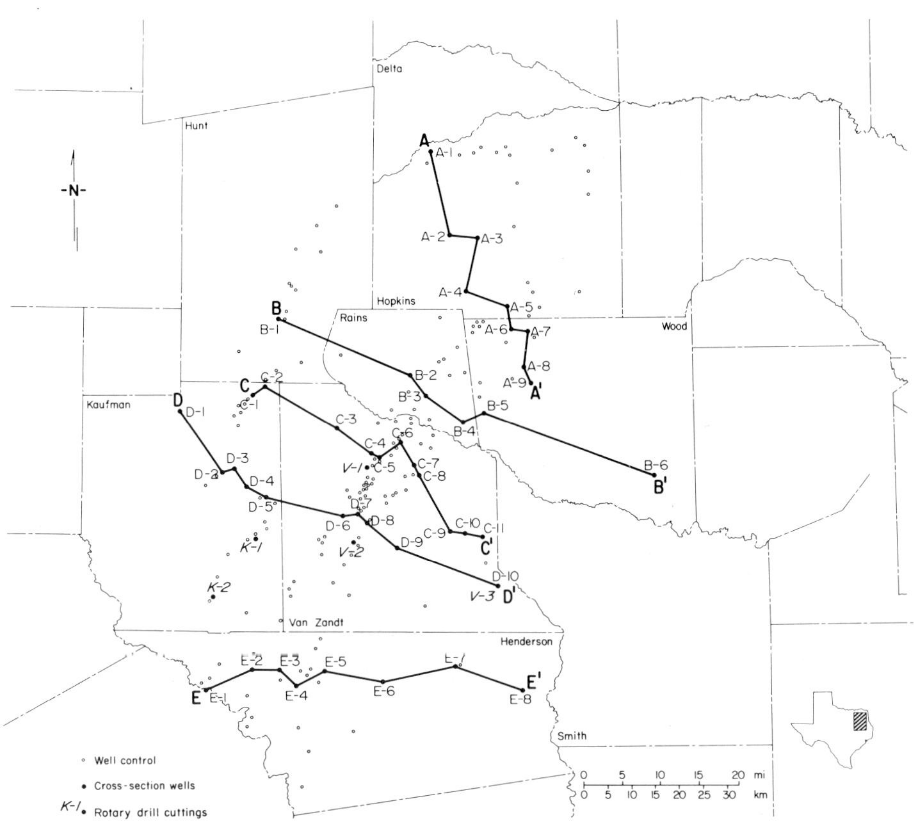

Figure 2. Index map showing well control, well cuttings, and location of stratigraphic cross sections.

the Louann Salt, the top of the Louann Salt, the top of the Gilmer Limestone (Cotton Valley Limestone) (Forgotson and Forgotson, 1976), and the top of the Pettet Limestone (Table 1). The fifth reflector, which we believe is the top of the Massive Anhydrite, was used in the northern part of the basin, where the Pettet Formation changes lithologically from a limestone facies to a sandy facies and thereby loses its character by prominent boundary reflections (Jackson and Harris, 1981). The reflector's inferred thickness, based on seismic interpretation, was correlated with gravity data. Zones of thicker salt generally coincide with gravity lows, whereas areas of thinner salt correspond to gravity highs (Jackson and Harris, 1981) (Fig. 3).

Table 1. Seismic Reflectors and Seismic Units in the Northwestern Part of the East Texas Basin

SEISMIC REFLECTOR	SEISMIC UNIT
Upper Navarro Marl	
Top of the Pecan Gap Chalk	
Top of the Austin Chalk	
Top of the Buda Limestone	
*Top of the Massive Anhydrite?	
*Top of the Pettet Limestone	D
*Top of the Gilmer Limestone	C
*Top of the Louann Salt	B
*Base of the Louann Salt	A

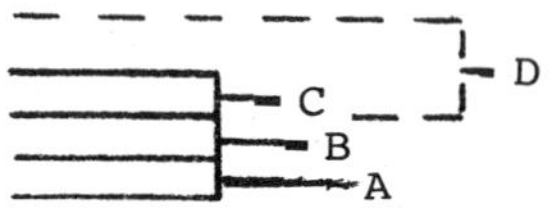

*Seismic reflectors used in this study.

Isopach, net-sandstone, and sandstone-percent maps of the Cotton Valley Group and the Hosston Formation were prepared. The boundary between the two was based on scout card information and regional correlations within the East Texas Basin. Using the Pettet Limestone as a datum, nine stratigraphic cross sections were constructed within the study area; selected sections are included in this report (Fig. 2).

Limitations of this data base include the following: First, although well spacing within individual oil and gas fields is good, overall spacing is poor, precluding detailed mapping of the Cotton Valley Group and Hosston Formation on a regional scale. Second, because conventional-core data were not available to verify environmental interpretations, facies designations were based entirely on electric log response and on sand-body geometry determined from net-sandstone maps, sandstone-percent maps, and textural and compositional features observed in well cuttings. And third, because Jurassic formations in northeast Texas are restricted to the subsurface, facies relationships of surface exposures could not be examined.

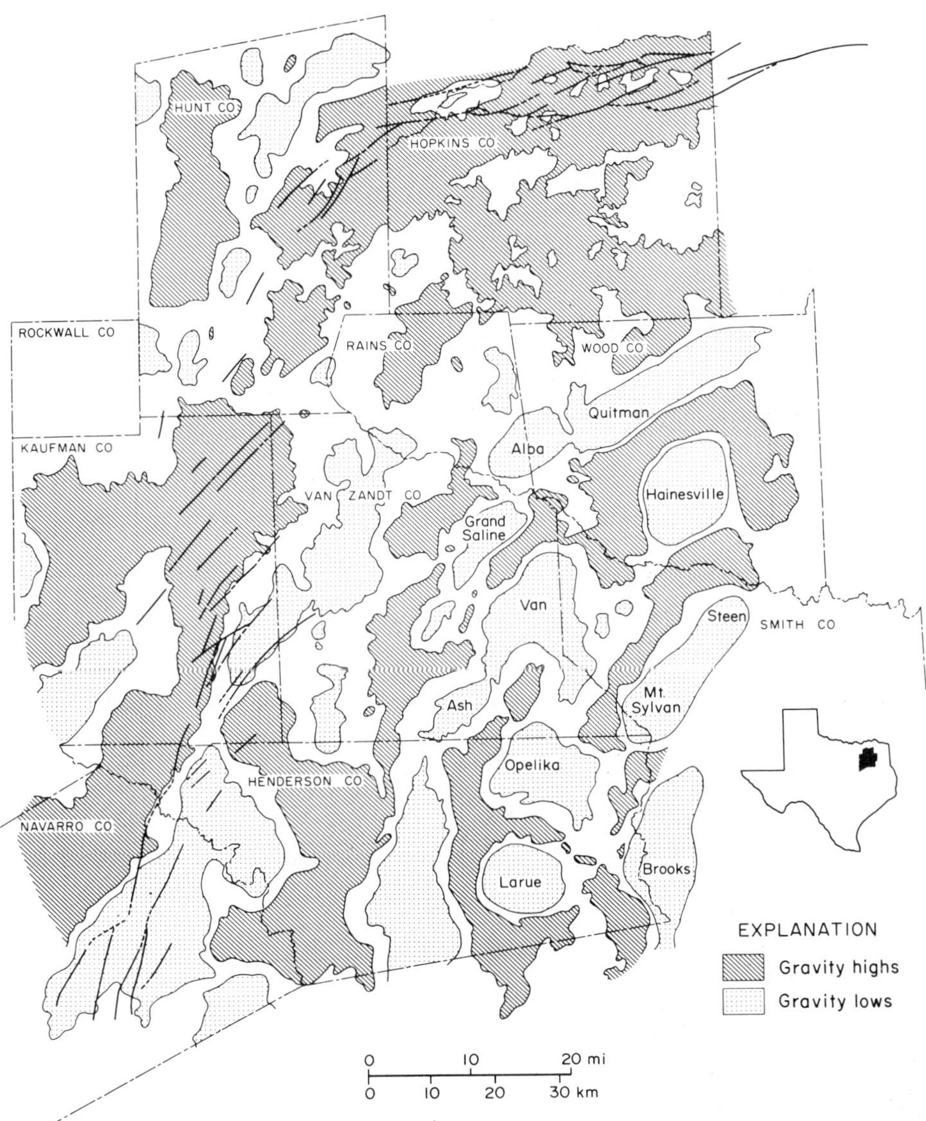

Figure 3. Generalized residual gravity map (modified from map done by Exploration Technique, Inc.).

II. TECTONIC FRAMEWORK

The East Texas Basin is recognized as a subbasin, or re-entrant, of the larger Gulf Coast Basin (Wood and Walper, 1974; Walper, 1980). Most researchers agree that the East Texas Basin was formed from one of either the megashear zones, the rift grabens, or the aulacogens that formed along the margins of the Gulf of Mexico, probably coincident with the breakup of Pangea and the separation of North and South America during the Triassic (Kehle, 1971; Burke and Dewey, 1973; Moore and Del Castillo, 1974; Wood and Walper, 1974; Beall, 1975; Salvador and Green, 1980). Kehle (1971) suggested that the interior salt basins of Mississippi and northern Louisiana also are foundered grabens, marginal to the ancestral Gulf Coast Basin. Major tectonic elements in and around the East Texas Basin are shown in Figure 4.

A. Tectonic History

Kehle (1971) and Wood and Walper (1974) maintained that the development of the interior salt basins resulted from the opening of the Gulf of Mexico. They described the history of these interior salt basins as follows: The interior salt basins (Mississippi, North Louisiana, East Texas, and Salinas Basins) represent the most marginal grabens that are associated with continental rifting. These grabens were initially filled with alluvial-fan deposits of the Eagle Mills Formation. With prolonged spreading, however, the interior grabens continued to founder. The southern margin of the East Texas Basin may have been elevated, thereby restricting circulation of sea water between the basin and the Gulf of Mexico. Evaporites of the Werner Anhydrite and Louann Salt were precipitated, possibly by the brine-mixing process (Raup, 1970).

Continued subsidence resulted in open marine conditions; this is evidenced in the widespread occurrence of carbonates in

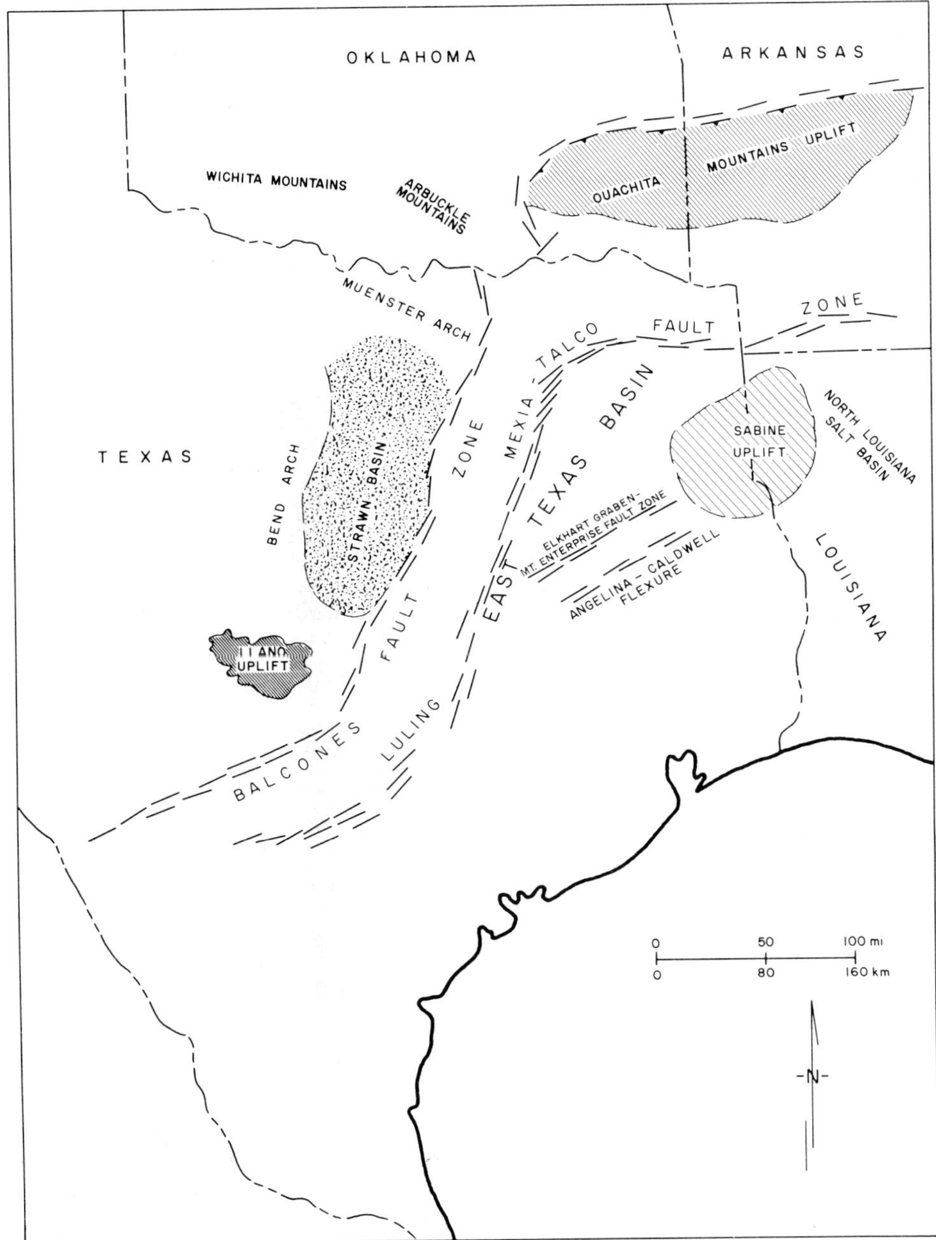

Figure 4. Tectonic elements in and around the East Texas Basin.

the Smackover and Gilmer Formations. One of the characteristics of rifting is that during the early stages the bounding crustal blocks tilt away from the incipient rift, allowing large quantities of terrigenous clastics to enter the basin only when the dip of the rift margin reverses. The major influx of terrigenous clastics into the East Texas Basin during the Late Jurassic (Cotton Valley) and Early Cretaceous (Hosston) may reflect this dip reversal.

B. Salt Tectonics

Two general observations about salt structures in the East Texas Basin can be made: First, the size and type of salt-related structures seem to be directly controlled by the thickness of the underlying salt. This relationship was also observed in the interior Mississippi salt basin by Hughes (1968). Second, salt apparently migrated at different times in different parts of the East Texas Basin through several mechanisms: faulting (common in the marginal areas of the basin) (Parker and McDowell, 1955; Rosenkrans and Marr, 1967; Hughes, 1968); gravity gliding of post-Louann strata, which was caused by basinward tilting (Kehle, 1971; Jackson and Harris, 1981); and mass imbalance caused by uneven sediment loading (Rogers, 1967; Turk, Kehle, and Associates, 1978; McGowen and Harris, 1981, 1984).

1. Dome Growth Mechanisms

Kehle (1971) maintained that uneven sediment loading is the dominant mechanism responsible for salt dome initiation. He observed that the density inversion caused by uneven sediment loading is accentuated in the resulting salt flow because the viscosity of salt is highly sensitive to shear stress. Kehle concluded that the unequal pressure gradient set up within a salt mass because of uneven sediment loading is dependent on the slope of the overlying sediment, which in the present study is a deltaic lobe.

Loocke (1978) applied this principle to the East Texas Basin, suggesting that the growth of the Hainesville salt dome (south-central Wood County) was initiated by mass imbalance caused by the progradation of Cotton Valley deltas into that part of the basin. By the end of Hosston (Early Cretaceous) sedimentation, the structure had grown into a topographically high salt pillow, or anticline (Loocke, 1978). The evolution of initial salt anticlines into salt domes depends on continued sediment loading (Kehle, 1971) as well as on the thickness of the underlying salt, which controls the supply of salt for continued growth.

2. Timing of Initial Salt Movement

Two seismic profiles available in the study area document the timing of salt mobilization (Figs. 5 and 6). Profile S-1 extends northwest to southeast through Kaufman, Van Zandt, northeastern Henderson, and western Smith Counties, profile S-2 trends north to south through Delta, Hopkins, Wood, and Smith Counties.

Seismic data suggest that salt movement was both pre-Gilmer and coeval with Cotton Valley-Hosston deposition (Jackson and Harris, 1981; McGowen and Harris, 1981, 1984).

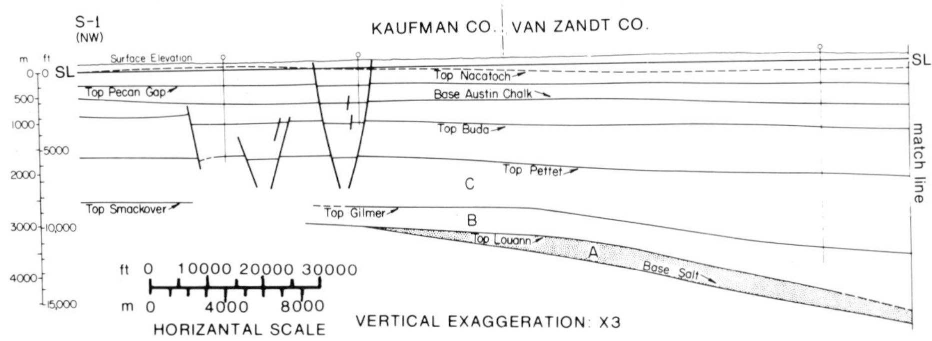

A

Figure 5A-C. Time-to-depth conversion section S-1. Seismic units A, B, and C refer to Table 1. Southeastern end of line shows salt-withdrawal feature associated with Mount Sylvan salt dome.

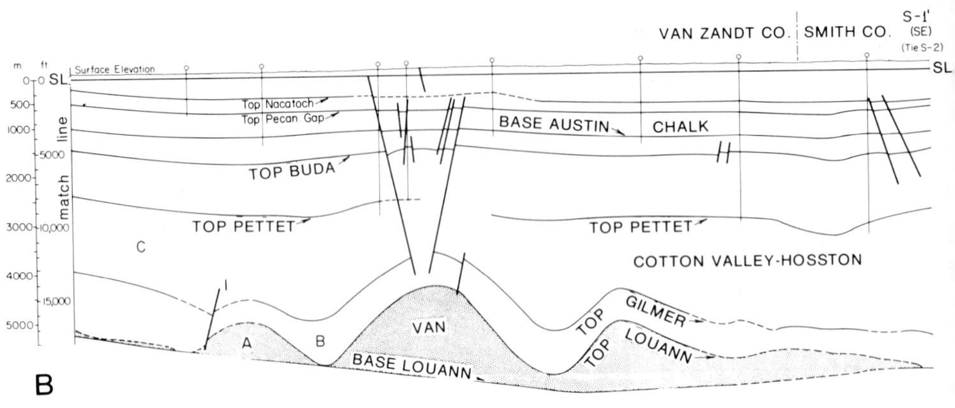

Figure 5B.

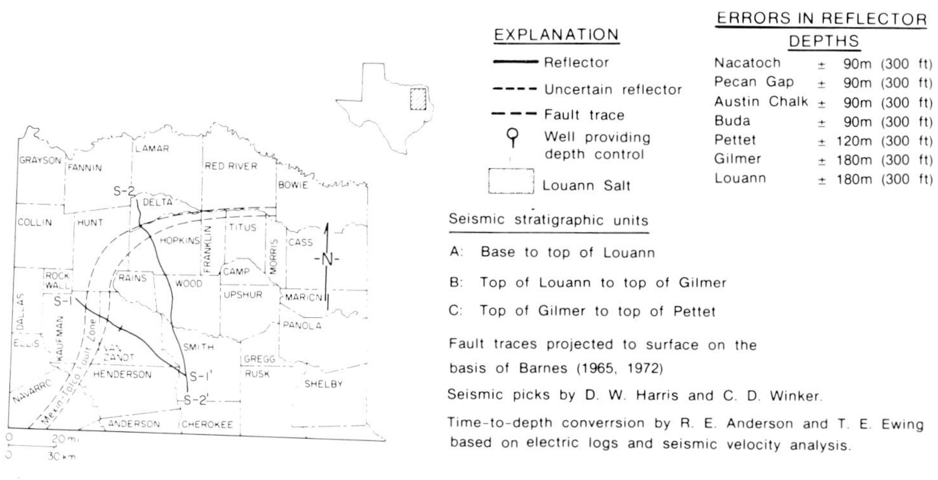

Figure 5C.

Pre-Gilmer salt migration occurred in an area north and west of a line through central Wood, eastern Rains, central Van Zandt, and central Henderson Counties (Figs. 5 and 6). Time-thickness variations are apparent between reflections at the top of the Louann Salt and the top of the Gilmer Limestone (seismic unit B,

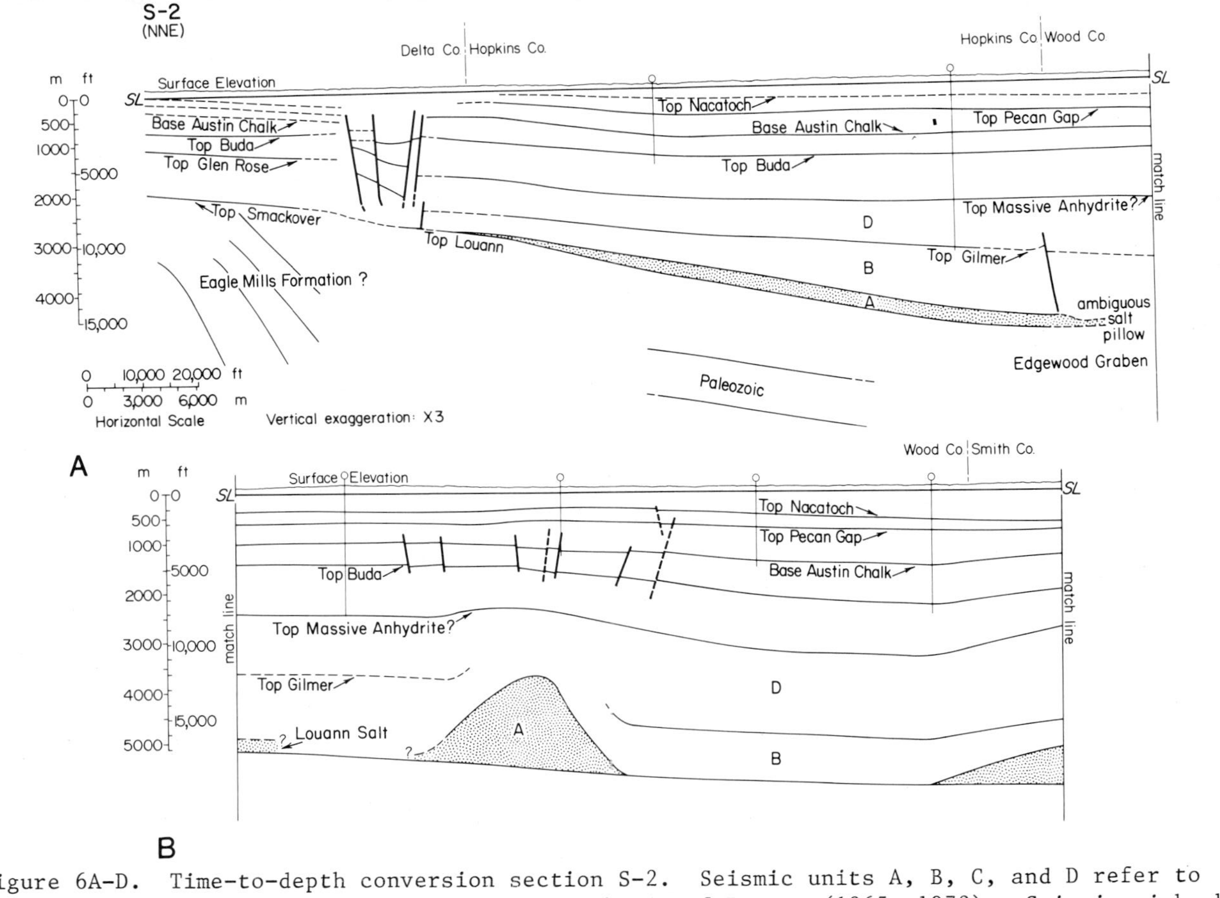

Figure 6A-D. Time-to-depth conversion section S-2. Seismic units A, B, C, and D refer to Table 1. Fault traces projected to surface on the basis of Barnes (1965, 1972). Seismic picks by D. W. Harris and C. D. Winker. Time-to-depth conversion by R. E. Anderson and T. W. Ewing based on electric logs and seismic velocity analysis.

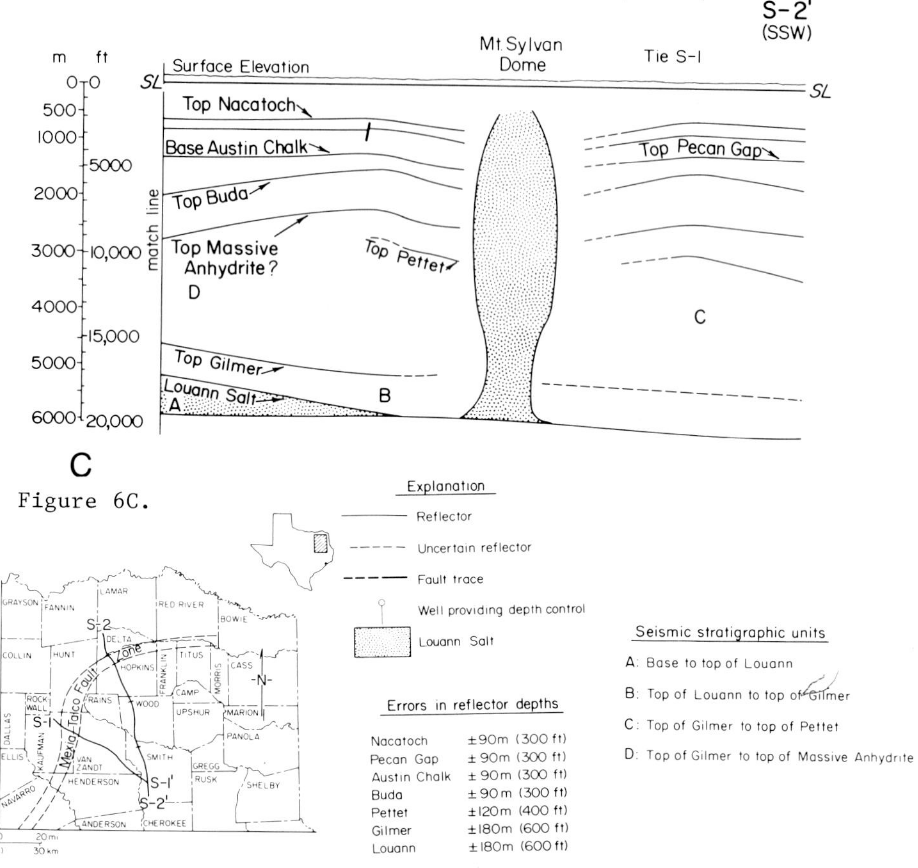

Figure 6C.

D

Figure 6D.

Table 1). No thickness variations caused by salt movement were observed in the seismic interval between the top of the Gilmer Limestone and the top of the Pettet Limestone (seimsic Unit C, Fig. 5) or between the top of the Gilmer Limestone and the top of the Massive Anhydrite (seismic unit D, Fig. 6). Thus, it can be concluded that salt moved during Smackover deposition, which preceded Cotton Valley-Hosston deposition.

During Cotton Valley-Hosston deposition, salt moved south and east of a line through central Wood, eastern Rains, central Van Zandt, and central Henderson Counties. The seismic interval between the top of the Gilmer Limestone and the top of the Pettet

Limestone (seismic unit C, Fig. 5B) apparently thins over salt structures and thickens within synclines caused by salt withdrawal. In contrast, the seismic interval between the reflections at the top of the Louann Salt and top of the Gilmer Limestone (seismic unit B, Fig. 5B) appears to be planar. This relationship is not as discernable on profile S-2 (Fig. 6) as on profile S-1 (Fig. 5) because seismic resolution is poor on the deep stratigraphic horizons in southern Wood County.

C. Structural Styles

Structural styles within the East Texas Basin can be grouped into four general categories: a peripheral graben system; a deformation-free zone; low- to intermediate-amplitude salt structures, which show pre-Cotton Valley salt movement; and salt anticlines and domes, which show movement coincident with Cotton Valley-Hosston deposition. These features have been discussed by Kehle (1971) and Jackson and Harris (1981). Similar structures have been observed in the Mississippi (Hughes, 1968) and the North Louisiana (Kehle, 1971) salt basins.

1. Peripheral Graben System

The Mexia-Talco fault zone bounds the study area on the north and west. The position of the fault system marks the updip depositional limit of the Louann Salt (Kehle, 1971; Agagu et al., 1980; Jackson and Harris, 1981). The fault zone is a series of en echelon normal faults and grabens that formed early in the history of the basin. However, there is no evidence on seismic lines observed during this study that the peripheral faults extend into the basement. Rather, evidence suggests that the grabens are based in the Louann Salt (Jackson, 1982). Turk, Kehle, and Associates (1978) maintained that major displacement along the fault zone was largely caused by downdip creep of the Louann Salt. Jackson and Harris (1981) suggested that fault displacement was also caused, in part, by basinward creep of clastic strata above the Louann decollement zone. Movement along the fault affected strata of Mesozoic through early Tertiary age.

Basinward of the Mexia-Talco fault zone, a second graben system (the Edgewood Graben) developed parallel to the Mexia-Talco trend (Rosenkrans and Marr, 1967) (Fig. 7). These faults were active during the Jurassic and became inactive during the Early Cretaceous before deposition of the Pettet Limestone (Fig. 6). These faults and related structures are discussed in more detail in Section III, C, 3.

2. Deformation-Free Zone

On both seismic profiles S-1 and S-2, a deformation-free zone between the peripheral graben system and the first occurrence of low-amplitude salt swells (Figs. 5 and 6) underlies Delta and Hopkins Counties to the north and Kaufman and western Van Zandt Counties to the west. The Louann Salt is recognized by prominent boundary reflections and lack of internal reflections (Jackson and Harris, 1981). Within the deformation-free zone, boundary reflections are planar and diverge basinward, indicating a thickening of the salt wedge (approximately 1,020 to 1,920 feet, or 340 to 640 m) (Jackson et al., 1982). Existence of the deformation-free zone suggests that a critical thickness of salt (approximately 1,500 feet, or 500 m) must be present to initiate flow (Jackson et al., 1982). Kehle (1971) maintained that the deformation-free zone is discontinuous along the full length of the basin margin.

3. Low- to Intermediate-Amplitude Salt Structures

Low-amplitude salt structures occur basinward of the deformation-free zone in southern Hopkins, southern Rains, and western Van Zandt Counties. Amplitudes, indicated by relief on the top of the salt anticlines, increase basinward as the salt wedge thickens. Basinward, folds are commonly tighter and salt has pierced some anticlines (Hughes, 1968; Kehle, 1971). Seismic profiles indicate that these low- to intermediate-amplitude structures formed early in the development of the basin. Salt movement began during deposition of the Smackover Formation; this is evidenced by time-thickness variations between reflections on the top of the Louann Salt and the top of the Gilmer Limestone (Figs. 5 and 6). Salt migration was probably initiated by downward creep that was induced by sedimentary loading of carbonate deposits (Rogers, 1967) and was enhanced by basinward tilting. Residual gravity maps suggest that these initial salt structures are aligned roughly parallel to the Mexia-Talco fault zone (Figs. 3 and 4). Hughes (1968) deduced that salt movement occurred as early as deposition of the Norphlet in the Mississippi salt basin. He also observed that low-amplitude salt structures in marginal areas of the basin are commonly arranged in ridges that parallel the peripheral faults.

Most growth of low-amplitude salt structures apparently occurred before deposition of Cotton Valley clastic sediment. This can be seen on seismic profiles, which show little variation in time-thickness between the reflections at the top of the Gilmer Limestone and the top of the overlying Pettet Limestone. The low-amplitude salt features exhibit little if any structural expression above the Gilmer Limestone. The thin underlying salt wedge appears to have largely controlled the size of the structures that formed in this part of the basin.

Near the Edgewood Graben (Fig. 7), salt structures of intermediate size are truncated by the fault system. Rosenkrans and Marr (1967) maintained that the salt structures were generated by faulting, and Jackson and Harris (1981) concluded that salt withdrawal by downdip creep, coincident with sediment loading within the graben, may have caused the faults. Fault movement along the system displaced strata of Late Jurassic and Early Cretaceous age but ended in the Early Cretaceous. Seismic profile S-2 (Fig. 6) indicates that faulting was contemporaneous with deposition of the Cotton Valley clastics.

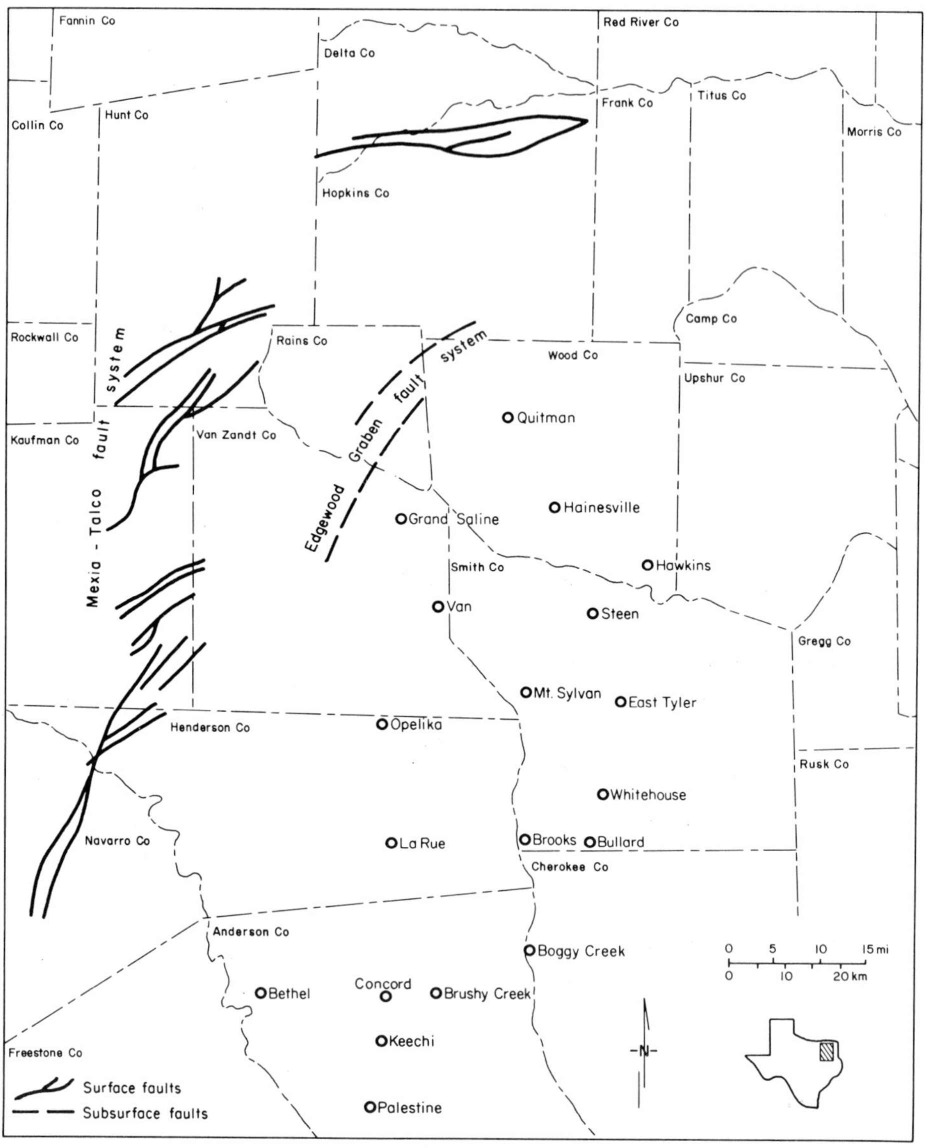

Figure 7. Salt domes and major fault systems, northwestern part of the Texas Basin.

Seismic profile S-2 (Fig. 6) also shows evidence of displaced reflections along the Edgewood graben during deposition of the Cotton Valley-Hosston strata. The following relationships were observed: First, a salt structure underlies the Edgewood fault, but the exact relationship of the structure to the fault is not clear on the seismic profile; Second, reflection of the Gilmer Limestone is offset down to the basin without appreciable time-thickness variations within seismic unit B (Table 1), indicating that most of the movement occurred after deposition of the Gilmer Limestone; Third, seismic unit D thickens on the downthrown side of the fault, not on the upthrown side, indicating that movement along the fault was contemporaneous with deposition of the Cotton Valley Group. The fault terminates within seismic unit D (Hosston Formation).

Evidence of faults that were active northwest of Van Dome during Cotton Valley-Hosston deposition is shown on seismic profiles S-1 (top structure, Fig. 5). First, updip of the fault, seismic unit A (Table 1) is distorted, and the configuration of the Louann Salt cannot be discerned. Second, seismic unit B thickens slightly basinward of the fault, suggesting that salt movement occurred during Smackover deposition. Third, the Gilmer Limestone reflection is clearly offset at the fault. Fourth, seismic unit C thickens appreciably on the downthrown side of the fault, indicating movement contemporaneous with deposition of Cotton Valley clastic sediment. Fifth, the fault terminates either in the upper part of the Cotton Valley or in the Hosston below the Pettet Reflection, which is planar.

4. Salt Anticlines and Domes

Within the study area, salt anticlines and domes occur south and east of a line through southern Hopkins, eastern Rains, central Van Zandt, and central Henderson Counties (Fig. 8). The location of salt anticlines and domes in the basin appears to have been controlled largely by the location of the Smackover-Gilmer carbonate platform, particularly where Gilmer carbonate shelf-edge strata overlie Smackover deposits. Rogers (1967)

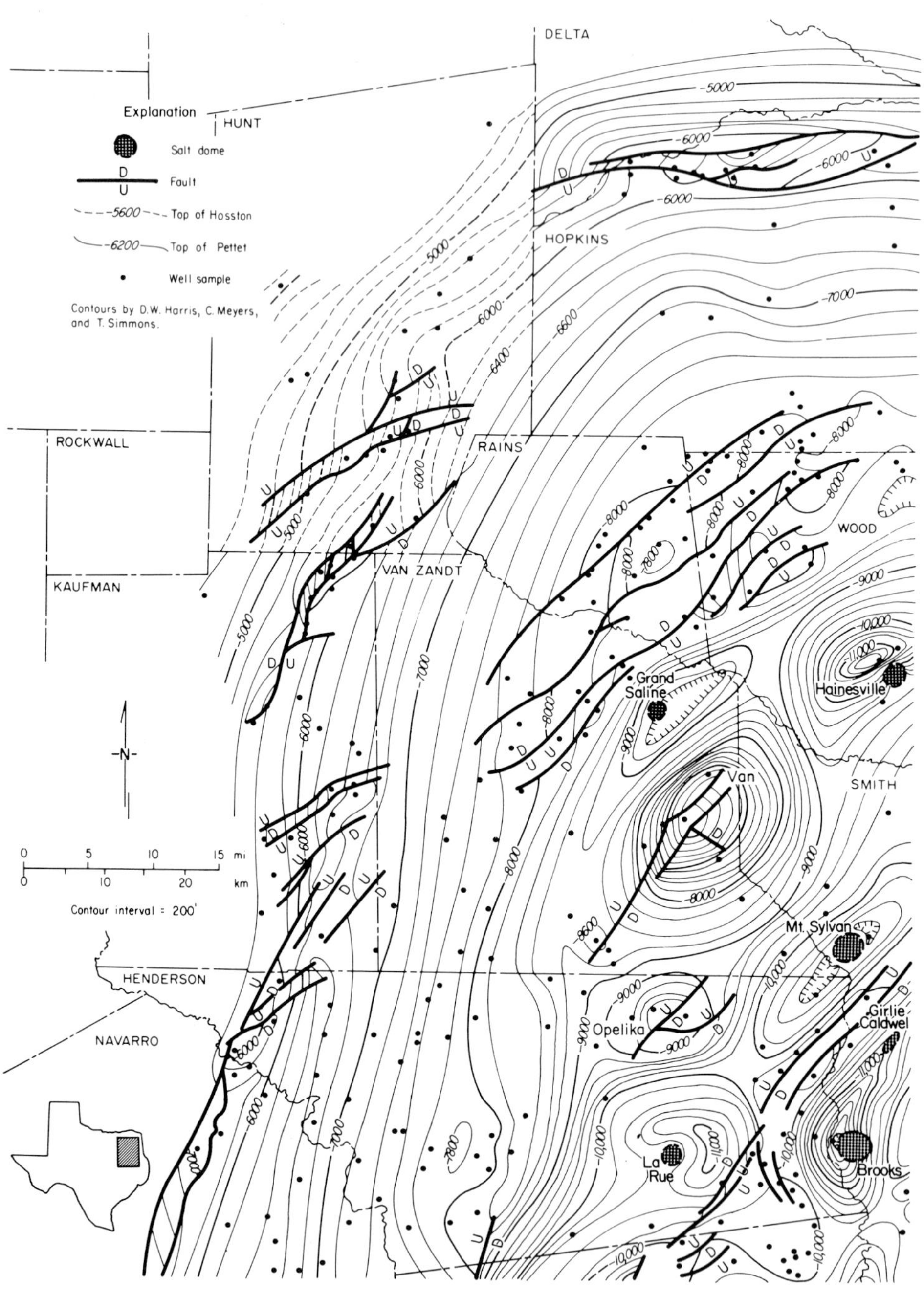
Explanation
Salt dome
D
U
Fault
-5600 Top of Hosston
-6200 Top of Pettet
Well sample
Contours by D.W. Harris, C. Meyers, and T. Simmons.
N
0 5 10 15 mi
0 10 20 km
Contour interval = 200'
DELTA
HUNT
HOPKINS
ROCKWALL
RAINS
KAUFMAN
VAN ZANDT
WOOD
SMITH
HENDERSON
NAVARRO
Grand Saline
Hainesville
Van
Mt. Sylvan
Opelika
Girlie Caldwel
La Rue
Brooks

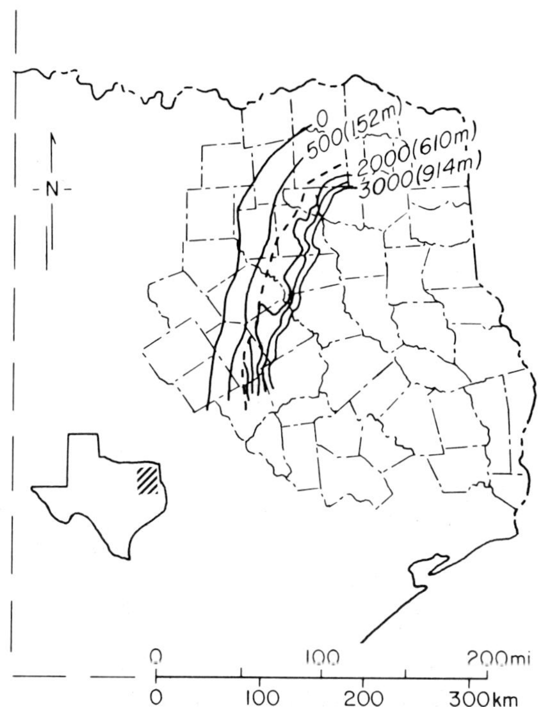

Figure 9. Isopach map of Smackover-Gilmer carbonate shelf facies (Rogers, 1967). Isopach of Smackover northwest of dashed line, combined isopach of Smackover and Gilmer southeast of dashed line.

maintained that a carbonate shelf-edge facies began to form during Smackover deposition and continued in some areas during Gilmer deposition, reaching a maximum combined thickness of 3,500 feet (1,166 m) (Fig. 9).

A carbonate shelf of this thickness could provide a stable platform, upon which fan-delta sediments of the Cotton Valley Formation would tend to spread laterally rather than to stack vertically. The isopach map of the Cotton Valley Group indicates a close correlation between the position of the Smackover carbonate shelf edge, as mapped by Eaton (1961) and Rogers

Figure 8. Structure map of the top of the Pettet Formation and the top of the Hosston Formation (Hunt County only) (facing page).

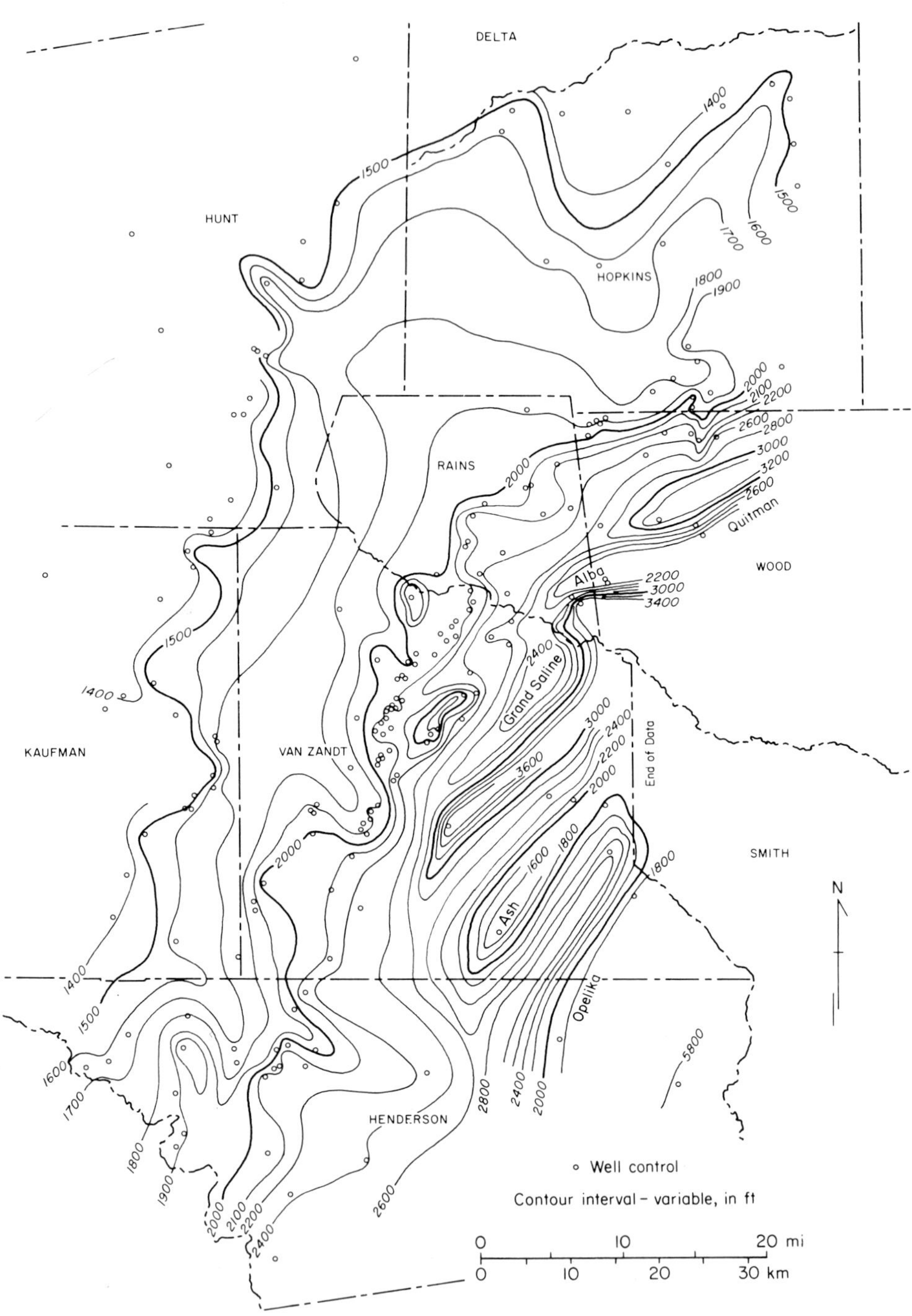
DELTA
HUNT
HOPKINS
RAINS
WOOD
KAUFMAN
VAN ZANDT
SMITH
HENDERSON
Quitman
Alba
Grand Saline
Ash
Opelika
End of Data
N
○ Well control
Contour interval – variable, in ft
0 10 20 mi
0 10 20 30 km

(1967), and the occurrence of Cotton Valley deltaic depocenters (Fig. 10). Superposition of deltaic sediments occurred in synclines that were located between salt ridges and were immediately basinward of the older carbonate wedge. Salt mobilization was initiated by basinward migration of the salt into ridges that fronted the progradational fan-delta system. The salt ridges probably were bathymetric highs that acted as effective sediment dams to perpetuate the synclines as depocenters until either the underlying salt was depleted or the rate of sedimentation was greater than the rate of subsidence. This allowed the delta to overrun the salt ridge.

During continued progradation, the fan-delta complex probably overran the initial syncline and salt ridge to establish a new depocenter basinward of and parallel to the original salt ridges. Consequently, sediments in depocenters should be progressively younger toward the center of the basin. Initial salt ridges that developed during the Late Jurassic and Early Cretaceous were later modified by continued salt flow, which resulted from uneven sediment loading by younger sedimentary units. Johnson (1980) showed that a pre-existing salt ridge on the outer continental shelf and upper slope of the Gulf Basin evolved into individual salt domes when buried by an influx of fluvial-deltaic sediments during the Pliocene.

A residual gravity map of the study area indicates a parallel arrangement of lows and highs (Fig. 3). A comparison of the Cotton Valley isopach map (Fig. 10) with the residual gravity map shows that sediment thicks generally coincide with gravity highs and salt thins, whereas sediment thins coincide with gravity lows and areas of thicker salt. The approximate parallel alignment of residual gravity lows again suggests that the original salt structures may have been a series of parallel salt ridges that evolved through time into salt anticlines and domes.

Figure 10. Isopach map of the Cotton Valley Group (facing page).

Seismic lines across the salt anticlines and domes in the study area indicate that, in most cases, salt moved after deposition of the Gilmer Limestone, whereas low-amplitude structures located in the marginal areas indicate pre-Gilmer movement. Deeper seismic reflections near Quitman Dome (Wood County) are distorted, and thus obscure the relationship between the salt structure and the overlying sedimentary units. However, a seismic survey profile across Mount Sylvan Dome (Smith County) shows virtually no evidence of pre-Cotton Valley salt movement (Fig. 6). West Tyler, to the southeast of Mount Sylvan Dome, and Boynton Field, to the northwest, are turtle structures created by early salt withdrawal that started coeval with Cotton Valley deposition (Jackson et al., 1982). Turtle structures are anticlinal structures without salt cores that were created by sediment thicks in primary withdrawal basins formed by early salt withdrawal (Trusheim, 1960; Kehle, 1971; Wood, 1981; Seni and Jackson, 1983). Turtle structures probably were depocenters during deltaic sedimentation of the Late Jurassic.

Seismic profile S-1 passes to the south of Grand Saline and Van Domes but crosses two parallel, lateral salt ridges that extend southeastward from Van Dome (Figs. 3 and 5) (Jackson and Harris, 1981). The seismic survey across the western ridge indicates that some pre-Gilmer salt movement occurred (seismic unit B, Fig. 5). Major movement, however, was coincident with deposition of Cotton Valley deltaic sediments (seismic unit C, Fig. 5). Basinward of the western ridge, no pre-Gilmer salt movement apparently occurred.

D. Summary

Seismic profiles indicate that salt movement occurred at different times in different parts of the East Texas Basin and demonstrate that more than one mechanism caused movement. Within marginal parts of the basin, low-amplitude salt structures formed during Smackover (pre-Gilmer) time. Salt migration was caused by loading of the Smackover carbonate deposits in conjunction with basinward tilting. Time-thickness variations caused by low-amplitude salt structures are not indicated on Cotton Valley-Hosston isopach maps. The sizes of salt structures increase basinward, coincident with thickening of the underlying salt wedge.

Basinward of a line running through central Henderson, Van Zandt, and Wood Counties, salt movement occurred later than in the marginal parts of the basin. Salt migrated during deposition of Cotton Valley-Hosston strata as a result of mass imbalance, which was caused by uneven sediment loading. Time-thickness variations (divergence) are evident between seismic reflections at the top of the Gilmer Limestone and the top of the Pettet Limestone. In contrast, reflections at the top of the Louann are parallel, suggesting that salt did not move in this part of the basin until Cotton Valley time.

III. SEDIMENTOLOGIC FRAMEWORK

A. Methodology

Early Jurassic sedimentation was dominated by the deposition of carbonate, evaporite, and mudstone facies. The first major influx of terrigenous clastic sediment into the East Texas Basin occurred during the Late Jurassic (Cotton Valley) and continued into the Early Cretaceous (Hosston).

In this study, the boundary between the Cotton Valley Group and the Hosston Formation is arbitrary, based on regional correlations and information of geologic picks from scout cards. In the northwestern part of the East Texas Basin, the top of the upper Cotton Valley Group is difficult to pick on electric logs and seismic profiles because Hosston sandstones overlie upper Cotton Valley sandstones. Nichols et al. (1968) maintained that deposition was continuous through Late Jurassic and Early Cretaceous time except around the western and northern margins of the basin, where an erosional contact was recognized. In these marginal areas, a basal Hosston conglomerate, ranging in thickness from 50 to 100 feet (16 to 33 m), separates the two formations (Nichols et al., 1968). It has not been determined whether this conglomerate bed occurs regionally or whether it consists of unrelated local facies, such as proximal alluvial-fan deposits or braided-stream channel-fill deposits. On the basis of seismic interpretations, Todd and Mitchum (1977) recognized a major unconformity between the Cotton Valley Group (Upper Jurassic) and the Hosston Formation (Lower Cretaceous) in the East Texas basin. However, we did not recognize a major unconformity on seismic sections in the northwestern part of the basin. The depositional sequence between either the Gilmer Limestone (Cotton Valley Limestone) or the base of the Buckner Anhydrite and the top of the Hosston Formation is thought to represent one regressive sequence with minor interruptions. The regression was terminated by a major transgression, which is evidenced by the transition of the uppermost Hosston Formation into the Pettet Limestone.

The volume, texture, and composition of sediment, the nature of the sediment-dispersal system, and the geometry of sands of the Cotton Valley Group and Hosston Formation were studied using sandstone-percent and net-sandstone maps. These were supplemented by regional stratigraphic cross sections, electric logs, isopach maps, and well cuttings. A sandstone-percent map is an effective way to reveal sandbody geometry because it minimizes the effect of thickness variations within a mappable unit. Stratigraphic markers are virtually absent in the Cotton Valley Group and Hosston Formation within the northwestern part of the basin; therefore, sandstone maps represent the composite sandstone thickness of superposed facies.

B. Source Area

The same source area supplied terrigenous clastics to the East Texas basin during both the Late Jurassic (Cotton Valley) and the Early Cretaceous (Hosston). Deposition of Cotton Valley terrigenous clastics in the East Texas Basin suggests that a reactivation of this source area along the northern and western margins of the basin began as early as the Late Jurassic. The Central Mineral Region and the Ouachita, Arbuckle, and Wichita Mountains were all highlands during the Late Jurassic and Early Cretaceous (Imlay, 1943) (Fig. 4).

Patterns of the net-sandstone and sandstone-percent maps suggest that during the Late Jurassic, terrigenous clastics were delivered to the East Texas Basin by many small streams and rivers. This differs from the major river and tributary system in Louisiana and Mississippi, where an ancestral Mississippi River was the principal fluvial system entering the North Louisiana Basin (Thomas and Mann, 1966).

Tongues of quartz conglomerate along the northwestern and northern margins of the basin in the Cotton Valley Group and throughout the Hosston Formation were probably derived from the Ouachita, Arbuckle, and Wichita highlands. Textural maturity and the dominance of very fine grained to fine-grained sandstone suggest that older sedimentary rock, surrounding the basin during the Late Jurassic and Early Cretaceous, was also an important source of terrigenous clastic sediment for the East Texas Basin. Most of the sandstone is quartzarenite and subarkose.

C. Depositional Systems

1. Cotton Valley-Hosston Fan-Delta System

Deposits of the Cotton Valley Group and Hosston Formation are interpreted to be a system of coalescing fan deltas that prograded from the west, northwest, and north. Facies interpretations are based on sandtone-percent values, sandstone distribution, electric log response patterns, descriptions of

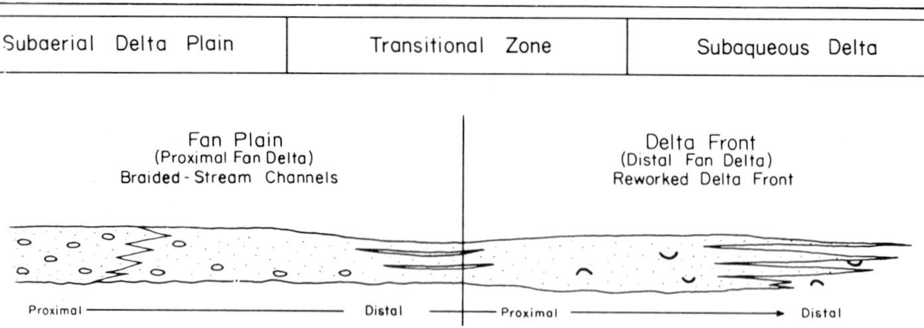

Figure 11. Diagrammatic cross section of fan-delta subenvironments (after Lucchi et al., 1981).

well cuttings and lithologic descriptions from other studies (McGowen and Harris, 1984). The morphology of fan-delta and braided stream systems used in this report is based on studies of recent depositional systems by McGowen, 1970; Miall, 1977; Rust, 1978; and Lucchi et al., 1981; (Fig. 11).

2. Cotton Valley Group

a. Facies. Facies of the Cotton Valley Group can be divided into three general categories: prodelta deposits, delta-front deposits, and braided stream deposits, all components of a fan-delta system (Figs. 2, 12 through 16). Prodelta deposits compose a thin subordinate facies in the study area; however, the facies thickens basinward. It consists of black or green calcareous, fossiliferous mudstone interbedded with light-gray to light-brown crystalline limestone and shelly, sandy limestone. Minor amounts of very fine-grained sandstone and siltstone occur within the facies.

Delta-front deposits consists of alternating beds of sandstone and mudstone, and a few thin beds of sandy limestone. Sandstone, the dominant lithology, is white to light gray and very fine-grained to fine-grained, containing glauconite, shells, and finely disseminated carbonized plant

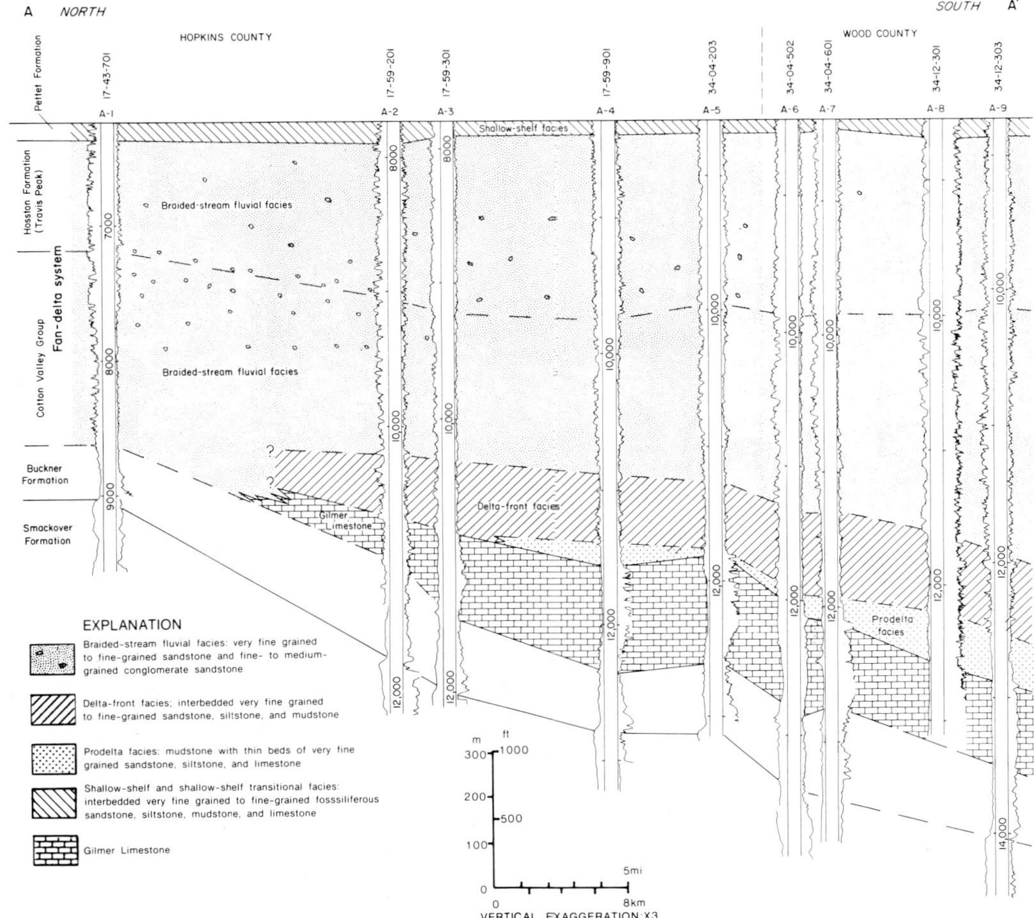

Figure 12. Stratigraphic cross section A-A' through Hopkins and Wood Counties.

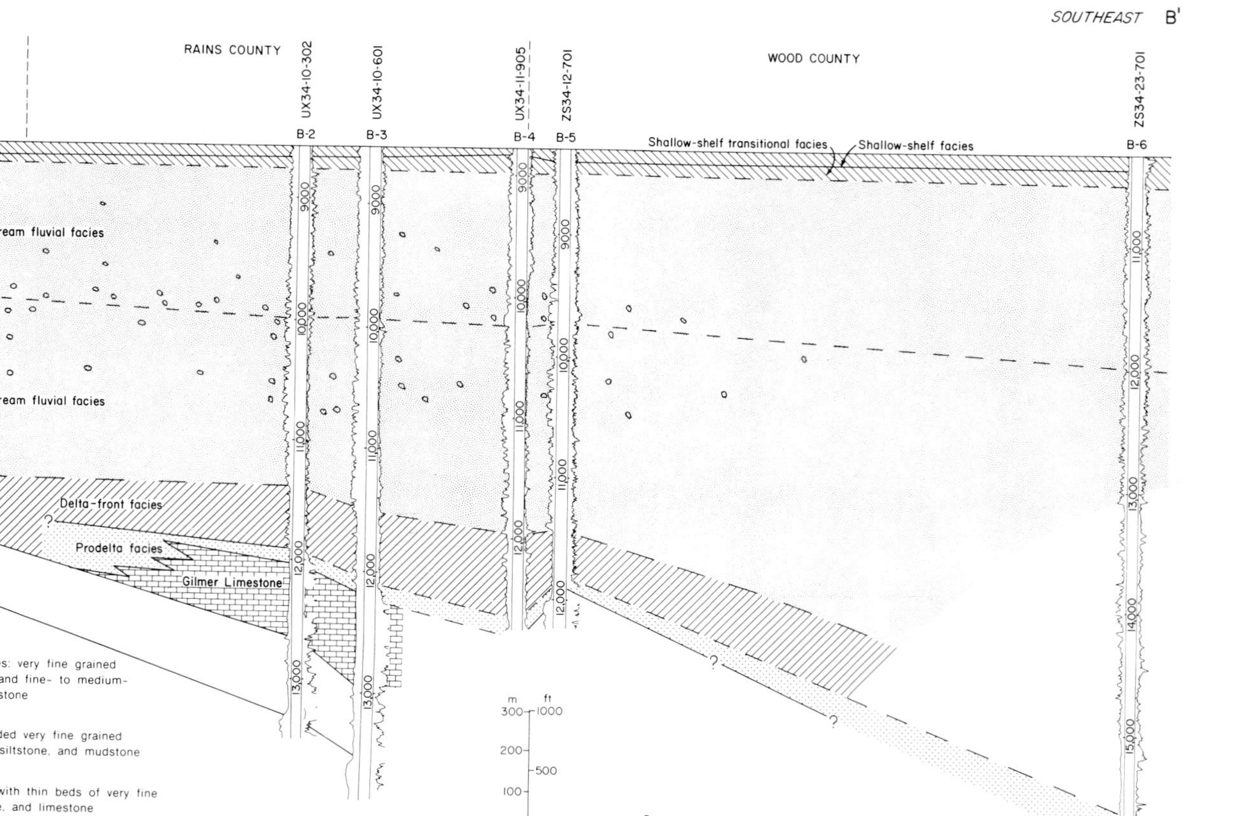

Figure 13. Stratigraphic cross section B-B′ through Hunt, Rains, and Wood Counties.

fragments. Interbedded mudstone is dark gray to black to green. Thick beds of conglomeratic sandstone occur in the proximal areas.

Braided-stream fluvial deposits consist of white to light red, very fine-grained to fine-grained sandstone, conglomerate sandstone, and chert-gravel and quartz conglomerate. Thin beds of red and green mudstone occur as a subordinate lithology. Conglomerate is more common updip, extending basinward as tongues (Newkirk, 1971). This facies exhibits blocky electric log patterns with sharp (erosional) bases and tops, characteristic of braided-stream deposits (Galloway et al., 1979). Well-developed upward-coarsening or upward-fining sections are rare to absent.

b. Sandstone distribution. Sandstone distribution in the Cotton Valley Group is indicated on the sandstone-percent map (Fig. 17). Higher sandstone-percent values are restricted to the basin margin, and values decrease basinward. Many dip-oriented sandstone-rich belts extend across Kaufman, Hunt, Hopkins, and northwestern Van Zandt Counties, suggesting the presence of many smaller streams in this part of the study area. Basinward, the dip-oriented sandstone geometry appears to change into a northeast to southwest strike-oriented trend. The change in orientation of sandstone trends marks the fluvial-marine interface. Seaward of this area, dominant marine processes have reworked sandstones into strike-oriented facies. This change in orientation seems to coincide with the position of an older Smackover-Gilmer carbonate shelf described by Eaton (1961) and Rogers (1967) (Fig. 9).

The Cotton Valley Group gradually thickens basinward toward a line that runs through central Henderson, central Van Zandt, and north-central Wood Counties (Fig. 10). Basinward of this line, the isopach contours outline parallel-aligned, strike-

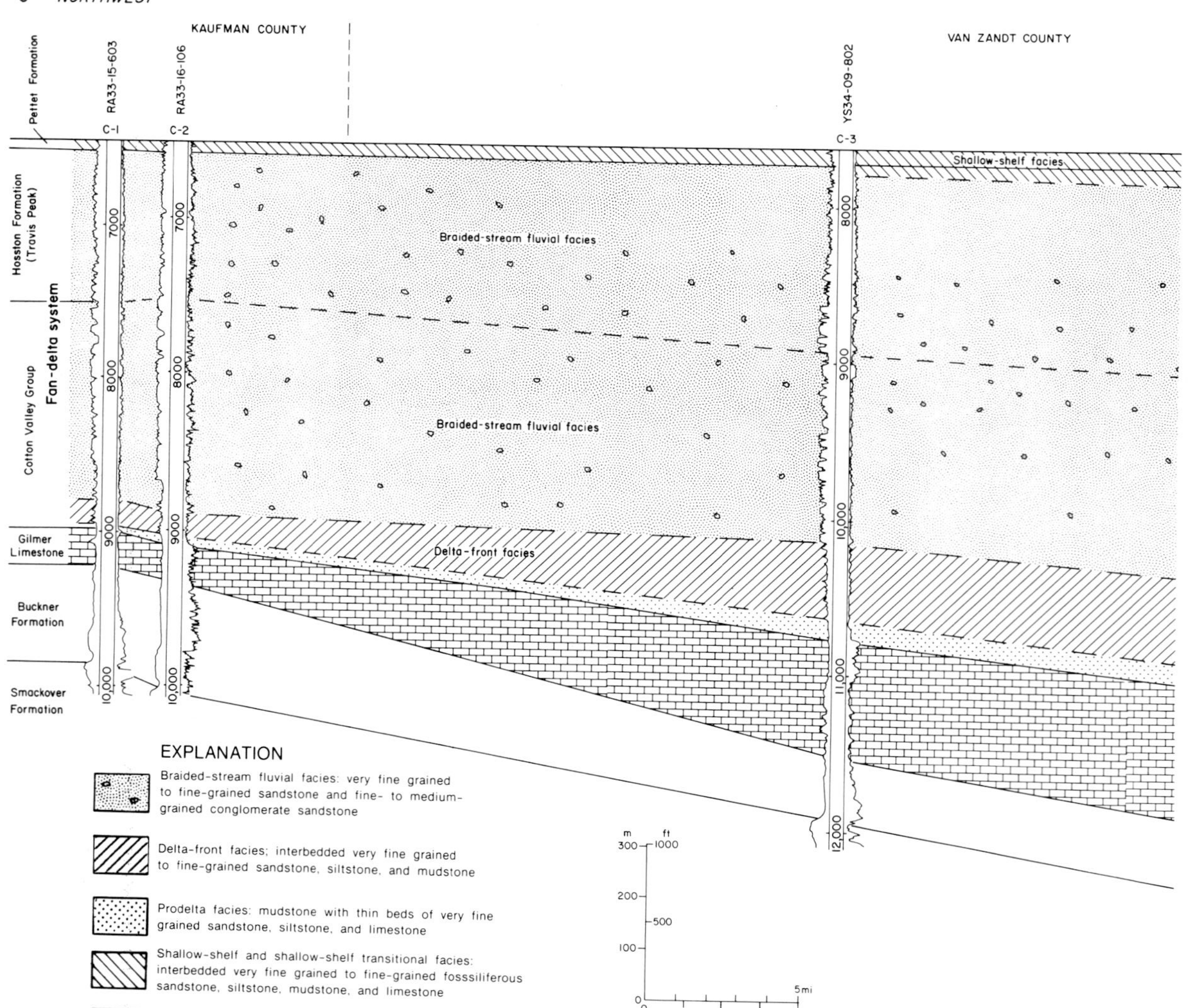

C NORTHWEST
KAUFMAN COUNTY
VAN ZANDT COUNTY
Pettet Formation
RA33-15-603
C-1
RA33-16-106
C-2
YS34-09-802
C-3
Hosston Formation (Travis Peak)
Cotton Valley Group
Fan-delta system
Gilmer Limestone
Buckner Formation
Smackover Formation
7000
8000
9000
10,000
11,000
12,000
Shallow-shelf facies
Braided-stream fluvial facies
Braided-stream fluvial facies
Delta-front facies
EXPLANATION
Braided-stream fluvial facies: very fine grained to fine-grained sandstone and fine- to medium-grained conglomerate sandstone
Delta-front facies; interbedded very fine grained to fine-grained sandstone, siltstone, and mudstone
Prodelta facies: mudstone with thin beds of very fine grained sandstone, siltstone, and limestone
Shallow-shelf and shallow-shelf transitional facies: interbedded very fine grained to fine-grained fosssiliferous sandstone, siltstone, mudstone, and limestone
m
ft
300
1000
200
500
100
0
0
5mi
0
8km

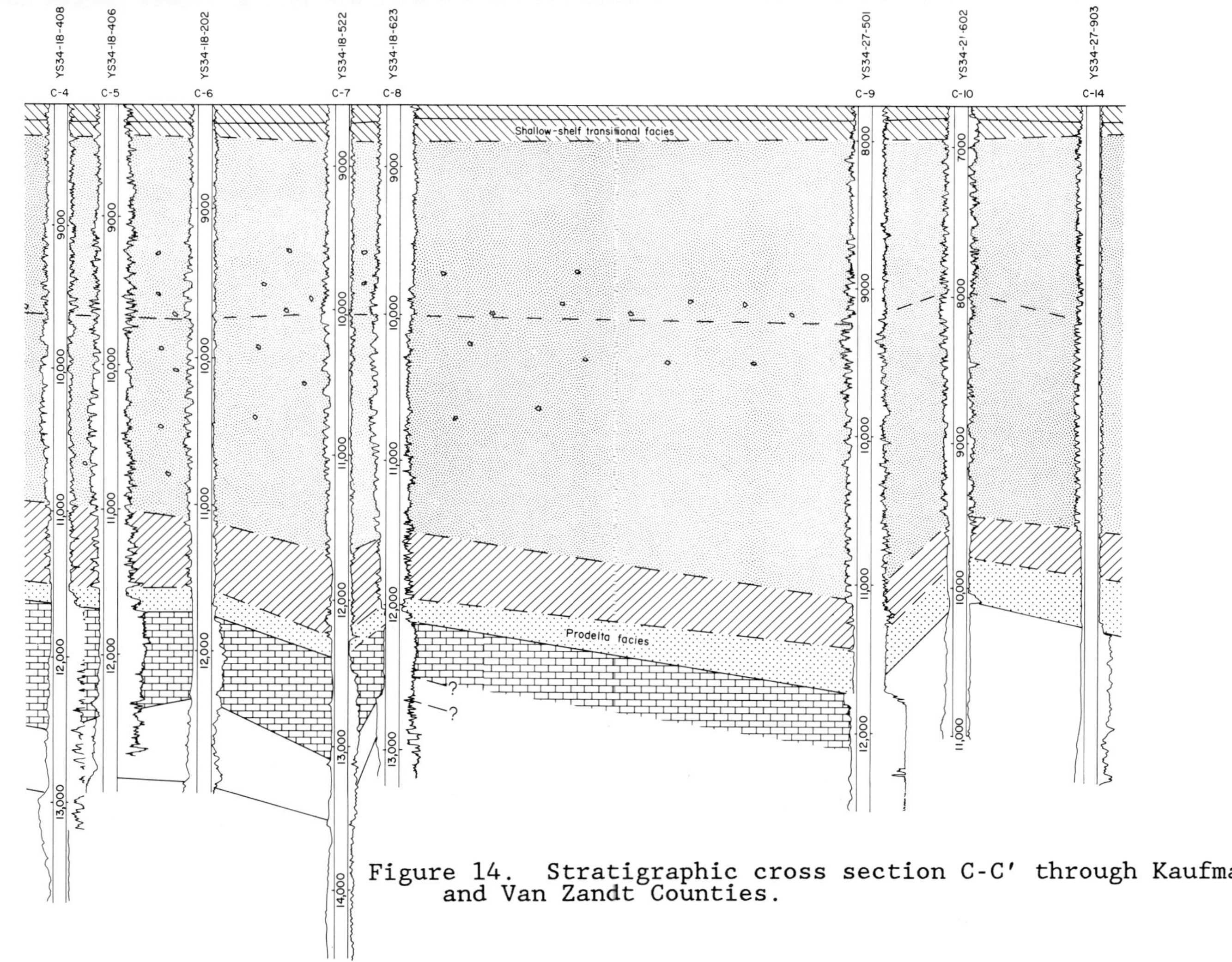

Figure 14. Stratigraphic cross section C-C′ through Kaufman and Van Zandt Counties.

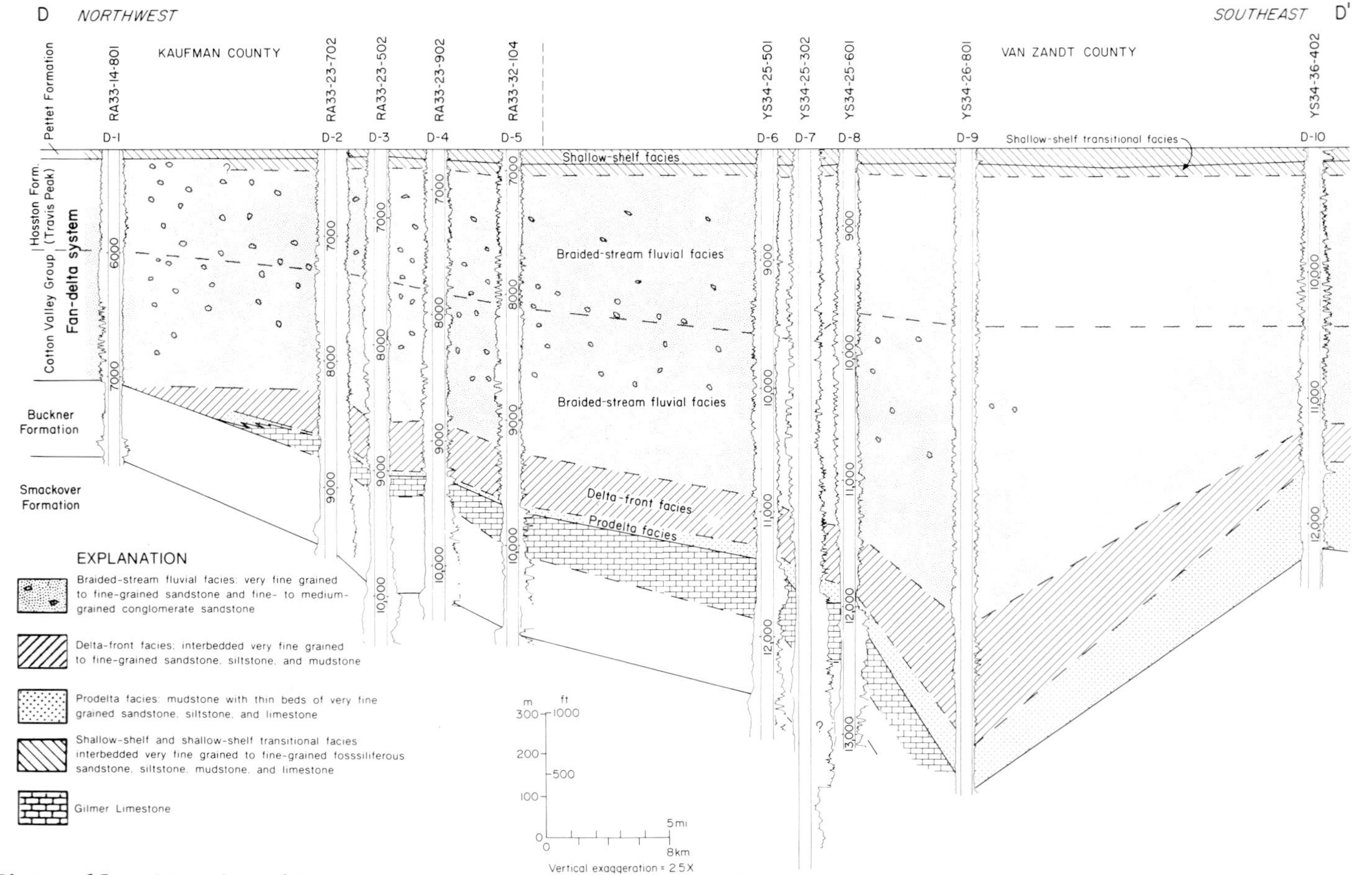

Figure 15. Stratigraphic cross section D-D′ through Kaufman and Van Zandt Counties.

oriented thicks and thins. When compared with the residual gravity map (Fig. 3), sediment thicks correspond to salt-poor (gravity highs), and sediment thins overlie salt structures (gravity lows). Parallel sandstone thicks indicate successive seaward depocenters of prograding fan deltas.

The area where the Cotton Valley Group gradually thickens basinward is coincident with subjacent Smackover-Gilmer carbonate shelf facies. The combined thickness of the Smackover-Gilmer deposits is approximately 3,500 feet (1,067 m). The isopach map

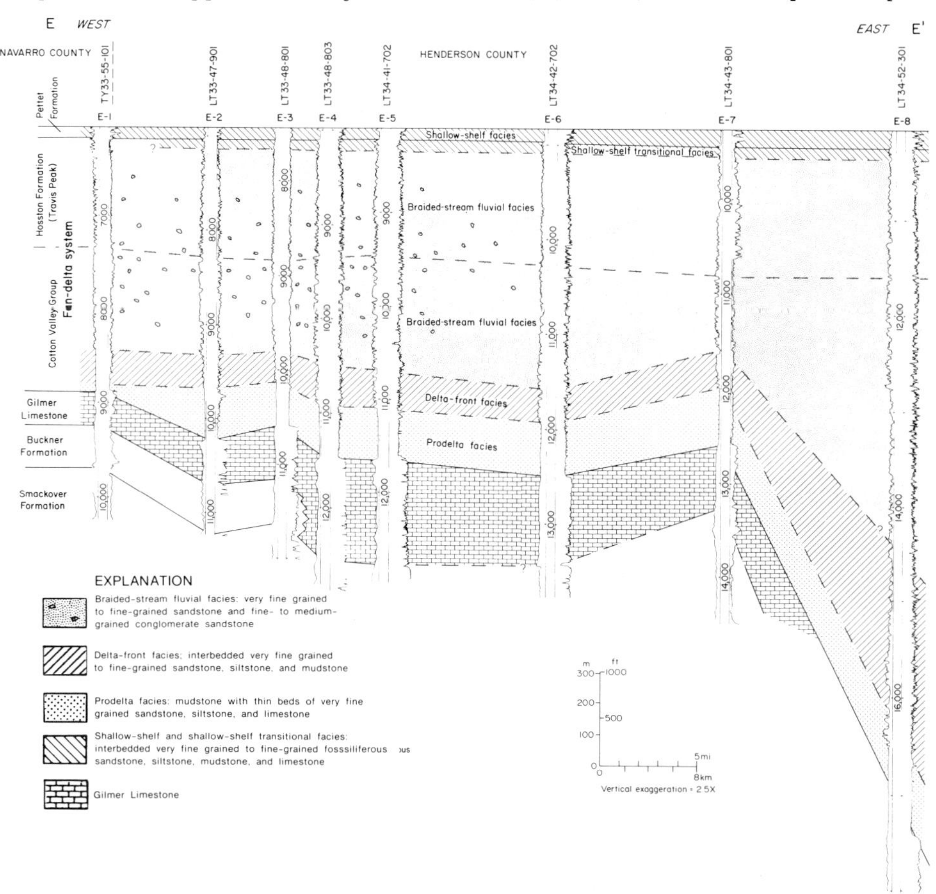

Figure 16. Stratigraphic cross section E-E' through Henderson County.

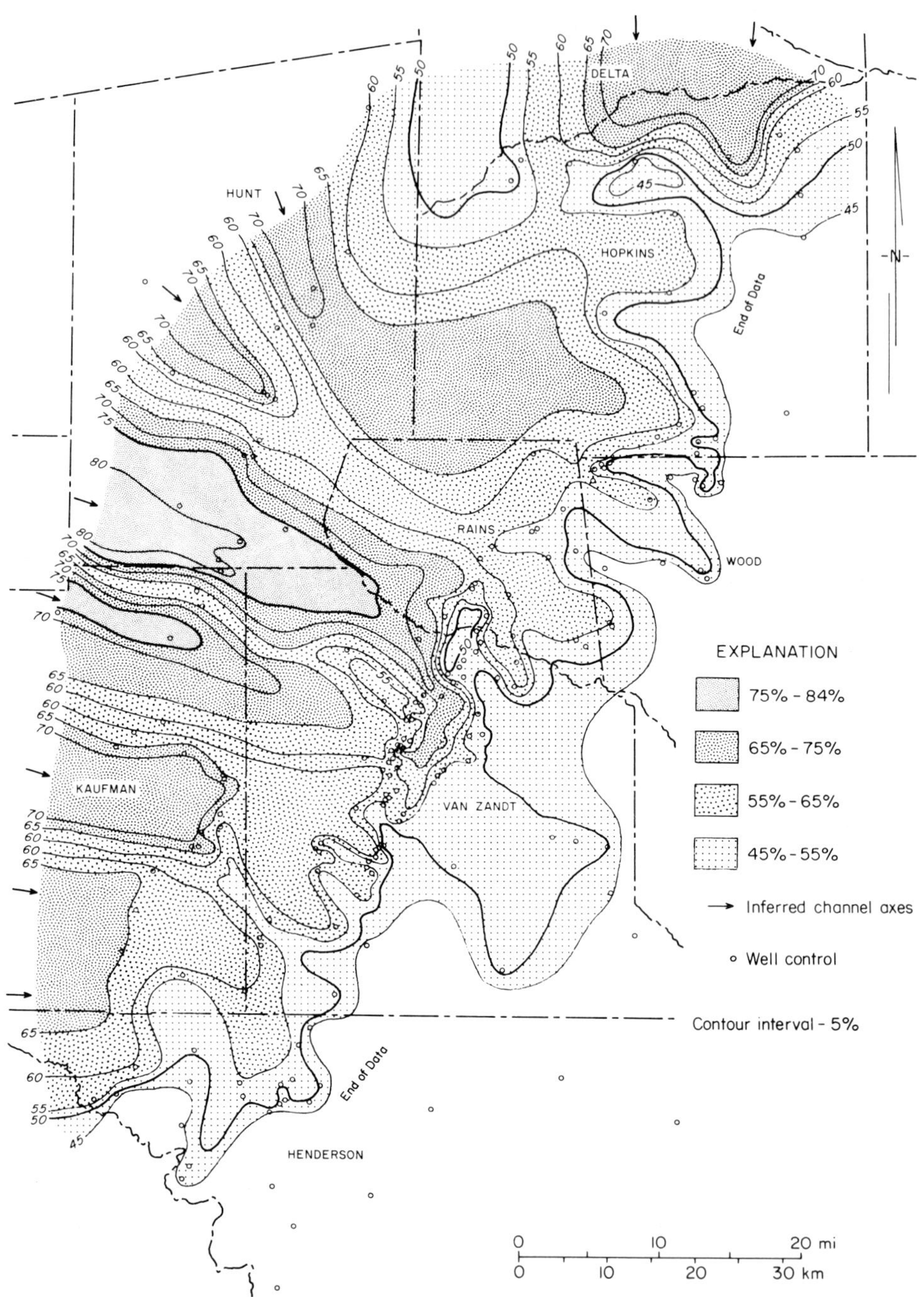
DELTA
HUNT
HOPKINS
RAINS
WOOD
KAUFMAN
VAN ZANDT
HENDERSON
End of Data
End of Data
-N-
EXPLANATION
75% - 84%
65% - 75%
55% - 65%
45% - 55%
Inferred channel axes
Well control
Contour interval - 5%
0 10 20 mi
0 10 20 30 km

of the Cotton Valley suggests that the Smackover-Gilmer shelf provided a stable platform over which the advancing Cotton Valley fan deltas prograded both laterally and basinward. Superpositions of deltaic sediments occurred in synclines that were located between salt ridges immediately basinward of the Smackover-Gilmer carbonate shelf edge. The thickness of the Louann Salt, which directly determined the amount of salt available for the formation of salt structures, also controlled the rate and amount of subsidence that occurred coeval with sedimentation.

The net-sandstone map indicates that sandstone-rich belts coincide with thick isopach trends of the Cotton Valley Group (Fig. 18). Net-sandstone values increase gradually across the Smackover-Gilmer shelf; beyond the shelf edge, net-sandstone highs correspond to sediment thicks indicated on the Cotton Valley isopach map (Fig. 10). Net-sandstone thicks coincide with salt-poor areas (Fig. 3).

Sediment accumulation around the northern and western margins of the basin appears to have been controlled, in part, by contemporaneous faulting in the Mexia-Talco fault zone. In Kaufman County, for example, massive sandstone units in the upper Cotton Valley Group are up to 600 feet (183 m) thick. These units are thought to be superposed braided-stream deposits that accumulated early in grabens of the Mexia-Talco fault zone. Subsidence was accelerated by accumulation of thick braided-stream deposits composed of quartz pebble conglomerate and fine- to medium-grained sandstone.

3. Hosston Formation

a. Facies. Two facies constitute the Hosston Formation: a braided-stream facies, which is dominant, and a transitional shallow marine facies, which is subordinate. The braided-stream

Figure 17. Sandstone-percent map of the Cotton Valley Group (facing page).

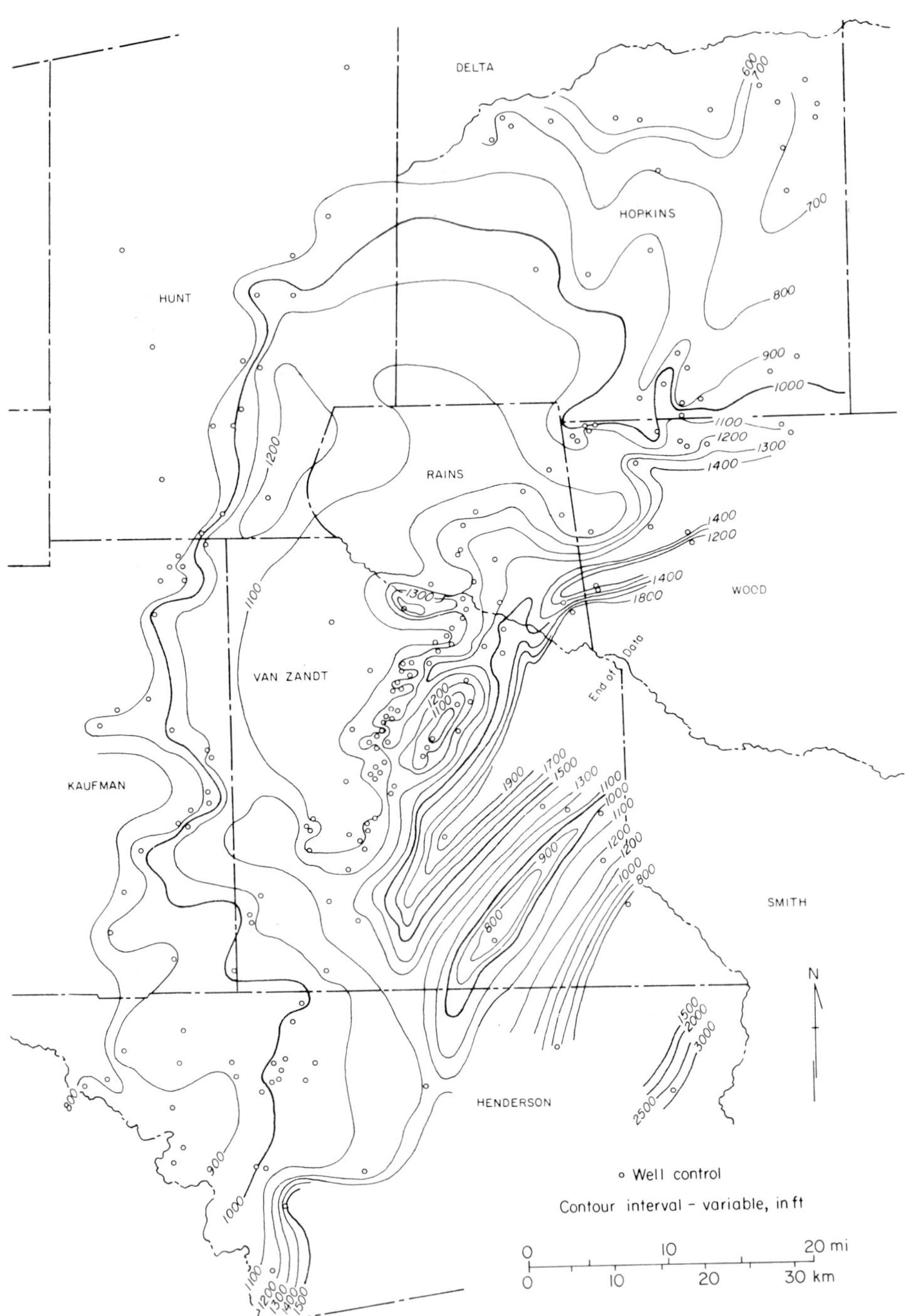
DELTA
HOPKINS
HUNT
RAINS
WOOD
VAN ZANDT
KAUFMAN
SMITH
HENDERSON
End of Data
N
Well control
Contour interval - variable, in ft
0 10 20 mi
0 10 20 30 km

deposits represent the subaerial fan-plain facies, a component facies of the larger fan-delta system (Fig. 11). Fluvial sedimentation ended with a major transgression reflected by a transitional shallow marine facies restricted to the upper 100 to 200 feet (33 to 67 m) of the Hosston Formation. The transitional deposits grade into the overlying Pettet Formation (Limestone).

The fluvial facies is composed of conglomeratic sandstone and white to light-red, fine- to medium-grained sandstone; light-gray to red muddy sandstone; white to pale-red siltstone; and thin beds of red and grayish-green mudstone. Carbonaceous and lignitic material occurs in some sandstone units. Quartz pebble conglomerate beds occur throughout the section but are more common in the lower part of the formation (Bushaw, 1968; Nichols et al., 1968). Electric log patterns of individual sand beds are blocky and have sharp (erosional) lower and upper boundaries. Upward-fining and upward-coarsening log responses are rare. Superposition of braided-stream deposits, occurring contemporaneous with faulting, is not as apparent here as it is in the Cotton Valley Group.

The upper 100 to 200 feet (33 to 67 m) of the Hosston Formation consists of interbedded white to light-red, very fine-grained to fine-grained sandstone, gray mudstone, and gray to light-tan sandy, fossiliferous, oolitic limestone. Uppermost Hosston deposits reflect a decrease in the sediment supply to the East Texas Basin because of a shift from dominantly fluvial deposits to dominantly marine deposits.

b. Sandstone distribution. Sandstone distribution within the Hosston Formation is characterized by dip-oriented sandstone rich belts in proximal areas (Fig. 19). There is, however, a change from many dip-oriented high-sandstone-percent trends in the Cotton Valley to a single dominant dip-oriented sandstone

Figure 18. Net-sandstone map of the Cotton Valley Group (facing page).

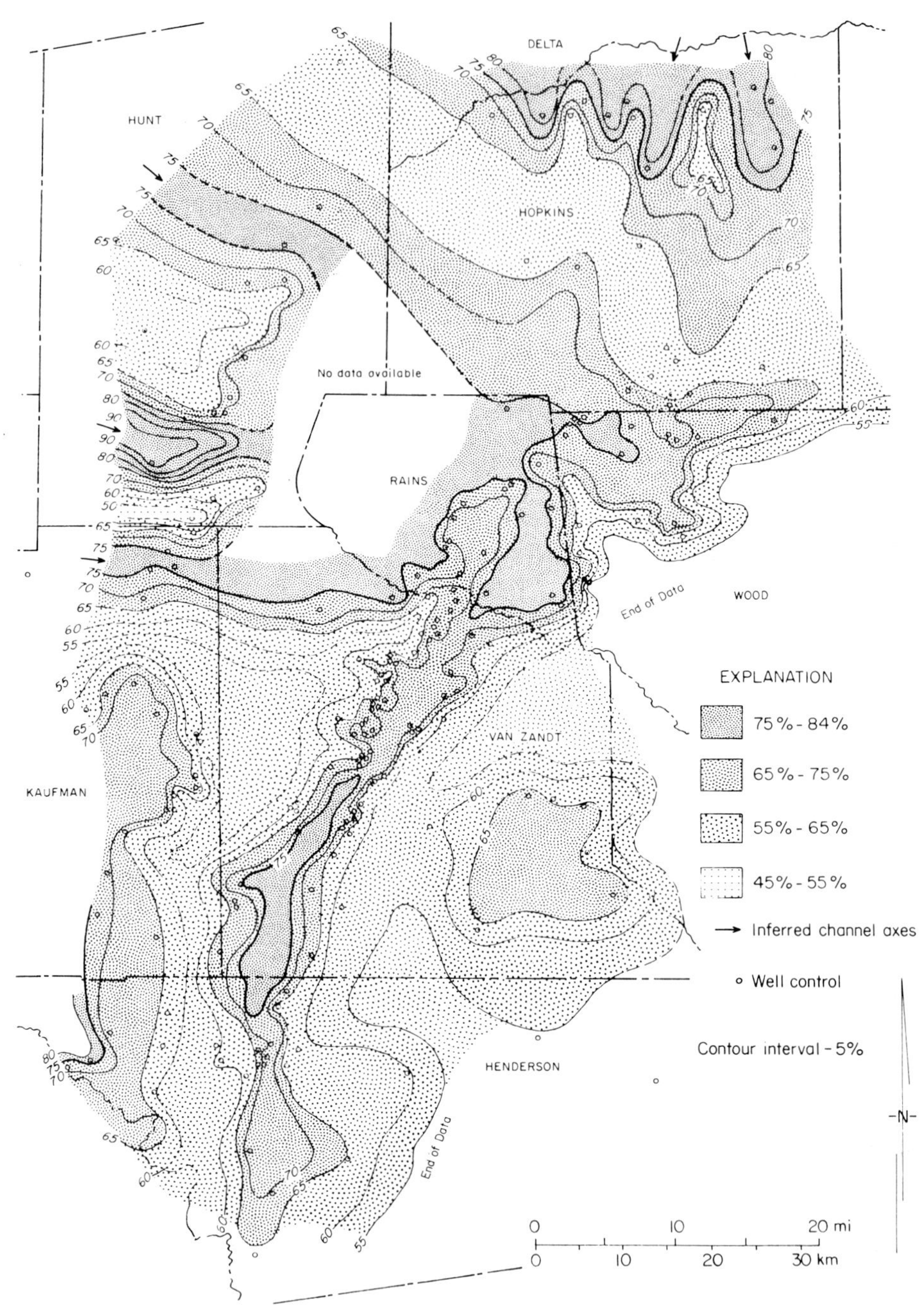
DELTA
HUNT
HOPKINS
No data available
RAINS
WOOD
End of Data
EXPLANATION
75%-84%
65%-75%
55%-65%
45%-55%
Inferred channel axes
Well control
Contour interval - 5%
VAN ZANDT
KAUFMAN
HENDERSON
End of Data
0 10 20 mi
0 10 20 30 km
N

trend in the Hosston centered in Hunt County (Figs. 19 and 21). Narrower sandstone-percent trends in the Hosston occur around the northern margin of the basin (Fig. 20). The sandstone-percent map suggests that the Cotton Valley drainage system had evolved, by Hosston time, from an immature fluvial complex made up of many smaller streams into a more mature system, the principal drainage system being located in the northwestern part of the study area.

In general, net-sandstone and isopach trends in the Hosston Formation (Figs. 20 and 21) resemble those of the Cotton Valley Group. Thickness and net-sandstone values of the Hosston Formation gradually increase basinward across the subjacent Smackover-Gilmer stable carbonate platform. Basinward of the platform, the parallel high-net-sandstone trends and sediment thicks in the Cotton Valley Group coincide with those of the Hosston Formation. Thickness variations, however, are not as great within the Hosston, suggesting that most of the Hosston deltaic sedimentation was basinward of the study area. A map showing depositional environments during middle Hosston time indicates a similar trend (Fig. 8 of Bushaw, 1968). Basinward, the Hosston Formation increases in thickness from 800 to 1,350 feet (244 to 412 m); the Cotton Valley Group exhibits a similar increase, from 1,400 to 5,800 feet (427 to 1,700 m).

IV. DEPOSITIONAL AND STRUCTURAL MODEL

The model selected for use in this study was proposed by Kehle (1971) to explain the initiation of salt movement on an originally flat salt surface. Kehle suggested that mass imbalance, resulting from uneven sediment loading, has greater impact on initial salt migration than any other mechanism. He attributed uneven sediment loading to several sedimentary processes, such as reef development, formation of submarine fans

Figure 19. Sandstone-percent map of the Hosston Formation (facing page).

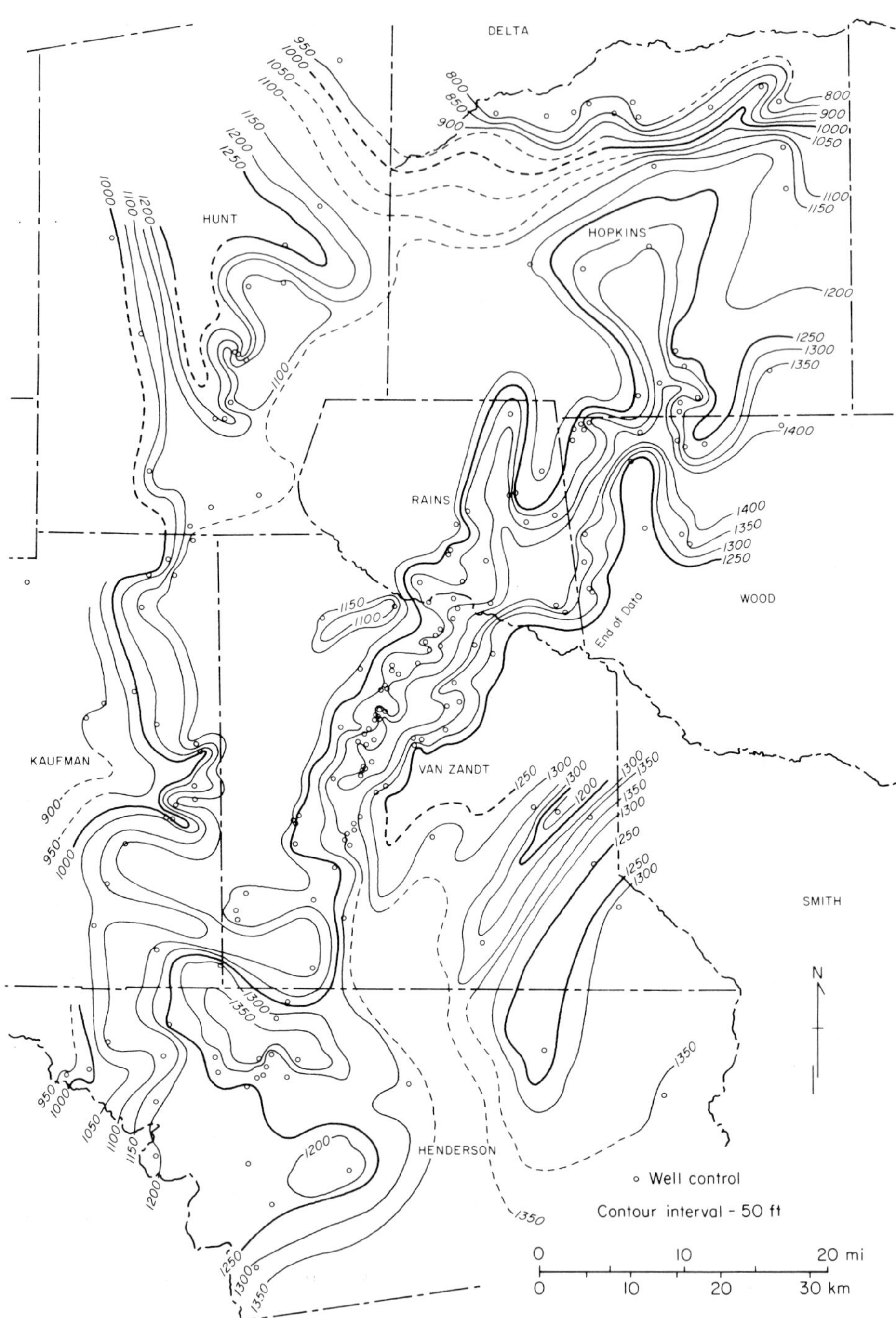
DELTA
HUNT
HOPKINS
RAINS
WOOD
End of Data
KAUFMAN
VAN ZANDT
SMITH
HENDERSON
N
Well control
Contour interval - 50 ft
0 10 20 mi
0 10 20 30 km

at the base of continental slope, and progradation of deltaic and strandline systems. According to this model, the underlying salt flows laterally away from the sedimentary anomaly, thereby forming an initial withdrawal basin. With continued accumulation of sediment, the withdrawal basin enlarges so that salt ridges are formed basinward of the sediment thicks. Lehner (1969) and Martin (1973) documented the presence of analogous features on the abyssal plain and lower continental slope in the northern Gulf of Mexico.

Bishop (1978) proposed a similar model to explain the emplacement of piercement diapirs. His model emphasized the importance of uneven sediment loading but also suggested additional factors that might contribute to the formation of incipient salt structures; these include regional dip, sedimentation rate, progradation rate, sediment density, thickness of overburden, and thickness of the underlying salt.

Fisher (1973) proposed a direct relationship between sedimentation and salt tectonics in the western Gulf Basin. He distinguished two main types of sedimentation styles and their related salt structures. In the first type, high-constructive lobate deltaic systems initiate mobilization of salt. The salt migrates laterally into interdeltaic areas, resulting in the formation of diapirs between the major deltaic lobes. In the second type, strike depositional systems generally induce broad salt ridges, rather than discrete salt structures, in the incipient stages. The second type is closer than the first to the salt structures that are thought to have formed during deposition of the Late Jurassic fan delta system.

In using these two models to explain how Late Jurassic depositional systems influenced salt migration, the following sequence of events is suggested: Reversal of dip along the rift margin allowed an influx of terrigenous clastics into the East

Figure 20. Isopach map of the Hosston Formation (facing page).

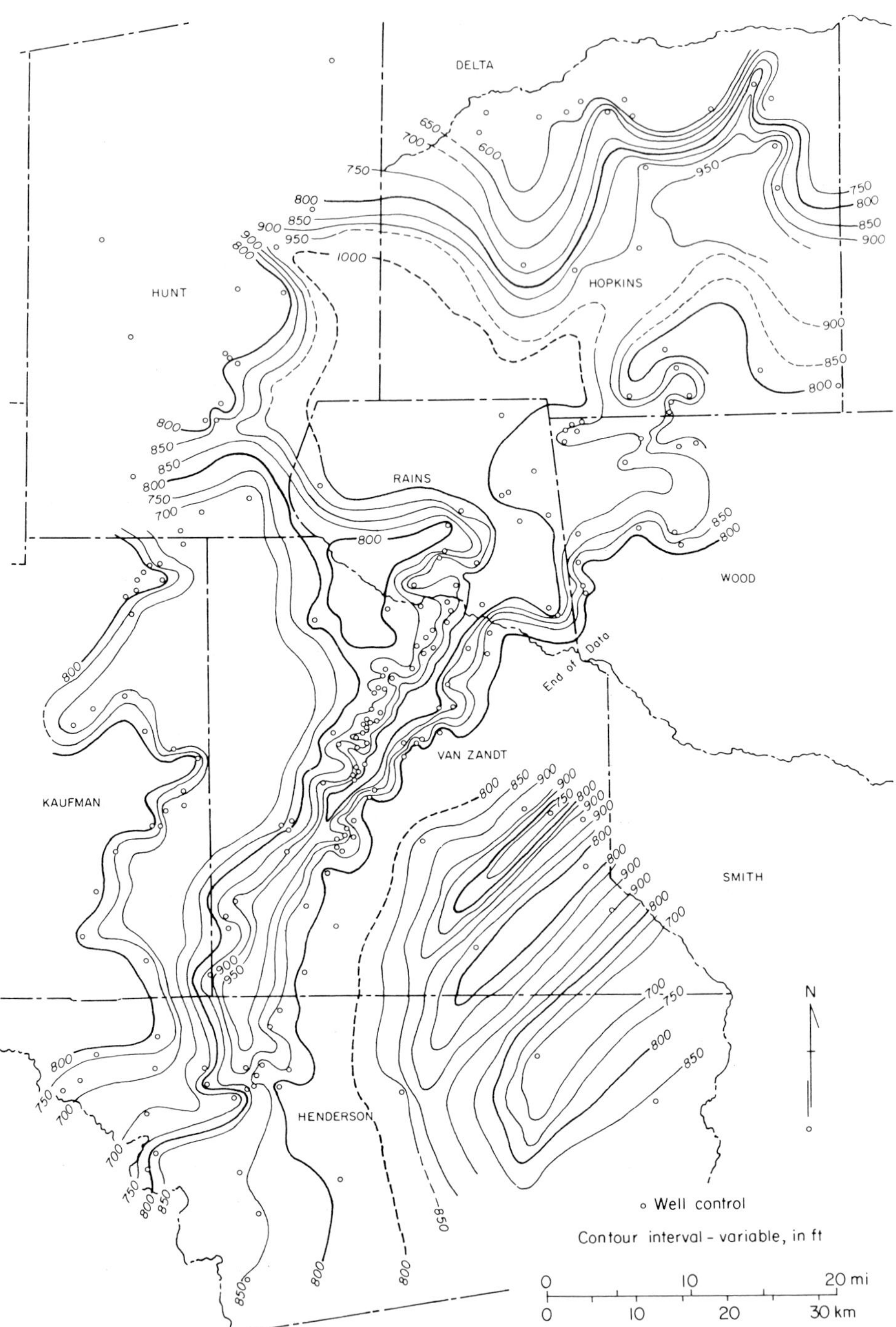
DELTA
HOPKINS
HUNT
RAINS
WOOD
KAUFMAN
VAN ZANDT
SMITH
HENDERSON
End of Data
N
o Well control
Contour interval - variable, in ft
0 10 20 mi
0 10 20 30 km

Texas Basin during the Late Jurassic and Early Cretaceous. Before that time, main drainage along the western margins of the basin flowed into the Triassic basins to the west. The Ouachita Mountains to the north were low-lying and thus supplied only a small quantity of terrigenous clastic sediment, which was limited mostly to the northern periphery of the basin.

As the supply of terrigenous clastic sediment increased, Cotton Valley and Hosston fan-delta systems, which were advancing southward and eastward and were supplied by a braided-stream complex, began prograding across the Smackover-Gilmer carbonate shelf in the East Texas Basin (Fig. 22). The shelf was a stable platform, and although regional subsidence occurred, it impeded the formation of local depocenters. Most likely, the seas over the carbonate platform were shallow and the rate of progradation of the advancing deltaic complex was rapid. Low-amplitude salt structures occur under the Smackover-Gilmer shelf complex, apparently the result of downward creep that was caused by loading of carbonate deposits (Rogers, 1967; Jackson and Harris, 1981). These low-amplitude salt swells are not evident in the overlying Cotton Valley-Hosston sediments.

Basinward of the ancient Jurassic shelf edge, the rate of progradation slowed, while subsidence caused by salt migration and increased water depth allowed the fan delta complex to stack up, forming elongate depocenters. Salt migrated basinward, forming a salt ridge in front of the sediment wedge. This ridge acted as a dam and thereby enhanced the effectiveness of the syncline as a sediment trap. As the Louann Salt became depleted through basinward migration, subsidence slowed and allowed aggradation to surpass it. Consequently, each subsequent depocenter and associated salt ridge shifted basinward, resulting in a series of parallel salt ridges and sediment thicks. As the

Figure 21. Net-sandstone map of the Hosston Formation (facing page).

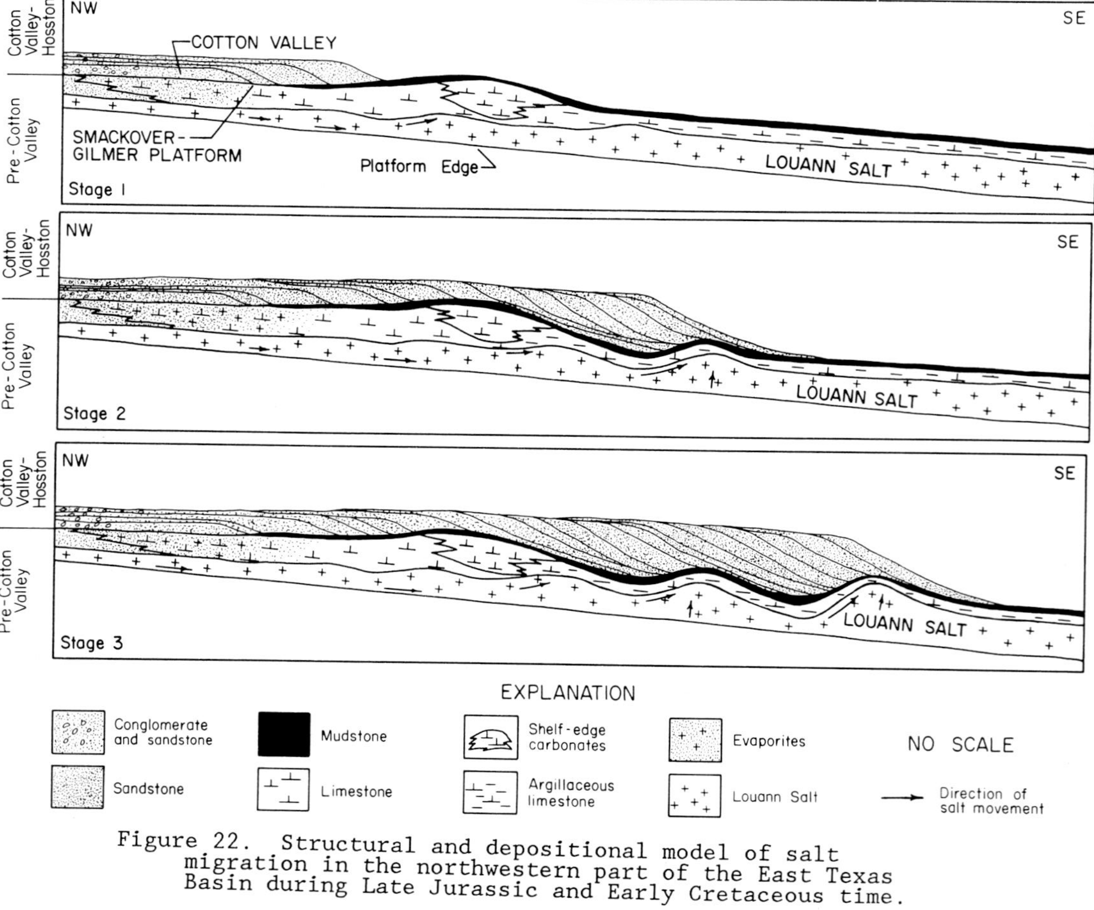

Figure 22. Structural and depositional model of salt migration in the northwestern part of the East Texas Basin during Late Jurassic and Early Cretaceous time.

older salt ridges became buried, continued salt migration allowed the original salt ridges to evolve into separate salt structures, such as salt anticlines and domes.

Toward the end of Early Cretaceous time, the volume of terrigenous clastic sediment entering the East Texas Basin decreased. The transitional nature of the uppermost Hosston and marine deposits in overlying units indicate that a marine transgression occurred.

V. CONCLUSIONS

Subsurface mapping of the Cotton Valley and Hosston depositional systems and subjacent salt structures leads to four major conclusions regarding the role of basin infilling in the initiation of salt movement within the northwestern part of the East Texas Basin. Absence of data in the central part of the basin, however, precludes delineation of the basinward extent of the proposed depositional systems.

First, a prograding fan delta system comprises the Cotton Valley-Hosston stratigraphic units (McGowen and Harris, 1984). The strata between the Gilmer and Pettet Limestones were deposited during a period of regression that was ended by a major post-Hosston transgression. The transgression is evidenced by transitional shallow marine strata in the upper 100 to 200 feet (33 to 66 m) of the Hosston Formation and by upward gradation into the overlying Pettet Limestone.

Second, the initial movement of salt that occurred in the proximal parts of the East Texas Basin during Smackover deposition was the result of downward creep that was induced by basinward tilting (Jackson and Harris, 1981). The resulting salt movement produced small salt structures. Relief on the salt structures increased basinward, coincident with thickening of the salt wedge.

Third, salt mobilization by mass imbalance was induced by Cotton Valley and Hosston deltaic deposition. Uneven sediment loading was controlled by the position of the underlying Smackover-Gilmer carbonate shelf. The Smackover-Gilmer carbonate shelf comprised a stable platform that impeded local subsidence and vertical aggradation of deltaic deposits. Therefore, advancing Cotton Valley-Hosston fan deltas spread laterally and basinward, depositing a fairly uniform wedge of sediments that gradually thickened basinward. Mass imbalance became a viable mechanism to initiate salt movement only when the advancing fan deltas prograded basinward of the stable carbonate platform. Salt migrated basinward of the prograding sediment wedge, forming a proximal incipient withdrawal basin and distal salt ridge. Subsequent depocenters and salt ridges shifted basinward, forming parallel sediment thicks and salt ridges.

Fourth, parallel salt ridges that formed during deposition of the Cotton Valley and Hosston apparently were the initial stage in the development of salt anticlines and domes. Through continued loading, the salt ridges evolved into discrete salt structures. Domal growth appears to depend directly on continued sediment loading, which occurs during periods of major deltaic deposition where the underlying salt wedge is adequately thick.

VI. ACKNOWLEDGEMENTS

Funding of this study was provided by the U.S. Department of Energy under Contract No. DE-AC97-80ET46617. Disscussions with M.P.A. Jackson, C.W. Kreitler, J.H. McGowen and S. J. Seni were very helpful. L.F. Brown, Jr., R.J. Finley, W.E. Galloway, M.P.A. Jackson, C.W. Kreitler, S.J. Seni and Noel Tyler reviewed the manuscript and made many helpful suggestions. Cynthia Lopez and Keith Pollman served as research assistants. The report was edited by Jean Trimble. Figures were drafted under the direction of Dan F. Scranton by John T. Ames, Thomas M. Byrd, Micheline R. Davis, Margaret L. Evans, Richard P. Flores, Byron P. Holbert, Jeffrey Horowitz, and Jamie McClelland.

REFERENCES

Agagu, O.K., McGowen, M.K., Wood, D.H., Basciano, J.M., and Harris, D.W. (1980). Regional tectonic framework of the East Texas Basin in Kreitler, C.W., and others, Geology and geohydrology of the East Texas Basin--a report on the progress of nuclear waste isolation feasibility studies (1979): The University of Texas at Austin, Bureau of Economic Geology Geological Circular 80-12, p. 11-19.

Barnes, V.E. (1965). Tyler sheet: University of Texas, Austin, Bureau of Economic Geology, Geologic Atlas of Texas, scale 1:250,000.

______, (1972). Dallas sheet: The University of Texas at Austin, Bureau of Economic Geology, Geologic Atlas of Texas, scale 1:250,000.

Beall, R. (1975). Plate tectonics and the origin of the Gulf of Mexico: Gulf Coast Association of Geological Societies Transactions 23, 109-114.

Bishop, R.S. (1978). Mechanism for emplacement of piercement diapirs: American Association of Geological Societies Bulletin 62, 1561-1583.

Burke, K., and Dewey, J.F. (1973). Plume-generated triple junctions: key indicators in applying plate tectonics to old rocks: Journal of Ccology 81, 406-433.

Bushaw, D.J. (1968). Environmental synthesis of the East Texas Lower Cretaceous: Gulf Coast Association of Geological Societies Transactions 18, 416-438.

Eaton, R.W. (1961). Development and future possibilities of the Jurassic in northeast Texas: South Texas Geological Society Publication, 15 p., reprinted in 1964, South Texas Geological Society Bulletin 4, 4-13.

Fisher, W.L. (1973). Deltaic sedimentation, salt mobilization, and growth faulting in Gulf Coast Basin (abs.): American Association of Petroleum Geologists Bulletin 54, 779.

Forgotson, and Forgotson, J.M., Jr. (1976). Definition of Gilmer Limestone Upper Jurassic formation: American Associatiion of Petroleum Geologists Bulletin 60, 1119-1123.

Galloway, W.E., Kreitler, W.W. and McGowen, J.H. (1979). Depositional and ground-water flow systems in the exploration for uranium: The University of Texas at Austin, Bureau of Economic Geology Research Colloquium, 1978, 267 p.

Hughes, D.J. (1968). Salt tectonics as related to several Smackover fields along the northeast rim of the Gulf of Mexico Basin: Gulf Coast Association of Geological Societies Transactions 18, 320-330.

Imlay, R.W. (1943). Jurassic formations of the Gulf region: American Association of Petroleum Geologists Bulletin 27, 1407-1533.

Jackson, M.P.A. (1982). Fault tectonics of the East Texas Basin: The University of Texas at Austin, Bureau of Economic Geology Geological Circular 82-4, 31 p.

_________, and Harris, D.W. (1981). Seismic stratigraphy and salt mobilization along the northwestern margin of the East Texas Basin, in Kreitler, C.W., and others, Geology and geohydrology of the East Texas Basin: a report of the progress of nuclear waste isolation feasibility studies (1980): The University of Texas at Austin, Bureau of Economic Geology Geological Circular 81-7, p. 28-32.

_________, Seni, S.J., and McGowen, M.K., (1982). Initiation of salt flow in the East Texas Basin (abs.): American Association of Petroleum Geologists Bulletin 66, 584-585.

Johnson, L.C. (1980). Structure and stratigraphy of an evolving salt ridge and basin complex, Louisiana continental shelf: The University of Texas at Austin, Master's Thesis, 120 p.

Kehle, R.O. (1971). Origin of the Gulf of Mexico: The University of Texas at Austin, Geology Library, call no. q557K260, unpublished report, unpaginated.

Lehner, P (1969). Salt tectonics and Pleistocene stratigraphy on continental slope of northern Gulf of Mexico: American Association of Petroleum Geologists Bulletin 53, 2431-2482.

Loocke, J.E. (1978). Growth history of the Hainesville salt dome, Wood County, Texas: The University of Texas at Austin, Master's Thesis, 95 p.

Lucchi, F.R., Colella, A., Ori, G.G. and Oglian, F., 1981, Pliocene fan deltas of the Intra-Apenninic Basin, Bologna, in Lucchi, F.R., ed. International Association of Sedimentologists Excursion Guidebook 4, 162 p.

Martin, R.G. (1973). Salt structure and sediment thickness, Texas-Louisiana continental slope, northwestern Gulf of Mexico: U.S. Geological Survey Open-File Report, 21 p.

McGowen, J.H. (1970). Gun Hollow fan delta, Nueces Bay, Texas: The University of Texas at Austin, Bureau of Economic Geology Report of Investigations 69, 91 p.

McGowen, M.K., and Harris, D.W. (1981). Preliminary study of the Upper Jurassic (Cotton Valley) and Lower Cretaceous (Hosston-/Travis Peak) Formations of the East Texas Basin, in Kreitler, C.W., and others, Geology and geohydrology of the East Texas Basin--a report on the progress of nuclear waste isolation feasibility studies (1980): The University of Texas at Austin, Bureau of Economic Geology Geological Circular 81-7, p. 43-47.

_________, and Harris, D.W. (1984). Cotton Valley (Upper Jurassic) and Hosston (Lower Cretaceous) Depositional Systems and their Influence on Salt Tectonics in the East Texas Basin in Ventress, W.P.S., Bebout, D.G., Perkin, B.F., and Moore, C.H., eds., The Jurassic of the Gulf Rim, GCS SEPM Third Annual Research Conference Proceedings, p. 213-253.

Miall, A.D. (1977). A review of the braided-river depositional environment, in Earth Science Reviews 13, 1-62: Amsterdam, Elsevier Scientific Publishing Co.

Moore, R.W. and Del Castillo, Luis (1974). Tectonic evolution of the southern Gulf of Mexico: Geological Society of America Bulletin 85, 607-618.

Newkirk, T.F. (1971). Possible future petroleum provinces of Jurassic, western Gulf Basin, in Cram, I.H., ed., Future petroleum provinces of the United States--their geology and potential: American Association of Petroleum Geologists Memoir 15, 927-953.

Nichols, P.H., Peterson, G.E., and Wuestner, C.E. (1968). Summary of subsurface geology of northeast Texas: American Association of Petroleum Geologists Memoir 9, 982-1004.

Parker, T.J. and McDowell, A.M. (1955). Model studies of salt dome tectonics: American Association of Petroleum Geologists Bulletin 39, 2384-2470.

Raup, O.B. (1970). Brine mining--an additional mechanism for formation of basin evaporites: American Association of Petroleum Geologists Bulletin 54, 2246-2259.
Rogers, J.K. (1967). Comparison of some Gulf Coast Mesozoic carbonate shelves: Gulf Coast Association of Geological Societies Transactions 17, 49-60.
Rosenkrans, R.R. and Marr, J.D. (1967). Modern seismic exploration of the Gulf Coast Smackover trend: Geophysics 32, 184-206.
Rust, B.R. (1978). Depositional models for braided alluvium, in Miall, A.D., ed., Fluvial sedimentology: Canadian Society of Petroleum Geologists Memoir 5, 605-626.
Salvador, A. and Green, A.R. (1980). Opening of the Caribbean Tethys (origin and development of the Caribbean and the Gulf of Mexico), in Colloque C5: Geologie de chaines alpines issues de la Tethys: Bureau de Recherches Geologiques et Minieres Memoire 115, 224-229.
Seni, S.J. and Jackson, M.P.A. (1983). Evolution of Salt Structures, East Texas Diapir Province, part 2: Patterns and Rates of Halokinesis: American Association of Petroleum Geologists 67, 1245-1274.
Thomas, W.A. and Mann, C.J. (1966). Late Jurassic depositional environments, Louisiana and Arkansas: American Association of Petroleum Geologists Bulletin 50, 178-182.
Todd, R.G. and Mitchum, R.M., Jr. (1977). Seismic stratigraphy and global changes of sea level, part 8: identification of Upper Triassic, Jurassic, and Lower Cretaceous seismic sequences in Gulf of Mexico and offshore West Africa, in Payton, C.E., ed., Seismic stratigraphy--applications to hydrocarbon exploration: American Association of Petroleum Geologists Memoir 26, 145-163.
Trusheim, F. (1960). Mechanisms of salt migration in northern Germany: American Association of Petroleum Geologists Bulletin 44, 1519-1540.
Turk, Kehle, and Associates (1978). Tectonic framework and history, Gulf of Mexico region: Report for Law Engineering Testing Co., Marietta, Georgia, 28 p.
Walper, J.L. (1980). Tectonic evolution of the Gulf of Mexico, in Pilger, R.H., Jr., ed., The origin of the Gulf of Mexico and early opening of the central North Atlantic Ocean: Louisiana State University Symposium Proceedings, Baton Rouge, 1980, p. 87-98.
Wood, D.H. (1981). Structural effects of salt movement in the East Texas Basin, in Kreitler, C.W. and others, Geology and geohydrology of the East Texas Basin--a report on the progress of nuclear wast isolation feasibility studies: The University of Texas at Austin, Bureau of Economic Geology Geological Circular 81-7, p. 21-27.
Wood, M.L., and Walper, J.L (1974). The evolution of the interior western basins and the Gulf of Mexico: Gulf Coast Association of Geological Societies Transactions 24, 31-41.

APPENDIX A

Cross-section wells

Well No.	Well Name	County
Line A-A′		
A-1 LZ 17-43-701	Humble Oil & Refining Co. No. 1 Dunham	Hopkins
A-2 LZ 17-59-201	Sunray DX Oil Co. No. 1 Seaman	Hopkins
A-3 LZ 17-59-301	North Central Oil Co. No. 1 Moseley	Hopkins
A-4 LZ 17-59-901	Hinton No. 1 Walker	Hopkins
A-5 ZS 34-04-203	Forest Oil Corp. No. 1 Asher	Hopkins
A-6 ZS 34-04-502	Hughley Operating Co. and No. Am. Expl. Co.	Wood
A-7 ZS 34-04-601	Humble Oil & Refining Co. No. 1. Allen	Wood
A-8 ZS 34-12-301	Getty Oil No. 1 Blalock	Wood
A-9 ZS 34-12-303	Shell Oil Co. No. 1 Wright	Wood
Line B-B′		
B-1 PH 33-08-204	Ohio Oil Co. No. 1 Popper	Hunt
B-2 UX 34-10-302	Texaco, Inc. Irvine Gas Unit No. 1	Rains
B-3 UX 34-10-601	Delta Drilling No. 1 Hare	Rains
B-4 UX 34-11-905	Caraway & Smith No. 1 Gilley	Rains
B-5 ZS 34-12-701	Samedan Oil Cor. Buchanan No. 1	Wood
B-6 ZS 34-23-701	Humble Oil & Refining Co. NW Hawkin No. 1	Wood
Line C-C′		
C-1 RA 33-15-603	Schneider and Murray No. 1 Jones Estate	Kaufman
C-2 RA 33-16-106	Santa Fe Minerals, Inc. No. 1 Barrow Estate	Kaufman
C-3 YS 34-09-802	E.C. Johnston Co. No. 1 Martin Gas Unit	Van Zandt
C-4 YS 34-18-408	Pan American Petr. Corp. Brown Gas Unit B-1	Van Zandt
C-5 YS 34-18-406	Pan American Petr. Corp. No. 1 Nichols Gas U.	Van Zandt

APPENDIX A (cont'd)

Cross-section wells

Well No.	Well Name	County
Line C-C' (cont'd)		
C-6 YS 34-18-202	Caraway and Smith Parker Gas Unit No. 2	Van Zandt
C-7 YS 34-18-522	R.J. Caraway No. 1 Stone (Fruitvale GU)	Van Zandt
C-8 YS 34-18-623	Continental Oil Co. No. 1 Elliot	Van Zandt
C-9 YS 34-27-501	Halbouty No. 1 Bowan	Van Zandt
C-10 YS 34-27-602	Midwest et al. No. 1 Clark	Van Zandt
C-11 YS 34-27-903	Pure Oil Co. D-8 Swain No. 11	Van Zandt
Line D-D'		
D-1 RA 33-14-801	Rockwall Exploration Co. No. 1 Wallace	Kaufman
D-2 RA 33-23-702	W. M. Hughes No. 1 Billings	Kaufman
D-3 RA 33-23-502	The T x 1 Oil Corp. No. 1 Liston	Kaufman
D-4 RA 33-23-902	Southland Royalty Co. No. 1 Frosch Unit	Kaufman
D-5 RA 33-32-104	Union Tex. Pet.-Lacal Pet. No. 1 Phillips	Kaufman
D-6 YS 34-25-501	R.J. Caraway No. 1 Yates	Van Zandt
D-7 YS 34-25-302	Superior Oil Co. No. 1 Porter	Van Zandt
D-8 YS 34-25-601	R.J. Caraway No. 1 Parker	Van Zandt
D-9 YS 34-26-801	R.J. Caraway No. 1 Gilmore	Van Zandt
D-10 YS 34-36-402	Pan American Petr. Corp. No. 1 Hobbs	Van Zandt
E-1 TY 33-55-101	Humble Oil & Refining Co.	Navarro
E-2 LT 33-47-0-1	Max Pray No. 1 Carter	Henderson
E-3 LT 33-48-801	Pan American Petr. Corp. No. 1 Sorrell	Henderson
E-4 LT 33-48-803	Rudman Resources, Inc. No. 1 Hammock	Henderson
E-5 LT 34-41-702	Lake Ronel Oil Co.-Bill Ross No. 1 Shaver	Henderson

APPENDIX A (cont'd)

Cross-section wells

Well No.	Well Name	County
Line E-E'		
E-6 LT 34-42-702	B. Smith, G. Lehnertz, W. Perryman No. 1 Lee	Henderson
E-7 LT 34-43-801	Lone Star Producing Co. 1-B Allyn	Henderson
E-8 LT 34-57-301	Texas Interstate Oil & Gas Co. No. 1 Cotton	Henderson

MODELLING OF BUOYANT SALT DIAPIRISM

I. Lerche[1]
J.J. O'Brien[2]

[1]Dept. of Geology
University of South Carolina
Columbia, SC 29208

[2]Standard Petroleum of Alaska
900 East Benson Blvd.
Anchorage, Alaska 99519

I. INTRODUCTION

While an extensive literature exists on the geology of salt deposits and structures, the work to date has been predominantly descriptive in nature (see, for example, Nettleton, 1955; Biot and Ode, 1965; Berner et al., 1972; Bishop, 1978). Many physical processes related to salt diapirism have been proposed but, in the main, have been discussed only in a qualitative manner. As a result it is difficult to gauge the relative significance of the various processes to determine which are the dominant ones, which are the minor ones and under which conditions this is so. The goal of the present work is to examine many of the physical processes associated with the development of salt structures due to buoyant instability in a quantitative fashion. We find that several of these processes are extremely robust; their development does not depend on the fine details of the models we study. We infer that these processes must be associated with the development of salt structures which result from buoyant instability. Examples of these factors include the upward buoyancy force which is exerted on a deeply buried salt body, the dependence of the buoyancy force on the contrast in density between salt and surrounding sediments, the depth of burial at which a buoyancy force is first exerted, the deformation of sediments overlying the salt body and the resulting stress in the sedimentary layers.

The presence of massive salt deposits can have a major impact on both the structural development of, and the trapping of hydrocarbons in, a sedimentary basin. This results from (1) the ability of salt to flow plastically on a geologic time scale under the influence of a differential stress, unlike other lithologies which deform basically as solids. (Overpressured shales may also flow plastically and so structures related to the flow of such shales bear many resemblances to structures related to salt flow.); (2) salt density remains approximately constant upon burial, in contrast with terrigenous clastic material which loses porosity and increases in density upon burial. At a critical depth of burial of the salt deposit a gravitational instability may develop wherein a more dense material overlies a less dense material which has the ability to flow on a geologic time scale. A salt dome may then develop in response to this instability. Other mechanisms may also provide the driving force for salt flow and the development of salt structures, e.g. differential loading or tectonism, but these are not considered in this paper.

Because of the economic significance of salt domes, a great deal of work has been performed on the mechanics of salt deposition, timing of deposition, mapping of domes, physical properties of salt, mechanical modelling of salt dome formations, etc. and an extensive literature exists on these topics. Rather than document those references here we refer the interested reader to the text by Halbouty (1979) which has an extensive list of references published up to 1978. Halbouty (1979) also thoroughly documents the occurrence of explored salt structures in the Gulf of Mexico region.

In the present paper we present quantitative models of several processes related to the dynamic development of salt diapirs which result from buoyant instability, placing emphasis on the central concepts underlying each process modelled, and the predictions and implications of each model.

II. BUOYANT UPLIFT OF SALT

In this section, we consider the response of a salt mass of constant density as the surrounding sediments are compacted during burial. Salt maintains a constant density ~2.2 gm/cm^3 during burial. On the other hand, typical clastic sedimentary material has a lower density of 1.6-1.9 gm/cm^3 at deposition. This density results from the high porosity of clastic sediments when being deposited. The rock density, ρ_R, reflects the density of the matrix material, ρ_M, the density of the formation fluids, ρ_f, and the porosity ϕ, and may be expressed in the following manner:

$$\rho_R = \rho_M(1 - \phi) + \rho_f \phi \qquad (1)$$

For typical values of ρ_M (2.6-2.7 gm/cm^3) and ρ_f (1.0 gm/cm^3) the formation density ρ_R is less than that of rock salt, 2.2 gm/cm^3, for porosities greater than 27%-30%.

While on a local scale porosity is a function of many variables, on a large scale porosity of clastic sediments is dominated by compaction effects, with density progressively increasing as porosity decreases in response to the increasing weight of overburden. These effects have been studied empirically in various sedimentary basins by several authors (Athy, 1930; Hedberg, 1936; Dickinson, 1953; Atwater and Miller, 1965; Rhodehamel, 1977; Sclater and Christie, 1980). These

authors find an overall trend of porosity decreasing with depth of burial in a manner which is best described by an exponential function of depth (Figure 1).
Thus, we may write

$$\phi(z) = \phi_o e^{-z/a} \tag{2}$$

where ϕ_o is the porosity at the sediment surface ($z = 0$), $\phi(z)$ is the porosity at a depth of burial z, and the parameter a is a measure of how rapidly porosity decreases with depth; the porosity $\phi(z)$ falls to half the value of the surface porosity at a depth of burial $z = 0.693a$.

From Figure 1 we see that the values of the parameters ϕ_o and a depend on lithology and also are different for different sedimentary basins. For this reason, we will keep the following discussion on salt diapirism quite general by referring to a generic porosity law of the form given in Equation (2). A reader interested in salt dome development in a particular sedimentary

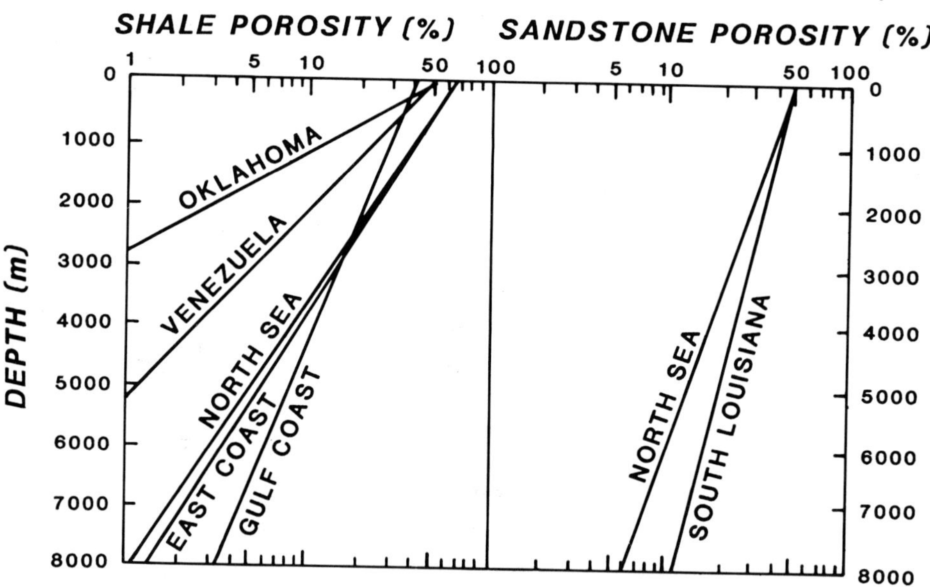

Figure 1. Porosity-depth relationship in various sedimentary basins.

basin may then apply these results simply by substituting the appropriate values of the parameters ϕ_o and a. Where we present numerical examples, the porosity model of Atwater and Miller (1965) for sandstones in South Louisiana (ϕ_o = 49%, a = 5500 m, ~18,000 ft.) or of Dickinson (1953) for Gulf Coast shales (ϕ_o = 40%, a = 3100 m, ~10,000 ft.) is used.

The point we emphasize is that with increasing depth rock density will eventually equal, and then exceed, the density of the salt. There is then an upward buoyancy force exerted on the salt. Due to its fluid characteristics, the salt will respond to this force by flowing upward and forming a domal structure. As a result of salt's ability to (i) flow on a geological time scale under the influence of an applied stress, and (ii) maintain an approximately constant density upon burial, salt deposits will develop into diapiric structures as they are buried deeper.

A. Quasi-equilibrium Model of Diapirism

We begin our discussion of the flow of salt in the subsurface by considering three factors which influence the flow. These are:

1. The buoyancy force exerted on the salt due to the contrast in density between the salt and the surrounding sediments.
2. The salt viscosity which in principle may limit the flow speed of salt. As we shall see presently, this effect is negligible on a geologic time scale.
3. The mechanical strength of the formations surrounding the salt which inhibits salt flow.

To understand the buoyancy force experienced by a salt body, consider the situation in which a large, laterally extensive, mother salt has developed an attached cylindrical dome structure (Figure 2). We compare the pressure at the top of the mother salt (Point A in Figure 2) with the pressure at a point at the same subsurface depth within the salt dome (Point B in Figure 2). At each point the pressure is the weight of the overburden, including the weight of the pore fluids. The difference in pressure between two points then reflects the density contrast

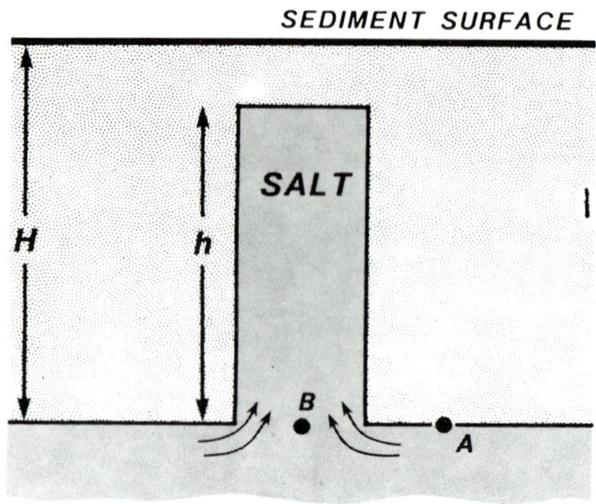

Figure 2. Geometry of model of salt dome growth resulting from gravitational instability.

between the two locations, integrated over the entire geologic section overlying these points. Thus it is not sufficient to consider only the density contrast between salt and sediments at the top of the mother salt; the density contrast must also be considered throughout all the overlying sediments. The pressure difference ($P_A - P_B$) may be expressed as:

$$P_A - P_B = \int_B^A (\rho_R(z) - \rho_s)\, g dz \tag{3}$$

where $\rho_R(z)$ is the formation density as a function of subsurface depth z, ρ_s is the salt density, g is the acceleration due to gravity, and the integration is performed over the height of the salt dome. The condition for equilibrium of the salt dome is that ($P_A - P_B$) be equal to $P_{resistance}$, the resistance to salt dome development provided by the surrounding formations:

$$P_A - P_B = P_{resistance} \tag{4}$$

As a first step let us consider the situation where $P_{resistance}$ is zero, i.e. the formations surrounding the salt dome provide negligible resistance to salt dome development. This will permit us to determine some minimum conditions for diapiric development. We will later consider the influence of non-zero resistance pressure. In the situation where $P_{resistance}$ is zero, if $(P_A - P_B)$ is positive then the integrated weight of overlying sediments at Point A in Figure 2 is greater than the weight of the overburden at Point B. In response to this differential pressure salt will flow from A to B and be added to the mass of the salt dome. Thus the salt dome will grow in height, provided that growth is not constrained, for example by the strength of the overlying sediments or by the supply of salt.

Modelling the diapiric motion of salt in a fully rigorous manner presents a formidable task indeed. In response to the upward buoyancy force, the salt experiences an upward acceleration. However, the flow of salt will be modified by viscous drag which retards the flow. Modelling of viscous flow is difficult in this case as the flow is not confined within defined boundaries; rather the boundaries are defined by the viscous salt flow which in turn depends on the boundaries defining the flow. Thus the problem is highly nonlinear. These difficulties are further compounded by lack of knowledge of the resistance provided by the sedimentary layers being deformed and breached by the rising salt.

Due to these complicating factors, and so as to gain physical insight into the processes associated with salt dome development, we examine individual aspects of the problem under simplifying assumptions.

The greatest simplification is obtained by assuming that the salt body is always in quasi-equilibrium, i.e. the response of salt is fast relative to the rate at which conditions develop which would tend to drive the system from equilibrium. Thus the salt dome is always very close to equilibrium. As the depth of burial of the mother salt is increased the salt dome evolves to the new equilibrium situation essentially instantaneously. This approximation, which we shall examine in detail later, implies that viscous drag has a negligible effect on the dynamical evolution of a salt diapir. The present calculation provides an estimate of the height, h, of a salt dome which can develop under these conditions when the mother salt is buried under a sedimentary column of height H.

Equation (3) implies that the formation immediately overlying the mother salt must be compacted at least to a critical porosity ϕ_{crit} (that porosity at which the sediment density equals the salt density), before a salt structure can be initiated. Using Equation (1), we obtain the following expression for ϕ_{crit}:

$$\phi_{crit} = \frac{\rho_M - \rho_s}{\rho_M - \rho_f} \qquad (5)$$

where ρ_M, ρ_f and ρ_s represent the densities of the rock matrix, the fluid saturating the rock, and the salt, respectively. For typical values of these parameters (ρ_M = 2.65 - 2.72 gm/cm^3, ρ_s= 2.2 gm/cm^3, ρ_f = 1.0 gm/cm^3) the sediments must compact from a surface porosity of ϕ_o to a porosity of less than 27%-30%, before a density inversion develops. Using Equation (2), the critical porosity at a critical overburden thickness H_{crit} is given by:

$$H_{crit} = a \ln (\phi_o/\phi_{crit}) \qquad (6)$$

The model of Dickinson (1953) for shale porosities in the Gulf Coast yields estimates of H_{crit} in the range 900-1200 m (2950-3950 ft.) while with the model of Atwater and Miller (1965) for sandstone porosity in South Louisiana, we obtain an estimate of H_{crit} in the range 2700-3300 m (8850-10800 ft.). Thus the critical depth of burial for salt dome initiation depends strongly on the dependence of rock density and porosity on depth and thus, among other things, on the shaliness of the overburden. These are minimum estimates; other factors, such as the strength of the surrounding formations, may delay the initiation of diapirism until the mother salt is buried to a greater depth (but see later).

As the thickness of the sedimentary column increases beyond H_{crit}, a diapiric structure develops which is fed by the mother salt. Now we ask: what is the height h of the salt dome when the thickness of the overburden overlying the mother salt is H? We can estimate the maximum salt dome height by neglecting the resistance provided by the mechanical strength of the surrounding formations and setting the excess pressure, $(P_A - P_B)$ in Equation (4), to zero. Using Equation (1), we find that the buoyancy pressure exerted on a salt body of thickness h when the mother salt is located at a depth H below the sediment surface is:

$$P_A - P_B = g\,(\rho_M - \rho_S)\,\{h - a\,\phi(H)\,\phi_{crit}^{-1}(e^{h/a} - 1)\} \qquad (7)$$

In this equation $\phi(H)$ is the formation porosity at a subsurface depth H, i.e., at the base of the salt dome. Setting $(P_A - P_B)$ equal to zero, Equation (7) yields the equilibrium salt dome height h in terms of the thickness of overburden H and of parameters describing the porosity-depth relationship:

$$h/a = \ln\{\,1 + \phi_{crit}\phi(H)^{-1}\,h/a\,\} \qquad (8)$$

We have solved Equation (8) for values of the parameters appropriate for Gulf Coast shales and South Louisiana sandstones.[1] These results are illustrated in Figure 3. A salt structure does not develop until the sediment density at the

---

[1]Equation (8) is a transcendental equation for h/a which can be solved iteratively using a hand calculator. Make an initial guess of the value of the quantity h/a. Evaluate the right-hand side of Equation (8) using this estimate of h/a. This yields a new estimate of h/a which is

depth of burial of the mother salt exceeds the salt density. This occurs at a depth of burial H_{crit}, which is approximately 900 m (2950 ft.) for 100% shale overburden and 2700 m (8850 ft.) for 100% sandstone overburden. For mixed lithologies H_{crit} should lie between these limits, being determined by the large scale dependence of rock density on depth. Beyond this critical depth of burial the height of the salt dome depends on the thickness of overburden in a manner which is approximately linear, the rate of growth of the salt column being approximately twice the rate of burial of the mother salt. From figure 3, we also see that an overburden thickness of ~1900 m (6230 ft.) is required over the mother salt before the top of the salt dome approaches the sediment surface for a 100% shale overburden while the corresponding thickness for a sandstone overburden is 5900 m (19350 ft.).

As we noted previously, this calculation provides an upper limit on the height of a salt dome for a given thick-ness of overburden overlying the mother salt. Several factors serve to inhibit the development of diapirs, including the lateral cohesive strength of the formations being penetrated, any significant variations in the density-depth dependence of the overburden, and the supply of salt from the mother layer.

Let us consider the effect of variations in the average density-depth relationship resulting from undercompaction of an overpressured formation at depth. (Overpressuring may also alter the lateral strength of the sediments and thereby modify diapir development, as discussed below). This change in the density law alters the integral in Equation (3). For simplicity, consider the case where the normal compaction law obtains down to a depth

---

[1] (Footnote continued) again used to evaluate the right-hand side of Equation (8). Iterate this procedure until the solution converges sufficiently well. Alternatively: choose h/a and solve Equation (8) for the corresponding value of H/a, plot h/a versus H/a as done in Figure 3. This is a linear procedure.

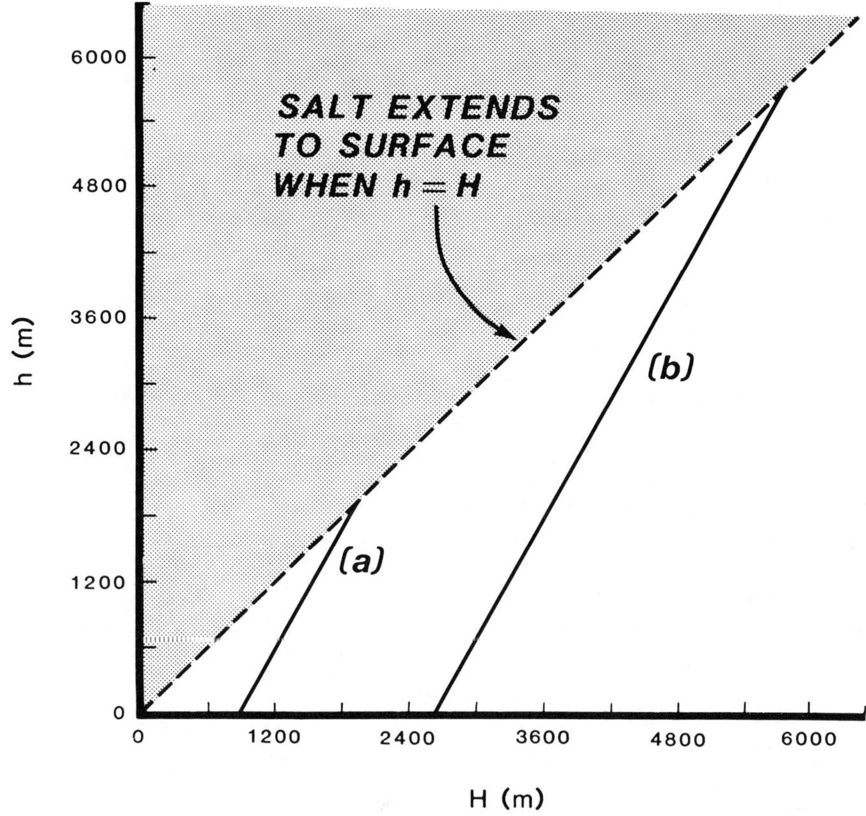

Figure 3. Equilibrium height (h) of a salt dome as a function of the depth of burial (H) of the mother salt. Curve (a) is the case where the overburden consists of shales which conform to the porosity/depth relationship of Dickinson (1953) and curve (b) refers to a sandstone overburden which follows the posority/depth law of Atwater and Miller (1965).

z_1, the top of the overpressured region, and that below z_1 the sediments do not compact any further, i.e., the rock density is constant below z_1 and has the value $\rho_R(z_1)$. In this case if z_1 is less than H_{crit}, the salt will not become buoyant if buried at any depth above or within this overpressured zone. If z_1 is greater than H_{crit}, salt dome development will proceed exactly along the lines we have described for thickness of overburden above the mother salt less than z_1. At this point, salt diapirism becomes inhibited. This is illustrated in Figure 4 in which we plot the height of a salt dome as a function of the thickness of the sedimentary section assuming the porosity-depth relationship of Atwater and Miller (1965) for the shallower section and where the deeper formations are undercompacted in accordance with the model described above. This indicates that the thickness of sediment required to induce a salt diapir to grow to the near surface may easily be greater than our first estimate. This example also shows that to model diapirism

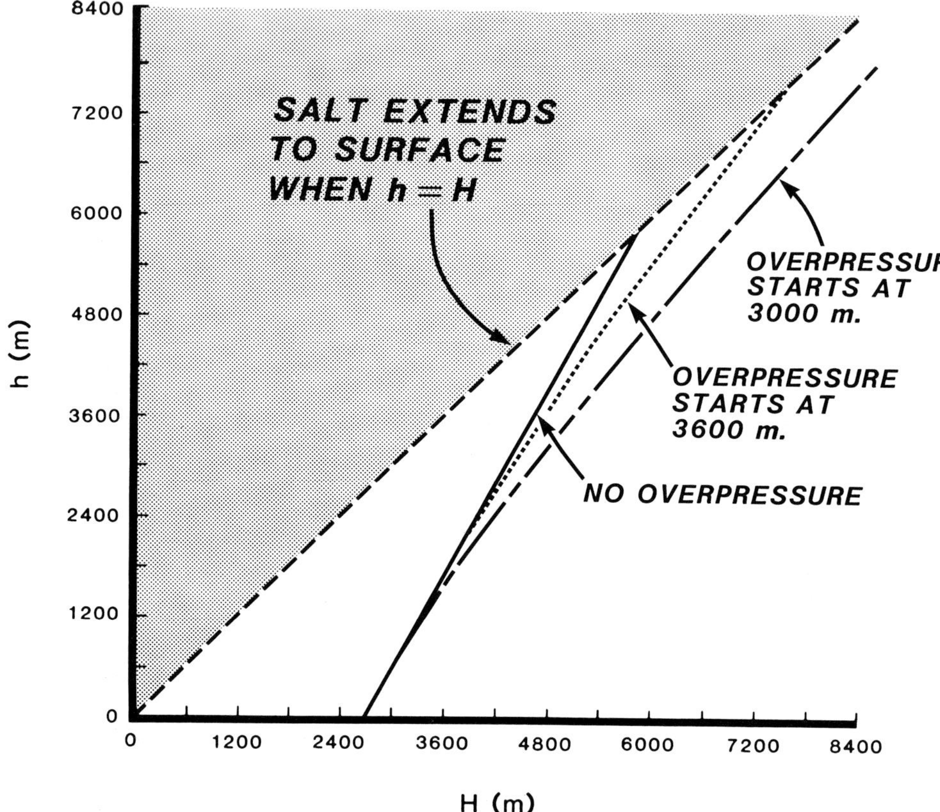

Figure 4. Equilibrium height of salt dome as a function of overburden thickness in overpressured region, assuming the porosity/depth relationship of Atwater and Miller (1965).

accurately, a good understanding is needed of the gross dependence of density on depth at all depths down to the mother salt.

B. Lateral Variations in Salt Diapirism Within a Basin

While there are many possible causes of lateral variability of salt dome structures two are of general occurrence:

1. The depth of overburden overlying a salt deposit will vary as one moves away from the axis of a sedimentary basin.
2. The progradation of a depositional surface across a marginal shelf also implies a laterally varying sediment load.

Here we construct two simple models to show the variations in salt diapirism resulting from basin shape and variable sediment load. In the first example a salt deposit occurs throughout the basin at a fixed height above basement (Figure 5). The thickness of overburden overlying the salt deposit depends on the distance from the axis of the basin. Thus, the isostatic equilibrium height of the salt column also depends on the radial location within the basin. In Figure 5, we show the development of salt diapirs due to buoyancy in such a basin. Diapirism is initiated once the overburden thickness exceeds a critical value. Salt domes are then best developed in the areas where there is the greatest thickness of overburden overlying the mother salt while salt domes are least developed, or even nonexistent, on the flanks of the basin.

A similar argument can be advanced concerning the development of salt structures underneath a prograding shelf. Salt structures resulting from buoyancy are expected to be best developed underneath the thicker landward section of the shelf, with less developed salt domes underneath the outer slopes of the shelf, as illustrated in Figure 6. This developmental pattern may be modified if the lateral variation in weight of overburden is significant relative to the horizontal confining forces. In

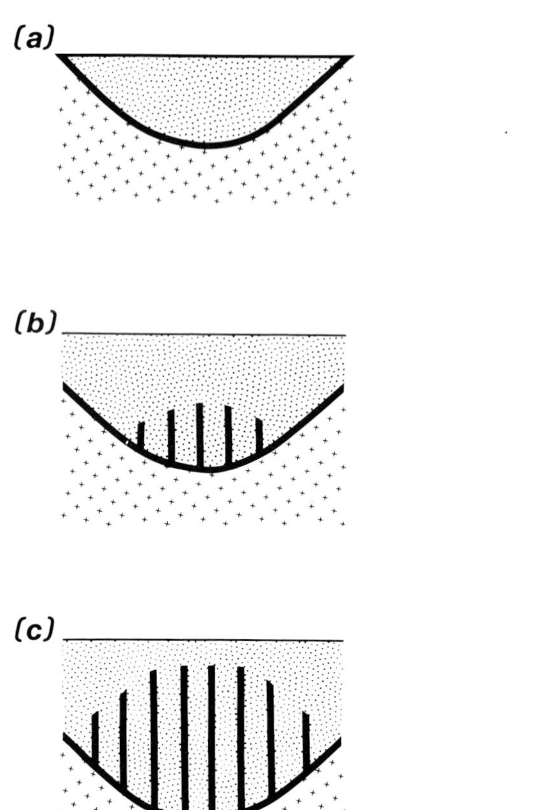

Figure 5. Influence of basin development on buoyancy diapirism. (a) Pre-diapir phase, maximum thickness of overburden less than H_{crit}, (b) diapiric development in deepest part of the basin, and (c) mature phase: diapirs occur throughout the basin and penetrate to the near surface where there is a sufficient thickness of overburden.

this situation the salt flow may have a horizontal component of motion, flowing from the thicker landward section of the shelf toward the shelf slope. Lateral flow does not depend on salt

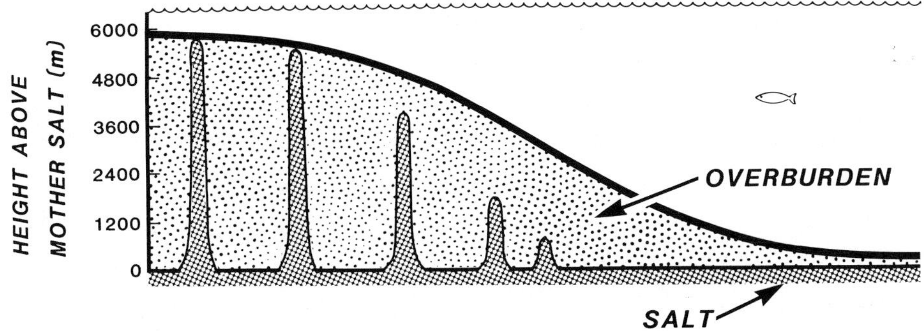

Figure 6. Salt dome development due to salt buoyancy under a prograding shelf, assuming the porosity/depth relationship of Atwater and Miller (1965).

buoyancy and so a critical depth of burial of the salt is not required to initiate lateral flow. Rather lateral variations in overburden and lateral confining forces are the dominant factors. This horizontal flow is most likely to occur for massive salt bodies buried at shallow depths under steeply dipping slopes.

III. GRAVITY ANOMALIES

In addition to salt diapirism, the characteristic contrast in density between rock salt and sedimentary formations is also reflected in the local gravitational field and may generate observable surface gravity anomalies. In this section we derive an estimate of the surface gravity anomaly associated with the quasi-static model of salt diapirism resulting from buoyancy. This anomaly consists of two major components. The first component results from the lower reaches of the salt dome where the density of salt is lower than that of the surrounding formations. Since more dense material has been replaced by lighter material, this results in a decrease in the gravitational force measured at the surface, i.e., a negative surface gravity anomaly. Because the density contrast which yields this feature occurs at depth, the resulting surface gravity anomaly is a relatively wide one.

The second component of the gravity anomaly results from the upper portion of the salt dome where the salt density is greater than that of the surrounding sediments. If a cap rock is present the dense cap rock material also contributes to this component. These materials, having a higher density than their surroundings, will contribute to a positive surface gravity anomaly. As this density contrast lies closer to the surface, the profile of this component of the gravity anomaly will be sharper than that of the negative component discussed above. Thus the net effect is a broad negative anomaly with a narrower positive anomaly superimposed at the apex. This response is illustrated in Figure 7 taken from Nettleton (1971). Typical anomaly magnitudes are also provided by Nettleton (1971), indicating that negative anomalies range from less than 1 mgal for a small dome with a salt column less than 1 mile in diameter to as much as 15 mgal for very large domes. The positive component may range from zero for domes which are buried deeply and have no cap rock to as much as 6 mgal for domes which penetrate to the near surface and have very thick cap rock.

While the surface gravity anomaly asociated with a salt dome of arbitrary shape must be calculated numerically, one may easily calculate the anomaly along the axis of a cylindrically symmetric structure. The mass difference in an annulus of a radius r at depth z is given by

$$dM = \Delta\rho(z)\, 2\pi\, r\, dr\, dz \tag{9}$$

where $\Delta\rho(z)$ is the density anomaly at depth z, and dr and dz are the radial thickness and height of the annulus respectively.

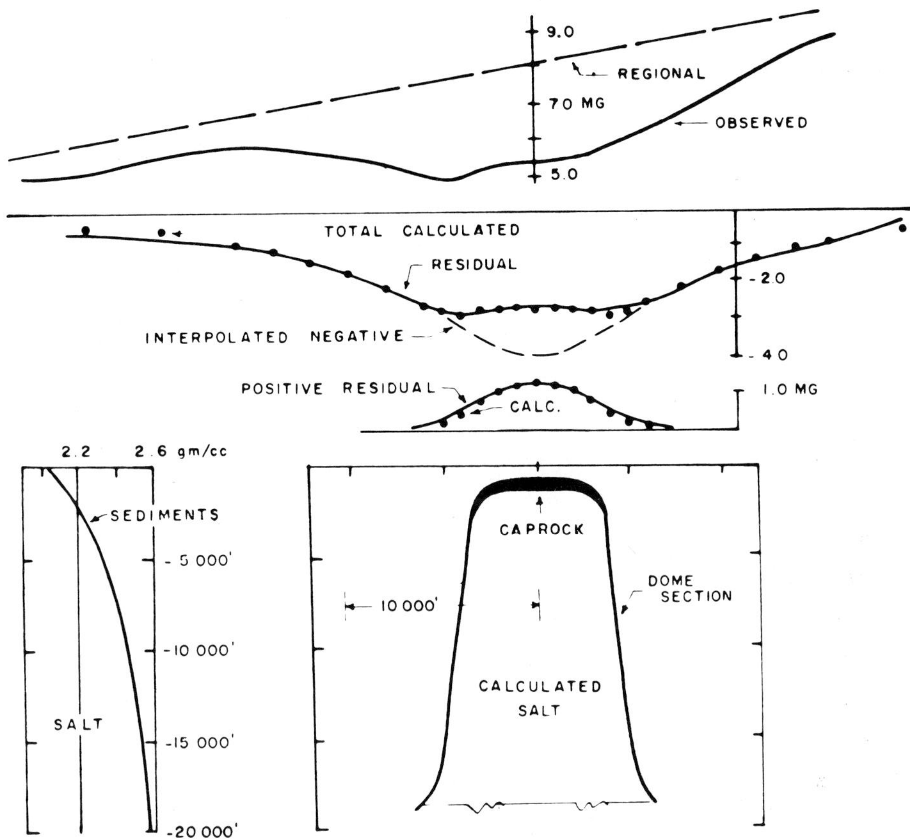

Figure 7. Illustration of the surface gravity anomaly over a salt dome showing how the negative anomaly from the lower salt and the positive anomaly from the upper salt and the caprock may be derived from the measured data. The salt dome model deduced from this data is also shown. (Adapted from Nettleton, 1971).

The surface gravity anomaly on the axis associated with this annulus is given by the following equation:

$$dg\ (z = 0,\ r = 0) = \frac{GdM}{r^2 + z^2} \cos \alpha \tag{10}$$

where G is the gravitational constant and 2α is the angle subtended at the surface location (z = 0, r = 0) by the annulus. The net surface gravity anomaly $\Delta g(z = 0,\ r = 0)$ is then obtained by integrating Equation (10) over the volume of the diapir:

$$\Delta g\ (z = 0,\ r = 0) = 2\pi g \int_{H-h}^{H} \Delta\rho(z)\{1 - z\,(R^2+z^2)^{-1/2}\}\,dz \tag{11}$$

In Equation (11) R(z) is the radius of the salt dome at depth z, h is the height of the salt column and the base of the salt is located at depth H.

For a uniform salt column of radius 800 m the equilibrium model predicts a positive anomaly component on-axis of ~3 mgal when the salt has penetrated to within 100 m of the surface and a negative component of ~-.4 mgal. The magnitude of this anomaly is in good agreement with the typical range of observed values for salt domes of these dimensions, as given by Nettleton (1971) (see Figure 7). However within the present model it is difficult to generate negative surface gravity anomalies of magnitude 15 mgal as quoted by Nettleton (1971) for very large salt structures; according to this model the salt should breach the surface before a dome of this magnitude is formed. The fact that such large domes are observed implies that in these instances other mechanisms must be operative which impede salt dome development, i.e. the resistance to salt dome development provided by the surrounding formations, denoted by $P_{resistance}$ in Eq. (4), must be included in the analysis. This point is addressed in the following sections.

IV. INFLUENCE OF FORMATION STRENGTH ON DIAPIRISM

The simple model described above predicts that the salt structure will penetrate to the sediment surface for a sufficient thickness of sediment overlying the mother salt. This thickness is approximately 1900 m (6230 ft.) for 100% shale overburden and 5900 m (19350 ft.) for 100% sandstone overburden. In this model increasing the depth of burial beyond this point would force all the salt from the mother salt layer up through the salt diapir to the surface, in obvious contradiction with observation. It is clear that some mechanism impedes the development of salt domes.

Let us consider the impact of the lateral cohesive strength of formations on salt dome growth. Consider a salt deposit of constant thickness h, and compute the buoyancy pressure exerted on the salt body as it is buried to increasingly greater depths (Equation 3). From Figure 8 we see that the buoyancy pressure increases monotonically with depth of burial, approaching an asymptotic value as the surrounding formations approach full compaction. The buoyancy pressure also increases with thickness of salt.

To determine the influence of lateral formation strength on diapir growth, we need to superimpose on Figure 8 a curve representing the formation cohesive strength. This strength depends on several parameters such as depth of burial, lithology, diagenetic history, formation pressures, regional stresses,

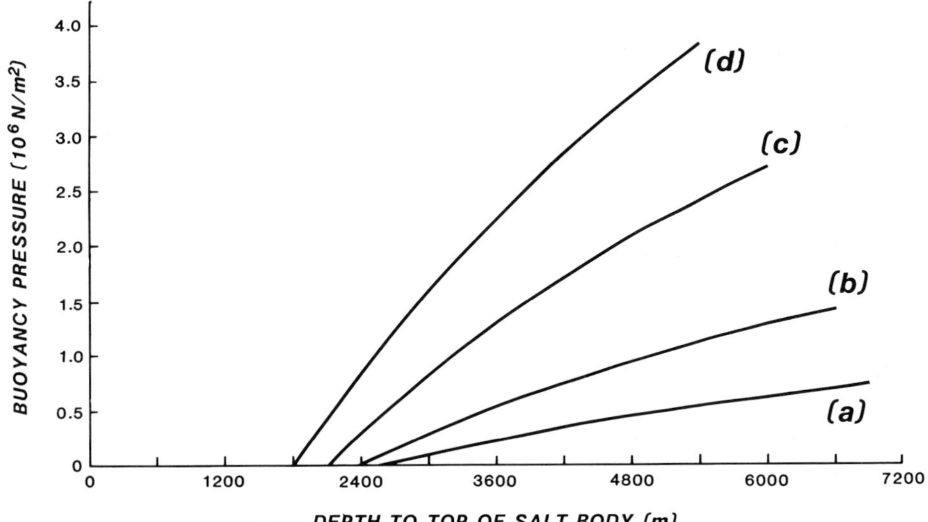

Figure 8. Buoyancy pressure on salt as a function of overburden thickness for salt bodies of thickness (a) 300 m, (b) 600 m, (c) 900 m and (d) 1200 m, assuming the porosity/depth relationship of Atwater and Miller (1965).

tectonic history, etc. For the sake of illustration we assume that the resistance provided by the lateral strength of sediments can be represented approximately by a constant, $P_{resistance}$, over the range of depths of interest. The following arguments may easily be made more specific if a more appropriate model of $P_{resistance}$ is available.

If the salt bed is thin, the buoyancy pressure exerted on the salt mass may never be large enough to overcome the resistance provided by the lateral strength of the surrounding formations. For example, if $P_{resistance} = 10^6$ N/m^2 and rock porosity follows the porosity-depth relationship of Atwater and Miller (1965), the pressure exerted by a 300 m thick salt body will never overcome the resistance of the sediments, while the pressure exerted by a 600 m thick salt will be just sufficient to breach the overlying sediments when the upper surface of the salt body is buried at 5000 m. Thus salt diapirs are not expected to be strongly associated with "thin" salt deposits unless the overlying sediments have very poor lateral strength. We require "thick" salt deposits to overcome the impedance provided by the overlying sediments. Thick beds, formed at deposition or resulting from post depositional salt deformation (e.g., tectonic activity or differential loading) may show a significant degree of diapirism.

Even for a thick salt bed, initiation of diapirism is inhibited by the lateral strength of the sediments and so the mother salt must be buried at a greater depth before a salt dome begins to grow. The critical depth of burial for salt dome initiation now depends on the mother salt thickness. For our previous example with $P_{resistance} = 10^6$ N/m^2, a salt layer 1800 m thick will begin to form a dome when its upper surface is buried at 2500 m and its base is at 4300 m. The delay becomes more extreme with decreasing salt thickness: a 600 m thick salt does not begin to form a diapir until the top of the salt is buried at

5000 m and the base of the salt is at 5600 m. Thus formation strength is seen to be an effective mechanism in inhibiting initiation of salt diapirism.

A similar argument shows that the cohesive strength of formations surrounding the salt will also inhibit salt dome growth during the later phases of diapirism, the overlying formations again providing some impedance to penetration by the uprising salt column. In this case, the strength of formations at the base of the dome may also be a contributing factor influencing the flow of salt from the mother layer into the salt column.

V. VISCOUS FLOW OF SALT

In responding to the buoyant force exerted by the surrounding sediments, the salt is restrained in its flow by viscous drag. This effect can be modelled by considering the flow of a viscous fluid through a pipe of given diameter. Due to the slow speed of salt flow and the small accelerations, inertial effect are negligible relative to viscosity effects in the development of salt domes. Reynolds' Number, which is a measure of the relative importance of inertial and viscous forces, is less than 10^{-22} for salt flow. This implies that the flow of salt is laminar rather than turbulent. (A value of Reynolds' Number less than 2,000 is the accepted criterion for laminar flow.)

For laminar flow through a cylinder, the fluid velocity u at a radial position r is given by the Hagen-Poiseuille law:

$$u = \frac{\Delta P}{4 \mu h} (R^2 - r^2) \tag{12}$$

where μ is the coefficient of viscosity and ΔP is the drop in pressure along a length h of the cylinder which has a radius R. Equation (12) indicates that the flow has a parabolic velocity distribution, the velocity being greatest along the axis of the cylinder and zero along the walls. The average velocity V is one half of the maximum velocity:

$$V = \frac{\Delta P \ R^2}{8 \ \mu \ h} \tag{13}$$

In the present case the excess pressure driving the flow is provided by the contrast in density between salt and the surrounding formations. Thus the speed V at which a salt dome of height h grows when the mother salt is buried under an overburden of thickness H can be determined by equating the buoyancy pressure $P_A - P_B$, given by Equation (3), with the back pressure ΔP provided by viscous drag. If we again assume an exponential porosity-depth relationship for the overlying sediments, we obtain:

$$V = gR^2(\rho_M - \rho_S) \ (8\mu h)^{-1} \ [h + \phi(H)\phi_{crit}^{-1} \ a(1-e^{-h/a})] \tag{14}$$

Thus if we know the thickness H of overburden overlying the mother salt at any given time and if we also know the height of the salt dome h at that time, we can determine the rate V at which the salt dome is growing, using Equation (14). In this way, the evolution of a salt dome may be traced for any arbitrary time-dependent overburden thickness H(t).

To assess the significance of this process we have carried out numerical calculations assuming a constant rate of burial, i.e., H increasing with time at a fixed rate. Carter and Heard (1970) have shown that while temperature and strain are influencing factors, salt viscosity in the subsurface falls in the range 10^{18} - 10^{19} Pa sec (10^{19} - 10^{20} poise). To bracket the possible range of the viscous effect, our calculations were performed over a wider range of values of viscosity.

From Equation (14), it is apparent that the speed of salt dome emplacement decreases as the radius decreases, which may occur as the height of the dome increases. To ensure that we do not under-estimate the effect of viscosity we permit the radius of the salt dome to decrease as the diapir develops. This is achieved by maintaining the volume of the salt dome at a constant value. In these model calculations the radius of the salt dome is approximately 170 m at a time when the salt has penetrated to the near surface. With this choice of the radius parameter R, we feel we have not under-estimated the impact of viscosity.

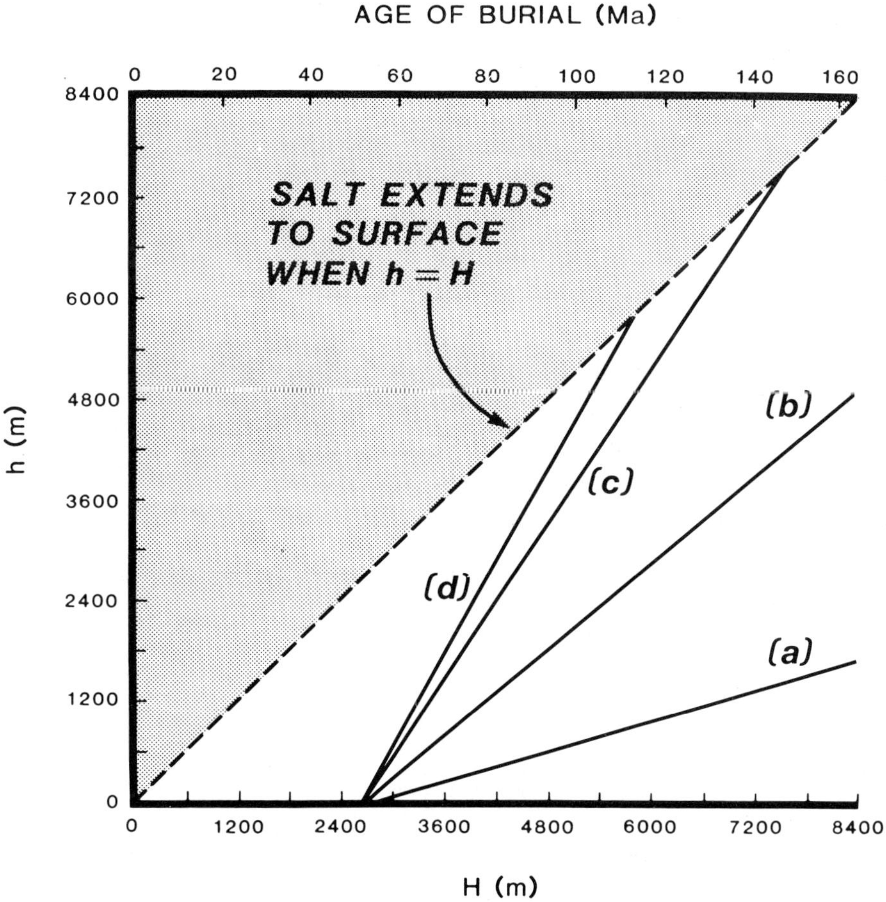

Figure 9. Height of salt dome as a function of age of burial, assuming viscous salt flow, a sedimentation rate of 50 m/Ma and the porosity/depth relation of Atwater and Miller (1965). The coefficient of friction μ for each case is (a) 10E22 Pa sec (=10E23 poise), (b) 10E21 Pa sec (=10E22 poise), 10E20 Pa sec (=10E21 poise), and (d) $\mu = 0$.

The results of these calculations are shown in Figure 9 for a moderate sedimentation rate (50 m/Ma) and in Figure 10 for a high sedimentation rate (300 m/Ma). Both figures show the height h of the salt dome as a function of thickness of overburden or, equivalently, of age of burial for a range of viscosities. For reference these figures also show the quasi-equilibrium evolution of a salt dome neglecting viscous drag.

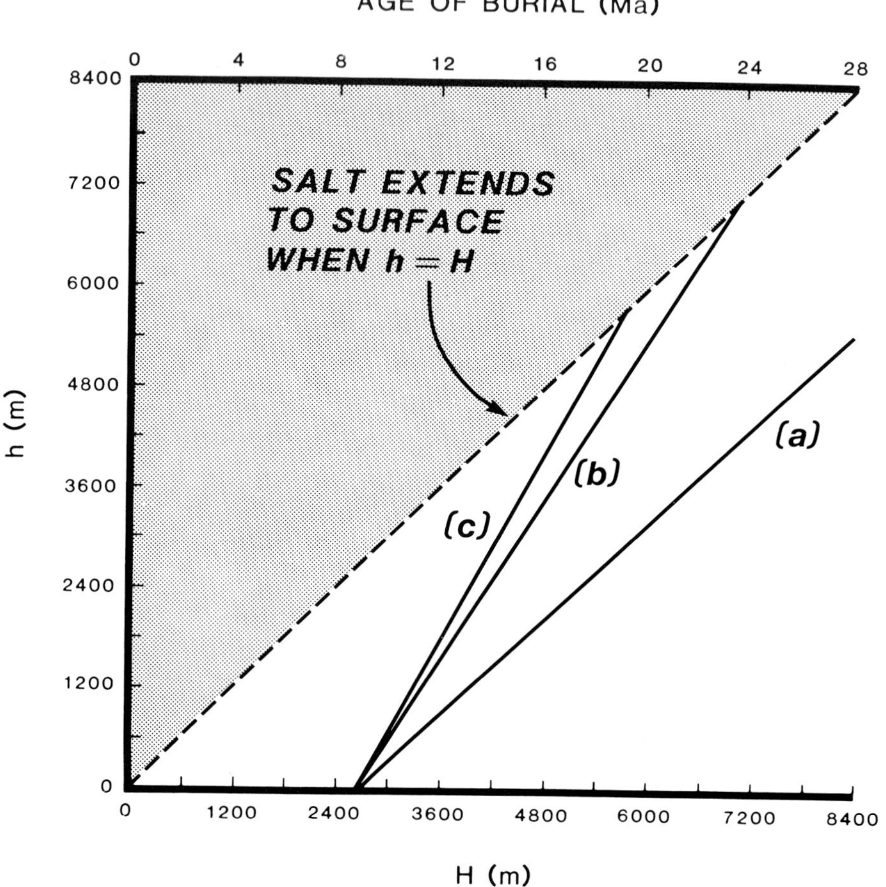

Figure 10. Height of salt dome as a function of age of burial, assuming viscous salt flow, a sedimentation rate of 300 m/Ma and the porosity/depth relationship of Atwater and Miller (1965). The coefficient of friction for each case is (a) 10E20 Pa sec (10E21 poise), (b) 10E19 Pa sec (10E20 poise), and (c) $\mu = 0$.

From Figures 9 and 10 we see that the height of the salt dome lags behind that predicted by the quasi-equilibrium calculation. The degree of lag increases with increasing sedimentation rate, with increasing time following diapir initiation during which constant sedimentation occurs, and with increasing viscosity.

However, Figures 9 and 10 show that even for a sedimentation rate as high as 300 m/Ma, a viscosity of 10^{20} Pa sec (10^{21} poise) is required to delay the salt dome development significantly with respect to the equilibrium prediction. For values of viscosity typical of rock salt in the subsurface, 10^{18} - 10^{19} Pa sec (10^{19} - 10^{20} poise), the impact of viscosity on diapir development is seen to be negligible. A quasi-equilibrium calculation is then an excellent approximation of salt dome development, i.e. salt viscosity does not significantly delay the response of a salt diapir, on a geologic time scale.

VI. MUSHROOM CAPS ON SALT DOMES

While we have assumed that salt domes can be modelled as having uniform cross-sections at all depths, this description is not quite accurate. Drilling experience indicates that salt domes oftentimes develop a mushroom-type top, or overhang, which may be drilled through to reach underlying sedimentary formations. Several factors may contribute to the development of this profile.

The first factor we consider is the dependence of salt viscosity on temperature. The data of Carter and Heard (1970) imply that salt viscosity decreases by a factor of two for an increase in temperature of ~190°C, i.e., the dependence of viscosity on temperature is relatively small. As viscous drag has a negligible impact on the flow of salt on a geologic time scale, we believe that such variations in viscosity due to temperature differences are of little consequence.

The next factor we consider is the impedance to salt motion provided by the surrounding sediments. This is expected to be greatest near the top of the salt dome, where overlying sedimentary formations are being breached. As we shall see in a later section, the stress concentration in the surrounding sediments is greatest near the top of the salt dome. This is the region in which the major deformation of the sedimentary formations occurs and so, correspondingly, this is the region in which the overburden exerts the greatest impedance to salt motion. Having once breached a sedimentary layer the salt may then flow more easily past this layer, the resistance to motion at this point being essentially sliding friction. The result of this differential resistance is to deform the top of the salt dome so as to enlarge the top of the dome in the form of a mushroom cap. The extent of this effect will depend on the strength of the formations being breached. A dome which has penetrated a sedimentary section composed of predominantly poorly consolidated shales is not expected to have a well-developed mushroom cap, in comparison with a dome which has penetrated highly competent formations, e.g. a thick carbonate sequence. This process is the same as that describing the deformation of the nose of a bullet fired into a solid material.

A further mechanism which promotes the development of a mushroom cap is that, while a net upward buoyancy force is experienced by the salt column as a whole, the upper reaches of the salt dome are in fact more dense than the surrounding formations. Locally the salt at the top of the dome tends to flow downwards, although this is counterbalanced by the upward flow of salt from the lower sections of the dome. The resultant effect is an increase in cross-section at the top of the dome. Thus differential impedance provided by the formations being pierced and differential buoyancy are believed to be the mechanisms which lead to the development of a mushroom cap.

However, the relative importance of these two mechanisms is not known at present.

VII. DEFORMATION OF FORMATIONS IN THE VICINITY OF A SALT DOME

Thus far we have concerned ourselves with the development of a buoyant salt column under the influence of sediment deposition and compaction. We now consider the structural deformation of formations surrounding a developing salt dome.

For simplicity let us neglect the lateral strength of the formations for the present. (We return to this point later in the discussion). Structural deformation then can be modelled as the response of a perfect fluid being penetrated by an uprising salt dome.

Calculation of fluid flow past a salt dome is facilitated by assuming that the salt has the shape shown in Figure 11, and described mathematically through the parametric form.

$$Z_{salt} = h(t) + R_o \{1 + 2 \cos\theta/2 - \sec\theta/2\} \quad (15a)$$

$$r_{salt} = 2 R_o \sin\theta/2 \quad (15b)$$

Here θ is a parameter which lies in the range $0 < \theta < \pi$, Z_{salt} and r_{salt} describe the surface of the salt dome for any value of the parameter θ, and $h(t)$ is the height of a reference point which is fixed within the salt column. In the limit of an extremely tall dome ($\theta \rightarrow \pi$) the radius of the dome at its base is $2R_o$. The speed of uplift of the salt column is $w = dh/dt$ so that we treat the salt column as an invariant form moving upwards at speed $w(t)$. At large distances above the salt, we suppose that the sediment is being deposited at a speed $R(t)$ which, in the interests of simplicity, we take to be independent of lateral position.

If we treat the surrounding formations as a perfect fluid we can describe the vertical velocity, u_z, and the radial velocity, u_r, of the sedimentary formations relative to the salt column using a stream function $\Psi(r,z)$:

$$u_z = \frac{-1}{r}\frac{\partial\Psi}{\partial r} \tag{16a}$$

$$u_r = \frac{1}{r}\frac{\partial\Psi}{\partial z} \tag{16b}$$

The convenience of the salt dome shape described by Equation (15) is that an analytical expression can be derived easily for the stream function of fluid flowing past this shape. We find (Lamb, 1879)

$$\Psi = R_o^2 w\ (z-h-R_o)/[(z-h-R_o)^2 + r^2]^{1/2} \tag{17}$$

$$\text{for } r > r_{salt} \text{ and } z > z_{salt}$$

Writing out the explicit dependence of u_z and u_r on vertical and lateral position we have:

$$u_z = R_o^2 w(z-h-R_o)[(z-h-R_o)^2 + r^2]^{-3/2} \tag{18a}$$

$$u_r' = R_o^2 w\ r\ [(z-h-R_o)^2 + r^2]^{-3/2} \tag{18b}$$

To determine the influence of a rising salt dome on sedimentary structure, we map the path followed by an element of sediment during burial. Consider a horizontal sedimentary formation located sufficiently far above the salt dome that it is initially uninfluenced by the motion of the salt. The location and shape of this horizon may be determined at any later time by tracing out the path followed by each element of the horizon. This can easily be achieved numerically since if we know the location of an element of sedimentary material at time t we can determine both the radial and vertical components of velocity using Equations (18a) and (18b). This allows us to determine the location of that element at time (t + dt). By repeating this process we map the path followed by a sedimentary element during burial.

The results of such a calculation are illustrated in Figure 11 where we show the paths followed by elements of sediment during burial. The rate of salt dome emplacement is taken to be 120 m/Ma. The radius of the salt at the base of the dome is 1500 m. We trace out the paths followed by elements of sediment which are initially positioned at radial intervals of 300 m far above the salt. Figure 11 illustrates the manner in which the overburden is pierced by a rising salt column. By superimposing isotime curves, we see the deformation of the sedimentary strata induced by the rising salt body. Strata overlying the dome are thinned, while on the flanks of the salt the formations are draped over the uprising dome. The angle of drape is predicted by this model to become quite large close to the flanks of the salt dome (> 60°). Because of this severe angle of drape, reflection seismology is expected to encounter problems in defining the salt-sediment interface on the flanks of a salt dome.

In this discussion we have neglected the strength of the sediments and assumed that sediments can be deformed freely under the influence of a rising salt body. However, the deformation implied by the present model is quite severe. This may be seen by considering cells of sedimentary formations in Figure 11 where the boundaries of a given cell are defined by adjacent flow paths and by adjacent isotime curves. By comparing the shapes of cells between two flow paths we trace the deformation of a cell as a function of time. This deformation is seen to be complex and to extend some distance above the top of the salt dome. Near the top of the salt this involves extension in the radial direction, contraction in the vertical direction, in addition to a shear

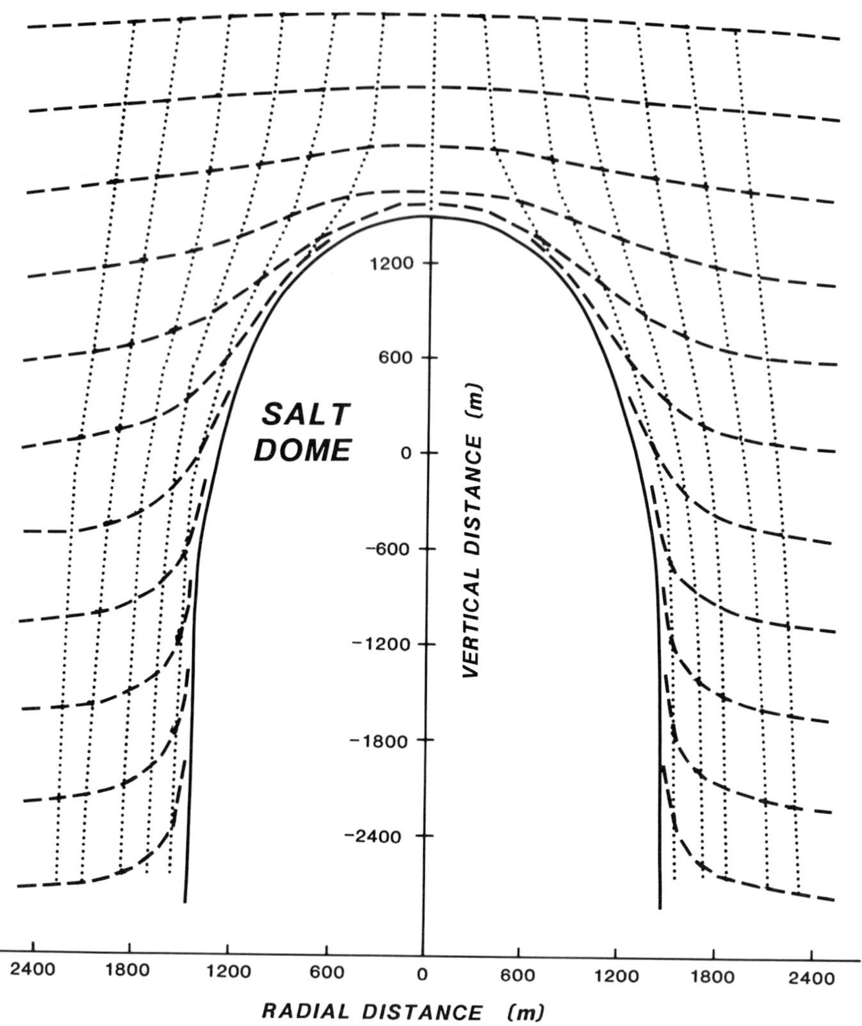

Figure 11. Model of sediment draping on the flanks of a salt dome. Dotted curves represent burial paths of elements of sediment and dashed curves show the deformation of bedding planes. Vertical distances are referenced to a point which is fixed within the salt dome.

deformation. There is also a component of azimuthal extension which is not apparent on this figure. The implication is that these formations are subject to significant deformational stress.

For modest levels of stress, the sediments respond by deforming elastically. However at the point at which the streses first exceed the lateral strength of competent formations surrounding the dome, faulting will occur, thus relieving the stress. As the deformational stresses, as implied by Figure 11, are quite significant we expect the critical point for fault initiation to lie some distance above the salt dome. By the time the formations are located on the flanks of the salt dome faulting should certainly have developed. (However, if the formations have low lateral strength, e.g., overpressured shales, they may tend to deform in a ductile manner under the applied stress rather than through faulting. This behavior would be similar to that of the perfect fluid which we have modelled above).

Once faulting has been initiated, the strength profile of the sediments is altered. The fault plane represents a region of low lateral formation strength. As the salt dome continues to rise the accompanying stress is most easily relieved by propagation of pre-existing faults. From Figure 11, we see that the dominant deformation of the formations near the top of the salt dome consists of lateral extension induced by the rising salt. The vertical contraction and shearing of strata then follow from mass balance considerations. The faulting pattern which develops under these extensional conditions consists of a system of normal faults which are initiated somewhat above the top of the salt, accompanied by antithetic faults. The fault development associated with a given salt dome is expected to depend strongly on any variations in sediment strength which, in turn, has a dependence on previous faulting.

A more complete model of post-depositional deformation and faulting could be developed by considering the deformation of a thick slab under the influence of an applied external force. While quite feasible, such a model is much more complicated mathematically than the model presented in this section due both to the fact that (i) we are considering the deformation of a thick rather than a thin body, and (ii) the deformation is large. Further it is then necessary to have a model of the lateral strength of the sediments overlying the salt dome which should reflect the impact of any previous faulting. We have chosen for the present to present a simpler, less exact, model which yields a clear insight into the physical processes involved. We will present a more rigorous analysis of the problem in a later paper in this volume.

In addition to the deformation of sedimentary strata near the top of the salt dome induced by the penetration of the rising salt body, a further component of deformation on the lower flanks of the salt column results from the flow of the salt from the mother layer into the vertical salt structure. Mass balance considerations imply a deformation of the sediments overlying the mother salt to compensate for this outflow of salt. A synclinal feature results surrounding the salt column, commonly referred to as a rim syncline. This syncline is not necessarily cylindrically symmetric; if salt flow to the salt column is preferentially from one direction, this fact will be reflected in the structure of the rim syncline. From mass balance arguments, we see that the volume deformation in the rim syncline must equal the volume of salt which has flowed into the salt column. This may yield a useful method of estimating the total volume of salt involved in salt dome development if, as implied by some theories of cap rock formation, a substantial portion of the salt column has been dissolved during or following salt dome emplacement. In such an application, one must discriminate between the rim syncline and other possible structural features.

From this discussion we see that two mechanisms contribute to the deformation of the formations surrounding a salt diapir:

1. Penetration of the salt through the overlying formations results in draping of these formations over the flanks of the dome and the development of an extensional fault system.

2. Flow of salt from the mother layer into the salt column results in the development of a rim syncline on the lower flanks of the dome.

At any location along the flanks of a salt dome, the structural style will then consist of a superposition of these two components. Draping of beds and faulting is expected to be dominant on the upper flanks, while rim syncline development is expected to be the dominant structural control on the lower flanks of an intrusive diapir.

VIII. CONCLUSIONS

In this paper we have examined quantitative mathematical models for several processes associated with the buoyant uplift of buried salt deposits. Using measured values of salt viscosity, we find that viscous drag has a negligible effect on salt diapirism on a geologic time scale and hence a quasi-equilibrium model is an excellent approximation for modelling salt dome growth. Using this model we obtain minimum conditions which must be met before diapirism can be initiated, as well as a formalism for estimating the height of a salt dome at any time during its subsequent develop ment. This model indicates that the factors which dominate buoyant diapirism are (a) depth of burial of the mother-salt, (b) the quantitative relationship between density (or, equivalently, porosity) and depth of burial, (c) the lateral cohesive strength of the formations being breached by the salt structure, and (d) the thickness of the mother-salt. This latter factor may represent both depositional and post-depositional processes, such as lateral salt flow in response to tectonic activity or to differential overburden load. Thus through the application of such mathematical models it is possible to obtain quantitative constraints which aid in the interpretation of salt diapirism.

We have also constructed a model of the deformation of formation overlying an uprising salt dome. This model reproduces the draping of bedding planes over the salt dome and the development of a rim syncline. However a limitation in this model is introduced by the difficulty of modelling the bulk elastic properties and stresses induced in the formations are severely deformed and subject to faulting. These features will be treated in greater detail in a later paper.

REFERENCES

Athy, L.F., (1930). Density, porosity and compaction of sedimentary rocks, A.A.P.G. Bull. 14, 1-24.
Atwater, G.I. and Miller, E.E., (1965). The effect of decrease in porosity with depth on future development of oil and gas reserves in South Louisiana, A.A.P.G. Bull., 49, 334-339.
Berner, H., Ramberg, H. and Stephansson, O., (1972). Diapirism in theory and experiment, Tectonophysics, 15, 197-218.
Biot, M.A. and Ode, H., (1965). Theory of gravity instability with variable overburden and compaction, Geophys. 30, 213-227.
Bishop, R.S., (1978). Mechanism for emplacement of piercement diapirs, A.A.P.G. Bull 62, 1561-1583.
Carter, N.I. and Heard, H.C., (1970). Temperature and rate dependent deformation of halite, Am. J. Sci. 269, 193-249.
Dickinson, G., (1953). Geological aspects of abnormal reservoir pressures in Gulf Coast Louisiana, A.A.P.G. Bull. 37, 410-432.
Halbouty, M.T., (1979). Salt Domes, Gulf Region, United States and Mexico (Second Edition), Gulf Publishing Company.
Hedberg, H.D., (1936). Gravitational compaction of clays and shales, Amer. J. Sci. 31, 241-287.
Lamb, H., (1879). Treatise on the Mathematical Theory of the Motion of Fluids, Cambridge University Press.
Nettleton, L.L., (1955). History of concepts of Gulf Coast salt dome formation, A.A.P.G. Bull. 39, 2373-2383.
Nettleton, L.L., (1971). Elementary gravity and magnetics for geologists and seismologists, S.E.G. Monograph Series, No. 1 (P.C. Wuenschel, ed.).
Rhodehamel, E.C., (1977). Sandstone porosities, geological studies on the Cost B-2 well, U.S. Mid-Atlantic Outer Continental Shelf Area: U.S.G.S. Circ. 750, 23.
Sclater, J.G. and Christie, P.A.F., (1980). Continental stretching: An explanation of the Post-Mid-Cretaceous subsidence of the central North Sea Basin, J. Geophys. Res. 85, 3711-3739.

STEPWISE CENTRIFUGE MODELING OF THE EFFECTS OF DIFFERENTIAL SEDIMENTARY LOADING ON THE FORMATION OF SALT STRUCTURES

M. P. A. Jackson
R. R. Cornelius

Bureau of Economic Geology
The University of Texas at Austin
Austin, Texas

I. INTRODUCTION

Salt structures of the northern Gulf of Mexico basin grew syndepositionally while the basin filled with Mesozoic and Cenozoic sediments. Not only did the sedimentary overburden influence the growth of salt structures, but this influence must have changed over time as the cover thickened and compacted, depocenters shifted, facies distributions changed, and the provenance varied with changes in source-area tectonics. In this changing environment the cover is clearly not laterally uniform, but varies laterally as well as vertically in thickness, density, effective viscosity, strength, and ductility. Growing salt structures will therefore be differentially loaded. Examples of differential loads are prograding deltas, alluvial or turbiditic fans, coral reefs, oolitic or terrigenous clastic shoals, volcanic accumulations, and thrust and ice sheets. Where sedimentary facies are stacked vertically on the margin of a deep, starved basin like the Gulf of Mexico, particularly in depositional compartments confined by salt structures or growth faults, the effects of differential loading are greatly accentuated and are propagated to great depth (Jackson and Talbot, 1986).

Sedimentologists recognize two processes by which sedimentary basins are filled laterally from the margin inward, as opposed to vertical filling by aggradation. The first is progradation, in which upward-coarsening deposits accumulate by sediment washing in from the basin margin across the clastic wedge. The second is lateral accretion, in which upward-fining deposits accumulate against the clastic wedge from detritus already within the basin. Both produce depositional units having the roughly sigmoidal geometry known as offlap. These structures are known as clinoforms in seismic stratigraphy (Mitchum et al., 1977) and are grouped as progradational reflection configurations. For dynamic modeling it is convenient and sufficient to follow this broad definition of progradation rather than the narrow one used by sedimentologists. Large-scale progradation structures are typical either of slopes fronting prograding shelves or of prodelta environments or slopes fronting prograding shelf deltas or shelf-margin deltas (Brown and Fisher, 1982). These environments are the most likely zones for the initiation of salt structures in Gulf Coast settings.

The effects of prograding deltas on salt tectonics are commonly invoked qualitatively in the literature (Ewing and Antoine, 1966; Lehner, 1969; Wilhelm and Ewing, 1972; Fisher, 1973; Bishop, 1978; Watkins et al., 1978; Humphris, 1979; Martin, 1980; Woodbury et al., 1980). However, our knowledge of the mechanical effects of pro-grading overburdens on the initiation and structural maturation of diapiric structures is extremely sparse. Mathematical and experimental modeling of these effects is complicated because loading, growth, density, effective viscosity, and strength are all time dependent and interactive in a subsiding clastic basin. Mathematical modeling has rarely been attempted. The work of Biot and Ode (1965) is an exception, but it addressed only the time-dependent effects of a thickening and compacting but otherwise uniform overburden. Apart from the transfer of otherwise uniform cover from above growing structures to synclinal areas between, differential loading was not simulated. Furthermore, their analysis applies only to the two-dimensional configuration of very low amplitude anticlines. Limited experimental modeling, referred to in the present paper where appropriate, has demonstrated qualitatively some of the effects of differential loading in three dimensions. Some of these experiments were based on the behavior of water-saturated incohesive sediment, scaled only for mesoscopic diapirs. Others were centrifuge experiments using material analogs like silicone putty.

The experiments described here are an initial attempt to understand the time-dependent effects of irregularly accumulating cover on syndepositional diapirism using scaled centrifuged models in which diapiric analogs grew spontaneously without artificial initiation. A stepwise centrifuge technique allowed progradation to be simulated, a new approach. Although some quantitative strain data were recorded, the results are intended for qualitative comparison with real-world examples. Experimental modeling was carried out in 1984 at the Hans Ramberg Tectonic Laboratory, Institute of Geology, University of Uppsala, Sweden.

II. EXPERIMENTAL METHODS

A. The Centrifuge

The raison d'etre of centrifuge modeling is as follows: by using centrifugal force as a body force equivalent to, but orders of magnitude greater than, the force of gravity, gravity tectonics can be simulated using much stiffer materials in a much shorter time than would be practical under normal gravity. Stiff materials enable intricately layered systems to be constructed before deformation and allow models to be sectioned after deformation for three-dimensional examination. The centrifuge technique has been comprehensively described by Ramberg (1981). The specific medium-capacity centrifuge and its peripheral equipment used in the present study were illustrated by Ramberg and Stephansson (1965). The centrifuge has a 10-cm-diameter cup for holding the model with a maximum capacity of 1.5 kg. Gearing allows any rate of spin up to 320 rad s^{-2} (3,050 rpm), equivalent to a centrifugal acceleration of 3,000 *g* (where *g* is the free-fall acceleration due to gravity equal to 9.807 m s^{-2}). Mirrors and a stroboscopic light allow the model to be observed during centrifuging. Refrigerated coils within the centrifuge prevent frictional heating of the spinning rotor and model by air.

B. Scaling

The principles and applications of geologic model scaling--one of the many applications of dimensional analysis (Langhaar, 1951; Kline, 1965)--have been comprehensively discussed by Hubbert (1937) and Ramberg (1981). Scaling calculations are required both in model design and in applying the experimental results to the natural prototype. The model ratio (subscript r) is equal to the model quantity (subscript m) divided by the corresponding prototype quantity (subscript p).

Geometric scaling was represented by a model length ratio equal to about 10^{-6}. Thus, a model with a 3-mm-thick source layer and a 6-mm-thick cover simulated thicknesses of 3,000 m and 6,000 m, respectively, in nature. As in all experimental or numerical models, the geometric scaling was a simplified version of the prototype. We designed some of our models to elucidate a general principle rather than simulate nature. For example, the model with a static differential load contained a tabular half-layer whose abrupt edge would be unlikely to be formed by sedimentary processes. Other models were designed to simulate prototypes more accurately, particularly in the dip of clinoforms. This dip reflects the maximum angle of repose of incohesive sediment, which is roughly equivalent to the angle of internal friction. Angles of repose range from 1.5° on the continental slope of the northern Gulf of Mexico to 34° in eolian sand dunes. For comparison, the clinoform prograding models had "foreset" dips ranging from 1.7° to 31°.

Kinematic scaling ensured that the model motions represented by strain and "sedimentation" were analogous to those in nature. We were mainly concerned with the balance between three different rates: (1) progradation of cover, (2) aggradation of cover, and (3) rise of buoyant struc tures. Table I compares rates in the three regularly prograding models with those in the prototype in the northwestern Gulf of Mexico. Comparative prototype rates are provided for the Quaternary, characterized by rapid rates of sedimentation and diapir rise, and for the whole of the Cenozoic, representing a period of average rates. Models were designed to simulate approximately the relative balance between the prototype rates of progradation and aggradation (*P/A* ratio) and between the rates of progradation and diapir rise (*P/R* ratio). These rates are discussed in more detail for specific models in Section IV.

Dynamic scaling ensures that the balance of relevant forces in the model is the same as that believed to occur in the prototype. Tables II and III are examples of model ratios used for dynamic scaling: the ratios of length, density, acceleration, and stress can be confidently derived; those of time and velocity are less exact. The least known natural parameter is that of the equivalent viscosity of terrigenous clastic rocks at various depths. However, because the equivalent viscosity of model materials can be directly measured, the same parameter in sediments can be estimated algebraically. Tables II and III show such estimates as 5×10^{19} and 3×10^{20} Pa s for rock salt, and 2×10^{20} and 1×10^{21} Pa s for the sedimentary cover. Scaling assumed that the cover is subaerial. A uniform thickness of

Table I: Comparison of Rates (ms^{-1}) and Ratios of Progradation (P), Aggradation (A), and Diapir Rise (R) in Prototypes and Models

	P	A	R	P/A	P/R
Prototype Cenozoic	[a] 1.4 E-10	[b] 3.7 E-12	[c] 3.0 E-12	38	47
Prototype Quaternary	[d] 3.6 E-10	[e] 3.2 E-11	[f] 2.4 E-11	11	15
840328	[g] 5.5 E-05	[g] 3.1 E-06	[h] 1.7 E-06	18	32
840413	[i] 8.3 E-04	[i] 7.4 E-05	[j] 2.7 E-05	11	30
840412	[k] 5.7 E-04	[k] 7.1 E-05	[k] 7.7 E-05	8	7.4
840402	[l] 0.0	[l] 1.6 E-08	[l] 2.4 E-08	0	0

Data Sources

[a] Mean Cenozoic progradation rate of northern Gulf Coast continental margin for the last 56 Ma from position at start of Wilcox time to present 200-m bathymetric contour (Winker, 1982)

[b] Mean Cenozoic aggradation rate of northern Gulf Coast continental margin for the last 56 Ma, based on mean thickness of 6,500 m from base of Wilcox Group to present surface (Salvador and Buffler, 1983)

[c] Mean Cenozoic rise rate of shallow diapirs for the last 56 Ma, based on mean thickness of 6,500 m from base of Wilcox Group to present surface and on mean depth of 1,250 m for 21 offshore Louisiana diapirs drilled to salt (New Orleans Geological Society, 1983)

[d] Mean Quaternary progradation rate of northern Gulf Coast continental margin for the last 1.8 Ma from position at start of Pleistocene (defined by extinction of Globoquadrina altispira at base of Lenticulina I fauna zone) to present 200-m bathymetric contour (Winker, 1982)

[e] Mean Late Quaternary aggradation rate of northern Gulf Coast continental margin for the last 1.8 Ma (defined by extinction of Discoasters at base of Trimosina B fauna zone), based on a mean Late Pleistocene thickness of 1,830 m (Woodbury et al., 1973). For comparison, mean aggradation rate of Pleistocene Sangamon faunal zone in same area (0.5-0.1 Ma ago) was 3.5 E-11; three-dimensional seismic reflection study by Wilson (1985)

[f] Late Pleistocene distortion of strata above rising salt diapir over the last 0.8 Ma on continental shelf of northern Gulf Coast; three-dimensional seismic reflection study by Wilson (1985)

[g] Steps 1 through 4 for 1603 s, representing period of progressive progradation and aggradation

[h] Steps 1 through 5 for 2367 s, representing entire period of diapiric growth

[i] Steps 1 through 19 for 121 s, representing period of progressive progradation and aggradation

[j] Steps 1 through 20 for 287 s, representing entire period of diapiric growth; based on mean of highest point of each wall

[k] Steps 1 through 18 for 170 s, representing period of steady progradation and aggradation and entire period of diapiric growth

[l] Static differential loading for 52 h

Table II: Dynamic scaling for model 840328[a]

Quantity	Model	Prototype	Model ratio
Thickness of cover	5 mm	6,500 m	$l_r = 7.69.10^{-7}$
Density of source layer	1,260 $kg.m^{-3}$	2,200(b) $kg.m^{-3}$	$\rho_r = 5.73.10^{-1}$
Density of cover	1,340(*) $kg.m^{-3}$	2,340 $kg.m^{-3}$	
Acceleration (body force/unit mass)	1,200 g	1 g	$a_r = 1.20.10^{3}$
Stress model ratio			$\sigma_r = \rho_r l_r a_r = 5.29.10^{-4}$
Velocity of dome rise(ms^{-1})	$1.7.10^{-5}$(c)	$2.97.10^{-12}$(d)	$v_r = 5.82.10^{5}$
Time model ratio			$t_r = l_r / v_r = 1.32.10^{-12}$
Viscosity model ratio			$\mu_r = t_r \sigma_r = 6.98.10^{-16}$
Viscosity of source	$3.6.10^{4}$(e) Pa.s	$5.16.10^{19}$(*)	
Viscosity of cover	$1.5.10^{5}$(e) Pa.s	$2.15.10^{20}$(*)	

Table II (cont.)

(a) Based on 21 offshore Louisiana diapirs (New Orleans Geological Society, 1983)

(b) Based on evaporite containing 95% halite and 5% anhydrite (Carmichael, 1984)

(c) Steps 1 through 5 for 2367 s, representing entire period of diapiric growth

(d) Mean Cenozoic rise rate of shallow diapirs offshore Louisiana, see [c], Table I

(e) Measured

(*) Values calculated by means of scale ratios

Table III: Dynamic scaling for model 840413[a]

Quantity	Model	Prototype	Model ratio
Thickness of cover	9 mm	6,500 m	$l_r = 1.38.10^{-6}$
Density of source layer	1,170 $kg.m^{-3}$	2,200[b] $kg.m^{-3}$	$\rho_r = 5.32.10^{-1}$
Density of cover	1,340[*] $kg.m^{-3}$	2,520 $kg.m^{-3}$	
Acceleration (body force/unit mass)	1,200 g	1 g	$a_r = 1.20.10^{3}$
Stress model ratio			$\sigma_r = \rho_r l_r a_r = 8.81.10^{-4}$
Velocity of dome rise(ms^{-1})	$2.7.10^{-5}$[c]	$2.97.10^{-12}$[d]	$v_r = 9.16.10^{6}$
Time model ratio			$t_r = l_r / v_r = 1.51.10^{-13}$
Viscosity model ratio			$\mu_r = t_r \sigma_r = 1.33.10^{-16}$
Viscosity of source	$3.6.10^{4}$[e] Pa.s	$2.71.10^{20}$[*]	
Viscosity of cover	$1.5.10^{5}$[e] Pa.s	$1.13.10^{21}$[*]	

Table III (cont.)

(a) Based on 21 offshore Louisiana diapirs (New Orleans Geological Society, 1983)

(b) Based on evaporite containing 95% halite and 5% anhydrite (Carmichael, 1984)

(c) Steps 1 through 20 for 287 s, representing entire period of diapiric growth; based on mean of highest point of each wall

(d) Mean Cenozoic rise rate of shallow diapirs offshore Lousiana, see [c], Table I

(e) Measured

(*) Values calculated by means of scale ratios

overlying water would not affect diapir rise rates materially, but a wedge of water above a sloping seafloor would increase rise rates below deeper water.

C. Model Materials and Their Rheology

One of the basic modeling materials was a pink silicone putty marketed under the brand name "Rhodorsil Gomme" (RG) by the Societe des Usines Chemiques Rhone-Poulenc (France). Like all bouncing putties, Rhodorsil Gomme is a mixture of a highly viscous polymer, polyborondimethylsiloxane (PBDMS), and solid filler (Weijermars, 1986a).

The other basic material was a variety of plasticine called "Plastilina" (PL), manufactured by Beckers Dekorima (Sweden). Slightly softer and weaker than Harbutts Plasticine (England), Plastilina consists mainly of quartz, kaolinite, hydrated halloysite, and K-feldspar, softened with interstitial fluid (Weijermars, 1986a). Red Plastilina is colored by liquid or dissolved pigments, whereas white Plastilina is colored by barite, wurtzite, and periclase.

Rhodorsil Gomme and red Plastilina were used separately or were mixed in two different proportions: 75RG/25PL was a mixture of 75 wt. % Rhodorsil Gomme and 25 wt. % Plastilina; 50RG/50PL

Figure 1. Rheology of model materials used, representing mixtures of Rhodorsil Gomme (RG) and Plastilina (PL), on a logarithmic graph of model shear stress, d (solid ordinate) against model shear strain rate, γ (solid abscissa). Slope of curves represents reciprocal of n, the power-law stress exponent (Table IV). Diagonal contours represent dynamic equivalent viscosity, μ. Data sources as follows: 1Hailemariam (1982, his Fig. 4, rotoviscometer, 21°C); and 7, Weijermars (1986a, his Figs. 7 and 9, Haake viscometer, 24°C); 3, 4, 5, and 6, this study (rotoviscometer, 21°C). Scaled values for prototype (natural equivalent) shear stress (dashed ordinate) and shear strain rate (dashed abscissa) were calculated from model ratios of stress and reciprocal time in model 840413 (Table III). Stippled area represents deformation field of natural salt diapirs, based on the following limits: a-b, buoyancy pressure at base of Oakwood Dome, East Texas, 5.5 km deep; c-d, maximum differential stress during last episode of strain recovery in shallow Gulf Coast salt

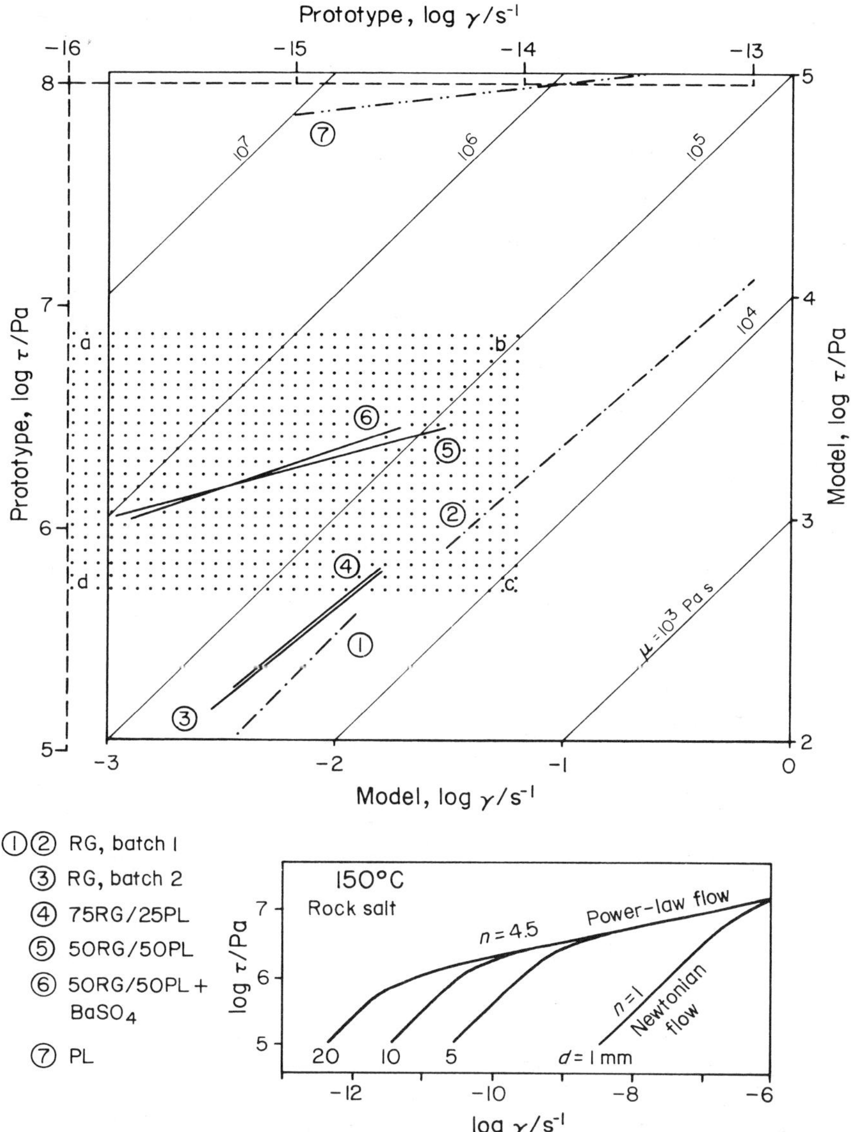

diapirs determined by average sub-grain size (Carter et al., 1982); b-c and a-d, approximate upper and lower strain-rate limits for East Texas and Gulf Coast salt diapirs (Seni and Jackson, 1983; Jackson and Talbot, 1986). Inset: straight lines show comparative creep behavior extrapolated from laboratory-derived flow laws for power-law creep predominant in dry rock salt (Carter and Hansen, 1983) and solution-transfer creep predominant in wet rock salt (Spiers et al., 1986); curved lines represent transitional behavior where both mechanisms operate; *d* = average grain size.

was a mixture of equal parts by weight of the two ingredients. Liquid mercury was used in the experiments on edge effects under normal gravity.

To allow layers to be visibly differentiated, modeling materials were colored by mixing in water-soluble powdered poster paints. Density was increased where necessary by adding powdered barite ($BaSO_4$) or magnetite ($[Fe^{2+}Fe^{3+}]_2O_4$).

A recent comprehensive review of the rheology of model materials (Weijermars, 1986a) stresses a point also made by Dixon and Summers (1985): the mechanical properties of silicone putties vary widely, not only among different brands, but also among different batches of the same brand from the same manufacturer. Varying contents of solid filler are the cause. All models described in the present report were made from the same RG batch 2.

The rheology of RG and its mixtures were tested in the centrifuge laboratory on a rotoviscometer by cylindrical Couette flow. Specifications and technique are described by Hailemariam (1982) and Weijermars (1986a). Strain rates during testing on this instrument bracket the typical shear strain rate of 10^{-2} s^{-1} during growth of the model diapirs. The creep behavior of the model materials is summarized in Table IV and in Figure 1. In addition to data from tests carried out by one of the authors (on RG batch 2, 75RG/25PL, and 50RG/50PL), Table IV also includes data from tests run by Hailemariam (1982) and Weijermars (1986a) on an earlier batch of Rhodorsil Gomme (RG batch 1) for comparison and on white Plastilina. Both batches of RG and the 75RG/25PL mixture exhibit a near-linear (Newtonian) relation between stress and strain rate; they have a stress exponent close to 1, a viscosity independent of strain rate, and a near-zero effective yield stress. In contrast, the 50RG/50PL mixtures and PL had markedly non-Newtonian power-law creep at strain rates typical of our models. On a logarithmic plot of strain rate versus stress (Fig. 1) these three shear-thinning have linear curves. The stress exponent, n, is given by the reciprocal slope of each curve, which varies from 2.9 to 9.5 (Table IV).

The creep behavior of RG and 75RG/25PL at the strain rates measured is virtually identical (Fig. 1, curves 3 and 4). However, increasing the proportion of PL above 25% increases the n value, the dynamic viscosity, and the threshold stress (discussed below)(Table IV and Fig. 1, curves 5 and 6). That this trend continues to a maximum with 100% PL (Fig. 1, curve 7) is confirmed by Weijermars' (1986a) data for mixtures of 33RG/67PL and 25RG/75PL. Increasing the percentage of barite powder added to RG from 20 to 45 causes similar increases in viscosity and n value at a typical model strain rate of 10^{-2} s^{-1} (Weijermars, 1986a). However, such differences are negligible with barite concentrations of less than 10%, the maximum used in our models (Fig. 1, curves 5 and 6).

Using a similar approach to that of Dixon and Summers (1985), the relevance of model creep behavior to that of the prototype can be shown using the model ratios of, for example, Table III. An alternative pair of abscissa and ordinate axes, representing the prototype strain rates and stresses, was calculated and added to the model data in Figure 1.

Strains produced at strain rates less than 1/100 of the typical model rates of 10^{-2} s^{-1} can be regarded as negligible on the time scale considered. This model threshold strain rate of 10^{-4} s^{-1} is equivalent to a prototype threshold strain rate of 1.5×10^{-17} s^{-1}. The actual mean strain rate calculated for salt diapirs of the East Texas Basin over 56 Ma is 6.7×10^{-16} s^{-1} This actual mean prototype strain rate is about 50 times the scaled prototype threshold strain rate, which is appropriate. The stress corresponding to this threshold strain rate is termed the threshold stress, which is calculated by linear extrapolation of the rheologic tests shown in Figure 1. The threshold stress can be regarded as an effective yield stress below which strain

Table IV: Creep rheology of model materials

Model material	Flow-law[a] $\gamma = C\,\tau^{n}$ s^{-1}	Threshold stress, τ at $\gamma = 10^{-4}\ s^{-1}$ Pa	Viscosity, μ at $\gamma = 10^{-2} s^{-1}$ Pa.s	N[b]
RG, batch 1[c]	$\gamma = 4.1.10^{-5}\ \tau$	3	$3.1.10^{4}$	6
RG, batch 2	$\gamma = 7.1.10^{-6}\ \tau^{1.2}$	9	$4.1.10^{4}$	8
75RG/25PL	$\gamma = 7.4.10^{-6}\ \tau^{1.2}$	9	$4.0.10^{4}$	6
50RG/50PL	$\gamma = 1.0.10^{-14}\ \tau^{3.7}$	530	$1.8.10^{5}$	8
50RG/50PL+ 9.3% $BaSO_4$	$\gamma = 2.5.10^{-12}\ \tau^{2.9}$	410	$2.0.10^{5}$	4
PL[d]	$\gamma = 1.5.10^{-48}\ \tau^{9.5}$	41,000	$6.9.10^{6}$	

Table IV (cont.)

(a) Flow law established by creep test using rotoviscometer. Data corrected for non-Newtonian Couette flow according to Krieger and Maron (1954)

(b) Number of data points in creep test

(c) Data from Hailemariam (1982, Fig. 4)

(d) Data from Weijermars (1986a, Fig. 9)

is trivial over the duration of the experiment or equivalent geologic time. A true yield stress is not actually present in the model materials, which are either Newtonian or pseudoplastic showing power-law behavior. It is difficult to derive meaningful yield stresses for tectonic strain rates by extrapolation of results from laboratory deformation, but the prototype threshold stresses calculated by scaling appear to be suitable. Model threshold stresses of 9 Pa (RG) to 530 Pa (50RG/50PL) correspond to prototype threshold stresses of 10 kPa to 600 kPa. The minimum relief in the top of the model source layer required to generate differential stresses that exceed the model threshold stress can be conservatively calculated--based on the minimum model density difference of 80 kg m^{-3} (model 840328) and on the standard centrifugal acceleration of 1,200 *g*--as 0.01 mm for RG and 0.6 mm for 50RG/50PL. Tiny undulations in the surface of the source layer, for example those produced by the rise of tiny random bubbles, were thus sufficient to initiate instability. With a typical length ratio of 10^{-6}, these values of minimum relief correspond to 10 m and 560 m in the prototype. By comparison, the threshold stress of rock salt, based on power-law creep of dry salt in the laboratory extrapolated to a threshold strain rate of 10^{-16} s^{-1}, corresponds to about 160 m of initial relief (Jackson and Talbot, 1986).

The stress-strain rate field of Gulf Coast salt diapirism in general is portrayed as the stippled field in Figure 1. These data were derived from geologic observations on both macroscopic and microscopic scales, not from dynamic scaling. The observed field corresponds well with the model data scaled to the prototype.

A final aspect of Figure 1 is the close simulation of the rheology of rock salt by the range of model materials used. The deformation micromechanisms are entirely different but their scaled stress-strain rate relations are similar. The inset in

Figure 1 shows experimentally derived creep behavior of dry salt deforming by crystal-plastic processes with $n = 4.5$ (after Carter and Hansen, 1983) and of wet salt deforming by solution-transfer processes with $n = 1$ (after Spiers et al., 1986). The power-law flow is simulated by the 50RG/50PL mixture and the Newtonian flow by RG or the 75RG/25PL mixture. We discuss the applicability of these results to behavior of prototype cover in Section II, E.

Elastic behavior at the strain rates used in material testing and modeling can be ignored. The Maxwell relaxation time (the quotient of viscosity and shear modulus) in elastoviscous materials is the time required for stresses to be dissipated by viscous flow to $1/e$ (where e = base of natural logarithm) of their original level. Consequently, loads imposed for durations much longer than the relaxation time should result in negligible elastic strain. We estimate the shear modulus of our model materials to be approximately $G = 130$ kPa by taking the mean of two published values for silicone putties ($G = 260$ kPa and 5.5 kPa for putties marketed by Imperial Chemical Industries [Cogswell et al., 1972] and Dow-Corning Chemical Company [Dixon and Summers, 1985]). Using the viscosities listed in Table IV, Maxwell relaxation times for RG and 50RG/50PL are 0.3 s and < 2 s, respectively. The actual (non-normalized) centrifuge run times of each step varied from 46 s to 514 s, and durations of rotoviscometer tests varied from 23 s to 324 s. The component of elastic strain in the models was therefore negligible. No signs of fracturing or faulting were visible in any of the silicone putty mixtures in any of the models.

D. Modeling Technique

Models were constructed in variously shaped layers that were added in increments between each centrifuge run. A lining of polythene on the outer surface of the base and walls of the model ensured easy removal from the 10-cm-diameter aluminum centrifuge cup. In the floor of the cup was a plastic disk designed to mold the base of the spinning model parallel to surfaces of equal centrifugal acceleration. "North" or "N" on the diagrams provides a reference direction but has no relation to actual geographic coordinates in model or prototype. At rest, N pointed horizontally toward the rotor axis; while the model was spinning, N pointed almost vertically downward. Acceleration increased downward in the model and radially outward from the axis of the spinning rotor. Models were kept as thin as possible to minimize the gradient in centrifugal acceleration. At the usual spin speed of 2,000 rpm (equivalent to about 1,250 *g*), the difference in centrifugal acceleration between the base and top of a 10-mm-thick model was only 4%.

To monitor surface strains during centrifuge runs, a millimeter-scale, mechanically passive, graphite grid was printed onto the upper surface of the model layers by means of an unfixed photostatic image of graph paper (Dixon and Summers, 1985).

Each centrifuge run, which was repeated up to 19 times in a single model, was termed a step. To study the kinematics of diapir growth, we cut vertical slices from most models during the experiment. For example, during a typical step an increment of cover was added to the model, the model was centrifuged, a vertical step slice was permanently removed to be photographed, and the two halves of the model were rejoined. Up to five step slices were removed during an experiment. Materials having a high proportion of silicone putty (RG and 75RG/25PL) sealed completely and irreversibly after cut faces were rejoined, so that the model always behaved as a continuum. The faces of material containing a high proportion of Plastilina (50RG/50PL) also were welded together by contact, but their welded contacts could be deliberately pried apart again. However, there was no sign of movement along these contacts during later centrifuging so that the joined cut could be considered as a passive material discontinuity that had no apparent mechanical influence on the strain history. Removal of step slices changed the mass distribution of the model because the earlier-deposited layers lost more volume than the younger layers. We do not think that the effect of this change on qualitative analysis is significant. Care was taken to avoid step or serial sectioning of the model through earlier cuts.

Each centrifuge step consisted of three phases: run-up (accelerating rpm typically lasting 40-50 s), steady (constant rpm), and run-down (decelerating rpm typically lasting 15 s). Maximum acceleration varied from 226 *g* to 1,988 *g*. The body force per unit mass during each step was calculated by integrating the centrifugal acceleration of the three phases over time. The duration of each step was normalized by dividing the integral by a standard acceleration of 1,200 *g*. Normalization allows durations of different steps in different models to be compared relative to a standard body force. Normalized durations are shown in the illustrations. Such calculations are only strictly valid where (1) rpm changed linearly with time, and (2) flow was Newtonian. The inaccuracy of these assumptions is trivial when compared with other scaling parameters, such as the effective viscosity of rocks and sediments.

Slices cut from the model after the final step are termed serial slices. Serial slices were both horizontal and vertical, except in models 840412 and 840423, where only vertical slices were cut. Horizontal slices were numbered from the top downward. Vertical slices were numbered from the centermost to the outermost; where both halves of the model were cut vertically, slices from one half were labeled with a prime mark as well as a serial number.

E. Assumptions and Limitations

Even the most sophisticated mathematical or experimental models require considerable simplification of the geometry and mechanics of natural systems either to make the mathematics tractable or for practical reasons of construction. The following simplifications apply to the centrifuged models described here.

(1) Compared with the intricate detail displayed within the model diapirs after they had grown, the initial geometry of the models was simplified. In the experiments simulating gradual accumulation of overburden by sedimentary aggradation, we added layers in increments larger than the periodic pulses of sediment with which basins are thought to fill. Accordingly, the strains within our models should be used only as a qualitative guide to natural processes.

(2) The experiments simulated only mechanical forces, which include forces associated with viscosity, pressure, gravity, and inertia. We did not model thermal or chemical processes. Thermal effects result from the strong temperature dependence of the rheology of both silicone putty mixtures and rock salt. Thermal processes might also include thermal convection within the salt structure, but this process is only likely to operate in areas of high geothermal gradient and in salt deforming extremely rapidly (strain rates of 10^{-11} s^{-1}) because of water softening (Jackson and Talbot, 1986). The most important chemical processes are the dissolution of rock salt and the formation of cap rock. These processes require that, among other things, the crest of the salt structure penetrate a nonsaturated aquifer, especially terrestrial fresh-water aquifers. The mechanical importance of dissolution is that the diapir's volume is reduced and its shape altered, typically with the formation of a planar dissolution table across the crest of the salt. Natural diapirs dissolved below ground typically have flatter crests than those of model structures. The cap rock that may form by dissolution represents a dense layer with higher equivalent viscosity and higher yield strength than salt. However, cap rock is a relatively thin skin in Gulf Coast diapirs and its mechanical

influences are probably slight. In the Sverdrup basin of Arctic Canada, virtually only gypsified anhydrite is exposed in scores of salt diapirs (Schwerdtner and Osadetz, 1983; Schwerdtner and van Kranendonk, 1984). These diapirs rose to the surface even with very thick, heavy cap rocks.

(3) The dimensions of length and time are independent, and inertial forces can be neglected (Ramberg, 1981, p. 39-40). The model ratios of acceleration, length, and time are rigorously related as follows: $a_r = l_r t_r^{-2}$. Strictly, this would constrain the duration of the experiment such that for typical model ratios of acceleration and length and simulated geologic durations (a_r = 1,200, $l_r = 10^{-5}$, t_p = 40 Ma), each centrifuge experiment should properly last three and a half millennia. Nevertheless, the velocity of rise of model diapiric structures was minute (Tables II and III), and inertial forces induced by changes of velocity in the models were correspondingly low. Measures of the importance of inertial forces relative to viscous and gravity forces are given by the nondimensional Reynolds and Froude numbers, respectively. For our centrifuged models, the Reynolds and Froude numbers were on the order of 10^{-8} and 10^{-11}, respectively. Inertial forces were therefore negligible. In real salt structures these numbers are many orders of magnitude lower still. For example, during its peak rate of growth 105 Ma ago, Hainesville dome in East Texas had a Reynolds number of 10^{-25} and a Froude number of 10^{-28} (based on data in Seni and Jackson, 1984, and the flow law for dry rock salt of Carter and Hansen, 1983). Consequently, inertial forces and inertial effects were vanishingly small in both models and prototypes, so length and time can be treated as independent variables in dynamic scaling.

(4) Both the buoyant layer and its overburden are assumed to behave like extremely viscous fluids in bulk in response to forces imposed over millions of years. We assume that over these long durations the rocks deform by steady-state creep--either

Newtonian or power-law, both of which can be modeled by materials previously described. As shown in Figure 1, the model materials chosen simulate the full range of steady-state creep behavior by wet or dry salt. Rheidity (Carey, 1954), equal to Maxwell relaxation time multiplied by 1,000, is the time required for the viscous component of strain to be 1,000 times the non-time-dependent elastic components under constant stress and varying strain. The rheidity of rock salt is 42,000 years, based on a shear modulus, $G = 1.5 \times 10^{10}$ Pa (Clark, 1966), and an equivalent viscosity, $\mu = 2 \times 10^{19}$ Pa s (at 100°C and strain rate of 10^{-14} s^{-1}; Carter and Hansen, 1983). For loads applied longer than the rheidity, the behavior of rock salt can thus be realistically modeled as effectively viscous. But our knowledge of the rheology of cover rocks is very sparse. We do not know the equivalent viscosity of the terrigenous clastic overburden, and hence cannot calculate its rheidity. Thus we do not know how unrealistic it is to model overburdens with viscous (linear or power-law) materials. Certainly the slower the natural strain rate, the more closely the material deforms like a highly viscous fluid. Faults and other fractures represent a component of the strain that is nonviscous and possibly related to temporary periods of higher than normal intrusion rates. However, more than 40 salt diapirs exposed in central Iran and currently under study clearly show that diapirs can pierce cover sediments, causing extremely high ductile strains in the cover but virtually no radial faulting. They represent spectacular proof of the proposition that under certain conditions both the salt and its country rock deform as viscous fluids on a macroscopic scale equivalent to that at which scale models are studied. Even where faults above and adjacent to diapirs are common, as in the Gulf Coast, their abundance may mean that the cover as a whole approximates a mechanical continuum like incohesive sand.

However, there is certainly ample scope for the modeling of brittle fracture in diapiric overburdens in future experiments.

F. General Theory and Nomenclature

A density inversion, in which a dense viscous layer, referred to as cover or overburden, overlies a less dense viscous layer, referred to as the source layer, is inherently unstable. This instability causes the system to overturn, which reduces gravity potential energy to a minimum. The gravitational instability of dense fluid overlying less dense fluid is known as Rayleigh-Taylor instability, the analytical theory of which has been applied to salt diapirs for more than 20 years (Danes, 1964; Biot and Ode, 1965; Selig, 1965; Biot, 1966; Ramberg, 1967, 1968; Hunsche, 1978; Turcotte and Schubert, 1982). Basically, an originally planar interface between the source and cover is initially deformed into sinusoidal waves of random wavelength (where wavelength is the horizontal distance between the crest points of adjacent waves). In a given system, waves of a certain wavelength grow fastest. Structures forming on this dominant wavelength eventually dominate the others in amplitude.

Analytical theory is valid only for the very early stages of deformation where amplitude/wavelength ratios are less than 0.015-0.09, depending on the viscosity contrast (Woidt, 1978). Numerical or material models (e.g., Parker and McDowell, 1955; Ramberg, 1967, 1981; Fletcher, 1972; Whitehead and Luther, 1975; Heye, 1978, 1979; Hunsche, 1978; Woidt, 1978, 1980) are required to simulate further deformation of the system. Numerical models have so far been restricted to two-dimensional vertical sections; the third dimension is built into the model by assuming plane strain or axisymmetry. Most of these models were run under normal gravity (Whitehead and Luther, 1975; Heye, 1978; Hunsche, 1978; Ramberg, 1981).

The structures that transfer the buoyant substratum to an overlying position in models closely resemble the shapes of natural diapirs known as salt stocks or salt walls. However, in many models even cylindrical "diapirs" do not pierce their greatly thinned overburden and are therefore not true diapirs (piercement by model diapirs is encouraged by high viscosity contrasts). Nor do we want to exclude from discussion low-amplitude model structures that resemble nondiapiric salt pillows and salt anticlines. For these reasons we refer to the model structures in terms used in fluid mechanics. Two types of rising structures are recognized on the basis of their shape aspect in plan view: those with low shape aspect (circular or slightly elliptical like pillows and stocks), known as fingers; and those with high shape aspect (like anticlines and walls), known as walls. This classification is independent of the amplitude (height) and maturity of these structures. Unless qualified otherwise, fingers and walls can be understood to be rising structures. Our comparatively meager knowledge of the evolution of walls into fingers is based largely on material models (but see Danes, 1964, and Whitehead and Luther, 1975, for aspects of mathematical theory).

G. Edge Effects

Experiments described here demonstrate the effect of particular geometries of loading on the planform of fingers and walls. To evaluate these loading influences, similar effects produced by the boundary walls of the experimental container are illustrated in Figures 2 and 3 and discussed in Appendix A.

Edge effects are imposed by lateral walls of the experimental container. These effects can be noted in most

Figure 2. Edge effects in a rectangular container shown by plan views of model 840410 at successive stages (time in seconds). Model consisted of 2-mm-thick liquid mercury cover (density = ρ = 13,550 kg m^{-3}) over 5-mm-thick tabular layer of silicone putty (ρ=1,090 kg m^{-3}), on the surface of which a 1-mm grid was printed to monitor surface strains. Model tray rests on mm-ruled graph paper, visible top and bottom of photographs. Reflection of camera from mercury surface visible at

Model 840410

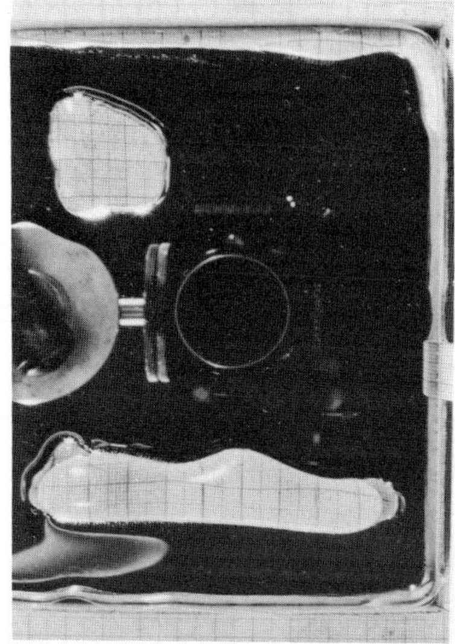

530 s

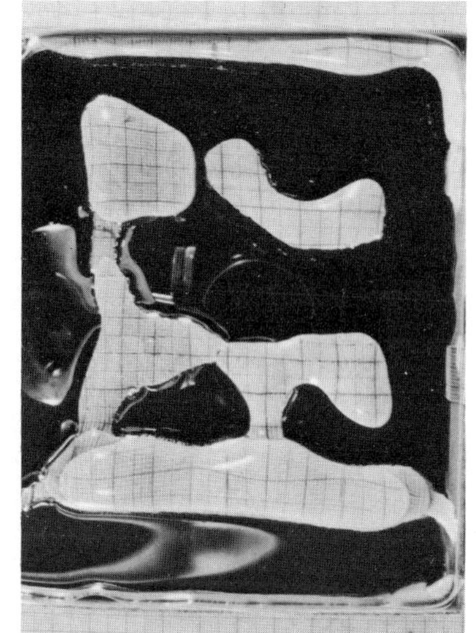

729 s

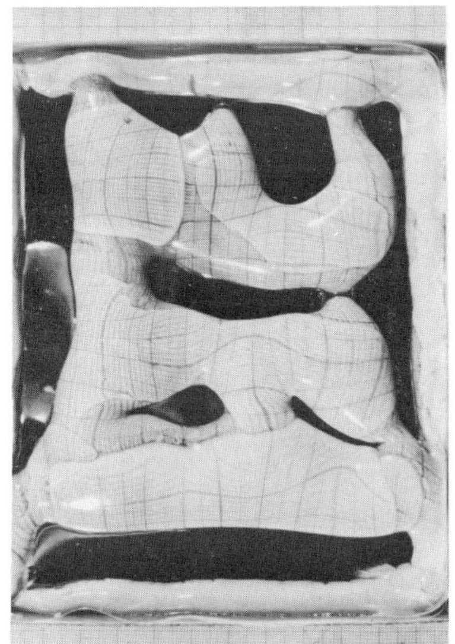

1440 s

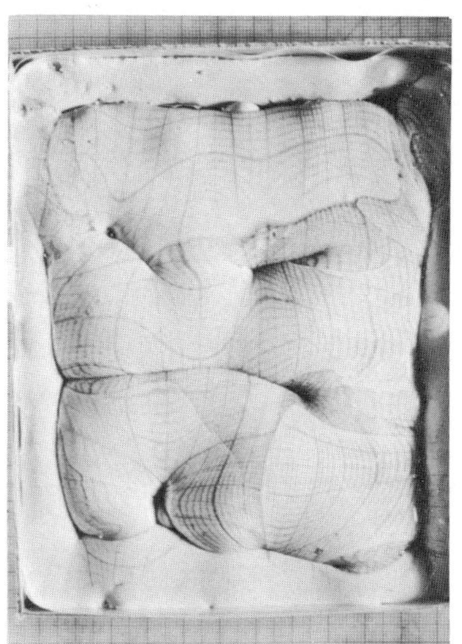

2955 s

530 s. First structure to breach the cover was peripheral wall on left before 500 s. Walls are orthogonal to model boundaries. Experiment by C. J. Talbot.

published multidome experiments run under normal gravity (e.g., Parker and McDowell, 1955; Ramberg, 1968, 1981; Berner et al., 1972; Talbot, 1974, 1977; Whitehead and Luther, 1975; Heye, 1978; Hunsche, 1978). Edge effects are less common in published centrifuged models (e.g., Talbot, 1977; Ramberg, 1981, p. 284; Dixon and Summers, 1983) for two reasons. First, most centrifuge experiments used initiators (built-in initial perturbations of shape) to focus finger or wall growth in desired sites, usually away from the model sides--a practice that reduces the influence of the lateral boundaries. Second, previous centrifuged models were almost invariably sectioned vertically rather than horizontally, which means that edge-effect walls may not have been recognized unless the horizontal dimension was graphically reconstructed using serial vertical sections. We routinely cut both horizontal and vertical sections, and no initiators were used to focus dome growth.

The geometry of edge effects described in Appendix A can be summarized as follows. A peripheral wall almost invariably forms in material models without deliberate initiation of walls or fingers. Not only is this peripheral wall remarkably complete and uniform along its circumference (before it segments upward into fingers), but its rate of growth is typically the highest in the model. The effects of this instability decrease away from the edges; they are typically visible one wavelength toward the

Figure 3. Edge effects in a circular container shown by plan views of model 840411 at successive stages (time in seconds). Model consisted of 3.5-mm-thick liquid mercury cover (ρ = 13,550 kg m^{-3}) over 4.5-mm-thick tabular layer of silicone putty (ρ = 1,090 kg m^{-3}), on the surface of which a 1-mm grid was printed to monitor surface strains. Model beaker rests on mm-ruled graph paper. Reflection of camera from mercury surface visible at 1,310 s. First structure to breach cover was peripheral wall at 360 s. Mercury cover was poured off at 2,578 s to show complete structure of source layer. Walls are parallel to model boundaries and therefore circular. Experiment by C. J. Talbot.

Model 840411

5 cm

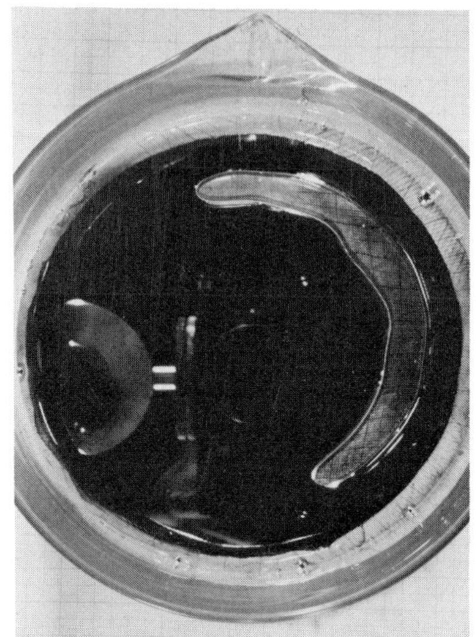

1310 s

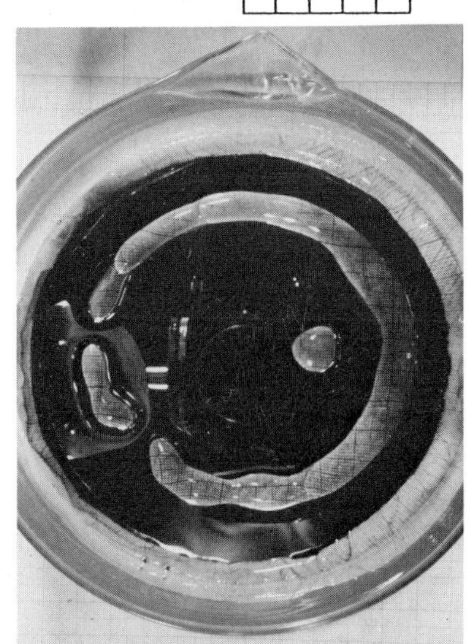

1615 s

2020 s

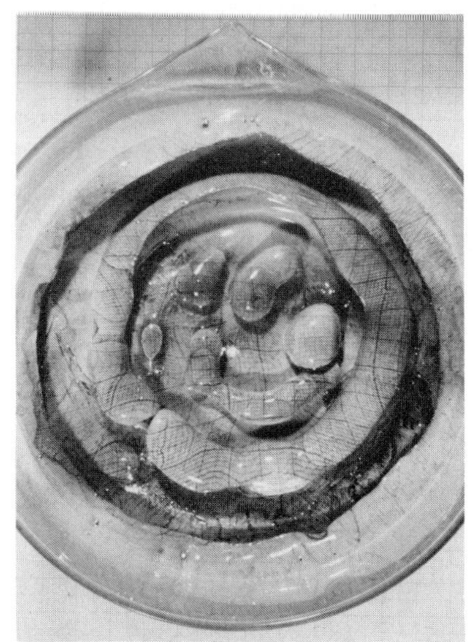

2578 s

center and may be partly apparent two wavelengths inward. This edge effect is a type of diapir family in which younger generations are propagated inward from the boundary. The classic salt dome families also show a laterally propagating instability that becomes progressively weaker with distance, but successive generations form outward, not inward, from a single strong perturbation.

Clearly, edge effects must be recognized and mentally removed while investigating the influence of particular types of loading on the pattern of finger and wall formation. If peripheral walls were caused by bubbles or surface tension, (as argued in Appendix A) they were peculiar to models and would not affect salt tectonics. Therefore we do not expect them to have their dynamic counterparts in natural salt tectonics, although geometric analogs may exist. For example, peripheral walls of salt are known in the North Sea Zechstein (Jenyon, 1985), where they were ascribed to updip flow of salt into natural barriers caused by facies changes in the evaporites.

One approach to filtering edge effects is to produce model structures of relatively small wavelength in a relatively large model container and to ignore those structures within one or two wavelengths of the boundary if they are visibly influenced by it. Another approach is to exploit container shape. If we expect to produce linear walls, we should experiment with a circular container. In this way any linear feature forming a chord across the circular model cannot be an edge effect. The converse holds true for a rectangular container. Both approaches were followed in our experiments: the first by using relatively thin layers so that the wavelength of the resulting structures was much smaller than the model diameter; the second by adding increments of cover with linear boundaries in a circular centrifuge cup.

III. STATIC DIFFERENTIAL LOADING OF SOURCE LAYER

A. Previous Experiments

Talbot (1977) demonstrated by centrifuge modeling that where the cover was laterally divided into more viscous and less viscous halves, fingers grew fastest beneath the faster-straining (less viscous) cover. He also studied the effect of lateral variations in density using abrupt and gradual lateral changes in density. In both cases the resulting fingers were tilted and they developed larger overhangs toward the side of denser cover, the side characterized by the greatest differential pressures.

The effect of laterally varying thickness (as opposed to density or viscosity) of cover on finger and wall shape has also been subjected to experiment. Rettger (1935) loaded a sand, bentonite, and clay sequence in a tank with bags of shot under normal gravity. The sequence became folded, with maximum amplitude just in front of the edge of the shot and two smaller folds equally spaced in front of the first fold. All folds verged away from the load and trended parallel to the distal edge of the load. McKee and Goldberg (1969) differentially loaded laminated clay with denser sand, producing upright, symmetrical walls at the edge of a static load; asymmetrical walls tilted away from the edges of gravity-spreading loads. Ramberg (1981, his Fig. 11.64) illustrated an analogous fold in a source layer of stitching wax in front of a static differential load of modeling clay bounded by a step. He termed this geometry an "edge of excess overburden" or "edge discontinuity." Across a step discontinuity where the overburden thickness or mass drops sharply, a pressure difference is created in the underlying source layer. Theoretically this surface-step effect (the influence of a stepped upper surface on model dynamics) causes the source to flow laterally toward the side with thinner overburden. According to Ramberg (1981, p. 283), the flow of source from beneath the heavier side of the surface step causes an anticline of thickened source to rise beneath the discontinuity. The experiment described below was designed to determine the manner in which this anticlinal frontal bulge acted to initiate fingers or walls.

B. Initial Configuration of Model 840402

A tabular, low-density source layer B, equivalent to salt (scaled density = 2,200 kg m^{-3}), rested on a rigid basement A (Fig. 4) and was overlain by a uniform cover layer C, whose density was greater than that of the source and equivalent to moderately compacted shale or sandstone (scaled density = 2,260 kg m^{-3}). A differential load was supplied by a dense, tabular half-layer D covering half the model; its scaled density of 2,320 kg m^{-3} was equivalent to well-compacted shale or slightly compacted marl. Viscosities of source and cover were identical, and flow was Newtonian. Both cover layers were added at the beginning of the experiment. The experiment was run under normal gravity for 52 hours (equivalent to 156 s at 1,200 *g*).

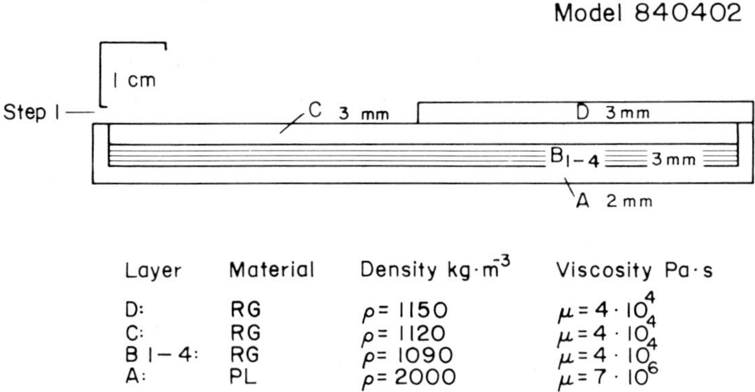

Figure 4. Initial configuration of static differential-loading model 840402 in vertical section.

C. Experimental Results

Adding the half-layer D created a surface step in the model. But in a few hours the dense half-layer sank, displacing the substratum sideways and raising the left half of the model to the same level as the right half, which contained the foundering half-layer. The upper surface therefore became planar by flow throughout the model container. Remaining deformation was restricted to partial overturn of layers B, C, and D because of the density inversion.

The material displaced by the sinking half-layer D was part of layers B and C, the source layer and uniform cover. Material was displaced from the loaded side (with respect to the half-layer) to the unloaded side (again, with respect to the half-layer). Analysis of 12 vertical slices shows that equal proportions of both layers B and C were displaced (Fig. 5). Layers B and C transferred an average of 11% and 12% of their volume from beneath the half-layer; there is no statistical difference between the two averages. The hydrostatic pressure difference created by a surface step is constant with increasing depth, even though the absolute pressures increase downward. Lateral flow of layer B is retarded by the top of the basement layer A. However, the top boundary is a frictionless free surface. Because the flow is Newtonian and laminar, the lateral flow velocities should increase upward from zero at the base of layer B to a maximum at the free surface. Both layers B and C have the same thickness and viscosity, so that a greater volume of layer C than layer B should have been transferred. Because the lateral flows were equal, some factor, perhaps related to the dynamics of a sinking plate, must counter the effect of a free surface above layer C. Where a differential load (such as a delta) overlies a weak substratum (like unconsolidated, overpressured prodelta mud), the differential load

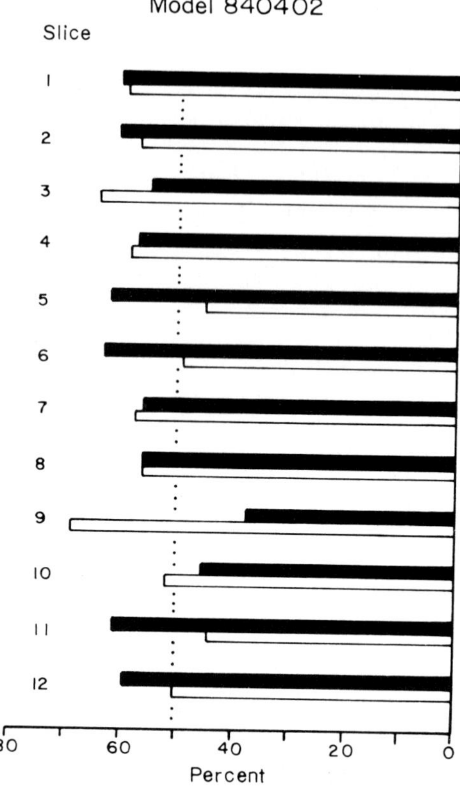

Figure 5. Mass-flow data for static differential loading model 840402, based on measured areas in serial vertical slices of final step after 52 h. Bars show percentages of layers B (source) and C (uniform cover) that accumulated within unloaded side of model by lateral flow from loaded side.

could be neutralized by lateral flow of the underlying weak fluid before a deeper buoyant layer of higher viscosity had time to react to differential loading.

Vertical slices through the model (Figs. 6 and 7) show that walls rather than fingers grew in both halves of the model. In the middle of the model the walls trended parallel to the surface step as a surface-step effect. At the model edges the usual

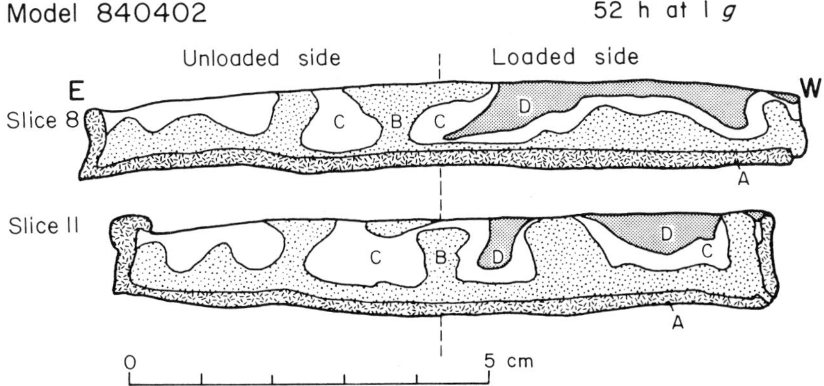

Figure 6. Two representative vertical slices of final step of differential-loading model 840402. Source layer B is stippled; D is half-layer exerting differential load.

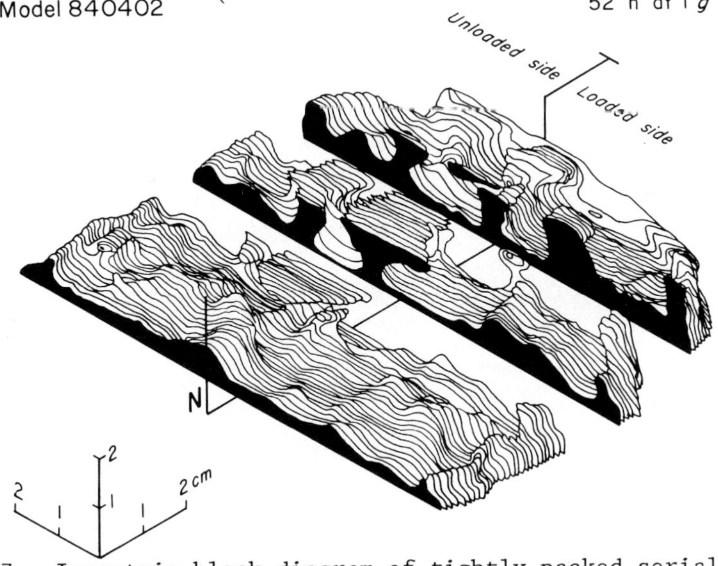

Figure 7. Isometric block diagram of tightly packed serial vertical slices of final step of static differential-loading model 840402 to show three-dimensional configuration of source layer. Cover has been excluded. Block diagram has been partly exploded to show slices illustrated in Figure 6.

peripheral walls formed because of the edge effect. In agreement with Ramberg's (1981) hypothesis, the surface step initiated growth of a frontal wall from the frontal bulge. The horizontal serial slices in Figure 8 show the frontal wall (slices 2 and 3) and a well-defined rim syncline around the largest finger on the loaded side (slices 1, 2, and 3).

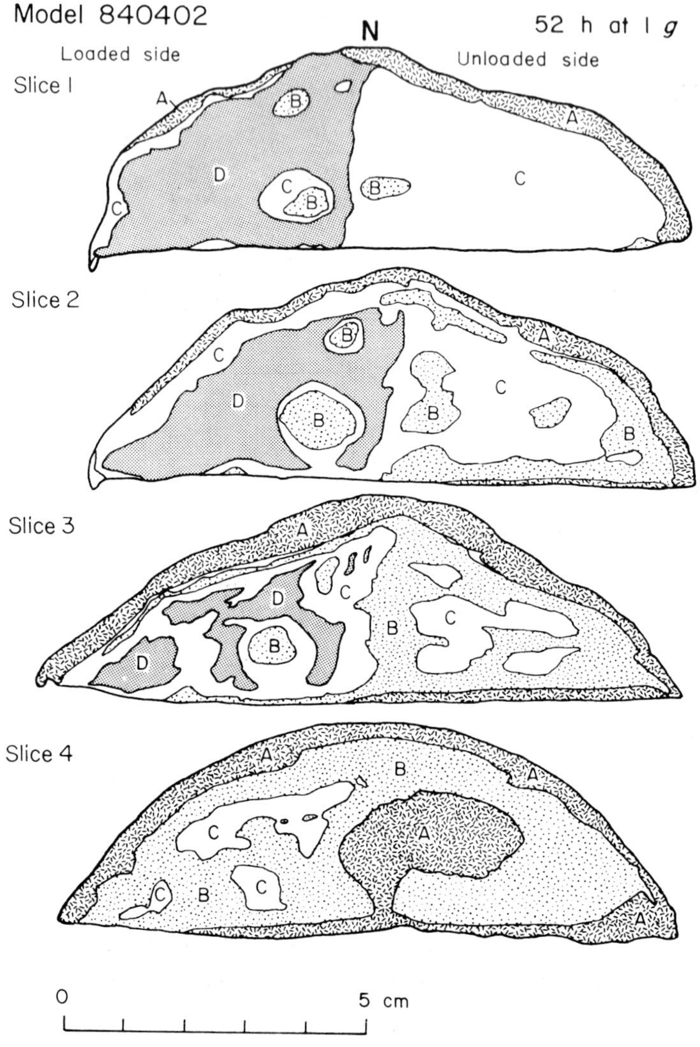

Figure 8. Serial horizontal slices of final step in differential-loading model 840402. Slice 1 is highest; slice 4 is lowest. Source layer is B, half-layer is D.

Data on structural tilt in the vertical slices are summarized in Figure 9. The datum for measuring tilt was the antiformal anticlinal core, or AAC (Jackson and Talbot, 1985), which extended up the stem and into the bulb. The trace of the AAC was defined by passive multilayers in the source layer. The curved AAC traces in Figure 9b were reduced to straight lines by drawing their chords in Figure 9c, which allowed tilt to be measured. Based on clustering of structures in the multiple vertical profiles shown in Figure 9a, the slices were divided into six domains, each domain containing a wall or finger. Three were on the loaded side (beneath the half-layer) and three on the unloaded side. The outermost domains were characterized by strong edge effects. The sides of the model exerted drag on the outer margins of the outermost movement cells. This meant faster flow of the inner sides of the movement cells, which in turn caused an inward tilt and asymmetry.

The mean tilts in each of the five inner domains (Figure 9d) showed that structures on each side tilted inward toward the surface step in the center. Tilt angles toward the loaded side were positive and those away from the loaded side were negative. The variation in tilt angle is shown in the histograms and cumulative frequency diagrams of Figure 9e. Structures on the unloaded side tilted more than those on the loaded side; this is probably a statistical artifact of the enhanced tilt of the peripheral wall farthest from the surface step. The distribution of tilt angles of structures on both sides is asymmetrical toward extreme values, with the skewness being greater on the unloaded side. Both sides have bimodal tilt distributions (Fig. 9e).

The variation in amount of tilt argues against simple inward propagation of an edge effect as the cause of tilt. Figure 9d shows that the mean tilt of the innermost domains, 3 and 4, is greater than that of the next outer domains, 2 and 5 (although their standard errors overlap). We interpret this to mean that

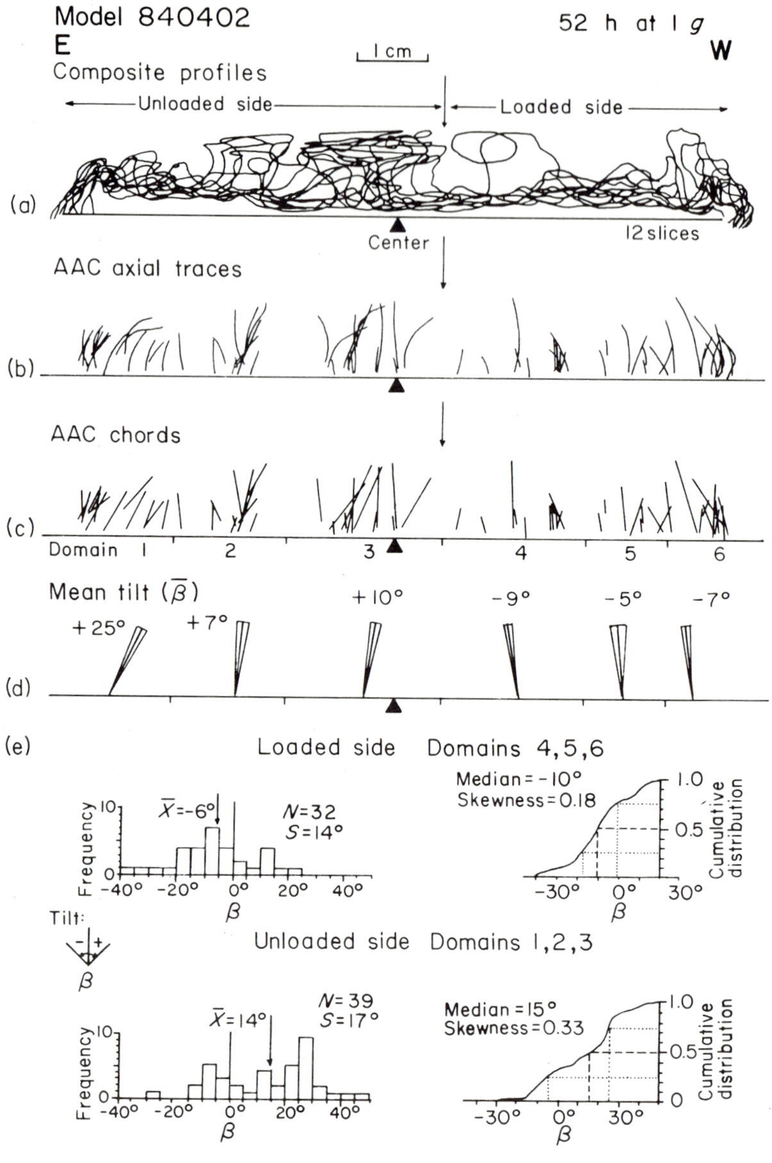

Figure 9. Tilt data for loaded and unloaded sides of final step of static differential-loading model 840402. (a) Superimposed profiles of source-layer structures traced from 12 serial vertical slices. Arrow marks edge of half-layer. (b) Axial traces of antiformal anticlinal core (AAC) of fingers and walls, defined by multilayers within source layer in the 12 slices. These traces indicate tilt and asymmetry of structures. (c) Chords

the foundered surface step in the center was responsible for the inward tilt on both the loaded and unloaded sides. The maximum lateral pressure gradients underlay the central edge discontinuity, so the frontal wall grew faster and higher than the others (Fig. 6a). Because the movement cell incorporating the frontal wall circulated the fastest, the sinking downwalls of cover within the cell increased the circulation speed of the neighboring movement cells by viscous traction. These neighboring upwalls therefore developed faster on their inward sides, causing a tilt toward the central surface step.

In summary, differential loading by a static half-layer produced walls parallel to the margin of the half-layer. On both the loaded and unloaded sides, structures tilted toward the surface step. Fingers and walls in front of the load were more tilted and more asymmetrical than those beneath the load.

IV. PROGRADING DIFFERENTIAL LOAD OVER TABULAR SOURCE LAYER

In Section III we described how a frontal wall was produced by differential loading parallel to and immediately in front of the edge of a static step. In this section we describe experiments demonstrating that this frontal wall was pushed ahead of the prograding load, but that its influence remained behind by the creation of linear initiators that were overridden by the load.

A. Previous Experiments

Rettger (1935) caused a small delta of sand to prograde over a thin layer of clay by suspension settling under normal gravity.

joining ends of axial traces in (b). (d) Six domains were defined on basis of clustering of structures in (a), (b), and (c). Elongated fan in each domain shows mean tilt (°) of AAC chords and standard error (= standard deviation/$\sqrt{N}$). (e) Histogram and cumulative distribution of tilt angles in loaded and unloaded sides of model. N = number of axial traces; x = arithmetic mean; S = standard deviation; skewness calculated on basis of upper and lower quartiles (facing page).

A train of as many as five folds built up in the clay ahead of the delta, the most pronounced being nearest the delta front. His folds moved ahead of the prograding delta as migrating waves. He did not examine the effects of this distortion and apparent retrodistortion on the substratum buried beneath the delta. McKee and Goldberg (1969) also produced model folds, both symmetrical and asymmetrical, in horizontally laminated mud beneath a prograding sand delta. Thrusts verging in the progradation direction were also produced at the delta front. They did not analyze the variation in vergence of the folds or thrusts.

B. Initial Configuration of Models

Model design was chiefly concerned with duplicating the natural balance between the rates of progradation, aggradation, and diapir rise in the prototype Gulf Coast (Section II, B and Table I). The relative dominance of these three processes has a powerful influence on structural evolution. The other main variable was the clinoform geometry.

Model 840328 simulated progradation by four thin strips of cover having elongated parallelogram-shaped vertical sections (Fig. 10). The prograding cover was deposited directly on the source layer. The ratio of progradation rate to aggradation rate (P/A) was intermediate between the long-term Cenozoic rate and the short-term Quaternary rate in the prototype (Table I). The same was true for the ratio of progradation rate to diapir rise rate (P/R). The source layer (scaled density = 2,200 kg m^{-3}) had Newtonian creep, whereas the cover (scaled density = 2,340 kg m^{-3}) had power-law creep; the equivalent dynamic viscosity ratio was 5.

The other two models (Fig. 10) simulated progradation by many narrow strips of cover, rather than merely four, directly onto a source layer. Model 840412 had 18 strips with rectangular vertical sections added in 18 centrifuge runs. Both the P/A and

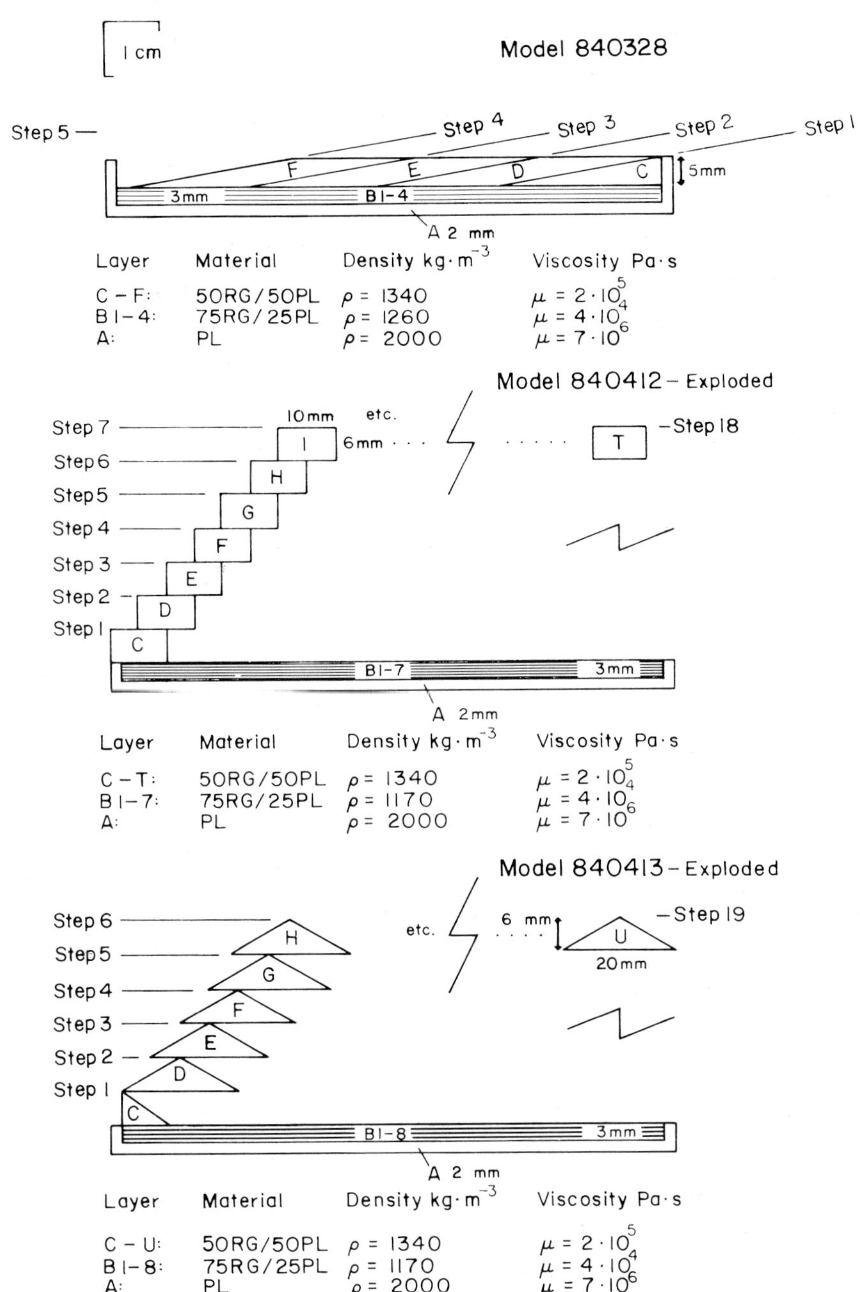

Figure 10. Initial configuration of prograding models before acceleration in centrifuge seen in vertical section. Not all cover units of 840412 and 840413 are shown to save space.

P/R ratios were slightly less than the Quaternary prototype ratios (Table I). Model 840413 had 19 strips with triangular vertical sections added in 20 centrifuge runs. The *P/A* ratio was the same as in the Quaternary prototype, whereas the *P/R* ratio was intermediate between the Cenozoic and Quaternary prototype ratios. The two models were otherwise of identical construction. The source layer (scaled density = 2,200 kg m^{-3}) underwent Newtonian flow, whereas the cover (scaled density = 2,520 kg m^{-3}) had power-law creep; the equivalent dynamic viscosity ratio was 5.

The side of the model from which progradation started is referred to as the proximal side; the side toward which the progradation advanced is the distal side.

C. Experimental Results

1. Model 840328

Figure 11 shows step vertical slices of model 840328. In step 1 the differential load of strip C caused considerable flow of the source from beneath it distally, raising all of the free surface of the source layer. The frontal wall was spread throughout the unloaded source layer. Bubble tracks, which acted as passive markers after the bubble had risen, were generally vertical but verged distally at the front of layer C.

Steps 2 and 3 show accentuated source transfer from beneath the prograding load after deposition of strip D. The horizontal width of the uncovered source layer narrowed as the source continued to thicken. Bubble tracks all verged distally away from the differential load. Some tracks were straight and others were listric (upwardly concave) to some degree. At the frontal tips (numbered 1 and 2 in Fig. 11) of cover layers C and D, the bubble tracks curved backward toward each frontal wall. The tracks were deformed by differential laminar flow across them. Their shapes therefore indicate the displacement profile of the flow. A proximal wall formed in step 3 as an edge effect.

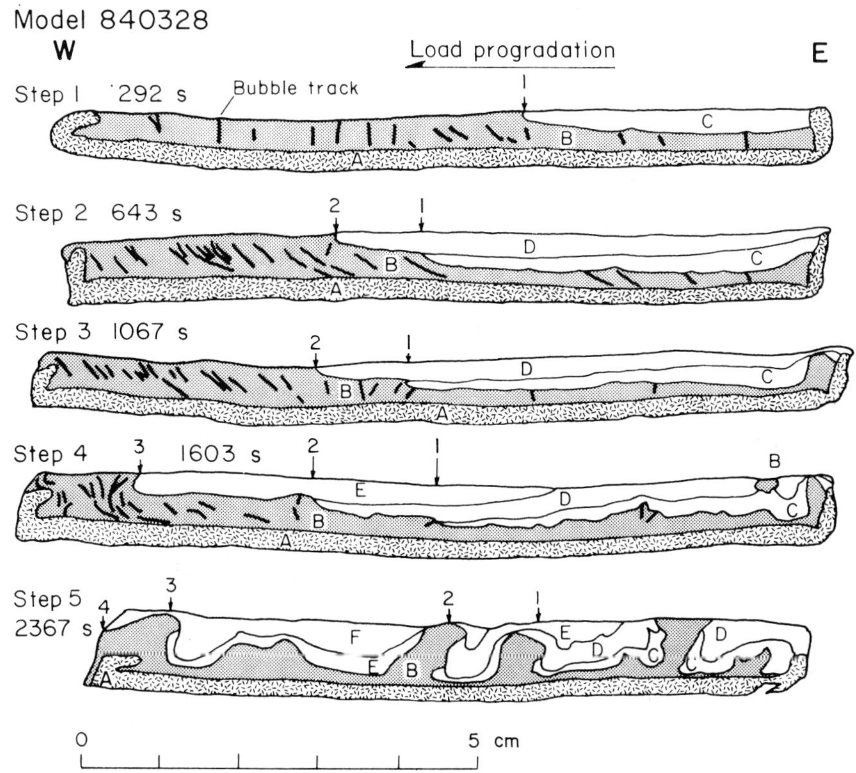

Figure 11. Step vertical slices of prograding model 840328. Bubble tracks were originally vertical before their distortion by lateral flow. Strips C through F are prograding cover directly overlying source layer.

In step 4, the first walls were initiated (excluding the earlier-formed peripheral wall) and some reached the surface and became extrusive (below the letter B). Bubble tracks indicate continued flow of the source toward the narrowing free surface. The source was being squeezed distally into an increasingly confined space, so the flow profiles curved strongly upward and back toward frontal tip 3.

By step 5 (Fig. 11) the prograding wedge completely covered the surface of the source layer after addition of strip F. The walls matured fully and their locations showed a strong control by the migrating frontal tips during progradation. Four mature walls formed; three were located at the three frontal tips, 1, 2, and 3. The rightmost, proximal wall grew beneath the middle of strip C, its position probably controlled by the dominant wavelength like that of the distal low wall beneath the middle of strip E.

Apart from controlling the location of structures in the source layer, the migrating frontal tips, which are equivalent to linear sedimentary depocenters, also controlled their form. Serial vertical slices yielded data that were processed into a structure-contour map of the source layer's upper surface, from which an isometric projection was prepared (Fig. 12). This shows deformation of the source layer into parallel walls, from which fingers rose in places. The walls trended perpendicular to the direction of progradation.

Figure 13 shows the changing mass balance within the source layer during steps 1 through 5 plotted against time. Up to step 2, source material continued to flow in the direction of progradation, a process we term distal incremental flow. After

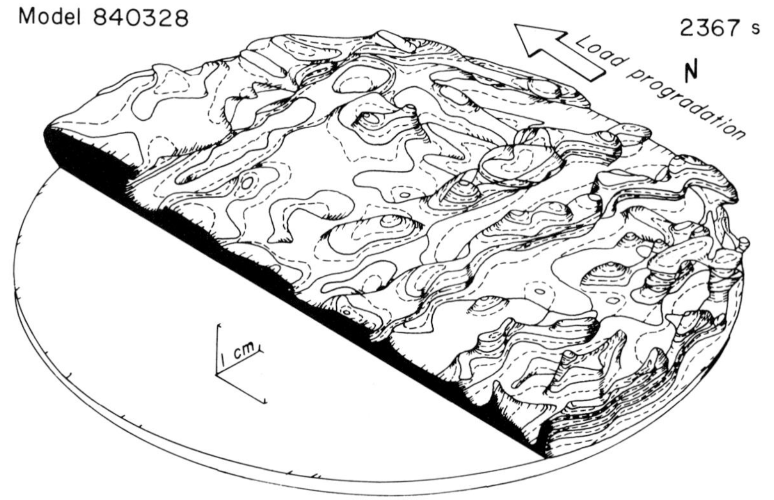

Figure 12. Isometric block diagram of upper surface of source layer in prograding model 840328 at final step. Cover not shown. Data derived from 22 serial vertical slices. Contour interval is 1 mm.

this, flow of the source layer reversed in direction, becoming proximal incremental flow. The word "incremental" emphasizes that although flow was proximal, the finite volume of layer B was still greater at the distal end because the reservoir of extra source material that had accumulated distally during early stages was still far from being exhausted. Flow was reversed to maintain balance after further cover was added distally and (less importantly, we think, based on model 840413) to supply the proximal walls that had begun to grow by this time.

The walls were strongly tilted proximally, opposite to the progradation direction. Clustering of AAC axial traces and chords (Fig. 14b,c) defined seven domains. Side domain 7 contained the large upwelling wall derived from distal flow of source material. The boxlike profile of the wall resulted in a bimodal distribution of AAC tilts: one for each shoulder, the inner one being strongly tilted in a proximal direction.

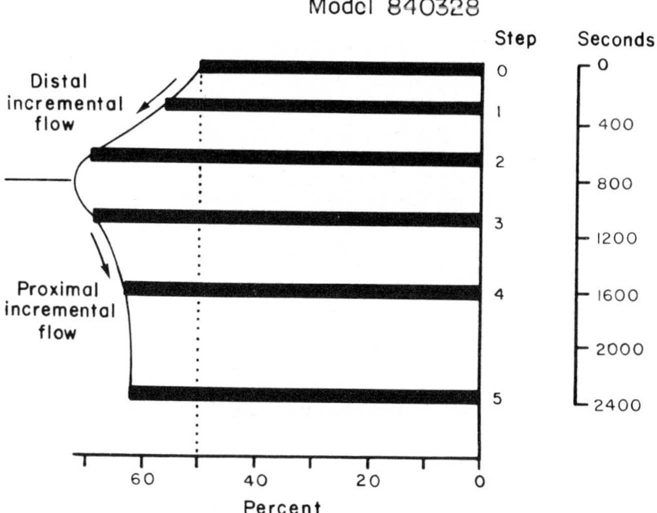

Figure 13. Mass-flow data from step slices of prograding model 840328, plotted against time. Bars show percentage of source layer in distal half of the model, which was covered by prograding cover late in experiment. Incremental flow reversed in direction between steps 2 and 3.

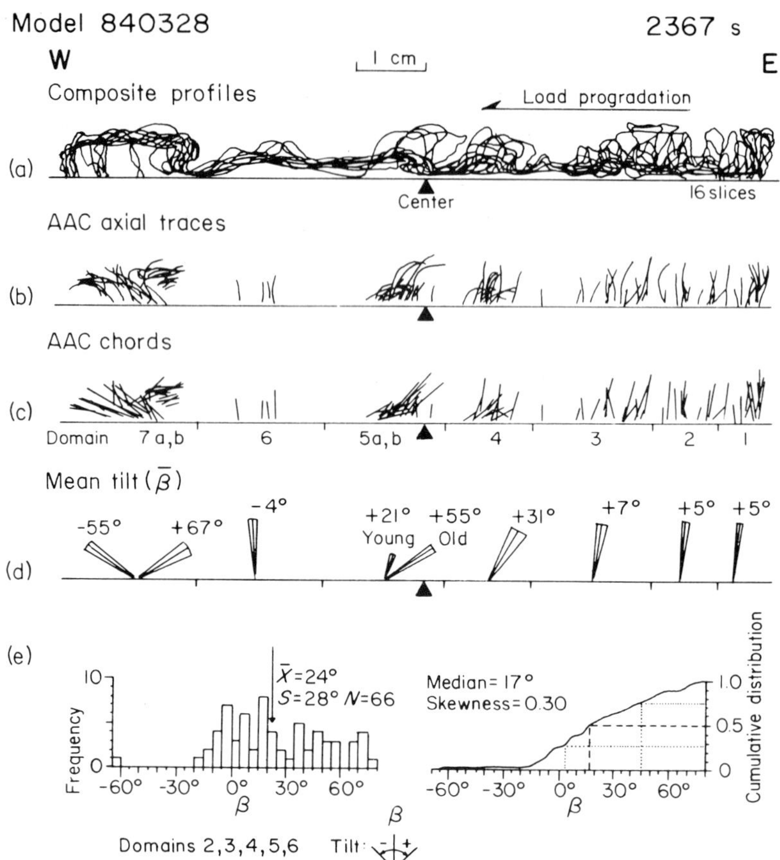

Figure 14. Tilt data for prograding model 840328. (a) Superimposed profiles of source-layer structures traced from 16 serial vertical slices. (b) Axial traces of antiformal anticlinal core of walls, defined by multilayers within source layer in the 16 slices. These traces indicate tilt and asymmetry of structures. (c) Chords joining ends of axial traces in (b). (d) Seven domains were defined on basis of clustering of structures in (a), (b), and (c). Elongated fan in each domain shows mean tilt (°) of AAC chords and standard error (= standard deviation/$\sqrt{N}$). Domains 5 and 7 are bimodal because of two generations of walls. (e) Histogram and cumulative distribution of tilt angles. N = number of axial traces; x = arithmetic mean; S = standard deviation; skewness was calculated on basis of upper and lower quartiles.

Disregarding domain 6, which contained only a few, low-amplitude, immature fingers, all the other domains were characterized by mean proximal tilts, that is, by positive β angles (Fig. 14d).

The most proximal domains, 1, 2, and 3, all had low positive tilts of 5° to 7°, whereas the distal domains had much greater positive tilts. The proximal walls were less affected by progradation than were the distal walls. Because proximal walls were soon overrun by the prograding load, the load seems to have acted as static cover for the remaining growth of the proximal walls. This hypothesis is supported by tilts in domain 5, which contained double walls. An older wall with a recumbent bulb was bowed upward by a younger wall growing beneath it. Significantly, the older structure was highly asymmetrical and tilted, whereas the younger structure was much less asymmetrical and tilted (Fig. 14d). Here the older structure grew in front of the prograding load. The younger structure grew much later, after the frontal instability had passed overhead and the load was uniform.

Excluding the outer peripheral walls, the distribution of tilt angles is shown in Figure 14e. Both the mean and median show a marked overall tilt opposite to the direction of progradation; the distribution is skewed away from extreme values of β.

2. Model 840412

Model 840412 (Fig. 10) simulated slow progradation (lower *P/A* and *P/R* ratios than in the prototype) as an intermediate between static loading and normal rates of progradation (Table I). No step slices were cut during the deposition of the 18 prograding strips, but the upper surface was photographed. Figure 15 shows the surface finite strains revealed by grid distortion at steps 2 and 7. The finite strain recorded two influences.

The first influence was an edge effect due to drag by the sides of the model during distal flow of the source away from the prograding load. Drag was shown in step 2 by distortion of the grid squares in the south to rhombs by right-lateral simple

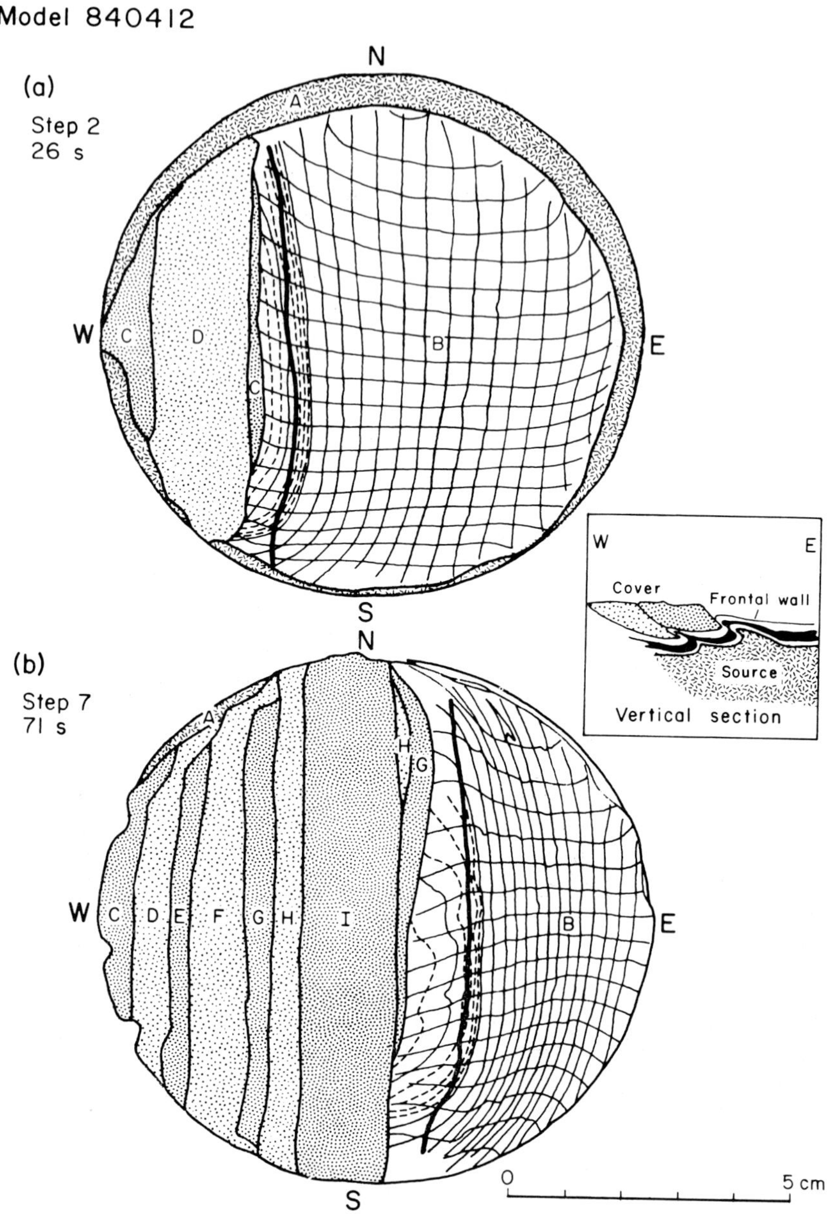

Figure 15. Surface strains ahead of prograding wedge of cover in model 840412 at two stages. Deformed 5-mm grid is based on 1-mm grid in actual model. Longitudi-

shear. Drag was more pronounced by step 7, when grid lines originally oriented north-south became distorted into arcs by drag against the northern and southern sides.

The second influence was the prograding frontal tip. Immediately in front of the tip in step 7, the upwelling frontal wall was marked by a zone of extension and area increase. Extension parallel to the linear front was negligible, but extension normal to the front was as much as 600%. Extension in the frontal wall was due to an increase in arc length as the source layer accommodated to the shape of the foundering frontal tip of cover (Fig. 15, inset). At the distal margin of the extended frontal wall was a line of no finite normal longitudinal strain, where the normal extension declined to zero. Ahead of this line a broad zone was shortened by 40% by step 7. This strain was less than in the frontal wall because it was distributed over a width approximately five times as great as the extension zone.

Figure 16 shows a representative vertical slice of the model after the final step. Because the progradation rate was slow, differential loading was marked, and almost all the source layer was squeezed distally into a confined space. The resulting source structure was bizarre. Its shape was controlled by the variation in age of the cover strips, which determined the degree of strain, and hence the final shape of each strip. The sequence of distortion can be seen by comparing the youngest, most distal, least deformed strip (T) with successively older, more proximal strips. Because strain was progressive, the proximal strips were the most distorted. The strips initially had rectangular vertical sections with dimensions of 6x10 mm.

nal normal strain is W-E change in length normal to prograding frontal tip. Proximal to (W of) heavy line is a zone of normal extension marking frontal wall squeezed up ahead of prograding wedge. Distal to (E of) the line is a zone of normal shortening compressed by frontal wall (facing page).

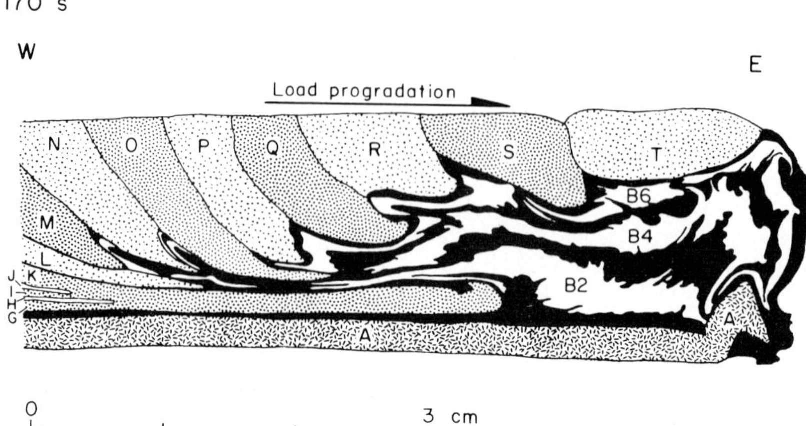

Figure 16. Representative vertical slice from final step of prograding model 840412. Detail within source layer B is shown by zebra-striped multilayers. Strips G through T are prograding cover.

The strips were successively distorted into listric sheets tapering downward and distally. The point of each tapering wedge of cover corresponded to the lower, distal corner of the rectangle before distortion. Arc lengths of these curved sheets increased to about 45 mm; most of this extension was taken up in the deep, gently dipping parts of each distorted sheet. Displacement vectors (not shown) indicate that the upper parts of the cover strips flowed proximally, whereas the lower parts of the older strips and the source layer flowed distally. These displacements indicate a movement cell on the scale of the entire model.

A series of appendages on the upper surface of the source layer changed proximally from triangular to hornlike as they matured by strain. Each appendage marked the junc tion of prograding strips, where minute flanges of the frontal wall of source spread proximally over distal tips of foundering cover strips. Each flange initiated growth of an appendage. At the top of the movement cell, the appendages and intervening sheets of cover deformed into curved horns by simple shear during distal laminar flow. The oldest, most proximal horn formed a recumbent isocline overlying a broad extent of flat-lying, foundered, and stretched cover (strips G through K). All the horns verged toward the oncoming prograding wedge.

3. Model 840413

The geometry and composition of model 840413 differed only slightly from the previous model, but its ratio of progradation rate to diapir rise rate was four times greater, and similar to the prototype ratio (Table I). Figure 17 shows growth stages at steps 5, 8, 11, 19, and 20. Features typical of previous prograding models were the transfer of source material from beneath the foundering, prograding wedge, the rise of the free surface of the source layer, and the distortion of cover strips into listric sheets.

However, several aspects of this model were different from the preceding model. First, the surface strains, which were monitored in the same way but are not reproduced here, were much lower because the *P/R* ratio was much greater; diapiric strains were low until the final step (20) after progradation had ceased. By step 3 the frontal wall was marked by a zone of normal extension only 1-2 mm wide; in all the later stages an extension zone was invisible at surface, though it might have been buried beneath the prograding frontal tips. The zone of normal shortening propagated ahead of the frontal wall; it increased in width and degree of shortening and reached the distal boundary by step 8.

Second, the cover strips do not show a regular proximal increase in strain; for example, strip S in step 19 is more strained than the oldest strips because of the rapid rise of the frontal wall in a confined space. Throughout much of the step 19

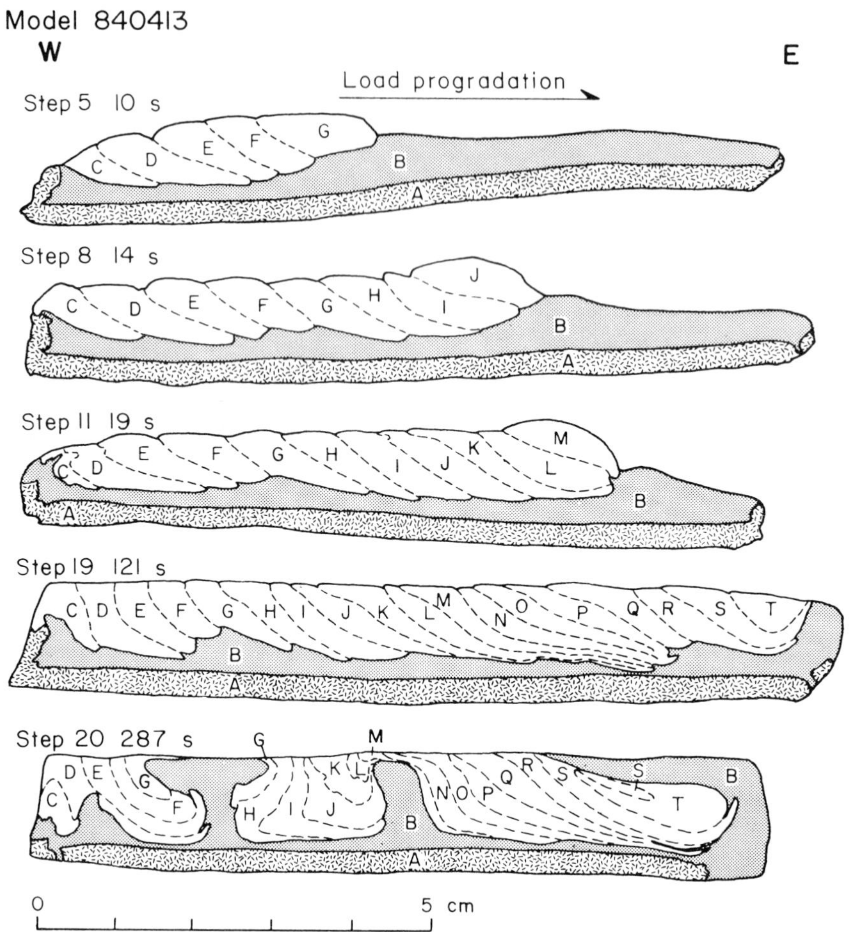

Figure 17. Step vertical slices of prograding model 840413. Source layer is B.

slice there appears to be a correlation between thickness of source layer and strain intensity of the cover strips: the most deformed strips overlie the thinnest, most drained, source.

By step 19 the top of the source layer showed peripheral walls and low irregularities at the junction of each cover strip. A 20th centrifuge run allowed structures to mature from some of these initiators (Fig. 18). Four parallel walls formed on a

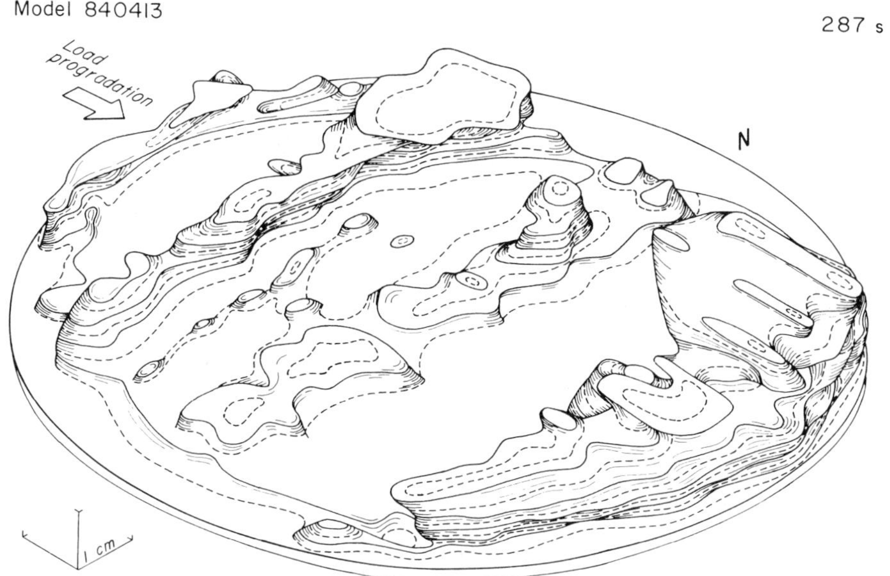

Figure 18. Isometric block diagram of upper surface of source layer in final step of model 840413 showing walls formed by progradation from W to E. Cover not shown. Data derived from 32 serial vertical slices. Solid contour interval is 2.1 mm.

dominant wavelength, trending normal to the direction of progradation. The dynamic control of the prograding differential load was so strong that the normally powerful edge effect was subdued in the north and south but amplified in the west and east, producing a large overhang on the distal side (Fig. 18).

Figure 19 summarizes mass transfer of source material as model 840413 evolved. Up to step 8 the source layer flowed distally away from the prograding wedge. Then flow reversed, ultimately completely draining the distal reservoir of extra source, so that by step 19 balance was restored. No interior walls had formed by step 19, so the reversal of flow cannot be ascribed to tapping of the source reservoir by growing proximal

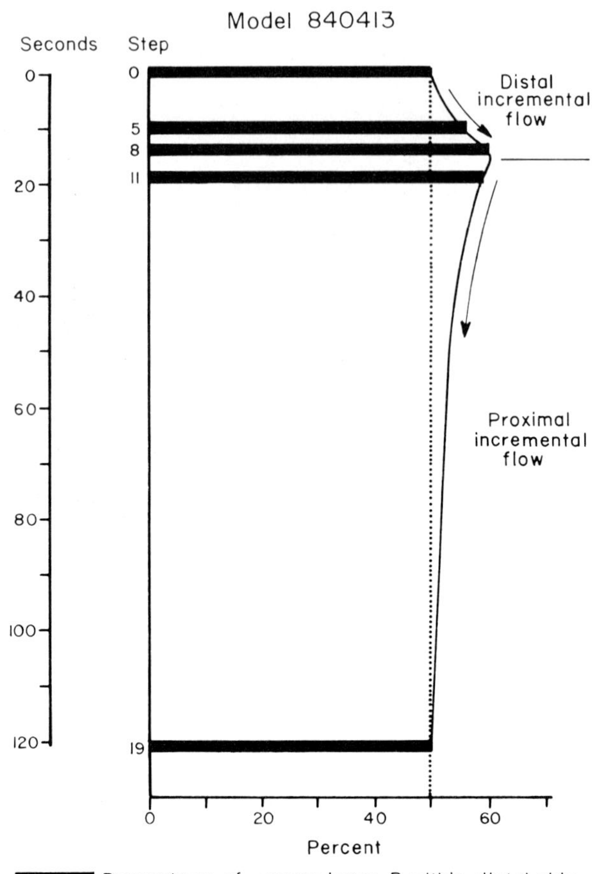

Figure 19. Mass-flow data for prograding model 840413, plotted against time. Bars show percentage of source layer in distal half of model, which was covered by prograding cover late in experiment. Incremental flow reversed in direction between steps 8 and 11.

walls (as was suggested for model 840328). The reversal resulted from progressive loading of the distal source layer by progradation.

Clustering of AAC axial traces and chords divided the model into six domains (Fig. 20b,c). The peripheral wall occupied domains 1 and 6. Because of the distal flow of source, the distal peripheral wall (domain 6) showed extreme development,

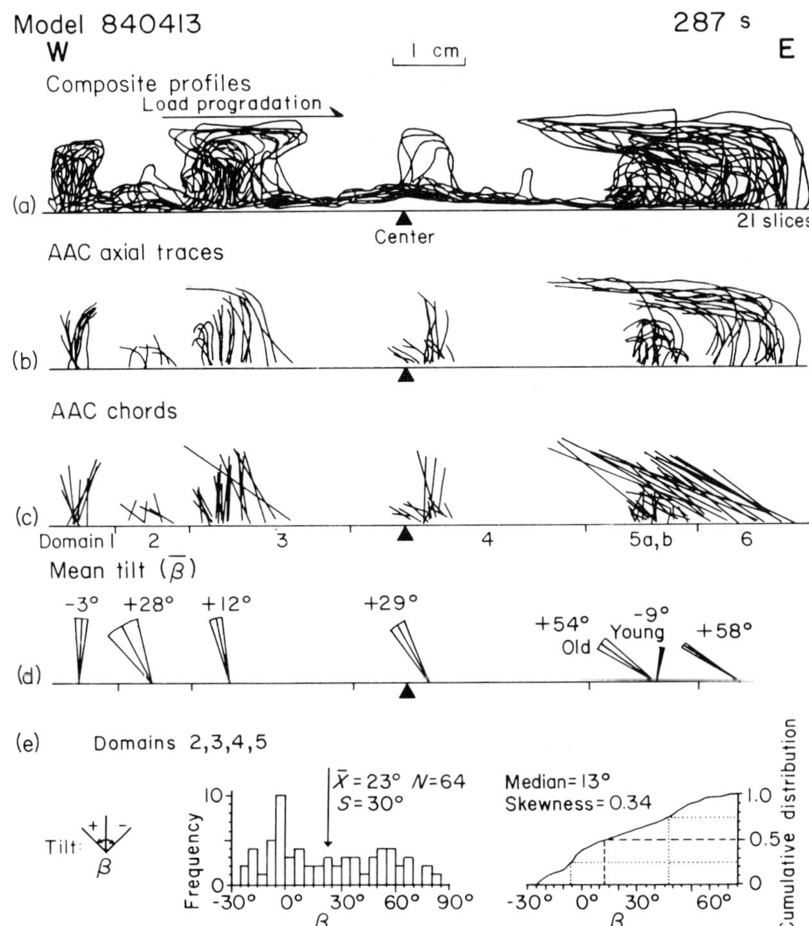

Figure 20. Tilt data for prograding model 840413. (a) Superimposed profiles of source-layer structures traced from 21 serial vertical slices. (b) Axial traces of antiformal anticlinal cores (AAC) of walls, defined by multilayers within source layer. These traces indicate tilt and asymmetry of structures. (c) Chords joining ends of axial traces in (b). (d) Six domains are defined on basis of clustering of structures in (a), (b), and (c). Elongated fan in each domain shows mean tilt (°) of AAC chords and standard error (= standard deviation/$\sqrt{N}$). Domain 5 is bimodal because of two generations of walls. (E) Histogram and cumulative distribution of tilt angles. N = number of axial traces; x = arithmetic mean; S = standard deviation; skewness was calculated on basis of upper and lower quartiles.

asymmetry, and tilt. Domain 5 contained the same type of double wall that was present in model 840328. The same comments apply: older walls were much more tilted than younger, lower walls because the former grew during active progradation. Walls were strongly tilted opposite to the direction of progradation; the distribution was broad and markedly skewed away from high β values (Fig. 20e). Proximal spreading of the extrusive bulb of the peripheral wall in domain 6 displaced cover both proximally and downward. However, data in Figure 20 do not support the argument that proximal displacement of cover tilted the interior walls by subhorizontal laminar flow. Ignoring the very low amplitude structures, which retain their primary asymmetry because of the shape of the prograding strips, the mature structures in domain 4 in the center actually lean toward the distal extrusion, whereas mature structures farther away in domain 3 lean in the opposite direction.

V. PROGRADING DIFFERENTIAL LOAD OVER BOUNDARY OF TABULAR SOURCE LAYER

We know of no previous experiments simulating the effects of loads prograding over the edge of a salt basin, away from the basin. Such effects are presumably minimal if the source layer wedges out very gradually because, despite thickening by outward flow of displaced source material from the basin center, the source would be too thin to generate diapiric structures at its margin. However, a salt margin could be abrupt if bounded by a syndepositional or, especially, postdepositional fault. The following two models show that the effects of progradation over an abrupt boundary are considerable.

A. Initial Configuration of Models

Each model contained a basement step defined by the rigid, dense layer A. The basement step confined a half-layer B of source material in one half of the model (Fig. 21); the other

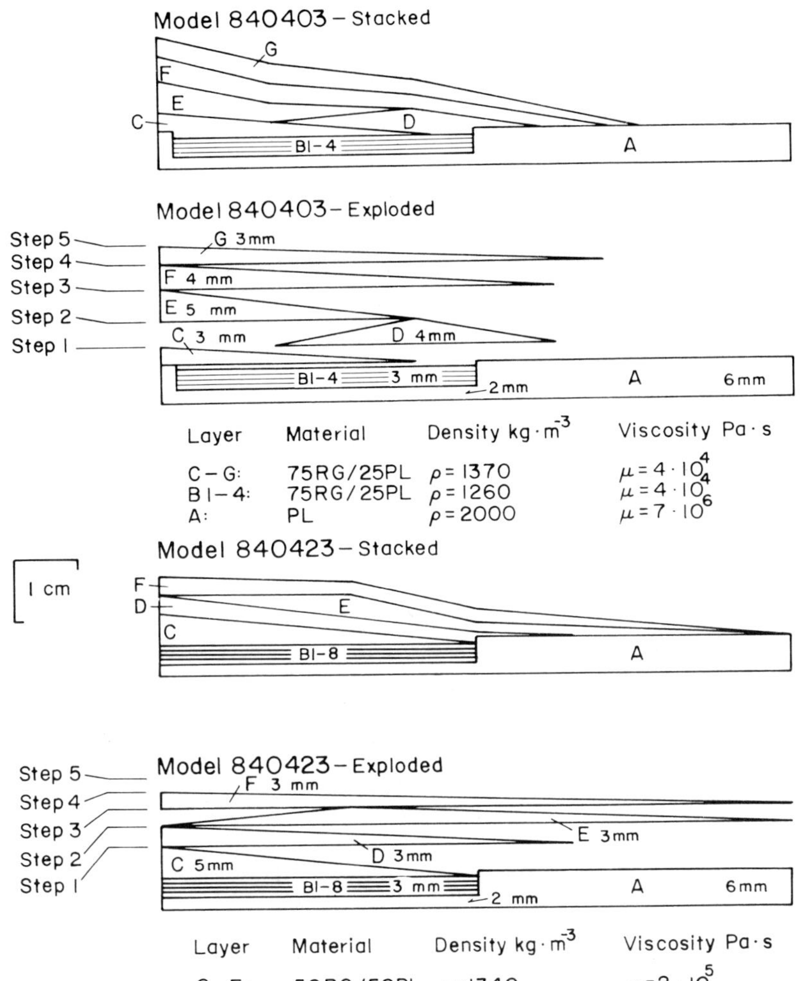

Figure 21. Initial configuration of prograding models 840403 and 840423 in vertical section before acceleration in centrifuge. Exploded views show actual shape of prograding strips added incrementally. Stacked views show hypothetical model geometry assuming no lateral flow. In actual experiment, each cover increment was added to an almost flat surface, which was an equilibrium profile after centrifuging.

half was originally bare. The models thus simulated a salt basin bounded by a linear basement high. Both models featured wide strips of cover that prograded across the source layer and its abrupt margin in the model center. Differential loading was more extreme in model 840403 than in model 840423, as indicated in the stacked profiles. In model 840403, the source (scaled density = 2,200 kg m^{-3}) and cover (scaled density = 2,390 kg m^{-3}) had the same viscosity, and both were Newtonian. In model 840423, the cover (scaled density = 2,520 kg m^{-3}) had five times the viscosity and deformed by power-law creep. Model 840403 had five increments of cover centrifuged in five steps. Model 840423 had four increments of cover centrifuged in five steps.

B. Experimental Results

1. Model 840403

The evolution of model 840403 is shown in step slices (Fig. 22). After the first cover strip was added in step 1, the source layer thinned beneath the load and thickened beneath the free surface because of distal flow. The free surface of the source rose above the basement step, causing an extrusive flange of source to spread over the lip of the plateau. The flange was buried by strip D in step 2 but continued to extend distally as a tongue. By step 3 this tongue had spread laterally to form a broad, isoclinal recumbent wall from which appendages branched. By steps 4 and 5 the recumbent wall had spread almost completely across the basement plateau (Fig. 22). All structures were asymmetrical in the distal direction. Asymmetry varied from slight extra growth of the distal side of overhanging bulbs to the extreme form shown in Figure 23, in which second-order diapirs streamed distally as if entrained by a current. Diapirs were also increasingly tilted in the same direction. In step 5 the structure just proximal to the basement step had tilted so much that its recumbent crest overhung the step (Fig. 22). Further maturation would have caused this to evolve into a second or third recumbent tongue above the principal recumbent wall (Fig. 24, slices 2 and 3). In slices not intersecting the lower parts

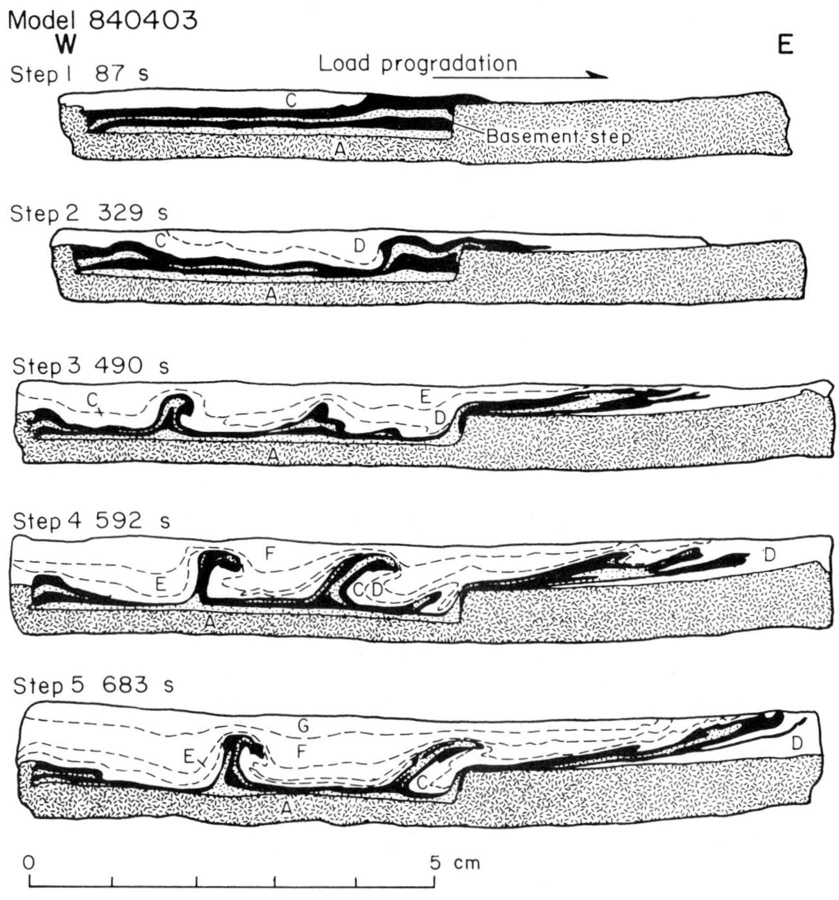

Figure 22. Step vertical slices of prograding model 840403. Multilayers in source layer B shown as zebra stripes. Strips C through G are prograding cover. Source layer and unit D on right define a recumbent isoclinal wall.

of the higher tongue connecting it to the source layer, the tongues appeared as "floating" detached sheets (Fig. 24, slice 4).

Unlike those in model 840412, these appendages were not nucleated by the junctions of cover strips; they formed spontaneously by amplification of small instabilities, some of which were tipped by a barb pointing distally. The appendages matured by strain into elongated triangles pointing distally in the flow direction, and finally formed long worms (Fig. 25), which were actually recumbent sheetlike isoclines in three dimensions. Cusps of similar appearance form on the margins of shortened, relatively incompetent dikes owing to viscosity contrasts with the country rock (Talbot, 1982). These cusps can reach flamelike proportions with very high strains (Jackson and Zelt, 1984). However, there is no viscosity contrast between source and cover in model 840403, and presumably the irregularities there result from Rayleigh-Taylor instability. Regardless of the mechanics of their initiation, these flamelike appendages are clearly features of high strain that evolved as flow folds by subhorizontal differential laminar flow across their initially upright axial surfaces.

Figure 26 shows successively deeper horizontal serial sections through half of the final model. The uppermost slice 2 shows the distal (eastern) front of the recumbent wall embedded in cover strip D. Because of drag by the sides of the centrifuge cup, the trace of the wall was arcuate. Proximal to the

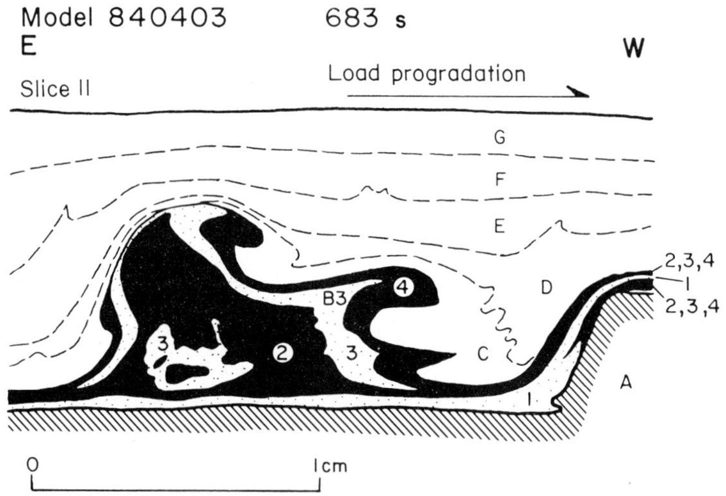

Figure 23. Extreme asymmetry of finger and distal appendages adjacent to basement step in vertical slice of prograding model 840403. Multilayers in source layer B shown as zebra stripes.

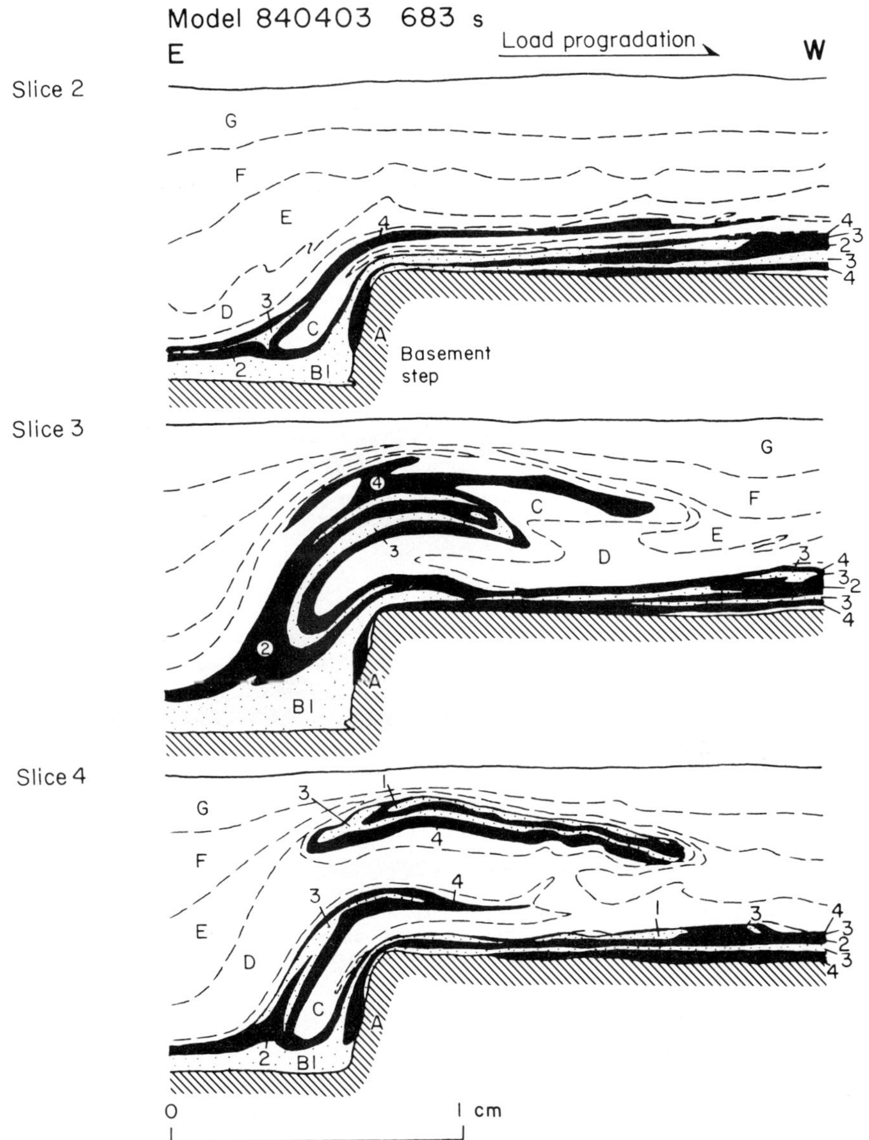

Figure 24. Second-order structures formed over the basement step in vertical slices of prograding model 840403. Multilayers in source layer B shown as zebra stripes.

recumbent wall in slice 2 are horizontal sections of several fingers. Three fingers partially or completely draped over the basement plateau, although rooted on the other side of the

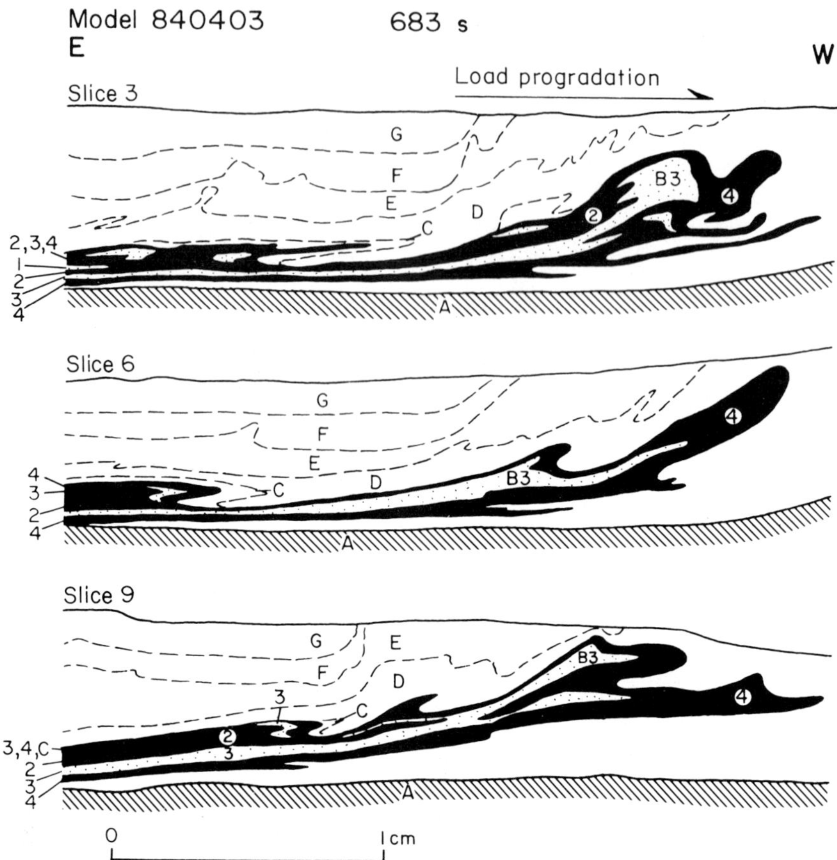

Figure 25. Second-order structures formed at leading edge of principal recumbent wall in vertical slices of prograding model 840403 after final step. Multilayers in source layer B shown as zebra stripes.

basement step, thus illustrating extreme tilt. The three fingers and their envelope of cover units C and D were highly elliptical in horizontal section. This could be due to distortion during distal flow (as monitored by surface strains and geometric considerations) or to the geometric effect of a horizontal section through a gently plunging cylinder with a circular cross section. Tilt lessened proximally, so the proximal fingers were not elliptical in slice 2.

The deepest horizontal slice 4 shows cellular walls of source below the fingers (Fig. 26). The cells were equidi-

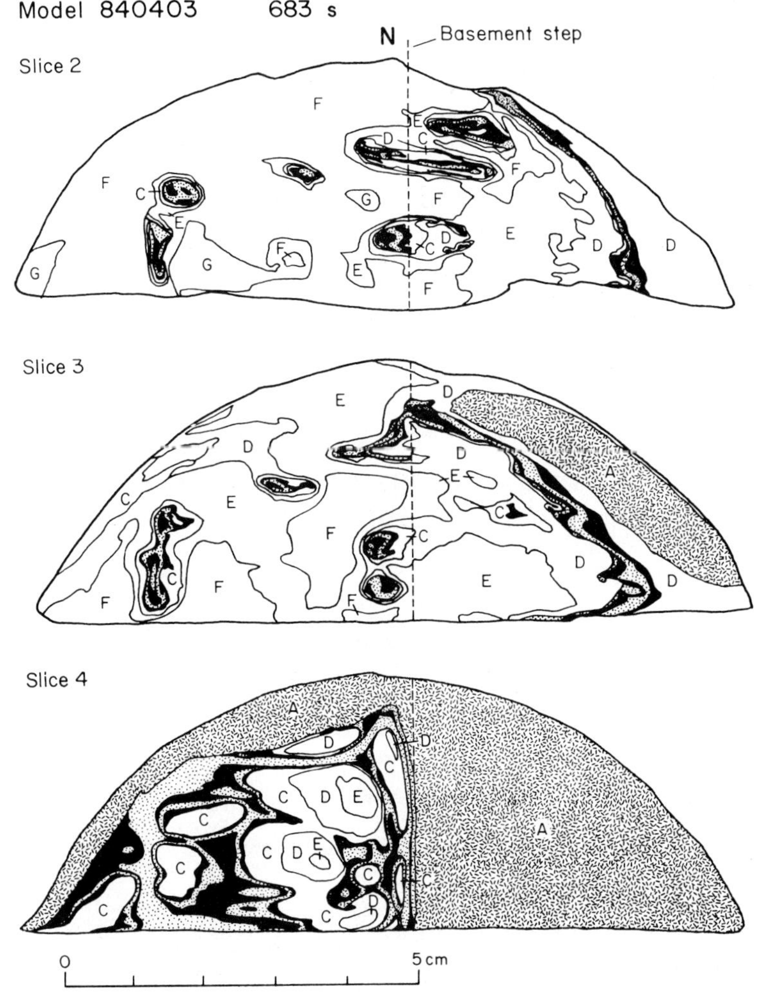

Figure 26. Serial horizontal slices of prograding model 840403 after final step. Multilayers in source layer B shown as zebra stripes. Slice 2 is highest; slice 4 is lowest. Arcuate source material on right is recumbent wall shown in Figures 22 and 25.

mensional away from the model sides, but toward the basement step they were successively flattened to ellipses against the step, ultimately forming a single straight wall that represented the root of the principal recumbent wall. We refer to the strong control of the basement step as a basement-step effect.

The geometry of the horizontal slices (Fig. 26), coupled with that of the serial vertical slices (not shown), indicates that over the original basin a mixture of fingers and walls formed. One of the walls was the usual peripheral wall; the other was a mature frontal wall along the proximal edge of strip D, which originally lay halfway across the source layer. This proximal frontal wall was unrelated to the basement step. Over the basement step, and strongly controlled by it, the source structure was a steep wall that became tipped on its side to form the principal recumbent wall overhanging the basement plateau.

2. Model 840423

Model 840423 had a stiffer, more regularly prograding cover than that of the previously described model 840403 (Fig. 21). The evolution of model 840423 was generally similar to the previous model, and we comment only on different features (Fig. 27): the extent of lateral flow, the style of the appendages, and the mode of deformation of the cover.

The principal recumbent wall in model 840423 began to develop at a similar rate to that in model 840403 (compare similar times in Figs. 27 and 22). But the lateral extent of the recumbent wall in Figure 27 after 501 s was much less than that of the recumbent wall in Figure 22 after 490 s. In contrast, the vertical relief of upright fingers and walls at comparable stages of each model was similar. These observations suggest that sideward growth of the 840423 recumbent wall was stunted not by the greater viscosity of the cover (which would have also stunted vertical growth) but by the milder differential loading, as shown in the stacked profiles of Figure 21.

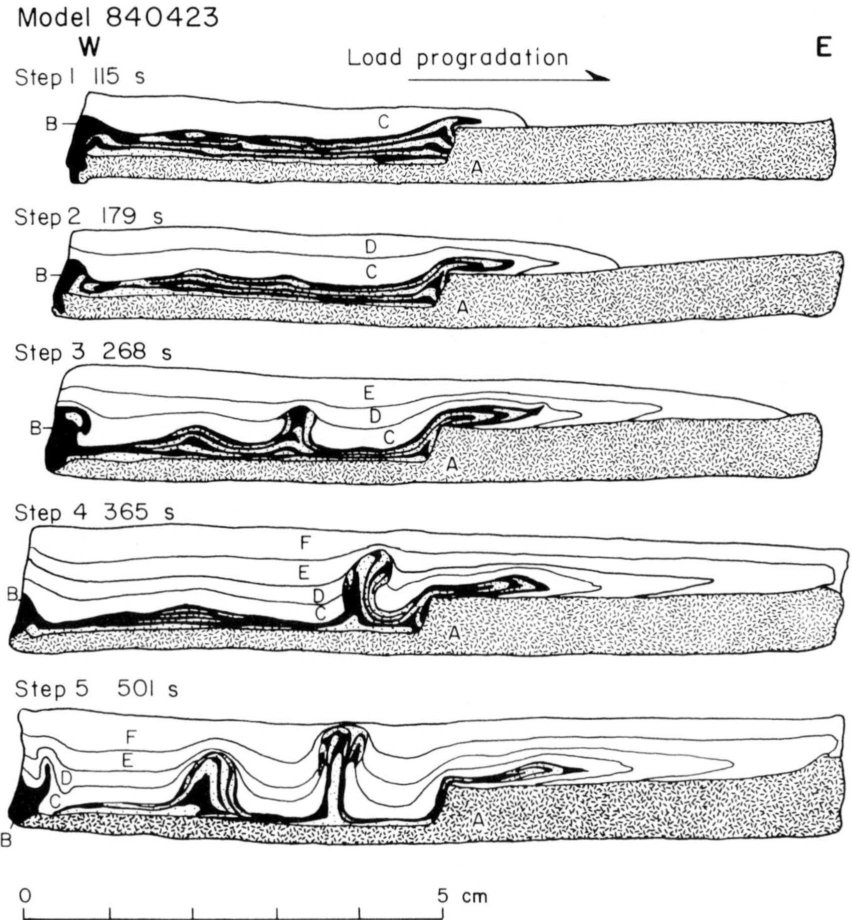

Figure 27. Step vertical slices of prograding model 840423. Multilayers in source layer B shown in zebra stripes. Strips C through F represent prograding cover.

Appendages on the recumbent wall were scarce in model 840423, although the model incorporated a viscosity contrast between source and cover unlike model 840403. Instead the recumbent wall was simply folded during steps 1 through 4 (Fig. 27). By step 5, smooth undulations formed in the upper surface (the non-inverted limb) of the recumbent wall. A second-order upright anticline invariably deformed the recumbent wall just above the basement step where the vertical source wall rotated to the horizontal (Fig. 28, all slices). Another second-order upright anticline generally formed in the more distal parts of the recumbent wall, either as a gentle fold or as a large, mature finger (Fig. 28, slices 13 and 3′). A pointed cusp was common on the leading tip of the recumbent wall, but where the wall was laterally short, the pointed tip was replaced by a balloonlike, mature structure representing a section through a secondary peripheral wall, an edge effect (Fig. 28, slice 16).

These results, together with those of model 840403, show that once formed, a recumbent wall can act like a horizontal source layer that initiates the growth of fingers or walls on its upper surface. As such it acts as a recycled allochthonous source layer perched high in the stratigraphic pile.

Model 840423 also illustrates the role of the cover during the gravity overturn and lateral flow of the system (the same mechanism operated in model 840403, but was less clearly demonstrated). Figure 27 shows that a recumbent sheath of cover lay ahead of and enclosed the recumbent source wall. All cover below the level of the front tip of the recumbent wall was overturned and younged downward. Kinematically, the folded cover represented a recumbent _growth anticline_. Cover units were added during its growth from steps 1 through 4; they therefore showed decreasing strain with decreasing age. The cover isocline was _open-cast_, in that it formed at surface and produced surface relief as it propagated ahead of the recumbent source wall.

In the overturned limb of the recumbent wall, all cover appeared to pinch out against the basement plateau (Fig. 27). The sites of apparent pinching out represent the original points of the cover wedges (Fig. 21). The part of the cover wedges that extended beyond the front of the recumbent source wall adhered to the basement plateau. During subsequent lateral flow of the composite structure, the adhered part of each cover wedge

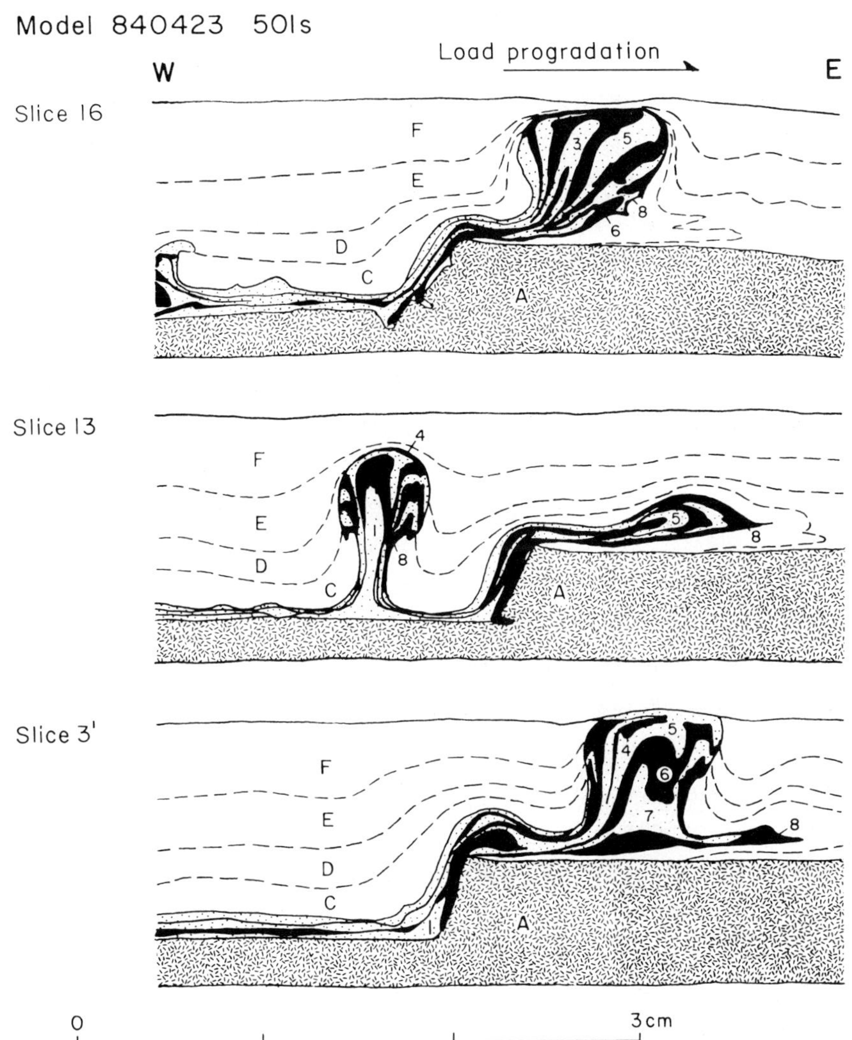

Figure 28. Vertical slices of final step in prograding model 840423. Multilayers in source layer B shown as zebra stripes. Strips C through F are prograding increments of cover.

remained anchored to the basement while the rest of the structure rolled overhead like a tank track. The motivating force for the lateral flow of both source and cover over the basement plateau is the shear stress exerted by a surface slope in a distal direction, well known to glaciologists (Paterson, 1981). This surface slope was built into the models from the onset (Fig. 21) to simulate a continental slope. Clearly, the lateral flow of the recumbent source wall in model 840423 was accomplished entirely by lateral spreading of cover, which rolled over its anchored parts. The source wall was carried along as a passive core. With free slip along the basement plateau, both the cover and source layer would have spread like a sliding thrust sheet or floating ice shelf.

Because of its original geometry, only cover strip D encloses the recumbent source wall in model 840403. Here the source wall spread faster than the enclosing cover. By step 2 (Fig. 22), strip D lay 19 mm ahead of the source basin edge; by step 5 this distance had shrunk to 2 mm. Because of the confining distal boundary of the centrifuge cup and because the system was overturning, cover unit D flowed beneath the leading edge of the recumbent source wall. The space vacated allowed the source tongue to rise and spread farther toward the upper, distal corner of the model. Nevertheless, the basic mechanism was the same as in model 840423 because cover unit D also was inverted below the recumbent wall.

A feature shown by both models was the boundary-layer separation within the source material as it surmounted the basement step (Figs. 22, 23, 24, 27, and 28). The basal boundary layer of the laminar flow could not follow the angular basement surface, so it separated in front of the step and ramped over a zone of stagnant or proximal-flowing source material. Similar flow separation has been observed in a salt glacier surmounting bedrock ridges and may also initiate folds in source layers being drawn inward to feed salt diapirs (Talbot and Jackson, in review).

In conclusion, a viscous source layer can demonstrably flow laterally as a recumbent wall for distances of at least five times the height of vertical fingers in the same system. During this process much of the source material can leave the confines of the original basin. Requirements for this flow are a surface slope and a cover that is effectively viscous or pseudoplastic and can spread under its own weight. This lateral squeezing of the source is similar to that described in previous prograding models where a frontal wall formed distal to a migrating depocenter and could migrate laterally like a ripple. But in models 840403 and 840423, the frontal wall was rooted at the basement step. It could only move sideways ahead of the prograding wedge by surmounting the step and rotating to form a recumbent wall. The wall then acted as a recycled, horizontal source layer for a second generation of steeper diapiric structures.

VI. DISCUSSION

A. Formation of Salt Walls in Nature

Our models show how source walls can be initiated as low-amplitude swells in front of a prograding load. If these frontal swells become overridden by the prograding load, they act as initiators for the growth of high-amplitude walls. The planform of these walls is directly controlled by the planform of the sedimentary depocenter that initiated them. Thick source layers tend to produce large diapirs that mirror the shape of the overall delta, whereas thinner source layers produce smaller diapirs that reflect smaller-scale morphology within the depocenter, such as channels or channel-mouth bars. Diapirs would be aligned concentrically or in a radial fan in fluvial-dominated and tide-dominated deltas, or parallel to the general shoreline in wave-dominated deltas. Such an origin has been suggested by Watkins et al. (1978) for Plio-Pleistocene elongated salt structures fronting the Mississippi fan (Fig. 29). Our models confirm that these salt walls probably originated as frontal walls pushed up by a prograding depocenter. However, there is no evidence that they formed by the coalescence of fingers, as thought by Watkins et al. (1978). Rather, fingers invariably evolve from walls instead of vice versa in all our experiments. This hypothesis of differential loading applies to the prototype Louann salt walls in the northern Gulf of Mexico. Salt walls elsewhere, such as the Zechstein walls in the North Sea and adjoining northwest Europe, may well owe their origin to subsalt faulting rather than differential loading because there the salt rests on a rugged, faulted basement, and major deltas were absent and sedimentation rates were much lower during diapirism.

B. Sigsbee Nappe Complex

Perhaps the least understood type of major salt structure is the Sigsbee nappe complex on the northern continental slope of the Gulf of Mexico (Fig. 30). The complex consists of many allochthonous sheets of Jurassic Louann Salt overlying and enclosed by Pleistocene and older strata.

Published descriptions have focused on the upper surface and leading edge of these allochthonous salt sheets because they are well imaged in reflection seismic profiles (DeJong, 1968; Amery, 1969; Lehner, 1969; Watkins et al., 1975, 1978; Buffler and Worzel, 1978; Humphris, 1979; Martin, 1980). The upper surfaces of the sheets vary from near planar to distorted into high-amplitude fingers and walls with conical vertical sections and irregular planforms.

The leading edge of salt underlies a steepening of the base of the continental slope, known as the Sigsbee Escarpment. Here the salt nappes wedge out seaward to thicknesses of a few meters. Reflectors in the Pleistocene are truncated by the base of the salt nappes, which dip gently landward toward the Cenozoic depocenter. This suggests that the nappes rest on thrust surfaces, narrow shear zones, or shallow-dipping intrusive contacts. Uncertainty exists because strong reflections from the

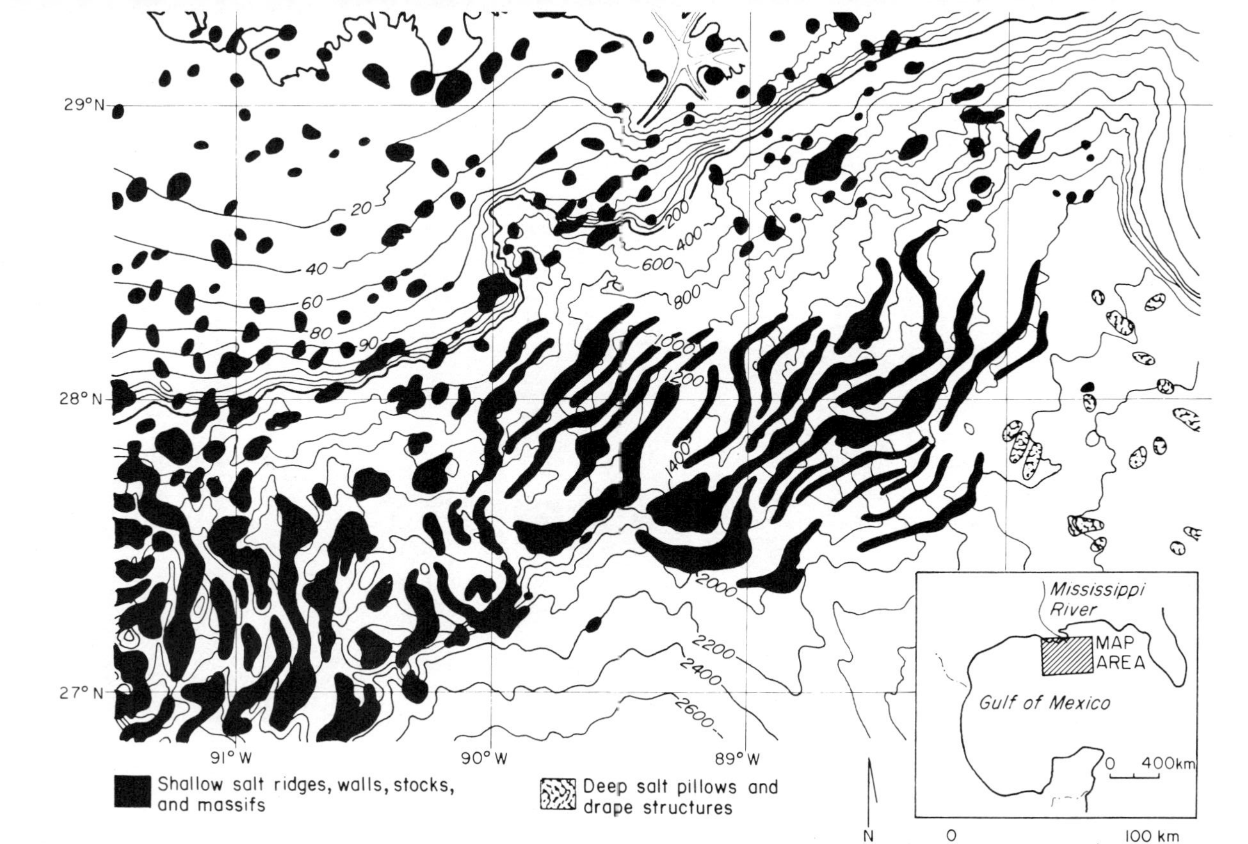

Figure 29. Elongated salt structures resulting from differential loading by progradation of distal Missisippi fan. The most seaward are non-diapiric salt anticlines; more landward structures are diapiric walls. On continental shelf, older salt diapirs have evolved to mature fingers with elliptical plan sections. Bathymetry in meters. (Adapted from Martin, 1980).

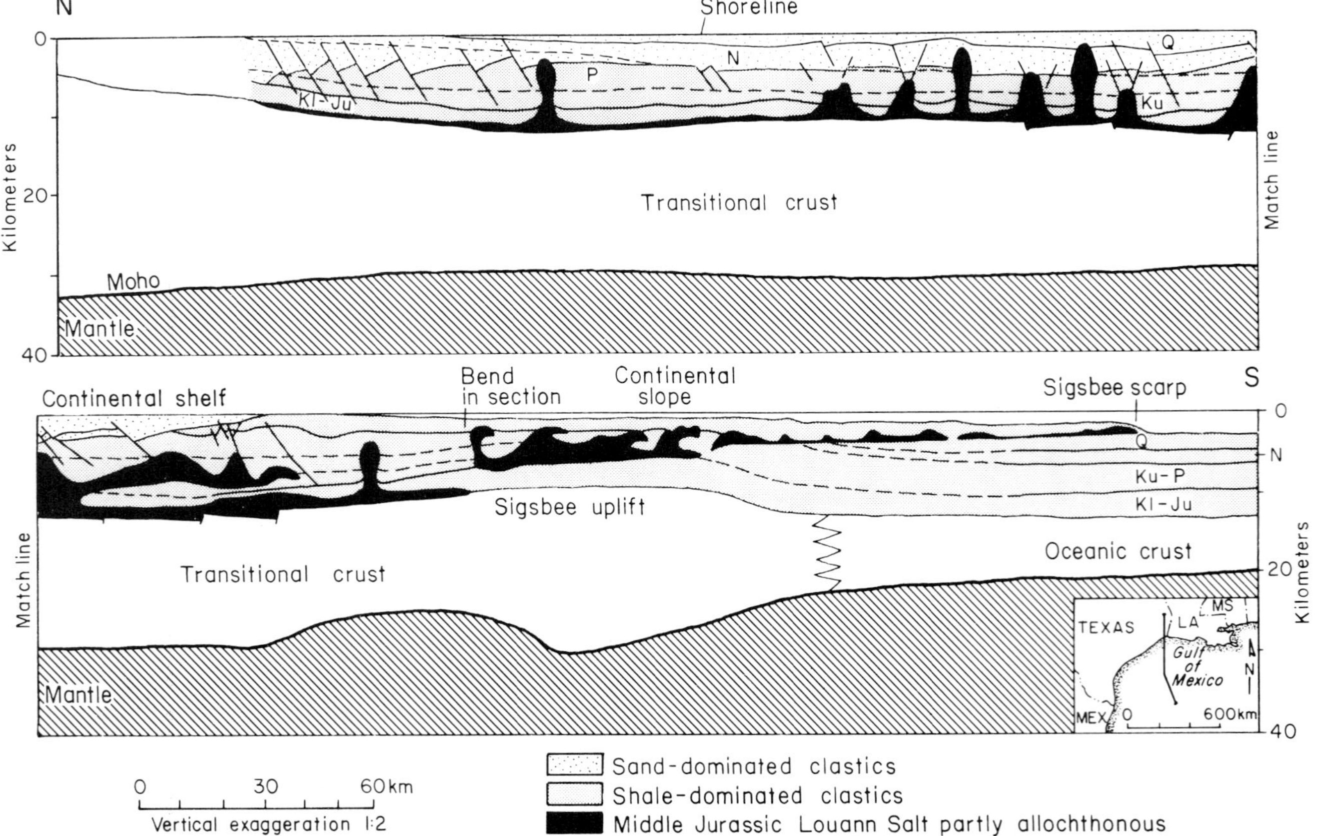

Figure 30. Vertical cross section across crust underlying the Gulf Coast coastal plain, continental shelf, slope, and abyssal plain in northern Gulf of Mexico. Inset shows line of section. Ju = Jurassic; Kl = Lower Cretaceous; Ku = Upper Cretaceous; P = Paleogene; N = Neogene; Q = Quaternary. (After Buffler et al., 1986).

generally shallow-dipping upper and lower contacts of the salt nappes return so much seismic energy that subsalt reflections are weak. Estimates for the subsalt extent of these basal discontinuities thus vary widely from 10-15 km (Martin, 1980) to as much as 170 km (Fig. 30). The latter estimate assumes that the entire southward bulge of the Sigsbee Escarpment is due to lateral intrusion of the nappe complex, a hypothesis supported by palinspastic plate reconstructions and subsidence modeling (Buffler and Sawyer, 1985).

The prevailing kinematic interpretation of the nappe complex appears to be that throughout most of their growth history these nappes remained close to the surface, moving ahead of the prograding continental slope. The usual mechanical explanation for movement is that rapid Cenozoic deposition squeezed underlying salt basinward in a series of laterally intruding tonguelike nappes (commonly described as extrusive, which incorrectly implies flow over the seafloor). During this lateral flow, the salt structures also climbed up stratigraphic section, so that they came to be underlain by much younger sediments (Fig. 30).

This mechanical hypothesis has only been described briefly and qualitatively in the literature, but clearly there is more to the process than merely lateral squeezing of salt. Additional features are illustrated by two of our models. Models 840403 and 840423 simulated the effects of a load prograding across and beyond the edge of a salt basin confined by a basement step. The prototype basement step is the Sigsbee Uplift (Buffler, 1984) below the northern Gulf of Mexico continental slope (Fig. 30). This uplift is probably a stranded horst or series of horsts of extended continental basement. Significantly it is thought to mark the seaward limit of the original Louann salt basin; ocean crust devoid of Louann Salt lies beyond.

Remarkable geometric similarities exist between the Sigsbee nappe complex and model 840403. The principal recumbent wall in the model simulates one of the many tonguelike salt nappes. Allochthonous, seemingly detached sheets of salt visible on seismic profiles have their counterpart in the higher, second-order recumbent walls in model 840403. The appendages also decorate the upper surface of the Sigsbee structures, where they are known as salt wedges. Similarly, the diapirs that corrugate the upper surface of the Sigsbee structures closely resemble the fingers of both models.

A major difference between models and prototypes is that in the Sigsbee nappe complex there is no obvious evidence of wholesale inversion of the Cenozoic and Mesozoic cover below the salt nappes. Even though our knowledge of the structure in this region is scanty, it seems intrinsically unlikely that Cenozoic strata of an equivalently large volume could be inverted. Nevertheless, we suggest that in the lower parts of the Sigsbee structures, strongly foliated basal salt overlies a thin zone of sheared, inverted, clastic sediments. On seismic profiles, this thin, basal shear zone with a thrust sense of movement would truncate reflectors of younger strata, as is known to be the case (Fig. 30). Such inversion would be difficult to recognize unless microfossils survive the penetrative strain. This prediction might one day be tested by drilling during exploration of the Sigsbee nappe complex, one of the largest known salt-related petroleum traps.

C. Controls on Tilt and Asymmetry

Because Rayleigh-Taylor instability is driven by gravity, any systems that are initially asymmetrical (in regard to orientation, thickness, density, effective viscosity, or yield stress) with respect to the pull of gravity will develop asymmetrical or tilted structures during overturn. Figure 31 summarizes schematic configurations known or thought to produce

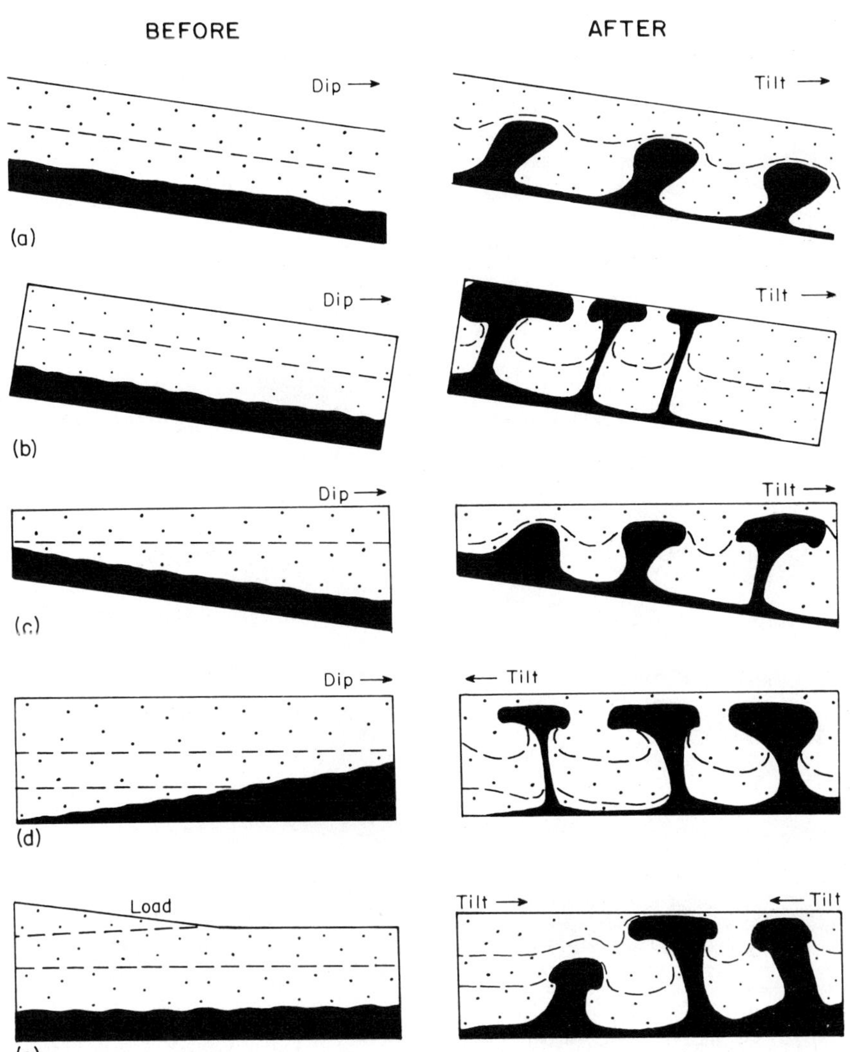

Figure 31. Schematic cross sections showing effects of various influences on tilt and asymmetry of diapirs formed from configurations shown in left columns. Source layer is black. (a) Source-dip effect without lateral boundaries (based on theoretical open-channel laminar flow). (b) Source-dip effect with lateral boundaries (based on oil-and-syrup 15° and 20° models of Talbot, 1977). (c) Cover-wedge effect (based on centrifuge model 9 of Talbot, (1977)). (d) Source-wedge effect (based on centrifuge model 20 and oil-and-syrup 7.5° model of Talbot, 1974). (e) Half-layer

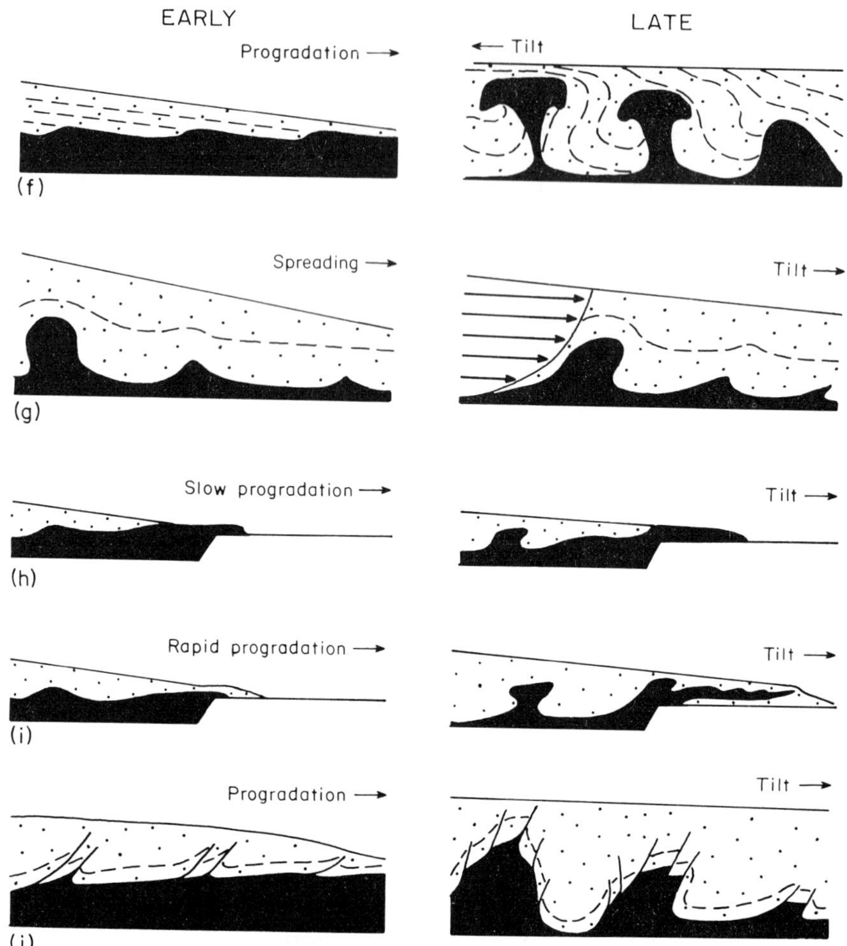

Figure 31 (continued) effect (based on model 840413).
(f) Inherited-asymmetry effect (based on models 840328 and 840413). (g) Gravity-spreading effect (based on theory and models by Ramberg, 1981, p. 195-226). (h and i) Progradation across a basin edge (based on models 840403 and 840423). (j) Combination of (f) and (g), (based on the Mississippi mudlumps of Morgan et al., 1968).

tilted or asymmetrical fingers or walls. Figures 31a-e apply to static or aggrading loads, whereas Figures 31f-j apply to prograding or spreading loads; with the detriment of slight inaccuracy, these contrasting conditions can be succinctly referred to as static loading and dynamic loading, respectively. Figure 31a is bounded laterally by hypothetical surfaces through which fluids of infinite volume can flow; Figures 31b-e have finite and constant volume; in Figures 31f-j, the volume of cover generally increases by deposition, and the configurations shown are representative "windows" of much larger systems.

For brevity we refer to all such structures of source material as diapirs. In each case a density inversion is present; that is, a dense cover overlies a less dense source layer. Tilt refers to the deflection from vertical (angle β) of the diapir stem in general and the axial trace of the AAC (antiformal anticlinal core defined by Jackson and Talbot, 1985) in particular. The tilt direction is the azimuth of displacement of the upper part of the diapir relative to th lower part. Asymmetrical diapirs are those with a bulb unequally developed on each side. A diapir with downdip asymmetry has most of the bulb volume on the downdip side.

1. Dipping tabular source/dipping tabular cover

According to fluid statics, where both source and cover are perfectly tabular and are unrestrained by lateral boundaries, a mobile source layer and its cover tend to flow downdip (down the elevation-head gradient), regardless of whether or not a density inversion is present (Fig. 31a, before). Diapirs form if initial perturbations of sufficient size are present. Laminar open-channel flow of both source and cover in a downdip direction shears the diapirs into asymmetry. However, in a closed, tilted box of finite volume (Fig. 31b, before), the source layer is displaced updip by downdip flow of the denser cover (Talbot, 1977). This circulation, comprising flow of the source and cover in opposite directions on the scale of the whole box was noted by Talbot (1977) and is termed first-order circulation (although both source and cover also flowed in Fig. 31a, flow was in the same direction so did not constitute circulation). Second-order circulation on the scale of individual diapir movement cells (three of which are present but not specifically shown in Fig. 31b) also takes place, as in all diapirism. First-order circulation causes the volume of diapirs to increase greatly updip; in contrast diapir relief increase downdip because of the downdip thinning of the source layer (Fig. 31b, after). Further influence of first-order circulation is the fact that the diapir rising nearest the lower end of the tilted box developed farther away from the end as the tilt of the box increased and first-order circulation intensified (Talbot, 1977). Diapirs initially rose perpendicularly to the tilted source layer, rather than vertically (Talbot, 1977). The reasons for this are not fully understood and may be connected with surface tension or first-order circulation. Regardless of the cause, we refer to the tendency of diapirs to grow perpendicularly to a dipping source layer as the source-dip effect. On nearing the upper boundary, the diapirs changed from being tilted and symmetrical (an intermediate stage not illustrated here) to being tilted and asymmetrical, spreading preferentially and being tilted 10° further from the vertical than the normal to the initial tilted interface in a downdip direction because of first-order circulation (Fig. 31b, after). Natural salt flow may be a case intermediate between Figures 31a and 31b because it is neither composed of infinite volumes of source and cover (like Fig. 31a) nor generally constrained by vertical walls (like Fig. 31b).

2. Dipping tabular source/horizontal wedgelike cover

With the configuration shown in Figure 31c, fluid statics indicates that the source layer will flow updip, regardless of whether the density of the dense cover is uniform or increases with depth. This is because the decreasing thickness of wedgelike cover updip causes the pressure-head gradient to be up the dip of the source layer. In contrast, the elevation-head gradient is downdip but is subordinate to the opposite pressure-head gradient because of the density inversion. The net hydraulic-head gradient is therefore updip, the direction in which the source therefore flows. This flow and the compensating downdip return flow of cover constitute first-order circulation. Diapirs initiated as perturbations initially grow symmetrically and perpendicularly to the dipping source layer by the source-dip effect. First-order downdip flow near the top of the system (whether or not confining lateral boundaries are present) causes asymmetrical growth preferentially on the downdip sides of the diapirs. The second-order flow centered on each diapir is also asymmetrical, favoring the downdip sides which are preferentially fed by updip flow of the source. In addition, the greater thickness of cover downdip means that rates of diapir rise increase downdip (because differential hydrostatic pressures increase downdip) and that at maturity the amplitude of diapirs increases downdip (Talbot, 1977, model 9). We refer to this combination of factors as the cover-wedge effect.

3. Horizontal wedgelike source/horizontal wedgelike cover

Previously modeled at 1 *g* and by centrifuge to simulate buoyant, diverging continental margins (Talbot, 1974), the configuration in Figure 31d also produces tilt and asymmetry of diapirs downdip (in relation to the top of the source layer). The cover-wedge effect previously described must operate here--and also the source-dip effect in the early stages--encouraging faster rise in downdip diapirs. These effects are overwhelmed by the source-wedge effect. The greater updip thickness of the source layer encourages increased rates of diapir rise updip because rise rate is proportional to the square of the absolute thickness (Ramberg, 1981, p. 107). But this updip increase in rise rate is countered by a subordinate opposite effect: rise rate is also proportional to the ratio of cover thickness to source thickness, which decreases updip. The amplitude of

Talbot's (1974) mature diapirs was uniform because the thicker end of the source wedge collapsed, leveling the top of the source layer between diapirs (Fig. 31d). The stalks of downdip diapirs, however, were thinner because less source material was available.

The geometry shown in Figure 31d was that of the initial configuration of Talbot's (1974) experiments. But the same geometry was present in our model 840328 by steps 3 and 4 (Fig. 11); the wedges of source and cover were dynamically produced by displacement of source from beneath prograding cover layers. Thus the strong tilt and asymmetry of diapirs in model 840328 in a direction opposite to that of progradation (Fig. 14) appears to be a combination of at least three effects already noted: source-dip effect, cover-wedge effect, and source-wedge effect.

4. Horizontal tabular source/horizontal half-layer cover

The geometry in Figure 31e corresponds to that of model 840402 apart from the fact that in the model the half layer was of uniform thickness rather than being wedge-shaped. The model showed that all diapirs tilted inward toward the edge of the half-layer, which we term the half-layer effect. Inward tilt of diapirs on the unloaded side is explained by first-order circulation caused by outward squeezing of the source from beneath the differential load. Inward tilt of diapirs on the loaded side may be caused by second-order circulation locally enhanced at the edge of the half-layer, as previously discussed in Section III, C.

A smaller-scale effect of static differential loading is the production of mesoscopic load casts, well known to sedimentologists. The differential load of dense sand is a narrow trough or plug in three dimensions rather than a half-layer as in model 840402. Thus diapirs on each side of the sagging trough or plug tilt inward. The diapir cores are commonly flame structures produced by sudden upward escape of overpressured pore fluids (McKee and Goldberg, 1969). The flame structures curve around the edge of the bulbous foundering load.

D. Controls On Tilt And Asymmetry Under Dynamic Loading

1. Kinetic considerations during progradation

The prograding models showed how lateral migration of a differential load could squeeze source material from beneath the load into a frontal bulge of variable width distal to the frontal tip of cover (Figs. 11, 15, and 17). Systems involving diapirism through prograding cover require at least five parameters to describe their kinetics, the first three of which were introduced and documented in Section 11< B and Table I: (1) P, the rate of cover progradation, or lateral migration of the frontal tip of the prograding wedge; (2) A, the rate of cover aggradation, or vertical accumulation; (3) R, the rate of diapir rise; (4) B, the rate of lateral migration of the frontal bulge of source material; and (5) I, the width of the clinoform cover increment or sediment pulse on the time scale being considered, which can influence diapir spacing as discussed below. The first four parameters have dimensions of velocity and the fifth has the dimension of length. Relations between P, R, B and I are schematically shown in Figure 32.

Under uniform loading, diapirs are spaced according to the dominant wavelength, λ_d, a function mainly of thickness and viscosity ratios. This spacing is perturbed during differential loading. The experimental importance of I (the width of the prograding increment) is emphasized by Figure 11, where frontal bulges initiated three of the four mature diapirs. On time scales of <1 Ma, natural progradation increments are much smaller than final wavelength of mature salt diapirs because of the brief interval of deposition. Small increments accumulated over brief durations cannot therefore influence the final wavelength of diapirism. However, on time scales of 1-10 Ma, pulses of sedimentation are large enough and last long enough to influence

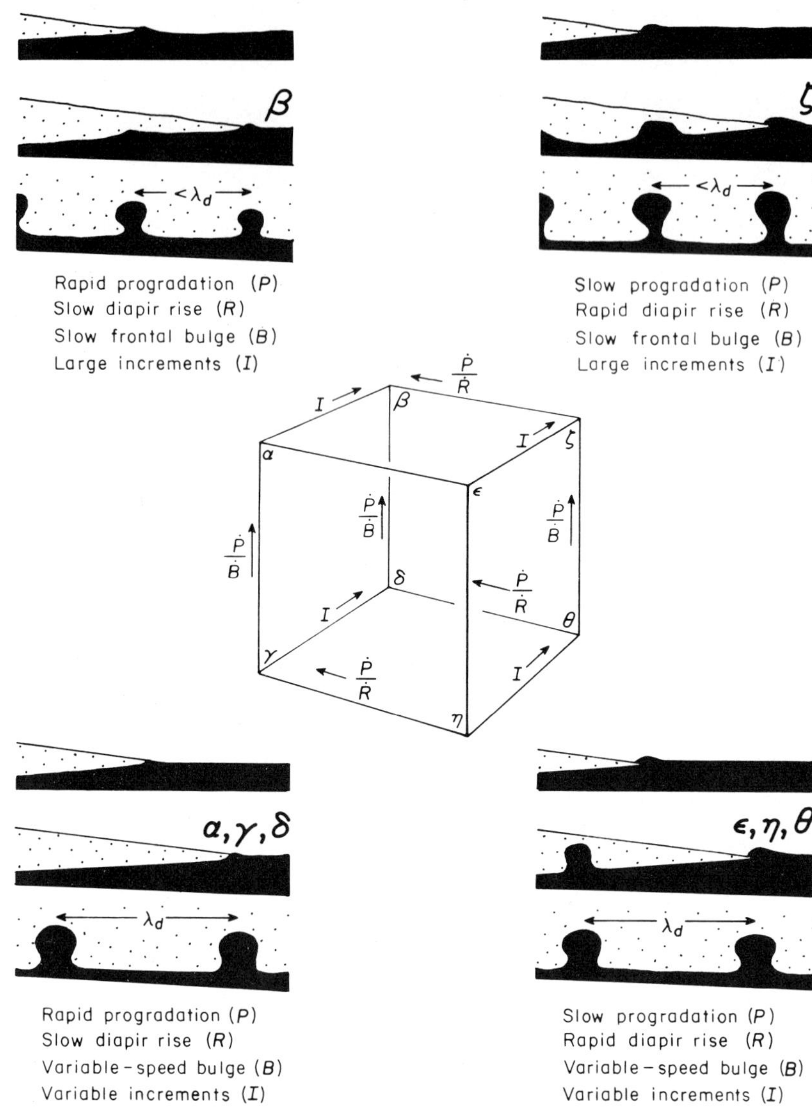

Figure 32. Schematic cartoon showing relations between cover progradation rate (*P*), migration rate of the frontal bulge of laterally squeezed source material (*R*), diapir rise rate (*R*), and the width of each increment of prograding cover (*I*). The eight corners of the cube, indicated by letters α through θ, correspond to different combinations of these four parameters, which have been selected for the convenience of illustration. The eight letters can be clustered in four principal kinematic groups shown in the sequential cross sections surrounding the cube.

diapirism. For example, in the northwest Gulf of Mexico major pulses with a mean frequency of 7 Ma and mean widths of about 50 km can be recognized in the Cenozoic sedimentary record (Winker, 1982; Jackson and Galloway, 1984). Where *I* is much greater than λ_d, the final wavelength is irregular, reflecting the combined influence of λ_d and the spacing of the initiating frontal bulges (e.g. Fig. 11). Where *I* is only slightly larger or smaller than λ_d, the influence of *I* can overwhelm that of λ_d. To illustrate this specific point in Figure 32, where *I* is referred to as "large" it is actually slightly smaller than λ_d, so the final diapir spacing is less than λ_d. Where *I* is much smaller than λ_d it has no influence on the final spacing, which is that of λ_d.

In nature and experiment, *P/R* ratios are generally greater than one (Table I). High *P/R* ratios mean that diapir growth is largely post-progradational, as in cases β and α, γ, δ; such diapirs would tend to develop on aggrading continental shelves landward of the continental slope where progradation is most active. Lower *P/R* ratios entail diapir growth through prograding cover, as in cases ζ and ϵ, η, θ, typically on the continental slope of a divergent margin.

A value of one for the *P/R* ratio signifies that the frontal bulge of source squeezed laterally by progradation advances at the same velocity as the frontal tip of cover, as in cases γ, δ and η, θ (Fig. 32). Because the frontal bulge is not buried during progradation, the size of the clinoform increment, *I*, is immaterial.

High values of *P/R* mean that the frontal bulge is overtaken and buried by progradation. Where *I* is large, it can influence the final diapir spacing, as in cases β and ζ. Where *I* is small, the resulting frontal bulge is negligible, so even though it is

Figure 32 caption continued. For the situations shown, *P/R* ranges from > 1 to >> 1, *P/B* ranges from 1 to >> 1, and *I* ranges from << λ_d to < λ_d where λ_d is the dominant wavelength.

buried, it does not influence the wavelength of the mature diapirs, so cases α and ϵ can be grouped with cases γ, δ and η, θ, respectively (Fig. 32).

2. Inherited asymmetry

Figure 31f illustrates an initial asymmetry produced by burial of frontal bulges, as discussed in the previous section and illustrated by models 840328 and 840413. In model 840413, when progradation had ceased by step 19, the interface between cover and source had returned to the horizontal (Fig. 17). Thus the dip and wedge effects previously described could not have operated during subsequent diapiric growth in step 20. Yet diapirs were strongly tilted opposite to the direction of progradation in the final step (20) (Fig. 20). In Section IV we found no evidence that this was an edge effect. An alternative but speculative explanation is that the asymmetry of the initial irregularities in the contact influenced the initial growth of the early diapirs, which in turn influenced the later growth stages, a process we term inherited asymmetry because the asymmetrical loading was only operative in the early stages of growth.

3. Distal gravity spreading

Stresses set up by the surface slope of a viscous or plastic mass cause it to spread like a glacier under its own weight until the wedge taper is reduced below a threshold angle determined by the yield strength of the spreading mass. Boundary-layer flow takes place during spreading if the base of the sequence is cohesive and the upper boundary is a free surface (Ramberg, 1981, p. 195-226). Where the sequence includes a diapiric source, spreading can distort originally upright and symmetrical diapirs and appendages and transform them to tilted, asymmetric structures (Fig. 31 g, late) as in the "end-member" case shown in Figure 31a for a tabular system. Because of the velocity profile, asymmetry is greatest for the smallest structures closest to the cohesive base of the spreading mass (Fig. 31g, late). This mechanism amplified and sheared irregularities in model 840403 to produce the bizarre appendages, which are asymmetrical and tilted distally (Figs. 22 through 25). De Boer (1979) described mesoscopic diapirs of coarse sand whose density was reduced by air trapped by rising tides; the diapirs became sheared to asymmetry by downslope gravity creep.

4. Progradation across a basin edge

Sections V and VI, B described how progradation across and beyond the abrupt edge of a basin of source material formed an upright wall at the basin edge. The wall remained rooted to the basin edge but was subsequently tipped on its side and squeezed or carried laterally to form a recumbent wall whose horizontal amplitude was much larger than its vertical dimension. The ratios in Figure 32 can also be applied to this case. In Figure 31h, slow progradation (low P/R, $P/B < 1$) causes the cover wedge to founder while the source extrudes as an exposed recumbent wall beyond the basin margin. If progradation is rapid (high P/R, $P/B > 1$), as in models 840403 and 840423, the recumbent wall is buried and attenuated laterally within the prograding, spreading cover (Fig. 31i).

5. Thrust-initiated asymmetry and gravity spreading

The Mississippi mudlumps of Morgan et al. (1968) illustrate a combination of mechanisms to produce asymmetry. Mudlumps are small shale diapirs, whose growth history in the Missippi delta has been monitored since 1876 because of their threat to shipping lanes once they reach surface. Their rapid appearance invariably coincides with times of river flood and rapid rates of sedimentation at the river mouth. Differential loading of clay containing silt intercalations (shelf, prodelta, delta front facies) by prograding denser sand and silt (channel and distributary-mouth-bar facies) provides the instability. Density contrast is about 250 kg m^{-3}. Diapiric intrusion of the clay is accompanied by reverse and thrust faults, commonly verging seaward, in the suprajacent cover (Fig. 31j, early). Further rise of the clay diapirs extends the cover by radial or parallel normal faulting, depending on diapir planform (Fig. 31j, late). The dynamics of growth are similar to those of models 840328 and 840413, apart from the brittle component of strain. At least two factors are likely to have controlled tilt and asymmetry during mudlump growth. Seaward-verging contraction faults provided the initial asymmetry. This was probably augmented by seaward downslope gravity creep of both the clay and overlying sand, which would have accentuated seaward asymmetry of the clay diapirs.

6. Summary

On the seaward-dipping flank of a salt basin down which a continental margin is prograding, most factors favor growth of diapirs that tilt and are asymmetrical toward the basin center. These factors are the source-dip effect, the cover-wedge effect, and the gravity spreading effect. In contrast, the source-wedge effect tends to produce landward tilt and asymmetry. The effect of inherited asymmetry depends on the nature of the initial asymmetry; we have illustrated both basinward and landward tilt and asymmetry. The half-layer effect also produces tilts and asymmetry in both directions. These effects are generally combined, even in our simplified models, and may be enhanced, retarded, or completely reversed by such combination. High-quality, deep-penetrating, 3-D reflection seismic data are required to determine the deep structure of Gulf Coast salt diapirs and establish whether these have preferred tilts and asymmetry in the specific environments discussed. Sufficient data are not yet available outside the petroleum companies and their service industries to confirm these hypotheses.

VII. ACKNOWLEDGEMENTS

Research was funded by the U.S. Department of Energy, Salt Repository Project Office under contract no. DE-AC97-83WM46651. The study could not have been done without experimental collaboration with, and enthusiastic support, advice, and countless hours and pages of discussion given by, Christopher Talbot, Director of the Hans Ramberg Tectonic Laboratory at the University of Uppsala. Ronald Arvidsson assisted MPAJ greatly in material preparation and testing and in model construction. Hans Ramberg, Ruud Weijermars, Peter Ronnlund, and Pierre Heeroma kindly provided practical advice and helpful comments during the experiments in Uppsala. Ruud Weijermars generously supplied a preprint of his paper on the rheology of model materials. We thank Christopher Talbot, Ruud Weijermars, Harro Schmeling, Ray Fletcher, John Dixon, Kelly Anne Bitner, Khosrow Bakhtar, and an anonymous referee for their critical comments of the manuscript, which was edited by Mary Ellen Johansen. T. B. Samsel III, Stephen Lawrence, Nan Minchow-Newman, Annie Kubert-Kearns, and Jamie McClelland drafted the figures. Half-tone printing was by David M. Stephens and diagram reproduction was by James A. Morgan.

APPENDIX A: EDGE EFFECTS

In a rectangular container the outermost source walls grow parallel to their nearest lateral boundary. If only a single internal wall is generated (because the source is thick relative to the width of the container), it is parallel to the longer side (Ramberg, 1981; Dixon and Summers, 1983). If walls are artificially initiated by low-amplitude bumps in the source layer, the effects of initiation can overpower those of the lateral boundary. For example, initiated walls can be made to grow parallel to the short side of a box without any development of the orthogonal direction (Dixon and Summers, 1983). All the edge effects described in this report are those operating without artificial initiation of walls or fingers.

Two experiments illustrate edge effects in rectangular and circular containers, each using a cover of mercury on a source layer of Rhodorsil Gomme. The high density contrast (12:1) caused complete overturn in a few hours under normal gravity outside the centrifuge, which allowed continual observation and photography. In model 840410 a few straight walls formed in the rectangular box (Fig. 2). Their orthogonal pattern was reflected in the geometry of the overhanging bulbs. In the final stage of 2,955 s, the sutured junctions between the spreading bulbs were marked by discon tinuities separating mismatched mm grids. Breaks that did not separate mismatched grids were merely "high-water marks" produced early in the experiment when mercury cover was in flood; after the mercury receded globules remained attached to the underlying RG and darkened its surface. The pattern of underlying walls that fed these spreading bulbs was one that changed spatially from orthogonal walls on the outside (an edge effect) to polygonal walls in the center. In other experiments where many of walls formed in a rectangular box, the pattern consisted entirely of polygonal walls except for a narrow fringe of orthogonal walls (P. Ronnlund, unpublished centrifuge modeling at the University of Uppsala, 1985).

In model 840411 a circular wall formed along the perimeter of the circular container (Fig. 3). Several internal structures were enclosed by the peripheral wall, and edge effects declined inward. The nearest wall to the peripheral wall was an entire circle, and the next wall inward was a semicircle, whereas the innermost structures were fingers. The final step at 2,578 s shows that the walls were not thick ridges as they appeared to be when surrounded by mercury, but were actually thin sheets that curved upward to the horizontal where they floated on mercury.

Here the cover was relatively thin, and a deep polygonal network of walls in the center was subtle or absent. In our other experiments not described here, the cover was thicker and the interior walls defined a polygonal network at depth similar to that in the center of a rectangular container.

Edge effects are common, but their origin has been either ignored or commented on only briefly in the literature. For a peripheral wall or series of fingers to form, the rising structures must overcome frictional resistance against the boundaries of the experimental container. Accordingly, there must be an outward decrease in hydrostatic pressure. Heye (1978) proposed that where the normally planar interface between syrup cover and oil source met the lateral boundary, differential surface tensions of the two fluids would cause flexure of the interface up against the boundary. After the box was inverted, such a flexure near the edges would provide the tiny perturbation necessary to trigger the rise of a wall.

A similar surface-tension effect explains the formation of peripheral walls in both experiments using a mercury cover, as shown in Figures 2 and 3. Liquid mercury has an upward-convex meniscus against glass or silicone putty because its internal cohesion is greater than its adhesion to glass or putty. Conversely, like water, silicone putty has an upward-concave meniscus against glass because its adhesion to glass is greater than its internal cohesion. At the edge of the container the two menisci combine, so that an extra-thick silicone source is overlain by an extra-thin mercury cover. This geometry, which was easily seen during the experiment, caused a decrease of pressure at the boundary, which would encourage the growth of the underlying perturbation of the top of the source layer. How thin would these layers have to be for surface tension to be the dominant force? The dimensionless Bond number, *Bo*, measures the force ratio of gravity to surface tension, and the Capillary number, *Ca*, the ratio of viscous force to surface tension (Weast, 1970):

$$Bo = \Delta\rho l^2 g/\sigma \quad (1)$$

$$Ca = \mu V/\sigma \quad (2)$$

where $\Delta\rho$ = density difference, l = thickness of sheet or radius of droplet, g = free-fall acceleration due to gravity, σ = surface tension, μ = equivalent dynamic viscosity, and V = velocity. For silicone putty against air, the following parameters apply: $\Delta\rho$ = 1,090 kg m^{-3}, g = 9.81 m s^{-2}, σ = 0.02 N m^{-1} (based on viscous PDMS of comparable viscosity, Weijermars, 1986b); for mercury against air, $\Delta\rho$ = 13,540 kg m^{-3} and σ = 0.471 N m^{-1}. Thicknesses corresponding to a Bond number of 1 are therefore 1.4 mm and 1.9 mm for silicone putty and mercury, respectively, against air. Below these thicknesses, at the edges of a fluid sheet, surface-tension forces will dominate gravity forces, thus allowing for the meniscus-type edge effects we have described. By comparison, the Capillary number for silicone putty (based on a viscosity of 4×10^4 Pa s for Rhodorsil Gomme [Table IV]) against air is 40, with a rise rate of 2×10^{-5} m s^{-1} (Tables II and III); the equivalent Capillary number for mercury (viscosity = 2×10^{-3} Pa s) against air is 9×10^{-8}. Thus surface tension forces are subordinate to viscous forces in the case of silicone putty against air, but are greatly dominant in the case of mercury against air.

Peripheral walls in centrifuged models composed entirely of silicone putty must have formed differently than those of the mercury models. The centrifuged models were constructed by adding thin layers on top of each other; surface tensions along their interfaces must have been similar. However, air bubbles were visibly trapped between these layers. Irregularities caused by the rise of bubbles during centrifuging acted as a multitude of random initiators for the growth of fingers and walls.

Because of the abundance and random distribution of bubbles, this type of initiation can be ignored; other factors, such as thickness and viscosity ratios, determined dominant wavelength. In contrast, along the vertical model boundaries, the bubbles were vertical sheets and were common because silicone putty abutted against a vertical, nonplanar surface of Plastilina or polythene sheeting around the model. The greater abundance of bubble sheets, together with their vertical orientation along the model boundaries, must have decreased locally the density and hydrostatic pressure of the overburden, thereby encouraging growth of a peripheral wall.

REFERENCES

Amery, G. B. (1969). Structure of Sigsbee Scarp, Gulf of Mexico. Am. Assoc. Petroleum Geologists Bull. 53, 2480-2482.

Berner, H., Ramberg, H., and Stephansson, O. (1972). Diapirism in theory and experiment. Tectonophysics 15, 197-218.

Biot, M. A. (1966). Three-dimensional gravity instability derived from two-dimensional solutions. Geophysics 31, 153-166.

Biot, M. A., and Ode, H. (1965). Theory of gravity instability with variable overburden and compaction. Geophysics 30, 213-227.

Bishop, R. S. (1978). Mechanism for emplacement of piercement diapirs. Am. Assoc. Petroleum Geologists Bull. 62, 1561-1583.

Brown, L. F., Jr., and Fisher, W. L. (1982). "Seismic stratigraphic interpretation and petroleum exploration." Am. Assoc. Petroleum Geologists Continuing Education Course Note Series No. 16, 56 p.

Buffler, R. T. (1984). Early history and structure of the deep Gulf of Mexico basin. Soc. Econ. Paleontologists and Mineralogists Gulf Coast Section, Fifth Annual Foundation Research Conference, Austin, Texas, 31-34.

Buffler, R. T., and Sawyer, D. S. (1985). Distribution of crust and early history, Gulf of Mexico basin. Gulf Coast Assoc. Geol. Socs. Trans. 35, 333-344.

Buffler, R. T., Winker, C. D., Rosenthal, D. B, Viele, G. W., Suter, Sherman, Lillie, R. J., Miles, A. E., Pilger, R. H., Jr., Nicholas, R. L., Watkins, J. S., Martin, R. G., and Sawyer, D. S. (1986). "Continent-Ocean Transect F1: Ouachitas to Yucatan." Geol. Soc. America Decade of North American Geology Program (in press).

Buffler, R. T., and Worzel, J. L. (1978). Deformation and origin of the Sigsbee Scarp--lower continental slope, northern Gulf of Mexico. 10th Annual Offshore Technology Conference, Houston, Texas, OTC; 3217, 1425-1439.

Carey, S. W. (1954). The rheid concept in geotectonics. Geol. Soc. Australia Jour. 1, 67-117.

Carmichael, R. S., (ed). (1984). "CRC handbook of physical properties of rocks, volume III." CRC Press, Boca Raton, Florida, 340 p.

Carter, N. L., and Hansen, F. D. (1983). Creep of rocksalt. Tectonophysics 92, 275-333.
Carter, N. L., Hansen, F. D., and Senseny, P. E. (1982). Stress magnitudes in natural rocksalt. Jour. Geophys. Research 87, 9289-9300.
Clark, S. P., Jr., (ed.) (1966). Handbook of physical constants (revised ed.). Geol. Soc. America Mem. 97, 587 p.
Cogswell, F. N., Gray, J. G. H., and Hubbard, D. A. (1972). Rheology of a fluid whose behaviour approximates to that of a Maxwell body. British Soc. Rheology Bull. 15, 29-31.
Danes, Z. F. (1964). Mathematical formulation of salt-dome dynamics. Geophysics 29, 414-424.
De Boer, P. L. (1979). Convolute lamination in modern sands of the estuary of the Oosterschelde, the Netherlands, formed as the result of entrapped air. Sedimentology 26, 283-294.
DeJong, A. (1968). Stratigraphy of the Sigsbee scarp from a reflection survey (abs.). Society of Exploration Geophysicists, Program, 21st Annual Meeting, Fort Worth, Texas, 51.
Dixon, J. M., and Summers, J. M. (1983). Patterns of total and incremental strain in subsiding troughs: experimental centrifuged models of inter-diapir synclines. Canadian Jour. Earth Sci. 20, 1843-1861.
Dixon, J. M., and Summers, J. M. (1985). Recent developments in centrifuge modelling of tectonic processes: equipment, model construction techniques and rheology of model materials. Jour. Structural Geol. 7, 83-102.
Ewing, M., and Antoine, J. (1966). New seismic data concerning sediments and diapiric structures in Sigsbee Deep and upper continental slope, Gulf of Mexico: Am. Assoc. Petroleum Geologists Bull. 50, 479-504.
Fisher, W. L. (1973). Deltaic sedimentation, salt mobilization and growth faulting in Gulf Coast Basin (abs.). Am. Assoc. Petroleum Geologists Bull. 57, 779.
Fletcher, R. C. (1972). Application of a mathematical model to the emplacement of mantled gneiss domes. Am. Jour. Sci. 272, 197-216.
Hailemariam, H. (1982). "Viscosity measurements of three different types of silicone putty by a co-axial cylinders viscometer." Uppsala University, C-Thesis, Uppsala, Sweden, 13 p.
Heye, D. (1978). Experimente mit viskosen Flussigkeiten zur Nachahmung von Salzstrukturen. Geol. Jahrb. E12, 31-15.
Heye, D. (1979). Modellversuche zum Salzdiapirismus mit zur Nachahmung von Salzstrukturen. Geol. Jahrb. E12, 31-15.
Heye, D. (1979). Modellversuche zum Salzdiapirismus mit viskosen Flussigkeiten. Geol. Jahrb. E16, 39-51.
Hubbert, M. K. (1937). Theory of scale models as applied to the study of geologic structures. Geol. Soc. America Bull. 48, 1459-1520.
Humphris, C. C., Jr. (1979). Salt movement on continental slope, northern Gulf of Mexico. Am. Assoc. Petroleum Geologists Bull. 63, 782-798.
Hunsche, U. (1978). Modellrechnungen zur Entstehung von Salzstockfamilien. Geol. Jahrb. E12, 53-107.
Jackson, M.P.A, and Galloway, W.E. (1984). Structural and depositional styles of Gulf Coast Tertiary continental margins: application to hydrocarbon exploration. Am. Assoc. Petroleum Geologists Continuing Educ. Course Note Ser. No. 25, 226 p.
Jackson, M. P. A., and Talbot, C. J. (1985). "The internal structure of model and natural salt domes." The University of Texas at Austin, Bureau of Economic Geology, final report prepared for U. S. Department of Energy under contract no. DE-AC97-83WM-46651, 57 p.

Jackson, M. P. A., and Talbot, C. J. (1986). External shapes, strain rates, and dynamics of salt structures. Geol. Soc. America Bull. 97, 305-323.
Jackson, M. P. A., and Zelt, G. A. D. (1984). Proterozoic crustal reworking and superposed deformation of metabasite dykes, layered intrusions, and lavas in Namaqualand, South Africa. In "Precambrian tectonics illustrated" (A. Kroner and R. Greiling, eds.), Internat. Union Geological Sciences/Schweizerbart'sche Verlagsbuchhandlung, Stuttgart, pp. 381-400.
Jenyon, M. K. (1985). Basin-edge diapirism and updip salt flow in Zechstein of southern North Sea. Am. Assoc. Petroleum Geologists Bull. 69, 53-64.
Kline, S. J. (1965). "Similitude and approximation theory." McGraw-Hill, New York, 229 p.
Krieger, I. M., and Maron, S. H. (1954) J. (1965). "Similitude and approximation theory." McGraw-Hill, New York, 229 p.
Krieger, I. M., and Maron, S. H. (1954). Direct determination of the flow curves of non-Newtonian fluids. III. Standardized treatment of viscometric data. Jour. Appl. Physics 25, 72-75.
Langhaar, H. L. (1951). "Dimensional analysis and theory of models." John Wiley, New York, 166 p.
Lehner, P. (1969). Salt tectonics and Pleistocene stratigraphy on continental slope of northern Gulf of Mexico. Am. Assoc. Petroleum Geologists Bull. 53, 2431-2479.
Martin, R. G. (1980). "Distribution of salt structures in the Gulf of Mexico: map and descriptive text, 1:2,500,000." U.S. Geol. Survey Misc. Field Studies Map MF-1213, 8 p.
McKee, E. D., and Goldberg, M. (1969). Experiments on formation of contorted structure in mud. Geol. Soc. America Bull. 80, 231-244.
Mitchum, R. M., Jr., Vail, P. R., and Sangree, J. B. (1977). Seismic stratigraphy and global changes of sea level, part 6: stratigraphic interpretation of seismic reflection patterns in depositional sequences. Am. Assoc. Petroleum Geologists Mem. 26, 117-133.
Morgan, J. P., Coleman, J. M., and Gagliano, S. M. (1968). Mudlumps: diapiric structures in Mississippi Delta sediments. Am. Assoc. Petroleum Geologists Mem. 8, 145-161.
New Orleans Geological Society (1983). "Salt domes of South Louisiana." Vol. III, 142 p.
Parker, T. J., and McDowell, A. N. (1955). Model studies of salt-dome tectonics. Am. Assoc. Petroleum Geologists Bull. 39, 2384-2470.
Paterson, W. S. B. (1981). "The physics of glaciers." 2nd ed. Pergamon Press, Oxford, 380 p.
Ramberg, H. (1967). "Gravity, deformation and the Earth's crust as studied by centrifuged models." 1st ed. Academic Press, London, 214 p.
Ramberg, H. (1968). Instability of layered systems in the field of gravity, I and II. Physics Earth and Planetary Interiors 1, 427-474.
Ramberg, H. (1981). "Gravity, deformation and the Earth's crust in theory, experiment and geological application." 2nd ed. Academic Press, London, 452 p.
Ramberg, H., and Stephansson, O. (1965). Note on centrifuged models of excavations in rocks. Tectonophysics 2, 281-298.
Rettger, R. E. (1935). Experiments on soft-rock deformation. Am. Assoc. Petroleum Geologists Bull. 19, 271-292.
Salvador, A., and Buffler, R. T. (1983). The Gulf of Mexico Basin. In "Perspectives in regional geological synthesis: Planning for the geology of North America" (A. R. Palmer, ed.), Geological Society of America, D-NAG Special Publication 1, p. 157-162.

Schwerdtner, W. M., and Osadetz, K. (1983). Evaporite diapirism in the Sverdrup basin: new insights and unsolved problems. Canadian Petroleum Geology Bull. 31, 27-36.
Schwerdtner, W. M., and van Kranendonk, M. (1984). Structure of Stolz diapir--a well-exposed salt dome on Axel Heiberg Island, Canadian Arctic Archipelago. Canadian Petroleum Geology Bull. 32, 237-241.
Selig, F. (1965). A theoretical prediction of salt dome patterns. Geophysics 30, 633-643.
Seni, S. J., and Jackson, M. P. A. (1983). Evolution of salt structures, East Texas diapir province, part 1: sedimentary record of halokinesis. Am. Assoc. Petroleum Geologists Bull. 67, 1219-1244.
Seni, S. J., and Jackson, M. P. A. (1984). "Sedimentary record of Cretaceous and Tertiary salt movement, East Texas Basin: Times, rates, and volumes of salt flow and their implications for nuclear waste isolation and petroleum exploration." The University of Texas at Austin, Bureau of Economic Geology Report of Investigations No. 139, Austin, 89 p.
Spiers, C. J., Urai, J. L., Lister, G. S., Boland, J. N., and Zwart, H. J. (1986). The influence of fluid-rock interaction on the rheology of salt rock and on ionic transport in the salt. Nuclear Sci. Technology CEC-EUR Ser. (in press).
Talbot, C. J. (1974). Fold nappes as asymmetrical mantled gneiss domes and ensialic orogeny. Tectonophysics 24, 259-276.
Talbot, C. J. (1977). Inclined and asymmetrical upward-moving gravity structures. Tectonophysics 42, 159-181.
Talbot, C. J. (1982). Obliquely foliated dikes as deformed incompetent single layers. Geol. Soc. America Bull. 93, 450-460.
Talbot, C.J. and Jackson, M.P.A. (in review). Internal dynamics and kinetics of salt structures. Am. Assoc. Petroleum Geologists Bull.
Turcotte, D. L., and Schubert, G. (1982). "Geodynamics: applications of continuum physics to geological problems." John Wiley, New York, 450 p.
Watkins, J. S., Worzel, J. L., Houston, M. H., et al. (1975). Deep seismic reflection results from the Gulf of Mexico, Part 1. Science 187, 834-836.
Watkins, J. S., Ladd, J. W., Buffler, R. T., Shaub, F. J., Houston, M. H., and Worzel, J. L. (1978). Occurrence and evolution of salt in deep Gulf of Mexico. Am. Assoc. Petroleum Geologists Studies in Geology 7, 43-65.
Weast, R. C. (ed.) (1970). "CRC handbook of chemistry and physics." 51st ed. Chemical Rubber Company, Cleveland, Ohio, 2364 p.
Weijermars, R. (1986a). Flow behaviour and physical chemistry of bouncing putties and related polymers in view of tectonic laboratory applications. Tectonophysics 124, 325-358.
Weijermars, R. (1986b). Finite strain of laminar flows can be visualized in SGM36-polymer. Naturwissenschaften 73, 33-34.
Whitehead, J. A., Jr., and Luther, D. S. (1975). Dynamics of laboratory diapir and plume models. Jour. Geophys. Research 80, 705-717.
Wilhelm, O., and Ewing, M. (1972). Geology and history of the Gulf of Mexico. Geol. Soc. America Bull. 83, 575-599.
Wilson, C. H. (1985). "Depositional, structural, and thermal evolution of a Pleistocene oil-productive area: Texas-Louisiana continental shelf." The University of Texas at Austin, Master's thesis, 93 p.
Winker, C. D. (1982). Cenozoic shelf margins, northwestern Gulf of Mexico. Gulf Coast Assoc. Geol. Socs. Trans. 32, 427-448.
Woidt, W.-D. (1978). Finite element calculations applied to salt-dome analysis. Tectonophysics 50, 369-386.

Woidt, W.-D. (1980). Analytische und numerische Modell-experimente zur Physik der Salzstockbildung. Institut fur Geophysik und Meteorologie der Technischen Universitat Braunschweig GAMMA 38, 151 p.

Woodbury, H. O., Murray, I. B., Jr., Pickford, P. J., and Akers, W. H. (1973). Pliocene and Pleistocene depocenters, outer continental shelf, Louisiana and Texas. Am. Assoc. Petroleum Geologists Bull. 57, 2428-2439.

Woodbury, H. O., Murray, I. B., Jr., and Osborne, R. E. (1980). Diapirs and their relation to hydrocarbon accumulation. In "Facts and principles of world petroleum occurrence" (A. D. Miall, ed.), Canadian Soc. Petroleum Geologists, Calgary, 119-142.

Table I: Comparison of Rates (m s^{-1}) and Ratios of Progradation (*P*), Aggradation (*A*), and Diapir Rise (*R*) in Prototypes and Models

		P		*A*		*R*	*P/A*	*P/R*
Prototype Cenozoic	[a]	1.4 E-10	[b]	3.7 E-12	[c]	3.0 E-12	38	47
Prototype Quaternary	[d]	3.6 E-10	[e]	3.2 E-11	[f]	2.4 E-11	11	15
840328	[g]	5.5 E-05	[g]	3.1 E-06	[h]	1.7 E-06	18	32
840413	[i]	8.3 E-04	[i]	7.4 E-05	[j]	2.7 E-05	11	30
840412	[k]	5.7 E-04	[k]	7.1 E-05	[k]	7.7 E-05	8	7.4
840402	[l]	0.0	[l]	1.6 E-08	[l]	2.4 E-08	0	0

Data Sources

[a] Mean Cenozoic progradation rate of northern Gulf Coast continental margin for the last 56 Ma from position at start of Wilcox time to present 200-m bathymetric contour (Winker, 1982)

[b] Mean Cenozoic aggradation rate of northern Gulf Coast continental margin for the last 56 Ma, based on mean thickness of 6,500 m from base of Wilcox Group to present surface (Salvador and Buffler, 1983)

[c] Mean Cenozoic rise rate of shallow diapirs for the last 56 Ma, based on mean thickness of 6,500 m from base of Wilcox Group to present surface and on mean depth of 1,250 m for 21 offshore Louisiana diapirs drilled to salt (New Orleans Geological Society, 1983)

[d] Mean Quaternary progradation rate of northern Gulf Coast continental margin for the last 2.8 Ma from position at start of Pleistocene (defined by extinction of Globoquadrina altispira at base of Lenticulina I fauna zone) to present 200-m bathymetric contour (Winker, 1982)

[e] Mean Late Quaternary aggradation rate of northern Gulf Coast continental margin for the last 1.8 Ma (defined by extinction of Discoasters at base of Trimosina B fauna zone), based on a mean Late Pleistocene thickness of 1,830 m (Woodbury et al., 1973). For comparison, mean aggradation rate of Pleistocene Sangamon faunal zone in same area (0.5-0.1 Ma ago) was 3.5 E-11; three-dimensional seismic reflection study by Wilson (1985)

[f] Late Pleistocene distortion of strata above rising salt diapir over the last 0.8 Ma on continental shelf of northern Gulf Coast; three-dimensional seismic reflection study by Wilson (1985)

[g] Steps 1 through 4 for 1603 s, representing period of progressive progradation and aggradation

[h] Steps 1 through 5 for 2367 s, representing entire period of diapiric growth

[i] Steps 1 through 19 for 121 s, representing period of progressive progradation and aggradation

[j] Steps 1 through 20 for 287 s, representing entire period of diapiric growth; based on mean of highest point of each wall

[k] Steps 1 through 18 for 170 s, representing period of steady progradation and aggradation and entire period of diapiric growth

[l] Static differential loading for 52 h

SECTION B

STRUCTURAL IMPACT OF SALT ON SURROUNDING FORMATIONS

ACTIVE SALT DOME DEVELOPMENT IN THE LEVANT BASIN, SOUTHEAST MEDITERRANEAN

Z. Garfunkel

Department of Geology
Hebrew University
Jerusalem, Israel

G. Almagor

Division of Marine Geology
Mapping and Tectonics
Geological Survey of Israel
Jerusalem, Israel

I. INTRODUCTION

Salt domes and diapirs occur in many parts of the world. These structures form by the upward flow of salt which is driven primarily by buoyancy forces. Such forces arise when a body of low density salt is covered by higher density sediments while the low effective viscosity of the salt allows it to creep at geologically significant rates (Nettleton, 1934; Parker and McDowell, 1955; Trusheim, 1960; Biot and Ode, 1965; Selig, 1956; O'Brien, 1968). Salt structures were studied extensively for several reasons. For one, they are associated with important hydrocarbon accumulations. Secondly, the mechanism of salt diapir formation is of great theoretical interest in itself and is also a good example of the general phenomenon of upward mass flowage driven by density inversion. The study of actively growing salt structures is, therefore, of special interest because in such cases the geologic conditions in which they are initiated and grow can be observed and studied in detail. One such place is the Mediterranean Sea where flowage of the salt-bearing Messinian series produced numerous diapiric structures of different types that are often still growing. The purpose of the present paper is to describe these structures in the Levant Basin, southeastern Mediterranean (Figs. 1 and 2), and to discuss the factors controlling their formation.

Flowage of the Messinian salt-bearing series and diapir formation in the Mediterranean Sea and in its eastern part in particular were described by many authors (Alla, 1970; Ryan et al., 1970; Finetti and Morelli, 1973; Biju-Duval et al., 1974; Neev et al., 1976; Ross and Uchupi, 1977; Woodside, 1977). The present work is part of an ongoing study by the authors on the Neogene-Quaternary structure of the Levant Basin and its margins. Some results have already been published (Almagor and Garfunkel, 1979; Garfunkel et al., 1979; Almagor, 1980, 1984; Garfunkel, 1984; Garfunkel and Almagor, 1985). This study is based primarily on a dense network of high-quality single-channel seismic reflection profiles, penetrating down to the base of the Messinian sequence, which were described in detail in the above-mentioned works. Also used were many multi-channel seismic reflection profiles from the Levant basin and its margins which penetrate into the Mesozoic sequence. Information about the western portion of the study area is derived from the profiles of Ross and Uchupi (1977) and Woodside and Williams (1977). The seismic profiles combined with information from oil wells from the coastal plain and continental shelf of Israel and the Nile Delta give an overall picture of the flowage and diapirism of the Messinian series in the Levant Basin.

II. GENERAL SETTING

A. Geologic Background

The Levant basin is located in the southeastern corner of the Mediterranean Sea (Fig. 1). The basinal water depth ranges from about 1.2 km at the base of the continental slope to more than 2 km at center. In the west the basin is delimited by the

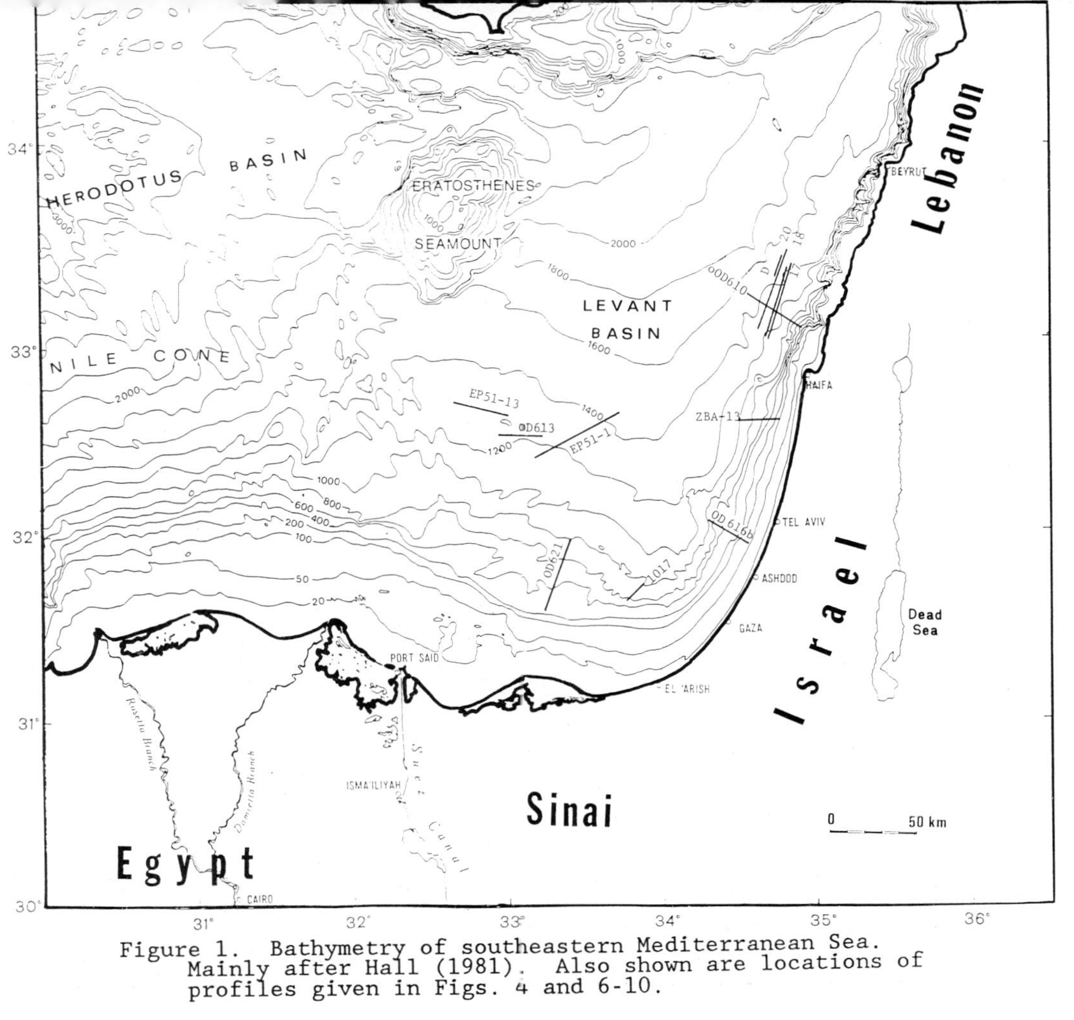

Figure 1. Bathymetry of southeastern Mediterranean Sea. Mainly after Hall (1981). Also shown are locations of profiles given in Figs. 4 and 6-10.

Eratosthenes Seamount, whereas to the southwest the seafloor deepens towards the Herodotus Basin where water depth exceeds 3 km. To the south and east these basins are bordered by the passive continental margins of the Arabo-African continent. This continental area was a structurally continuous unit during most

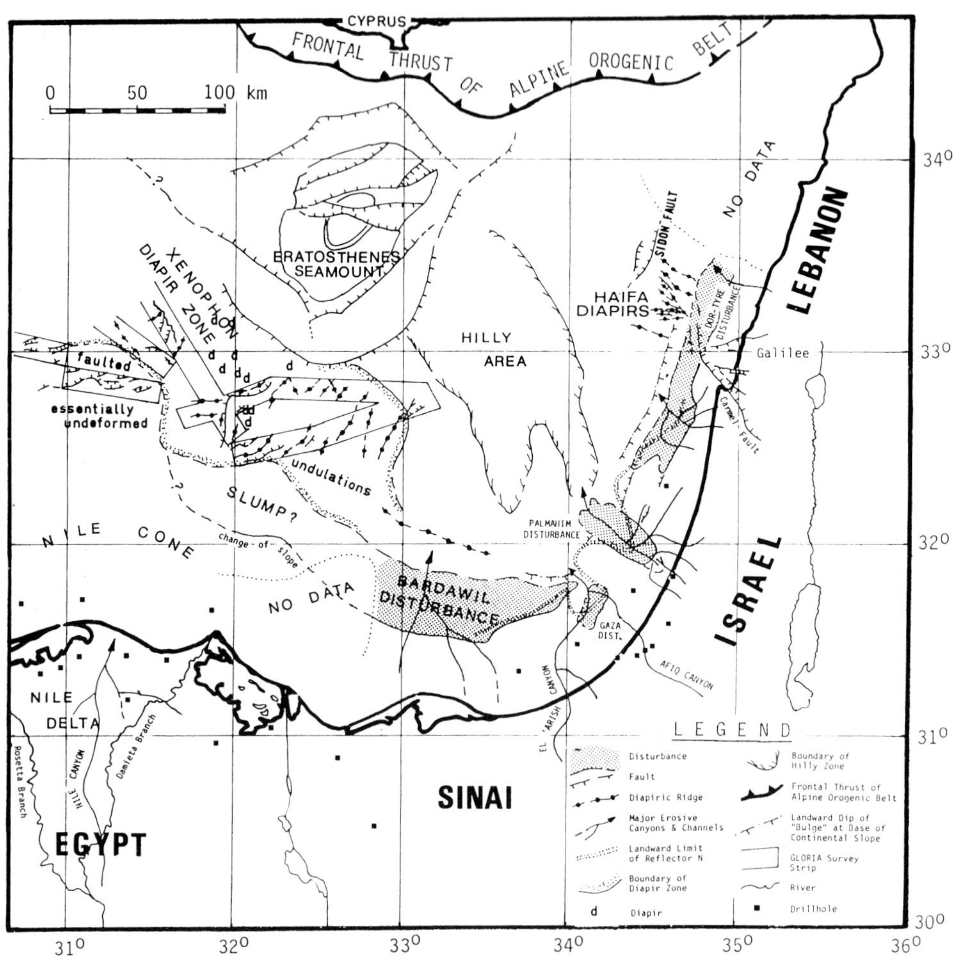

Figure 2. Main Pliocene-Quaternary structural elements of southeastern Mediterranean Sea. Modified after Garfunkel, 1984; Garfunkel and Almagor, 1985; Said, 1981. GLORIA survey data after Kenyon et al., 1975.

of the Phanerozoic, but in mid-Cenozoic times it broke into several independent plates that are now separated by the Red Sea and Gulf of Suez rifts, and by the Dead Sea transform. Results from deep drilling on the coast and offshore Israel demonstrated that the passive margins of the Levant Basin formed by Triassic-Liassic rifting and that deep-water conditions already existed there in Jurassic and Cretaceous times (Derin, 1974; Bein and Gvirtzman, 1977; Garfunkel and Derin, 1985). This shows that the Levant Basin was formed by the Early Mesozoic rifting event that affected the entire northern edge of Gondwanaland and shaped the Mesozoic Tethys area (Smith, 1970; Dewey et al., 1973; Biju-Duval and Dercourt, 1980; Smith and Woodcock, 1983). Seismic refraction studies (Lort et al., 1974; Ginzburg and Gvirtzman, 1979; Ginzburg and Folkman, 1980; Makris et al., 1983) showed that the crystalline crust of the Levant Basin is about 10 km thick, that it is overlain by about 10 km of sediments, and that the crustal thickness increases under the continental margin, reaching 35-40 km under Israel and Sinai. Correlation of the seismic profiles with the geologic data shows that the thick sediment fill of the basin is of Mesozoic and Cenozoic age. The heat flow of the eastern Mediterranean basin is lower than 40 mWm^{-2} (Cermak and Hurtig, 1979), as is to be expected in such an old basin.

Since Late Cretaceous times the northern edge of the Levant Basin was involved in plate convergence along the southern border of the Alpine orogenic belt (Fig. 2), but the area studied was only mildly deformed, with formation of broad folds that are sometimes faulted. This deformation practically ceased by Neogene times. When the nearby lands were rifted in Late Cenozoic times the continental margin, and perhaps the basin too, subsided considerably. This movement produced a basinward flexure along the continental margin with a structural relief of

1.5-2.5 km on which a 2-3 km thick prism of Neogene to Recent sediments accumulated (Ginzburg et al., 1975; Neev et al., 1976; Salem, 1976; Gvirtzman and Buchbinder, 1978).

A. Late Miocene to Recent History

The diapirs in the southeastern Mediterranean, like in the other parts of the Mediterranean Sea, formed as a result of flowage of the Messinian evaporitic series. Thus, diapirism characterizes the young stage of the basin's history, which followed the Messinian salinity crisis. This event involved a drop of the Mediterranean sea level, occasional desiccation of the basin, and deposition of a voluminous evaporitic series over the entire Mediterranean basin and its margins (Hsu et al., 1973, 1978; Ryan, 1978). On seismic reflection profiles the top and the base of this series appear as very prominent reflectors, called reflectors M and N, respectively (Ryan et al., 1970, 1973). Mapping of these two reflectors and results of drilling supply information about the diapirs of the Levant Basin and other parts of the Mediterranean Sea.

Two domains of erosion and sedimentation can be distinguished in the Levant Basin during the course of the Messinian salinity crisis:

(a) A basinal domain, extending under the Levant Basin and farther west, in which the base of the Messinian series (seismic reflector N) is a rather even surface that generally has an overall slope to the north and west, except for the elevated Eratosthenes Seamount (Fig. 3). Erosion of the underlying beds produced a mild relief in a few places, but in general the contact with the underlying beds is conformable. The predominance of evaporites - halite, gypsum and anhydrite - in the Messinian series between seismic reflectors M and N is proven by correlation with data from wells on the nearby coasts and continental shelves of Israel and Egypt (Fig. 2), correlation with DSDP drillholes and the seismic character of this interval

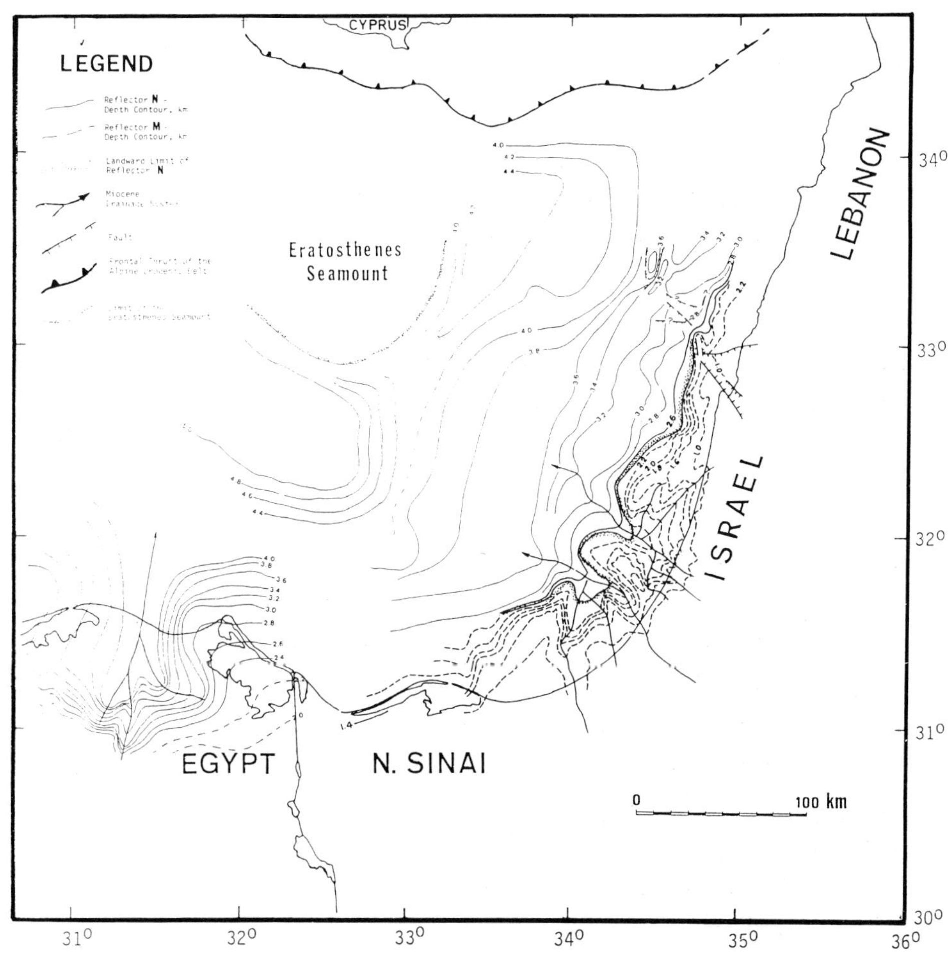

Figure 3. Structural map of the base of the Messinian sediments: In the basinal domain it is marked by reflector N; its depth in km was calculated with the following seismic velocities: 1.5 km/sec in water, 2.2 km/sec in the Pliocene-Quaternary series and 4.0 km/sec in the Messinian series. Under the continental slope of Israel and Sinai this is mostly an unconformity surface. Nile Delta area according to Said, 1981.

(Hsu et al., 1973, 1978). Summarizing seismic data, Montadert et al. (1978) showed that the upper part of the Messinian series in the Mediterranean is generally bedded and contains other sediments besides halite, whereas its lower part consists mostly of halite. In the eastern Mediterranean this lower

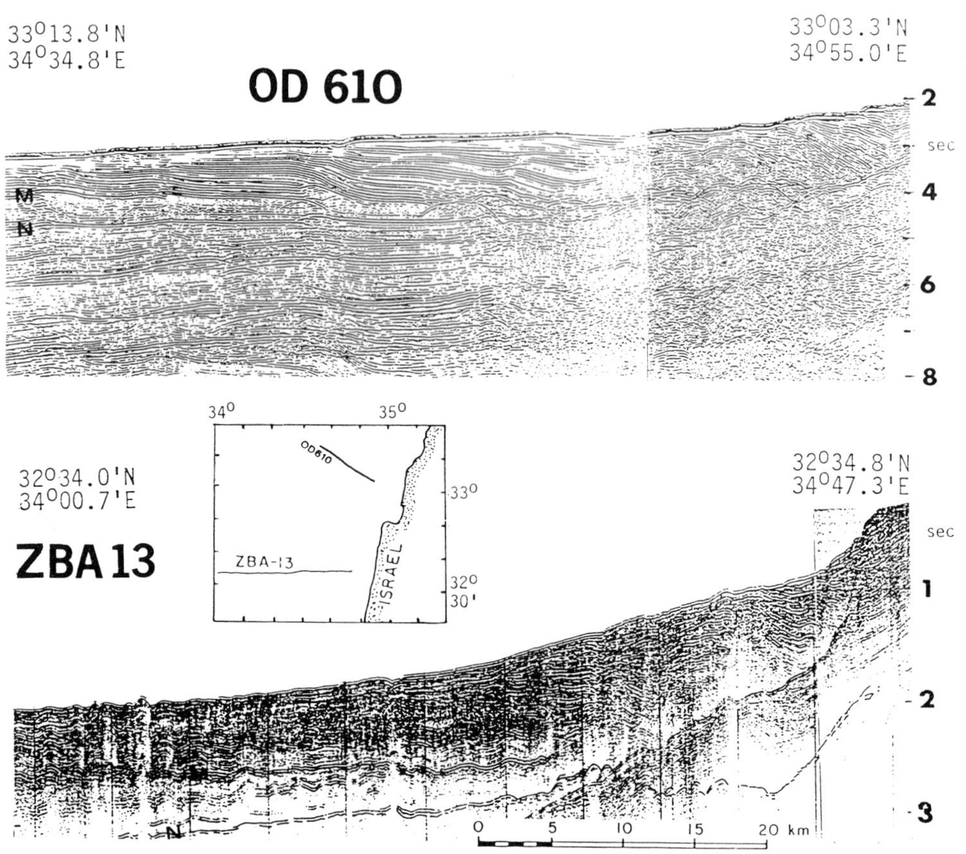

Figure 4. Deformation of base of continental slope and adjacent part of Levant basin. Note different vertical exaggerations in two profiles. Both profiles show the seismic character of the Messinian and Pliocene-Quaternary sequences, and the landward pinching out of the Messinian sequence, expressed by mergence of reflectors M and N. Profile OD 610 shows base of Tyre slump (disturbance). Note that the tilting and faulting of the Pliocene-Quaternary series do not affect the base of the Messinian series. Profile ZBA 13 extends from base of Dor slump (distrubance) and shows small-amplitude anticlinal undulations in nearby basin.

halite-rich part is best developed in the deep Herodotus Basin and thins towards the shallower Levant Basin. In the west of the Levant Basin the Messinian series is 1.5-2.0 km thick, but thins landward to a few hundred meters at the base of the continental slope. Several seismic reflectors are evident in this series and in places the series even contains bedded intervals (Fig. 4), which is quite unusual in pure halite series. Although the average seismic velocity in this series is about 4 km/sec, the velocity in the upper part is only 3.0-3.8 km/sec (Finetti and Morelli, 1973; Lort et al., 1974; Ross and Uchupi, 1977; Mart and Ben-Gai, 1982), whereas the seismic velocity in pure halite is 4.0-4.2 km/sec and in gypsum and anhydrite exceeds 4.7 km/sec (Schreiber et al., 1973; Gardner et al., 1974). Taken together these features show that in the Levent Basin the Messinian sequence, especially its upper part, contains not only evaporites (halite and gypsum), but also some non-evaporitic constituents. These are most likely marl or silty clay interbeds, similar to the sediments which were deposited in the basin before and after the Messinian.

(b) A marginal domain, which extends under the continental margins and coastal plains of mainland Egypt, Sinai, Israel and probably also Lebanon. This domain was actually the Late Miocene continental slope and shelf, and had a much steeper overall slope than the basinal area. During the Messinian the marginal domain was strongly affected by subaerial erosion which produced a drainage system characterized by deeply incised valleys, some of which extend far inland. Under the eastern margin of the basin the erosive valleys, up to 1 km deep, formed a well-defined drainage system (Fig. 3; Ginzburg et al., 1975; Neev et al., 1976; Gvirtzman and Buchbinder, 1978; Garfunkel et al., 1979). Even stronger was the erosion under the present Nile Delta, where the Messinian Nile gorge was more than 2 km deep (Rizzini et al., 1978; Said, 1981; Barber, 1981). Off Israel the transition between the marginal and basinal domains is now at a depth of 2.2-2.4 km below sea level, whereas off Egypt it is about 3.5-4 km below sea level.

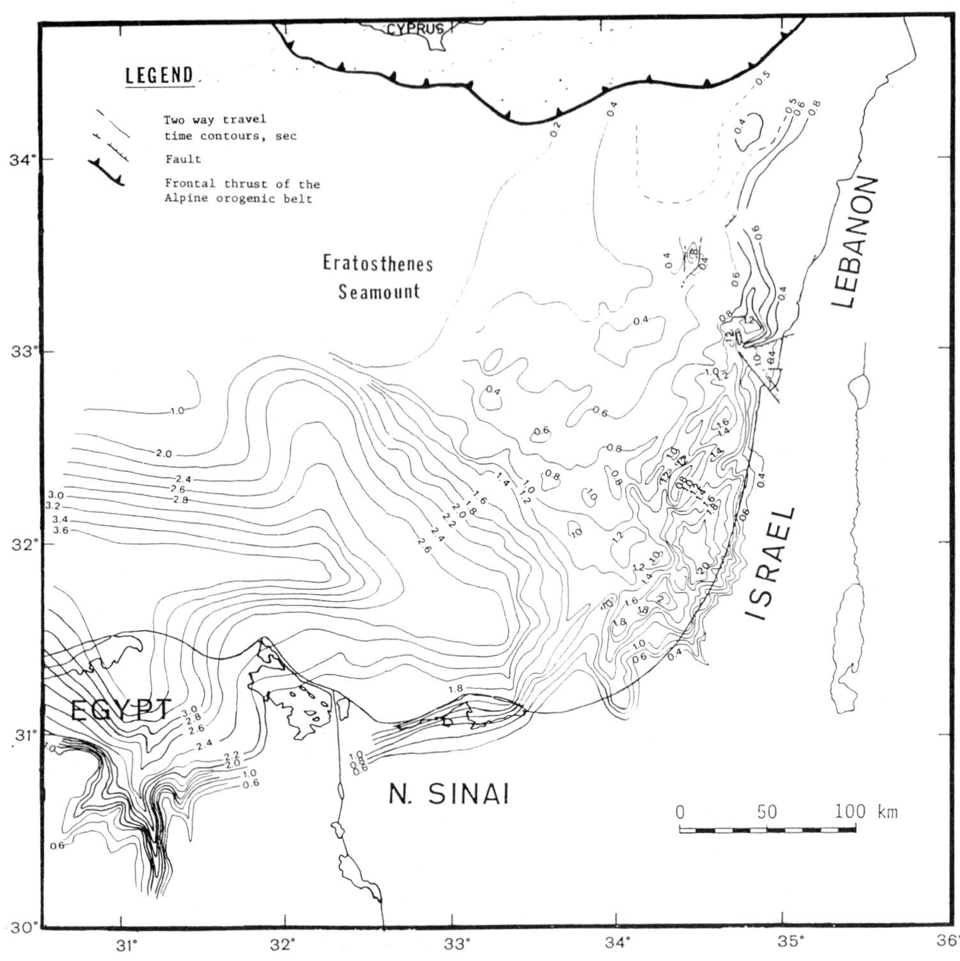

Figure 5. Two-way travel time of post-Messinian series. Western part modified after Ross and Uchupi, 1977, Woodside and Williams, 1977, Barber, 1981, Said, 1981.

This difference is in part due to subsidence caused by the large load of the Nile delta sediments. However, the much greater depth of the ancestral Nile valley compared to the valleys in Israel shows that already in the Late Miocene the Herodotus Basin was deeper than the Levant Basin.

On seismic reflection profiles of the Levant Basin reflectors M and N usually merge at the transition from the basinal to the marginal domains, recording the landward pinching out of the Messinian series (Figs. 4 and 5). However, lobes of the Messinian series (called Mavqi'im Formation in Israel), up to several hundred meters thick, extend from the basinal domain into the deeper parts of the valleys. Drillholes in these erosive valleys encountered halite and gypsum as well as land derived clastics (Derin and Gerry, 1971; Gvirtzman and Buchbinder, 1978). On the higher parts of the relief the seismic profiles do not have sufficient resolution to reveal the distribution of Messinian sediments in this area, but drilling showed that here dolomite, gypsum and shale, up to several tens of meters thick, occur discontinuously. At the mouth of the ancestral Nile gorge well over 1 km of Late Miocene coarse clastics (Qawasim Formation) accumulated. These are capped by a sequence of anhydrite interbedded with shales (Rosetta Formation), which is less than 100 m thick (Rizzini et al., 1978; Said, 1981). The seismic reflection profiles available to us do not cover the transition between the Messinian sections of the Nile Delta and that of the Levant Basin, so the behavior of reflectors M and N in this transition remains unknown. Therefore, on Figures 2 and 3 the merging of these reflectors at the base of the marginal domain was not shown west of long. 33°E.

With the renewed flooding of the Mediterranean in the Pliocene, normal marine sedimentation was resumed. Since the Pliocene the Nile River has been the main source of sediments for the eastern Mediterranean. A large part of the load of predominantly silty-clayey sediments built the large delta and submarine cone, but a part of this load was distributed over the eastern Mediterranean basin and its margins, where the Nilotic source can be recognized by the distinct clay mineral and heavy mineral assemblages (Nachmias, 1969; Ryan et al., 1970; Venkatarathnam and Ryan, 1971; Nir and Nathan, 1972; Biju-Duvai et al., 1974; Gvirtzman and Buchbinder, 1978; Rizzini et al.,

1978; Said, 1981). The Pliocene-Quaternary sediments have thicknesses exceeding 3.5 km under the Nile Delta, but they thin to 2-2.5 km under the continental shelf of Sinai and central Israel and to 1.5 km farther north (Fig. 5). In the Levant Basin the post-Messinian sediments thin from more than 1.5 km at the base of the continental slope to a mere 0.5 km under its northern part. On the seismic reflection profiles the Pliocene-Quaternary sequence appears to be mostly well bedded.

C. Young Structure

The Messinian and overlying beds are highly deformed because of flowage of the evaporitic series, but they are also affected by several structures of deeper origin. Besides long wavelength undulations, e.g., basinward flexing of the continental margins and flexing of the basin by the load of the Nile Delta, whose effects are difficult to measure, several fault zones have also recently been active (Fig. 2). One group of faults extends from northern Israel and Lebanon across the continental margin, the most important being the Carmel Fault which is an offshoot of the Dead Sea transform; in the deeper basin off this zone is the Sidon Fault which offsets reflector N by almost 1 second (Fig. 2; Garfunkel and Almagor, 1985). Another important fault system probably extends seaward from the Suez rift. However, though there is evidence for young tectonic activity along the isthmus of the Suez Canal (Garfunkel and Bartov, 1977), the seaward extension of the active structures into the Nile Cone is not clear. The transition between the Levant and Herodotus basins, which is in line with the Suez rift, is probably also tectonically controlled (Woodside, 1977), but the deep structure of this area is obscured by the thick sediments of the Nile Cone and by intensive diapirism. North of this zone young faulting affected the Eratosthenes Seamount as well as the area surrounding it. In addition to these well-defined zones, small active faults probably exist in the Levant Basin, as is indicated

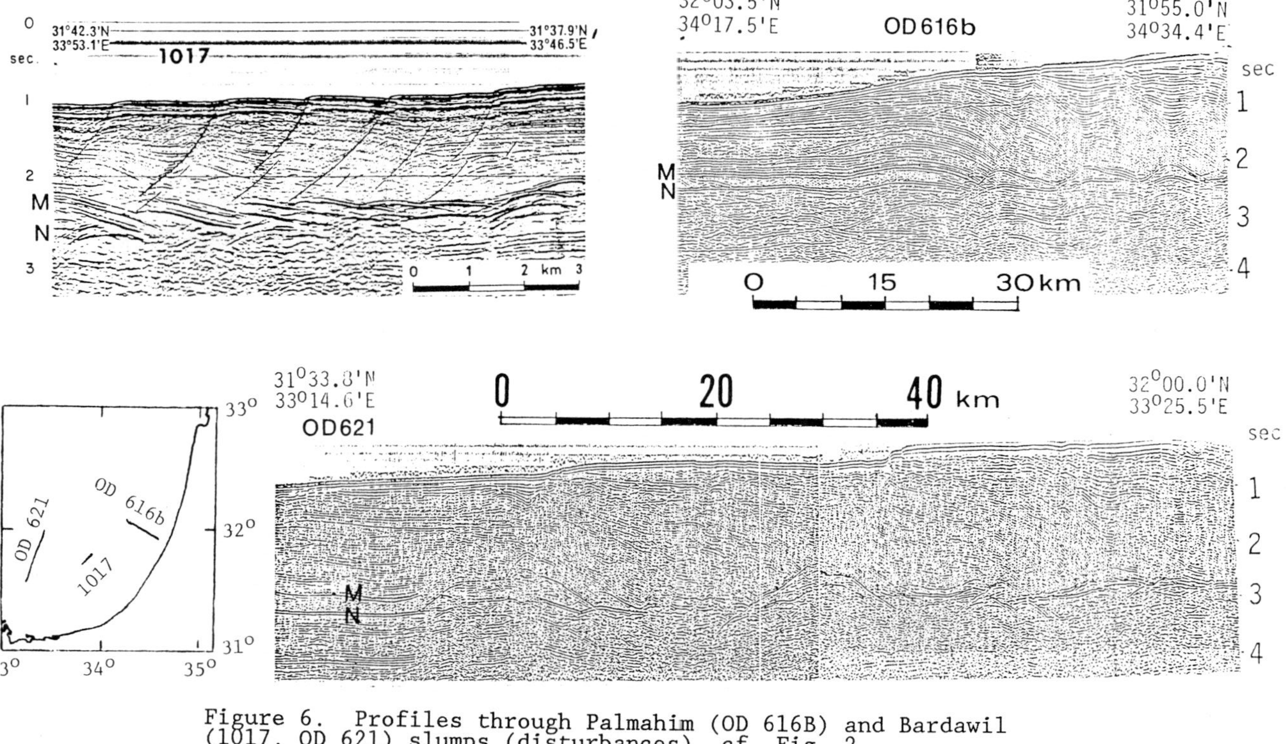

Figure 6. Profiles through Palmahim (OD 616B) and Bardawil (1017, OD 621) slumps (disturbances), cf. Fig. 2.

by the seismic activity of this area (Arieh et al., 1985), but hitherto such faults have not been positively identified.

III. FLOW OF THE MESSINIAN EVAPORITIC SERIES

A. General Features

The Messinian evaporite-bearing series flowed to various extents under much of the Levant Basin and its margins and in the Herodotus Basin, producing several types of structures (Figs. 4, 6-10). This is expressed by the varying degrees of deformation of reflector M and of the reflectors within the Messinian series, as well as by warping of the overlying Pliocene-Quaternary series.

On the seismic reflection profiles reflector N, marking the base of the Messinian series, mimics the undulations of reflector M, but with reduced amplitudes (Figs. 4, 6, 8-10). This is essentially a "pull-up" effect that results from the contrast between the seismic velocity in the Messinian series and the considerably lower seismic velocity in the overlying, incompletely compacted, clastic sediments. In many cases correction for this "pull-up" effect shows that the base of the Messinian series and the underlying beds are undeformed, or that their deformation is small and below the limit of resolution. The uncertainty arises only because the seismic velocities are not known with sufficient accuracy. Thus, the deformation of the Messinian series is largely rootless, i.e., it is a shallow phenomenon that does not directly reflect the structure in the underlying beds. However, though some of the large diapirs originating in the Messinian series seem to be developed over faults in deeper levels, in these cases, too, the shallow deformation is much stronger than, and different from, the deformation of the underlying beds.

The flowage of the Messinian sequence deformed the bedding of the overlying Pliocene-Quaternary series. In the lower part of the latter series the bedding is always parallel to the undulations of reflector M, there are no thickness variations, and onlap relations are never observed. In the upper 1/4-1/2 of the Pliocene-Quaternary series, however, thicknesses are always reduced over the structural highs (Fig. 4). This dates the beginning of deformation to the Late Pliocene or Early Quaternary, i.e., to the last 2-3 Ma. Thus, the Messinian series began to flow only after its overburden reached a certain critical thickness, and continued since then while the upper part of the Pliocene-Quaternary series was being deposited. The thickness variations are apparent even in the youngest sediments and their deformation is usually directly reflected in the physiography of the seafloor. These features show that deformation is still active.

The structures produced by flowage of the Messinian series are of several kinds that differ in size and in amplitude. Though gradational types occur, it is convenient to distinguish between relatively small-amplitude anticlinal undulations of reflector M and larger amplitude diapiric structures (Figs. 4, 7-10). The structures of the first type are found in much of the Levant Basin, whereas the diapirs appear in two groups: the Haifa Diapirs and the Xenophon Diapirs (Fig. 2). In addition, a distinct group of structures formed as a result of flow of the Messinian beds under the continental slope and under its base (Figs. 2 and 6).

B. Deformation under the Continental Slope and its Base

The Messinian series and the overlying beds are deformed under much of the continental slope fringing the Levant Basin and along the base of the slope (Figs. 2, 4 and 6). The structures comprise large-scale rotational slumps and large bulges of the Messinian series along the base of the continental slope (Garfunkel et al., 1979; Almagor, 1980, 1985; Garfunkel, 1984). In the slumps (Figs. 4 and 6), also called disturbances

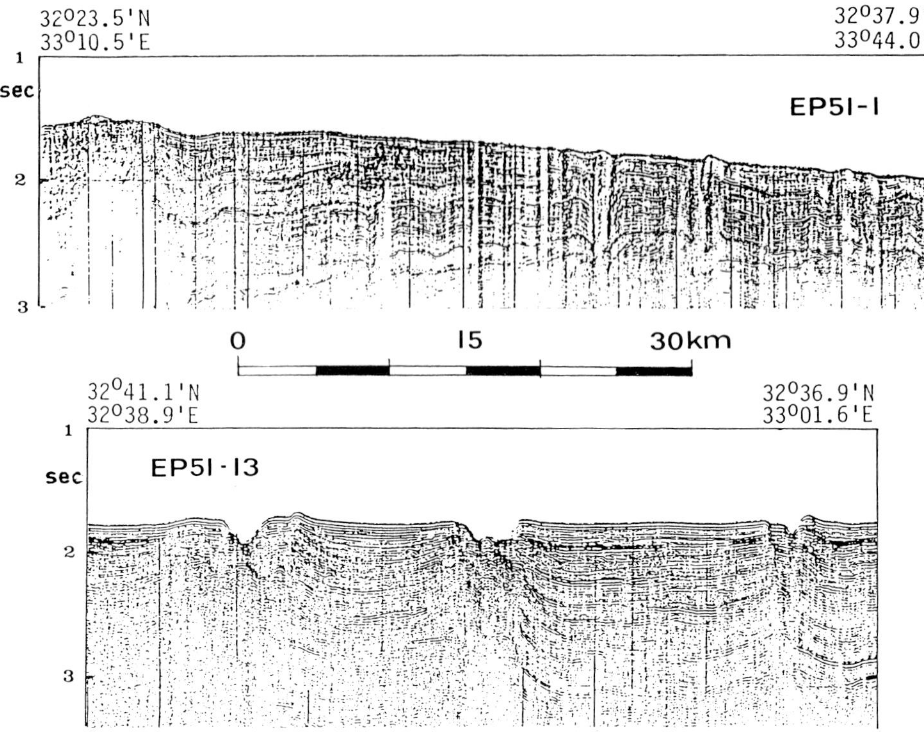

Figure 7. Profiles from southwestern part of Levant Basin, see Fig. 1 for location. EP51 shows small-amplitude anticlinal undulations and a diapir at the extremity of the Xenophon Diapir zone. EP51-13 shows diapirs with crestal grabens and valleys.

(Garfunkel et al., 1979), the Pliocene-Quaternary series glided basinward over the Messinian series and was broken by approximately shore-parallel listric growth faults into tilted blocks. Tilting is generally landward, amounting to 5° or more, but the dips of the strata decrease upward. This proves that tilting proceeded while deposition of the upper part of the Pliocene-Quaternary series was still in progresss. Thinning of recent sediments over the updip edges of the tilted blocks, and the formation of up to several tens of meters high scarps where the growth faults reach the seafloor, prove that slumping is still active.

The Messinian series beneath the slumps is strongly deformed, forming shore-parallel ridges up to a few hundred meters high, under the updip edges of the tilted blocks (Fig. 6). The growth faults flatten downward and are rooted in these ridges, and do not extend to deeper levels. It is these ridges of Messinian beds that absorb the shallow deformation and allow the shallow structure to be detached from deeper levels. Drilling in one slump and in several places along the coast (Derin and Gerry, 1971; Gvirtzman and Buchbinder, 1978) proved that the Messinian series beneath the slumps contains halite that is up to a few hundred meters thick. However, other lithologies - gypsum, anhydrite, dolomite and clastics - are also present. Neither the faulting nor the tilting affects the base of the Messinian series or the underlying beds in any identifiable way. The base of the Messinian series beneath the slumps slopes basinwards at angles of 1°-3°, the greater slopes being found closer to the coast. The structural relations are most obvious in sections perpendicular to the continental slope (Fig. 6), but in shore-parallel sections too the bedding of the post-Messinian beds is often at an angle to reflector M.

The slumps are developed only where the Pliocene-Quaternary series is 1.5-2.0 km thick. On higher parts of the continental slope and on the divides between the erosive valleys, where this sequence is thinner, slumping did not occur. Thickness changes in the sediments which record the beginning of block tilting, and by inference also the beginning of slumping, are discernible only about 0.7-1.0 km above reflector M. This was the critical thickness needed of the post-Messinian overburden to initiate basinward slumping. However, the presence of a thick enough overburden was not a sufficient condition for slumping, because

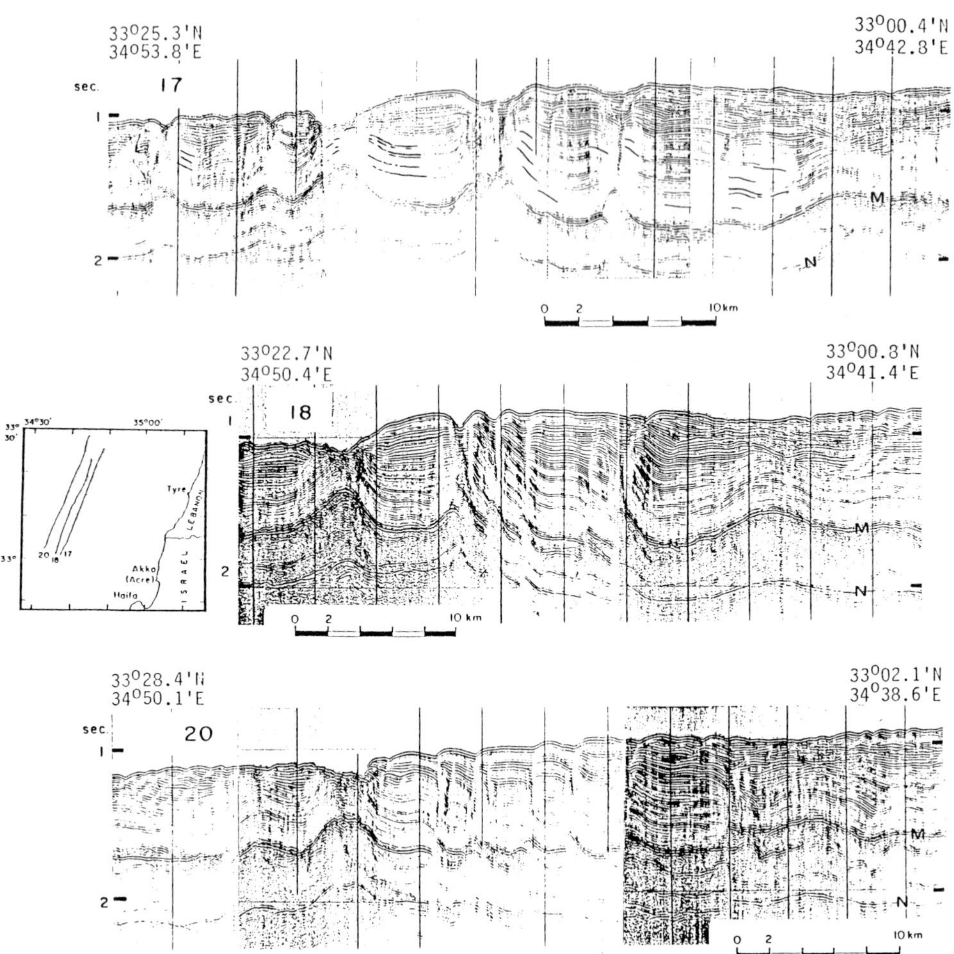

Figure 8. Profiles across the Haifa Diapirs. Note that reflector N extends continuously under some diapirs. The northernmost structure (on right end of profiles) does not deform the youngest sediments.

slumping occurred only where the Messinian series was also thick enough to be able to flow. The critical combination of a mobile substratum and an overburden thick enough to allow gravitational slumping occurs either on the lower part of the continental slope or in the lower parts of the Late Miocene erosive valleys into which thick lobes of the Messinian series extend from the basinal domain.

The westward continuation of the largest slump, the Bardawil Disturbance, is not clear because of lack of data. However, in this region the breakaway of the slump forms an accentuated slope of the seafloor at water depths of 100-500 m. This bathymetric feature extends to west of long. 32° E, where a few seismic reflection profiles (Ross and Uchupi, 1977) reveal a shallow faulted structure similar to that of the Bardawil Disturbance. Therefore this area is interpreted as the westward continuation of the slump (Fig. 2). Farther west, the part of the Nile Cone west of long. 31°30′ E appears to be essentially undeformed on both N-S and roughly E-W trending profiles (lines B, D of Woodside and Williams, 1977; lines 15, 17, 26, 32 of Ross and Uchupi, 1977). It is likely that under this area the Messinian

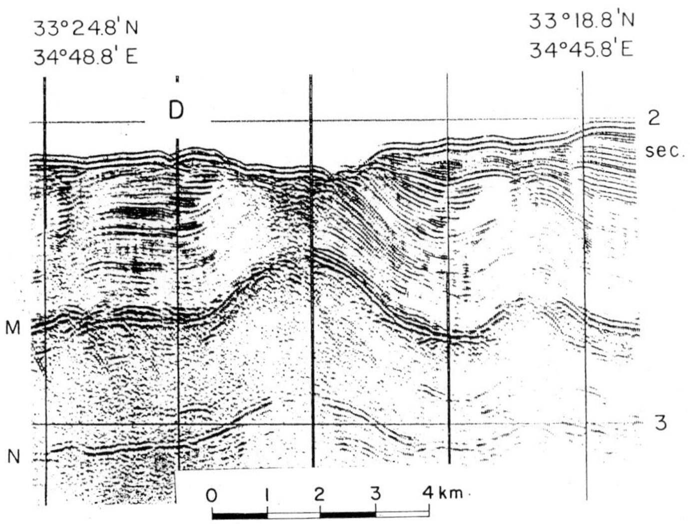

Figure 9. Profile through one of the Haifa Diapirs. For location see Fig. 1. Shows reflector N to be continuous beneath the diapirs.

evaporites did not flow because they are too thin and here the Messinian sequence mostly consists of clastics that were deposited in front of the ancestral Nile gorge.

Related to the slumps are large bulges of the Messinian evaporites that occur along their toes, just off the base of the continental slope. Most conspicuous are the bulges in front of the Dor Disturbance and in the lower part of the Palmahim Disturbance (Figs. 2, 4, 6). In these structures the Messinian series flowed basinwards and broke away from the marginal domain along a strongly faulted zone. The structure of the Pliocene-Quaternary series shows that deformation was contemporaneous with slump activity. These bulges seem to have formed where the Messinian series was pushed in front of the prograding sediments of the continental slope and the basinward moving slumps. Diapiric ridges in front of the western part of the Bardawil Disturbance, the largest of which is more than 50 km long (Fig. 2; Ben Avraham and Mart, 1981), are probably related features. Reflector N is not visibly offset across these diapirs. Since the diapirs are parallel to the growth faults of the slump, it is most likely that they formed as a result of pushing of the Messinian series in front of the slump.

C. Small-Amplitude Anticlinal Undulations

Included under this heading are structures in which reflector M forms undulations with amplitudes that do not exceed a few hundred meters and with wavelengths of 1-3 km (Figs. 4 and 7). The thickness of the Messinian series under the anticlinal undulations is 1-1.5 km, while the deformed Pliocene-Quaternary series is usually less than 1 km thick. The deformation of the Messinian beds up-arched the overlying Pliocene-Quaternary series to form anticlines of varying complexity. These structures occur in great numbers under much of the Levant Basin, but they are most common and more accentuated in a hilly area in its central part, where they deform the seafloor and produce rounded

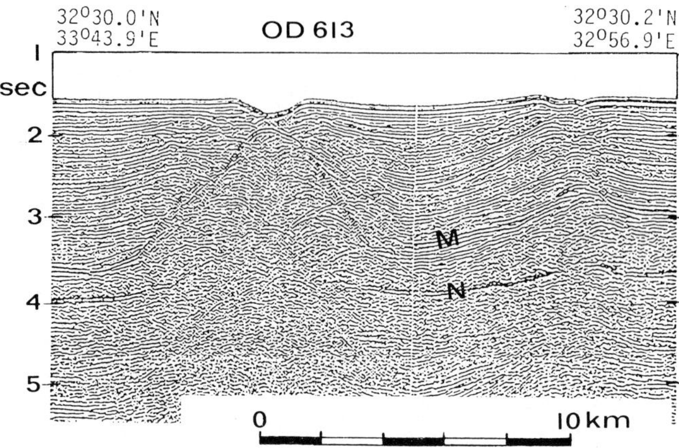

Figure 10. Typical appearance of Xenophon Diapirs.

hills that are up to a few tens of meters high and 1-2 km wide (Figs. 2, 4). Similar structures, but whose growth was slower than sedimentation, are found in places where the seafloor is smooth. In the northern part of the Levant Basin, where the post-Messinian overburden is thinnest, these structures are slightly developed.

While both the Messinian series and the overlying beds are deformed, reflector N is found to be practically undeformed when the pull-up effect is taken into consideration. Moreover, within the uncertainty limits of the seismic velocities the depth of reflector N does not change across these structures but it outlines a smooth surface. The reflectors within the Messinian series also participate in the deformation, but sometimes their shapes differ from reflector M (Fig. 4). Thicknesses tend to decrease towards the crests of the structures only in the higher parts of the Pliocene-Quaternary section and occur even in the youngest sediments. This, and the frequent reflection of the structure in the seafloor physiography, show that deformation is still active.

The intensity of deformation of the anticlinal undulations is quite variable (Figs. 4, 7). In the smaller structures the deformation of reflector M has amplitudes of 50-100 m and the overlying beds are either continuous or only slightly broken. Where deformation was stronger and amplitudes exceed 100-150 m the crests of the structures are faulted and reflector M either appears to be offset or is not continuously recorded. In some cases there are no coherent reflections from the cores of the anticlines, which suggests that here the Pliocene-Quaternary sequence is intruded by material rising from the Messinian series. In plan view these structures appear to be equidimensional, but additional data are needed to verify this inference.

D. Diapirs

The term diapir is applied here to describe structures with larger amplitudes and larger wavelengths than those described in the previous section. Such diapirs occur in two distinct groups in the study area: the Haifa Diapirs off northern Israel and the Xenophon Diapirs at the base of the Nile Cone (Fig. 2). In the diapir cores the Messinian sediments rose much more than under the anticlinal undulations described above, and they often break or pierce through the overlying beds, sometimes almost reaching the seafloor. Some structures in the western Levant basin are gradational between the two groups, but owing to their geographic separation a clear distinction can be made between the diapirs and the small-amplitude undulations.

1. The Haifa Diapir Group

This group comprises linear diapirs that are located close to the base of the continental slope of northern Israel (Figs. 2, 8, 9; Garfunkel and Almagor, 1985). The area of diapir development is a particularly deformed zone, whereas to the west and south similar structures are not found and the Messinian and overlying beds are conspicuously less deformed.

In the diapirs the top of the Messinian series (reflector M) forms E-W to WNW-ESE trending ridges, 200-400 m high, 1-5 km wide, up to 20 km long, and 2-7 km apart. Reflector M is continuous on the crests of some ridges, while in others, especially the narrower ones, it is faulted (Fig. 8). Over the wider ridges the Pliocene-Quaternary sediments are usually up-arched and form symmetrical anticlines whose crests are broken into complex grabens. Over the narrower ridges the post-Messinian sediments form flexures which are crossed by conspicuous faults that extend from discontinuities of reflector M (Fig. 8). Along the faults usually there are no coherent reflections, which suggests that material rising from the Messinian series intruded the Pliocene-Quaternary series along these faults. Deformation usually extends to the seafloor where the crestal grabens are reflected in the bathymetry as valleys up to a hundred meters deep. Over the wider diapiric ridges the valleys are up to 3 km wide, whereas over the narrower ones they form about 0.5 km wide furrows. Dissolution of the rising salt could have produced the valleys. However, in many structures the salt may be about 0.5 km below the seafloor, which suggests that other processes, such as submarine erosion of highly fractured sediments at the crests of diapirs, could have also contributed to the formation of the valleys.

In the Haifa Diapir zone the Messinian series is 0.7-1.0 km thick, but east of the diapir zone the thickness of this series decreases rapidly on approaching the Messinian marginal domain. Between the diapirs the Pliocene-Quaternary series is 0.8-1.3 km thick and thickness changes and unconformities associated with the diapirs usually occur throughout the upper half of this series. However, over the crest of the northernmost diapirs the thickness of the very shallow sediments does not vary, which shows that this diapir is no longer active (Fig. 8).

The Haifa Diapirs appear to be structurally controlled. This is indicated by their position along the submarine extension of the active Carmel Fault and the Galilee faults, and by their trends which are parallel to the fault trend. On the other hand, off the unfaulted continental margin farther south similar diapirs are absent, even though the Messinian series and its overburden have the same thicknesses. There only shore-parallel structures related to slumps and to the base of the continental slope are present. Moreover, the depth of reflector N changes by up to 200 m across several diapirs, which indicates that they are developed over faults offsetting the underlying beds. These faults probably triggered the development of these diapirs and caused them to be linear rather than equidimensional in plan view. Under other diapirs, however, reflector N is clearly continuous, showing that in these cases deformation was confined to the Messinian and overlying beds (Fig. 9). Basinward the diapirs die out towards a structural high east of the active Sidon Fault (Fig. 2). This prominent fault is not associated with any flowage of the Messinian series, which indicates that faulting does not necessarily trigger significant diapirism.

2. The Xenophon Diapir Group

The most extensive occurrence of diapirs in the eastern Mediterranean is in the region at the base of the Nile cone (Fig. 2), named the Xenophon Basin by Ryan (1978). The diapirs occur in a belt that extends from the western part of the Levant Basin into the deeper parts of the Herodotus Basin. These diapirs were mentioned by many authors (Finetti and Morelli, 1973; Kenyon et al., 1975; Neev et al., 1976; Ross and Uchupi, 1977; Woodside and

Williams, 1977; Woodside, 1977). Here we discuss only the eastern part of this belt, which is about 100 km wide and trends NW-SE.

In the western Levant Basin the base of the Messinian series slopes westward, descending about 1 km towards the diapir zone (Fig. 3) and the thickness of this series and its overburden increase in that direction to 1.5-2.0 km and 2-3 km, respectively. In the diapir zone itself reflector N is not very clear on the available profiles. According to Montadert et al. (1978) and Ryan (1978), however, the base of the evaporitic sequence deepens and its thickness increases northwestward along the Xenophon Diapir zone. In fact, this zone is developed over a lobe of thick evaporites which extends from the Herodotus Basin.

In the diapirs the Messinian series rose 0.5-2.0 km and even more, up-arching the overlying beds and often breaking them. In the smaller amplitude structures, which are developed in the southeastern corner of the diapir zone, the post-Messinian sediments are up-arched and form unbroken undulations above the rising evaporitic masses which form simple pillow-like bodies. In structures with amplitudes exceeding 1 km the rising masses pierced and broke the Pliocene-Quaternary beds and produced complex grabens over the crests of the structures. These grabens are expressed on the seafloor by valleys, a few kilometers wide and up to 250 m deep, whose flanks are often up-arched and elevated above the nearby seafloor (Figs. 7 and 10). Very young faulting in the crestal grabens and dissolution of salt were probably the main processes that formed the valleys, although erosion of highly fractured sediments may have also occurred. The spacing between the diapirs is 8-12 km, occasionally 15 km, and between them the post-Messinian beds were usually deformed into broad synclines. At the level of reflector M the diapirs are at least 5-7 km wide. Thus, even in the larger structures with summital grabens, the rising masses form protrusions whose sides have moderate slopes, often less than 45°. The stage of formation of diapiric columns or walls with vertical sides, such as those found in the North German Basin, around the Gulf of Mexico and elsewhere (Trusheim, 1960; Lehner, 1969; Garrison and Martin, 1973; Humphries, 1979), was not reached in the Xenophon Diapir Zone.

Important information about the shape of the diapirs in plan view is provided by a GLORIA (long range side-scan sonar) survey in the Xenophon Diapir zone which revealed a system of linear grabens, a few tens of kilometers long, that mark diapir crests (Fig. 2; Kenyon et al., 1975). Indeed, wherever crossed by seismic reflection profiles the grabens prove to be at the summits of piercing diapirs. The dominant diapir trends are NW and especially NNE to NE, both trends being at an angle to the overall elongation of the diapir belt, and both directions are at an angle to the overall slope of the seafloor. Where GLORIA data are not available, the shapes of diapirs crossed by seismic reflection lines are not known (these were shown as "d"s on Fig. 2), but it is likely that at least some of them are also linear.

The area of diapir development described here is sharply bounded. On the northeast is the high-standing and faulted area of the Eratosthenes Seamount, whereas on the southwest there is an abrupt transition to the slightly deformed portion of the Nile Cone (Fig. 2). Since the base of the Messinian series is relatively deep in the Xenophon Diapir zone, this zone can be interpreted as a structurally controlled depression in which the evaporitic series is particularly thick. It is noteworthy that young faulting is very conspicuous in the nearby Eratosthenes Seamount and that the diapir zone is approximately in line with the Suez rift, although a direct link between them has not been established to date. This raises the possibility that the linear diapir trends are fault controlled. Kenyon et al. (1975) interpreted the NW trending diapirs as being controlled by faults having the Suez trend. However, the dominant diapir trends, NE and NNE, are quite different; if they are fault controlled, a local structural pattern is suggested. However, non-tectonic processes could have also influenced the distribution of the thick evaporite series. Thus, the evaporites may have filled a depression that formed between the Eratosthenes high and the thick prograding pile of clastics, deposited in front of the ancestral Late Miocene Nile gorge (more than 1 km thick near the coast). More data is needed to establish the exact role of faulting in the formation of the Xenophon Diapirs.

IV. DISCUSSION: CONDITIONS OF FLOWAGE OF THE MESSINIAN SERIES

According to the accepted notions about diapirism, this process is a result from the combination of two conditions: (a) The low density salt is overlain by denser sediments, which produces buoyancy forces, (b) The salt is ductile enough to flow at geologically significant rates in response to the buoyancy forces. It is only the critical combination of both these conditions, not one of them alone, which causes salt (and shale) diapirism, allowing original irregularities of the upper surface of the low density layer to grow significantly (Nettleton, 1934; Parker and McDowell, 1955; Biot and Ode, 1965; Selig, 1965; O'Brien, 1968; Whitehead and Luther, 1975). We now examine these two factors in the studied part of southeastern Mediterranean within the general geologic framework of the diapirism and evaporite flowage outlined above.

In the southeastern Mediterranean the Pliocene-Quaternary overburden of the evaporitic series consists of fine clastics and is thus expected to become compacted upon burial. The published compaction curves (reviewed by Baldwin and Butler, 1985) show that shales will become denser than 2.16 g/cm^3, which is the density of pure halite, at depths of 0.4-0.8 km, while sandstones will compact to this density only about 1 km deeper. The compaction of the young fine clastic sediments of the eastern Mediterranean is better described by the shale curves than by the sandstone curves. Thus, the data on the thickness of the Pliocene-Quaternary series that existed at the onset of diapiric deformation in the southeastern Mediterranean shows that flowage of the Messinian series began when a density inversion relative to halite was just established, or a short time after that. When the burial depth increased to 1 km the density inversion can reach 0.05-0.1 g/cm^3 and twice as much when the overburden reached 2 km. Thus, the condition of density inversion is satisfied in the eastern Mediterranean, though the density contrast was quite small when diapiric flowage began.

The ductility of the Messinian series in the studied area can be assessed on the basis of experimental work done on the creep of halite (Heard, 1972; Carter and Hansen, 1983). These studies showed that at temperatures of 100 °C, and probably less, the creep of halite is governed by a power law and thus its strain rate depends critically on both the stress difference and on the temperature. Since the heat flow in the eastern Mediterranean is quite low (Cermak and Hurtig, 1979), the normal thermal gradient of 30°C/km is likely an overestimate of the gradient in the Pliocene-Quaternary series. In the Messinian series the thermal gradient will be considerably lower because of the high thermal conductivity of halite (Clark, 1966). Thus, where a 2 km thick evaporitic series has a 2 km thick overburden, which are maximum figures for the studied area, the temperatures at the base of the evaporites will hardly reach 100-200°C. This upper bound on the temperature shows that in the eastern Mediterranean flowage of the Messinian series occurred mostly at lower temperatures, probably as low as 50-70°C. To achieve geologically significant strain rates of 10^{-14}-10^{-16}/sec at temperatures of 60°C-100°C, stress differences of 7.8-2.8 bars and 4.4-1.6 bars, respectively, are needed according to Carter and Hansen's (1983) results on natural halite. At 150°C the required stress differences are 2.5-0.9 bars. For deformation of synthetic halite aggregates (Heard, 1972) the necessary stresses are more than twice as large. Such stress magnitudes are indeed recorded by the textures of salt in well developed diapirs (Carter et al., 1982). In the eastern Mediterranean, however, it is not certain that such stresses existed when the Messinian series began to flow.

Thus, to produce a buoyancy pressure difference of 1 bar, the undulations of the halite-overburden contact must be 100-200 m high when the density contrast is 0.05-0.1 g/cm^3. This means that until the swells at the top of the Messinian series grew to amplitudes of several tens of meters at least, buoyancy could not have provided stress differences that are large enough to cause halite to flow at geologically significant rates. In other words, according to the available experimental data halite is too strong to allow diapirism to begin spontaneously in the conditions prevailing in the eastern Mediterranean. On the other hand, the observation that diapirism began only when a thick enough overburden accumulated indicates that buoyancy forces were important. This is also indicated by the fact that the deformation of the Messinian series is much stronger than that of the underlying beds. Therefore, it must be concluded that either the Messinian series as a whole was much weaker than the halite samples that were experimentally investigated, or that another process besides buoyancy initiated the flowage of the Messinian series.

Several causes for weakening the Messinian series can be proposed. One possibility is the presence of beds of fine clastics interbedded with the evaporites. This is suggested by the presence of seismic reflectors in the Messinian series and is also compatible with the occasional low seismic interval velocities. Such sediments are expected to be deposited in the basinal domain at times of low stands of the Mediterranean sea level, which are known to have occurred during the Messinian (Hsu et al., 1973; 1978). The enclosing impervious evaporites can hinder the dissipation of pore pressure in such interbeds, so that abnormal pressures could have built up in them as the Pliocene-Quaternary overburden accumulated. Such overpressured interbeds are intrinsically weak, and may be mobilized by small stress differences and also by seismic shaking, which occurs repeatedly in the study area. Another possible cause that could weaken the Messinian series is the presence of interstitial water that remained trapped in the halite and in other interbedded sediments. Such water could greatly facilitate the creep of halite in the early stages of deformation. As deformation proceeds and the halite recrystallizes, the trapped water may be expelled, but if this happens when the deformation of the top of the flowing series reaches sufficiently large amplitudes, the buoyancy forces at that time can be large enough to sustain further creep at significant rates.

If these were indeed the factors weakening the Messinian series compared with the experimentally studied rocks, then diapirism would have begun spontaneously once the density of the overburden was high enough if the top of the Messinian series was irregular. This certainly must have been the case because, as was revealed by deep-sea drilling (Hsu et al., 1973, 1978), the top of this series was exposed and eroded to some extent during the terminal Miocene desiccation event. Once spontaneous diapirism driven by buoyancy begins, swells with a characteristic wavelength will form (Biot and Ode, 1965; Selig, 1965; Whitehead and Luther, 1975). Indeed the small-amplitude undulations of the Levant Basin and the diapirs tend to have a regular spacing, and the spacing is largest in the Xenophon Diapir zone where the Messinian series is thickest, as expected theoretically.

Moreover, neighboring structures are expected to be of rather similar sizes and amplitudes because they grew under similar conditions, which is indeed the case in the area described in this work.

The low ductility of halite can be overcome also if large enough perturbations of the top of the Messinian series are created by an external agent, such as faulting (Parker and McDowell, 1955; Jenyon, 1985). By itself faulting need not cause flowage of the evaporitic series and growth of diapirs unless the proper conditions for that to happen exist already. In that case, if faulting offsets the top of the evaporitic series by a few tens of meters the buoyant forces will be large enough to cause the halite to flow at geologically significant rates.

As noted above, faulting probably triggered some of the Haifa Diapirs and may have controlled their shapes. The absence of similar diapirs in other parts of the Levant Basin where the Messinian series and its overburden have similar thicknesses also shows that the Haifa Diapirs were triggered by a unique local mechanism. However, even in this diapir group, reflector N is clearly continuous below some structures (Figs. 8 and 9). This shows that the growth of fault-triggered diapirs could enhance the flowage of the Messinian series and cause the formation of additional diapirs in nearby areas, probably by creating large enough buoyant stresses.

The influence of faulting on the Xenophon Diapirs is less clear. However, since it is difficult to see any relation between the trends of these linear diapirs and the shape of the underlying evaporite body or of the Nile Cone (cf. Fig. 2), and since the existence of active faults is likely in this area, as shown above, it is very likely that faulting did influence the formation of these diapirs. Faults could trigger the growth of some diapirs and also could have caused the linear shapes of the diapirs. However, the data available to us do not enable identification of the location or trend of the faults. The fact that the diapir trends are not easily related to a simple tectonic pattern and are at an angle to the overall trend of the diapir zone also makes tectonic interpretation difficult. It is also possible that only a few diapirs developed above faults and they triggered the formation of the other diapirs, in analogy with the Haifa diapirs.

In contrast to the Haifa Diapirs, important faulting does not appear to exist beneath the small-amplitude anticlinal undulations of the Levant Basin. As noted, in most cases the base of the Messinian series does not appear to be offset beneath these structures, although hitherto unidentified faults could have triggered the growth of some structures. On the other hand, the tendency of the structures to have a characteristic wavelength and especially their rather uniform size suggest that their growth was dominated by buoyancy forces rather than by faults, which are expected to have varying offsets and not to be evenly spaced.

In the Mediterranean another factor that could have enhanced diapirism should be considered: deep-sea drilling showed that gypsum and anhydrite (densities of 2.32 g/cm^3 and 2.96 g/cm^3, respectively, Clark, 1966) occur in the upper part of the Messinian series. If these components are abundant, they could have provided a density inversion already at the end of the Messinian. Moreover, upon burial the original gypsum will tend to pass into the high density anhydrite, which will greatly increase the density contrast. However, to enhance salt flowage the gypsum- and anhydrite-bearing layer must be deformed so that its thickness changes, or the salt must penetrate the gypsum-anhydrite layer. Thus the influence of such components on the initial stages of deformation is doubtful.

In the continental slope gravitational instability, rather than a buoyancy controlled instability, was the important factor that caused flowage of the Messinian series. Here this series acted as a decollement layer and a mobile substrate beneath the basinward-gliding Pliocene-Quaternary sediment pile. This role shows that the Messinian series was overall much weaker and more ductile than its overburden. However, since this overburden consists of unconsolidated silty clays, it is difficult to understand how it could be more ductile than halite (or gypsum and anhydrite) which are at temperatures of 60°C at most. It was already suggested (Garfunkel et al., 1979; Garfunkel, 1984) that beds of fine clastics (which were actually drilled under the continental shelf) are interbedded with the Messinian evaporites and that they are overpressured. Such beds could have controlled the location of the decollement and could also have facilitated the deformation of the Messinian series. Slumping down the continental slope and the pressure in front of the slump are the likely causes of the flowage at the base of the slopes of Israel and northern Sinai. The structures in these areas are characteristically parallel to the continental slope. As flowage of the evaporite sequence in these structures was significant, it must have also affected the nearby basinal area, but it is difficult to assess the role of this process. In contrast to these structures the Haifa Diapirs and the Xenophon Diapirs are not related to the regional slope of the seafloor, which indicates that their formation was not directly influenced by the adjacent continental slope and Nile Cone, respectively.

V. CONCLUSIONS

The foregoing description shows that the Messinian series of the eastern Mediterranean, and of the Levant Basin in particular, flowed to varying extents, forming a variety of structures. The deformation is essentially a shallow feature that was usually independent of any deformation at deeper levels; even where such deformation occurred in post-Messinian times, the shallow deformation was much stronger and of a different type. Flowage of the Messinian series began a short time, not more than 2-3 Ma, after the end of its deposition about 5 Ma ago. At that time the thickness of the overburden varied from slightly more than 0.5 km to about 1.0-1.3 km. By now well-developed salt pillows and diapirs have grown where the overburden is 0.8-2.0 km thick, the larger structures having formed where the overburden is thickest. This, and the tendency of neighboring structures to be of similar sizes and to be evenly spaced, are compatible with the interpretation that buoyancy forces caused the flowage of the Messinian series. The fact that deformation began only when a sufficient density contrast was established also indicates the importance of buoyancy forces. However, for diapirism to begin spontaneously, the evaporitic sequence in the eastern Mediterranean must have been considerably more ductile than experimentally studied halite. In addition, faulting was probably an important factor in triggering the larger linear diapir groups.

These features characterize, in fact, all the diapir occurrences in the Mediterranean basin, and are not unique to the area studied. Compared with other diapir provinces such as those of northern Germany or the Gulf of Mexico (Trusheim, 1960; Lehner, 1969; Garrison and Martin, 1973; Humphries, 1979) several differences can be noted. In the Mediterranean Sea diapirs occur in many relatively small areas which reflect the complex structure of this tectonically active area, whereas in other regions diapirs occur in larger areas with simpler structures, and there is no evidence for tectonic activity at the time of diapir growth. In addition, in other regions the overburden is thicker, diapirism was active for much longer periods and has reached more advanced stages than in the Mediterranean Sea. Also in places, e.g., the Gulf of Mexico and the Niger Delta (Evamy et al., 1978) basinward motion of blocks delimited by growth faults is much more important, and must have exerted a greater influence on diapirism than in the Mediterranean. On the other hand there is much similarity in size and shape of the diapiric structures, so it is likely that the Mediterranean can serve as a good model for youthful diapirism. Therefore, the conclusions reached above regarding the conditions of initiation of diapirism are probably of general applicability. In particular, it appears that evaporitic series may generally be much weaker than laboratory work suggests. Otherwise, in the absence of faulting that could trigger diapirism, which appears to be the rule in most places, diapirism can begin only under a larger density contrast (i.e. thicker overburden) and when the heat flow is higher than in the eastern Mediterranean.

VI. ACKNOWLEDGEMENTS

We are very grateful to the captain and crew of R/V Shiqmona of the Israel Oceanographic and Limnologic Research, Ltd., for their help in obtaining seismic profiles. We are indebted to Z. Ben Avraham and to D. Ross for letting us use the profiles which they obtained, and to L. Montadert and J. Fisher for permission to use multi-channel profiles. D. Ross gave us access to multi-channel profiles. This research was in part done within the framework of the Geological Survey of Israel, project 22000 - seismic profiling. Last but not least we want to thank Mrs. B. Katz for help with editing the English manuscript.

REFERENCES

Alla, G. (1970). Etude sismique de la plaine abyssale au sud de Toulon, Revue Institut Francais Petrole 25, 291-304.

Almagor, G. (1980). Halokinetic deep-seated slumping on the Mediterranean slope of northern Sinai and southern Israel, Mar. Geotechnol. 4, 83-105.

Almagor, G. (1984). Salt tectonics on the Mediterranean slope of central Israel, Mar. Geophys. Res. 6, 227-243.

Almagor, G. and Garfunkel, Z. (1979). Submarine slumping in the continental margin of Israel and northern Sinai, Amer. Assoc. Petrol. Geol. Bull. 63, 324-340.

Arieh, E. Artzi, D., Benedik, N., Eckstein (Shapira), A., Issakow, R., Reich, B., and Shapira, A. (1985). Revised and updated catalog of earthquakes in Israel and adjacent area, 1900-1980. Report of Inst. for Petroleum Research and Geophysics, Holon, Israel, 38 p.

Baldwin, B., and Butler, C.O. (1985). Compaction curves, Bull. Amer. Assoc. Petrol. Geol. 69, 622-626.

Barber, P.M. (1981). Messinian subaerial erosion of the Proto-Nile Delta, Mar Geol. 44, 253-272.

Bein, A. and Gvirtzman, G. (1977). A Mesozoic fossil edge of the Arabian plate along the Levant coastline and its bearing on the evolution of the Eastern Mediterranean. In "Structural History of the Mediterranean Basins" (B. Biju-Duval and L. Montadert, eds.), Editions Technip, Paris, pp. 95-109.

Ben Avraham, Z., and Mart, Y. (1981). Late Tertiary structure and stratigraphy of northern Sinai continental margin, Amer. Assoc. Petrol. Geol. Bull. 65, 1135-1145.

Biju-Duval, B., and Dercourt, J. (1980). Les bassins de la Mediterranee orientale representent-ils les restes d'un domain oceanique, La Mesogee, ouvert au Mesozoique et distinct de la Tethys? Bull. Soc. Geol. France (7), 22, 43-60.

Biju-Duval, B. Letouzey, J., Montadert, L., Courrier, P., Mugniot, J.F., and Sancho, J., (1974). Geology of the Mediterranean Basin, In "The Geology of Continental Margins" (C.E. Burk and C. Drake, eds.), Springer Verlag, p. 695-712.

Biot, M.A., and Ode, H. (1965). Theory of gravity instability with variable overburden and compaction, Geophysics 30, 213-227.

Carter, N.L., and Hansen, F.D. (1983). Creep of rocksalt, Tectonophysics 92, 275-333.

Carter, N.L., Hansen, F.D. and Senseny, P.E. (1982). Stress magnitudes in natural rocksalt, J. Geophys. Res. 87, 9289-9300.

Cermak, V., and Hurtig, E. (1979). Heat flow map of Europe. In "Terrestrial Heat Flow in Europe" (V. Cermak and L. Rybach, eds.) Springer-Verlag.

Clark, S.P. (ed.) (1966). "Handbook of physical constants, revised edition", Geol. Soc. Amer. Memoir 97, 587 p.

Derin, B. (1974). The Jurassic of central and northern Israel. Ph.D. Thesis Hebrew Univ., Jerusalem, Israel, 152 p. (in Hebrew, English abstract).

Derin, B., and Gerry, E. (1971). Micropaleontological report, Joshua no. 2. Isr. Inst. Petrol., Micropaleontological Lab., Report 9/71, 18 p.

Dewey, J.F., Pitman, III, W.C., Ryan, W.B.F. and Bonin, J., (1973). Plate tectonics and the evolution of the Alpine System, Geol. Soc. Amer. Bull. 84, 3137-3180.

Evamy, B.D, Haremboure, J., Kamerling, P., Knaap, W.A., Malloy, F.A. and Rowlands, P.H. (1978). Hydrocarbon habitat of Tertiary Niger Delta, Amer. Assoc. Petroleum Geol. Bull 62, 1-39.

Finetti, I., and Morelli, C. (1973). Geophysical exploration of the Mediterranean Sea, Boll. Geofiz. Teor. Applic. 15(60), 263-344.

Gardner, G.H.F., Gardner, L.W., and Gregory, A.R. (1974). Formation velocity and density - the diagnostic basics of stratigraphic traps, Geophysics 39, 770-780.

Garfunkel, Z. (1984). Large-scale submarine rotational slumps and growth faults in the Eastern Mediterranean, Mar. Geol. 55, 305-324.

Garfunkel, Z. and Almagor, G. (1985). Geology and structure of the continental margin off northern Israel and the adjacent part of the Levantine Basin, Mar. Geol. 62, 105-131.

Garfunkel, Z., Arad, A., and Almagor, G. (1979). The Palmahim Disturbance and its regional setting. Bull. Geol. Surv.Israel 72, 56 p.
Garfunkel, Z. and Bartov, Y. (1977). Tectonics of the Suez Rift. Bull. Geol. Surv. Israel 71, 44 p.
Garfunkel, Z., and Derin, B., (1985). Permian-Early Mesozoic tectonism and continental margin formation in Israel and its implications for the history of the eastern Mediterranean, Spec. Publ. Geol. Soc. London 17, 187-201.
Garrison, L.E. and Martin, R.G., Jr. (1973). Geologic structures in the Gulf of Mexico Basin, U.S. Geol. Surv. Prof. Paper 773, 85 p.
Ginzburg, A., Cohen, S.S., Hay-Roe, H., and Rosenzweig, A. (1975). Geology of the Mediterranean shelf of Israel, Amer. Assoc. Petrol. Geol. Bull. 59, 2142-2160.
Ginzburg, A., and Folkman, Y., (1980). The crustal structure between the Dead Sea Rift and the Mediterranean Sea, Earth Planet. Sci. Lett 51, 181-188.
Ginzburg, A., and Gvirtzman, G. (1979). Changes in the crust and in the sedimentary cover across the transition from the Arabian platform to the Mediterranean basin: evidence from seismic refraction and sedimentary studies in Israel and in Sinai, Sediment. Geol 23, 19-36.
Gvirtzman, G., and Buchbinder, B. (1978). The Late Tertiary of the coastal plain and continental shelf of Israel and its bearing on the history of the eastern Mediterranean, Initial Reports DSDP 52(2), 1195-222. U.S. Gov't Printing Office, Washington, D.C.
Hall, J.K. (1981). The MEDMAP project: a systematic analysis of Eastern Mediterranean geophysical data, Isr. Geol. Surv., Current Research during 1980, pp. 89-94.
Heard, H.C. (1972). Stead-state flow in polycrystalline halite at pressure of 2 kilobars. In "Flow and fracture of rocks" (Heard, H.C. Borg, I.Y., Carter, N.L. and Raleigh, C.B., eds.), Amer. Geophys. Union Monogr. 16, 191-210.
Hsu, K.J., Cita, M.B., and Ryan, W.B.F. (1973). The origin of the Mediterranean evaporites, Initial Reports DSDP 13(2), 1203-1231. U.S. Gov't. Printing Office, Washington, D.C.
Hsu, K.J., Montadert, L., Bernoulli, D., Cita, M.B., Erickson, A., Garrison, R.E., Kidd, R.B., Meliers, F., Muller, C., and Wright, R. (1978). History of the Mediterranean salinity crisis, Initial Reports DSDP 42(1), 1053-1078. U.S. Gov't. Printing Office, Washington, D.C.
Humphries, C.C., Jr. (1979). Salt movement on continental slope, northern Gulf of Mexico, Amer. Assoc. Petrol. Geol. Bull. 63, 782-798.
Jenyon, M.K. (1985). Fault-associated salt flow and mass movement, J. Geol. Soc. London 142, 547-553.
Kenyon, N.H., Stride, A.H. and Belderson, R.H. (1975). Plan views of active faults and other features on the lower Nile Cone, Geol. Soc. Amer. Bull. 86, 1733-1739.
Lehner, P., (1969). Salt tectonics and Pelistocene stratigraphy on continental slope of northern Gulf of Mexico, Amer. Assoc. Petrol. Geol. Bull. 53, 2431-2476.
Lort, J.M., Limond, W.Q., and Gray, F. (1974). Preliminary seismic studies in the eastern Mediterranean, Earth Planet. Sci. Lett. 21, 355-366.
Makris, J., Ben Avraham, Z., Behle, A., Ginzburg, A., Giese, P., Steinmetz, L., Whitmarsh, R.B. and Eleftheriou (1983). Seismic refraction profiles between Cyprus and Israel and their interpretation, Geophys. J. R. Astr. Soc. 85, 575-591.
Mart, Y., and Ben-Gai, Y. (1982). Some depositional patterns at continental margin of southeastern Mediterranean Sea, Amer. Assoc. Petrol. Geol. Bull. 66, 460-470.

Montadert, L., Letouzey, J., and Mauffert, A. (1978). Messinian Event: Seismic evidence, Initial Reports DSDP 42(1), 1032-1050. U.S. Gov't. Printing Office, Washington, D.C.

Nachnias, J. (1969). Source rocks of the Saqiya Group sediments in the coastal plain of Israel - a heavy mineral study, Isr. J. Earth Sci. 18, 1-16.

Neev, D., Almagor, G., Arad, A., Ginzburg, A., and Hall, J.K. (1976). The geology of the southeastern Mediterranean, Bull. Geol. Surv. Israel 68, 51 p.

Nettleton, L.L. (1934). Fluid mechanics of salt domes, Amer. Assoc. Petrol. Geol. Bull. 18, 1175-1204.

Nir, Y., and Nathan, Y. (1972). Clay mineral assemblages in recent sediments of the Levantine Basin, Mediterranean Sea, Bull. Gr. Franc. Argiles 24, 187-195.

O'Brien, G.D. (1968). Survey of diapirs and diapirism. In "Diapirs and diapirism" (Braunstein, J., and O'Brien, G.D., eds.), Amer. Assoc. Petrol. Geol. Mem 8, 1-9.

Parker, T.J. and McDowell, A.N. (1955). Model studies of salt-dome tectonics, Amer. Assoc. Petrol. Geol. Bull 39, 2384-2470.

Rizzini, A., Vezzani, F., Cococcetta, V., and Milad, G. (1978). Stratigraphy and sedimentation of the Neogene-Quaternary section in the Nile Delta area (A.R.E.) Mar. Geol. 27, 327-348.

Ross, D.A. and Uchupi, E. (1977). Structure and sedimentary history of southeastern Mediterranean Sea - Nile Cone area, Amer. Assoc. Petrol. Geol. Bull. 61, 872-902.

Ryan, W.B.F. (1978). Messinian badlands on the southeastern margin of the Mediterranean Sea, Mar. Geol 27, 349-363.

Ryan, W.B.B., Cita, M.B., Hsu, K.J. and others (1973). Initial Reports DSDP, 13(1), 514 p. U.S. Gov't. Printing Office, Washington, D.C.

Ryan, W.B.F., Stanley, D.S., Hersey, J.B., Fahlquist, D.A. and Allan, T.D. (1970). The tectonics and geology of the Mediterranean Sea. In "The Sea" (A. Maxwell, ed.), Said, R. (1976). Evolution of Eocene-Miocene sedimentation patterns in parts of northern Egypt, Amer. Assoc. Petrol. Geol. Bull. 60, 34-64.

Schreiber, E., Fox, P.J., and Peterson, J.J. (1973). Compressional wave velocities in selected samples of gabbro, schist, limestone, anhydrite, gypsum and halite, Initial Reports DSDP 13, 595-598. U.S. Gov't. Printing Office, Washington, D.C.

Selig, F. (1965). Theoretical prediction of salt dome patterns. Geophysics 30, 633-643.

Smith, A.G. (1970). Alpine deformation and the oceanic areas of the Tethys, Mediterranean and Atlantic, Geol. Soc. Amer. Bull. 82, 2039-2070.

Smith, A.G., and Woodcock, N.H. (1982). Tectonic syntheses of the Alpine-Mediterranean: a review, Amer. Geophys. Union Geodynamics Series 7, 15-38.

Trusheim, F. (1960). Mechanism of salt migration in northern Germany, Amer. Assoc. Petrol. Geol. Bull. 44, 1519-1540.

Venkatanathnam, K. and Ryan, W.B.F. (1971). Dispersal patterns of clay minerals in the sediments of the eastern Mediterranean, Mar. Geol. 11, 261-282.

Whitehead, J.A., Jr. and Luther, D.S. (1975). Dynamics of laboratory diapir and plume models, Jour. Geophys. Res. 80, 705-717.

Woodside, J.M. (1977). Tectonic elements and crust of the eastern Mediterranean, Mar. Geophys. Res. 3, 317-354.

Woodside, J.M. and Williams, S.A. (1977). Geophysical Data Report of the Eastern Mediterranean Sea: RRS Shackleton Cruises 3/72, 5/72, 1/74. Dept. of Geodesy and Geophysics, Cambridge Univ., UK, 225 p.

THIN-SKINNED DEFORMATION OVER SALT

Dan M. Davis(1)
Terry Engelder(2)

[1]Lamont-Doherty Geological Observatory
of Columbia University
Palisades, NY 10964

[2]Department of Geosciences
Pennsylvania State University
University Park, PA 16802

I. INTRODUCTION - SALT STRUCTURES AND FOLD BELTS

There are more than a dozen fold-and-thrust belts that have at some time, and at least in places, moved atop a layer of evaporites, usually including salt. The Triassic evaporites beneath both the Jura (Laubscher, 1972; Gwinner, 1978) and Pyrenees (Brinkman and Logters, 1968; Liechti, 1968) are intimately involved in the deformation resulting from horizontal compression. The same is true of the Cambrian Saline River formation beneath the Franklin Mountains in northwestern Canada (e.g., Cook and Aitken, 1973; Aitken et al., 1982) and the Ordovician Bay Fiord and Otto Fiord formations beneath the Canadian Arctic fold belts (e.g. Price and Douglas, 1972; Fox, 1984). The Cambrian salt beneath the Salt Range of Pakistan provides a weak detachment upon which thin-skinned deformation can advance far beyond that in adjacent, apparently salt-less, areas along strike (e.g., Sarwar and DeJong, 1979; Seeber et al., 1981; Burbank, 1983). The tectonic development of the Zagros belt is dominated by the horizontal decoupling and vertical diapirism of massive salt beds of Late Precambrian and Miocene age (e.g., Farhoudi, 1978; Colman-Sadd, 1978). Other mountain belts whose tectonics are strongly influenced by evaporites include the

Dynamical Geology of Salt
and Related Structures

Sierra Madre Oriental of Mexico (Rogers et al., 1962; de Cserna, 1971), the Cordillera Oriental of Colombia (Campbell and Burgl, 1965; McLaughlin, 1972), the Atlas Mountains of Tunisia and Algeria (Tortochaux, 1978), the southern Urals (Nalivkin, 1973; Kazantsev and Kamaletdinov, 1977), the Tadjik fold-and-thrust belt (Leith, 1984), and the anticlinal province of the Amadeus Basin in Australia (McNaughton et al., 1968).

"Thin-skinned" mountain belts result from horizontal compression (e.g. Chapple, 1978). Their structures are predominantly oriented parallel to the intermediate stress axis (Anderson, 1942), in the horizontal direction perpendicular to the maximum shortening. Deformation of the overlying rocks can produce both the overburden inhomogeneity and stress instability required to initiate diapiric growth, and the geometric ordering required to initiate systematically elongate salt structures. Even where there is no development of salt domes, such as in the fold belts of the Appalachian Plateau (e.g. Gwinn, 1964) and the Jura (e.g., Laubscher, 1977), the folds contain considerable anticlinal thickening of the salt, associated with synclinal thinning of salt.

Salt diapirism is usually associated with faulting in overlying rock. The growth of salt domes requires a density inversion and leads to faulting of the displaced younger strata (e.g. Nettleton, 1934). In the absence of significant horizontal differential stresses, such faulting and the associated salt domes, should be expected to have no strongly preferred orientation in map-view.

Conversely, deformation of the overlying sediments and the tectonic forces that cause it can be major factors in controlling the development of salt diapirs. When folding and faulting alters the thickness of the overlying section or the depth to the top of the potentially mobile evaporite, then it enhances the mechanical instability required to initiate diapiric movement.

Such deformation of the overlying rocks will tend to have a geometry controlled by the regional state of stress, and that geometry will be inherited by the resultant diapiric structures, whether of a piercement or non-piercement type. The very close correspondence between pre-existing faults and the elongated salt structures of the southern North Sea and northern Germany is a particularly well known example of this relationship.

There are numerous examples of prominent elongate salt-related structures associated with fold-and-thrust belts (for example, see Figure 1) where salt is intruded along thrust faults or at anticlines (e.g., Harding and Lowell, 1979). In addition to the Zagros (e.g., Stocklin and Nabavi, 1973), these include the Carpathian (e.g., Paraschiv and Olteanu, 1970), Sierra Madre Oriental (e.g. Weide and Martinez, 1970), Pyrenees (e.g., Liechti, 1968), Amadeus Basin (e.g., McNaughton et al., 1968) and Atlas (e.g., Tortochaux, 1978) fold belts.

II. IDEAS ABOUT THIN-SKINNED DEFORMATION

A. Historical development

Over a century ago it was recognized that the Glarus Thrust in Switzerland represented a major overthrust (Heim, 1871). Permian rocks in the upper plate of the thrust had "traveled" many km north by northwest over the Mesozoic rocks of the lower plate. This observation sparked a debate among Alpine geologists that was to last many decades: were these rocks pushed from the rear or did they slide down hill? It was soon recognized that frictional resistance along the basal decollement and the failure strength of rock were two major parameters controlling the length of overthrust sheets. By early in this century, the brittle strength of rocks was well documented in the laboratory. Smoluchowski (1909) recognized that for an overthrust to move over a horizontal surface, it would have to be pushed from the side with a force equal to the total frictional resistance along its base (Figure 2). He assumed a reasonable coefficient of friction and calculated how strong the granite of the overthrust must be in order to permit a long thrust sheet to be pushed from behind. Laboratory tests suggested that granite was not strong enough for a push of this magnitude. Instead, the granite sheet would crush at the rear before slip could be initiated. Smoluchowski concluded that either all long thrust sheets must glide down hill with the help of gravity or that the assumed coefficient of friction was too large. To date, there is no evidence for widespread, downhill sliding of large thrust sheets.

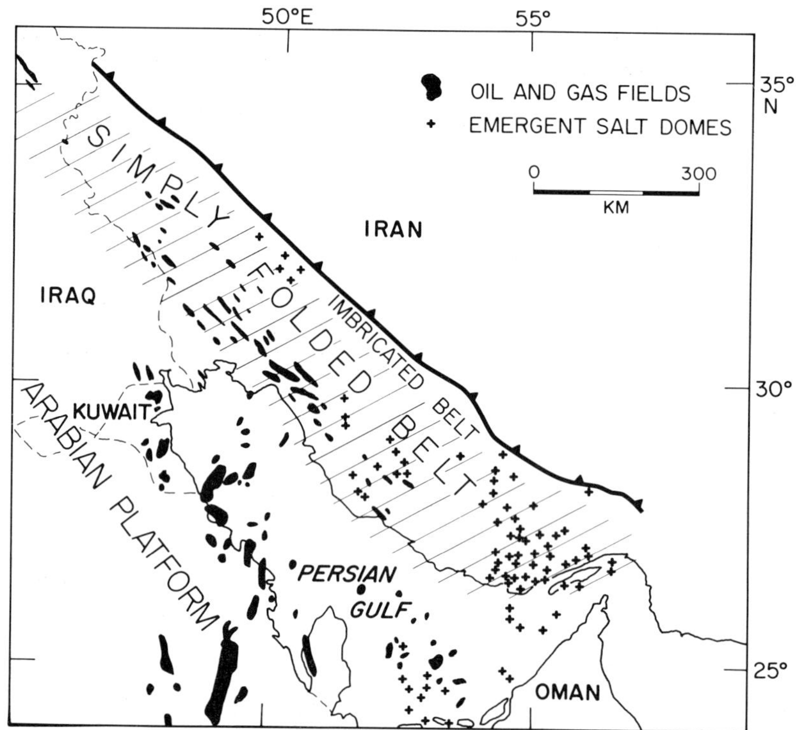

Figure 1. (a) Map (after Colman-Sadd, 1978) illustrating the relationship between salt domes, elongate oil and gas fields, and the Simply Folded Belt of the Zagros, which overthrusts salt.
(b) Highly schematic cross-section through the Zagros (after Farhoudi, 1978), illustrating the complicated relationship between diapirism of salt (shaded) and folding.

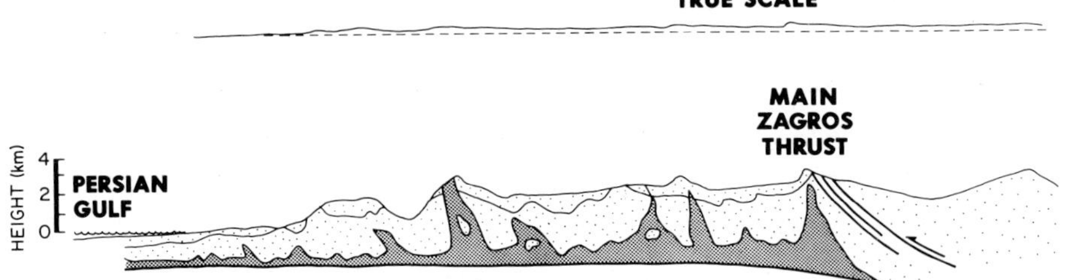

Figure 1 continued.

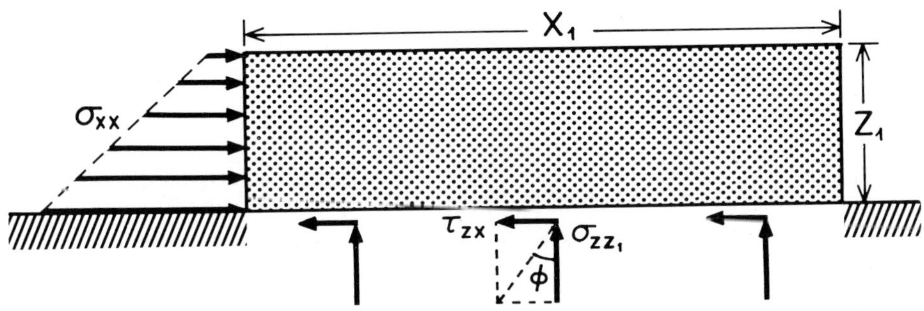

Figure 2. The balance of forces on a simple overthrust block. The frictional resistance to sliding must be balanced by a horizontal push.

This problem was partially resolved by the recognition (Hubbert and Rubey, 1959) that frictional resistance to sliding along a thrust fault is reduced in the presence of elevated pore-fluid pressures. Pore-fluid pressures support part of the normal stress but none of the shear stress, thus bringing the total stress state closer to failure. The general Coulomb criterion for the shear traction τ at failure is

$$\tau = S_0 + \mu\sigma_n(1-\lambda) \qquad (1a)$$

where $\mu = \tan \Phi$ is the coefficient of internal friction, σ_n is the normal traction, S_0 is the cohesive strength, and λ, the

Hubbert-Rubey pore-fluid pressure coefficient, is given by the relation

$$\lambda = P_p/\rho gz \tag{1b}$$

Here, P_p is the pore fluid pressure, ρ is the mean density of the sedimentary overburden, g is the gravitational acceleration, and z is the depth. The pore fluid pressure is termed 'lithostatic' (λ=1) if it is equal to the total overburden pressure of the overlying sediments. With unimpeded pore space connectivity to the water table (assumed to be at the surface), the pore fluid pressure is termed 'hydrostatic'. In that case $\lambda = (\rho_w/\rho) \approx 0.4$, where ρ_w is the density of water.

Data from laboratory experiments enabled the calculation of the size and length of an overthrust sheet that could be pushed from the rear with elevated pore pressures along its base. Assuming reasonable rock strengths, Hubbert and Rubey (1959) showed that known overthrusts could have been pushed horizontally from the rear if sliding was facilitated by nearly lithostatic pore-fluid pressures.

B. Critical wedge theory

Chapple (1978) listed several characteristics that are shared by foreland fold-and-thrust belts and that are essential to consider in modeling their mechanics. These are: 1) the basal surface of detachment or decollement, below which there is little deformation, dips toward the interior of the mountain belt. This dip direction is usually away from the craton. In other words, most fold-and-thrust belts were pushed up hill and did not benefit from the help of gravity gliding. 2) a large horizontal compression occurs in the rock above the basal flattening in response to this large compression. 3) in cross section, the deformed rock mass has a characteristic wedge shape which tapers toward the margin of the mountain belt. 4) the basal layer beneath the fold belt commonly consists of rock that is, for some reason, relatively weak (e.g. commonly a shale).

The interiors of mountain belts are in fact deformed in compression, as was predicted by Smoluchowski (1909) and Hubbert and Rubey (1959). Thus, a taper is produced in the direction of the craton, with a topographic surface that slopes gently toward the craton and a decollement surface that slopes away from the craton (Figure 3). Chapple (1978) showed that a perfectly plastic fold-and-thrust belt over a very weak basal layer can overthrust its base if there is sufficient topographic slope.

Davis et al. (1983) presented a simple model for a cohesionless, time-independent, homogeneous and isotropic Coulomb wedge that deforms in a manner analogous to soil being pushed by a bulldozer. For most fold-and-thrust belts, this model does not require an extremely weak basal layer. Despite its simplicity, this model proved successful in predicting the gross geometry of the active fold-and-thrust belt of western Taiwan, particularly when the role of cohesion was taken into account (Dahlen et al., 1984). This wedge deforms until it attains a steady state or critical taper and then slides stably, continuing to grow self-similarly as additional material is accreted at the toe (Davis et al., 1983; Dahlen et al., 1984). The thick end of a fold-and-thrust belt wedge is at the core of the mountain range. In the

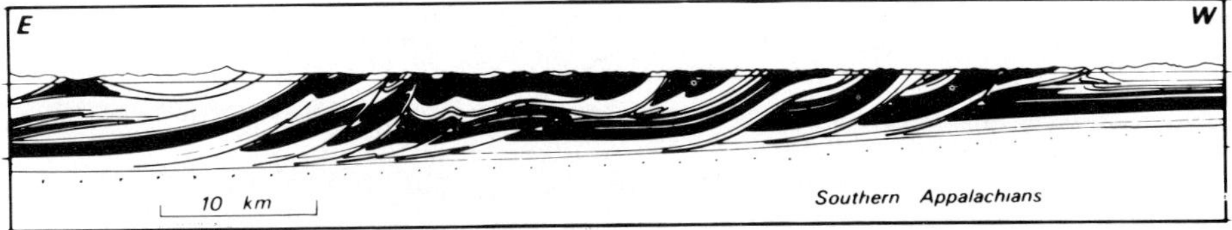

Figure 3. A cross-section through the southern Appalachians (Roeder et al., 1978), modified after Davis et al. (1983). The intense folding and thrusting is concentrated in the wedge-shaped overthrust belt, which overlies a layer of decollement. Note that structures verge predominantly toward the craton.

Appalachians this core is presently eroded to the crystalline rocks that acted as a bulldozer pushing the clastic wedge of the Valley and Ridge and the Appalachian Plateau.

Davis et al. (1983) calculated the critical taper for the fold-and-thrust wedge of western Taiwan, making the assumption that the rock within the wedge is everywhere on the verge of shear failure according to the Coulomb failure criterion. The taper is simply the sum of the mean local slope of topography, α, and the local dip of the basal decollement, β. The magnitude of the critical taper at which the wedge can overthrust its base is governed by the balance of forces in the up-dip direction of the decollement. It is a function of the internal friction angle Φ of the wedge, the friction coefficient at the base of the wedge μ_b, and λ, S_0, ρ, g, and z, all of which were defined above for Eq. (1). In the limit as decollement strength becomes much less than that of the overlying sediments (which, as we shall see, is appropriate for thrusting over salt) the critical Coulomb wedge taper can be found iteratively using a relation adapted from Dahlen et al. (1984).

$$\alpha+\beta = \frac{\beta\rho gz + (1-\lambda)\mu_b - \left[2S_0 \cot\Phi(\alpha+\beta)/\rho gz(\operatorname{cosec}\Phi-1)\right]}{1 + 2(1-\lambda)/(\operatorname{cosec}\Phi-1)} \qquad (2)$$

Knowing the wedge geometry and having measured the pore-fluid pressure ratio $\lambda = 0.67$ (pore pressure $\approx$ 67% of lithostatic) for the Taiwan fold-and-thrust belt, Davis et al. (1983) find that the frictional strength of the decollement does not have to be exceptionally low: it appears to be 80% to 85% of the strength of the overlying rocks. If the decollement in Taiwan were much weaker than 80% of the strength of the overlying rocks, then the critical taper would be smaller than its actual value of 9°. Of course, the decollement cannot be stronger than the overlying rocks, or else some other, weaker, horizon in the overlying rocks would become the layer of decollement.

III. WEAK DETACHMENT MECHANICS

A. Rock strengths

Strain in the crust takes place very slowly by human standards: a strain rate of 10^{-10} sec^{-1} represents very rapid deformation, and 10^{-14} sec^{-1} is more typical. In this range of strain rates and at temperatures less than about 250°C, almost all common crustal rocks deform in a brittle manner that is essentially independent of time (eqn. (1)). Byerlee (1978) pointed out that an extraordinarily wide range of crustal rocks have a very similar friction coefficient of 0.85. The only major exception that he noted was for clay minerals (some of which are often found in fault gouges), which were typically 3 or 4 times weaker. At the high temperatures of mid-crustal depths, quartz-rich rocks start to become ductile, because of thermally activated flow mechanisms. Below this brittle-ductile transition, rock strengths decrease rapidly with depth (Figure 4).

The results of experiments on dry natural rocksalt (Carter and Hansen, 1983) in the strain rate range 10^{-6} $sec^{-1} > \epsilon > 10^{-9}$ sec^{-1} can be fit very well by a steady-state flow equation of the form

$$\epsilon = A\ (\sigma_1 - \sigma_3)^n \exp[-B/RT] \qquad (3)$$

where $A = 7.6\ 10^{-4} \pm 2.8\ 10^{-3}$ sec^{-1}, the activation energy $B = (66.5 \pm 17.5)\ 10^3$ kJ/mol, $\sigma_1 - \sigma_3$ is the stress difference, $n = 4.5 \pm 1.3$ is a constant, R is the gas constant, and T is the temperature (in °K). Note that the strain rate is dependent upon the stress difference, $\sigma_1 - \sigma_3$, but not upon the absolute magnitude of the confining stress. In this way, ductile flow is fundamentally different than frictional strength, which is proportional to confining pressure.

B. Mechanisms for reduction of friction

There are at least two common classes of thin-skinned deformation in which overthrusting is met with very weak resistance to sliding. The first of these classes consists of thin-skinned wedges with very high pore-fluid pressures. Although near-lithostatic pore pressures in fold-and-thrust belts appear to be very rare (e.g., Fertl, 1976; Suppe and Wittke, 1977), there are many accretionary prisms in which the accreting sediments are very porous and must dewater as part of the diagenetic process (e.g., von Huene and Lee, 1982; Moore and Biju-Duval, 1984). There is a growing body of data suggesting that deformation at the frontal toe of at least some accretionary prisms takes place under conditions of extremely high pore pressures. As demonstrated by Hubbert and Rubey (1959), this permits overthrusting (or, in this case, subduction) at very low shear stresses.

The other special class of thin-skinned deformation includes those fold-and-thrust belts that overthrust, at least in part, a detachment zone in an evaporite, and it is that class with which this paper is concerned. There are more than a dozen fold-and-thrust belts that have formed atop an evaporitic layer.

For moderate geothermal gradients (15° to 25°C km^{-1}), a temperature of 100°C is found at 3 to 5 km below the surface. Let us assume that at this depth, there is a detachment along a 100 m thick salt layer, where 1 cm/yr of slip takes place. Equation (3) indicates that at this strain rate ($\approx 3.10^{-12} sec^{-1}$) the differential stress required to drive this slip is only 1 MPa, which corresponds to a maximum shear stress of 500 kPa (5 bars). Varying the strain rate by two orders of magnitude changes this result by factor of less than 3, with higher strain rates requiring higher differential stresses.

Evaporites in general (and rock salt in particular) are much weaker than any other common rock type, including shales.

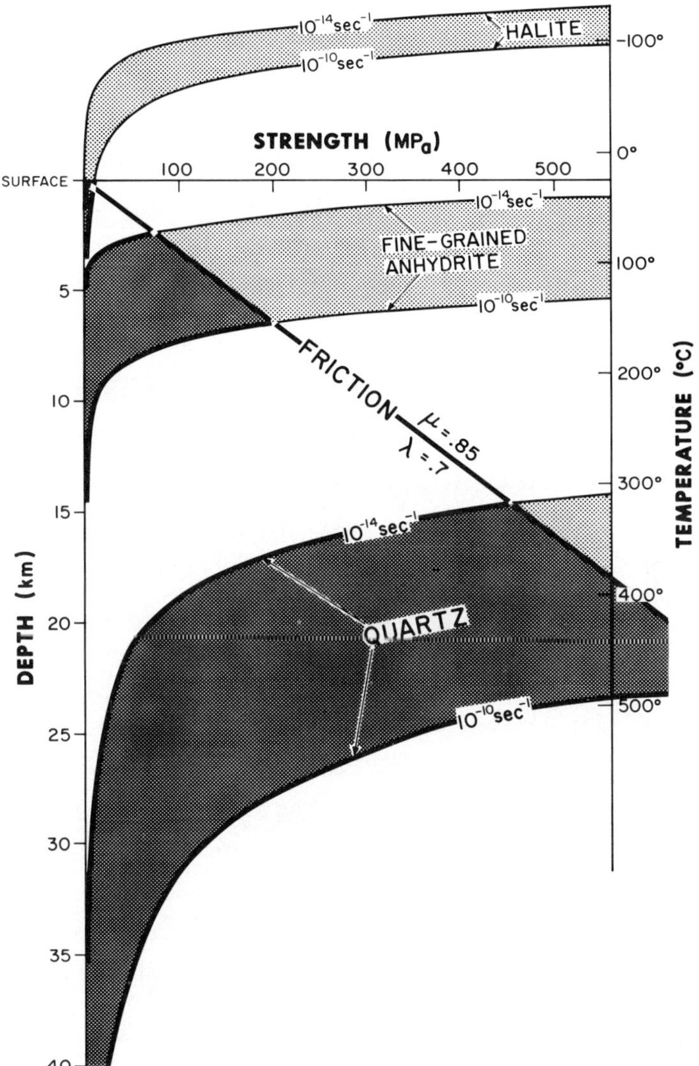

Figure 4. Strengths of common rock types versus depth and temperature, showing the relative weakness of halite. The frictional strength assumes that the coefficient of friction μ=0.85 (Byerlee, 1978), density = 2.75 g/cc, and pore fluid pressure ratio λ=0.7. The flow strengths are as follows: quartz from Brace and Kohlstedt (1980), halite from Carter and Hansen, (1983), and anhydrite from Muller et al., (1981).

Laboratory measurements of the deformation of salt (e.g. Carter and Hansen, 1983) indicate that for reasonable strain rates salt typically reaches its brittle-ductile transition at a depth of less than 1 km (Figure 4). Indeed, in the presence of small amounts of water, dynamic recrystallization (e.g. Spiers et al., 1986) may make salt even weaker at geologic strain rates than the low strength suggested by most laboratory measurements. Such extreme weakness seems reasonable in light of the intense shallow deformation that has been observed within salt domes and salt glaciers.

At typical depths for a basal detachment (2 to 8 km), the shear strength of salt is commonly less that 1 MPa for typical strain rates and geothermal gradients, making it between 1 and 2 orders of magnitude weaker than most other rocks (Figure 4). Anhydrite is also weak at very moderate temperatures, although not as weak as is halite. Muller et al. (1981) found that a fine-grained anhydrite from the Table Jura of Switzerland (Wandflue) is weak (<10 MPa) at low strain rates and temperatures over 100°C (Figure 4), but at a strain rate of 10^{-10} sec^{-1} it is weak only at temperatures of over 200°C. Furthermore, they found that a less fine-grained anhydrite from Riburg required roughly 100°C higher temperatures in order to become as weak as its fine-grained counterpart at similar strain rates. The grain size appropriate for natural deformation will, of course, vary with both initial depositional character and the degree of recrystallization under strain.

It is the extreme weakness of salt that makes the structural style of overthrusting on salt so distinct from that over other types of rock (Davis and Engelder, 1985). The two mechanisms for weak-coupling decollement (overpressures and evaporites) are not mutually exclusive. Indeed, salt can form an impermeable lid for the maintenance of excess pore pressures (e.g., Fertl, 1976).

C. Effect of salt on fold belts

1. Foldbelt taper and width

A detachment in salt has a dramatic effect upon the cross-sectional taper and width in map-view of a fold-and-thrust belt. The critical taper equation (eqn. (2)) is, to a large extent, a reflection of the relative strengths of the overthrusting wedge and the stratum over which it rides. Thus, a weak detachment zone should not require a large wedge taper in order to permit overthrusting.

As we have seen in Section III.B and in Figure 4, the basal salt layer in fold-and-thrust belts which overthrust salt is typically a few km deep and is likely to be well below the depth of brittle-ductile transition in salt. For reasonable strain rates, the shear strength of the salt layer is not much greater than 1 MPa, and may be considerably less. Furthermore, for a constant strain rate, that strength is expected to be independent of the normal traction exerted by the overburden. We thus express the shear strength of the basal detachment in salt as a constant, τ_0, whose value is likely to be less than 1 MPa. The critical Coulomb taper of such an overthrusting wedge is calculated by replacing μ_b in eqn. (2) with $\mu_b = \tau_0/\rho gz$, giving

$$\alpha+\beta = \frac{\beta\rho gz + (1-\lambda)\tau_0 - \left[2S_0 \cot\Phi(\alpha+\beta)/(\operatorname{cosec}\Phi-1)\right]}{\rho gz[1 + 2(1-\lambda)/(\operatorname{cosec}\Phi-1)]} . \tag{4}$$

Thus, a foldbelt needs a taper of only a few tenths of 1° in order to ride over a salt layer, if salt is as weak as eqn. (4) suggests and if the rocks of the foldbelt deform by brittle failure. Non-brittle failure in the wedge will be considered further (in Section IV.C).

Given a certain thickness of the sedimentary section, a narrower wedge taper implies that it is possible to build a broader fold belt. In other words, a reduced resistance to sliding should make it possible for the "bulldozer" to push a foldbelt that is longer as well as more narrowly tapered. Thus, the part of a foldbelt that overthrusts a salt basin should be broader in map-view and more narrowly tapered in cross-section over the salt than at points along strike that are not over the salt. As for all of the arguments made in this paper, the same should be true, though to a somewhat lesser degree, for overthrusting on anhydrite, which has a brittle-ductile transition somewhat deeper than that of halite (Figure 4).

2. Stress orientations and vergence

Another parameter controlled by the presence of a very weak basal detachment is the dip ψ_b at which the axis of maximum compressive stress σ_1 dips toward the foreland with respect to the basal detachment. The orientation of the principal stress axes is important because of the following relation to deformation. The Coulomb failure criterion is satisfied most readily along the two planes containing the σ_2 (intermediate) stress axis and inclined about the σ_1 axis at an angle θ defined by the simple relation

$$\theta = \pm (45° - \Phi/2) \,. \tag{5}$$

In order to have the proper sense of shear traction along the basal decollement, the σ_1 (maximum compression) axis must dip toward the foreland, as shown by Hafner (1951) for a rectangularly shaped overthrust. The angle ψ_b at which the σ_1 axis dips with respect to the basal decollement for a thrust wedge has been calculated by Davis et al. (1983) and Dahlen et al. (1984). The cohesive strength can be neglected in this calculation if it is much smaller than the component of strength due to rock sliding friction, such that $S_0 \ll \rho g z \mu (1-\lambda)$. Here, S_0, ρ, g, z, μ, and λ are the cohesive strength, rock density, gravitational acceleration, depth, rock friction coeffieicnt, and Hubbert-Rubey pore-fluid pressure coefficient, respectively. In the limit of negligible cohesion

$$\psi_b = \tfrac{1}{2}\arcsin(\sin\Phi_b/\sin\Phi) - \Phi_b/2 \quad , \tag{6a}$$

where Φ and Φ_b are the internal friction angle of the wedge and the sliding friction angle for its base, respectively. This relation can be demonstrated by using a Mohr-Coulomb diagram (Davis et al., 1983). Dahlen et al. (1984) show that in the presence of significant wedge cohesion, it is necessary to solve iteratively for ψ_b using the relation

$$[\mu(1-\lambda)+(S_0/\rho zg)]\sin 2\psi_b = \mu_b(1-\lambda)[(1+\mu^2)^{1/2}-\mu\cos 2\psi_b] \tag{6b}$$

where $\mu_b=\tan\Phi_b$ is the coefficient of sliding friction along the basal detachment layer. The magnitude of ψ_b at the base of the deforming wedge is very strongly dependent upon shear traction which can be supported across the basal detachment. Dahlen et al. (1984) have calculated that beneath the toe of the Taiwan fold-and-thrust belt, $\psi_b = 12°$. If the coupling were weakened by replacing the Talu Shale with a thick layer of salt, then the basal friction μ_b would be replaced by $\tau_0/\rho gz$, where τ_0 is a normal-stress independent strength, as for the critical taper calculation in Section III.C.1. In that case, with τ_0=1 MPa or less, then ψ_b would be considerably less than 1°.

Both sets of possible slip planes are symmetric about the σ_1 axis, which dips forward at an angle ψ_b. Therefore (Figure 5) the forward verging slip planes have an optimal dip at an angle $\delta_f=\theta-\psi_b=45°-(\phi/2)-\psi_b$ and the backward vergent ones dip at $\delta_b=\phi+\psi_b=45°-(\phi/2)+\psi_b$. The fact that the forward verging slip planes dip more shallowly (by a margin of $2\psi_b$) probably explains why they are the more common of the two in most thrust belts (Davis and Engelder, 1985). The reason for this is simple. The shallower dip of the forward verging slip planes permits a greater amount of horizontal shortening for the same increase in gravitational potential energy. Forward verging thrusts are also favored because of stratigraphic strength anisotropy, which should favor thrusting closer to the orientation of bedding. In addition, the section behind the frontal thrust zone has typically been thickened (and therefore strengthened) by earlier thrusting. However, if the basal detachment is extremely weak (i.e., in salt), ψ_b is very small, so the two candidate slip planes have more nearly equal dips, the wedge is much more subtly tapered, and forward-vergent thrusts should be less predominant. If folds in salt-basal fold belts are related to blind thrusts, which should face both directions, then the folds should also have relatively symmetrical form.

IV. APPALACHIAN PLATEAU DEFORMATION

A. Contrast with the Central Appalachian Valley and Ridge

Perhaps the single most distinctive aspect of the Appalachian Plateau Province is seen in large-scale map view. In the Plateau of New York and Pennsylvania, Appalachian deformation proceeds about 200 km further onto the craton than at other points along strike. This deformation is thin-skinned, and has a decollement in salt. At present, the post-salt rocks thin from 3.7 km at the southeastern end of the Plateau to 600 m to the northwest, giving a present-day wedge taper only 1.15° (Frey, 1973). It is not certain exactly how much thicker the taper was when the foldbelt was active, but it was clearly always quite narrow.

The narrow wedge taper and broad foldbelt of the Appalachian Plateau are highly suggestive of the theoretical discussion of the previous section. Several other salt-basal foldbelts share these attributes of narrow taper and unusually great breadth, including the Salt Range of Pakistan (e.g., Burbank, 1983; Jaume et al., 1985), the Franklin Mountains and the Colville Hills in front of them (e.g., Cook and Aitken, 1973), the Zagros Simply

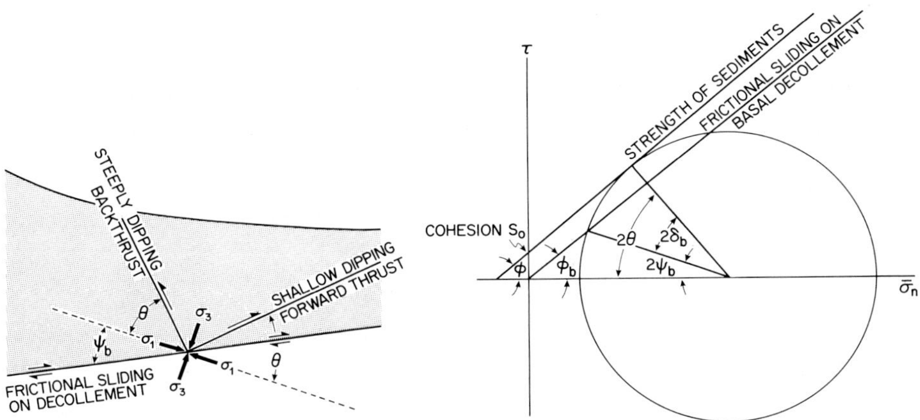

Figure 5. (a) An illustration of why, in the presence of significant decollement strength, forward vergent slip planes dip more shallowly than do the backward vergent planes. This dip difference is a direct result of the fact that ψ_b is positive. For a salt-basal fold belt, ψ_b is very small (<1°), so thrusts and folds are relatively symmetric.
(b) A Mohr-Coulomb diagram illustrating graphically the relationship between the stress orientation angle ψ_b and the strengths of the wedge and the basal detachment (after Davis and Engelder, 1985).

Folded Belt (e.g., Farhoudi, 1978; Colman-Sadd, 1978), the Sierra Madre Oriental (e.g., de Cserna, 1971), and the Parry Islands Fold Belt (e.g., Price and Douglas, 1972; Fox, 1984).

The Appalachian Plateau of Pennsylvania and western New York State lies in front (to the northwest) of the Valley and Ridge Province of the Central Appalachians (Figure 6). The folds of both provinces are involved in the northwestward thin-skinned transport of allochthonous Paleozoic rocks. Along the boundary between the two provinces, the level of decollement steps up from Cambrian shales beneath the Valley and Ridge Province to eventually reach the Silurian Salina salt beneath the Appalachian Plateau (Figure 6).

These wavelengths are comparable in the two provinces; 8 to 16 km in the Valley and Ridge and a bit greater (8 to 32 km) in

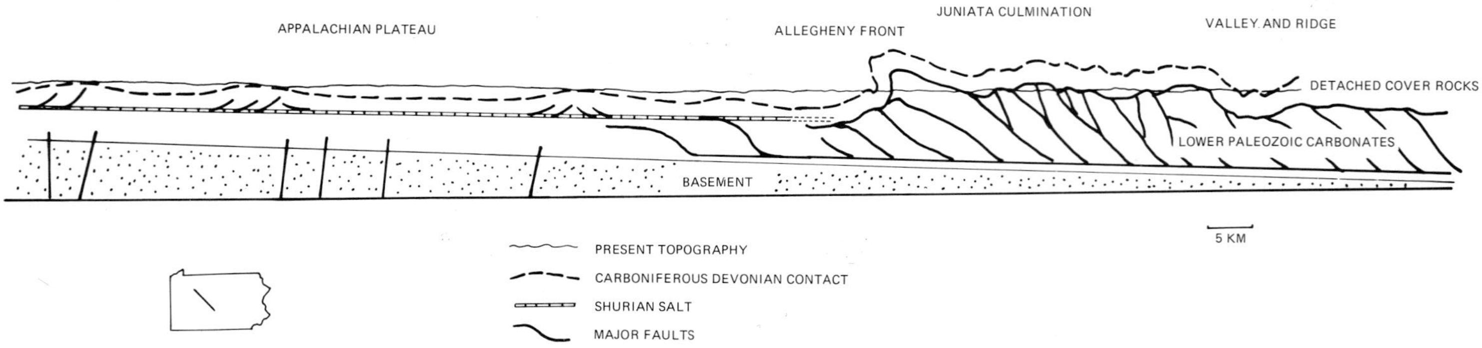

Figure 6. Interpretive cross-sections of parts of the Appalachian Plateau area and the Valley and Ridge Province. (after the Geologic Map of Pennsylvania of the Geological Survey of Pennsylvania and after P. Geiser, personal communication, 1986).

the Plateau (Rodgers, 1963; Wiltschko and Chapple, 1977). However, the styles of folding are quite different. Only one Plateau fold has structural relief at the surface exceeding 800 m or a flank dip greater than 10°, but the first fold in the Valley and Ridge Province has relief of 8 km and is overturned (Rodgers, 1963). In addition, all Valley and Ridge folds (Figure 6b) are much steeper on their northwestern sides, but Plateau folds are much less asymmetric, with only a slight preference toward a greater steepness on the southeastern side (Figure 6a). This tendency toward slightly oversteepened southwest limbs is most prominent in the northwestern part of the Plateau, where amplitudes are the smallest (Sherrill, 1934; Gwinn, 1964). The contrast between the two provinces extends to faulting as well. Although there are many surface thrust faults (all dipping to the southeast and verging to the northwest) in the Valley and Ridge, there are extremely few surface faults in the Plateau, and those that have been found do not seem to have any marked preference in vergence (Rodgers, 1963).

This lack of a strongly preferred direction of structural vergence is a very common attribute of fold-and-thrust belts over salt. For example, both the folds and the thrust faults of the Franklin Mountains are inconsistent in their direction of asymmetry. Indeed, there are several places in the northern Franklin Mountains (Figure 7) where thrust faults with opposite senses of transport occur near to, and along strike from, each other (Cook and Aitken, 1973).

B. Fold geometry

A cross section through the Appalachian Plateau (Figure 6a) shows that Coulomb-like failure is localized in some of the more rigid formations (sandstone and limestone) just above the salt decollement, in the core of widely spaced anticlines. Engelder and Geiser (1979) show that above those rigid layers the

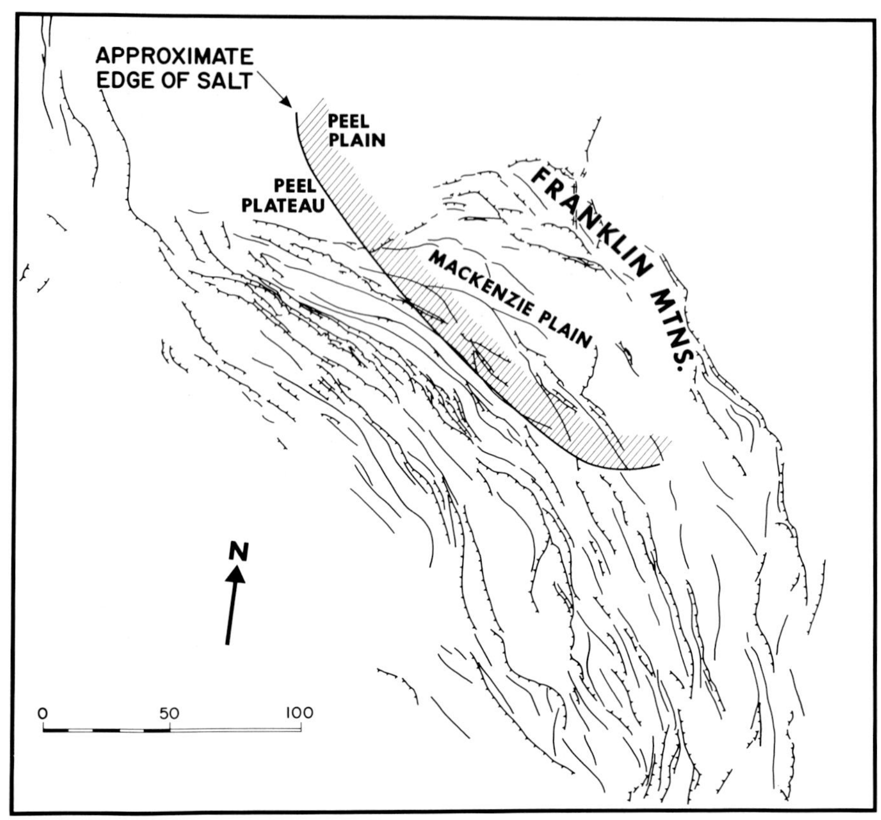

Figure 7. Map of the Franklin Mountains area (after Davis and Engelder, 1985), showing the relationship between structures and the edge of the Cambrian Saline River Formation salt.

predominantly shale section of the Plateau was affected by a penetrative strain probably reflecting a state of time-dependent viscous failure.

We noted previously that plateau structures are more symmetric than are those of the Valley and Ridge. This observation is consistent with the arguments of Section III.C.2. However, in detail many of the folds on the Appalachian Plateau are cored with blind thrust systems where the majority of the thrusts follow backward planes (Gwinn, 1964). This geometry

(Figure 8) is opposite that for most fold-and-thrust belts with forward stepping thrusts. A consequence of this thrusting is that many of the folds of the Appalachian Plateau have a subtle asymmetry with the steep limb dipping toward the interior of the mountain belt. The initial Davis and Engelder (1985) assumption that the Appalachian Plateau was in a state of homogeneous Coulomb failure cannot be used to explain the asymmetry.

If Coulomb failure occurs locally and behaves according to the Davis and Engelder (1985) model, then the Appalachian Plateau asymmetry and back thrusting can be explained using the following hypothetical sequence of events. First, initial Coulomb failure and subsequent thrusting is symmetrical on both forward and backward thrust planes (Figure 9a). This thickens the section in the vicinity of the blind thrusting. Thickening acts to increase the strength of the rocks by virtue of increased normal stress across Colomb-failure planes. The next point of failure must then be backward or forward of the thickened section

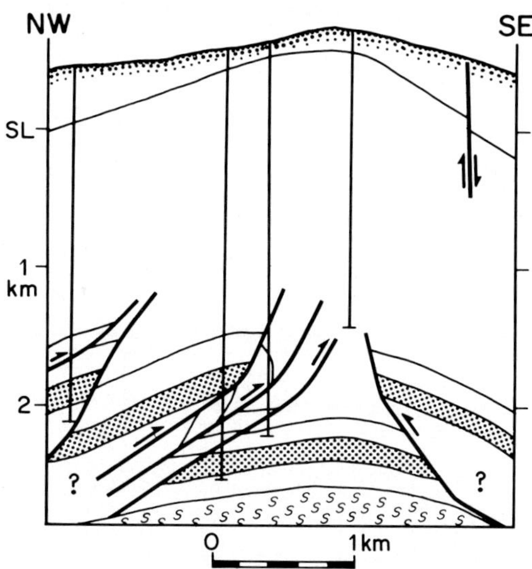

Figure 8. A cross-section through part of the Chestnut Ridge anticline in Pennsylvania (after Gwinn, 1964).

but not in it. Furthermore, the higher normal stresses of the thickened section repress additional Coulomb-like failure in that area.

The location of additional Coulomb failure is a function of frictional dissipation of the traction pushing the Appalachian Plateau toward the foreland. This dissipation is low. But, with everything else equal, the traction on the interior side of the thickened section is larger than the traction on the exterior (cratonward side).

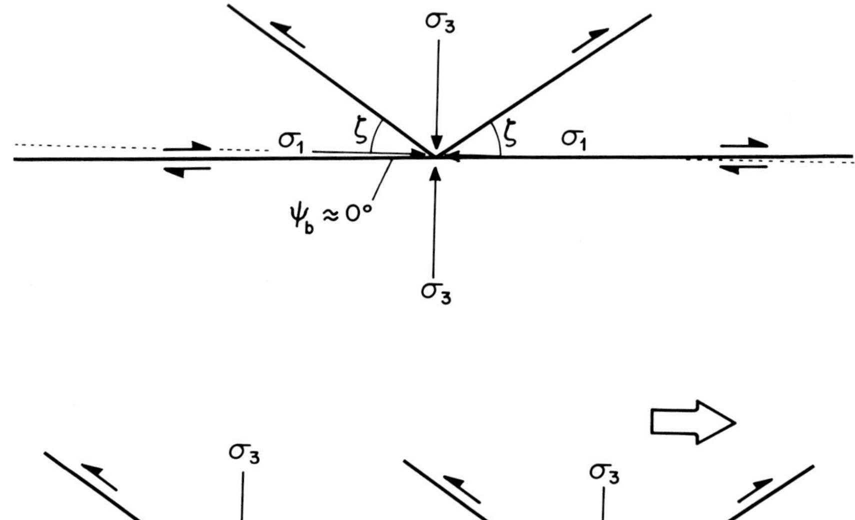

Figure 9. Proposed mechanism for the subtle asymmetry of Appalachian Plateau folds, with the steeper limb facing toward the interior of the mountain belt.
(a) Initial, symmetric deformation.
(b) Development of subsequent, additional backward vergent thrust(s) in response to greater traction immediately behind the original thrusts.

Failure is then favored on the interior side of the thickened section by virtue of a slightly larger traction there.

For Coulomb failure on the interior side of the thickened section, failure along back thrusts is favored where the fracture propagates back into a thinner section. Here the normal stress across the failure planes for back thrusts is lower than normal stress for forward thrusts. This gives the section a second back thrust and further steepens the asymmetry of the folds on the Appalachian Plateau (Figure 9b). Several asymmetric folds have apparently developed across the Appalachian Plateau by this process.

Plateau folds within 50 km of the structural front with the Valley and Ridge Province have structural relief close to the local thickness of the salt. Wiltschko and Chapple (1977) suggest that this relief is produced as a result of synclinal thinning of salt. Near the structural front the synclines appear to be essentially depleted, but further away from the structural front the synclines are relatively less depleted. This is taken to be an explanation for the reduction in structural relief of the folds towards the northwest. In some areas, anhydrite apparently takes part in the flowage along with the halite. Wiltschko and Chapple (1977) have also pointed out that as a syncline becomes depleted of salt, it must become relatively more difficult to continue the thinning of the synclinal salt because the stress required for a constant rate of thinning is proportional to the inverse cube of the layer thickness.

The strength of the basal decollement in the salt layer also increases with synclinal thinning. A thinner salt layer must deform at a proportionally higher strain rate in order to attain a given rate of overthrusting slip. However, as we noted in Section III.B. the shear stress is relatively insensitive to the strain rate (eqn. (3)). Nevertheless, forward oversteepening of folds in the Zagros fold belt, where the Cambrian salt is exceptionally mobile, has been attributed to increased basal resistance in areas of synclinal salt depletion (Colman-Sadd, 1978).

C. Strain and Paleostress

The sediments cratonward (to the northwest) of frontal folds in the Appalachian Plateau have undergone layer-parallel shortening by pressure solution. Measurements by Engelder (1979) and by Engelder and Geiser (1979) indicate that the amount of horizontal shortening by pressure solution and intergranular mechanisms is closely related to the distribution of the Silurian salt. Along each of the three traverses in Figure 10, compressive strain increases from very low levels (⩽2%) where there is no significant Silurian salt to 15% over areas with thick salt. Apparently, in the absence of salt, strata younger than Silurian are strongly coupled to the rocks below and did not deform, but where the Silurian includes significant salt deformation was significant.

The driving stresses of the Alleghanian Orogeny in the Appalachian Plateau have been estimated using several different techniques (e.g., Prucha, 1968; Rutter, 1976; Engelder, 1982). Engelder (1982) inferred that the level of differential stress was 6.2 MPa by observing the axial canals of crinoid columnals within the wedge, which acted as stress concentrators and caused twinning adjacent to the canals. Residual stress measured in rocks from more deeply buried portions of the Appalachian Plateau wedge averages 11 MPa (Engelder and Geiser, 1984). These data indicate that shortening took place at relatively low differential stresses within the wedge, and provide a constraint on the mechanism for the development of the Appalachian Plateau structures over the Silurian salt.

One conclusion that can be drawn from these data is that after the initiation of the initial blind thrusting and splay faults, much of the deformation took place in a more distributed,

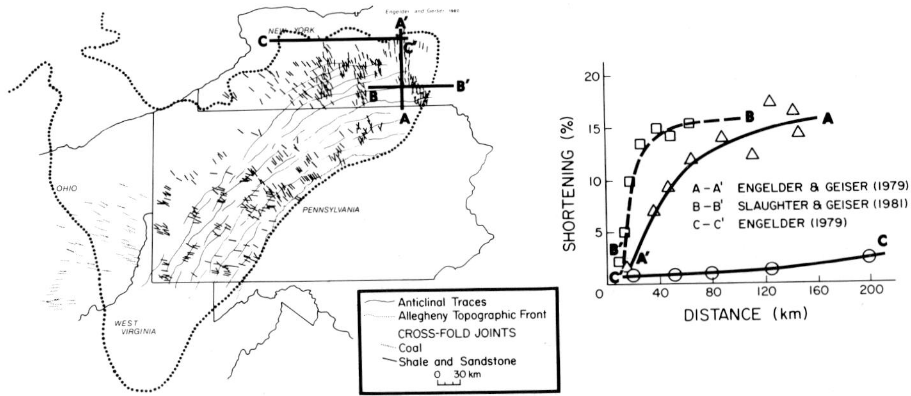

Figure 10. (a) A map of the Appalachian Plateau area, showing the limit of the Silurian salt basin and the locations of three lines, along which measured shortening has been plotted in part (b) (after Davis and Engelder, 1985).

long-term manner, at a relatively low strain rate and a relatively low differential stress. The reasoning is as follows: in the narrowly tapered Plateau, there is no marked deformation front or frontal thrust at which deformation is concentrated. Therefore, at any given time the locus of shortening is quite large (a sizeable fraction of the whole Plateau, as opposed to merely a single frontal thrust). With shortening distributed over a wide area, the strain rate at any given point in space and time must be quite low, apparently low enough to make layer-parallel shortening a viable mechanism for accommodating most of the shortening in the northwestern part of the Plateau.

Given the present taper ($\approx 1.5°$) of the post-Silurian strata (Section 4.A.), it appears that the original, pre-orogenic basinal taper of the Appalachian Plateau may well have been greater than the Coulomb critical taper (eqn. 4) at which high-stress Coulomb failure will take place within the wedge. If the original taper of the Appalachian Plateau wedge exceeded the critical taper, then there would have been no need for further thickening by widespread Coulomb failure. This would explain the lack of major thrust faults everywhere in the Plateau except occasionally just above the salt and the predominance of the observed homogeneous, lower stress, lower strain rate deformation mechanism, layer-parallel shortening.

D. Relation of folding salt basin

There is a very close relation between Appalachian Plateau deformation and the distribution of the Salina salt, as is clear from the distribution of layer-parallel shortening (Figure 10). The transition from Valley and Ridge folding and faulting to Appalachian Plateau folding occurs at the southeastern edge of the salt basin, where the decollement steps up section until it reaches the mechanically favored weak Silurian salt (Figure 6). A very similar transfer of slip is seen to occur in other fold belts that encroach upon a salt basin for part of their lengths.

One particularly interesting example of this process is found in the Franklin Mountains of northwestern Canada. Where there are no Saline River Formation evaporites, the part of the section younger than Late Cambrian is effectively pinned to basement. However, where the frontal thrusts of the Mackenzie Arc reach part of the basin containing the salt, there is a transfer of slip to that level. There, deformation does not stop with the frontal Mackenzie thrust fault, but continues another 100 km or so onto the craton (Figure 9). The front of significant folding does not reach the Peel Plain even though there is salt beneath it, because there is no salt to the southwest along the Mackenzie front into which slip could be transferred (Aitken and Cook, 1975; Aitken et al., 1982).

The Burning Springs Anticline at the southwestern edge of the Appalachian Plateau is an anomalously trending, large amplitude structure. This is not a normal Plateau fold. It appears to be related to a transfer of slip to the surface where there was no longer sufficient salt in which to form a weak decollement (Rodgers, 1963). The Burning Springs Anticline follows the trend of the edge of the salt, and might be likened to the piling up of a rug adjacent to where it is pinned down.

V. SALT-RELATED STRUCTURES IN OTHER FOLDBELTS

A. Box Folds

The folds of the Appalachian Plateau (Figures 6,8) are typically characterized by splay faults off the basal decollement which produce salt-cored anticlinal uplift (Gwinn, 1964). These folds are much like the drape folds described by Stearns (1971). Salt has thickened under these regularly spaced anticlines (Wiltschko and Chapple, 1977). Another style of folding observed above salt is that found in the Jura, where non-sinusoidal folds in a considerably thinner overthrusting section have been described as a superposition of different kinds of instabilities in space and time (Laubscher, 1972, 1977). They are most realistically modeled as box fold structures with very large (hundreds of meters across) kink bands with rounded edges (Figure 11).

It is possible that the presence of either Appalachian Plateau or Jura type folding may depend upon whether there is a major 'rigid' structural member low in the section as is the case in the Appalachians (the Onondaga Limestone and Oriskany Sandstone). The difference between fold style in the Appalachian Plateau and the Jura may also be related to differences in the thickness and the mechanical homogeneity of the sections, with the Jura comparatively thin and homogeneous.

B. Effects of basement structure

Basement faults and warps are known to influence deformation in many fold-and-thrust belts. Wiltschko and Eastman (1983) have demonstrated a number of mechanisms for the generation of stress concentrations due to basement structure. They point out that a

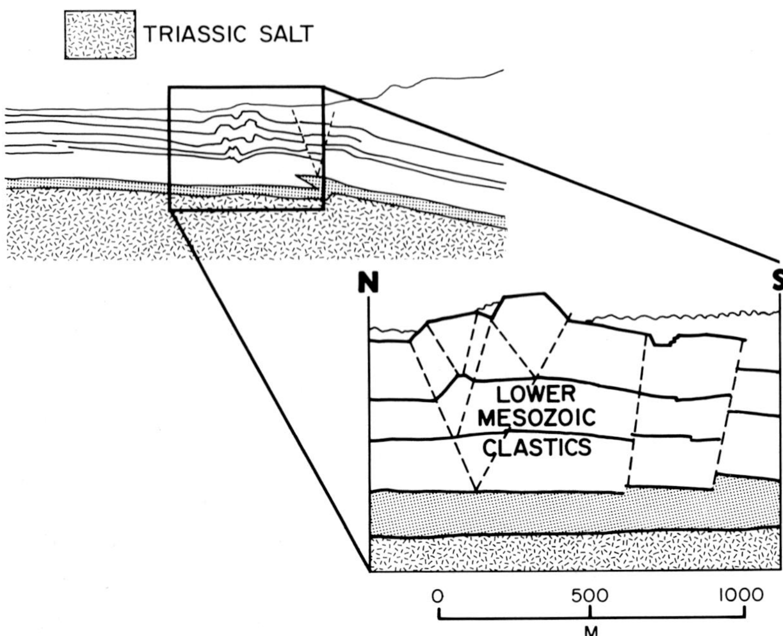

Figure 11. A cross-section through a Jura box fold (after Laubscher, 1977).

small number of stress concentrations, generated by an equally small number of basement structures, can control the gross geometry of structure in an entire fold-and-thrust belt. We suggest that this conclusion may be particularly applicable to salt-basal foldbelts which, as we have seen, commonly lack a consistent preferred direction of structural vergence.

The basal decollement of the Tadjik (Vakhsh) foldbelt in the Soviet Union is in a thick Jurassic salt (Leith et al., 1981). Recent studies have determined that deformation in the Cretaceous and Tertiary strata of the fold-and-thrust belt is, in the central Depression, conspicuously absent north of a buried basement fault that marks both the hinge-zone of the late Mesozoic passive margin and the northern boundary of the evaporite (decollement) horizon. However, in the eastern part of the Depression, thrusting has moved coherent thrust sheets north, over the block fault. These sheets now lie flat atop the autochthon to the north. Thus, the crustal structure inherited from a Mesozoic extensional phase has strongly influenced the Cenozoic pattern of deformation, producing a fold-and-thrust belt in which structures are strongly concentrated toward the northern (toe) end, near the basement fault. Apparently, the development of the thrust system has included the progressive overlapping of thrusts, with the later thrusts having formed internal to the older thrusts which they subsequently overrode. This implies that thrusts in front of a basement offset in at least this salt-based fold-and-thrust belt develop in a reverse sequence, away from the craton (Leith and Alvarez, 1986).

Unlike foldbelts without basal evaporites, salt-basal foldbelts are likely to display stress-guide behavior, because they come close to having a zero shear-stress boundary on the bottom (in salt) as well as on top (in air). When confined and compressed (as is the case in the Tadjik foldbelt), a stress guide (the sedimentary pile) will break at its thinnest (and therefore weakest) point. That thinnest point is located at the outer, marginal edge of the basin; in the case of the Tadjik foldbelt that is immediately adjacent to the Illiac fault. Thrusting leads to a buildup of taper at and a strengthening of the toe in front of the basement fault. Eventually, the frontal thrust must lock, leading to fracture of the adjacent sheet. The process continues: displacement of a sheet on a new ramp; buildup of topography and strength at the toe; and formation of a new internal thrust. The result (Figure 12) is the anomalous, backward migrating, thrust sequencing over the basement normal fault at the front of the Tadjik foldbelt (Leith and Alvarez, 1986).

The original depositional geometry of the Tadjik Basin apparently exceeds the negligible critical taper needed for it to overthrust the Jurassic salt. This allows the thrust sheet to

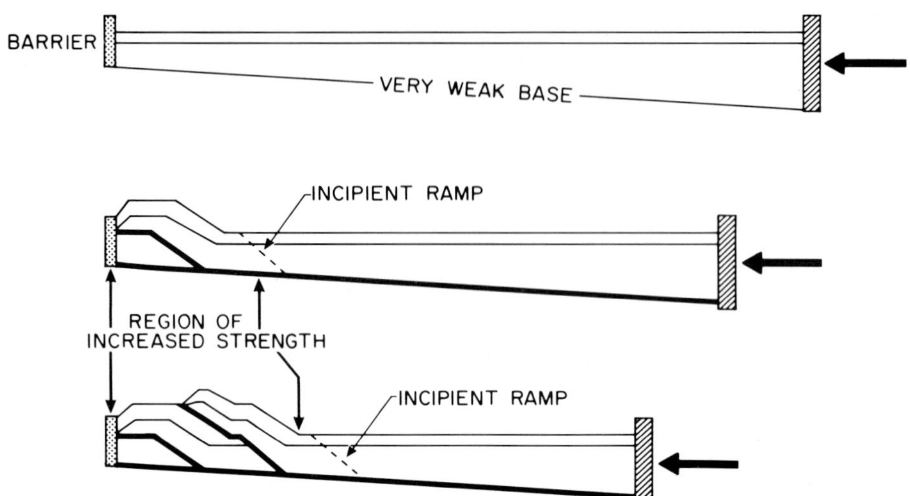

Figure 12. Schematic cross-section through a weak-basal foldbelt with a large basement offset at its front, (such as the Tadjik foldbelt), showing a backward migration of structures (after Leith and Alvarez, 1986).

slide well below its failure strength everywhere except near its northern end where there is a large mechanical barrier (the Illiac fault, which offsets basement by 3 km). This barrier causes localized failure and buildup of topography in much the same way as the depletion of the Salina salt in West Virginia caused the development of the Burning Springs Anticline of the Appalachian Plateau.

In summary, foldbelts over salt can act almost like stress guides, with relatively little shear traction exerted on its bottom (in salt), as well as none exerted on its top (in air). This is the essential reason why they can respond in such a complicated manner to a major basement uplift, as in the Salt Range (e.g., Yeats and Lawrence, 1985) or the Tadjik fold and thrust belt (e.g., Leith, 1984), and why they generally do not show the strong structural discrimination between their forward and backward directions typically found in strong-basal mountain belts.

C. Effects of the distribution of salt

Oroclines are curved mountain belts, which are usually assumed to have started out straight. Because salt facilitates overthrusting, the presence of a salt basin provides an obvious thin-skinned mechanism for the development of oroclines. However, oroclines can form in other ways as well. The original shape of a continental margin is likely to be ziz-zag in form because of the effect of transform faulting. It has been suggested (Thomas, 1977) that margin re-entrants become structural salients and margin promontories become structural recesses. Pre-existing basement structure is the obvious explanation for the Mackenzie Arc of northwestern Canada, but it cannot explain the outward extension of the Franklin Mountains and the Colville Hills (Figure 7), which is a thin-skinned phenomenon related to the distribution of the Cambrian salt.

The distribution of thin-skinned thrusting is one of the many aspects of Andean tectonics that appears to be very closely related to the angle of subduction of the Nazca plate (e.g., Jordan et al., 1983). However, Jensen (1984) has pointed out that there is a nearly one-to-one correspondence between the distribution of thin-skinned thrusting in Chile and Argentina and that of Oxfordian (Upper Jurassic) gypsum.

As the deformation in an evaporite-basal foldbelt moves far ahead of the position of the structural front in areas where there is no salt, there are inevitably going to be drag and rotational features at the edges of the salt layer along strike. The Burning Springs Anticline of the Appalachian Plateau (Section III) and the wrapping of folds around the Coahuila Peninsula in the Sierra Madre Oriental (Kleist, 1984) are two probable examples of this phenomenon. Such rotations, if they exist, should be reflected in paleomagnetic declinations. The results to date are mixed. Paleodeclinations from the Salt Range (Opdyke et al., 1982) and Sierra Madre Oriental (Kleist et al., 1984) are consistent with the expected degree of rotation due to the uneven advance of the overthrust, but at least some data from the Jura and the Appalachian Plateau are not (Elderedge et al., 1985). A recent study in the Appalachian Plateau (Kent and Miller, 1986) finds evidence for oroclinal bending and suggests that previous negative results there can be attributed to post-folding remagnetization. However, even if this is the case, such bending is probably more a response to the basement structure of the marginal salient than to along-strike variations in thin-skinned deformation.

The Verkhoyansk fold belt in Siberia is a north-trending Mesozoic foldbelt that separates the flat-lying upper Precambrian and Paleozoic rocks of the Siberian platform from the Paleozoic Chersky foldbelt and Kolyma massif to the east. Permian clastic rocks over 7000 m thick are deformed into large, linear folds and are bordered on the west by a thick section of syn-orogenic Cretaceous sediments derived from the Verkhoyansk foldbelt (Churkin, 1972). The Verkhoyansk complex probably represents the Paleozoic eastern continental margin of the Siberian plate, deformed in the Mesozoic by collision with the North American plate (Fujita, 1978; Fujita and Newberry, 1982). The Verkhoyansk foldbelt matches the characteristics of a "typical" evaporite-based fold-and-thrust belt so well that we consider the existence of undetected evaporites there to be a distinct possibility (Davis and Engelder, 1985). Published data indicate that some salt and a greater amount of other evaporites have been found in the area, but little is known of their extent (Zarkhov, 1981).

VI. SUMMARY

Although there is probably not a very great difference in strength between the basal detachment zone and the overthrusting rocks in many thin-skinned mountain belts, this is not true of overthrusting over salt. The style of deformation in thin-skinned fold-and-thrust belts is critically dependent upon the resistance to sliding along the detachment between the mass of deforming sediments and the underlying rocks. Evaporites can provide an extremely weak horizon within which a basal detachment can form and along which only a relatively small shear traction can be supported. Fold-and-thrust belts that form atop a salt layer share several readily observable characteristics (Davis and Engelder, 1985). The weakness of a detachment in salt permits a fold-and-thrust belt to maintain a much narrower cross-sectional taper (as little as a few tenths of 1°) than is commonly observed in the absence of salt (8° or more). This is reflected in the great width of many salt-basal fold belts, which project far outward toward the craton map view. Along strike, at the boundary between the parts of the fold belt that do and do not ride on salt, this difference in basal resistance often produces complex structures related to the accommodation of the different amounts of overthrusting in the two areas. Rodgers (1963) explained the anomalously trending Burning Springs anticline near the southwestern edge of the Silurian salt beneath the Appalachian Plateau as just this sort of salt-termination differential-drag structure. En echelon folding (e.g. Norris, 1972) and rotation about a salt-less hinge area (e.g., Seeber and Jacob, 1977; Crawford, 1974) appear to be other ways in which overthrusting rocks accommodate a large contrast in resistance along the detachment between areas with and without salt. The predicted subhorizontal orientation of the maximum principal stress axis over a detachment in salt is also consistent with the commonly observed lack of consistent vergence in structures within salt-basal overthrust belts.

VII. ACKNOWLEDGEMENTS

We would like to thank Hannes Brueckner, Charlotte Schreiber, Bill Leith, and Gordon Lister for their reviews and helpful comments. This work was supported by NSF grant EAR 84-07816 (DMD) and EPRI contract #RP2556-24 (TE).

REFERENCES

Andersen, E.M. (1942). The dynamics of faulting, Oliver and Boyd, London.
Aitken, J.D. and Cook, D.G. (1975). Upper Ramparts River and Sans Sault Rapids map areas, District of Mackenzie (106G, 106H), Geol. Surv. Canada, Open file 272.
Aitken, J.D., Cook, D.G., and Yorath, C.J. (1982). Upper Ramparts River (106G) and Sans Sault Rapids (106H) map areas, District of MacKenzie, Geol. Surv. Canada Memoir 388.
Brinkman, R., and Logters, H. (1968). Diapirs in western Pyrenees and foreland, Spain, In: Diapirism and Diapirs: A Symposium, edited by J. Braunstein and G.D. O'Brien, Am. Assoc. Petrol. Geol. Memoir 8, 275-293.
Burbank, D.W. (1983). The chronology of intermontane-basin development in the northwestern Himalaya and the evolution of the Northwest Syntaxis. Earth Planet. Sci. Lett. 64, 77-92.
Brace, W.F. and Kohlstedt, D.L. (1980). Limits on lithospheric stress imposed by laboratory experiments. J. Geophys. Res. 85, 6248-6252.
Byerlee, J. (1978). Friction of rocks. Pure Appl. Geophys. 116, 615-626.
Campbell, C.J. and Burgl, H. (1965). Section through the Eastern Cordillera of Colombia, South America. Geol. Soc. Am. Bull. 76, 567-590.
Carter, N.L. and Hansen, F.D. (1983). Creep of rocksalt. Tectonophysics 92, 275-333.
Chapple, W.M. (1978). Mechanics of thin-skinned fold-and-thrust belts. Geol. Soc. Am. Bull. 89, 1189-1198.
Churkin, M., Jr. (1972). Western boundary of the North American plate in Asia. Bull. Geol. Soc. Am. 83, 1027-1036.
Colman-Sadd, S.P. (1978). Fold development in Zagros Simply Folded Belt, southwest Iran. Geol. Soc. Am. Bull. 62, 984-1003.
Cook, D.G. and Aitken, J.D. (1973). Tectonics of northern Franklin Mountains and Colville Hills, District of MacKenzie, Canada, In: Arctic Geology, edited by M.J. Pitcher. Am. Assoc. Petrol. Geol. Memoir 19, 13-22.
Crawford, A.R. (1974). The Salt Range, the Kashmir syntaxis, and the Pamir arc. Earth Planet. Sci. Lett 22, 371-379.
Dahlen, F.A., Suppe, J., and Davis, D. (1984). Mechanics of fold-and-thrust belts and accretionary wedges: Cohesive Coulomb theory, J. Geophys. Res. 89, 10087-10101.

Davis, D.M. and Engelder, T. (1985). The role of salt in fold-and-thrust belts. Tectonophysics 119, 67-88.
Davis, D. et al. (1983). Mechanics of fold-and-thrust belts and accretionary wedges, J. Geophys. Res. 88, 1153-1172.
De Cserna, Z. (1971). Development and structure of the Sierra Madre Oriental of Mexico, Abstracts with Programs, Geol. Soc. Am., 377-378.
Elderedge, S. et al. (1985). Paleomagnetism and the orocline hypothesis. Tectonophysics 119, 153-179.
Engelder, T. (1979). The nature of deformation within the outer limits of the Central Appalachian foreland fold-and-thrust belt in New York State. Tectonophysics 55, 189-210.
Engelder, T. (1982). A natural example of the simultaneous operation of a free-face dissolution and a pressure solution. Geochim. Acta. 46, 69-74.
Engelder, T., and Geiser, P. (1979). The relationship between pencil cleavage and lateral shortening within the Devonian section of the Appalachian Plateau, New York. Geology 7, 460-464.
Engelder, T. and Geiser, P. (1984). Near-surface in-situ stress 4. Residual stress in the Tully Limestone - Appalachian Plateau, New York. J. Geophys. Res. 89, 9365-9379.
Farhoudi, G. (1978). A comparison of Zagros geology to island arcs. J. Geology 86, 323-334.
Fertl, W.H. (1976). Abnormal Formation Pressures. Developments in Petroleum Science, No. 2.
Fox, F.G. (1984). Structure sections across Parry Islands fold belt and Vesey Hamiltion Salt Wall, Arctic Archipelago, Canada, In Seismic Expression of Structural Styles, A Picture and Work Atlas, edited by A.W. Bally, AAPG Studies in Geology Series No. 15, vol. 3, Am. Assoc. Petrol. Geol., Tulsa, 3.4.1.54-3.4.1.72.
Frey, M.G. (1973). Influence of Salina Salt on structure in New York-Pennsylvania part of Appalachian Plateau. Am. Assoc. Pet. Geol. Bull. 57, 1027-1037.
Fujita, K. (1978). Pre-Cenozoic tectonic evolution of northeast Siberia. J. Geol. 86, 159-172.
Fujita, K. and Newberry, J.T. (1982). Tectonic evolution of northeastern siberia and adjacent regions. Tectonophysics 89, 337-357.
Gwinn, V.E. (1964). Thin-skinned tectonics in the Plateau and northwestern Valley and Ridge Province of the Central Appalachians. Am. Assoc. Pet. Geol. Bull. 75, 863-900.
Gwinner, M.P. (1978). Geologie der Alpen: Stratigraphie, Palaogeographie, Tektonik, 2. Auflage, Stuttgart, 480 pp.
Hafner, W. (1951). Stress distributions and faulting. Geol. Soc. Am. Bull. 62, 373-398.
Harding, T.P. and Lowell, J.D. (1979). Structural Styles, Their Plate-Tectonic Habitats, and Hydrocarbon Traps in Petroleum Provinces. Am. Assoc. Petrol. Geol. Bull. 63, 1016-1058.
Heim, A. (1871). Untersuchungen uber den Mechanismus der Gebirgsbildung. Basel.
Hubbert, M.K. and Rubey, W.W. (1959). Role of fluid pressure in mechanics of overthrust faulting, I, Mechanics of fluid-filled solids and its application to overthrust faulting. Geol. Soc. Am. Bull. 70, 115-166.
Jaume, S.C. et al. (1985). The mechanics of the Salt Range-Potwar Plateau: Pakistan. EOS Trans. AGU 66, 1090.
Jensen, O.L. (1984). Andean Tectonics related to geometry of subducted Nazca plate: Discussion and Reply. Discussion, Geol. Soc. Am. Bull. 95, 877-879.
Jordan, T.E. et al. (1983). Andean tectonics related to geometry of subducted Nazca plate. Geol. Soc. Am. Bull. 94, 341-361.
Kazantsev, Yu.V., and Kamaletdinov, M.A. (1977). Salt tectonics in the southern part of the Cis-Uralian foredeep and its connection with thrusts. Geotectonics 11, 294-298.

Kleist, R. et al. (1984). A paleomagnetic study of the Lower Cretaceous Cupido Limestone, northeast Mexico: Evidence for local rotation within the Sierra Madre Oriental, GSA Bull. 95, 55-60.

Laubscher, H.P. (1972). Some overall aspects of Jura dynamics. Am. J. Sci. 272, 293-304.

Laubscher, H.P. (1977). Fold development in the Jura. Tectonophysics 37, 337-362.

Leith, W. (1984). The Tadjik Depression, USSR - Geology, Seismicity and Tectonics. Ph.D. thesis, Columbia University, New York, 122 pp.

Leith, W. and Alvarez, W. (1985). Structure of the Vakhsh fold-and-thrust belt, Tadjik SSR: Geologic mapping on a Landsat image base. Bull. Geol. Soc. Am. 96, 875-885.

Leith, W. and Alvarez, W. (1986). Structure of the Vakhsh fold-and-thrust belt, Tadjik SSR: Geologic mapping on a Landsat image base (Reply). Bull. Geol. Soc. Am. 97, no. 6.

Leith, W. et al. (1981). Structure and permeability: Geologic controls on induced seismicity at Nurek Reservoir, Tadjikstan, USSR. Geology 9, 440-444.

Liechti, P. (1968). Salt features of France. Geol. Soc. Am. Special Paper 88, 83-106.

McLaughlin, D.H., Jr. (1972). Evaporite deposits of Bogota area, Cordillera Oriental, Colombia. Geol. Soc. Am. Bull. 56, 2240-2259.

McNaughton, D.A., Quinlan, T., Hopkins, R.M., and Wells, A.T. (1968). Evolution of salt anticlines and salt domes in the Amadeus Basin, Central Australia. Geol. Soc. Am. Special Paper 88, 229-247.

Miller, J.D. and Kent, D.V. (1986). Paleomagnetism of the Upper Devonian Catskill Formation from the southern limb of the Pennsylvania salient: Possible evidence of oroclinal rotation. Submitted to Geophys. Res. Letts.

Moore, J.C. and Biju-Duval, B. (1984). Tectonic synthesis, Deep Sea Drilling Project Leg 78A: Structural evolution of offscraped and underthrust sediment, Northern Barbados Ridge Complex. In: B. Biju-DuQval and J.C. Moore et al. Init. Repts. DSDP 78A, 601-621: Washington (U.S. Govt. Printing Office).

Muller et al. (1981). Deformation experiments on Anhydrite rocks of different grain sizes, rheology, and microfabrics. Tectonophysics 78, 527-543.

Nalivkin, D.V. (1973). Geology of the U.S.S.R.. Univ. of Toronto Press, 855 pp.

Nettleton, L.L. (1934). Fluid mechanics of salt domes. Am. Assoc. Petrol. Geol. Bull. 18, 1125-1204.

Norris, D.K. (1972). En echelon folding in the northern cordillera of Canada. Bull. Can. Pet. Geol. 20, 634-642.

Opdyke, N.D., Johnson, N.M., Johnson, G.D., Liny, E.H., and Tahirkheli, R.A.K. (1982). Paleomagnetism of the middle Siwalik formations of northern Pakistan and rotation of the Salt Range decollement. Paleogeogr., Paleoclimatol., Paleoecol. 37, 1-15.

Paraschiv, D. and Olteanu, Gh. (1970). Oilfields in the Mio-Pliocene zone of eastern Carpathians (district of Ploiesti), In Geology of Giant Petroleum Fields, edited by M.T. Halbouty. Am. Soc. Pet. Geol. Memoir 14, 399-427.

Price, R.A. and Douglas, R.J.W. (1972). (editors), Variations in tectonic styles in Canada. Geol. Soc. Canada Spec. Paper 11.

Prucha, J.J. (1968). Salt deformation and decollement in the Firtree Point anticline of central New York. Tectonophysics 6, 273-299.

Rodgers, J. (1963). Mechanics of Appalachian foreland folding in Pennsylvania and West Virginia. Am. Assoc. Pet. Geol. Bull. 47, 1527-1536.

Roeder, D. et al. (1978). Evolution and macroscopic structure of the Valley and Ridge thrust belt, Tennessee and Virginia. Stud. in Geol. 2, 25 pp., Dept. of Geol. Sci., Univ. of Tenn., Chattanooga.
Rodgers, C.L. et al. (1962). Tectonic framework of an area within the Sierra Madre Oriental and adjacent Mesa Central, north central Mexico. U.S. Geol. Surv. Prof. Paper 450C, Article 68, C21-C24.
Rutter, E.H. (1976). The kinetics of rock deformation by pressure solution. Phil. Trans. R. Soc. Lond. A 283, 203-219.
Seeber, L. and Jacob, K.H. (1977). Microearthquake survey of northern Pakistan: Preliminary results and tectonic implications. Proc. CNRS Symp. Geology Ecology Himalayas, Paris, Dec. 1976, 347-360.
Sarwar, G. and De Jong, K.A. (1979). Arcs, oroclines, syntaxes: The curvatures of mountain belts in Pakistan. In Geodynamics of Pakistan, edited by A. Farah and K.A. De Jong. Geol. Surv. Pakistan, Quetta, 341-349.
Seeber, L., Armbruster, J.G. and Quittmeyer, R.C. (1981). Seismicity and continental subduction in the Himalayan arc, in Zagros - Hindu Kush - Himalaya Geodynamic Evolution. Geodynamics Series v. 3, edited by H.K. Gupta and F.M. Delaney, 215-242.
Sherrill (1934). Symmetry of the Northern Appalachian Valley and Ridge Province, Ky. Geol. Surv. Special Pub. 1, 150-166.
Smoluchowski, M.S. (1909). Some remarks on the mechanics of overthrusts. Geol. Mag., N.S., Dec. 5, VI: 204-205.
Spiers, C.J., Urai, J.L., Lister, G.S. and Zwart, H.J. (1984). Water weakening and dynamic recrystallization in salt. Abstracts with Programs. Geol. Soc. Am. 16, 665 (abstr.).
Stearns, D.W. (1971). Mechanics of drape folding in the Wyoming Province. 23rd Annual Field Conference, Wyoming Geological Association Guidebook, p. 125-143.
Stocklin, J. and Nabavi, H.M. (1973). Tectonic map of Iran. Iran Geol. Survey, scale 1:2,500,000.
Suppe, J. and Wittke, J. (1977). Abnormal pore-fluid pressures in relation to stratigraphy and structure in the active fold-and-thrust belt of northwestern Taiwan. Pet. Geol. Taiwan 14, 11-24.
Thomas, W.A. (1977). Evolution of Appalachian-Ouachita salients and recesses from recesses and promontories in the continental margin. Am. J. Sci. 277, 1233-1278.
Tortochaux, F. (1978). Occurrence and structure of evaporites in North Africa. Geol. Soc. Am. Special Paper 88, 107-138.
von Huene, R. and Lee, H. (1982). The possible significance of pore fluid pressures in subduction zones, In Studies in Continental Margin Geology. Am. Assoc. Petrol. Geol. Memoir 34, 781-791.
Weide, A.E. and Martinez, J.D. (1970). Evidence for diapirism in Northeastern Mexico. Am. Assoc. Pet. Geol. Bull. 54, 655-661.
Wiltschko, D.V. and Chapple, W.M. (1977). Flow of weak rock in Appalachian Plateau faults. Am. Assoc. Pet. Geol. Bull. 61. 6535-6570.
Wiltschko, D. and Eastman, D. (1983). Role of basement warps and faults in localizing thrust fault ramps. Geol. Soc. Am. Memoir 159, 177-190.
Yeats, R.S. and Lawrence, R.D. (1986). Tectonics of the Himalaya thrust belt in northern Pakistan. Proceedings of the U.S.-Pakistan workshop on Marine Science in Pakistan, Karachi, Nov. 11-16, 1982, in press.
Zharkov, M.A. (1981). History of Paleozoic Salt Accumulation. Springer-Verlag, New York, 308 pp.

SALT CONTROL ON THRUST GEOMETRY, STRUCTURAL STYLE AND GRAVITATIONAL COLLAPSE ALONG THE HIMALAYAN MOUNTAIN FRONT IN THE SALT RANGE OF NORTHERN PAKISTAN

Robert W. H. Butler[1]
Michael P. Coward[2]

[1]Department of Geological Sciences
The University of Durham
Durham DH1 3LE UK

[2]Department of Geology
Royal School of Mines
Imperial College
London SW7 2BP UK

Gill M. Harwood

Department of Geology
The University
Newcastle-upon-Tyne NE1 7RU UK

Robert J. Knipe

Department of Earth Sciences
The University of Leeds
Leeds LS2 9JT UK

I. INTRODUCTION

The importance of evaporite sequences in the dynamic and kinematic evolution of thrust belts has long been suspected. The reasons are that evaporites have extremely low yield strengths compared to other lithologies. Evaporites are also relatively common constituents of present day continental margins such as those around the Red Sea and the Gulf of Mexico. Ancient continental margins are commonly believed to have been the precursors of many foreland fold and thrust belts which are generated during the later stages of continental collision. Mechanically weak and laterally persistent horizons, as provided by evaporite formations, play an important part in controlling the structural geometry of a thrust belt. Davis and Engelder (1985; following Davis et al., 1983) suggest that to overcome the shear resistence at the base of a thrust belt, a critical wedge taper must be exceeded. This taper will be derived from two components; a basal slope along the active thrust surface plus the synorogenic surface slope which will generally dip outwards towards the foreland. The basal slope, apart from local variations in thrust morphology, will be controlled by the magnitude of lithospheric flexure generated by loading of thrust sheets onto the foreland. Hence, close to the orogen, the basal slope will dip into the hinterland (e.g. Beaumont, 1981). The critical gradient of surface slope will be a function of the shear resistance at the base of the thrust pile (Davis et al., 1983). Therefore, upon encountering a weak horizon, the active thrust can propagate ahead of the main topographic expression of the thrust belt, thereby attaining a reduced mean surface gradient. The net result is to transfer displacement forward, from the main locus of crustal shortening in the orogenic interior, onto stratal shortening on the foreland. Recognition of this type of relationship (Fig. 1) is clearly important in understanding not only the overall dynamics of thrust sheet emplacement but also the mechanisms of large scale crustal deformation, and is critical for other aspects of mountain belt geology. The migration of stratal shortening far ahead of the main region of crustal shortening means that a significant proportion of the flexural depression, which rims any mountain belt, will be occupied by thrust structures and hence will not be available for the deposition of the unroofing products of the mountain belt (Fig. 1). Such thrust belts will severely disrupt the sediment pathways and drainage patterns in foreland basins, for understanding provenence and depositional facies within foreland basin successions.

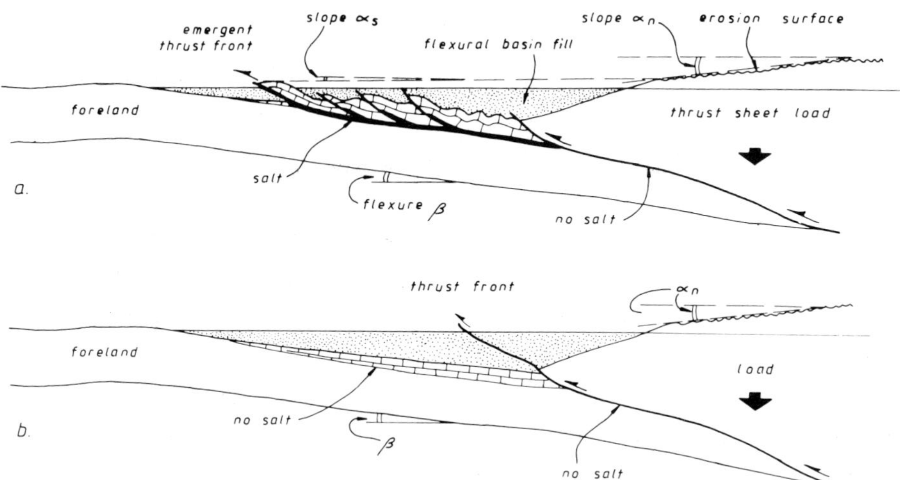

Figure 1. Diagrams illustrating the hypothetical contrasting foreland basin geometries for: (a) a thrust front substantially ahead of the main topographic load, this being caused by the decoupling of basement and cover along evaporite units; and (b) a thrust front lying immediately ahead of the main topographic load. The effect of enhanced forward thrust propagation is to reduce the surface slope, and hence the thrust wedge taper (α_s), over the salt. Steeper surface slope gradients (α_n) may be required to overcome the greater resistance to slip along the more internal portions of the basal thrust detachment beneath the main mountain belt. Note that the basal slope of both cases is controlled by the detailed thrust geometry and by the foreland lithospheric flexure (β) generated by loading.

Arguably the most famous example of the detachment of sedimentary rocks from basement during tectonic shortening are the Jura mountains of Europe. This belt of folds and thrusts lies about 100 km outboard of the NW Alps, the two separated by the Swiss molasse basin (Fig. 2). Gravity data show that basement has not been involved in folding beneath the Jura, although a combination of drill hole, gravity, seismic and geological data show that the basal decoupling surface for the Jura folds passes back (SE) into a zone of basement thrusting within the Alps (Menard and Thouvenot, 1984; Butler et al., 1986). The Triassic evaporite horizon lies at the base of a carbonate sequence several kilometers thick. In the NW this decoupling surface carried these carbonates out onto the Oligo-Miocene sediments of the Bresse graben, at which point the thrust front presumably became emergent. There the net result of the spatial separation between the thrust front and main crustal shortening site is that the last 25-30 km displacement within the orogen was transferred beneath the Swiss molasse basin. However, further S around the Alpine arc the separation between the thrust front and the locus of crustal shortening is much reduced; in the Grenoble area it is just 25 km (Fig. 2). The main foreland basin site remained in situ on its basement in Bas Dauphine, ahead of the thrust front, and hence was not subjected to the same degree of uplift as experienced by the Swiss molasse basin. A much more complete record of the upper Miocene molasse deposition history is preserved in this southern area. Vann et al. (1986; following Rigassi, 1977) interpret the variations in the separation between thrust front and crustal shortening sites as reflecting the distribution of Triassic evaporites on the foreland prior to the last increments of Alpine thrusting. It is likely that the evaporite formations are better developed to the northern part of the western Alpine foreland compared to in the south. This is not to argue that evaporite formations are absent directly beneath the Bas Dauphine basin. Figure 3 shows how a local paucity in evaporites would cause the active thrust surface to climb to a higher level, above the deeper evaporites. Clearly an important requirement of evaporite formations is lateral continuity if they are to develop into regional decoupling surfaces.

While the distribution of evaporites, and their control on the pattern of thrust belt development, are clear enough in the Alps, a number of problems remain unanswered. Since thrusting terminated in late Miocene times the thrust front has been eroded back several kilometers and the fold belt has been partially

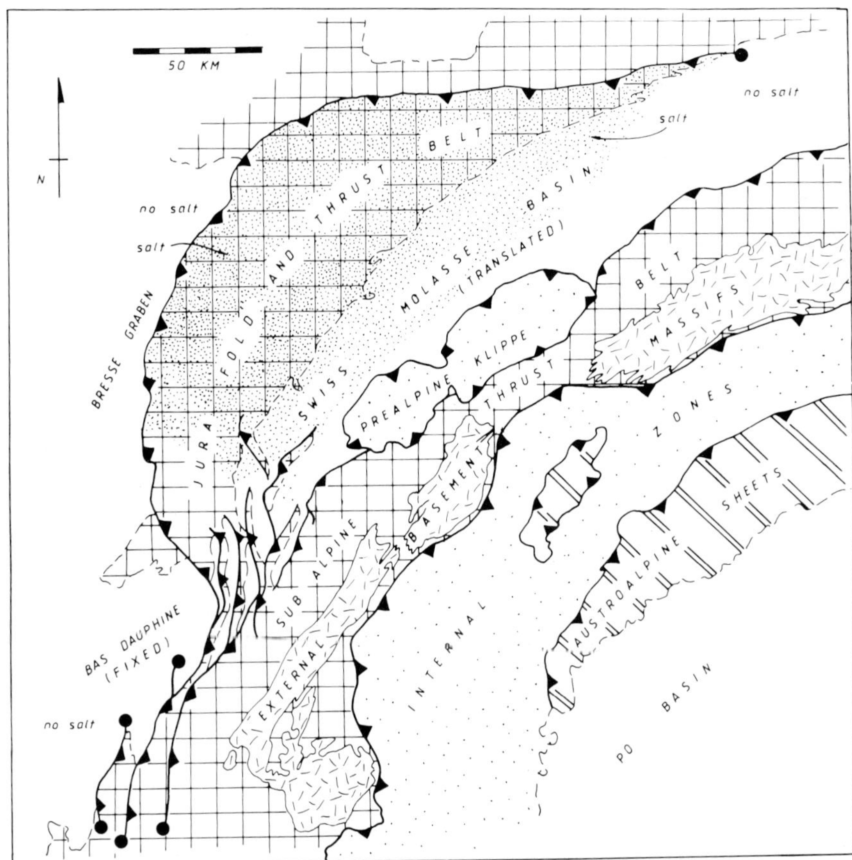

Figure 2. Simplified map of the NW Alps illustrating the outlying position of the Jura fold belt ahead of the main part of the mountain belt (the internal and Austroalpine zones are ornamented by spaced stipple and diagonal rule respectively). The presumed distribution of Triassic evaporites along the foreland basement (random flecks) and cover (cross-hatch ornament) is illustrated by dense stipple (after Vann et al. (1986)). The surviving portions of the molasse basins are unornamented.

unroofed. Furthermore, the mountain belt as a whole has experienced rapid uplift in Quaternary times so that the present day erosion level though the thrust pile probably bears few similarities with the synorogenic surface. Uplift will also have altered the basal slope by changing the conditions of lithospheric flexure, not only with respect to the load magnitude

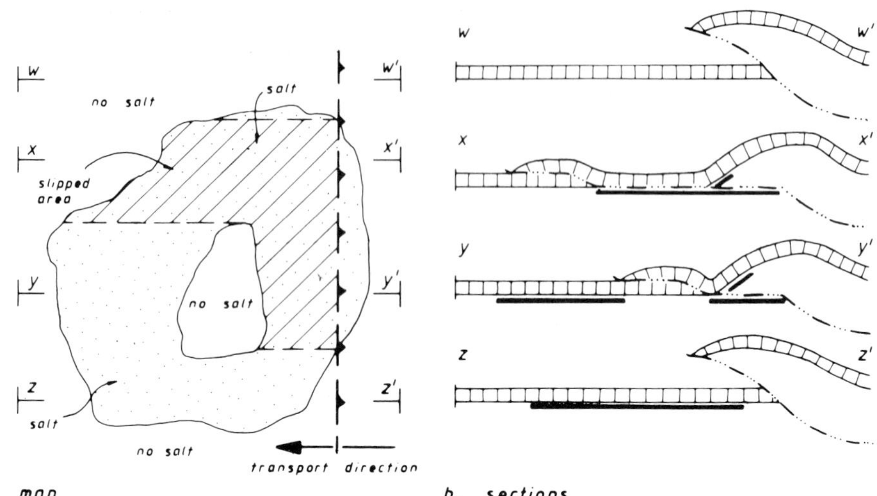

Figure 3. Map (a) and sections (b) illustrating the importance of salt continuity in generating thrust geometries. The thrust is considered to climb up to the salt-bearing horizon illustrated on (a) at the barbed line, where it can propagate forwards. At the scale of the diagram the thrust can propagate infinitely far forwards once it encounters salt. The thrust will climb to higher stratigraphic levels if the salt is either not encountered or lost from the detachment horizon. The diagrams also assume that forward thrust propagation is very much faster than it is laterally. The thrust geometry, resulting from the exploited area of the salt horizon, is controlled by the continuity of the salt for the particular thrust transport direction. This salt layer is depicted on the sections (b) by the thick line.

and morphology but also with respect to the thermal structure and hence rigidity of the foreland. Any intermontaine depocenters which existed during deformation have been eroded. Thus the detailed interactions between thrust tectonics, evaporites and molasse sedimentation during the deformation are obscure. To examine these aspects we must choose a modern thrust front in an evaporite province. The example chosen here is the Salt Range of northern Pakistan which lies outboard of the western Himalayas in the Punjab foreland basin. We discuss the structural geometry and evolution of the Salt Range from late Miocene to Recent times emphasizing the relationships between structural style and displacements with the distribution, mobilization and migration of evaporite deposits, together with the control of depocenter location and unroofing histories of molasse sediments.

II. TECTONIC SETTING

The mountain belts of N Pakistan lie at the western end of the main Himalayan chains (Fig. 4) in the apex of a major oroclinal deflection in structural trend. Further west the folds and thrusts trend more north-south and can be traced around

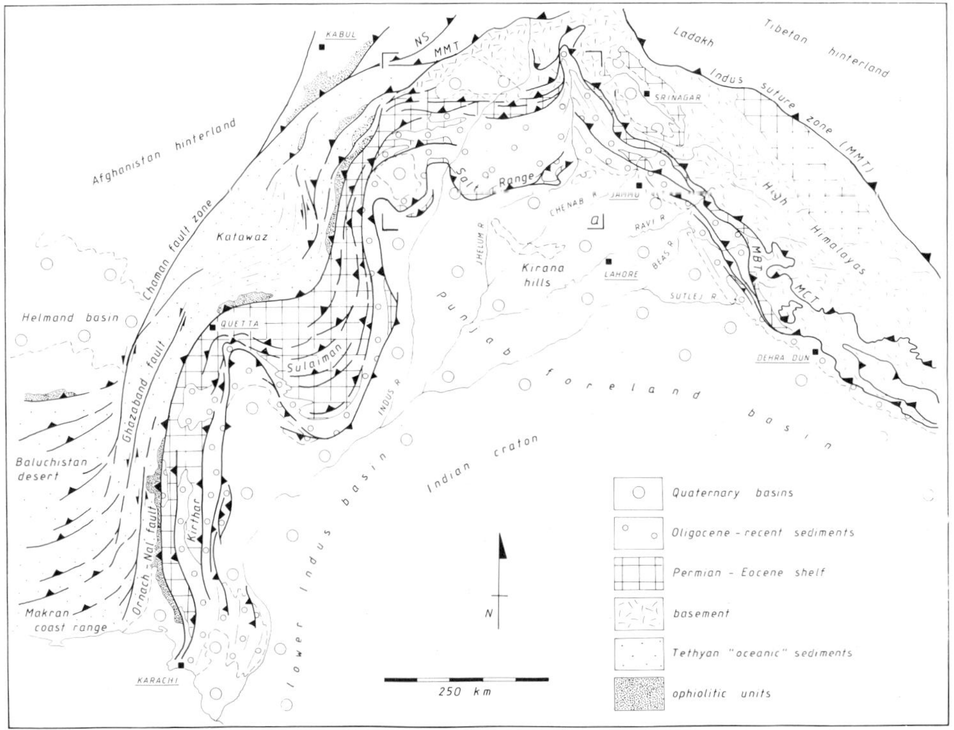

Figure 4. Tectonic sketch map of the NW Himalayas and surrounding thrust systems in western Pakistan, modified from Gansser (1964) and Butler and Coward (in press). NS - Northern (Kohistan) Suture, MMT - Main Mantle thrust, MCT - Main Central thrust, MBT - Main Boundary thrust, a - area of Figs. 5 and 6.

Table 1. Stratigraphic Framework of the Salt Range District (modified after Shah, 1977).

Era	Period	Epoch	Group	Formation
		QUATERNARY		QUATERNARY ALLUVIAL SEDIMENTS (no formation name)
CENOZOIC	TERTIARY	LOWER PLEISTOCENE	SIWALIK GROUP	LEI CONGLOMERATE
				SOAN FORMATION
				DHOK PATHAN FORMATION
		PLIOCENE		NAGRI FORMATION
				CHINJI FORMATION
		MIOCENE	RAWALPINDI GROUP	KAMLIAL FORMATION
		?UPPER OLIGOCENE		MUREE FORMATION
		EOCENE		CHORGALI FORMATION
				SAKESAR LIMESTONE FORMATION
				NAMMAL FORMATION
		PALEOCENE		PATALA FORMATION
				LOCKHART LIMESTONE FORMATION
				HANGU FORMATION
MESOZOIC		LOWER CRETACEOUS		LUMSHIWAL FORMATION
				CHICHALI FORMATION
		UPPER JURASSIC		
		MIDDLE JURASSIC		SAMANA SUK FORMATION

Table 1 (continued)

Era	Period	Group	Formation
MESOZOIC	LOWER JURASSIC		SHINAWARI FORMATION DHATTA FORMATION
	UPPER TRIASSIC		KINGRIALI FORMATION
	MIDDLE TRIASSIC		TREDIAN FORMATION
	LOWER TRIASSIC		MIANWALI FORMATION
PALEOZOIC	UPPER PERMIAN	ZALUCH GROUP	CHHIDRU FORMATION WARGAL LIMESTONE FORMATION AMB FORMATION
	LOWER PERMIAN	NILAWAHAN GROUP	SARDHAI FORMATION WARCHHA SANDSTONE FORMATION DANDOT FORMATION TOBRA FORMATION
	MIDDLE CAMBRIAN		BAGHANAWALA FORMATION JUTANA FORMATION KUSSAK FORMATION
	LOWER CAMBRIAN		KHEWRA SANDSTONE FORMATION
	/EOCAMBRIAN		SALT RANGE FORMATION

after Shah 1977.

festoon-like subsidiary arcs to the Makran coast range, marginal to the Indian Ocean. In the main Himalayan belt of India and Nepal (e.g. Gansser, 1964) the distribution of topography is relatively simple with the high ground (of about 6000 m) lying to the north of the regionally extensive Main Boundary Thrust (MBT) zone. There is only a narrow belt of thrusts and folds ahead of the MBT before the main foreland basin is encountered (Fig. 4). However, the apex between north-south and NW-SE trending belts shows a more complex distribution of topography, and the tectonic analog of the High Himalayas has an average elevation of only about 2500 m. The Salt Range lies in this apex, about 100 km south, outboard from the main part of the western Himalayan foothills, but well within the flexural depression which moats the entire Himalayan range from Karachi to the Ganges delta.

Figure 5. LANDSAT mosaic of the ground between the Punjab foreland basin and the Peshawar basin in the Himalayan thrust belt, locality (a) on Fig. 4. See Fig. 6 for tectonic overlay of the same area.

Figures 5 and 6 illustrate this setting in more detail together with the distribution of the major present day sedimentary depocenters for the alluvial molasse which are being fed by the major river systems of the Indus and Jhelum draining the Himalayan range. The largest depocenter lies ahead of the Salt Range and is the main Indus-Punjab foreland basin. A region of molasse outcrop, north of the Salt Range but ahead of the Main Boundary Thrust Zone, has experienced recent uplift and now forms the Potwar Plateau. This area probably formed part of the main Indus-Punjab basin but has been uplifted during the deformation which also produced the Salt Range (Burbank and Raynolds, 1984). To the west this uplift was less pronounced locally, so that the Marwat and Khisor (Trans-Indus) ranges merely ponded molasse behind them producing the enclosed, intermontaine depocenter of the Bannu basin similarly, to the north the Peshawar and Campbellpore basins act as a transitory reservoir for alluvial sediments from the Kabul and Indus rivers, ponded behind the Attock-Cherat and Kala Chitta ranges (Burbank and Tahirkheli, 1985).

It has long been suspected that the Salt Range represents a Himalayan analog for the Jura of the NW Alps; uplift was caused by thin-skinned detachment and shortening of the Indian continental shelf sediments above a decoupling surface within or at the base of these sediments. The best independent evidence to support this model comes from gravity data (Farah et al., 1977; Fig. 7). A gradually increasing negative Bouguer anomaly on a south to north section trends obliquely to the main surface expression of the Salt Range, although the range does not coincide with any deep structural element which might change the gravity pattern. Thus basement structure is preserved from the foreland back beneath the Salt Range and southern Potwar plateau. The systematic increase in the negative Bouguer anomaly is characteristic of flexural subsidence of foreland continental

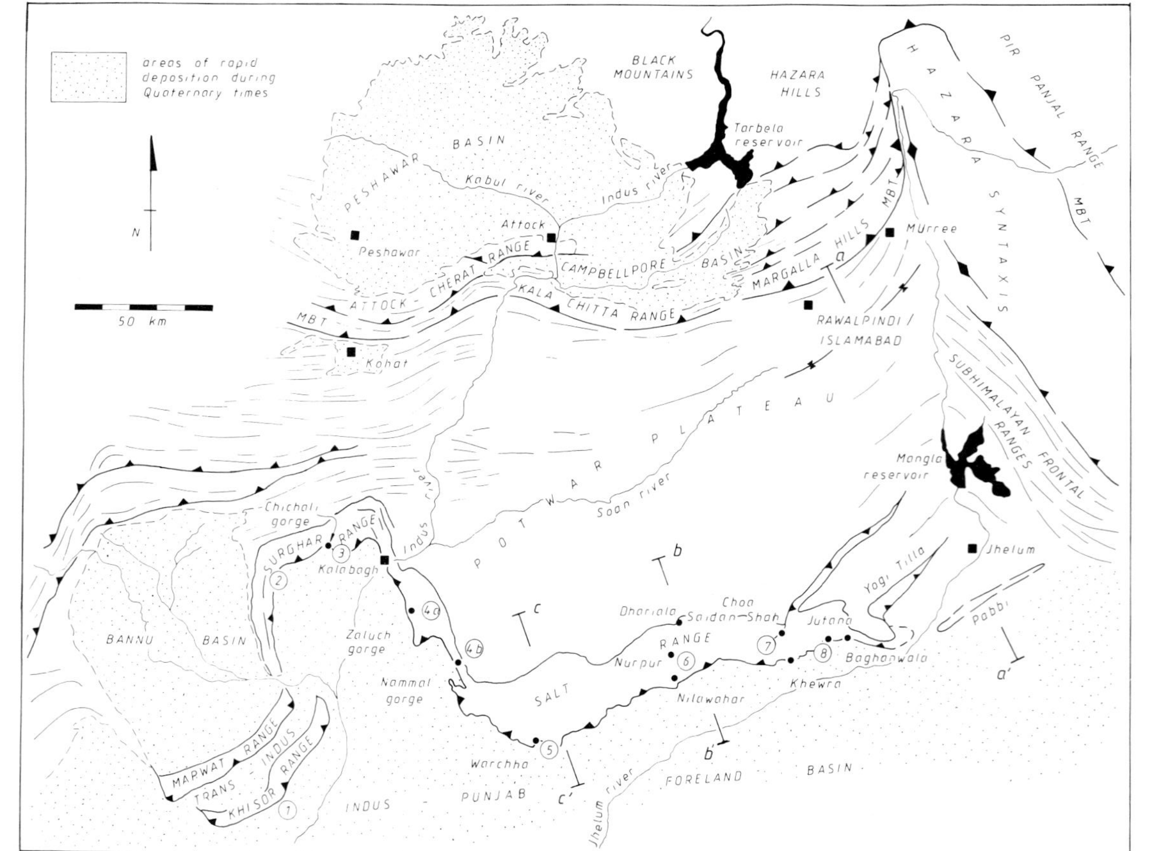

Figure 6. Tectonic overlay to the LANDSAT mosaic of Fig. 5, locality illustrated on Fig. 4. MBT - Main Boundary thrust. The section lines of Fig. 10 and 11a,b are depicted a-a', b-b' and c-c' respectively. The locations of the measured stratigraphic sections in Fig. 9 are indicated by circled numbers.

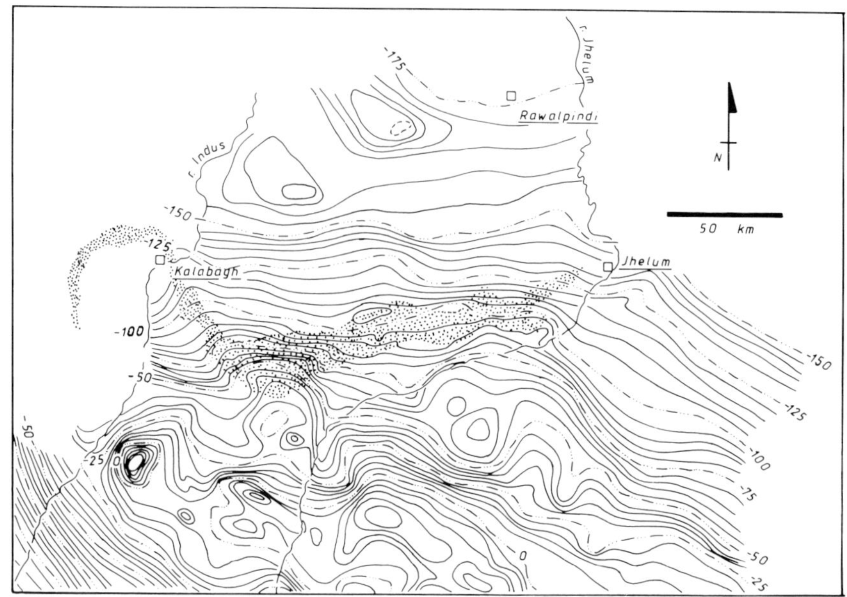

Figure 7. Map of the Bouguer gravity anomaly (contours in mgals) in the northern Punjab foreland and Potwar plateau, after Farah et al. (1977). The positions of the Salt and Surghar ranges are stippled.

crust under an adjacent load (Karner and Watts, 1983) (Fig. 1). Proterozoic basement rocks outcrop in the Kirana Hills (Fig. 4) and coincide with the ridge of positive Bouguer anomalies (Fig. 7) to the south of the flexural basin. These hills may represent a low amplitude peripheral bulge, although this is probably superimposed on a pre-existing ridge within the basement.

The relationship between the Salt Range and the main Himalayan thrust belts to the north is illustrated on two parallel cross-sections (Fig. 8). These show contrasting structural geometry, illustrated by the different levels of detachment within the sedimentary pile for the regional sole thrust. In the east this thrust has climbed up into the molasse sediments just south of the Main Boundary Thrust zone and has shortened these units on a series of hinterland and foreland directed thrusts. About 50 km west the sole thrust has run along the base of the sediment pile so that the thrust front brings up the pre-molasse shelf rocks to outcrop. Variations on this second type of thrust profile, while retaining the essential feature of basement-cover detachment, characterize the structure along the Salt Range. However, both regional sections (Fig. 8) display the same spatial relationship between thrust front and the main location of imbrication. Figure 8a is based on a regional balanced cross-section across the entire Indian continental thrust belt in northern Pakistan (Coward and Butler, 1985). Almost 500 km shortening of Indian continental cover and upper crust are implied, of which only the last 45 km becomes emergent ahead of the Main Boundary Thrust zone. Thus the deformation within the Salt Range represents a very small proportion of the whole thrusting process within the Himalayas. The initial collision between the Indian continent and the Asian (Tibetan) block occurred in middle Eocene times. There has been about 2000 km of later convergence across the whole mountain belt, as evidenced by palaeomagnetic and Indian ocean floor magnetic anomaly studies (Patriat and Achache, 1984). In northern Pakistan these data suggest that the convergence direction had a NNW-SSE axis throughout the collision process, coincident with the direction of thrust transport deduced from small scale structures within the northern Pakistan thrust belts (Coward et al., in press). Although these thrusts represent a large amount of displacement, far more (1500 km) must have occurred to the north in Tibet and the Pamirs. Broader relationships between the Salt Range and regional tectonics will be considered later. but first we introduce the stratigraphy exposed within the Salt Range before proceeding to examine the structural geometry, its evolution and interactions with salt dynamics associated with the frontal tens of kilometers displacement in northern Pakistan.

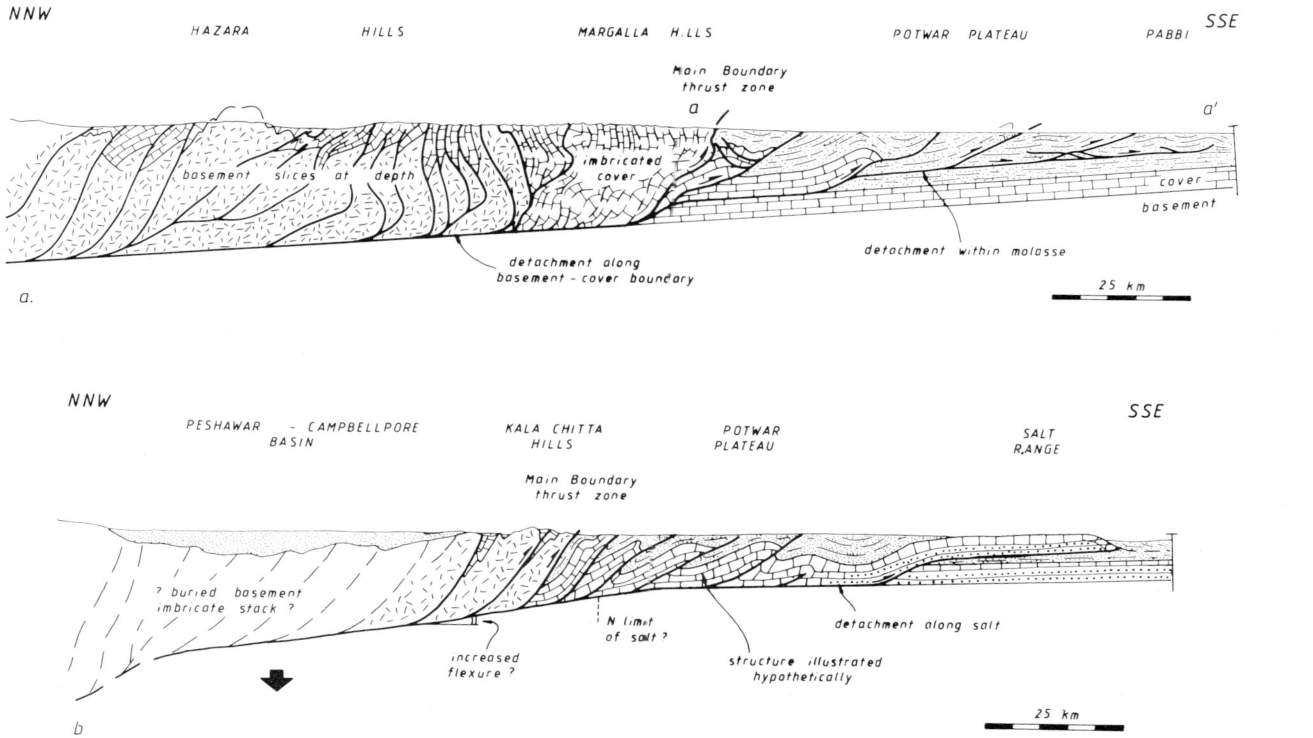

Figure 8. Two regional cross-sections across the southern part of the Himalayan thrust systems in northern Pakistan illustrating the contrasting geometry of thrusts, the outlying nature of the thrust front, and the disruption of foreland basin and intermontaine basin sediments (stippled). a) lies along a projection of the line a-a' on Fig. 6 and is simplified from the regional balanced section of Coward and Butler (1985). b) lies approximately 50 km to the west of, and parallel to (a), approximately along the projected line b-b' on Fig. 6.

III. STRATIGRAPHIC FRAMEWORK

The stratigraphy of the Salt Range can be considered as three major units: the salt, the overlying carapace sediments, and the molasse. Stratigraphic details are summarized in Table 1 and the stratigraphic relations depicted in Fig. 9. The oldest sediments are the Eocambrian - Cambrian Salt Range Formation (Asrarullah, 1967), an evaporite sequence overlain by Early Cambrian sediments. In turn these are unconformably overlain by Permian glacial and peri-glacial sediments which pass upwards into Permian carbonates, and then into a thick Mesozoic sequence of shallow water carbonates and clastics. There is evidence for deepening water conditions in the Upper Jurassic and Lower Cretaceous with subsequent rapid shallowing in the Aptian-Albian. Paleocene and Eocene sediments are fluvio-deltaic clastics, with some local coals, interbedded with carbonates. These sediments are unconformably overlain by syn-tectonic molasse sediments, which are involved in the latest tectonic movements. The major formations are briefly described below, a more detailed description can be found in Shah (1977).

A. Salt Range Formation

The Salt Range Formation is the major evaporite-bearing formation in the area and comprises red marls, gypsum and dolomite, passing upwards into gypsum and, above, thick beds of red-stained and translucent halite interbedded with thinner red marls and some poor grade oil shales. Minor potash salts are also present. It represents a thick series of playa and distal alluvial fan evaporite sediments, with paleocurrent directions indicating transport from the south (Asrarullah, 1967). In the type section in the Khewra Gorge (Fig. 9) the formation attains a thickness of over 830 m, of which more than 630 m is halite. In the southern Potwar at Dhariala (Fig. 6), some 2000 m of salt has been drilled. Regional variations in the stratigraphic thickness

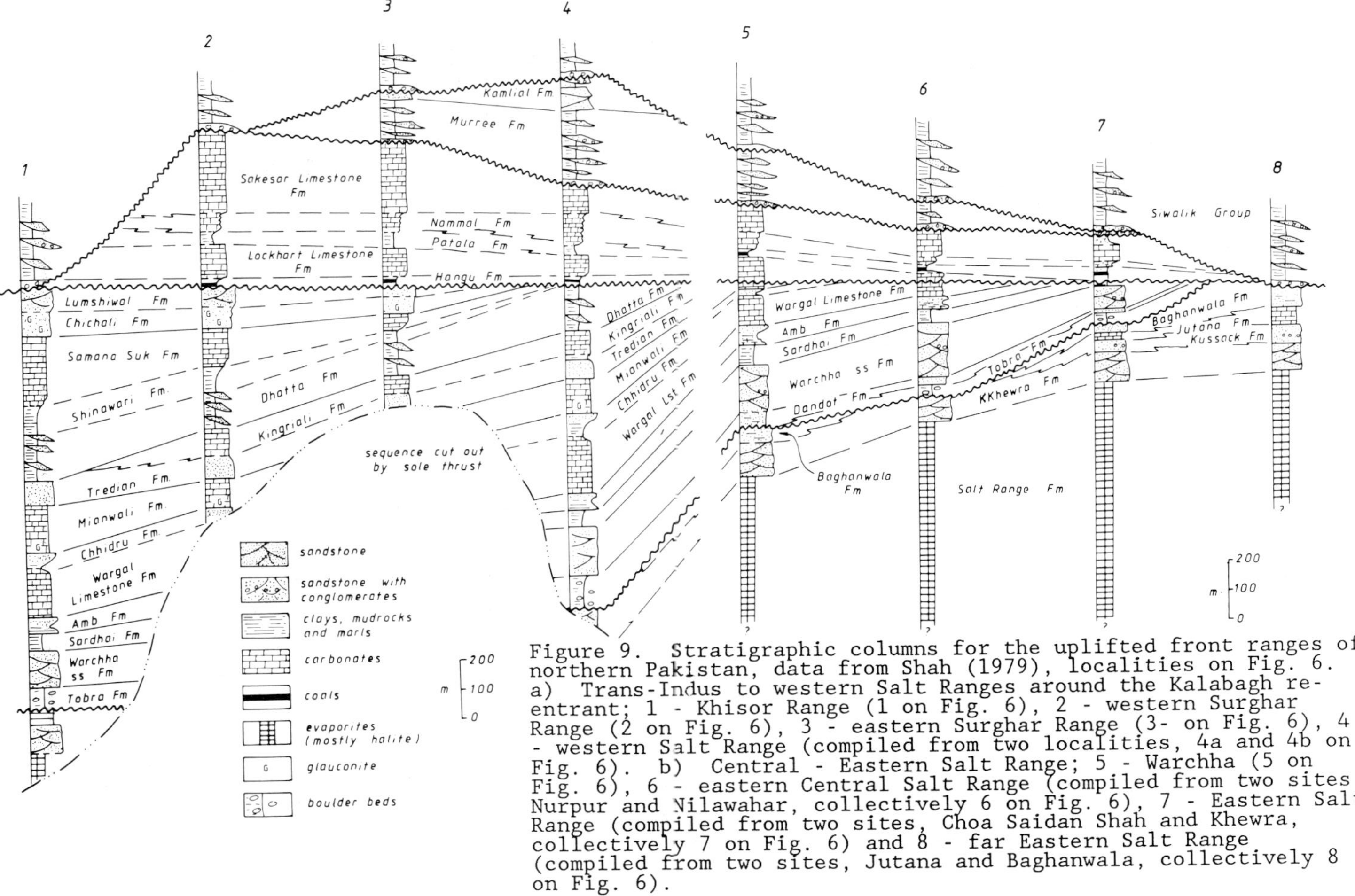

Figure 9. Stratigraphic columns for the uplifted front ranges of northern Pakistan, data from Shah (1979), localities on Fig. 6. a) Trans-Indus to western Salt Ranges around the Kalabagh re-entrant; 1 - Khisor Range (1 on Fig. 6), 2 - western Surghar Range (2 on Fig. 6), 3 - eastern Surghar Range (3- on Fig. 6), 4 - western Salt Range (compiled from two localities, 4a and 4b on Fig. 6). b) Central - Eastern Salt Range; 5 - Warchha (5 on Fig. 6), 6 - eastern Central Salt Range (compiled from two sites, Nurpur and Nilawahar, collectively 6 on Fig. 6), 7 - Eastern Salt Range (compiled from two sites, Choa Saidan Shah and Khewra, collectively 7 on Fig. 6) and 8 - far Eastern Salt Range (compiled from two sites, Jutana and Baghanwala, collectively 8 on Fig. 6).

have been modified by tectonics, salt diapirism and extrusion. The formation is exposed throughout the main Salt Range from east of Baghanwala to Kalabagh in the west (Fig. 6), but is locally absent along the western Salt Range and is not present in the Surghar Range (see Fig. 6 for locations). Drill holes have encountered the formation beneath the Punjab foreland basin (several tens of kilometers south of the Salt Range where the formation rests on metamorphic basement), and also beneath the Potwar plateau (Shah, 1977).

B. Carapace sediments

The carapace sediments can be divided into three main units, the lowest being of Cambrian sediments, the next involving rocks of Permian and Mesozoic age, and the youngest of Lower Tertiary age. The Cambrian sediments directly overlie the Salt Range Formation, a contact considered to be conformable by Gee (1945) but tectonic by Sahni (1947). Field traverses by the authors have shown numerous small-scale thrusts and associated folds near the contact, but we consider that these are associated with Himalayan collision tectonics and that, prior to this deformation, the contact between the Salt Range Formation and its carapace was conformable.

The Cambrian sediments are dominantly reddened micaceous sandstones containing cross-stratification and ripple lamination with evidence for sub-areal exposure from mudcracks and halite psuedomorphs. Collectively these sediments comprise the Khewra Sandstone and Baghanawala Formations and probably represent distal alluvial fan deposits. They are overlain by the Kussak and Jutana Formations, present only in the eastern Salt Range (Fig. 9), which represent a partial return to an inland drainage basin with playa facies displaying less dominant sandstones with dolomites and gypsum. All these formations have been dated as Early to Middle Cambrian in age (Schindewolf and Seilacher, 1955; Teichert, 1964).

Permian sediments overlie the Cambrian formations with a slight angular unconformity (Fig. 9). They can be divided into the Nilawahan and Zaluch Groups (Table 1). The oldest sediments in the Nilawahan Group, the Tobra Formation, comprise tillites with diamicrites, boulder beds, sandstones and siltstones while, in the Central Salt Range, a series of lacustrine freshwater shales and siltstones contain an Early Permian fauna of freshwater bivalves. In the Eastern Salt Range the Tobra Formation grades upward into marine sandstones of the Dandot Formation, but in the Warchha area (Fig. 6) the Dandot Formation lies directly on Cambrian sediment. The overlying Warchha Formation has an abrupt, albeit conformable, lower contact with the Dandot and Tobra Formations. The Warchha is composed of cross-stratified medium-to-coarse grained reddened sandstones which are commonly arkosic and contain evidence for fluviatile deposition over a broad alluvial plain. This Warchha Formation forms a distinct, sandy sequence of up to 180 m thickness over much of the Salt Range (Fig. 9). In contrast, the overlying Sardhai Formation contains clays and, locally, carbonaceous shales with subordinate sandstones. In the Khisor Range (Fig. 6) the formation contains dark, argillaceous limestones, the whole unit probably representing flooding of the alluvial plain.

The overlying Zaluch Group, of Early through Late Permian age, shows less siliciclastic input and represents a change to warmer depositional conditions. The lowest unit is the Amb Formation which has a sharp basal contact with the Nilawahan Group and contains a good marine fauna of plentiful fusilinids, productids, bryozoa, bivalves and gastropods. The top contact of the Amb with the Wargal Limestone Formation is abrupt. The Wargal Formation is of Late Permian (Guadalupian) age and contains limestones, dolomites with local cherts and contains a rich, shallow, marine fauna of abundant bryozoa, brachiopods, bivalves, gastropods, nautiloids, trilobites and crinoids.

Increasing upwardshale content forms a gradation into the Chhidru Formation, which contains shales with rare phosphatic nodules grading up into calcareous sandstones and rare sandy limestones (Kummel and Teichert, 1970). Both the sandstones and the carbonates are fossiliferous, indicating Late Permian deposition, and probably have an offshore marine origin.

Fossil evidence indicates the absence of the uppermost Permian stage (Kummel and Teichert, 1970) and the Triassic (Scythian) Mianwali Formation contains some reworked Permian faunas. The Mianwali Formation incorporates carbonates with subordinate sandstones, siltstones and marls; limestones and dolomites in the lowermost member are glauconitic, indicating moderate water depths, but the formation shallows upwards and becomes very fossiliferous with the uppermost beds being dolomitic. There is a sharp junction with the sandstones and minor shales of the Tredian Formation, which pass up into distinctive massive white sandstones of probable shallow marine origin. An increasing dolomitic component occurs up stratigraphic section, grading into the Kingriali Formation of Late Triassic age. This formation, in turn, contains thinly bedded to massive dolomites with dolomitic limestones, and terminates upwards at an irregular iron-rich dolomitic layer which may be a hardground. Dolomitization has destroyed detailed fossil evidence but Shah (1977) records brachiopods, bivalves and crinoids.

The Jurassic sediments of the Salt Range are shallow marine to non-marine carbonates, shales and sandstones. Emergence took place at the end of Triassic times, and the oldest Jurassic rocks of the Dhatta Formation rest disconformably on the Kingriali and Tredian Formations. The Dhatta Formation contains alluvial and deltaic red, grey and white sandstones with siltstones, shales, mudstones and fire clays with carbonaceous horizons. The Dhatta is widely developed in the Salt and Trans-Indus Ranges and grades upwards into thin to well-bedded limestones, nodular marls, shales and calcareous sandstones of the Shinawari Formation. Locally oolitic carbonates, and cross-stratified and ripple laminated sandstones with iron-rich horizons within the formation, show a return to marine, shallow water, conditions. Bivalves, corals and gastropods indicate a Lower Jurassic (Toarcian) age with possible middle Jurassic components. Open marine depositional conditions prevailed in the overlying Samana Suk Formation where grey limestones with thin shales contain rich skeletal fragments with local ooids. The formation is present in the Western Salt Range and thickens westwards into the Surghar and Trans-Indus Ranges (Fig. 9). The upper contact of the Samana Suk is abrupt and represents a disconformity with the Upper Jurassic to Lower Cretaceous Chichali Formation. These latter are glauconitic sandstones, nodular carbonates, phosphatic and iron-rich sandstones with many probable non-sequences, indicative of a change to a deeper water, offshore depositional environment. Abundant belemnites are present within the glauconitic sandstones. The Chichali Formation passes up into thick-to-massive bedded, cross-stratified sandstones with local shale horizons which collectively characterize the overlying Lumshiwal Formation. This Aptian to Early Albian sequence is preserved only in the far Western Salt and Trans-Indus Ranges, and represents a rapid shallowing upwards in the area. The Lumshiwal in the Khisor Range contains fluvio-deltaic sandstones with tree fragments.

Emergence in Upper Cretaceous times with gentle tilting led to a slight angular unconformity with the overlying Paleocene and Eocene sediments. The oldest of these sediments is the Lower Paleocene Hangu Formation, which comprises deltaic sandstones and shales, locally iron-stained, with carbonaceous shales and coals in the Western Salt and Surghar Ranges. The overlying Lockhart Limestone Formation contains foraminifera which are the major faunal constituent but corals, gastropods, echinoids, algae and bivalves are recorded by Shah (1977) yielding a Paleocene age. An increased siliciclastic input, combined with a shallowing depositional environment, marks the transition into the Late Paleocene Patala Formation. The Patala comprises shales, marls and thin limestones with calcareous sandstones towards the top of the formation, coals are found in the central and eastern Salt Range. The overlying Nammal Formation contains abundant foraminifera and sparse molluscs within a sequence of alternating shales, marls and limestones, of Early Eocene age, increasingly dominated by limestones towards the upper, transitional contact with the Sakesar Limestone Formation. This Early Eocene carbonate is prominent through much of the area although eroded from the far eastern Salt Range and western Surghar Range. The carbonate is grey or cream in color, commonly nodular, with cherts abundant in the upper parts and yields foraminifera, molluscs and echinoids which suggest deep shelf to upper slope facies. In the Eastern Salt Ranges the Salecsav Limestone is overlain by the Chorgali Formation whose alternating sequence of shales and limestones have yielded an Early Eocene fauna. All the Eocene sediments are overlain unconformably by molasse sediments atop a weathered palaesol surface (Fig. 9).

It should be emphasized that these stratigraphic notes are only appropriate to the mountain front area and that large scale variations in stratigraphy have been recognized in the Kala Chitta and Hazara hills (e.g. Calkins et al., 1975). There is also evidence for late Precambrian - early Palaeozoic deformations in these districts which are not represented in the Salt Range.

C. Molasse

The present mountain belt from Karachi to the Ganges delta is rimmed by a moat filled with clastic sediments derived from the Himalayas and associated regions of uplift. Karner and Watts

(1983) show that this moat represents flexing of the Indian continental lithosphere caused by loading from the adjacent mountain belts. As such the moat forms a classic foreland basin providing a depository for the eroded detritus from the mountain belt. Work within the mountain belts has shown that the deformation which caused their uplift, and hence the load, has migrated southward across the Indian continent with time (e.g. Coward and Butler, 1985). Therefore the flexural basin itself will have migrated into the foreland, the basin sediments forming a transgressive sequence onto the underlying Cambrian-Eocene cover rocks. This transgressive package of molasse sediments has been divided into two. The lower group is the Rawalpindi Group, which, in turn, is split into the Murree and, overlying, Kamlial Formations (Shah, 1977). The Murree Formation comprises a distinctive sequence of dark red and purple clays with sandstones and local intraformational conglomerates. The Kamlial Formation is found overlying the Murree in the western Salt and Surghar Ranges (Fig. 9), and is essentially a coarser unit with subordinate shales. In the northern Potwar plateau, the Rawalpindi Group attains a thickness of over 3000 m but is only 200 m thick in the Nammal district of the western Salt Range.

The Rawalpindi Group is overlain by the Siwalik Group which is much greyer than the underlying molasse. The Siwaliks have been subdivided into a further five formations, listed in Table 1, and can exceed a thickness of 4000 m. The Siwaliks have a much greater sandstone content, with conglomerates, and represent the unroofed detritus from the Himalayas (Pilgrim, 1910; Gill, 1951).

The ages of the various molasse sediments have been given by vertebrate fauna and plant fragments (e.g. Pilgrim, 1910). Extremely detailed dating has been possible in the Siwaliks of the Potwar and Peshawar basins using the detailed magnetic reversal stratigraphy preserved in the sediments. Excellent studies by Johnson and coworkers (Johnson et al., 1979), supported by detailed sedimentological work (Burbank, 1983; Burbank and Raynolds, 1985), show the southward transgression of foreland basin sedimentation, and can provide absolute ages for the growth of folds and thrusts in the Potwar and Salt Range districts. The Murree Formation sediments, which are the oldest foreland basin fill in the Potwar area, have yielded Early Miocene ages, whereas the Siwaliks in this district date from Middle Miocene to Recent times. Alluvial sedimentation continues to the present day as the Indus and Jhelum rivers, with tributaries, discharge Himalayan detritus into the foreland flexural and intermontaine basins.

IV. STRUCTURE OF THE THRUST FRONT

In plan form (Figs. 5 and 6) the Salt Ranges have a markedly irregular pattern. In the east, around Jhelum, the frontal structures are distributed with an anticline in the south and widely spaced thrusts behind (Johnson et al., 1979), illustrated by the cross-section through the eastern Potwar and the edge of the Salt Ranges (Fig. 10a). The frontal anticline is considered to be detached above a blind thrust which locally shows very little displacement. This thrust has run along a horizon within the molasse. The remaining section, northward to Rawalpindi, is based on a road section supported by high quality commerical seismic reflection data. These data confirm that, north of Jhelum, thrust structures break the present erosion surface and were probably emergent during displacement. An opposed dip complex has developed within the Siwalik molasse and can be traced using the two prominent reflective markers (Fig. 10a). Most of the hinterland-directed displacement at depth is transferred upwards onto a single, foreland-directed thrust, which outcrops just north of Jhelum and carries the Jogi Tilla anticline (Fig. 6). Further north another thrust carries an anticline cored by

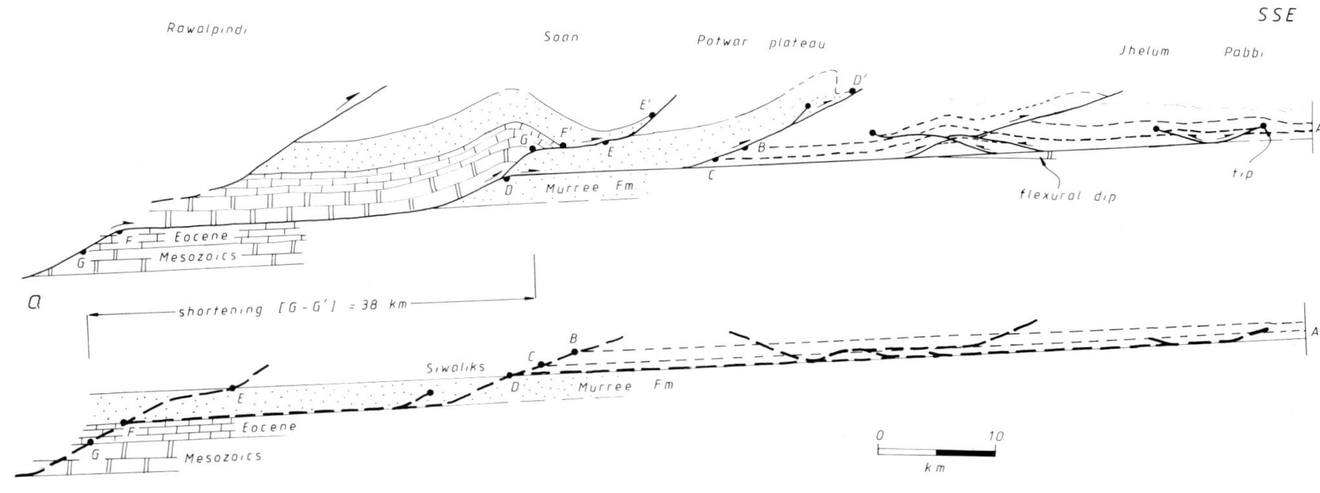

Figure 10. Balanced (a) and restored (b) cross-section couplet through the eastern Potwar and front folds around Jhelum, based on surface data and commercial seismic reflection profiles. Section line a-a′ on Fig. 6, after Butler and Coward (in press).

Murree Formation molasse of the Rawalpindi Group, suggesting that the regional basal detachment, or sole thrust, has cut deeper into the sedimentary pile towards the north. Thus, in common with most thrusts, it has cut upsection in its direction of transport. This geometry is illustrated on the restored section (Fig. 10b) derived from the cross-section (Fig. 10a). This restoration confirms the validity of the structural model illustrated on Fig. 10a, particularly that the frontal folds and thrusts have decoupled along a up-cutting detachment within the molasse sediment. Any salt within the stratigraphic section must have remained untaped at depth. For this eastern section, therefore, the basal detachment only cuts back down into Eocene and older rocks in the northern Potwar.

A major advantage in constructing balanced cross-sections, apart from testing the validity of structural models, is that they provide minimum estimates of horizontal shortening. The section through the eastern Potwar and Salt Ranges is no exception; those thrusts and folds south of the Soan river (Fig. 6) can be traced into the main part of the Salt Range to the west. Thus, a restoration of these structures allows estimation of the amount of convergence across the frontal structures between 35 and 40 km.

The structure of the main Salt Range is markedly different to that described in the east. Thrusts and folds detach within, or at, the base of the Salt Range Formation evaporites, so that the full stratigraphic section is involved in the deformation. Yet, there are still marked structural variations along the range, illustrated on the simplified sections through the western and eastern parts of the central Salt Range (Fig. 11, see Fig. 6 for locations). These sections are based on surface geology supported by good quality commercial seismic reflection profiles in the southern Potwar. The westernmost of these sections shows a simple fold train underlain by a greatly thickened sequence of

Salt Range Formation evaporites. However, in the east, the structure is dominated by a simple ramp-type uplift, morphologically similar to the classic Appalachian examples (Rich, 1934). These variations, and how they relate, will be discussed in more detail now. The following account is based on field studies by the authors, but is also heavily dependent on the excellent 1:50,000 geological map coverage of the Salt Range from Kalabagh to Jogi Tilla (Fig. 6) produced by Gee (1980), which has been modified only locally.

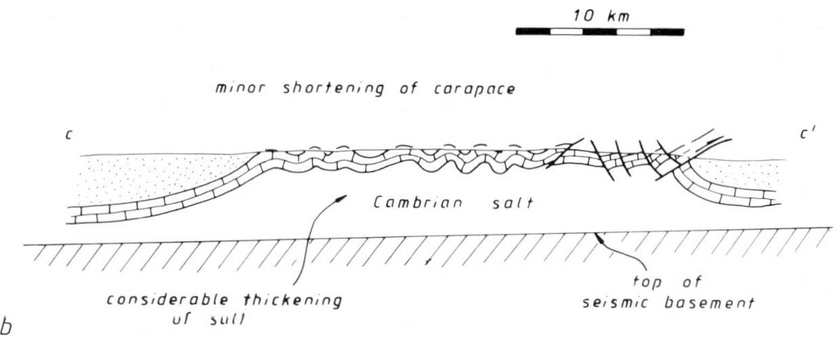

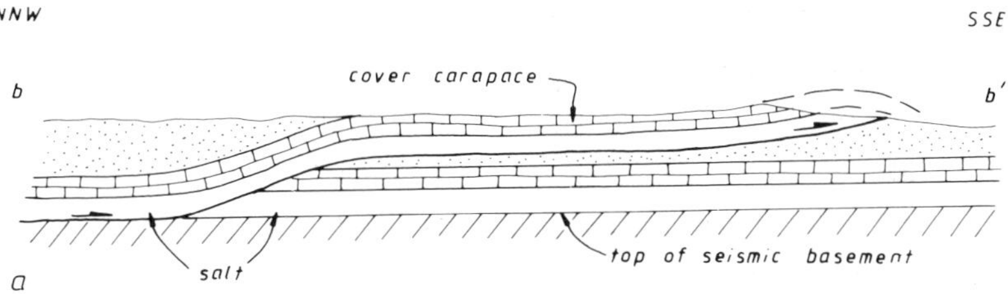

Figure 11. Generalized cross-sections through the eastern Central Salt Range (a, section line b-b' on Fig. 6) and western Central Salt Range (b, section line c-c' on Fig. 6) based on surface data and commercial seismic reflection data. The cover carapace is ornamented by bricks, salt is unornamented and the molasse sandstones are stippled.

A. Central Salt Range

The distribution of folds and thrusts in the western Central Salt Range is illustrated in Fig. 12. The major splay from the Range Front Thrust can be traced entering the region from the west, just north of Golewali. In this sector there are relatively few folds and these trend between NW-SE and E-W, locally mimicking the outcrop pattern of the thrust. To the east, these folds swing into an east-west trend and become far more common, notably around Sakesar (Fig. 12). A major fold belt continues south of Uchhali, gradually losing amplitude to the east. A further fold belt trends WSW-ENE from Sakesar, north of San Sakesar Kahar, to bound a tract of Quaternary sediments. Older parts of this molasse sediment are deformed, suggesting that much of the folding has occurred during Quaternary times.

Within the major fold belt there are local thrusts (e.g. SE of Uchhali, Fig. 12) which terminate laterally. To the west (south of Sakesar) other thrusts can be traced up from the Range Front to terminate in the fold trains. This geometry can be best understood by considering three parallel cross-sections (Fig. 13a,b,c; see Fig. 12 for locations). The western most of these (x-x', Fig. 13a) demonstrates that the fold train is dominated by upright to SSE-overturning structures, which pass downwards into an imbricate splay of thrusts emanating from the Salt Range Formation evaporites. As these thrusts lose displacement upwards they generate folds, generally an anticline in the hanging-wall and a syncline in the footwall. Further east (section y-y', Fig. 13b), the folds have lost amplitude and probably originate as simple buckles detaching directly from the salt. Exceptions occur in the center of the belt where foreland and hinterland-directional thrusts break surface. Further east again (section z-z', Fig. 13c), the buckle folds lose amplitude and the thrusts

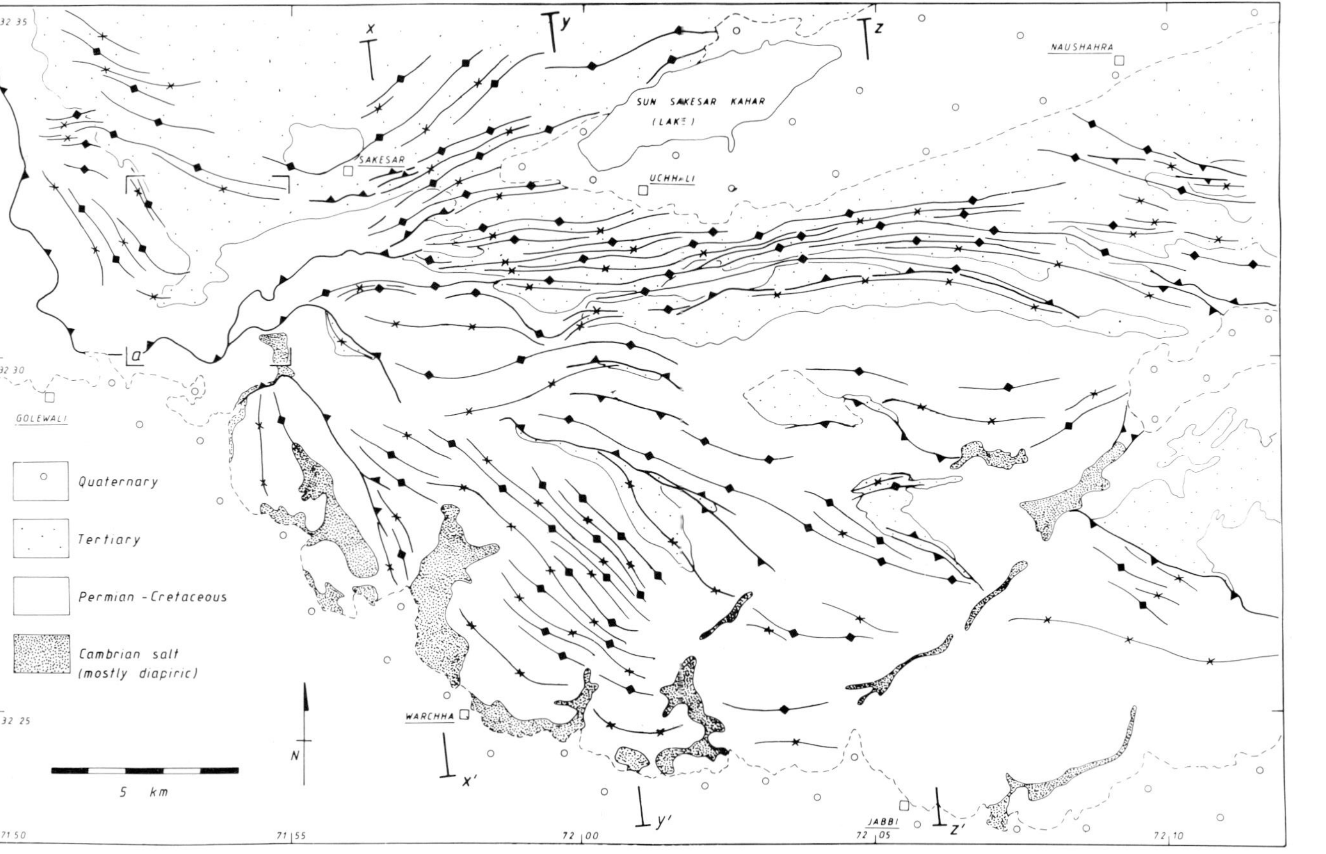

Figure 12. Detailed map of the western Central Salt Range, derived from a compilation of the maps of Gee (1980) and the authors' own observations. Major folds and thrusts are indicated, as are the section lines (x-x', y-y' and z-z') of Fig. 13 (a, b, c respectively) and the location (a) of Fig. 36.

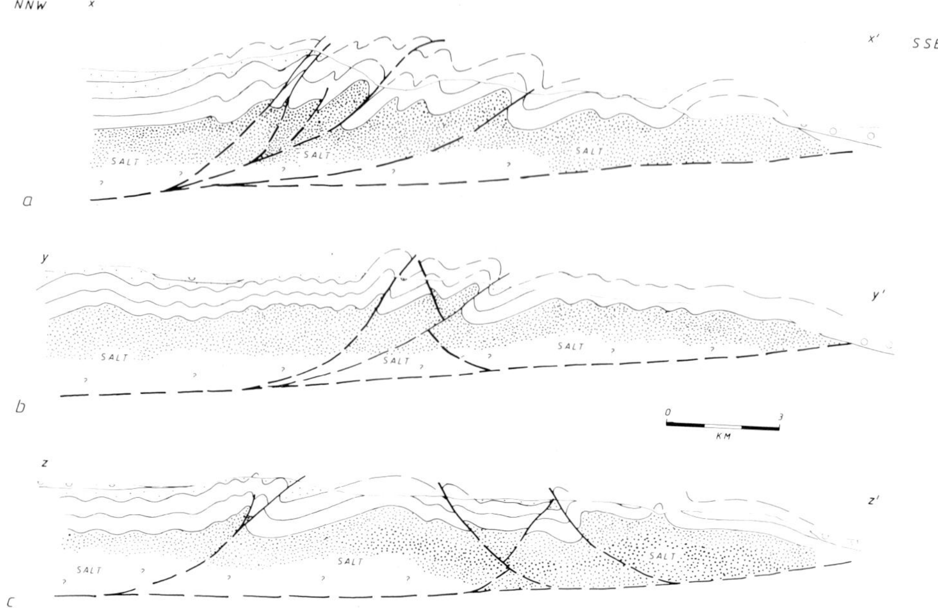

Figure 13. Three serial cross-sections (x-x', y-y' and z-z' on Fig. 12) through the western Central Salt Range, key as Fig. 12.

which do reach outcrop have reduced displacements. The structure is dominated by long wavelength, low amplitude folds. Figure 12 shows that very few folds are developed further east.

This pattern of folds and thrusts essentially represents a reduction in the amount of deformation and shortening across the Salt Range from west to east. There is no indication that salt thicknesses vary systematically beneath the fold belt. Indeed, there are outcrops of Salt Range Formation evaporites all along this sector of the thrust front (Fig. 12). This loss of shortening across the Salt Range, however, does not include displacement on the Range Front thrust. The simplified section across the Central Salt Range to the east (Fig. 11b) shows a minimum of about 30 km offset between hanging-wall and footwall cut-offs of the cover carapace above the salt. While the footwall cut-off can be located from the monoclinal warp at outcrop, and appears on seismic sections, the matching hanging-wall ramp has been eroded at the thrust front. Thus the shortening partitioned within the range is probably less than a quarter of the total convergence between the southern Potwar and the foreland. It is likely that this shortening is transferred out onto the thrust front towards the east.

Using these geometric arguments it is possible to predict the geometry of the Range Front thrust at depth as the structural syyle varies from west to east. Returning to the general section through the western Central Salt Ranges (Fig. 11a), the convergence between the Potwar and the foreland has been accommodated at depth by thickening within the salt. This presumably occurred by ductile deformation and flowage so that the overlying cover carapace deformed independently. The salt has thickened about 2.5 times, creating the uplift within the Salt Ranges. This amount of thickening is equivalent to about 30 km displacement along the basal decoupling surface beneath the salt. However, the folds in the cover carapace, together with the minor thrusts which breach it, probably represent about 5 km shortening. A further length of carapace, probably about 25 km, must have existed prior to deformation. This missing length of carapace must have been eroded at the thrust front. Thus, in situ ductile shortening within the Salt Range Formation passed upwards onto an upper detachment at the base of the carapace. The cover was translated out across the foreland land surface to the south, from where it has since been eroded. Further east the cover and salt did not decouple, but rather formed a simple sheet which was carried out onto the foreland.

A final problem here is why this eastern sheet was not eroded back while erosion did occur in the carapace to the west. The solution is implicit in the cross-sections. Whereas in the west the thrust front climbed up at least 5 km into the molasse sediment and probably moved across the paleo-land surface, in the east the thrust has just one kilometre of molasse in its footwall. Thus the thrust itself remained buried beneath part of the foreland. The front would have become emergent only some distance ahead of the advancing carapace. Not only is the uplift on the carapace reduced by several kilometers in this model, but the leading edge is protected by a mantle of molasse sediment. Using this reasoning, Figure 14 is an illustration of the predicted position of the cover carapace to the Salt Range Formation at depth beneath the Salt Range, and of the estimated position of the leading edge of this carapace in the thrust sheet, neglecting the effects of erosion.

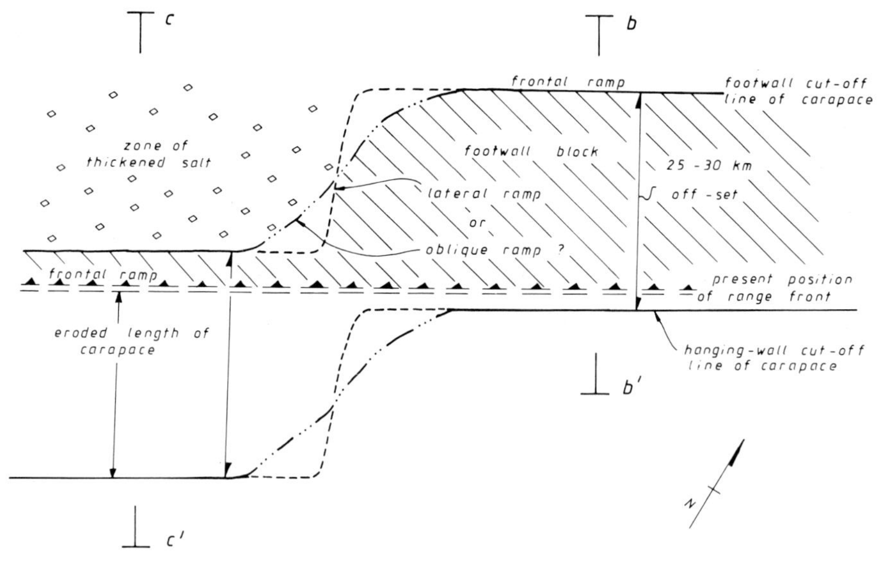

Figure 14. Cut-off line map of Central Salt Range, illustrating the predicted continuity of thrust front structure with depth between the section lines of Fig. 11 (b-b' and c-c' on this figure and Fig. 6).

B. Eastern Salt Ranges

Although the structural geometry of the eastern Central Salt Range is relatively simple, this picture does not continue for any great distance further east. On the section between Jhelum and Rawalpindi (Fig. 10) we have previously seen that the equivalent structures include both foreland and hinterland-directed thrusts and folds. This pattern emerges within the eastern Salt Range, shown by an area (Fig. 15, modified after Gee (1980)), between the section lines of the two contrasting sections (Figs. 10 and 11b). In the west the Permian-Eocene shelf sediments of the carapace, together with Cambrian clastics of the Khewra and Baghanwala Formations, are not disrupted by thrusts and are only folded gentle structures (e.g. in the Dulmial area on Fig. 15). However, running NE from the town of Choa Saidan Shah, is a thrust which carries the carapace up onto middle Siwalik (Pliocene and younger) molasse. This thrust is directed towards the Potwar plateau, away from the foreland, and is associated with a series of SW-NE trending folds which either overturn towards the NW or are upright structures. There are relatively few thrusts and folds within the carapace between this hinterland directed belt and the present thrust front. The front is presently buried beneath thick fanglomerate deposits between Khewra and Jalalpur. Eastwards, the Salt Range becomes very narrow, with Quaternary alluvium lying in the hinge of a major syncline just a few kilometers behind the thrust front. The syncline trends NW-SE behind the range, with which it converges eastwards, near the village of Jalalpur. The Salt Range essentially terminates here in a remarkable S-bend, to become replaced further to the north by the Jogi Tilla ridge. Note that this change in structure coincides both with the loss of Permian to Eocene shelf sediments, cut out by the sub-molasse unconformity (Fig. 9b), and with the apparent loss of salt along

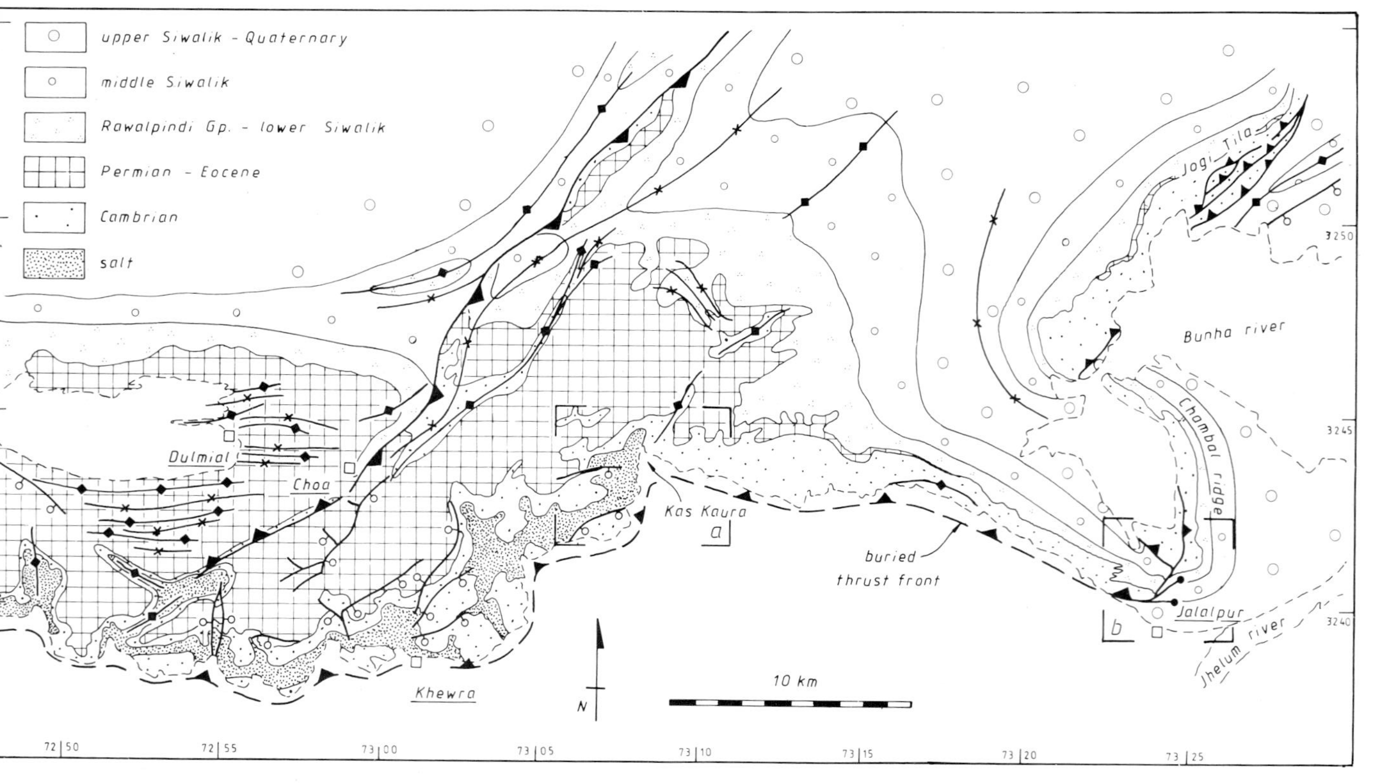

Figure 15. Simplified map of the Eastern Salt Range, modified from Gee (1980), illustrating the detailed localities of Figs. 19 (b) and 34 (a). Recent alluvium of the foreland basin is unornamented.

the thrust front to the east of Kas Kaura (Fig. 15). However, thin streaks of salt are brought up on the thrusts and faults along the Chambal ridge (Gee, 1980).

Vann et al. (1986) suggest that back thrusts develop due to the reduced ability of a basal detachment to accommodate displacement. This type of behavior might be expected if the Salt Range Formation lost definition. Davis and Engelder (1985) proposed that the lateral terminations of salt-based thrust belts should be characterized by rotations of thin sheets, and this can be accomplished by alternating displacement on a pair of foreland hinterland-directed faults. Indeed, paleomagnetic results in the southern Potwar (Opdyke et al., 1982; Fig. 16) show a 20° counterclockwise rotation of the Miocene molasse on the thrust sheet between the range front and the back thrust. Elsewhere, such as behind the western and central Salt Range, the paleomagnetic pole directions are little different to the rest of the thrust belt, suggesting that the rotations in the eastern Salt Range are due to local thrust effects. Provided that the back thrust displacements die in the Choa Saidan Shah district

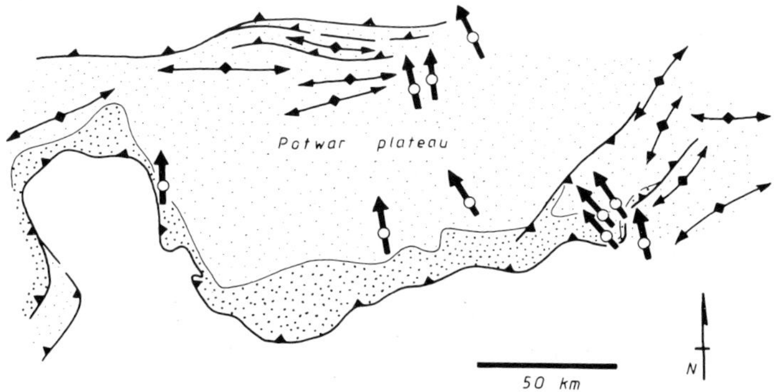

Figure 16. Sketch map of the southern Potwar (fine stipple) and frontal ranges (coarse stipple) illustrating the palaeomagnetic rotations recognized by Opdyke et al. (1981).

(Fig. 15) and do not pass forward onto the range front nor are cut out by the thrust front, then the observed rotations in the eastern sheets can be generated (Fig. 17). The preferential wedging of the Potwar plateau beneath the eastern end of this structure, while pinned at the western end, must either deform the entire hinterland or generate rotations in the back thrust sheet. An interesting possibility occurs if displacements alternate between the forethrust at the front of the range and the back thrust. The rotation can then occur during the overall displacement of the Potwar onto the foreland, as envisaged by Davis and Engelder (1985). Alternatively, the back thrust may predate displacements on the eastern end of the thrust front, so

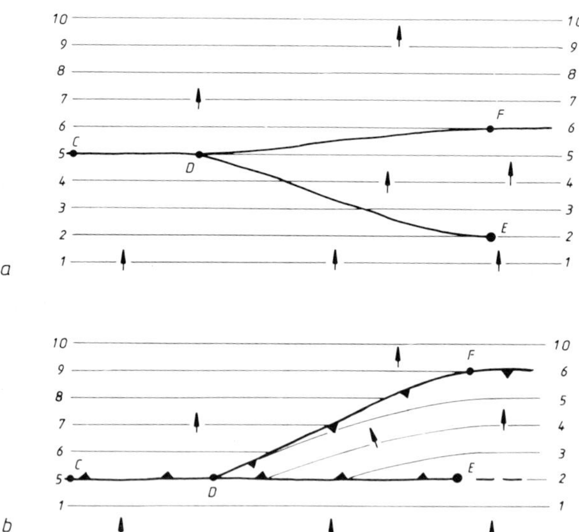

Figure 17. A model for the rotation of thrust sheets by the alternation of displacement across a laterally limited forethrust - backthrust couplet, depicted in plan form. a) illustrates the pre-thrust situation with parallel palaeomagnetic poles and lines of latitude (numbered). b) illustrates the geometry after displacement. A consistent convergence of four units is preserved along strike so that the boundary conditions are plane strain. Fore-thrust displacements are conserved along fault trace C-D but are lost along trace D-E. Backthrust displacements increase from D-F so that, to the right of F, the back thrust contains all the convergence. Thrust sheet EDF experiences a counterclockwise rotation.

that the rotations occurred while the back thrust sheet was part of the foreland. It might be possible to distinguish between these models after considering the structure of the Chambal ridge and Jalalpur district.

One question remains open regarding the origin of the back thrust: does its presence reflect the local loss of salt and, hence, can back thrusts, in general, be used diagnostically in tracing salt distributions (as implied by Davis and Engelder, 1985 and Vann et al., 1986)? To answer this we present a hypothetical model (Fig. 18) which shows the evolution of a single back thrust above a detachment. This model illustrates that, although, like any other fault orientation, a backthrust can initiate by instantaneous propagation following inhibited displacement on part of the thrust surface, for movements to be transferred onto the new backthrust the basal detachment must continue to propagate into the foreland (Butler, in press). This model can be contrasted with two other options: (a) the generation of a fore-thrust ramp bypasses the frontal portion of a level of detachment by transferring displacement up to higher horizons, (b) folding requires a basal detachment to progressively decrease displacement towards the foreland. Thus, backthrusting is the only case actually favored by uninhibited forward propagation of the basal detachment. In the other two options this property will either inhibit, or have little direct bearing on, the evolution of the structure. However, the present

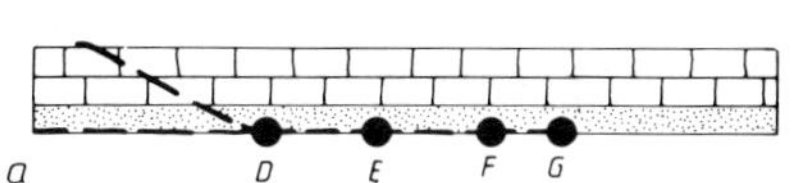

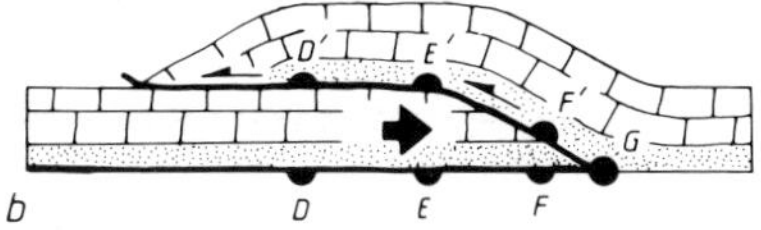

Figure 18. The requirement of continued foreland propagation of a basal forethrust detachment during backthrust movements (modified after Butler, in press). a) illustrates the pre-deformational geometry with the future thrust location pecked. b) illustrates the post-deformational geometry. Note that the leading edge of the thrust complex buried in the section must migrate from D to G as the thrust wedge drives into the foreland.

location of trailing edge of the backthrust on the nose of the advancing thrust wedge (G on Fig. 18) may indicate the limit of uninhibited basal detachment propagation for ancient thrust belts, provided subsequent displacements are partitioned elsewhere. In the case of the eastern Salt Range, the later movements may have occurred on fore-thrusts along the present thrust front.

The second problem with the eastern Salt Range concerns the lateral termination around the village of Jalalpur (Fig. 15). To the east of the village lies the Jhelum river which drains a large part of the western Himalayas. River gravels from the Jhelum have been tilted along its northern bank so that they now dip at up to 60° SSW towards the foreland. These gravels have yielded ages of less than 700,000 yr. (Yeats et al., 1984). The mountain front in this locality, rather than being emergent as it is ahead of the central Salt Range, is a simple monocline. There is no indication that the thrust front becomes emergent further to the south, since some topographic expression would be preserved in such a young thrust belt. To the west of Jalalpur village, however, the mountain front emerges as a forethrust bringing Cambrian clastics and Salt Range Formation onto Siwaliks. These contrasting geometries are illustrated on a sketch map of the Jalalpur district (Fig. 19) and sketch sections (Fig. 20).

Consider a traverse across the mountain front from foreland to hinterland up Kahan Kas (x-x′ on Fig. 19; Fig. 20a). The frontal monocline in the Quaternary-Recent river gravels is complicated by some minor back thrusts which can be detected from offsets in the alluvial molasse stratigraphy mapped by Gee (1980). The major thrust front itself is very steep, running NE from the southern slopes of the hill Mangal Dev, and contains fault-bounded slices of Cambrian clastics (Jhelum Group), molasse (Rawalpindi and Siwalik Groups), and salt. To the NW lies an

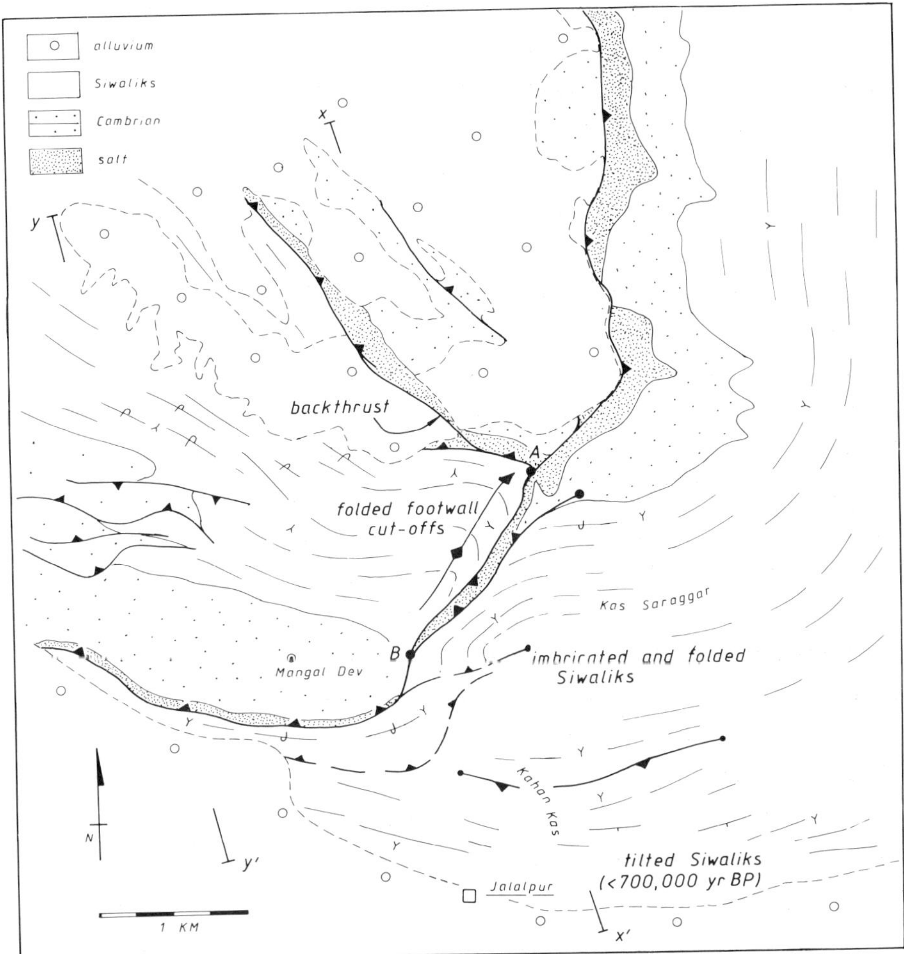

Figure 19. Sketch map of the Jalalpur district (b on Fig. 15), modified after Gee (1980) and Yeats et al. (1984). Section lines of Figs. 20a and 20b,c are depicted x-x' and y-y' respectively. A and B represent critical points of thrust intersection and are indicated on Fig. 20a. The Rawalpindi Group is omitted from the map for clarity, it lies at the base of the Siwalik molasse.

anticline which overturns slightly to the SSE. The complementary syncline to the N contains klippen of Salt Range Formation and Jhelum Group rocks, but is largely covered by blown sand. We interpret these klippen as being the remnants of a backthrust sheet, which roots along the thrust front and is largely responsible for generating the frontal monocline on the banks of the Jhelum river. A series of splays in the hanging-wall to this back thrust can be recognized from Gee's (1980) mapping along the southern Chambal ridge (A on Figs. 19 and 20a). The backthrust has there been folded by an anticline which lies in the hanging-wall to the forethrust along the front of the Salt Range. In this locality it must show very minor displacements since forethrust imbricates terminate towards the east around Kas Saraggar (Fig. 19).

There is some support for greatly reduced forethrust displacements along the eastern range front near Mangal Dev. Figure 15 shows the greatly reduced width of the range outcrop as measured by the distance between Quaternary rocks in the hinterland and the undeformed foreland. It is unlikely that the range front has been eroded back a substantial distance, as was proposed earlier for the western Central Salt Range (Fig. 14), since Cambrian units are folded down onto the Range Front thrust between Kas Kaura and Jalalpur (Fig. 15). These geometric elements have been combined in a section (Fig. 20b) which indicates less than 5 km displacement on all foreland-directed structures. These structures may have over-ridden an earlier backthrust; they certainly deform the backthrust (Fig. 20a). Fig. 20c thus illustrates the proposed geometry of back and forethrusts in the Mangal Dev region. The model suggests that the eastern Salt Range mountain front initiated as a backthrust, but that this system eventually locked up to be bypassed by

Figure 20. Cross-sections through the thrust front in the Halalpur district (see Fig. 19 for locations). a) is section line x-x' on Fig. 19 and illustrates an early back thrust imbricate fan (A is a branch point in this fan, see also Fig. 19) being folded around a forethrust and fold complex with relatively subordinate displacements. B is the cut-off of the earlier backthrust against the forethrust. b) is the basic section implied by surface geology alone (see Fig. 19 and Gee, 1980) along the line y-y' on Fig. 19 (key as Fig. 20a). c) is a modified version of Fig. 20b illustrating the possible over-running of a back thrust

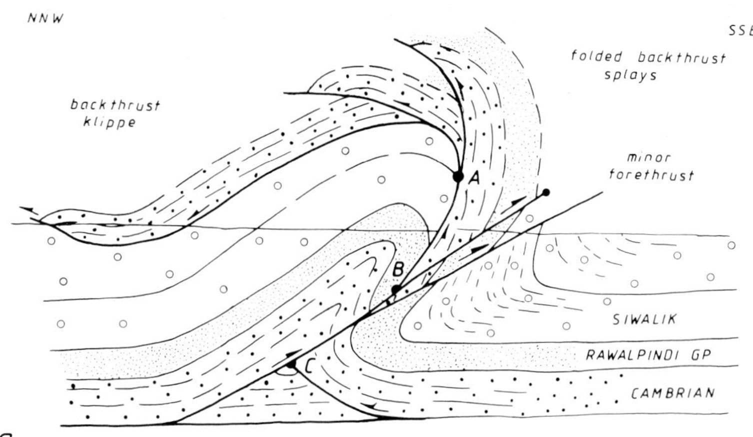

a.

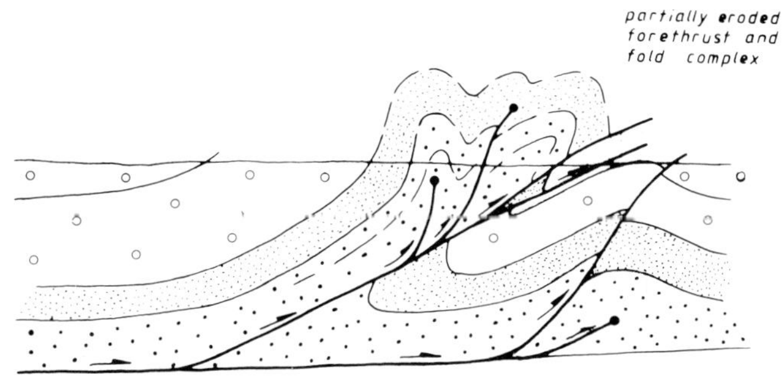

b.

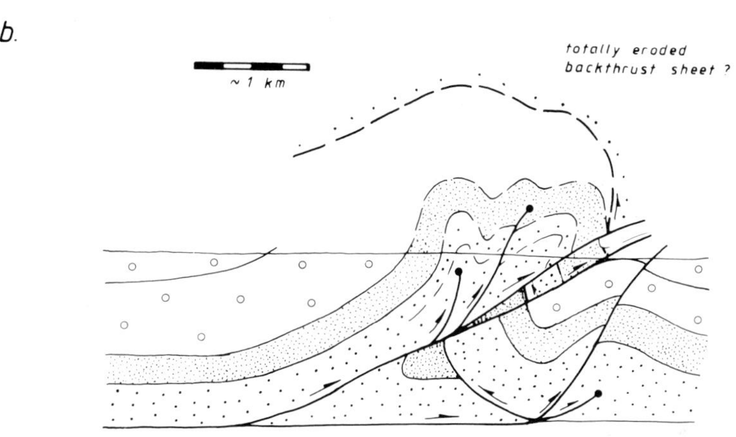

c.

complex by the simple forethrust structure, thereby maintaining displacement continuity around the mountain front.

foreland-directed thrusts and folds. These later structures have masked the root of the backthrust system in the Mangal Dev area, and it is only preserved to the east (beyond B, the cut-off point of the backthrust against the Range Front forethrust, on Fig. 19).

Unfortunately, the interactions between backthrusts and forethrusts are less well exposed at the northern end of the Chambal ridge, at the top of the S-bend. Gee (1980) mapped the thrust front to the east as a conventional emergent forethrust, with associated imbricates (Fig. 15) carrying the Jogi Tilla anticline. This front can be traced laterally onto the section line of Fig. 10 where it features as the opposed dip complex north of Jhelum.

C. Eastern Termination of the Salt Range

From the above discussion it should be clear that the displacement on the Range Front thrust, accommodating all the convergence between the southern Potwar and the foreland in the eastern Central Salt Range (e.g. Fig. 11b), is transferred over a complex system of backthrusts, forethrusts and related folds towards the east. It is by no means certain how these structures relate to the main sub-Himalayan thrust systems with which they converge in this region (Fig. 4). However, since these folds and thrusts developed in a foreland basin, which experienced active deposition of alluvial sediments, it is possible to gain estimates of the relative and absolute ages of fold initiation and growth. Excellent, detailed studies of reversal magneto-stratigraphy have been calibrated against known successions, and supported by paleontological data (Johnson et al., 1979; 1982). These provide minimum ages for deformation (Fig. 21).

The first major difference is the convergence of thrust transport direction of the two systems (Fig. 21). The Salt Range moved towards the SSE, whereas the movement direction in the western sub-Himalayan frontal ranges was towards the SW. This difference is also followed by the transport directions of more internal Himalayan thrust systems (e.g. Brunel, 1986), those in the Pakistan sector lying parallel to the relative plate convergence vector (i.e., towards SSE), while those in the main Himalayan belt (Fig. 4) are radial to the arc and trend NE-SW.

The most outlying structure of the sub-Himalayan fold trends is the Mangla-Samwall anticline (Fig. 21; Johnson et al., 1979). This anticline deforms alluvium, which has reversal stratigraphies indicative of deposition from about 2.7 Ma, and controls its deposition until about 1.5 Ma (Johnson et al., 1979). As such the anticline immediately predates the eastern folds of the Salt Range, namely the Rhotas and the Pabbi anticlines. These formed between 1.7 and 0.4 Ma (Rhotas) and 1.2 and 0.4 Ma (Pabbi), marginally post-dating the Chambal ridge which grew between 2.4 and 0.7 Ma (Johnson et al., 1979).

With regard to the model proposed earlier for the evolution of the eastern termination of the Salt Range around Jalalpur (Fig. 20), these dates support the notion of fore-thrusting deforming the marginally older backthrusts. The Chambal ridge probably initiated as a backthrust frontal monocline (Vann et al., 1986), but was enhanced by forethrusting and associated folding during the last 1.5 Ma. This forethrusting process migrated out into the foreland to develop the Pabbi anticline. These rather limited data suggest that, at any instant, the deformation could occur across several folds, but that this diffuse deformation tended to migrate outwards towards the foreland with time.

The present distribution of folds (Fig. 21) suggests that the basal detachment for the sub-Himalayan folds reached the Jhelum district significantly earlier than that for the Salt Range. At present, the lateral propagation of the Salt Range detachment would be seriously inhibited by the sub-Himalayan folds. Thus the lateral termination of the Salt Range system may

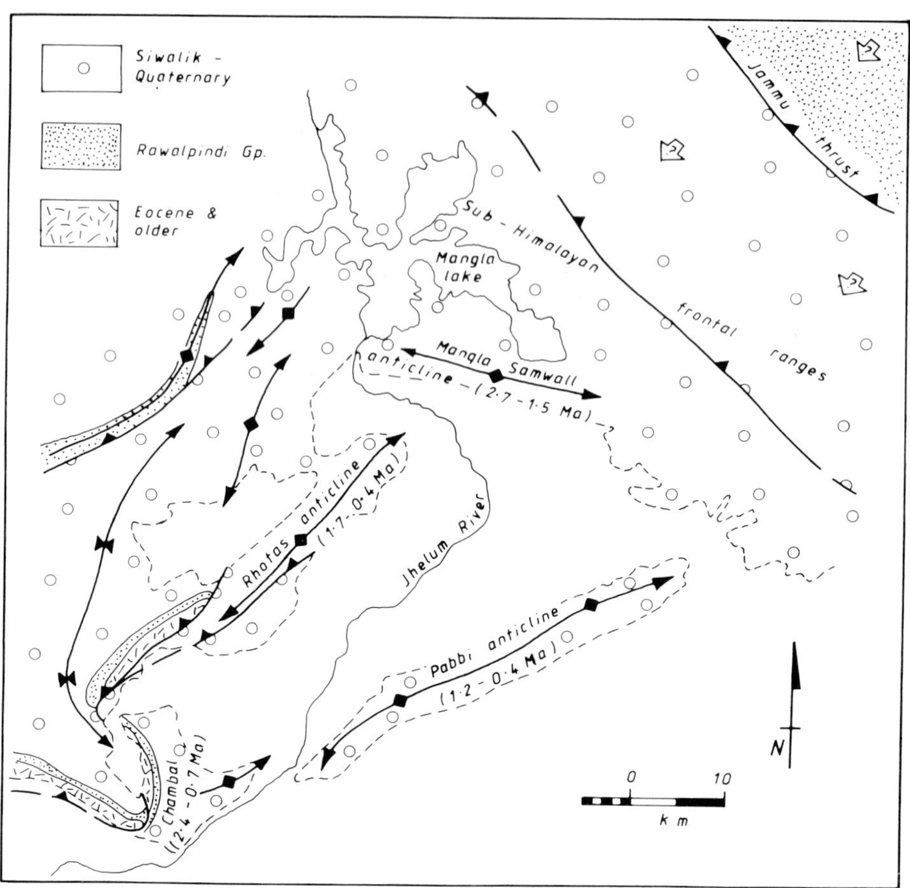

Figure 21. Simplified map of the distribution of folds and thrusts at the eastern termination of the Salt Range in the Jhelum district (modified after Johnson et al., 1979) illustrating the ages of folding for various structures determined by magnetic reversal stratigraphic dating of the Siwalik molasse.

reflect either (a) the proximity of the main Himalayan belts or, (b) the associated foreland flexural depression and its thicker pile of molasse sediments, rather than reflect the distribution of salt at depth (Burbank, 1983). Yet the distribution of salt may have controlled the geometry of the Salt Range detachment immediately to the west. We know that the structures along the section line of Figure 10 decouple within molasse sediments and not along the Salt Range Formation. This change in decoupling level, the basal detachment climbing laterally from salt in the west to within molasse in the east, must occur along a line running across the entire Potwar plateau, from between Kas Kaura and Jalalpur in the south to just west of Rawalpindi in the north. This lateral ramp structure could then transfer displacements across the adjacent sections illustrated schematically on Figure 8. Note that, as Figure 3 illustrated, this control on lateral ramp location could be produced by a very local loss of the Salt Range Formation beneath Rawalpindi. The salt could still exist beneath, say, the Jhelum district.

D. Surghar Range Front Structure

The eastern margin of the Salt Range is marked by a broad zone of folds and thrusts which essentially continue to transfer thrust displacement laterally towards the sub-Himalayan chains. The western margin is rather different. Figure 6 shows that the mountain front trends NNW-SSE, parallel to the relative plate convergence and thrust transport direction in northern Pakistan. This trend links up with the spatially offset lateral termination of the Salt Range, the Surghar Range, which then strikes perpendicular to the thrusting direction for about 40 km. The western end of the Surghar Range is marked by a further strike swing, back to NNW-SSE, to link with the mountain front in the Trans-Indus Ranges of Marwat and Khisor. In plan view, therefore the mountain front shows a sharp re-entrant, about 90 km in length (Fig. 6). Before examining the structure of the lateral, NNW-SSE trending segments we first discuss the nature of the thrust front in the Surghar Range.

Figure 22 is a sketch section through the frontal part of the Surghar Range at the Chichali gorge (see Fig. 6 for location). On a foreland to hinterland traverse, the first sign of deformation is a weakly developed train of low wavelength, moderate-amplitude, folds which bring up Rawalpindi Group molasse sediments. Wadia (1949, Fig. 26) shows that the northern end of these folds contain small-scale thrusts. These thrusts carry up an inverted stratigraphic contact between Eocene limestones and the younger molasse sandstones. Inverted, and steeply-dipping, rocks exist further north, passing down the stratigraphic section into the uppermost Cretaceous units of the Chichali Formation. This continuous passage into older rocks is terminated at a thrust which carries right-way-up Jurassic rocks. Minor, low-angle, thrusts repeat part of the Jurassic succession, but the hanging-wall contains very few other signs of deformation. The moderate hinterland dips of the Mesozoic and Tertiary units in the hanging-wall are considered to be the products of thrust sheet rotation above the footwall ramp.

The sketch section (Fig. 22) shows a relatively simple thrust front. The thrust which carries Jurassic rocks to outcrop has an unknown displacement since the hanging-wall ramps are not preserved on this section. This thrust will have developed through a previously folded sequence, propagating as a fore-limb thrust (Dahlstrom, 1970). This folding could represent a reduced

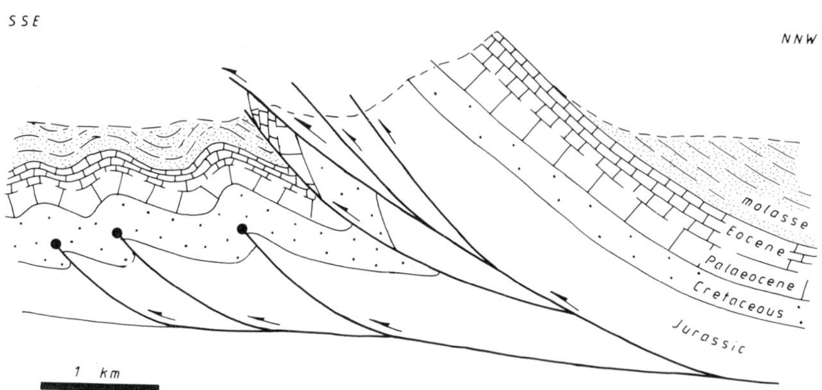

Figure 22. Simplified cross-section through the Surghar range front structure at the Chichali gorge, heavily modified after Wadia (1949) on the basis of the authors' observations. See Fig. 6 for the gorge location.

ability for the thrust front to propagate, a process which also explains the presence of smaller folds in the footwall (Fig. 22). An important point is the absence of any Palaeozoic rocks along the thrust front in this sector, indicating that the basal detachment propagated within the Mesozoic succession and was not located by any salt horizon.

E. The Eastern Margin of the Kalabagh Re-entrant

The structural relationship between the Surghar and western Central Salt Ranges is critical in unravelling the evolution of the Salt Range detachment. Several workers have tentatively proposed, or implied, that the eastern margin of the Kalabagh re-entrant contains a tear fault (Burbank and Raynolds, 1984). Various linear features on the LANDSAT compilations (Yeats et al., 1984; Fig. 5) have been proposed as candidates for this structure. The fault is envisaged to have offset the previously linear thrust front which linked the two WSW-ENE trending range segments, a displacement of about 90 km. As outlined on Fig. 23 such a model has particular consequences for displacement continuity around the mountain front. Since the lineaments do not continue into the foreland at the SW corner of the Salt Range, the tear fault displacements would have to be transferred onto the thrust front to the east. Thus this eastern front would display not only the displacement now seen along the Surghar range (at least 5 km, possibly a lot more) but also the additional 90 km transferred along the tear fault. Presumably these displacements would also be found as thrusts and folds developed behind the Surghar Range, the thrust front here being abandoned in favor of deformation in the hinterland (Fig. 23a).

While there may be few problems with this later argument (the geology of the ground north of the Surghar Range is poorly known) substantial problems arise in accommodating up to 100 km displacement in the western Central Salt Range. The cross-sections through the Central Salt Range (Fig. 11) show no

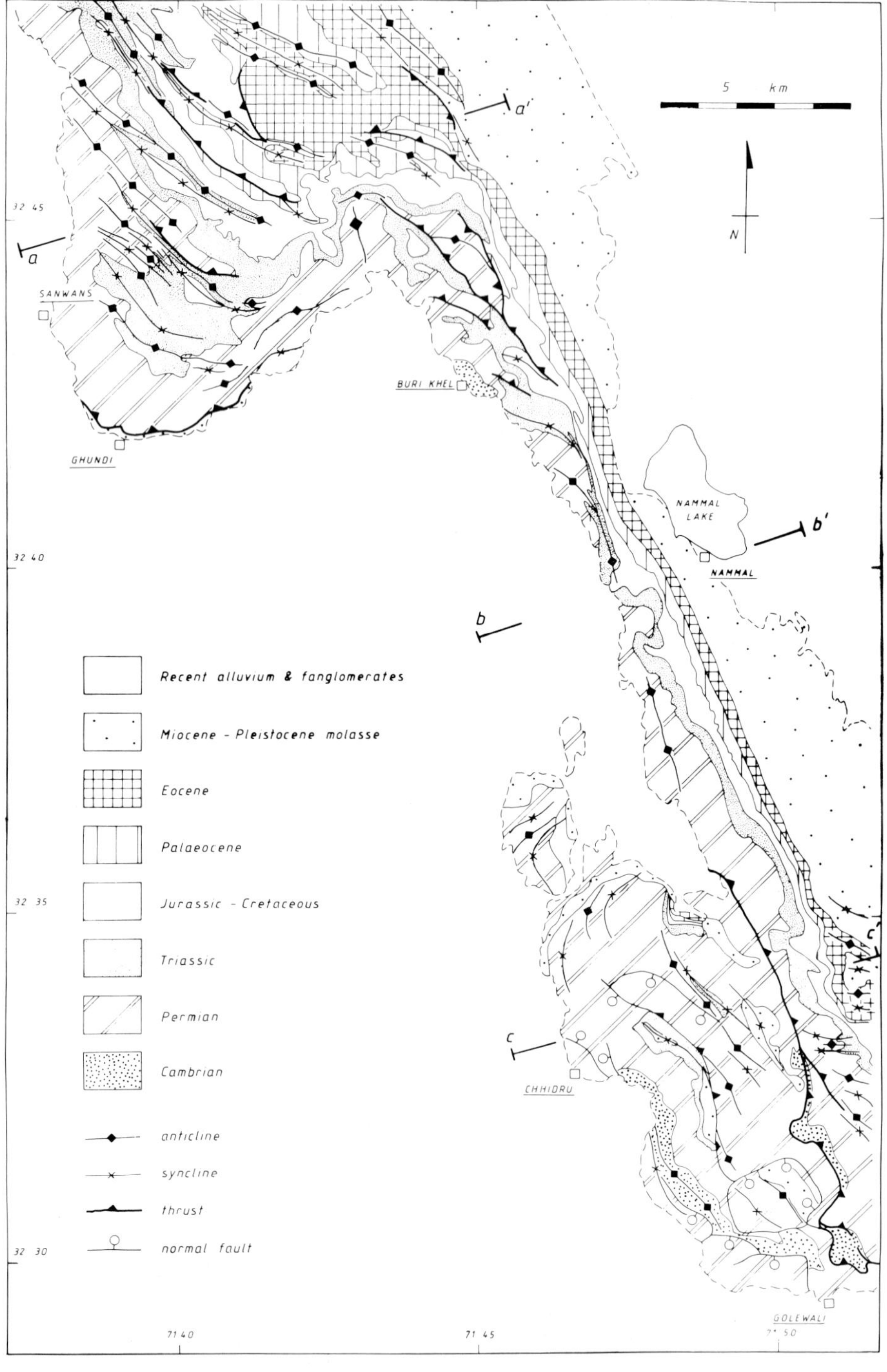

5 km
N
a
a'
b
b'
c
c'
SANWANS
GHUNDI
BURI KHEL
NAMMAL LAKE
NAMMAL
CHHIDRU
GOLEWALI
32 45
32 40
32 35
32 30
71 40
71 45
71 50
Recent alluvium & fanglomerates
Miocene - Pleistocene molasse
Eocene
Palaeocene
Jurassic - Cretaceous
Triassic
Permian
Cambrian
anticline
syncline
thrust
normal fault

evidence of these displacements, although they can accommodate around 30 km shortening. Certainly most of the circa 100 km displacement required would have to be lost west of the Jhelum district. It is most unlikely, however, that this eastern sector accommodated more than about 40 km. Hence the Salt Range detachment would show a lateral decrease in displacement of 60-70 km in a distance of about 150 km. This would generate bulk rotations of the Potwar plateau relative to the Indian foreland of 20-25° (Fig. 23 b). However, palaomagnetic data which (Fig. 16) show such large rotations are restricted to the sheet bounded by the back and forethrusts (Fig. 17), and to parts of the transect between Rawalpindi and Jhelum. Large rotations appear to be absent in the central and western Salt Ranges, although we admit that data are sparse. The counterclockwise rotations of up to 40°, described by Burbank and Raynolds (1984), south of Rawalpindi are too restricted in space, as presently defined, to be sufficient to accommodate the required differential movement component across the Potwar. We here propose a more appropriate model for the structure of the eastern margin of the Kalabagh re-entrant.

The structure of this critical area is illustrated by Figure 24. In this sector the thrust front is exposed in only one locality, at Ghundi, where Permian rocks are emplaced onto Quaternary fanglomerates (Gee, 1980; Yeats et al., 1984). The ranges behind contain a belt of folds and thrusts which are arcuate in plan. These can be traced around the promontory in the range structure from the NW to near Buri Khel (Fig. 24), where the fold belt is considerably narrower. A WSW-ENE cross-

Figure 23. Implications of siting a large-displacement tear fault down the eastern margin of the Kalabagh re-entrant (as proposed by Burbank and Raynolds, 1984, amongst others) for the bulk rotation of the Potwar plateau and the continuity of thrust displacements around the mountain front in northern Pakistan, illustrated in plan form. The thrust front is offset (x-x') by about 100 km but the eastern thrust front shows just 30 km displacement (facing page).

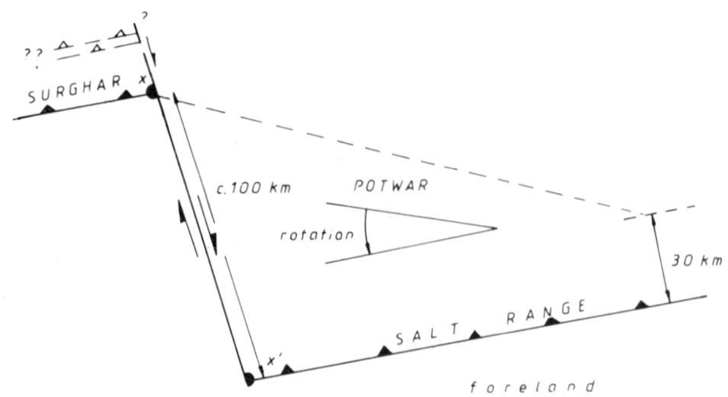

Figure 24. Geological map of the southern sector of the eastern margin of the Kalabagh re-entrant, modified from Gee (1980). The section lines of Fig. 25a, b, c are indicated a-a', b-b' and c-c' respectively. The SE corner of this map is continuous with Fig. 12.

section through the wide part of the fold belt (Fig. 25a) illustrates the simple structural style above the basal thrust, somewhat similar to the structure of the western Central Salt Range (Fig. 13b). The arcuate plan form of these structures is not disrupted by a tear fault and the only site for such a displacement lies along the range front itself where the basal thrust dips gently beneath the hills.

The structure of the Salt Range near Nammal lake (Fig. 24) is extremely simple, the fold belt being reduced to a single anticline trending NNW-SSE (Fig. 25b). The thrust front itself is masked by scree but presumably lies close to the range. This simple structure can be traced southward into a range of hills composed of folds and local thrusts. One of these thrusts appears to be the direct continuation of the buried detachment predicted for the Nammal lake sector, running out of the Punjab alluvial plain 5 km NE of Chhidru (Fig. 24). This thrust can be traced through the hills to near the village of Golewali where it changes to an E-W trend. The same thrust is illustrated on Fig. 12, outlined by thin outcrops of salt, and locally carries up the oldest rocks in the overlying cover carapace, the Permian Tobra

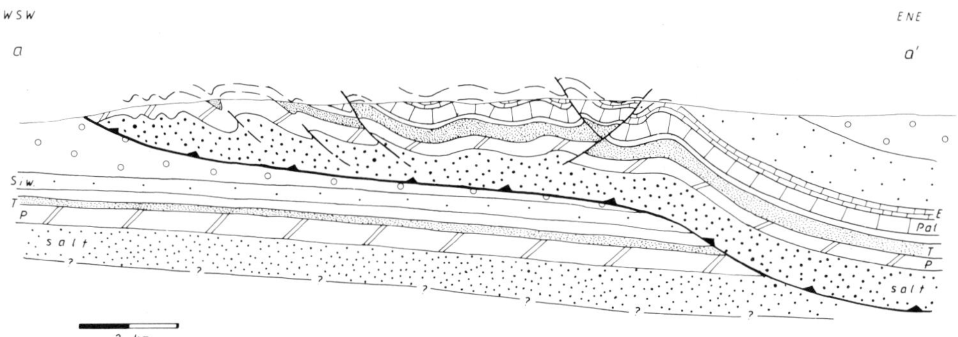

Figure 25. Serial WSW-ESE sections (a-a', b-b' and c-c') through the mountain front covered by the map of Fig. 24 with which the lines of section and the key are common. The major Range Front thrust is indicated by the barbed line.

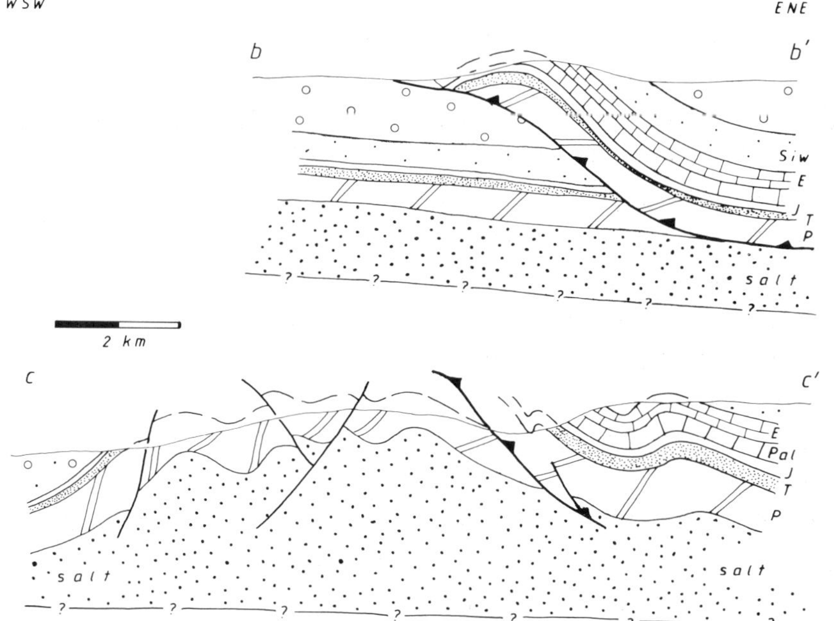

Figure 25 continued.

Formation. A WSW-ENE section through this sector (Fig. 25c) illustrates how the main range front detachment is folded by later structures developed in its footwall, so that the detachment has been uplifted above the level of the Punjab alluvial plain. As such, the range front structure is similar to that in the Surghar Range (Fig. 22), except that the folds and thrusts have detached within the salt rather than at the base of the Mesozoic succession.

Thus, the eastern margin of the Kalabagh re-entrant is not bounded by a steep tear fault, but rather by a simple thrust front, which displays many similarities to the range front structures elsewhere in the region. At present, the most likely model of displacement continuity transfers the SSE-directed thrust motion along the lateral ramp on the eastern margin of the re-entrant and onto the adjacent thrust front in the Surghar Range. If the lateral displacements continue north of the re-entrant, as is implied by the continuity of the tear fault lineament as recognized by Burbank and Raynolds (1984), then additional movements must be transferred forwards from the Surghar hinterland, branching onto the Range Front thrust at Kalabagh. However, in this district the obvious lineament visible on LANDSAT (Fig. 5) shows dominantly extensional offsets, downthrowing towards the WSW. This fault is not an integral part of the thrust network but has a related origin which will be discussed later (Figs. 37 and 38).

The thrusts and folds trend oblique to, and locally parallel, the thrust transport direction on the eastern side of the re-entrant (Fig. 24). A differential movement across a strike-slip segment of a thrust system will generate oblique folds (Coward and Potts, 1983) without involvement of the basement. It is thus possible to interpret the eastern margin of the Kalabagh re-entrant as a coherent part of the Salt Range front.

The three sections (Fig. 25) are constructed perpendicular to the presumed thrust direction, so that the ramp structures on

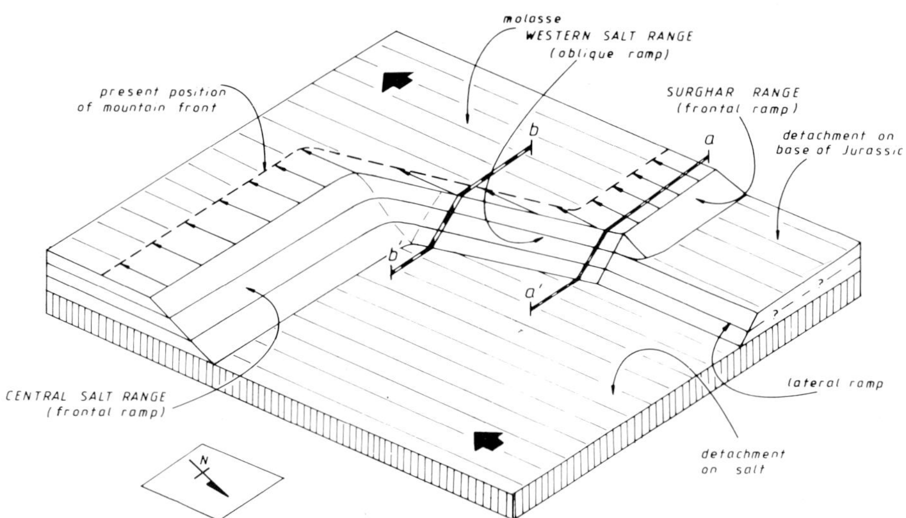

Figure 26. Block diagram illustrating the three dimensional thrust geometry of the eastern margin of the Kalabagh re-entrant. Section lines of Fig. 27 are depicted a-a' and b-b' The thrust transport direction is arrowed.

the basal thrust detachment must show substantial strike-slip components. Therefore they are lateral or oblique ramps. Figure 26 illustrates the three dimensional nature of the thrust front and its evolution. The displacement on the basal detachment must be locally transferred forwards at a late stage to uplift the foreland between Chhidru and Golewali (Fig. 24). We are not certain how the three dimensional ramp structure links through to the Surghar Range, except that the basal detachment must climb up stratigraphic section towards the west so that no Palaeozoic rocks are involved. The lack of regional uplift of the hanging-wall between Kalabagh and the eastern Surghar Range indicates that ramps trend almost parallel to the thrust transport direction. Thus the older molasse sandstones and underlying Eocene carbonates dip towards the ENE at the eastern Surghar Range (Fig. 27).

The above model assumes a consistent transport direction on the Salt Range basal detachment throughout the mountain front

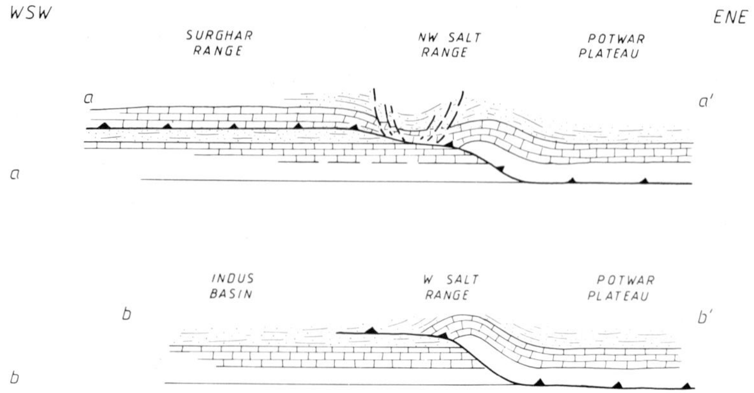

Figure 27. Longitudinal sections through the Surghar and NW Salt Ranges illustrating the variations in thrust geometry along the mountain front. Note that the movement direction is out of the page. Salt is unornamented, the carapace is brick-ornamented and the molasse is stippled. Section lines located on Fig. 26.

structure and is supported by the paleomagnetic data in the Potwar (Fig. 16). Thus the Potwar must have been emplaced in a single direction. The structural geometry and linked nature of the thrust front around the Kalabagh re-entrant provide a firm constraint on this transport direction since the re-entrant is bounded on both sides by lateral ramps. Such primary re-entrants, and the resulting castellated form to the thrust front, can result only from a transport direction towards the SSE which is constrained by the corrugated nature of the basal detachment. It is also pertinent that the trend of folds and the related thrusts in the eastern side of the re-entrant give misleading information about the direction of maximum compression. When taken in isolation, they would imply NE-SW stratal shortening. As part of a coherent thrust and fold belt they require regional thrust transport towards the SSE.

V. THRUST FRONT GEOMETRY AND ITS IMPLICATIONS FOR SALT DISTRIBUTION

The theoretical illustration of thrust geometry related to the distribution of a pre-existing salt horizon in Figure 3 provides a useful starting point for the following discussion. The propagated map pattern of a thrust, even assuming complete control on its form by the salt, is related more to the linked continuity of salt than to the simple distribution of salt directly beneath the thrust front. Clearly the hypothetical thrust (Fig. 3) could have a greater displacement to carry the leading edge of the thrust sheet out beyond the salt patch. Thus the pattern of salt distribution can be established only from restored maps and sections and not simply from the final geometry. Furthermore, this restored pattern will provide only a minimum distribution for the salt since areas could be in disadvantageous positions for thrust propagation (e.g. sections y-y′ and z-z′ on Fig. 3b). On this cautionary note, we can speculate as to the pre-existing distribution of the salt beneath the foreland basin immediately prior to the propagation of the Salt Range basal detachment.

The first aspect to consider is the control of salt on the actual position of the thrust front. In this respect, the location of the Himalayan mountain front in the Central Salt Range, some 120 km ahead of the main zone of shortening, has occurred because of easy slip above the basement beneath the Potwar plateau. Several deep drill holes confirm the presence of salt beneath much of the Potwar (e.g. Shah, 1979; Gee, 1980). However, it is less clear whether this salt continues into the east, onto the section line of Fig. 10. In this section the thrust has climbed up into molasse to the north of Rawalpindi, at the extreme end of the section (below G on Fig. 10b). Even if we assume that sites of thrusts climbing from the basement cover contact represent an absence of salt, salt could exist a short distance further south, where the thrust system has climbed up above the salt to use other detachment surfaces (notably at the top of the Eocene carbonates and within the molasse sediments). Any outlying salt would remain undisturbed and have no influence on thrust geometry until the footwall ramp (beneath G on Fig. 10a) failed, allowing displacements to transfer forwards. Similarly, to the north of the Kalabagh re-entrant, the thrusts have detached within Mesozoic rocks so any deeper level salt horizon would be unusable unless it connected directly back to the footwall basement-cover cut-off. At present we have not been able to locate this cut-off line as there is insufficient geological and seismic data available to construct balanced cross-sections into the hinterland. However, there is no structural evidence to discount the possibility of the Salt Range Formation continuing across the Kalabagh re-entrant. Indeed, in the southern part a deep drill hole has encountered the formation (Gee, 1980). The southern Khisor Range also includes the Salt Range Formation (Shah, 1977), indicating that the decoupling surface beneath the Bannu basin runs at this stratigraphic level, similar to beneath the Potwar plateau.

The next question concerns the actual location of the thrust front: could this location be controlled by the local absence of the salt at depth? In the west at Golewali (Fig. 24), in the footwall to the main frontal thrust, the folds in the foreland certainly decouple within salt which has extruded diapirically up some faults and is exposed in the cores of some anticlinal structures. The Salt Range Formation has been encountered in the few deep drill holes on the foreland. Commercial seismic reflection data suggest that the salt layer can be traced with a considerable (more than 1 km) thickness out to the foreland along the section lines of Fig. 11. These observations suggest that, although the salt is instrumental in allowing the Himalayan thrust front to propagate far into the foreland, the salt does not specifically locate the ramp across the overlying carapace which generates the mountain front structures. Neither does the salt specifically control the structural geometry of the thrust front, whose various forms are summarized on Fig. 28.

It is, perhaps, not surprizing that the thrust front geometry appears not to be controlled by a change in the rheological properties of the basal decoupling surface. For a thrust front to operate it must transfer shortening equally across all layers above a decoupling surface, provided the convergence is driven by plate tectonic forces. The foreland template in northern Pakistan might be considered as an infinite layer of salt overlain by a stronger carapace. It is implausible that the basal decoupling surface will propagate infinitely far, since the carapace will have to deform at some locality to accommodate the continued convergence. The problem then becomes one of the three dimensional rheological contrasts within the carapace, their spatial distribution, and the boundary conditions imposed upon the system. In essence the problem of deformation localization is identical to that in larger scale plate tectonics for which there is, as yet, no convincing explanation. A simple control on ramp location might be the local disruption of the salt layer by pre-Himalayan normal faults. Some of these faults are visible on seismic reflection profiles and, if they are Mesozoic or younger in age, could provide local weaknesses in the carapace sediments. However, an important control on the location of the ramp across the carapace might be imposed by the adjacent segments of the thrust system which are not influenced by salt decoupling. In these areas, in the Surghar and Rawalpindi-Jhelum transects, thrust propagation and displacement would have been inhibited by the relatively strong rheologies, as evidenced by folding preceding thrusting in both cases (Figs. 10

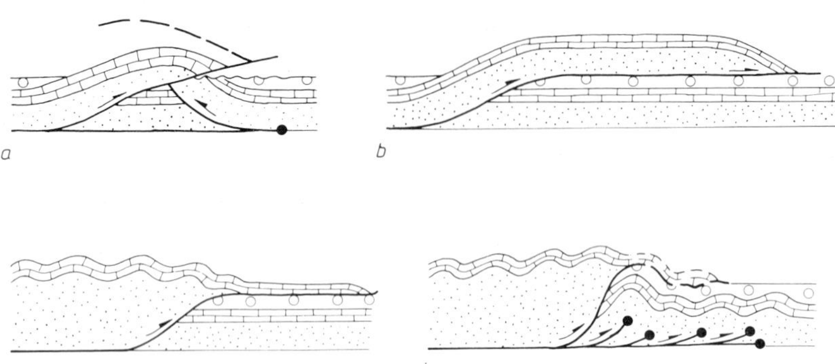

Figure 28. Four types of mountain front structure exhibited around the Salt Range. a) illustrates the masking of a backthrust front by a relatively subordinate forethrust, as defined for the Jalalpur district (see Fig. 20c). b) shows a simple Rich-type (1934) flat-topped anticline generated by displacement up a ramp-flat thrust profile, as is probable for the eastern Central Salt Range (See Fig. 11a). c) depicts the decoupling of the carapace from the salt so that the cover is thrust over the foreland while the salt thickens by distributed processes in the hinterland, as suggested for the western Central Salt Range (Fig. 11b). d) is a hybrid form of the last example where minor displacements are transferred onto the foreland, folding the thrust front, as appears the case in the SW corner of the Salt Range (Fig. 25c).

and 22). The much faster propagation rates possible in the salt-controlled sector beneath the Potwar could be achieved only if the linking lateral ramps could propagate at the faster rate.

Inability to do so would inhibit the otherwise rapid sectors, causing deformation within the carapace, possibly failure as a ramp. This third dimension in thrust propagation has yet to be explored by geologists to any great extent. However, the predicted line of the lateral ramp, between the salt detachment beneath the central Potwar and the intra-molasse detachment in the east, coincides with a zone of locally large (40°) counterclockwise rotations described by Burbank and Raynolds (1984). Rotations of this kind have been considered to be indicative of laterally inhibited thrust propagation (Coward and Potts, 1983), and are predicted to occur on the margins of salt-based thrusts systems (e.g. Davis and Engelder, 1985).

A final problem concerns the variety of thrust front morphologies along the Salt Ranges, summarized in Figure 28. For the sake of discussion these morphologies can be grouped into two types. The first is typified by Figure 28a and b, where the salt behaves geometrically similar to other units above the basal decoupling surface. The salt retains a broadly constant thickness, so that little or no flowage can have occurred. This behavior occurs primarily in the eastern Salt Range and can be contrasted with Figure 28c where the salt has been greatly thickened beneath the carapace. In this second case the principal decoupling occurs, not at the basement-cover contact, but along the base of the carapace, principally in the western Salt Range. Figure 28d illustrates a hybrid form, restricted to the far western sector around Golewali (see Fig. 25c). The upper detachment is favored initially, but the folding then appears to be controlled by lower detachment without carapace decoupling. The key to this behavior appears to be the activation of the upper detachment, an explanation possibly being offered by the lateral stratigraphic variations around the range (see Fig. 9).

In its type area in Khewra gorge (Fig. 6 and 15), the Salt Range Formation is conformably overlain by the Lower Cambrian Khewra Sandstone, its basal part being dominantly shaley (Shah, 1977), so that the rheological variation between salt and carapace is likely to be transitional. In the west, around Ghundi (Fig. 24), the Salt Range Formation is locally overlain directly by the Lower Permian Nilawahan Group, the Cambrian formations being cut out by the sub-Permian unconformity. The base of this Group is a poorly sorted tillite (Shah, 1977), which is likely to have had starkly contrasting rheological properties to the Salt Range Formation. Intuitively, decoupling is most likely to occur in the latter case. We propose that the tendency to develop a detachment surface along the sub-Permian unconformity allowed the salt to thicken beneath the carapace while, in the east, the lack of such a pronounced rheological change may have preserved the salt-carapace boundary intact. It is interesting that a continuous presence of Cambrian clastics above the salt does not occur in the western Salt Range. The exception lies in the foreland folds around Golewali and Chhidru where the upper detachment has not developed. It is possible that some of the minor thrusts which breach the carapace in the western Salt Range may nucleate at remnant patches of the conformable Cambrian cover. Clearly, complete decoupling did not occur because the carapace is buckled in this eastern sector.

VI. SALT FLOWAGE AND THRUST FRONT COLLAPSE

Most thrust belts exhibit the full range of structures described above, together with more complex ones relating to the interactions between arrays of thrusts. However, the Salt Range further exhibits a suite of extensional faults, particularly well developed in the western sector, which are unusual in salt-free terrains. Figure 29 documents the distribution of these normal faults together with the outcrop pattern of the salt. Let us first consider a detailed area about 10 km SE of Naushahara, illustrated by a sketch map (Fig. 30), a cross-section (Fig. 31) and a plate (Fig. 32). The high plateau of Tertiary carbonates has been dropped down towards the foreland on a suite of normal faults which bottom out in the salt. The geometry of bedding in the frontal fault blocks strongly suggests that the faults have a listric geometry, decoupling near the base of the salt along the range front. The fault blocks in the north have antithetic development, and appear to behave in a more rigid rotation manner, suggesting that the normal faults remain planar at depth. This second type of geometry requires substantial flowage of salt during displacement, suggesting a rather poorly defined zone of

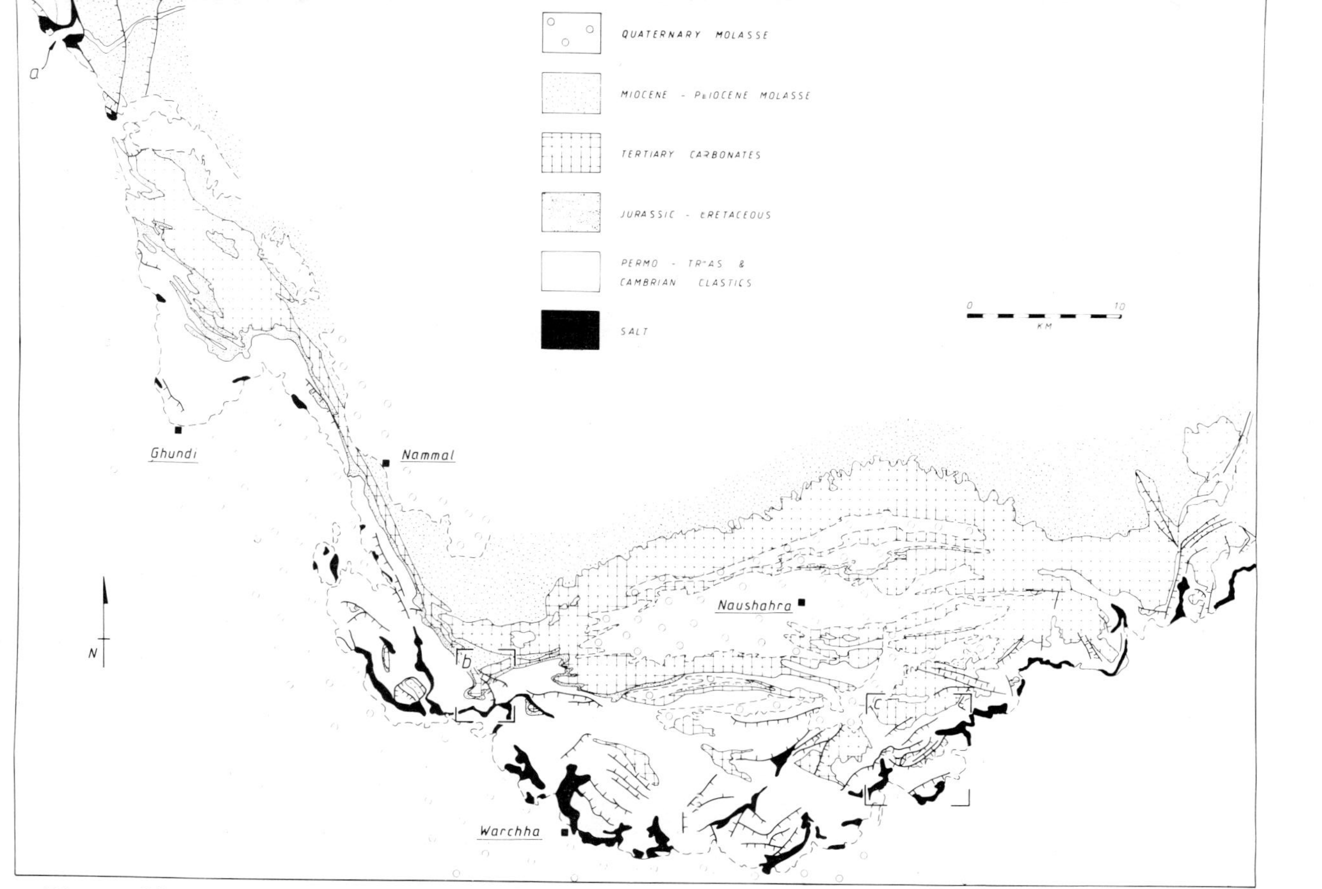

Figure 29. Simplified geological map of the Western and western Central Salt Range (modified after Gee, 1980) illustrating the distribution of normal faults and the detailed locations of; a- Fig. 37, b - Fig. 36, c - Fig. 30.

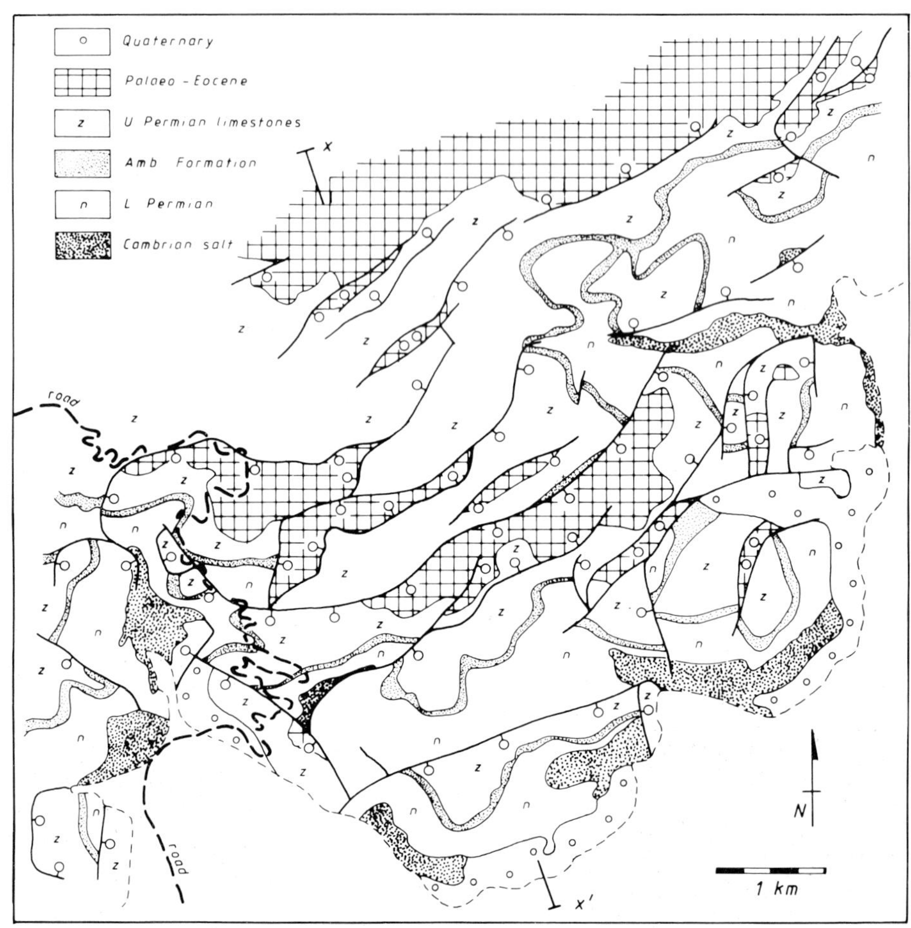

Figure 30. Sketch map (modified after Gee, 1980) of normal fault collapsing the frontal escarpment of part of the Central Salt Range, approximately 10 km SE of Naushahra, locality c on Fig. 29. The section line of Fig. 31 is depicted x-x'.

decoupling. The complete system of faults displays about 1 km extension, measured on offsets of the Permian Amb Formation. This displacement must pass onto the range front thrust detachment which, locally, gains extra movement.

The cross section (Fig. 31) illustrates that the normal faults are effectively large land-slide structures, and the regional map (Fig. 29) shows that they are restricted to lie on

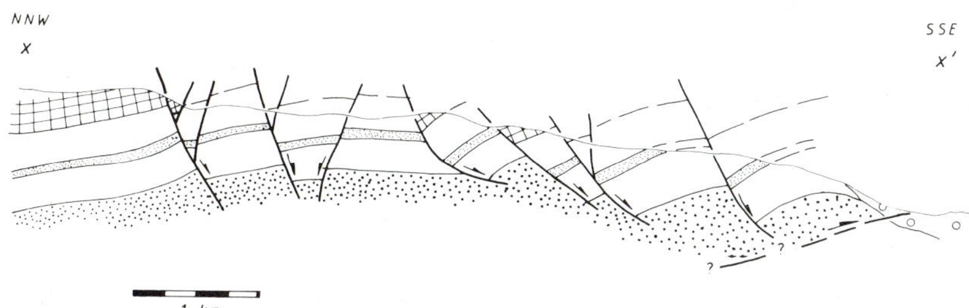

Figure 31. Sketch section x-x' on Fig. 30 across the normal faults responsible for the collapse of part of the Central Salt Range mountain front (key as Fig. 30).

Figure 32. Plate of the normal faults in Figs. 30 and 31.

the steep outboard flanks of the range. A gravitational driving force is clearly implicated. A problem, however, exists as to why some of the outer range slopes fail and others do not.

Davis and Engelder (1985) suggest that salt-based thrust belts need to maintain only a foreland wedge taper of about one

degree to continue movement, as opposed to 8° to 12° for belts with strong basal detachments. The frontal slopes of the Salt Ranges are generally about 30°, with a present-day topographic relief of up to 1500 m above the foreland, providing the available potential to drive gravitational collapse. Perhaps the best trigger for this collapse would be a dramatic weakening of the basal salt layer. Experimental data by Urai (1983) suggests that the presence of only small amounts of brine will greatly enhance rates of grain boundary migration, diffusional mass transfer, and subgrain development, leading to substantial loss of strength at low temperatures (60°C). Thus the introduction of water at the thrust front (Fig. 33) would lead to local developments of extremely weak zones within the salt. Should these zones link up the salt could begin to flow laterally under the topographic load of the uplifted carapace, leading to salt extrusion along the thrust front and, in extreme cases, large scale collapse of the range front slopes. This mechanism would be an efficient method of unroofing and eroding the carapace in the western Central Salt Range.

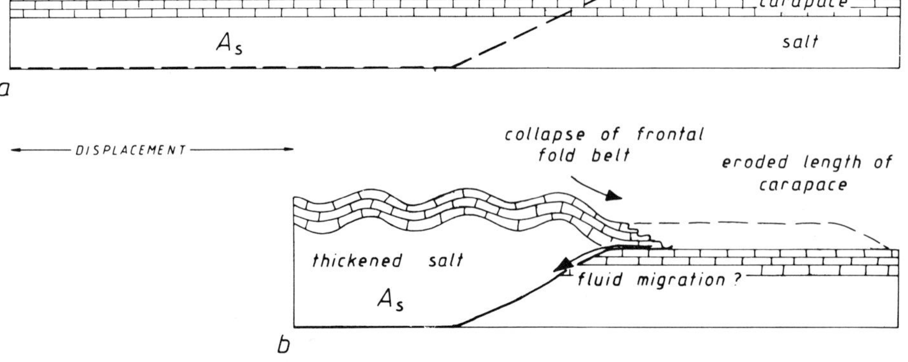

Figure 33. Cartoon of the structural history of the frontal structure of the western Central Salt Range. a) illustrates the pre-deformational geometry of salt and its carapace. b) shows the thickening of salt (area As), the concomitant thrusting of the carapace onto the foreland (subsequently eroded), and the collapse of the eroded thrust front, possibly facilitated by the greatly reduced strength of salt caused by the introduction of fluids from the foreland.

Unfortunately much of the range front in the Central Salt Range is masked by recent screes and fanglomerates so the possible salt extrusions are obscured. However, in the eastern sector, at Kas Kaura (Fig. 15), a section through part of the front is exposed in a stream section. This area contains several normal faults which drop higher parts of the range front structure down towards the Punjab alluvial plain. The Salt Range Formation is particularly thick along this part of the thrust front. In nearby Khewra mine (Fig. 15) the salt is at least 850 m thick and behaves diapirically in some localities. However, at Kas Kaura (Fig. 34) there is an erosional re-entrant in the range so that the mountain front locally trends N-S. Along this segment the salt has flowed out (Fig. 35) across the folded Cambrian sandstones and local outcrops of river gravels. Linear

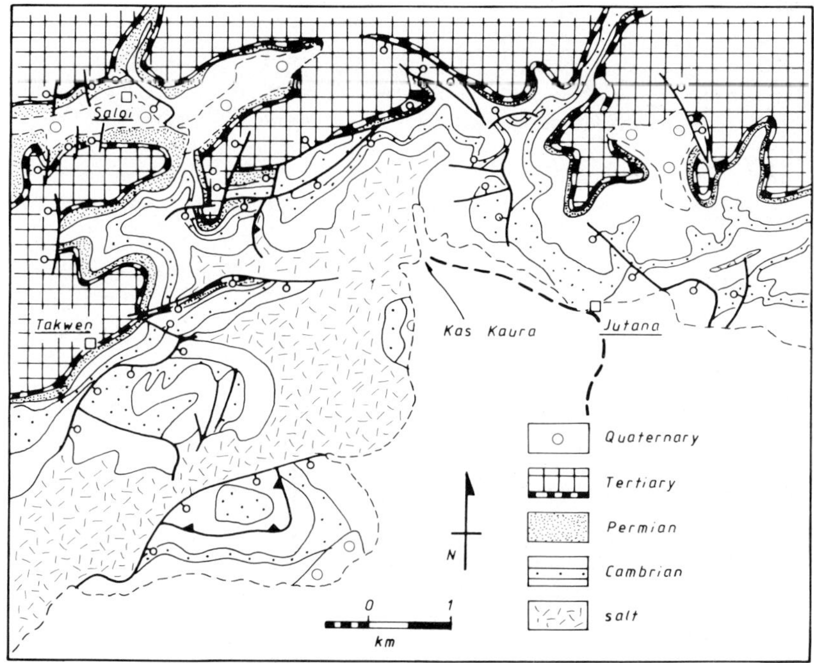

Figure 34. Sketch map of the Kas Kaura district of the Eastern Salt Range (a on Fig. 15) illustrating the distribution of normal faults and the location (arrowed) of Fig. 35.

fabrics within the salt, which has been extensively recrystallized, together with shear fibers on more discrete surfaces, imply a flow direction towards ESE, a direction parallel to the local slope and bearing no relationship to the regional direction of thrust transport.

Elsewhere in the Salt Range, Gee (1980) mapped arrays of normal faults converging downwards to join thrust faults (Fig. 36; see Figs. 12 and 29 for location). These faults drop the Tertiary carbonates and underlying Permian rocks towards the foreland. Salt along the thrust, which presumably formed a continuous layer prior to the extensional collapse, is now disrupted, suggesting both flowage and decoupling along the early

Figure 35. Plate of salt extrusion at Kas Kaura.

thrust contact. It is not certain whether this episode of gravitational collapse occurred significantly in the thrusting process, or whether the normal faults merely tapped into a convenient detachment surface during propagation.

Arguably, the most spectacular examples of large-scale collapse of the thrust front within associated salt flowage and migration occur in the Kalabagh area (Fig. 29), where the Indus river now crosses the range front. The northern banks of the river contain a large fault-bounded block, entirely underlain by salt. The structure is illustrated by a map (Fig. 37) and two cross-sections (Fig. 38).

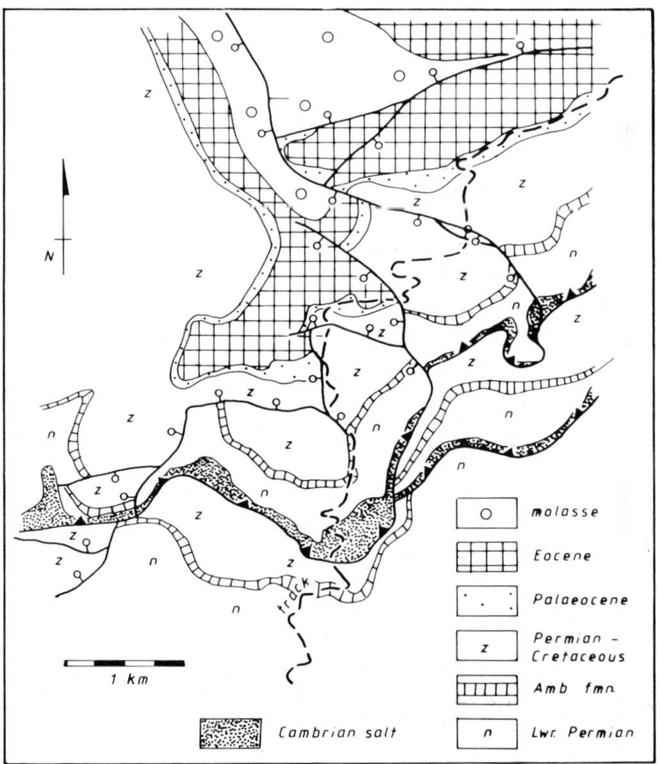

Figure 36. Map of part of the SW Salt Ranges (modified after Gee, 1980), and located (b) on Fig. 29. Note that the normal faults in the center of the map converge southwards (and downwards) with thrusts which are decorated by strips of salt.

When viewed on a cross-section constructed parallel to the thrusting direction (i.e. NNW-SSE, Fig. 38b), the structure of Kalabagh Hill is relatively simple. The thrust front is emergent onto molasse sediment as young as Quaternary age (Gee, 1980), and carries a thick wedge of salt with its carapace of Permian through to Eocene rocks with molasse on top. However, in a perpendicular section this simple geometry is heavily disrupted by normal faults which decouple in the salt wedge and down-throw towards the WSW. Further complications arise because the normal faults cut previously developed oblique folds. The resulting half-graben contains Quaternary alluvium and fanglomerates, probably related to the paleo-Indus river, and which show signs of being deposited during the extensional movements. Associated with this collapse there has been considerable flowage of salt, which has migrated both up the back of the normal fault blocks

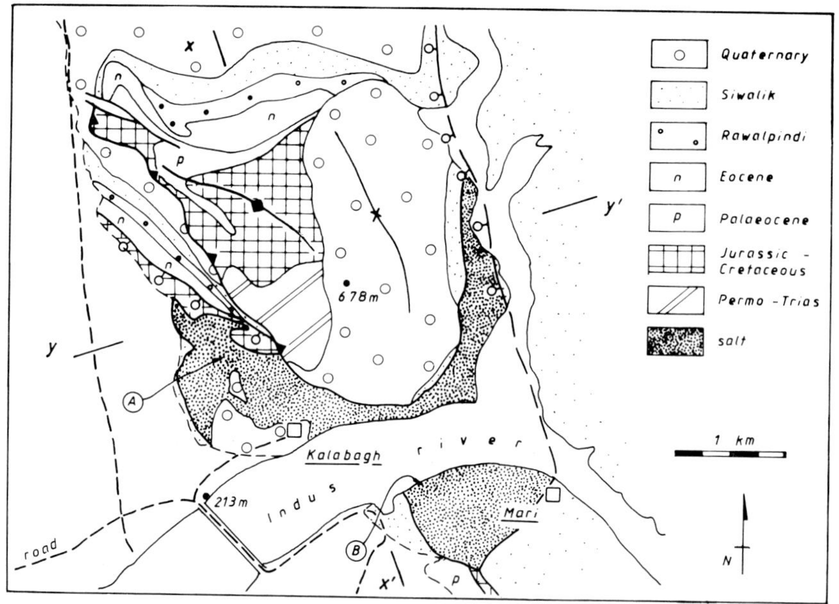

Figure 37. Sketch map of the Kalabagh Hill district of the NW Salt Range (locality (b) on Fig. 29) with the section lines of Fig. 38 (x-x' and y-y') together with the locations of plates in Fig. 39 indicated (modified after Gee, 1980).

and forward beneath them. The structural geometry in the perpendicular section (Fig. 38a) suggests substantial volume loss from the salt. The pathway for the loss of salt leads up through the Quaternary river deposits and is associated with spectacular small scale diapiric structures, some of which are illustrated (Fig. 39). Almost all the salt has been extensively recrystallized, and has well-developed planar and linear fabrics, isoclinal folds and mylonitic textures, which presumably relate to the mobilization. The relationships between thrust front morphology, extensional collapse, and salt migration are illustrated hypothetically on Figure 40.

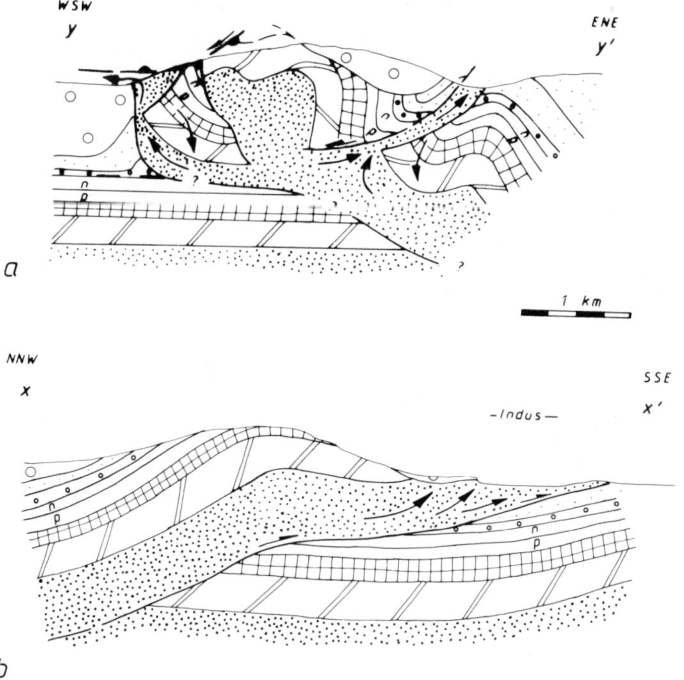

Figure 38. Perpendicular cross-sections through Kalabagh Hill, lines illustrated on Fig. 37. a) lies perpendicular to the thrust transport direction and shows the lateral collapse of the thrust front towards the foreland while (b) lies parallel to thrusting and shows a simple frontal geometry. Arrows indicate the inferred flowage direction of salt. Key as Fig. 37.

Figure 39. Plates of salt diapirism around Kalabagh Hill.

VII. DISCUSSION

We have demonstrated from the above account of the structure of the Salt and Surghar Ranges that, despite a relatively simple tectonic setting, mountain front structure can show rapid variations along strike (Fig. 41). These variations are not controlled simply by the mere presence or absence of the infra-Cambrian salt. The main variations in thrust front morphology, particularly in changing from a fold belt to a simple thrust ramp, have been tentatively linked to the degree of coupling between the salt and its cover carapace. However, we emphasize that, contrary to the predictions of Davis and Engelder (1985), it is probably unwise to draw up simple rules to define the location of salt from variations in structural style. This

Figure 39B.

advise arises because all the primary thrust structures we have found along the Salt and Surghar Ranges, together with those which are implicit on the published maps (Gee, 1980), have analogs in other, salt-free, thrust belts. Furthermore, much of a subcrop pattern of salt may be in a disadvantageous location to be used by thrust detachments (Fig. 3). Thus, much of the undisrupted foreland basin, together with parts of the thrust belt which decoupled at relatively high stratigraphic levels, may also be underlain by salt.

The distinctive feature of the mountain front in northern Pakistan is the presence of gravitational collapse structures, together with salt extrusion, which we consider to be a direct result of deforming the Cambro-Eocene carapace above the salt. The presence of a mechanically weak layer, at the base of thrust-

Figure 39C.

generated topography, was sufficient to permit the generation of essentially large scale land-slips, directed down topographic slope. The introduction of water, with its effect of greatly reducing the strength of salt, may have been critical in the timing, geometry and activation of these collapse structures and their associated salt extrusions (Fig. 41). In the Khewra area (Fig. 15), where these extrusions are particularly well developed, the thrust front may have actively over-ridden previous extruded salt, so that the great thickness of dominantly in situ salt beneath the frontal parts of the range may have been contaminated by post-Cambrian material. The mining records at the Khewra site (Pakistan Mineral Development Corporation, personal communication, 1984) apparently contain accounts of wood being found at deep levels within the salt. This contamination

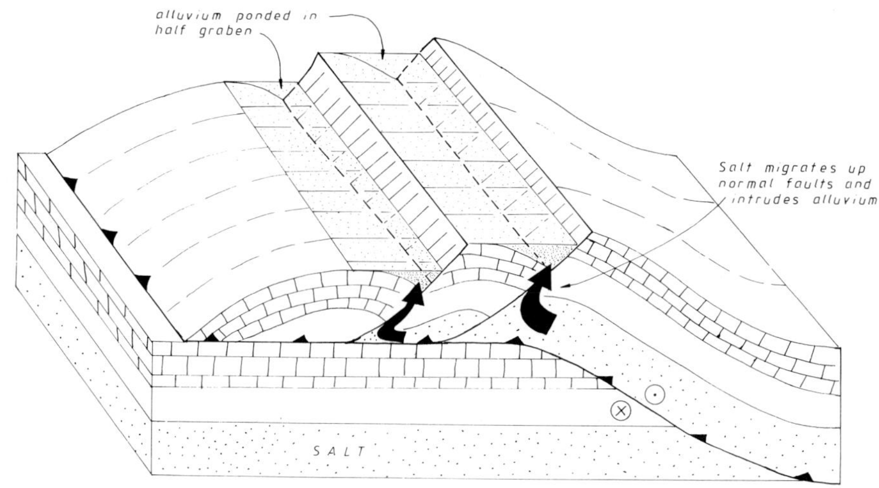

Figure 40. Hypothetical block diagram illustrating the collapse of a thrust front, and the mobilization of salt up the normal faults to intrude molasse sediments ponded in the resultant half-graben.

may explain some of the early confusion as to the age of the Salt Range Formation (Gee, 1945). The intimate mixing of in situ and contaminated salt would have been enhanced by diapirism.

Although we might predict that salt-based thrust fronts are characterized by gravitational collapse structures and by salt extrusion, this prediction will be of limited use in interpreting more ancient belts. Future erosion of the Salt Range would remove much of the evidence, and the mountain front could preserve merely the simple thrust structures. Furthermore, we do not know the extent or importance of gravitational collapse in salt-free active mountain fronts. Certainly similar features occur at the toes of accretionary prisms, structures which are essentially sub-marine thrust belts. Rather than list characteristic structures for particular tectono-stratigraphic regimes (e.g. salt-based thrust belts), it might be more fruitful, if more time-consuming, to make studies of thrust geometry in three dimensions in conjunction with stratigraphic and geophysical surveys for each particular region of interest.

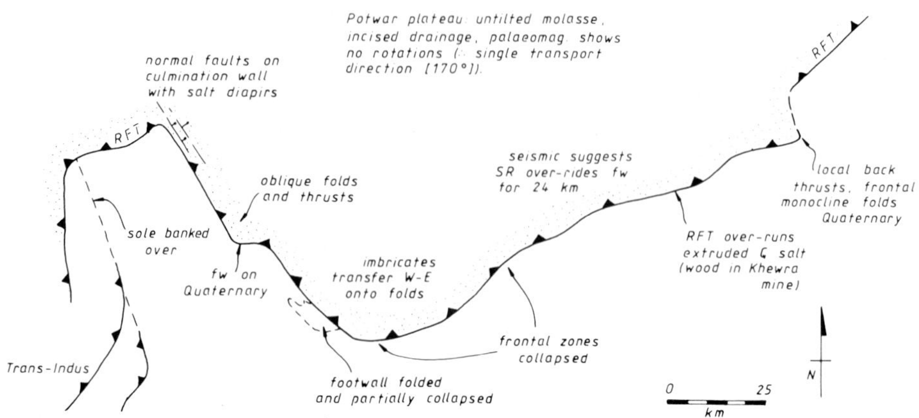

Figure 41. A summary of range-front structure.

There remains one attribute of the Salt Ranges which, perhaps, reflects least ambiguously the control of sub-carapace salt on thrust geometry. This attribute is the overall tectonic setting of the mountain front, some 100 km ahead of the main Himalayan thrust belts (Figs. 6 and 8). Using the reasoning of Davis and Engelder (1985), we might predict that the entire Potwar is underlain by salt. Thus, as the thrust belt encounters the salt beneath what is now the northern margin of the Potwar, it requires a reduced surface slope to generate a critical thrust wedge taper in order to move. The thrust detachment can then propagate, unhindered by the low resistance to slip within the salt, to climb out at the site of the Salt Ranges. The Potwar would then be simply transported over the salt with only minor shortening. Along strike to the east along the Rawalpindi-Jhelum transect (Fig. 8a), the thrust detachment apparently propagated above the salt (Fig. 10) at a stratigraphic level which had a greater mechanical resistance to slip. This required a greater critical wedge slope to propagate and move, so that a stack of folds and imbricates developed. The eastern Potwar is, therefore, far more disrupted, and the molasse sediments are more folded, than in the central Potwar district. This type of control was proposed by Burbank (1983) to explain the eastern termination of the Salt Range. This model is essentially dynamic (Davis et al., 1983), with the propagating thrust geometry generating its own critical wedge taper, as illustrated in Fig. 1.

It is tempting to relate the Peshawar and Campbellpore basins to this model of critical thrust wedge generation. These basins lie directly behind, and some 100-150 km north of, the Salt Range mountain front (Figs. 5 and 6), and represent a greatly reduced hinterland topography compared with the adjacent Hazara hills (Fig. 6). Detailed stratigraphic studies of the molasse deposits (Burbank and Tahirkheli, 1985) show that the Peshawar basin developed between 3 and 5 Ma and is still acting as a depocenter. The Salt Range thrust can be similarly dated, using detailed magneto-stratigraphy, as forming at about 3-5 Ma and moving within the last 1 Ma (Burbank, 1983; Yeats et al., 1984). These data suggest that the actual thrust displacement rate was about 1 $cm.yr^{-1}$ (30-40 km movement in 3-4 Myr), but the thrust propagation rate was about an order of magnitude faster beneath the Potwar plateau. The Main Boundary Thrust zone probably ceased being active in the northern Potwar at about 4-5 Ma (Burbank, 1983), so that the thrust detachment propagated to the Salt Ranges (about 80 km) in about 1 Myr. During, and following, this period of rapid thrust propagation, the intermontaine Peshawar and Campbellpore basins developed.

One link between the Peshawar-Campbellpore basins and the generation of the Salt Range mountain front might be that the thrust wedge reduced critical taper upon reaching the sub-carapace salt. However, the intermontaine basins in the north are built on previously stacked, and already deeply eroded, thrust structures. The adjacent Hazara hills comprise stacked basement and cover rocks (Calkins et al., 1975; Coward and Butler, 1985) which plunge beneath the Peshawar basin. Thus, for one thrust wedge model, the internal part of the thrust belt would have actively subsided to generate a reduced critical taper (Fig. 42). This model is distinct from that in Fig. 1, requiring the thrust wedge to have been super-critical upon reaching the salt, and for simple accelerated forward thrust migration to have been insufficient to reduce the active taper. This model is essentially identical to the gravitational collapse of the mountain front along the Salt Range (Fig. 40), discussed earlier, but on a much larger scale. The model generates extra thrust displacements at the mountain front to accommodate the thrust sheet extension (Fig. 42).

The super-critical thrust wedge has profound implications for the lateral continuity of thrust displacement along the mountain front. Since the presumed subsidence is limited to the

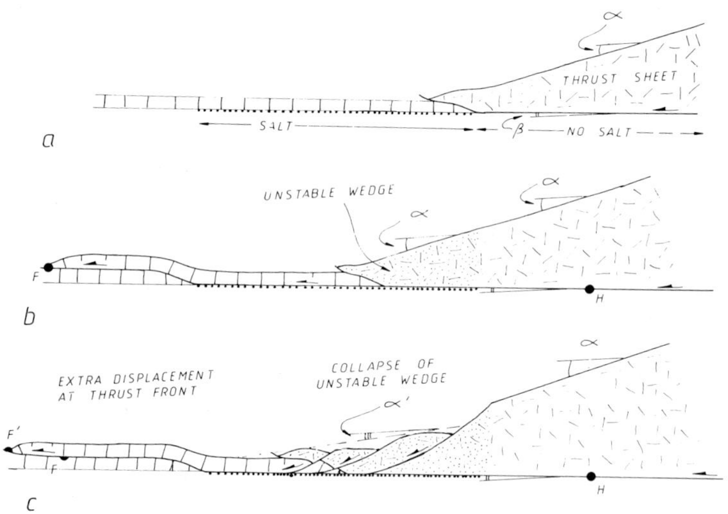

Figure 42. Generation of the Peshawar basin by the collapse of a thrust wedge upon encountering the salt horizon with greatly reduced resistance to shear. Note that this generates displacements (F-F') in addition to those provided by the regional thrust transport.

present site of the Peshawar basin and this subsidence would, for this model, drive extra thrust displacement, the mountain front should show variable displacements. The central Salt Range should show more shortening compared to the Surghar and eastern Salt Ranges, which lie ahead of apparently intact parts of the Rthrust wedge. The Potwar plateau should be crossed by tear faults linking the zones of differential subsidence, which should post-date many of the folds and thrusts behind the Salt Range. No such faults have been recognized. Furthermore, it is difficult to envisage greatly increased displacements in the central Salt Range.

An alternative model might be that the internal part of the thrust belt in the Peshwar basin area has always been low. The present thrust wedge geometry along this section line (Fig. 8a) is merely a critical (rather than super-critical) wedge model (Fig. 1). The Hazara Hills section might rather have been uplifted to generate a greater wedge thickness in the hinterland, accentuating the frontal shortening south of Rawalpindi. This behavior could occur by either late thrusts restacking the Hazara thrust belt, or by a lower level duplex structure beneath. The lower duplex model would avoid many of the problems of displacement continuity introduced by the super-critical wedge model, since the deep level displacements can be equivalent to the shortening on the thrust structures at outcrop. But the duplex model would, however, require modification of Coward and Butler's (1985) cross-section. It would require both the Peshawar basin to originate prior to Salt Range thrusting and the thrust belt to have had a very low angle taper even before it encountered the salt. There is little supporting evidence for this scenario from the molasse stratigraphy studies (Burbank, 1983; Burbank and Raynolds, 1984; Burbank and Tahirkheli, 1985). These studies show that the early molasse deposits (Miocene - Pliocene) are intensely folded, and predate the present

VIII. ACKNOWLEDGEMENTS

We thank personnel at the National Centre of Excellence in Geology at Peshawar University, particularly M. Qasim Jan and R.A.K. Tahirkheli for discussions on Himalayan geology, their hospitality and logistic support, together with personnel at the Rawalpindi office of the Pakistan Mineral Development Corporation for discussions on the structure of the Salt Range and for arranging accommodation in PMDC houses. We are also grateful to staff at the Hydrocarbon Development Institute for allowing access to commercial seismic lines and for discussion. Field work was funded by Natural Environment Research Council grants and travel funds were kindly made available to RWHB from the Royal Society and to GMH from the Research Fund of Newcastle University. A Royal Society (Jaffe Donation) Research Fellowship is gratefully acknowledged by RWHB.

REFERENCES

Asrarullah (1967). Geology of the Khewra Dome. Pakistan Sci. Conf. 18th/19th Jamshoro Proc. III. Abs., F3-F4.

Beaumont, C. (1981). Foreland Basins. Geophys. J.R. Astr. Soc. 65, 291-329.

Brunel, M. (1986). Ductile thrusting in the Himalayas; shear sense criteria and stretching lineations. Tectonics 5, 247-265.

Burbank, D.W. (1983). The chronology of intermontane basin development in the northwestern Himalaya and the evolution of the Northwest syntaxis. Earth Planet. Sci. Lett 64, 77-92.

Burbank, D.W. and Raynolds, R.G.H. (1984). Sequential late Cenozoic disruption of the northern Himalayan foredeep. Nature 311, 114-118.

Burbank, D. and Tahirkheli, R.A.K. (1985). The magnetostratigraphy fission-track dating and stratigraphic evolution of the Peshawar intermontane basin, northern Pakistan. Bull. Geol. Soc. Am. 96, 539-552.

Butler, R.W.H. (in press). Thrust sequences. J. Geol. Soc. London.

Butler, R.W.H., Matthews, S.J. and Parish, M. (1986). The NW external Alpine thrust belt and implications for the geometry of the Western Alpine origin. In "Collision Tectonics" (M.P. Coward and A.C. Ries, eds.), Spec. Publ. Geol. Soc. London 19, 245-261.

Calkins, J.A., Oldfield, T.Q., Abdullah, S.K.M. and Tayab Ali, S. (1975). Geology of the Southern Himalaya in Hazara, Pakistan and adjacent areas. Prof. Pap. U.S. Geol. Surv. 716C, 1-20.

Coward, M.P. and Butler, R.W.H. (1985). Thrust tectonics and the deep structure of the Pakistan Himalaya. Geology 13, 417-420.
Coward, M.P., Butler, R.W.H., Khan, M.A. and Knipe, R.J. (in press). The tectonic history of Kohistan and its implications for Himalayan structure. J. Geol. Soc. London.
Coward, M.P. and Potts, G.J. (1983). Complex strain patterns developed at the frontal and lateral tips to shear zones and thrust zones. J. Struct. Geol. 5, 383-399.
Dahlstrom, C.D.A. (1970). Structural Geology at the eastern margin of the Canadian Rocky Mountains. Bull. Can. Pet. Geol. 18, 332-406.
Davis, D.M. and Engelder, T. (1985). The role of salt in fold- and thrust- belts. Tectonophysics 119, 67-88.
Davis, D.M., Suppe, J. and Dahlen, F.A. (1983). Mechanics of Fold- and Thrust-Belts and Accretionary Wedges. J. Geophys. Res. 88, 1153-1172.
Farah, A., Mirza, M.A., Ahmad, M.A. and Butt, M.H. (1977). Gravity field of the buried shield in the Punjab-Plain, Pakistan. Bull. Geol. Soc. Am. 88, 1147-1155.
Gansser, A. (1964). "Geology of the Himalayas." Wiley, Interscience, London, 289 pp.
Gee, E.R. (1945). The age of the saline series of the Punjab and of Kohat, India. Nat. Acad. Sci. Proc., Sec. B, 14, 269-310.
Gee, E.R. (1980). Salt Range Series, Pakistan Geological Maps at 1:50000, six sheets, Directorate of Overseas Surveys, UK.
Gill, W.D. (1951). The stratigraphy of the Siwalik Series in Northern Potwar, Punjab, Pakistan. Q. J. Geol. Soc. London 107, 375-394.
Graham, R.H. (1978). Quantitative deformation studies in the Permian rocks of the Alpes Maritimes. Proc. Geol. Assoc. 89, 125-142.
Johnson, G.D., Johnson, N.M., Opdyke, N.D. and Tahirkheli, R.A.K. (1979). Magnetic reversal stratigraphy and sedimentary tectonic history of the Upper Siwalik Group, eastern Salt Range and southwestern Kashmir. In "Geodynamics of Pakistan" (A. Farah and K. DeJong, eds.), 149-165, Geol. Suv. Pakistan, Quetta.
Johnson, G.D., Zeitler, P., Naeser, C.W., Johnson, N.M., Summers, D.M., Frost, C.D., Opdyke, N.D. and Tahirkheli, R.A.K. (1982). Fission-track ages of late Neogene and Quaternary volcanic sediments, Siwalik Group, northern Pakistan. Palaeogeogr. Palaeoclimatol. Palaeoecol. 37, 63-93.
Karner, G.D. and Watts, A.D. (1983). Gravity anomalies and flexure of the lithosphere at mountain ranges. J. Geophys. Res. 88, 10449-10477.
Kummel, B. and Teichert, C. (1970). Relation between the Permian and Triassic formations in the Salt Range and Trans-Indus Ranges, West Pakistan. Neues Jahrab. Geol. Palaeont. Abh. 125, 297-333.
Menard, G. and Thouvenot, F. (1984). Ecaillage de la lithosphere europeene sous les Alpes Occidentales: arguments gravimetriques et sismiques lies a l'anomalie d'Ivrea. Bull. Soc. Geol. France 26, 875-884.
Opdyke, N.D. Johnson, N.M., Johnson, G.D., Lindsay, E.H. and Tahirkheli, R.A.K. (1982). Paleomagnetism of the middle Siwalik formations of northern Pakistan and rotation of the Salt Range decollément. Palaeogeogr. Palaeoclimatol. Palaeoecol. 37, 1-15.
Patriat, P. and Achache, J. (1984). Collision chronology and its implications for crustal shortening and the driving mechanism of plates. India-Eurasia Nature 311, 615-621.
Pilgrim, G.E. (1910). Preliminary notes on a revised classification of the Tertiary fresh-water deposits of India. Rec. Geol. Surv. India 40, 185-205.

Rich, J.L. (1934). Mechanics of low-angle overthrust faulting illustrated by Cumberland thrust block, Virginia, Kentucky and Tennessee. Bull. Geol. Soc. Am. 18, 1584-1596.
Rigassi, D. (1977). Genese tectonique du Jura: une nouvelle hypothese. Palaeolab News 2, 1-27..
Sahni, B. (1947). Micro-fossils and the Salt Range Thrust. India Natl. Acad. Sci. Proc. B. 16, 1-44.
Schindewolf, D.H. and Seilacher, A. (1955). Beitrage zur Kenntis des Kambriums in der Salt Range, (Pakistan). Abh. Akad. Wiss. Lit. Mainz. Abh. Math-Nat. 10, 446.
Shah, S.H.I. (1977). "Stratigraphy of Pakistan." Mem. Geol. Surv. Pakistan, v. 12, Quetta.
Teichert, C. (1964). Recent German work on the Cambrian and Saline Series of the Salt Range, West Pakistan. Rec. Geol. Surv. Pakistan 11, 1-2.
Urai, J. (1983). Deformation of wet salt rocks: an investigation into the interaction between mechanical properties and microstructural processes during deformation of polycrystalline carnallite and bischofite in the presence of a pore fluid. Unpubl. Ph.D. thesis, Univ. Utrecht, Utrecht, 221 pp.
Vann, I.R., Graham, R.H. and Hayward, A.B. (1986). The structure of mountain fronts. J. Struct. Geol. 8, 215-227.
Wadia, D.N. (1949). "Geology of India" MacMillan and Co., London, 536 pp.
Yeats, R.S. Khan, S.H. and Akhtar, M. (984). Late Quaternary deformation of the Salt Range of Pakistan. Bull. Geol. Soc. Am. 95, 958-966.
Zeitler, P.K. (1985). Cooling history of the NW Himalaya, Pakistan. Tectonics 4, 127-151.

MODELLING OF THE DEFORMATION AND FAULTING OF FORMATIONS OVERLYING AN UPRISING SALT DOME

John J. O'Brien[1]
Ian Lerche[2]

[1]Standard Petroleum of Alaska
900 East Benson Blvd.
Anchorage, Alaska 99519

[2]Dept. of Geology
University of South Carolina
Columbia, SC 29208

I. INTRODUCTION

Emplacement of diapiric masses can have a profound effect on the structure of the formations overlying the diapir and of the formations being intruded. This influence may be manifested in many ways, such as the deformation of these formations to accommodate the intrusion, development of faulting in the supradomal formations, modification of the depositional pattern resulting in thinning of beds as one approaches the dome, and the development of a rim syncline at depth. These features may, in turn, also impact the further development of the salt diapir and may also create an environment which is favorable for the trapping of hydrocarbons.

Several authors have discussed the structural development of the formations overlying a salt dome. Halbouty (1979) has presented a comprehensive discussion of the salt domes of the Gulf of Mexico, including an analysis of the structures observed in the overlying formations and their implications for the trapping of hydrocarbons. Halbouty identified several structural elements which can contribute to the development of hydrocarbon traps around salt domes as follows:

1. Simple domal anticlines.
2. Graben fault trap over dome.
3. Porous caprock.
4. Flank sand pinchout and sand lens.
5. Trap beneath overhang.
5. Trap uplifted and buttressed against salt plug.
7. Unconformity.
8. Fault trap downthrown away from dome.
9. Fault trap downthrown toward dome.

Seismic examples of such supradomal structural elements have recently been presented by Inderwiesen (1983), Owen and Taylor (1983), Sunwall et al. (1983), and Wanslow (1983). In addition to the structural development associated with the emplacement of salt diapirs, the formations overlying a massive salt deposit may also be impacted by other aspects of the behavior of salt. Larberg (1983) has documented the development of listric normal faults compensating for the withdrawal of salt at depth, while Jenyon (1984, 1987), Jenyon et al. (1984) and Lohmann (1972) have identified a number of structural features associated with salt dissolution in North Sea seismic examples.

In this paper we present a mathematical model of the initiation of faulting in the formations overlying a salt diapir which is being emplaced at depth. We first propose a technique to estimate the stress induced in the formations surrounding the uprising dome. This technique is based on analysis of the deformation induced in these formations and is expected to be valid for small to moderate levels of deformation. To underscore the physical processes which are operating, in the present work we invoke a model of the deformation of the sedimentary formations surrounding a salt dome; however this analysis may also be extended to the case where the deformation of the formations is estimated from the available observational data. Having specified the deformation of the sedimentary formations, we then infer the stresses which generated these deformations through application of the theory of elasticity. In this way we can model the state of stress in the formations surrounding a salt dome during the time in which the dome is being emplaced.

This yields information on the nature of the induced stresses, and on the magnitude and distribution of the stresses in formations overlying the uprising salt dome. By comparison with the strength of the formations we can infer the timing of initial rock failure which can be related to the initiation of faulting in the supradomal region, and we can also predict many of the characteristics of the fault patterns which follow.

A. Analysis of Stress and Strain

Before analysing the deformation generated by a salt dome we first review some basic concepts related to stress and strain of thick bodies. For a fuller description we refer to such texts as Ramsay (1967), and Jaeger and Cook (1976). In the present work we employ a cylindrical coordinate system with the positive z axis pointing upwards, radial distance from the z-axis denoted by r and azimuthal angle denoted by θ. We adopt the sign convention of Jaeger and Cook (1976) whereby compressive stresses and strains are positive in sign.

To characterize the deformation of a solid we must specify nine components of the strain tensor. The condition that the body is in equilibrium and is not rotating reduces the number of independent components to six, while if we require that the body deformation is symmetric about the z-axis we reduce the number of independent strain components to four. If we denote the displacement vector upon deformation by ξ then the four independent strain components are as follows:

$$\epsilon_r = - \frac{\partial \xi_r}{\partial r} \quad , \qquad (1)$$

$$\epsilon_z = - \frac{\partial \xi_z}{\partial z} \quad , \qquad (2)$$

$$\epsilon_\theta = - \frac{1}{r} \; (\xi_r - r) \quad , \qquad (3)$$

$$\Gamma_{rz} = \frac{1}{2} \left(\frac{\partial \xi_r}{\partial z} + \frac{\partial \xi_z}{\partial r} \right) , \qquad (4)$$

Here ϵ_r, ϵ_z and ϵ_θ denote compressional strains in the r,z and θ directions respectively, and Γ_{rz} represents shearing strain in the (r,z) plane. In addition it is also convenient to define the volumetric strain as follows:

$$\Delta = \epsilon_r + \epsilon_\theta + \epsilon_z \qquad (5)$$

Similarly, to characterize the state of stress of a body we must, in the general case, specify nine components of the stress tensor. Again, if the body is in equilibrium, is not rotating, and if the stress field is axially symmetric, these nine components reduce to four independent stress components which we denote as σ_{rr}, σ_{zz}, and $\sigma_{\theta\theta}$, the normal stress components in the r,z and θ directions respectively, and τ_{rz} which is the shear stress in the (r,z) plane. If the materials display an isotropic linear elastic response then the connection between stress and strain for the axially symmetric case is provided through the following:

$$\sigma_{rr} = \lambda\Delta + 2\mu\epsilon_r \quad , \qquad (6)$$

$$\sigma_{\theta\theta} = \lambda\Delta + 2\mu\epsilon_\theta \quad , \qquad (7)$$

$$\sigma_{zz} = \lambda\Delta + 2\mu\epsilon_z \quad , \qquad (8)$$

$$\tau_{rz} = 2\mu\Gamma_{rz} \quad . \qquad (9)$$

Here λ and μ are the Lame constants.

While stress is usually considered as the independent variable and equations (6-9) are used to predict the resulting strain, the central point of our analysis is that these equations can equally well be applied to infer stresses from information concerning the strains. Thus if we can estimate the deformation of a body and we possess the required information concerning the elastic properties of the body, we can then deduce the stress field which generated this deformation. This approach is valid provided that the response of the material is elastic, i.e. the material has not been stressed beyond its elastic limit, it has not failed, and creep is not significant.

II. PRINCIPAL AXES

The discussion of stress and strain is greatly simplified by introducing the concept of principal axes. To this end, we first consider the stresses acting on a surface which has some arbitrary orientation. In general these stresses can be resolved into a shearing stress (τ) and a normal stress (σ). If the orientation of the surface is varied then the magnitude of τ and σ also varies. For certain orientations, referred to as principal axes, we find that the shearing stress vanishes and the stress acting on the surface consists solely of a normal stress. For the cylindrically symmetric case considered here, the magnitude of the two principal stresses in the (r,z) plane are given by the following (Jaeger and Cook, 1976):

$$\sigma_A(z,r) = \frac{1}{2}\left\{(\sigma_{zz} + \sigma_{rr}) + \left[(\sigma_{zz} - \sigma_{rr})^2 + 4\tau_{zr}^2\right]^{1/2}\right\} \tag{10}$$

$$\sigma_B(z,r) = \frac{1}{2}\left\{(\sigma_{zz} + \sigma_{rr}) - \left[(\sigma_{zz} - \sigma_{rr})^2 + 4\tau_{zr}^2\right]^{1/2}\right\} \tag{11}$$

The directions of the principal stress axes in the (r,z) plane are given as follows:

$$\left.\frac{dr}{dz}\right|_A = -2\tau_{rz}\left\{(\sigma_{rr} - \sigma_{zz}) - \left[(\sigma_{rr} - \sigma_{zz})^2 + 4\tau_{rz}^2\right]^{1/2}\right\}^{-1} \tag{12}$$

$$\left.\frac{dr}{dz}\right|_B = - 1/ \left.\frac{dr}{dz}\right|_A . \tag{13}$$

In the case of cylindrical symmetry the stress in the azimuthal direction is not coupled to the stresses in the (r,z) plane. Thus the third principal stress component is identically equal to the azimuthal stress and the third principal axis is in the azimuthal direction. In our analysis it will prove convenient as an intermediate step to rotate our coordinate system to coincide with the three principal stress directions.

Having performed this rotation, we can now determine the normal stress (σ) and shear stress (τ) across a plane whose normal lies in the (r,z) plane and which is inclined at an arbitrary angle (β) to (σ_A):

$$\sigma = \frac{\sigma_A + \sigma_B}{2} + \frac{\sigma_A - \sigma_B}{2} \cos 2\beta \quad , \tag{14}$$

$$\tau = -\tfrac{1}{2} (\sigma_A - \sigma_B) \sin 2\beta \quad . \tag{15}$$

From equations (14) and (15) we see that the normal stress is greatest across a plane whose normal lies along the direction of (σ_A), smallest across the plane whose normal lies along the direction of (σ_B), while the shearing stress is greatest in magnitude along a surface whose normal is at 45° to the directions of (σ_A) and (σ_B). This may be illustrated graphically using the Mohr diagram which shows the relationship between normal stress and the magnitude of the shearing stress as a function of the angle (β), as in Figure 1. Equations (14) and (15) and Figure 1 also show that as the angle (β) is varied, the state of stress traces out a circle in the normal stress/shearing stress plane.

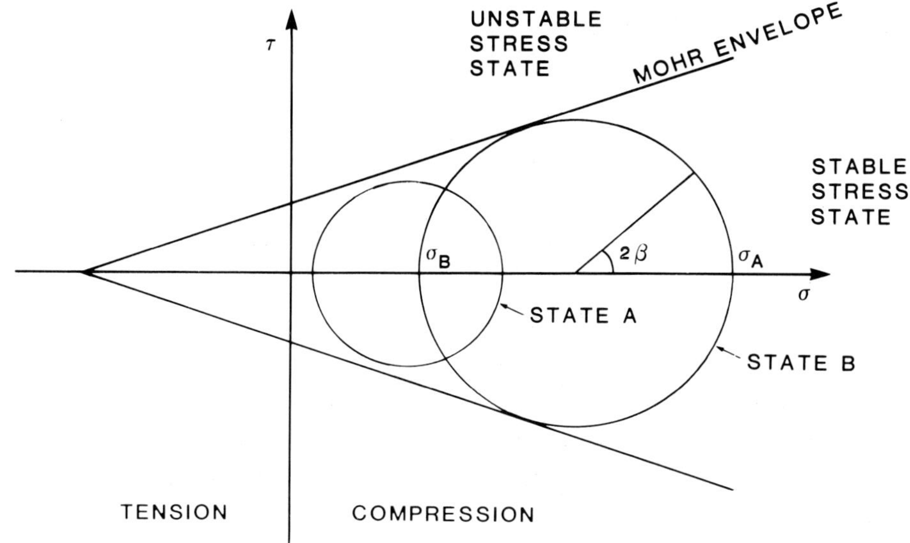

Figure 1. The Mohr diagram showing the relationship between the normal stress (σ) and the shearing stress (τ) for a plane whose normal is at an angle (β) to the direction of the maximum principal stress. The envelope corresponding to the linear Mohr failure criterion is also shown, separating stable and unstable stress states. Stress state A is stable along all planes, while stress state B satisfies the Mohr criterion along two planes.

A. Shearing Failure

For a body in an arbitrary state of stress, the normal stress and shearing stress components across a plane oriented at an angle (β) to the principal axes can be represented by a circle in the Mohr diagram. However, all the states which can be represented by points on the Mohr diagram are not physically realizable; in particular if the shearing stress along a plane exceeds some critical value, the material is found to undergo shear failure. Mohr (1900) showed that the critical shearing stress is functionally related to the stress normal to that surface. In the simplest case where the critical shearing stress depends linearly on the normal stress, the criterion for shear failure can be expressed as follows:

$$|\tau| = S_o + \nu\sigma \quad , \tag{16}$$

where S_o is a constant which is characteristic of the material and which may be regarded as its inherent shear strength. This linear equation is sometimes referred to as the Coulomb criterion and is analogous to the equation relating the frictional force opposing motion along an inclined plane to the force acting normal to the plane. In analogy with sliding along a plane, the coefficient (ν) in Equation (16) is referred to as the coefficient of internal friction. As the sign of the shearing stress only determines the direction of sliding, we consider only the magnitude of the shearing stress in Equation (16). The Coulomb criterion (16) yields a cone shaped envelope in the Mohr diagram separating stable and unstable stress states; stress states within this envelope are stable while stress states which lie on the envelope satisfy the Coulomb criterion and are liable to shear failure.

To determine conditions under which shear failure is initiated, let us consider the quantity $|\tau| - \sigma$. The value of this quantity attains its maximum value for a plane whose normal is oriented at an angle β to the axis of the maximum principal stress where β is given by

$$\tan 2\beta = \pm 1/\nu \quad . \tag{17}$$

Combining Equations (14), (15), and (17) with (16), we find that for this plane the Coulomb criterion can be expressed as follows:

$$\sigma_A = C_o + q\,\sigma_B \quad , \tag{18}$$

where

$$q = [(\nu^2 + 1)^{1/2} + \nu]^2 \quad , \tag{19}$$

and

$$C_o = 2q^{1/2}S_o \quad . \tag{20}$$

The quantity C_o represents the value of σ_A at which rock failure occurs when σ_B is zero, and so can be identified with the uniaxial compressive strength. Thus when σ_A, the maximum compressive principal stress, meets the criterion stated in Equation (18), shear failure is predicted to occur along either of the two planes whose normals lie in the plane of σ_A and σ_B and which make an angle of $\pm\beta$ with σ_A.

This formation of the Mohr criterion incorporates an assumption concerning the nature of the stress field, i.e. it assumes that the normal stress σ is a compressive stress. This implies the following constraint on our solution:

$$\sigma_A > \tfrac{1}{2} C_o \quad . \tag{21}$$

B. Deformation of Formations overlying a Salt Dome

In order to estimate the strain induced in the overlying formations by an uprising salt dome, in this section we introduce a model of the prefaulting deformation associated with a salt dome. The basic assumptions of this model are, first, that the deformation does not result in any change in volume of the formations but, instead, to a change in shape which may include extensional, compressional and shearing components. Second, we do not consider any stress relief associated with faulting and so this model is applicable only to prefaulting deformation. Third, we assume that all formations deform conformably and that the deformation can be modelled as resulting from an upthrusting diapir which maintains its shape throughout.

Construction of such a model is facilitated by considering a salt structure whose upper surface can be described mathematically by the following parametric form:

$$Z_{salt} = h(t) + R_o\{1 + 2\cos(\theta/2) - \sec(\theta/2)\} \; , \tag{22}$$

$$r_{salt} = 2R_o \sin(\theta/2) \qquad (23)$$

Here θ is a parameter which lies in the range $0<\theta<\pi$; $\theta = 0$ represents the apex of the dome while $\theta \rightarrow \pi$ represents, asymptotically, the flanks of the salt dome near the base of the structure, at which point the radius of the dome approaches a value of $2R_o$. The parameter h(t) describes the location of a reference point within the salt relative to some datum, such as basement, which is uninfluenced by salt flow, sedimentation, or structural development related to salt flow. The speed of uplift of the salt column is given by $w = dh(t)/dt$ so that we treat the salt column as an invariant form rising upwards at speed w(t).

At the sedimentation surface, which is taken to be at a height H(t) above the reference datum, we suppose that sedimentary material is being deposited at a rate R(t) which, in the interests of simplicity, we take to be independent of lateral position. In the absence of diapiric effects or other tectonic influences the deposited material should maintain its structure unaltered upon burial which, in this instance, represents a layercake geometry. We seek the modifications to this geometry produced by the uprising salt diapir.

If we require that the bulk density of rock is not affected by the emplacement of diapirs, then the deformation of the formations overlying an uprising diapir can equally well be treated as the deformation of material which is moving past a stationary diapir in a flow which can be described by a stream function. The convenience of the salt dome shape described by Equations (22) and (23) is that it represents a surface on which the stream function for flow of a perfect fluid is constant. Letting the elastic sedimentary formations flow around the salt according to a perfect fluid the stream function then provides that no change in volume of the sedimentary material takes places while significant deformation can yet occur. The vertical velocity, $u_z(r,z)$, and the radial velocity, $u_r(r,z)$, of the sediments are then described in terms of a stream function $\psi(r,z)$ where,

$$u_z(r,z) = -\frac{1}{r}\frac{\partial\psi}{\partial r} , \qquad (24)$$

$$u_r(r,z) = \frac{1}{r}\frac{\partial\psi}{\partial z} . \qquad (25)$$

With the salt dome shape described by equations (22) and (23), an analytical expression can be derived easily for the stream function of material flowing past the diapir. Following Lamb (1879) we obtain the following expression for the stream function:

$$\psi(r,z) = R_o^2\, w(z-h-R_o)\, [(z-h-R_o)^2 + r^2]^{-\frac{1}{2}} . \qquad (26)$$

This expression is valid only for those regions outside the salt body. Writing out the explicit dependence of u_z and u_r on vertical and lateral position, we obtain the following expressions:

$$u_z(r,z) = \frac{R_o^2\, w(z-h-R_o)}{[(z-h-R_o)^2 + r^2]^{3/2}} , \qquad (27)$$

$$u_r(r,z) = \frac{R_o^2\, rw}{[(z-h-R_o)^2 + r^2]^{3/2}} . \qquad (28)$$

In Equations (27) and (28) we see that u_r is non-negative everywhere, representing a radial outward component of motion of material being diverted by the uprising salt dome. We also note that at large radial distances from the salt dome ($r \to \infty$) and at large vertical distances above the salt ($z \to \infty$), both the radial and vertical components of velocity approach zero, indicating that deformation is limited to the vicinity of the salt dome.

To determine the deformation of formations close to the salt dome, we map the path followed by an elementary volume of rock through the period in which deformation is occurring. First, we consider a bedding plane located sufficiently far above the salt mass that it is, initially, uninfluenced by the motion of the salt. Taking this bedding plane as our reference, the salt dome is seen to be rising upwards at a speed w(t) while the bedding plane is being buried below the sediment surface at a rate of R(t). However, taking the salt diapir as our reference the bedding plane is seen to be moving downwards towards the salt dome at an initial speed of w(t) while the sediment surface is approaching at a speed of (w(t) - R(t)). The downward moving material is diverted by the salt diapir, causing deformation of the bedding surfaces, which were initially horizontal. The location and shape of this reference horizon may be determined at any later time by tracing out the paths followed by elements of rock which characterize the horizon. This can easily be achieved numerically since, if we know the location of the element of rock at time t, we can determine both the radial and the vertical components of its velocity, relative to basement, through application of equations (27) and (28). This then allows us to determine the location of the element of rock at time $t+\Delta t$ and so, by repeating this process, we can map out the path followed by the element of rock. The results of such a calculation are illustrated in Figure 2, where we show the paths followed by elements of rock during emplacement of a diapiric mass. In this example the rate of sedimentation is taken to be 50 m/Myr. (164 ft./Myr.), the rate of emplacement of the diapir is 100 m/Myr. (328 ft./Myr.), while the asymptotic radius of the salt at the base of the dome is 1500 m (4920 ft.), i.e. $2R_0$. In this model, we trace out the paths followed by elements of rock which were initially positioned at radial intervals of 1000 ft. when the bedding plane with which they are associated was far above

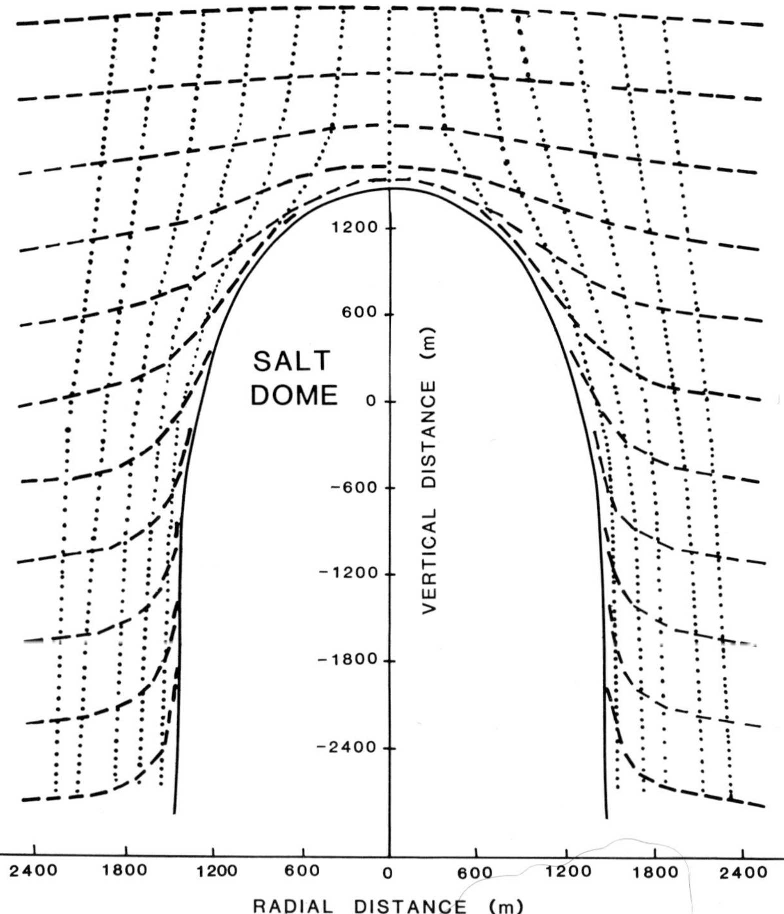

Figure 2. Deformation of formations, excluding faulting, induced by emplacement of a salt diapir. Dotted curves indicate displacement in the (r,z) plane of formation elements while dashed curves indicate bedding surfaces.

the top of the salt. Figure 2 illustrates the manner in which the individual elements of rock are diverted by the rising salt diapir. By superimposing isotime curves, we observe the progressive deformation of the sedimentary strata. Inspection of this structure indicates that strata overlying the dome are thinned, due solely to the postdepositional deformation induced by the rising salt dome. Several authors have observed thinning of formations overlying diapirs (see, for example, Jenyon, 1987) but this is usually attributed to the impact of diapirism on sedimentation patterns. This model suggests that postdepositional deformation may also contribute to the thinning and draping of formations overlying the diapir. The present model also predicts that the strata on the flanks of the salt are draped over the uprising dome. The angle of drape predicted by this model becomes quite large close to the flanks of the dome ($>60^\circ$). Because of this severe angle of drape, reflection seismology may be expected to encounter difficulties in defining the salt/formation interface on the flanks of a salt dome.

To analyse the strain induced by emplacement of the salt, let us consider cells of sedimentary formation in Figure 2 where the boundaries of a cell are defined by adjacent flow paths and by adjacent isotime curves. By comparing the shapes of cells between two flow paths we can trace the progressive deformation of the cell as a function of time. This deformation is complex; in the supradomal region it involves extension in the radial direction, compression in the vertical direction, in addition to shearing deformation in the (r,z) plane and axial extension, which is not shown on this Figure. On the flanks of the salt structure the strain predicted by this model becomes progressively greater. However at this point the induced stresses may have exceeded the strength properties of the sedimentary formations and a complete model for deformation along the flanks of a dome should include the impact of faulting.

C. Faulting of Formations Overlying a Salt Dome

In the previous section, we have seen how the progressive deformation of the formations overlying a salt diapir may be modelled. This model proceeds by computing the displacement of a representative set of points which characterize the bedding planes as these bedding planes are deformed. By comparing the displacement of adjacent points on a given bedding surface and by comparing adjacent bedding surfaces we can map the development of strain in the formations induced by the rising salt dome, through application of equations (1-4).

Having determined the components of the strain tensor, referenced to a cylindrical polar coordinate system, the components of the stress tensor can be calculated in this same coordinate system through application of the stress/strain relationships, as in equations (6-9). The principal stress components can be determined using equations (10) and (11), and the Coulomb criterion, equation (18), can be tested to determine whether stress conditions have developed which would induce shear failure in the formation. Analysis of the orientation of the principal stress axes also provides information concerning the planes along which shear failure occurs.

To evaluate this procedure numerically, we must specify a number of rock properties which characterize the geologiic column overlying the salt on a gross scale. These parameters include the elastic properties, strength of the formations against shear failure, and the coefficient of internal friction of the formations. These parameters are expected to depend on a number of independent variables, notably lithology, depth of burial, age, porosity, degree of consolidation, cementation, possible overpressuring, etc. For a rigorous model of a given salt dome, these parameters should be specified as accurately as possible.

In the present study our interest lies in gaining insight into the physical processes which underlie the deformation and faulting of formations, and so we attempt to characterize the physical properties of the geologic column in a more general way.

The values of the Lame constants, (λ) and (μ) which are effective under isothermal conditions are related to the seismic compressional wave and shear wave velocities, V_p, and V_s, as follows:

$$V_p = \left(\frac{\lambda + 2\mu}{\rho} \right)^{1/2} , \tag{29}$$

$$V_s = \left(\frac{\mu}{\rho} \right)^{1/2} , \tag{30}$$

where ρ represents the formation density.
The ratio of the compressional velocity to the shear wave velocity can be expressed as follows:

$$(V_p/V_s)^2 = \frac{\lambda + 2\mu}{\mu} \tag{31}$$

Several authors have reported measurements of the V_p/V_s ratio for a variety of sedimentary formations. Ultrasonic laboratory measurements based on core data indicate that the V_p/V_s ratio for consolidated sandstones is in the range of 1.6 to 1.8 (Castagna et al., 1985; Domenico, 1984; Tatham, 1982) and in the range 1.8 to 2.0 for carbonates (Rafaevich et al., 1984; Wilkins et al., 1984). Accordingly, in numerical evaluation of the present model we have assumed that the two Lame constants λ and μ are equal in magnitude, which corresponds to a V_p/V_s ratio of 1.73 and a value of Poisson ration of .25.

If we can use seismic compressional and shear wave velocities to estimate the Lame constants, then the gross trend of increasing seismic velocity with increasing depth of burial, associated primarily with increasing compaction, implies a corresponding trend in the dependence of the Lame constants on depth of burial. This gross trend is modelled through the following analytic formulation:

$$\lambda(z) = \lambda_o + (\lambda_{max} - \lambda_o) \tanh(z/L) \tag{32}$$

$$\mu(z) = \mu_o + (\mu_{max} - \mu_o) \tanh(z/L) \tag{33}$$

In this formulation, λ_o, λ_{max}, μ_o and μ_{max} represent the values of λ and μ at the sediment surface and for fully consolidated rocks respectively. The mathematical function tanh(z/L) then assures a smooth transition from the near surface values of the rock properties to those of the fully consolidated formations at depth, where the depth at which these fully consolidated values are attained can be specified through choice of the scale parameter L. In the examples presented below we assume values of λ_o and μ_o of 2.2 x 10^9 N/m^2 (3.2 x 10^5 psi), values of λ_{max} and μ_{max} of 3.1 x 10^{10} N/m^2 (4.5 x 10^6 psi) and a depth scale parameter L of 4000 m (13120 ft.). Under adiabatic conditions, these elasticity parameters correspond to a seismic compressional wave velocity of 1850 m/sec. (6070 ft./sec.) and a density of 2.0 g/cm^3 at the sediment surface, a velocity of 6100 m/sec. (20000 ft./sec.) and a density of 2.5 g/cm^3 for fully consolidated formations, and a value of Poisson ratio of 0.25 which is independent of depth.

In modelling the shear strength of sedimentary formations, we note that uniaxial and triaxial laboratory measurements yield a broad range of estimates of the uniaxial compressive strength of sedimentary rocks; for example, Jaeger and Cook (1976) quote a range of values from 1.4 x 10^7 N/m^2 (2000 psi) to 3.5 x 10^8 N/m^2 (50,000 psi), independent of the depth of burial. We shall return to the problem of specifying the strength parameter later.

According to Jaeger and Cook (1976), estimates of the coefficient of internal friction ν range in value between 0.51 (sandstone) and 0.75 (marble), with the majority of values clustered around 0.65. As the orientation of the plane along which shear failure occurs, relative to the maximum principal compressive stress, is determined by ν (see equation (17)), the effective value of this parameter may be determined from an analysis of the fault orientation. In this paper we assume a value of the coefficient of internal friction of 0.65, which corresponds to orientation of the faulting planes at an angle of approximately 30^0 to the maximum compressive stress axis.

D. Numerical Evaluation

In this section we present some numerical examples of the initiation of faulting in the formations overlying a salt dome. In the initial models we assume that the asymptotic radius of the salt dome, denoted by $2R_0$, has a value of 300 m (985 ft.) and that the top of the salt dome is at a subsurface depth of 4000 m (13120 ft.) when the model formation is deposited. We also assume that the rate of emplacement of the salt is twice the rate of deposition at the surface, in accordance with the approximate rule of thumb given by O'Brien and Lerche (1987) for salt emplacement driven by gravitational instability.

In Figure 3 we illustrate the progressive deformation of a formation induced by the emplacement of a salt mass at depth. This formation is modelled as being deposited as a planar surface which, as it is being buried, is concurrently being deformed in response to the motion of the salt dome. This deformation is accompanied by an increase in the stresss field such that, for the present scenario of a deeply buried salt dome and our present choice of rock parameters, the Mohr criterion, equation (18), is satisfied along the axis of the dome when the formation is buried at a depth of approximately 500 m (1640 ft.). An additional 1000 m (330 ft.) of salt has been emplaced since the model formation was deposited, resulting in the deformation of this bedding plane into a domal structure having vertical closure of approximately 4 m. (13 ft.). We note also that the radial extent of the deformation at the level of the model layer is much greater than the lateral extent of salt dome, this deformation being of significant size out to a distance which is an order of magnitude greater than the radius of the deeply buried salt dome.

In Figure 4 we illustrate the development of strain in the model formation as the salt is emplaced at depth. The vertical strain, which is compressive in nature, is seen to be the largest component of strain. Thus, according to the present model, the

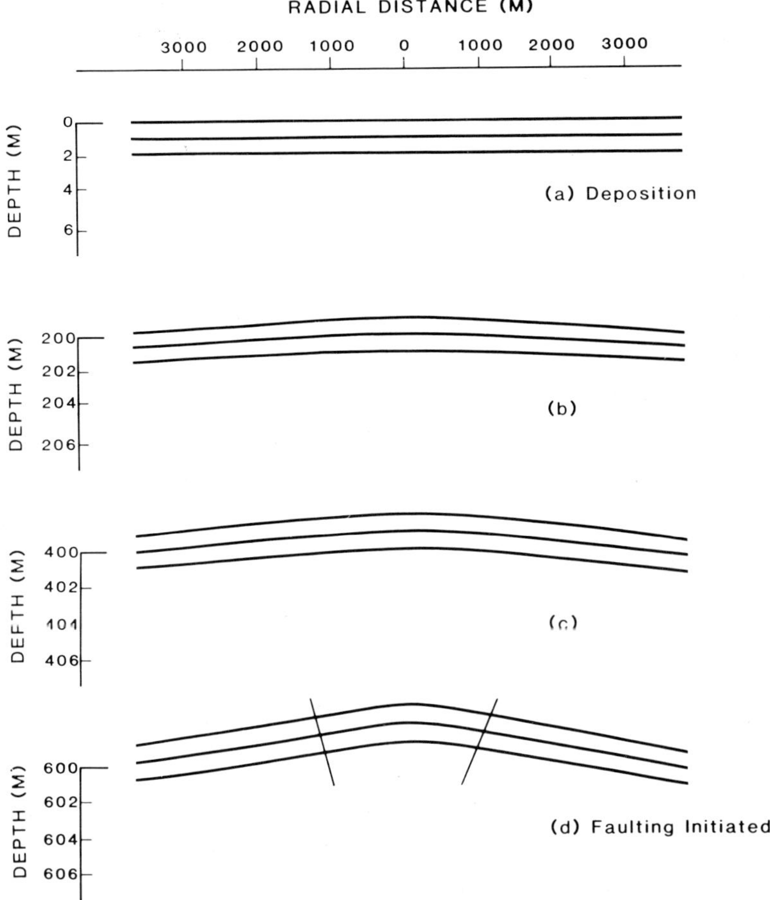

Figure 3. Progressive deformation of a formation induced by emplacement of a salt dome, assuming that the top of the diapir is at a depth of 4000 m (13120 ft.) when the formation is deposited, has an asymptotic radius of 300 m (985 ft.), and that the rate of emplacement of the diapir is twice the rate of burial of the formation. (a) Formation is deposited. (b) Formation is buried at a depth of 200 m (656 ft.). (c) Formation is buried at a depth of 400 m (1312 ft.). (d) Formation is buried at a depth of 600 m (1968 ft.), and salt has moved upwards a distance of 1200 m (3936 ft.) since the formation was deposited. Mohr's criterion is satisfied for azimuthal faulting when the model formation is buried at a depth of 500 m (1640 ft.). Indication of faulting is schematic.

deformation in the supradomal region consists primarily of vertical compression, induced by the rising salt body, accompanied by an outward motion of material as required by mass

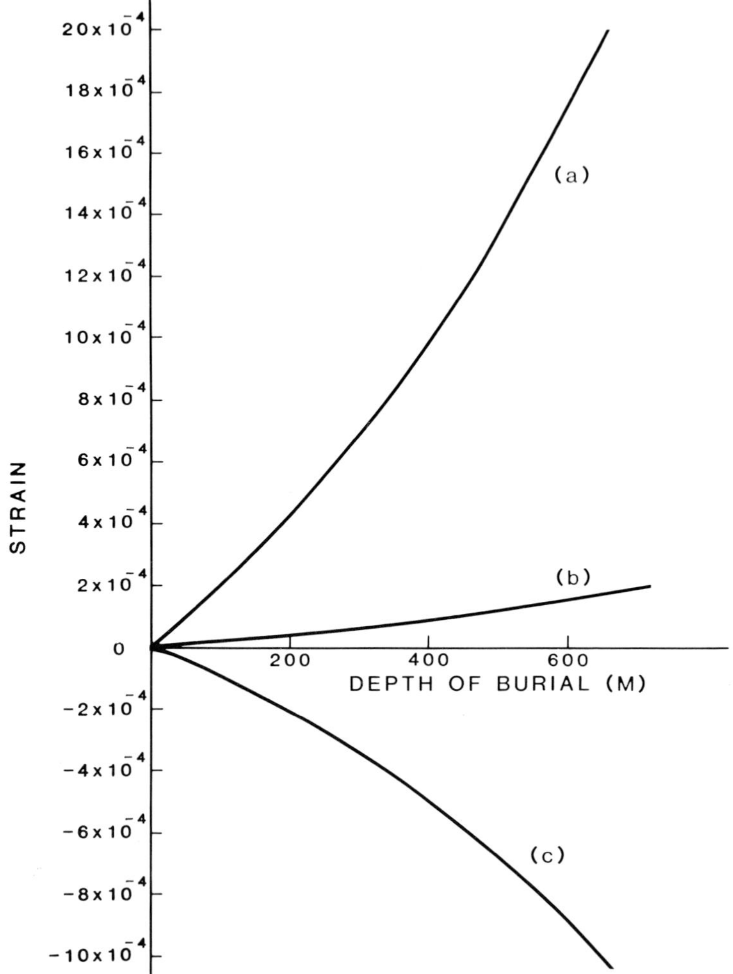

Figure 4. The development of strain during the burial and concurrent deformation of a lithologic unit induced by an uprising diapir for the model illustrated in Figure 3. This plot refers to strains along the axis of the salt dome. (a) The compressive vertical strain. (b) The shear strain in the (r,z) plane. (c) The extensive radial strain and the azimuthal strain which are essentially equal along the axis of the structure.

balance considerations. This implies that the strain in both the radial and azimuthal directions is extensive in nature. As Figure 4 shows, these strains are approximately equal in the supradomal region and are approximately half the magnitude of the vertical compressive strain. The shearing strain in the (r,z) plane is the smallest component of strain in the supradomal region, although it can be of greater significance along the flanks of the salt structure. From Figure 4 we further note that at a depth of 500 m, the depth at which the present model predicts the initiation of shear failure, the value of the vertical strain component is approximately 14×10^{-4} while the radial and azimuthal strain components are approximately 7×10^{-4}.

Figure 5 shows the development of the principal stress components in the shallow subsurface due to the emplacement of salt at depth. This shows that the greatest principal stress component is the compressive stress in the (r,z) plane. The second principal stress in the (r,z) plane and the principal stress component which lies in the azimuthal direction are both extensional and approximately equal in magnitude. Thus we may expect to see two groups of faults: a first group with fault planes oriented in the azimuthal direction and which are controlled by the values of the two principal stresses which lie in the (r,z) plane, and a second group with fault planes oriented radially and which are controlled by the values of the maximum principal stress, which lies in the (r,z) plane, and the principal stress component which lies in the azimuthal direction. This Figure indicates our predictions for the location of shearing failure will be very similar for both azimuthal and radial faulting.

Figure 5 shows the critical value of the maximum principal stress, given by equation (18), at which shear failure is

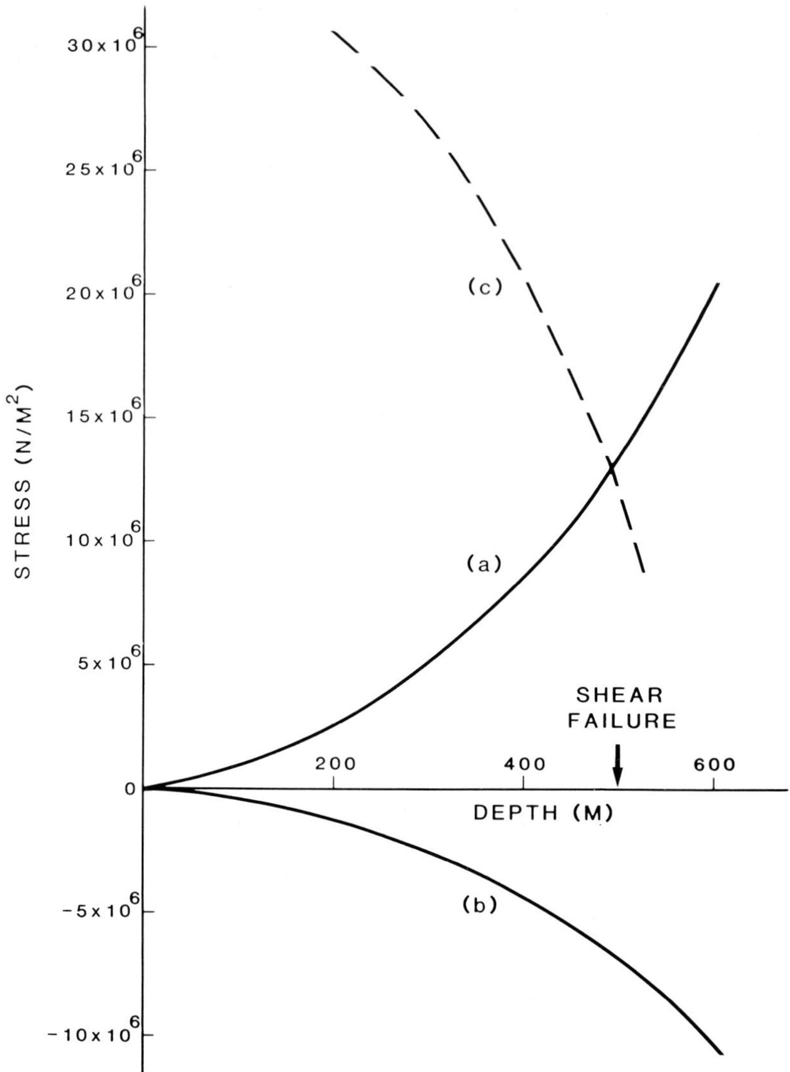

Figure 5. Development of the principal stresses along axis in the shallow subsurface, induced by an uprising diapir for the model illustrated in Figure 3. (a) The maximum principal stress component, which is a compressive stress and which lies in the (r,z) plane. (b) The second principal stress component in the (r,z) plane and the azimuthal principal stress component. Both are extensive and essentially equal in value. (c) The critical value of the maximum principal stress at which the Mohr criterion predicts shearing failure in the (r,z) plane.

predicted to be initiated in the (r,z) plane. In the present model, initiation of shear failure is predicted to occur at a depth of burial of approximately 500 m (1640 ft.).

As noted previously, the shallow structure resulting from emplacement of a salt body at depth is not confined laterally to within the radius of the salt dome itself. This is illustrated in Figure 6, which shows the development of the maximum principal stress at a number of radial locations. Principal stress development is essentially identical at all locations out to a radial distance of 1000 m (3280 ft.) from the axis of a deeply buried salt having an asymptotic radius of 300 m (985 ft.).

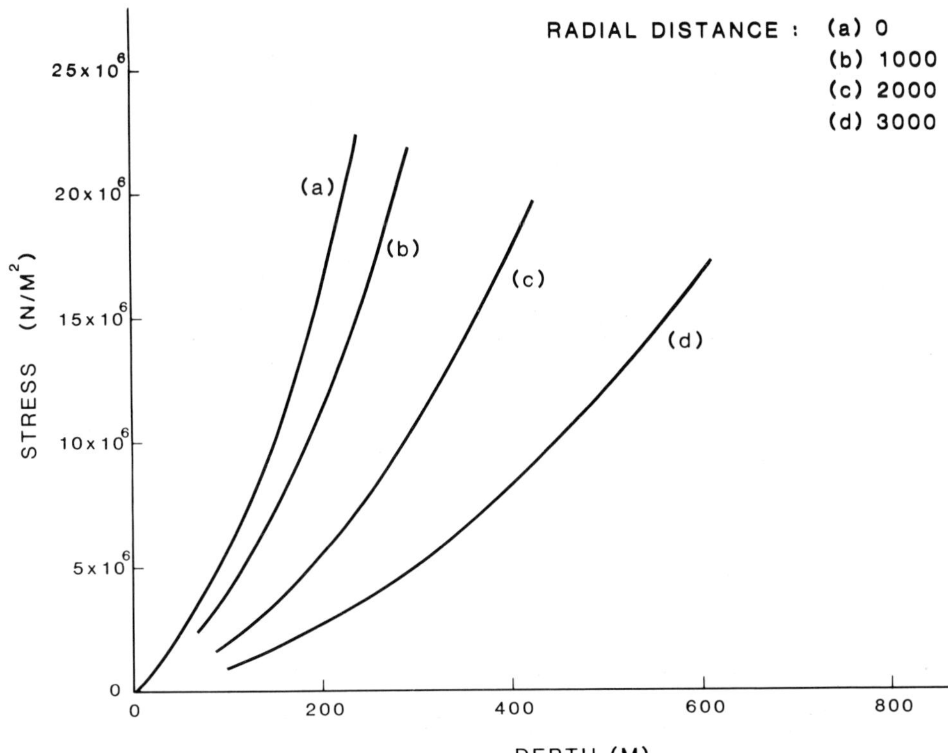

Figure 6. Development of the maximum principal stress at shallow subsurface depths induced by an uprising salt dome, at various radial distances from the axis of the salt structure. This stress is calculated using the model illustrated in Figure 3.

Beyond approximately 1000 m (3280 ft.) stress development is somewhat retarded; the stress field may develop to the point where shearing failure is initiated, but the model layer must be buried to a greater depth before the Mohr criterion is satisfied.

III. ANGLE OF FAULT PLANE

In Figure 7 we illustrate the angle of the two fault planes with normals in the (r,z) plane, along which shear failure is predicted to occur. We recall that the Mohr theory predicts that shear failure can occur along either of two planes, the axis of the greatest principal stress bisecting the normals to these two planes. In the present deformation model, the axis of the largest principal stress is vertical along the axis of the salt, which implies that the two possible azimuthal fault planes are oriented at approximately $+30^\circ$ and -30° to the vertical. The actual angle of the fault planes depends on the coefficient of internal friction, which we have assumed to be 0.65. In this analysis, positive fault planes angles imply normal faulting which is downthrown away from the salt. Thus this model predicts that we can have graben structures overlying a salt dome, faulting which is downthrown towards the salt mass, or a combination of these two.

Moving radially away from the axis of the salt, the axis of the largest principal stress component becomes inclined to the vertical, and the angles of the fault planes rotate accordingly. Beyond a certain radius both solutions predict faulting which is downthrown towards the salt. Thus azimuthal faulting which is downthrown away from the salt is restricted to a region near the axis of the salt. Faulting which is downthrown towards the salt is not subject to this restriction.

We also note in Figure 7 that at large radial distances one solution for the azimuthal fault plane angles becomes quite large, approaching 90°. While this solution is valid

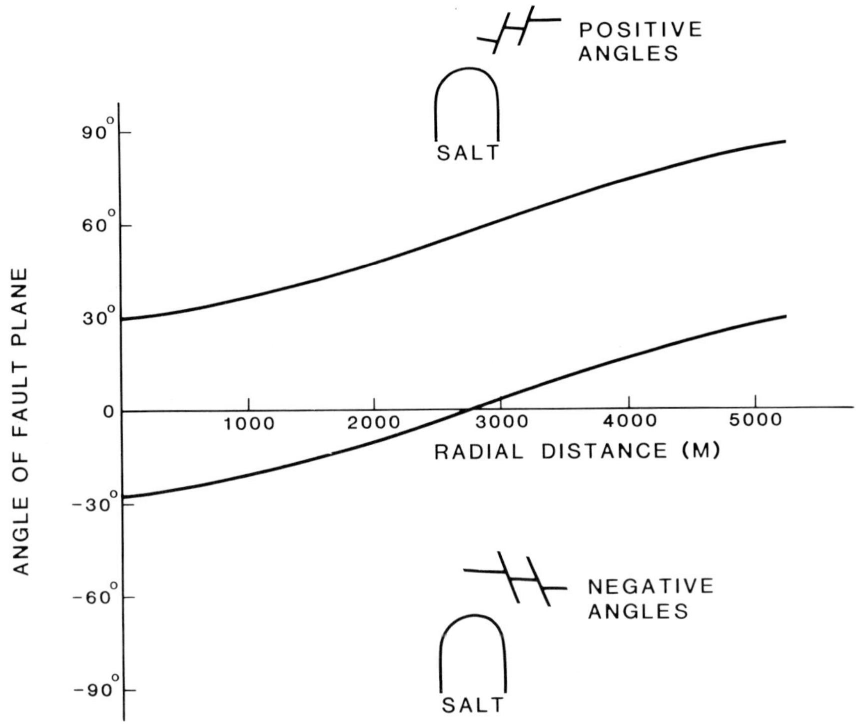

Figure 7. Angle of the two planes along which azimuthal faulting is predicted, as a function of radial distance from the axis of the salt dome, for the model illustrated in Figure 3. A positive angle implies faulting which is downthrown towards the axis of the salt, while a negative angle implies faulting which is downthrown away from the salt structure (see inserts).

mathematically, it appears unfeasible on physical grounds, and so we probably have just one set of fault planes in any radial direction in this region, representing faulting which is downthrown towards the salt.

IV. RADIAL FAULTING

While our discussion thus far has been concerned primarily with the initiation of azimuthal faults having normals in the (r,z) plane, Mohr's criterion can equally be applied to the initiation of radial faults, i.e. faults oriented such that a radial vector lies in the fault plane. When considering radial faulting the minor principal stress component, denoted by σ_B in Equation (18), should be identified with the azimuthal principal stress component. As seen in Figure 5, the present model of deformation indicates that the second principal stress component in the (r,z) plane and the azimuthal stress component are essentially equal in magnitude. Thus the predictions presented in this paper regarding the depths at which azimuthal faulting is initiated apply equally to the initiation of radial faulting; both azimuthal and radial faulting are predicted to be initiated initially at the same subsurface depths. Near the axis of the salt dome the axis of the largest principal stress component is vertical. Thus radial faulting in this region represents normal faults which are downthrown in either the positive or negative azimuthal directions. Thus two sets of radial faults are predicted in any radial direction, which, for a coefficient of internal friction of 0.65, are oriented at angles of $+30^\circ$ and -30° to the vertical. Several such sets of radial faults may be associated with a given salt dome, radiating outwards in different directions from the axis of the domal structure. At larger radial offsets from th axis of the salt, the axis of the largest principal stress becomes inclined to the vertical. The normal to the radial fault plane is no longer solely in the (θ,z) plane but also contains a component in the (r,z) plane. In this region these faults no longer represent simple normal motion with strata downthrown in either the positive or negative azimuthal directions; they also combine a radial inward component of motion toward the axis of the salt and a change in orientation of the fault planes. Thus while the nature of the fault planes is quite simple near the axis of the structure, these faults become significantly more complex off-axis.

V. INFLUENCE OF SALT DOME SIZE

In this section we consider the dependence of shear failure on the parameters which describe the salt dome. In Figure 8 we plot the development of the maximum principal stress at several radial positions for the case where the top of the salt is located at a depth of 2000 m (6560 ft.) when the model formation is deposited. Comparing this figure with Figure 6, we see an accelerated development of stress associated with the shallower salt dome. This results in the formations reaching the critical value of stress for shear failure and the initiation of faulting at shallower depths. Figure 9 illustrates this point further, showing the location at which the Mohr criterion for shear failure in the (r,z) plane is first satisfied as a function of the depth of the salt dome when the model formation was deposited. This shows that formations deposited when the top of the salt is at 4000 m (13120 ft.) will satisfy Mohr's criterion when buried, on axis, at a depth of 500 m (1640 ft.) whereas formations deposited when the top of the salt is at a depth of 2000 m (6560 ft.) are expected to fail at a depth of 170 m (560 ft.).

The radius of the salt dome is a second factor which strongly influences the locations at which shear failure can occur. This is illustrated in Figure 10, which shows the locations at which the Mohr criterion for shear failure in the (r,z) plane is first satisfied as a function of the asymptotic radius of the salt dome. For each of these cases we assume that the top of the salt dome is at a depth of 4000 m (13120 ft.) when the model formation is deposited.

Figure 10 shows that salt domes of large radii induce shear failure at much shallower depths than domes of smaller radii. As one example, we note that formations deposited over a salt dome buried at 4000 m (13120 ft.) and having an asymptotic radius of 1000 m (3280 ft.) will satisfy Mohr's criterion when its depth of

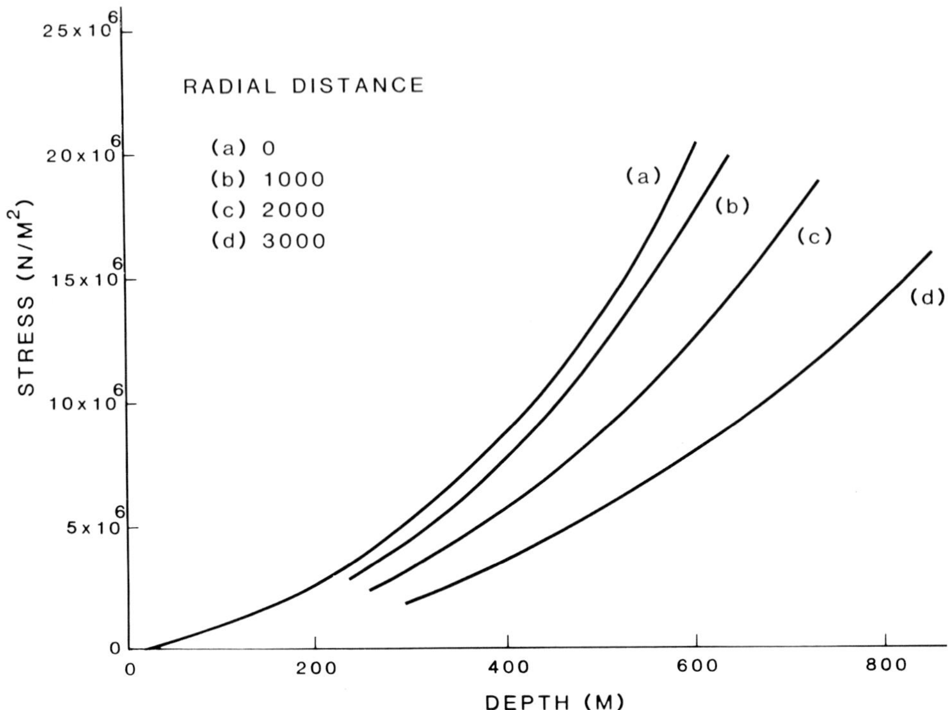

Figure 8. Development of the maximum principal stress in the near subsurface induced by emplacement of a salt dome at depth, as a function of the radial distance from the axis of the salt structure. This assumes that the top of the salt structure is at a depth of 2000 m (6560 ft.) when the model formation is deposited and that the salt dome has an asymptotic radius of 300 m (985 ft.).

burial, on axis, reaches 160 m (525 ft.), whereas formations deposited over a similar dome but which has an asymptotic radius of 200 m (656 ft.) are not expected to fail until buried at a depth of 760 m (2500 ft.). Figure 10 also shows that the stress field associated with a salt dome of large radius propagates a greater radial distance from the axis of the structure than the stess field induced by a dome of smaller radius. Thus we expect faulting over a dome having a smaller radius to be more localized to the structural axis than faulting over broader diapirs.

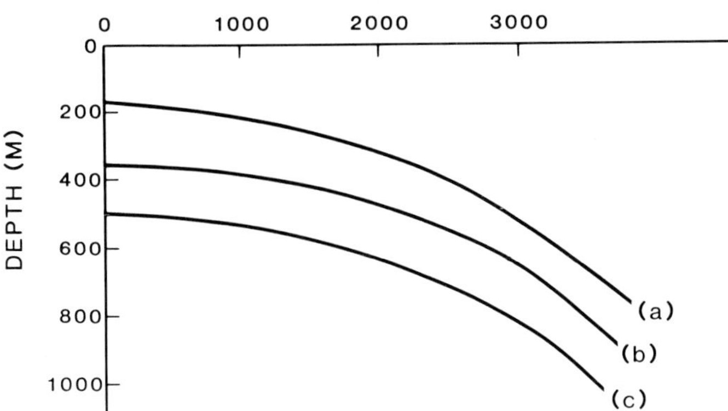

Figure 9. Location at which shearing failure in the (r,z) plane is initiated as a function of the depth of burial of the salt dome. Depth of the top of the salt dome when the model layer is deposited is (a) 4000 m (13120 ft.), (b) 3000 m (9840 ft.), and (c) 2000 m (6560 ft.). The asymptotic radius of the salt dome is 300 m (985 ft.).

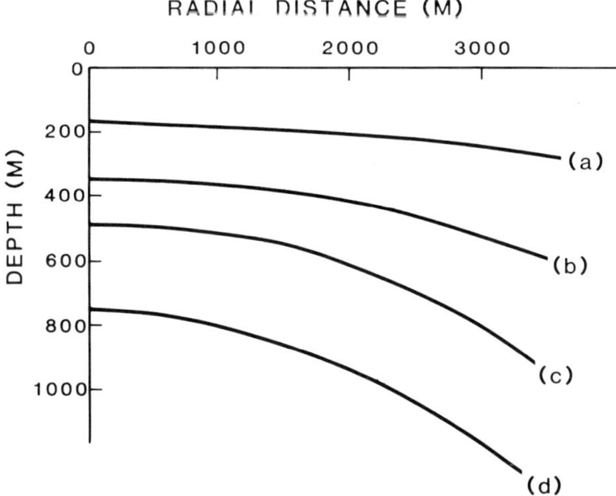

Figure 10. Location at which shearing failure in the (r,z) plane is initiated as a function of the radius of the salt dome. Salt dome asymptotic radius is (a) 1000 m (3280 ft.), (b) 500 m (1640 ft.), (c) 300 m (984 ft.), and (d) 200 m (656 ft.). Depth to the top of the salt when the model layer is deposited is 4000 m (13120 ft.) in each case.

VI. STRESS RELAXATION

A fundamental tenet of this paper thus far has been that there is a one-to-one relationship between stress and strain. We model stress as developing in lockstep with strain until the Mohr criterion is satisfied, at which time the formation fails and the stress is relieved. However it is possible that other mechanisms of stress relief may also be active. One example of an inelastic response to stress on a geologic scale is provided by the compaction of clastic formations during burial in response to the increasing overburden load.

A second inherent assumption of our model thus far is that stress is independent of the rate of geologic processes, but instead depends solely on the total strain which has developed since deposition. This assumption may not always be valid as, for example, rapid deposition of a shale facies may lead to subsequent development of an overpressured condition which may influence the rock property parameters in a time-dependent fashion.

Effects such as these can readily be incorporated into our model by introducing a stress relaxation factor. This is most easily effected by assuming that, neglecting any other geologic processes which would alter the stress tensor, stresses will tend to relax back to their equilibrium values on a timescale which can be characterized by a parameter (τ). This can be expressed mathematically as follows:

$$\sigma_i(t) = \sigma_i(t{=}0)e^{-t/\tau} \quad , \tag{34}$$

where $\sigma_i(t)$ denotes the value of any component of the stress tensor at time t. Thus stress relaxation can be incorporated into our model by permitting the stress in the model formation to decay according to equation (34) during each timestep, in conjunction with the increase in stress associated with the increase in deformation during the same time interval.

To illustrate the impact of stress relaxation in our model, we show the locations at which the Mohr criterion is first satisfied for a variety of timescale parameters (τ) in Figure 11. In these calculations we assume that the top of the salt is at a depth of 4000 m (13120 ft.) when the model formation is deposited and that the asymptotic radius of the salt dome is 300 m (985 ft.).

Inspection of Figure 11 shows that including stress relaxation effects in our model tends to retard the development of stress. Thus for a given choice of rock property parameters, Mohr's criterion is not satisfied until a greater deformation is attained, the timing of initial failure now depending on the rate of deformation, relative to the relaxation time constant (τ), in addition to the dependences already discussed. For the rock

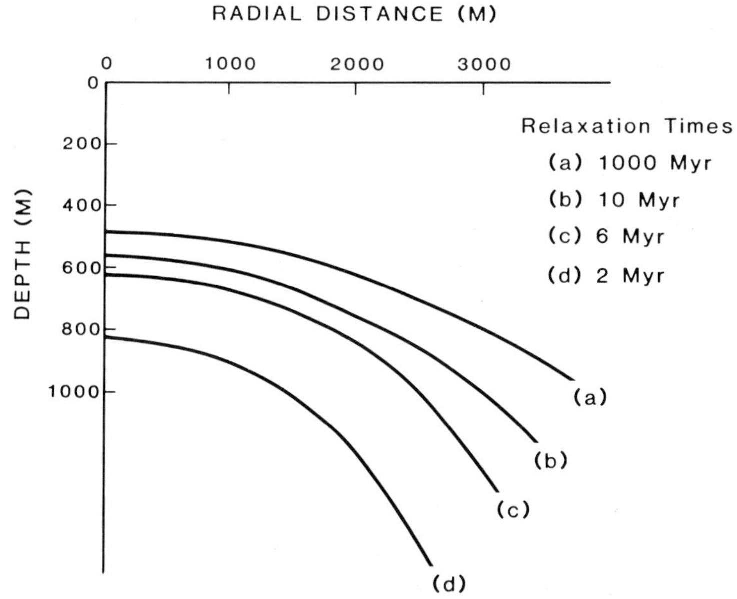

Figure 11. The locations at which the Mohr criterion is first satisfied for a range of values of the relaxation time constant (τ). The depth of the top of the salt is 4000 m (13120 ft.) when the model layer is deposited and the asymptotic radius of the diapir is 300 m (985 ft.).

property values chosen here, time constants of 5 Myr. or greater have a moderate effect on the initiation of faulting, tending to delay initiation until a somewhat greater depth of burial and degree of deformation, and to cause faulting to be more localized to the structural axis. For the same rock property values and geological process rates, time constants less than approximately 5 Myr. have a greater impact, retarding fault initiationn to a significant degree.

The extent to which stress development on a geologic scale is dependent on the rate of strain and the appropriate time constants which should be used when analyzing salt induced faulting are unclear at this time. In future case studies we hope to address these and other questions related to the values of the physical properties which control the initiation of faulting over salt diapirs.

VII. ROCK PROPERTY VALUES

In evaluating the model of fault initiation presented in this paper, it is necessary to specify a number of physical parameters which characterize the formations overlying the salt diapir in a gross sense. These parameters refer to the intrinsic strength of the formations, the elastic properties as given by the Lame constants, and the coefficient of internal friction. In estimating the numerical values of these parameters, the information which is available to us refers to the strength and elastic properties of core samples as measured in the laboratory, and estimates of the Lame constants as derived from seismic velocity data. However the relationship between these measurements and the values of these properties which control faulting is not entirely clear.

First, laboratory measurements are performed on a limited population of small samples, which may or may not adequately represent the geologic section overlying the salt. In compositing these individual measurements to estimate the properties of the larger geologic section, one must consider both the range of values and also the spatial distribution of these values. Clearly, an alignment of low strength units will have an impact on the failure characteristics of the larger body.

A second source of uncertainty lies in the fact that the measured values describe the response of the formations over a short time period; for example, seismic propagation times can be measured in milliseconds. These measurements are performed under adiabatic conditions rather than the isothermal conditions which are appropriate to geologic processes. The relationship between the rock property values which we can measure and the values which apply on a scale of hundreds of metres over geologic times is not obvious.

One approach to this problem is to infer the rock property values from information on the faulting pattern in the supradomal region. As the material strength and elasticity parameters all enter the linear Mohr criterion in a linear fashion, the characterization of initial shearing failure depends on the ratio of the unaxial compressive strength, C_0, to the Lame constants, λ and μ. If the Lame constants can be related to each other, such as through the equality relationship which we have postulated in this paper, and if a relaxation time constant can be specified, then initiation of rock shear failure can be described by just one ratio of rock property parameters, such as the ratio C_0/λ, for example. To illustrate this point, Figure 12 plots the locations at which the linear Mohr criterion is satisfied for a range of values of the ratio C_0/λ. In this Figure the depth of the top of the salt is 4000 m (13120 ft.) when the model layer is deposited, the asymptotic radius of the diapir is 300 m (985 ft.), and we have neglected relaxation effects.

From Figure 12 we see that formations having low strength, relative to the elasticity parameters, are expected to fail at shallow depths. In addition, shear failure may extend a significant radial distance from the structural axis. Stronger formations are not expected to fail until buried more deeply and, when it does occur, failure is expected to be confined more closely to the axis of the structure.

Conversely, if we know the depths and radial locations at which shear failure is initiated, we may be able to infer the ratio of the strength and elasticity parameters at the time of failure. This approach has potential in case studies where, using seismic and well log data to deduce fault history, one may be able to infer the effective ratio of compressive strength to Lame constants for the geologic section.

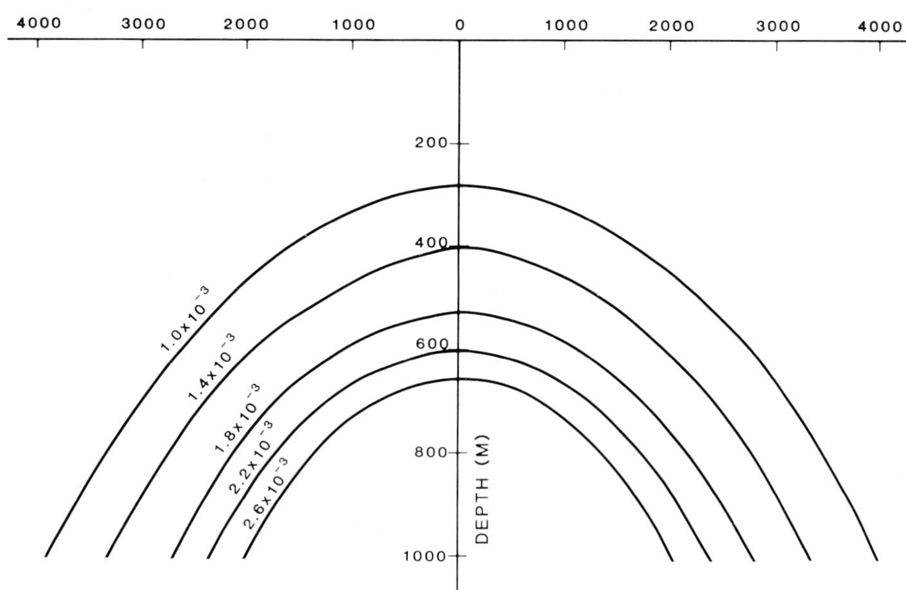

Figure 12. The locations at which the linear Mohr criterion is first satisfied, neglecting relaxation effects, for a range of values of the ratio C_o/λ. The depth of the salt is 4000 m (13120 ft.) when the model layer is deposited and the asymptotic radius of the diapir is 300 m (985 ft.).

VIII. CONCLUSIONS

In this paper we have presented a technique for analyzing the stress distribution in the formations overlying a salt dome which is being emplaced at depth. The input which is required by this technique consists of the deformation induced by the diapir. This information may be provided by observational data or, as in the present paper, through a mathematical model. The latter approach has the advantages that any uncertainties concerning the size, shape or history of the salt mass are removed and that the response of the system can be modelled for a broad range of rock property values.

Having developed a model of the stress distribution, we can then study the initiation of shear failure and subsequent fault development through application of the Mohr criterion. This model predicts the occurrence of radial faults, of azimuthal faults downthrown towards the salt, and of azimuthal faults downthrown away from the salt. It also predicts the angle of the faults and the range of locations at which faulting may be initiated. The dominant factors which influence fault initiation on a gross scale are as follows:

1. Depth of the salt dome.
2. Radius of the salt dome.
3. Ratio of the compressive strength of the formations undergoing failure to the elastic constants of the formations.
4. Stress relaxation may possibly be a factor, but the full significance of this effect is not known, as yet.

With reasonable choices of rock property parameters this model is successful in predicting the faulting patterns which develop at shallow depths above salt diapirs which are being emplaced at depth. With this formalism in hand, we are now faced with the enticing possibility of being able to analyze the deformation of formations overlying a salt diapir to infer the history of stress development leading to fault initiation, with obvious implications for the modelling of the development of hydrocarbon traps. Such modelling may aid significantly in understanding and exploring for hydrocarbon traps associated with salt domes.

REFERENCES

Castagna, J.P., Batzle, M.L., and Eastwood, R.L. (1985). Relationships between compressional-wave and shear-wave velocities in clastic silicate rocks. Geophysics 50, 571.

Domenico, S.N. (1984). Rock lithology and porosity determination from shear and compressional wave velocity. Geophysics 49, 1188.

Halbouty, M.T. (1979). Salt Domes, Gulf Region, United States and Mexico, Gulf Publishing Company, Houston.

Inderwiesen, P.L. (1983). Salt anticline - East Texas Basin, in "Seismic expressions of structural styles - a picture and work atlas", A. W. Bally (ed.), AAPG Studies in Geology Series No. 15, vol. 2.

Jaeger, J.C., and Cook, N.G.W. (1976). "Fundamentals of Rock Mechanics", Chapman and Hall, London.

Jenyon, M.K. (1984). Seismic response to collapse structures in the Southern North Sea. Mar. Petrol. Geol. 1, 27.

Jenyon, M.K. (1987). The development by salt diapirs of superficial overhang features, and effects on associated sediments, in "Dynamical Geology of Salt and Related Structures", I. Lerche and J. J. O'Brien (eds.), Academic Press.

Jenyon, M.K., Cresswell, P.M., Taylor, J.C.M. (1984). Nature of the connection between northern and southern Zechstein basins across the mid North Sea high. Mar. Petrol. Geol. 1, p. 355.

Lamb, H. (1879). Treatise on the Mathematical Theory of the Motion of Fluids, Cambridge University Press.

Larberg, G.M. Byrd (1983). Contra-regional faulting: salt withdrawal compensation, offshore Louisiana, in "Seismic expressions of structural styles - a picture and work atlas", A.W. Ballt (ed.), AAPG Studies in Geology Series No. 15, Vol. 2.

Lohmann, H.H. (1972). Salt dissolution in subsurface of British North Sea as interpreted from seosmograms, AAPG Bull. 56, 472.

Mohr, O. (1900). Welche Umstande bedingen die Elastizitatsgrenze und den Bruch eines Materials? Z. Ver. dt. Ing. 44: 1524, 1572.

O'Brien, J.J. and Lerche, I. (1987). Modelling of Buoyant Salt Diapirism , in "Dynamical Geology of Salt and Related Structures", I. Lerche and J. J. O'Brien (eds.), Academic Press.

Owen, P.F. and Taylor, N.G. (1983). A salt pillow structure in the southern North Sea, in "Seismic expressions of structural styles - a picture and work atlas", A.W. Bally (ed.), AAPG Studies in Geology Series No. 15, Vol. 2.

Rafaevich, F., Kendall, C.G.St.C., and Todd, T.P. (1984). The relationship between acoustic properties and the petrographic character of carbonate rocks. Geophysics 49, 1622.

Ramsay, J.G. (1967). "Folding and Fracturing of Rocks", McGraw-Hill, New York.

Sunwall, M.T., McQuillan, K.A., and Nick, C.J. (1983). Salt Diapir - Gulf of Mexico, in "Seismic expressions of structural styles - a picture and work atlas", A.W. Bally (ed.), AAPG Studies in Geology Series No. 15, vol. 2.

Tatham, R.H. (1982). Ratio of seismic compressional and shearwave velocities and lithology. Geophysics 47, 336.

Wanslow, J.B. (1983). Piercement salt dome - Upper Continental Slope, in "Seismic expressions of structural styles - a picture and work atlas", A.W. Bally (ed.), AAPG Studies in Geology Series No. 15, Vol. 2.

Wilkens, R., Simmons, G. and Caruso, L. (1984). The ratio V_p/V_s as a discriminant of composition for siliceous limestones. Geophysics 49, 1850.

FRACTURED CHALK OVERBURDEN OF A SALT DIAPIR, LAEGERDORF, NW GERMANY - EXPOSED EXAMPLE OF A POSSIBLE HYDROCARBON RESERVOIR

Andreas G. Koestler
Werner U. Ehrmann

GEO-RECON A.S.
Bernhard Herres vei 3,
0376 Oslo Norway

I. INTRODUCTION

Although fractured carbonate reservoirs associated with salt diapirs are important sources of hydrocarbons in NW Europe (Ekofisk, see van den Bark & Thomas, 1981; Albuskjell Field, see Watts, 1983), geological situations of this type can be studied at the surface at only a very few localities. This paper describes one of these, Laegerdorf (NW Germany, some 45 km NW of Hamburg), from the point of view of its potential reservoir characteristics. At Laegerdorf, typical Upper Cretaceous chalks have been pushed upwards through thick Tertiary and Quaternary sediments by an underlying salt ridge, and are now exposed at the surface (Fig. 1). The chalks have reacted to the superimposed stresses by brittle fracturing, with strain concentrated on distinct faults and fault zones arranged in complicated conjugate sets (Koestler & Ehrmann, in press). The chalks are industrially exploited in several large quantities, where excellent exposures allow the deformation features to be studied in three dimensions at different scales.

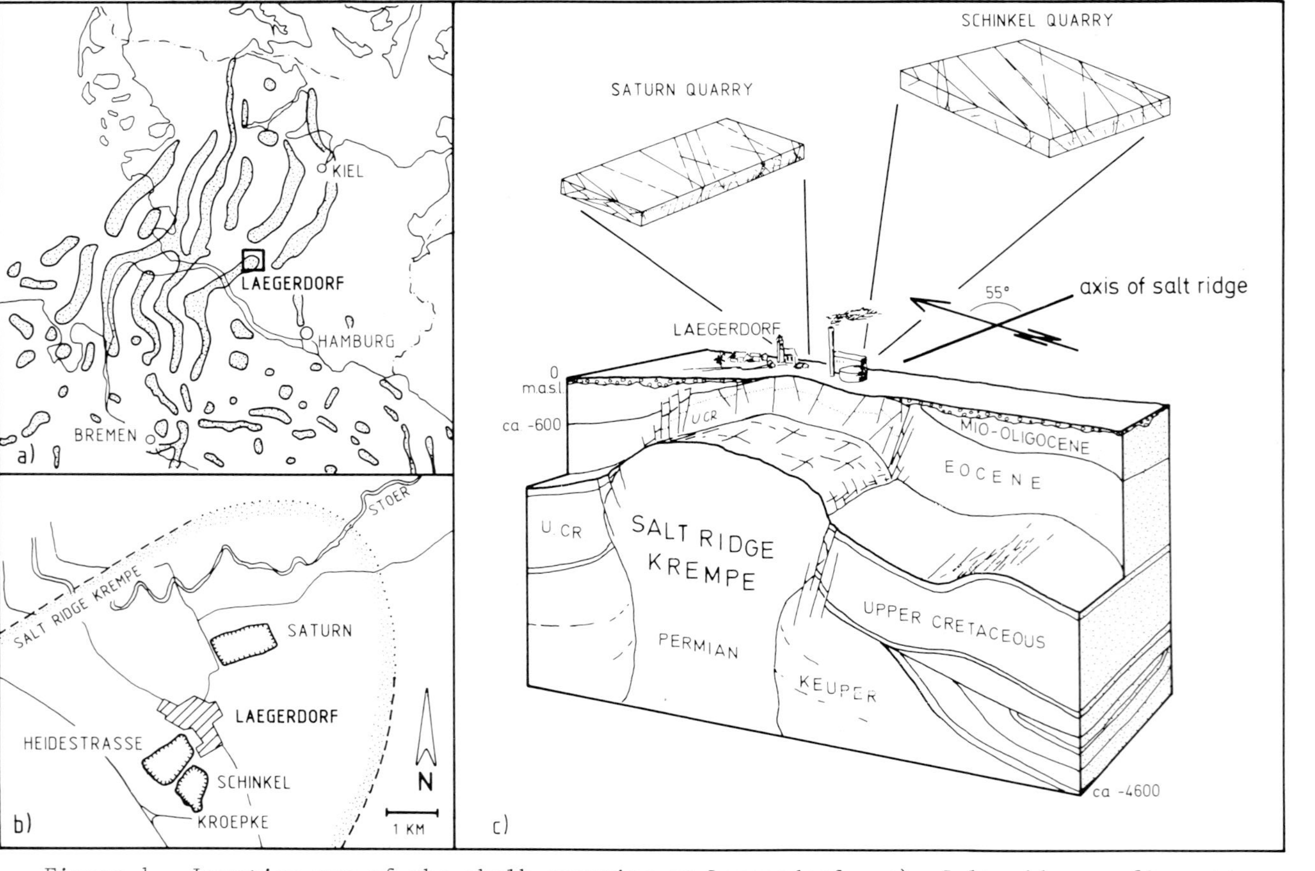

Figure 1. Location map of the chalk quarries at Laegerdorf. a) Salt ridge configuration in the subsurface of Northern Germany (From Jaritz, 1973). b) Location of quarries Saturn, Heidestrasse, Schinkel and Kroepke. c) Schematic block diagram of the tectonic situation of the quarries on salt ridge Krempe (geologic section redrawn from Grube, 1955).

II. GEOLOGIC SETTING

In the North German Lowlands, subparallel ridges of Permian salt are typical features of the diapiric tectonics in the subsurface (Jaritz, 1973). Locally, they bring geological formations close to the surface which are normally hidden below a thick sediment cover. In the area of Laegerdorf the NE-SW striking salt ridge Krempe lifts Upper Cretaceous chalks by more than 1000 m, and makes them accessible to industrial exploitation and geological investigations. The Laegerdorf quarries are situated near the NE end of the Krempe salt ridge, which may be a depression in the crest only, and close to the salt ridge axis (Fig. 1). They are up to 1000 m long, several hundred meters wide and up to 60 m deep.

The undeformed Upper Cretaceous chalk is a very homogeneous and almost monomineralic sediment (cf. Hardman, 1982). The exposed sedimentary sequence ranges stratigraphically from Middle Coniacian to Lower Maastrichtian (Schulz et al., 1984) and has a thickness of about 420 m. The carbonate content varies between 90 and 98%, and the acid insoluble residue of the chalks consists mainly of montmorillonite and quartz.

Illite, kaolinite, chlorite and clinoptilolite may occur in smaller quantities (Ehrmann, 1986). The porosity of this extremely finegrained sediment (grain size usually <5 μm) is about 40-50%, the matrix permeability about 2-10mD (cf. Koestler & Ehrmann, in press; Hardman, 1982).

The chalks of Laegerdorf are extremely bioturbated (Schulz et al., 1984) and hardly any bedding planes are visible. However, numerous bedding-parallel layers of nodular chert, some marly horizons and occasional pyrite-impregnated beds indicate a general bedding dip of about 10 degrees to the NW as a result of the doming. These layers serve as markers for displacement measurements on the individual faults.

The structural features in the chalks could be mapped on different scales and in several subareas covering the whole outcrop area. The general tectonic style on top of the salt diapir is given by the orientation of fault zones, the amount of displacements on the faults and the tilting of the fault blocks. In addition, detailed mapping revealed relationships between the structural patterns of joints and fault zones and mesoscopical features such as brecciation, slickenside striations and clay enrichment on the faults (Koestler & Ehrmann, in press). Microscopical studies have enabled a variety of textural changes due to deformation close to and on fracture surfaces to be described.

In the area of Laegerdorf, the chalk has also been deformed by ice loading during the Quaternary. However, the corresponding features are always restricted to the uppermost 10 m of the outcrops and can be clearly distinguished from the features caused by salt diapirism. Only the latter are described here.

From the point of view of potential reservoir characteristics, the deformational features in the Laegerdorf chalk will be considered under two different headings: geometrical characteristics and physico-chemical characteristics. Geometrical characteristics include fracture morphology, definitions of fracture sets and their geometrical relations, and discussion of fracture spacing and the shape of interfracture blocks. Physico-chemical characteristics encompass transformations suffered by the original chalk during and after fracturing (mechanical grinding, pressure solution, mineral growth on fractures, etc.), resulting in zones of brecciation, slickensides, stylolites, clay enrichment, neomineralization, etc.. In a qualitative way, the sealing potential of faults, the fracture permeability, and the interconnection of faults will be evaluated.

III. FRACTURING: GEOMETRICAL CHARACTERISTICS

The general fault patterns and deformation features in the Laegerdorf quarries have been described in detail elsewhere (Koestler and Ehrmann, in press); here we give a brief summary of these results. This paper will discuss the deformation features in the overburden of a salt diapir more in the light of a potential hydrocarbon reservoir in fractured chalk. The chalks are deformed by fracturing giving a range of deformation features roughly classified as joints, faults and fault zones. Figure 2 shows the inhomogeneous deformation in the chalk as exemplified by two several hundred meter long quarry walls (Saturn, S-wall and W-wall), where large fault zones, partly in conjugated sets, displace the slightly visible bedding. The area between the fault zones contains irregularly distributed small faults and joints, but the underlying pattern is an heterogeneous strain, with large volumes of minor deformation bordered by high strain zones.

A. Joints

Joints (single fractures without visible displacement) are observed in a wide range of lengths, from centimeters to several tens of meters. They occur as planar, curved and strongly irregular features. Planar fractures in the range of meters to tens of meters are the most common type of deformation feature in the chalk (Fig. 2, 3). They can show slight morphologic variations, but on the macroscale and mesoscale they intersect the quarry walls as straight traces indicating a planar shape in three dimensions. The termination of planar joints can be either a complex zone of minor planar fractures, an abrupt termination without other visible deformation features in the region of the tip, or a transition into a curved joint inducing a complex pattern of further fractures. Curved fractures are mainly found in connection with lenses filled with argillaceous material (Fig.

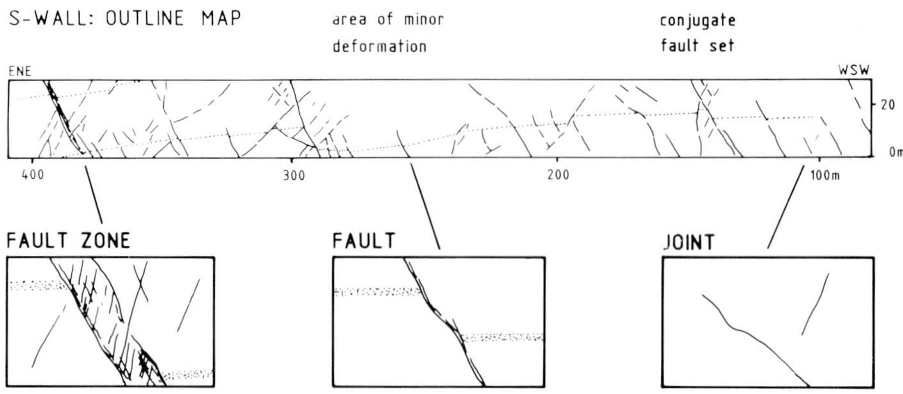

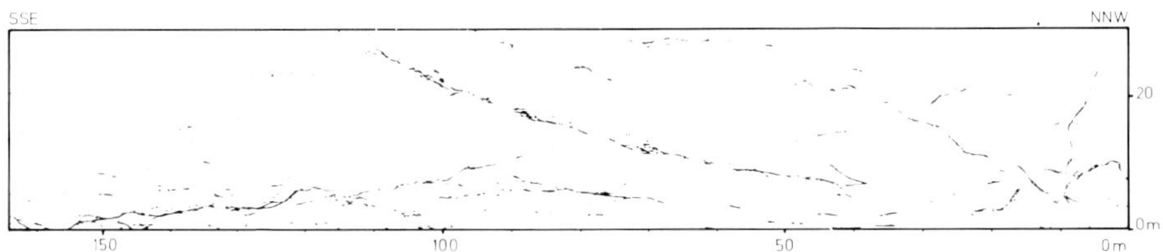

Figure 2. Deformation features of the chalk overburden shown with examples of S-wall and W-wall in Saturn quarry, Laegerdorf.

3). These lenses show the characteristics of a pull-apart opening or tension joint in a vertical orientation. The lenses are also observed on faults which converge in places into one single fracture. The curved fractures in the surrounding seem to resemble stress trajectories and are interpreted to be the result of stress concentrations during the propagation of the main fault. The enhanced stresses may also be responsible for the enrichment of argillaceous material in these lenses due to pressure solution. Irregular fracture surfaces are characterized by morphologic lows and highs on a scale of centimeters. They are mainly observed with a subhorizontal orientation and usually lack an argillaceous coating and slickenside striations. The surface shape, the lateral extension and the lack of surface features suggest that these joints are the result of late unloading, due to regression of the ice sheet in the Quaternary and to erosion of the overlying sediments.

Joints are not only discontinuities in the less deformed part of the chalk, but also elements in the structurally more complex fault zones (Fig. 3).

B. Faults and Fault Zones

The most obvious deformation features in the chalks are faults and fault zones displacing the bedding in the range of meters (maximum 12 m) and bordering the tilted fault blocks. They occur mainly in conjugate sets and are composed of a complex configuration of minor faults and joints. They usually cut the

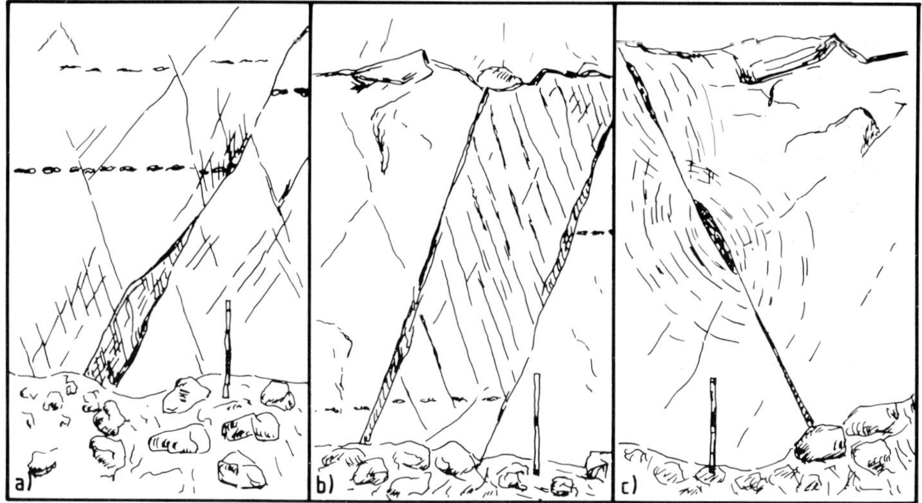

Figure 3. Fault zones in Heidestrasse quarry, Laegerdorf. a) Major fault with a complex pattern of conjugate fracture sets. Displacement is indicated by a chert layer. b) Zone bordered by two major faults with an oblique internal pattern of shear fractures. Displacement is concentrated on one bordering fault. c) Lens-shaped pockets with enrichment of argillaceous material, surrounded by curved joints. Scale bar is 1 m.

quarry wall from top to bottom, and can be identified again on the facing quarry walls, i.e. over distances of a couple of hundred meters (Fig. 4).

Fault surfaces are planar or curved and always show an argillaceous coating and slickenside striations. Striations not parallel to the dip line of the faults indicate that displacement was not always vertical, although dip-slip was the most common movement, and curved striations indicate that rotational components during deformation were locally active. The width of faults is quite variable in the range of millimeters to centimeters, and the fracture infilling is mainly argillaceous material partly containing small chalk particles and pyrite concretions. In places, faults are fine fractures without any filling and with widths of less than one millimeter (Fig. 3), but with a displacement of several meters. Many faults contain lensoid-shaped pockets filled with argillaceous material (Fig. 3). The overlap of the two bordering faults usually equals the displacement on the fault. The lenses show the same configuration as pull-apart openings, however, in a vertical orientation.

Fault zones occur with two different characteristics depending on whether the main displacement occurred on one or two major discontinuities. The internal structures show quite different patterns in these two types. The fault zones with one major plane of displacement (Fig. 3a) show a wide variation in width of fractured rock along the fault zone. In areas where the surroundings are highly fractured, the major fault and the related fractures show the constellation of conjugated sets. The orientation of the bisectrix of the acute angle in this case is usually vertical. Fault zones of this type often contain relatively narrow zones of highly brecciated rock always bordered

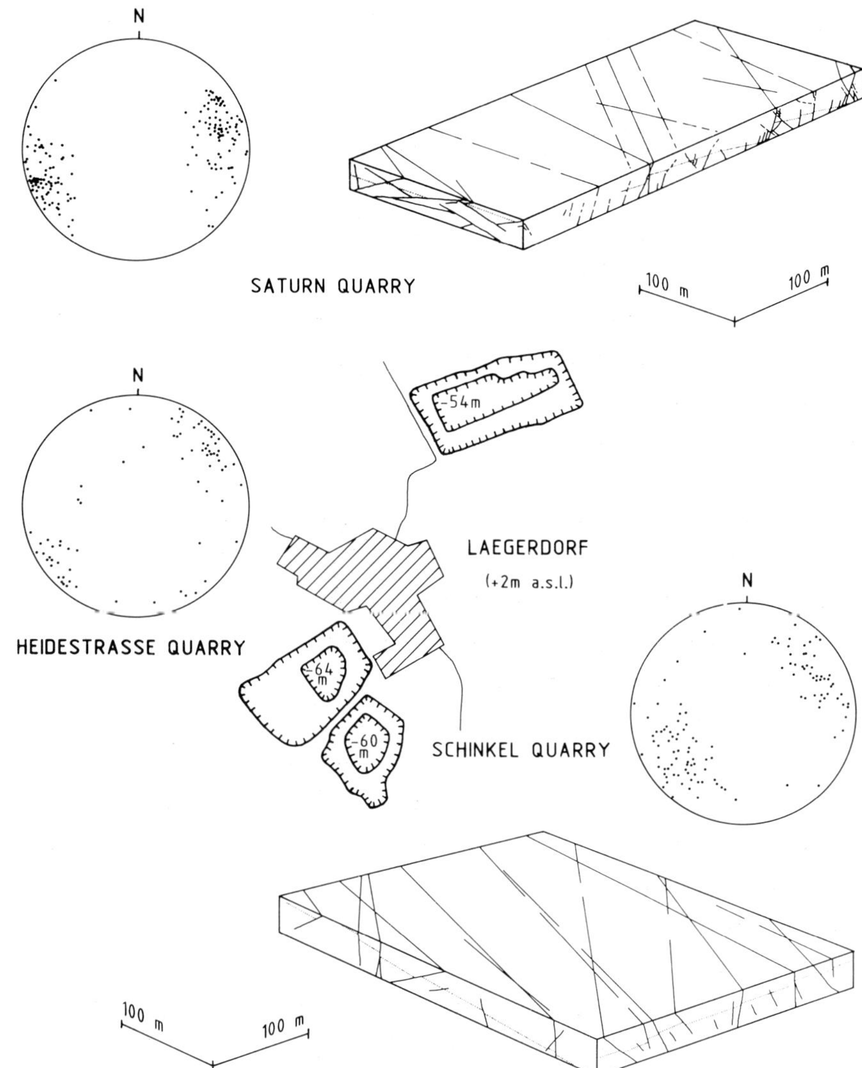

Figure 4. Pattern of faults and fault zones in the quarries Saturn, Schinkel and Heidestrasse. The block diagrams illustrate the fault sets conjugated in the horizontal and vertical plane. The stereographic projections of structural elements (poles to fracture planes) show the variation in dip and strike. (Block diagram Schnikel quarry redrawn from Grube, 1955).

by at least one distinct fault plane. In some places, the breccia zones contain an internal fault plane highly enriched in argillaceous material.

The fault zones bordered by two distinct faults show a more regular pattern of internal structures, and brecciation is seldom observed (Fig. 3b). The internal fractures in such zones constitute a parallel set of structures oblique to the bordering faults in a staggered pattern. The oblique fractures indicate the normal shear sense in the zone with the trace of a main stress axis parallel to them. The displacement observed by the use of chert layers as markers is usually inhomogeneously distributed across the zone, however, and is mostly accomodated by one or both of the bordering faults (Fig. 4c).

C. Geometrical Pattern

Some nine hundred orientations of joints and faults were measured to establish a three dimensional understanding of the fracture patterns. All the structural elements in the overburden of the salt ridge occur in conjugate sets easily seen in the more or less vertical quarry walls (e.g. Fig. 2). The reconstruction of the fault pattern over the large quarried area reveals conjugate sets in a horizontal plane as well (Fig. 4). The stereographic plots of the poles of joints and faults within a large range of extensions (sizes) indicate that the main strike orientation of the structural features is around NW-SE to NNW-SSE. The dip of joints and faults vary in the range of 45 to 85 degrees. Since all the observed faults are extensional with respect to the subhorizontal bedding, this statistical pattern indicates a main extension direction roughly parallel to the salt ridge axis, and a minimum extensional direction in the subvertical which may be compressional due to the rising salt diapir. The intermediate bulk strain axis has a NW-SE orientation in the horizontal plane and is clearly also extensional. The overall structure shows many analogies to the theoretical and experimental constellations of fractures resulting from deformation in a 3-D stress field (Fig. 5; see Reches, 1978, 1983; Reches & Dietrich, 1983).

The interpretation of the fracture pattern in the chalk overburden of Laegerdorf must take into account both local and regional tectonics. The rising diapir induced a compressional stress in a vertical direction resulting in conjugate sets of normal faults with a main extension parallel to the salt ridge axis. That the main extension is parallel to and not perpendicular to the ridge axis (as expected) suggests the influence of a regional stress field overprinting the diapirism. The reason for this stress field has to be looked for in the tectonic situation of NW-Europe during Late Cretaceous and Cenozoic times. Large strike-slip zones dominate the northern border of central Europe (e.g. Ziegler, 1982; Ziegler, 1985; Pegrum, 1984). The Tornquist Zone, extending from the central North Sea into Poland, and parallel zones, record dextral movements during Late Cretaceous and Early Tertiary (e.g. Pegrum, 1984). This dextral shear regime on NW-SE striking tectonic zones suggests a regional compressive stress regime with a NW-SE to N-S orientation. It seems that this regime could have been responsible for the fault pattern observed at Laegerdorf. The regional stress overprint over a doming ridge was studied by Withjack & Scheiner (1982) in clay experiments. They showed that a regional compressive stress field oriented perpendicular to the ridge axis induces fracture sets mainly parallel to this stress direction, partly in conjugate sets with an acute angle pointing in the direction of maximal compressive stress. In these experiments the main extension is parallel to the ridge axis, against the expected result of extension connected to a rising salt ridge, but closely comparable to the situation at Laegerdorf.

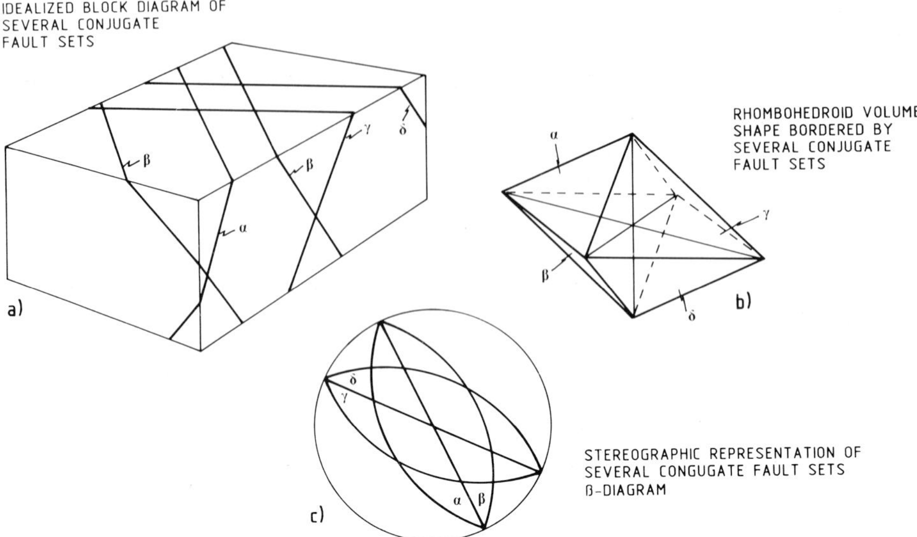

Figure 5. Several conjugate fault sets. a) Block diagram, b) rhombohedroid volume shape, c) stereographic representation. The fault sets are indicated with **α**, **β**, , **δ**, respectively.

Due to the conjugate sets of faults in the horizontal and vertical plane, the faults border volumes of only slightly deformed chalk with a special shape (Fig. 5). These rhombohedroid rock masses have dimensions in the range of tens to several hundred meters. From the point of view of the chalk as a potential hydrocarbon reservoir, these rock masses represent the reservoir volumes which are not further subdivided by faults and which can be accessed by enhanced faulting permeability only along the zone faces. Additional fracturing due to enhanced effective pressure during hydrocarbon production will enlarge the fracture permeability in these fault-bordered volumes.

IV. FRACTURING: PHYSICO-CHEMICAL CHARACTERISTICS

Changes in the Laegerdorf chalk on and around fractures and fracture zones have been studied on two different scales: in outcrop and under the microscope (Scanning Electron Microscope). Outcrop scale features include clay enrichment, stylolites, slickensides and zones of brecciation, whereas microscopic study reveals details of the mechanisms of denser physical packing, pressure solution and new mineral growth.

A. Outcrop-scale features

Clay enrichment occurs on most fracture planes and can be the effect of two different mechanisms. In profile sections, where the chalk is interbedded with marly layers (Middle Coniacian), clay smearing along minor faults is observed, even when the displacement is only a few centimeters to decimeters. However, pressure solution is probably the most important mechanism for clay enrichment. Due to stress concentration in the chalk during the development of fractures and existing fractures, the calcareous components of the sediment are dissolved and removed, so that an insoluble residue is concentrated on the fracture surfaces. Clay mineral analyses at four different fractures in quarry Saturn (unpublished data) show that the composition of the argillaceous fracture fill in Laegerdorf is equivalent to the clay composition of the surrounding chalk. Stylolites, with the typical features of interfingering cones covered with a seam of clay minerals, also indicate widespread pressure solution. They are mainly observed in the lower part of the stratigraphic column (Middle Coniacian). Usually, the stylolite surfaces are parallel to the bedding, and sometimes to fractures. In both cases the stylolitic cones show subvertical orientations. The widespread solution of the carbonates is confirmed by observations on a microscale level (see below).

The slickenside striations vary quite a lot in their intensity, from slight striations showing almost no argillaceous coating to sets of sharp grooves clearly covered by argillaceous material. Usually, the degree of clay enrichment between the striated fault planes increases with increasing intensity of striation. This observation, and the similarity of fracture surfaces and stylolite surfaces under the microscope, support the idea that slickenside striations are not only the result of mechanical movements of the adjacent fault blocks, but also of solution of the calcareous sediment components along fracture surfaces (cf. Hancock, 1985). Slickenside striations in the chalk seem to represent a special type of stylolite in places, but they are also clear indicators of relative movements of adjacent fault blocks. Fibrous growth of minerals on fracture surfaces indicating displacement direction and amount was not observed. On some fracture surfaces, superimposed slickenside striations are observed, indicating various directions of relative displacement. Some of them are curved due to morphological variations of the fracture surfaces, some of them indicate different generations. These generations of striations can be interpreted in terms of changes in local stress conditions.

Brecciation is widespread along major fault zones. The chalk is crushed to angular fragments in the size range millimeters to centimeters in a fine-grained matrix. The matrix is argillaceous material quite similar to the coating of the fractures. The surfaces of breccia fragments show slickensides and argillaceous coating as observed on fracture surfaces. Some pyrite concretions seem to be generated in the clay matrix of the breccias. There are layers of highly enriched clay material internal in the breccia zones running parallel to their borders. These clay layers seem to represent fault planes accommodating a great part of the displacement, and seem to have developed during the brecciation of the chalk. In areas where the fragment content exceeds the amount of matrix, the fragmentation follows fracture sets corresponding to the conjugated sets belonging to the fault zone outside the breccia zone. Therefore, the breccias are thought to be developed during the same deformation event as the fault zone, perhaps during a phase of enhanced strain rate.

B. Microscopic-submicroscopic Features

Scanning electron microscopy reveals a number of deformation features which can be correlated with the mesoscopic observations. The special microtextural configuration of the original chalk enables it to react to the deformation by different mechanisms. The undeformed chalk is a highly porous medium composed of minute particles (usually less than 5 μm in diameter) which are mainly fragments of calcareous nannofossils (Fig. 6a). The high porosity (40-50%, cf. Hardman, 1982) makes it possible for the chalk to react physically under stress by a denser packing of the microcomponents (Fig. 6b). This process, with its concomitant volume loss, is most effective close to fracture surfaces. Due to the small particle size and the high porosity, chemical processes are unusually active along fractures and both solution and redeposition of carbonate can be observed. Along fractures, the coccoliths show dissolved edges and are reduced in size. Close to fracture surfaces an increased neomineralization of calcite as fibrous and blocky crystals can be observed (Fig. 6a). Clay enrichment is observed on the fracture surfaces in the form of irregular seams composed of montmorillonite, illite and kaolinite. These seams cover the surfaces with variable thicknesses usually forming a continues film (Fig. 6c). Therefore, the fracture surfaces themselves tend to act as impervious screens. On a few fractures, new growth of barytes has been observed.

The surfaces of breccia fragments and of fractures coated by argillaceous material are usually cut by micro-cracks, with a

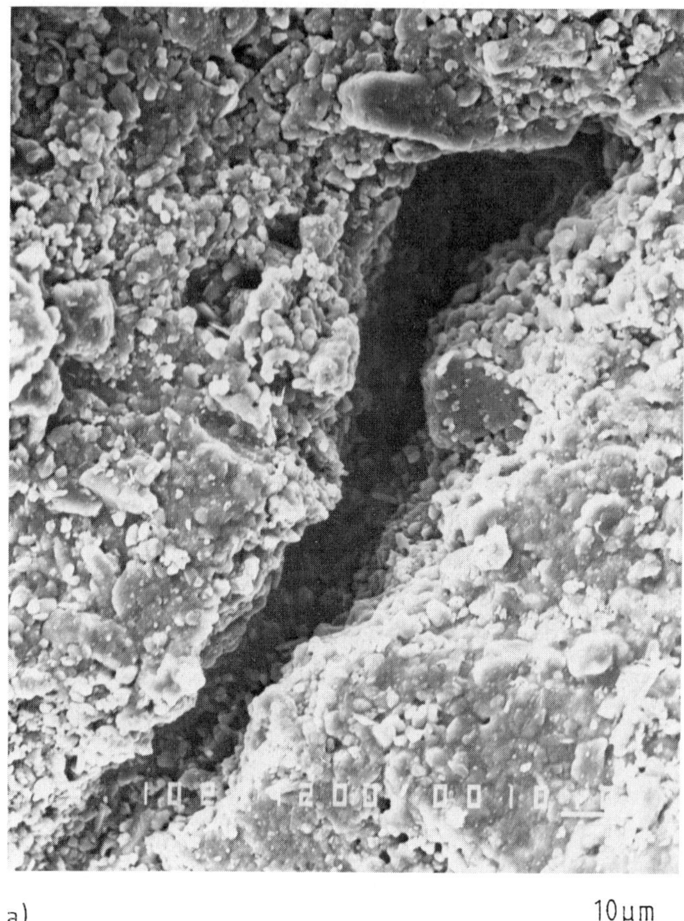

Figure 6. SEM-pictures from deformed chalk. a) Microfissure cutting the coated fracture surface. Denser packing can be observed just beneath the fracture surface. b) Blocky calcite growth close to a fracture surface. Note the texture of undeformed chalk in the right upper corner. c) Fracture surface sealed with clay minerals to give an impermeable screen.

width of less than 10 μm, and which show a slightly preferred orientation. They probably have some importance for the permeability of the rock masses by maintaining communication between fractured and undeformed chalk. Microscopically, the surfaces of breccia fragments show the same features as the surfaces of single fractures. Mostly, the surface geometry is

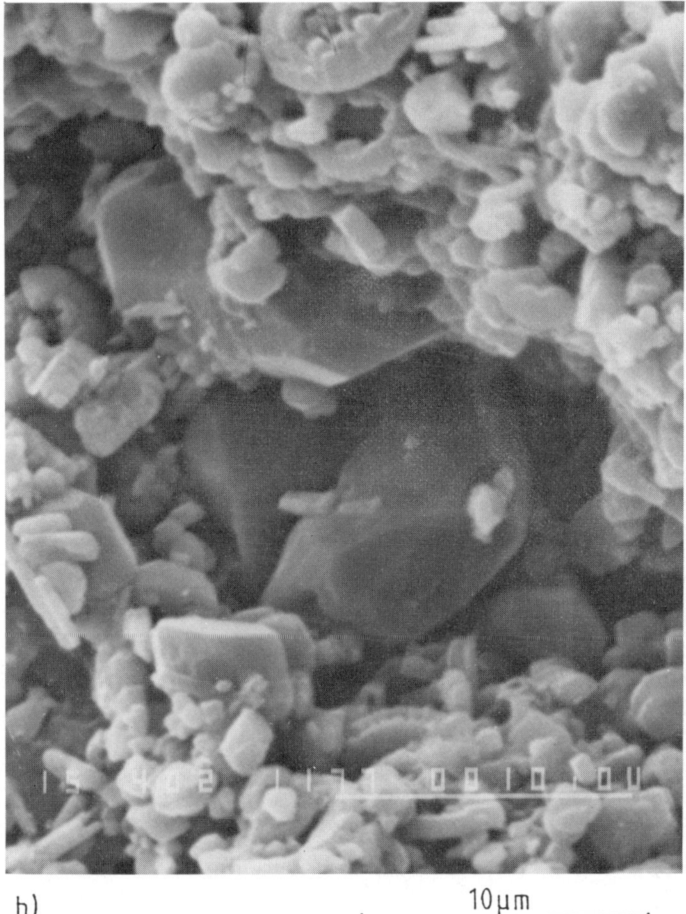

Figure 6B.

highly irregular and the fragments are floating in an argillaceous matrix. This matrix is composed of the same clay minerals as the seams on single fractures and as the argillaceous components of the undeformed chalk.

The observed mesoscopic and microscopic effects due to deformation of the chalk are summarized in Table 1.

Figure 6C.

IV. CONCLUSIONS

The chalk of Laegerdorf is heterogeneously deformed due to the diapirism of the underlying salt ridge. A main extension direction parallel to the ridge axis is shown by the fault orientation with mainly NW-SE trends. The pattern of conjugated fault sets both in the vertical and in the horizontal plane is interpreted in terms of deformation in a three-dimensional strain

Table 1:

PROPERTIES OF UNDEFORMED CHALK	DEFORMATION PROCESSES	EFFECTS
Porosity: 40–50% Permeability: 2–10 mD Size of particles:<5 μm Type of authigenic mineralization: calcite, plagioclase, zeolites	Pressure Solution	Stylolites Concentration of clay minerals Sealing
Clay minerals: montmorillonite, illite, kaolinite/chlorite	Compaction	Change in microtexture Volume decrease Loss of porosity
	Neomineralization (fibrous calcite, blocky calcite, baryte)	Reduction of porosity
	Fracturing	Brecciation Microfissures Increased permeability Corrosion of particles
	Faulting	Clay smearing Sealed surfaces Slickenside striations

Tab. 1: Mesoscopic and microsopic deformation features in the chalks.

field. The local strain induced by the rising salt crest was overprinted by a regional stress field which produced a strike-slip regime during Upper Cretaceous and Early Tertiary along major NW-SE trends (Tornquist Zone).

Viewed as a potential hydrocarbon reservoir, the chalk in the overburden of the salt diapir is subdivided into volumes of low deformation bordered by large fault zones. These volumes bordered by the four conjugated fault sets have dimensions of order tens to a hundred meters and constitute the basic reservoir units.

Observations on fracture surfaces surrounding them indicate a high sealing potential. Clay smearing, clay enrichment and neomineralization on and close to fracture surfaces reduce the porosity of the chalk, which normally has a high porosity but a low permeability. However, observed micro-cracks on fault surface and the high density of fractures close to fault zones increase the possibility of enhanced permeability on the fault zones, and facilitate communication between undeformed chalk and fracture zones.

The large quarries at Laegerdorf are exceptionally favorable for investigating in detail the deformed overburden of a salt diapir. The outcrop situation enables one to get a three-dimensional understanding of the fracture pattern. The study of deformation features on all scales gives a unique insight into a fractured chalk reservoir and the complex relationships which affect the permeability and sealing potential of subsurface fracture systems.

VI. ACKNOWLEDGEMENTS

The authors want to express their special gratitude to A.G. Milnes (Trondheim) for his constructive criticism of the manuscript and for correction of the English text. This study is an extension of a research project which was partly supported by

Saga Petroleum a.s. (Hovik, Norway). The Alsen-Breitenburg Zement- und Kalkwerke GmbH and the Vereinigte Kreidewerke-Dammann KG kindly permitted the investigations in the quarries near Laegerdorf.

REFERENCES

Ehrmann, W.U. (1986). Zum Sedimenteintrag in das zentrale nordwesteuropaische Oberkreidemeer. - 163 pp., Diss. Univ. Kiel.

Grube, F. (1955). Tektonische Untersuchungen in der Oberkreide von Lagerdorf (Holstein). - Mitt. Geol. Staatsinst. Hamburg 24, 5-32.

Hancock, P.L. (1985). Brittle mocrotectonics: Principles and practice. J. Struct. Geol 7, 437-457.

Hardman, R.F.P. (1982). Chalk reservoir of the North Sea. Bull. Geol. Soc. Denmark 30, 119-137.

Jaritz, W. (1973). Zur Entstehung der Salzstrukturen Nordwestdeutschlands. Geol. Jb. A10, 1-77.

Koestler, A.G. and Ehrmann, W.U. (in press). Fault patterns in the calcareous overburden of a salt diapir: Laegerdorf, NW Germany. N. Jb. Geol. Palaont. Mh., 1986.

Pegrum, R.M. (1984). The extension of the Tornquist Zone in the Norwegian North Sea. Norg. Geol. Tidssk. 64, 39-68.

Reches, Z. (1978). Analysis of faulting in three-dimensional strain field. Tectonophysics 95, 111-132.

Reches, Z. (1983). Faulting of rocks in three-dimensional strain fields. II. Theoretical analysis. Tectonophysics 95, 133-156.

Reches, Z. and Dietrich, J.H. (1983). Faulting of rocks in three-dimensional strain fields. I. Failure of rocks in polyaxial, servo-control experiments. Tectonophysics 95, 111-132.

Schulz, M. -G., Ernst., G., Ernst, H. and Schmid, F. (1984). Coniacian to Maastrichtian stage boundaries in the Standard Section for the Upper Cretaceous white chalk of NW Germany (Lagerdorf-Kronsmoor-Hemmoor): Definitions and proposals. Bull. Geol. Soc. Denmark 33, 203-215.

Van den Bark, E. and Thomas, O.D. (1981). Ekofisk: First of the giant oil fields in Western Europe. Amer. Assoc. Petrol. Geol. Bull. 65, 2341-2363.

Watts, N.L. (1983). Microfractures in chalks of Albuskjell Field, Norwegian sector, North Sea: Possible origin and distribution. Amer. Assoc. Petrol. Geol. Bull. 67, 201-234.

Withjack, M. and Scheiner, C. (1982). Fault pattern associated with domes - an experimental and analytical study. Amer. Assoc. Petrol. Geol. Bull. 66, 302-316.

Ziegler, P.A. (1982). Geological atlas of Western and Central Europe. 130 pp., 40 encl., The Hague (Shell Internationale Petroleum Maatschappij B.V.).

Ziegler, P.A. (1985). Late Cretaceous and Tertiary compressional deformation in the Alpine Foreland - a geodynamic model. Terra cognita 5, 108.

THE ORIGIN AND DEVELOPMENT OF SALT STRUCTURES IN NORTHWEST GERMANY

W. Jaritz

Federal Institute for Geosciences and Natural Resources
D-3000 Hannover 51, Germany

I. INTRODUCTION

North-west Germany is a classic salt structure area (see Fig. 1). As a result of the intensive seismic investigations for hydrocarbon exploration the layering of all the beds down to the saliferous base are nowadays well known. The saliferous base is generally encountered at a depth of 3 to 6 km, occasionally however down to 10 km depth.

Significant progress regarding salt structures was brought about by the method of analysing the development of salt structures by considering the development of the thicknesses of the surrounding rock (Trusheim, 1957, 1960 and Sanneman, 1963, 1968). Since its introduction this method has been well proven and forms the basis of this work.

II. SALIFEROUS SEDIMENTS IN NORTH-WEST GERMANY

The oldest saliferous formation that is associated with the north-west German salt structures is the Oberrotliegendes (Upper New Red). It occurs in the coastal areas of the North Sea, around the Lower Elbe and in Schleswig-Holstein (Trusheim, 1971; Plein, 1978). It is made up of rock salt and numerous interstratified clay bands; carbonate rock and anhydrite are generally missing. Its thickness is more than 1,000 m. The salt formation of the Oberrotliegendes (Upper New Red) is nowadays found above all at the centre of the NNE-SSW striking salt walls on either side of the Lower Elbe.

The most important saliferous formation in north-west Germany is the Zechstein. It occurs over the whole of north-west Germany and comprises seven cycles (Table 1).

Higher saliferous sediments in the Werra sequence occur only in the southern borderlands. Cycles 2 to 4, Staßfurt to Aller sequence, are to be found over the whole of north-west Germany and occur in all salt stocks. Cycles 5 to 7 do not appear to be involved in the salt stock structures. The complete cycle in each case consists of mudstone,

Table 1.

Cycles		Authors
7th Cycle	Mollin sequence	Best (1986)
6th Cycle	Friesland sequence	Kading (1977)
5th Cycle	Ohre sequence	Reichenbach (1970)
4th Cycle	Aller sequence	Richter-Bernburg (1955)
3rd Cycle	Leine sequence	ibid
2nd Cycle	Staßfurt sequence	ibid
1st Cycle	Werra sequence	ibid

carbonate, anhydrite, rock salt and potassium-magnesium salt, corresponding to the evaporation process. This sequence is followed by a very thin retrograde sequence.

The total thickness of the Zechstein in the north-west German basin is approximately 1000 to 1500 m. About 70% of this is made up of the Staßfurt sequence.

Other saliferous sediments are to be found in north-west Germany in the Triassic and Jurassic. Extended salt beds occur in the Upper Bunter and in the Middle Muschelkalk (Anisian-Ladinian boundary). They are mainly made up of rock salt and

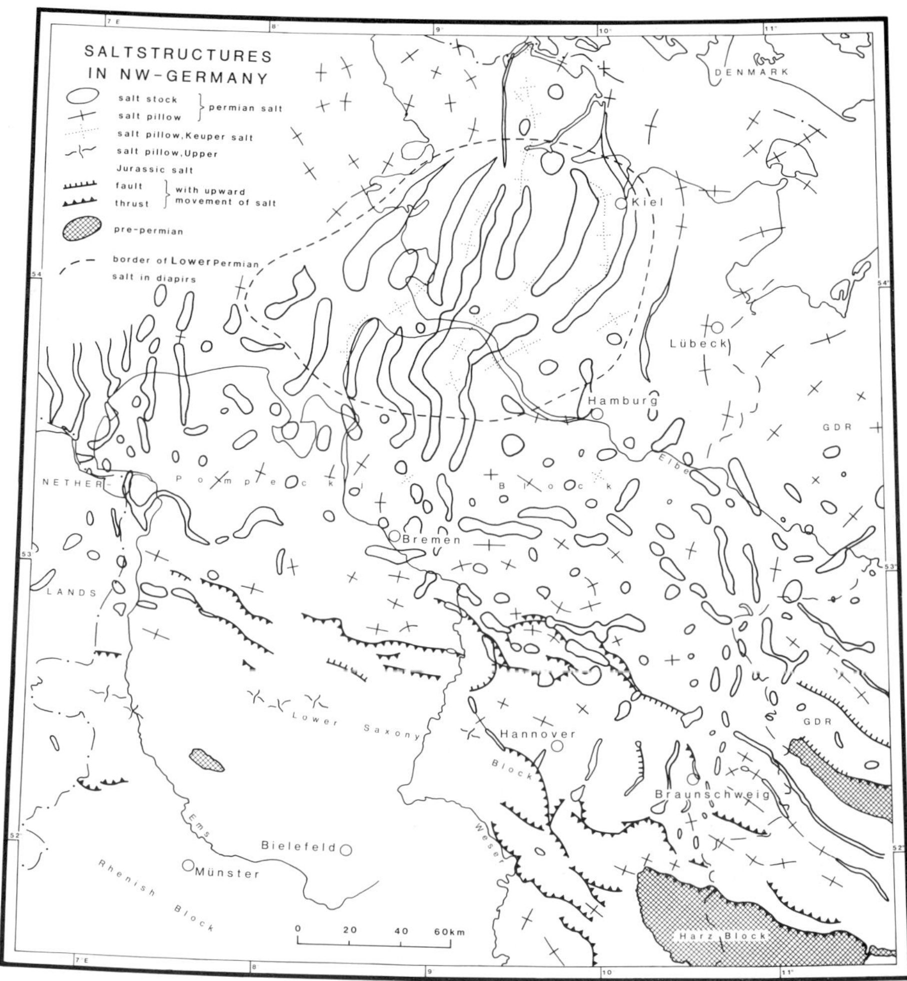

Figure 1. Map of salt structures in north-west Germany and neighboring areas (after Jaritz, 1973 and Kockel et al., 1985).

have clayey or clayey-marly intercalations. The thickness is about 100 m and varies little. These salts do not form independent structures.

Further saliferous intercalations are to found at various levels of the Middle Keuper (corresponds to about Karnian to

Norian). These are restricted to the rim synclines around salt stocks which passed through the diapir stage in the Triassic (Trusheim, 1971; Jaritz, 1973). Consequently they are limited in extent, with a maximum thickness generally of several hundred meters and in places over 1000 m. Independent salt pillows have been created from the particularly thick deposits of Keuper salt. Moreover, the addition of Keuper salt has resulted in places in a widening of the Permian salt walls.

Salt sedimentation took place in the Munder Mergel (corresponds to about Lower Purbeck) in the Upper Jurassic; this is associated with troughs and semi-troughs of the Lower Saxony Block. Pelite, limestone and anhydrite is intercalated between the rock salt sequences. The maximum rock salt thickness is about 1000 m. Independent salt pillows have developed locally from the salt beds.

III. REQUIREMENTS FOR THE FORMATION OF SALT STRUCTURES

The ability of rock salt and other salt minerals to creep is that particular property which encourages the formation of salt structures, in the first place as salt pillows. To a great extent creep is dependent on the temperature and on the pressure gradient. A lower limit for these two variables, however, is not known. Results from laboratory tests, for instance, cannot be directly related to the halokinetic processes as the time available for halokinesis is approximately 10^7 times greater than that available for laboratory tests.

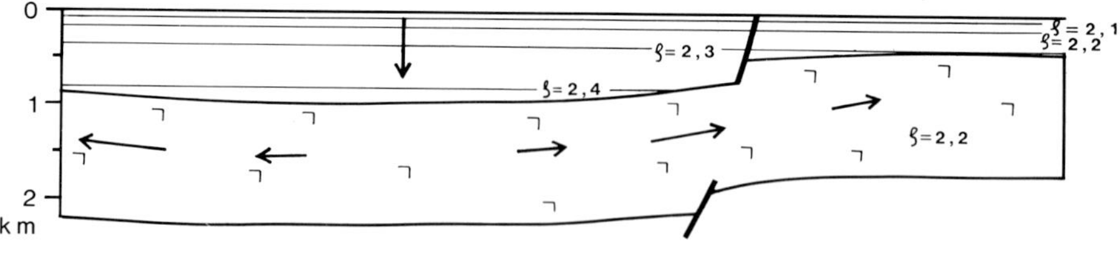

Figure 2. Schematic section. Semi-trough as initial disturbance for salt migration.

Local variations in the overburden load may develop as formations are being buried. Consequently the lithostatic pressure on the salt formation is dependent on the locality. A pressure gradient builds up accordingly and the salt begins to creep. Salt pillows are then formed. Figure 2 shows schematically how irregular loading of the subsurface creates a pressure gradient. The density values shown in the figure originate from Matthesius (1974) and refer to clastic Tertiary sediments. A distinct density inversion results leading to the Rayleigh-Taylor instability (Hunsche, 1976, etc.). During the Lower and Middle Bunter the density instability was more likely greater because the considerable carbonate and anhydrite content of the upper sequence of the Lower Bunter must have caused a rapid increase of the average rock density (see, however, Brink, 1984, 1986).

It can be seen at various places that troughs and semi-troughs have found during the Lower and Middle Bunter. The clearest example of this is the Gluckstadt Trough which was the impetus for the formation of the salt stock family of Schleswig-Holstein. The Horn Trough in the North Sea, just outside the area under discussion, is similarly significant for halokinesis (Best et al., 1983). Further, generally smaller faulting is to be found at various localities in north-west Germany. These faults, too, were important for the beginning of halokinesis. The lifting of the region can cause continental conditions to arise. The resulting morphology with peaks and valleys then likewise leads to a pressure gradient in the salt beds (Fig. 3). Examples of this process are probably provided by some salt structures in which Triassic fault tectonics are not observed. In such cases the morphological development prior to the transgression of the Solling sequence (uppermost section of the Middle Bunter) may have acted as the initial disturbance for the structural development. Details have yet to be explained.

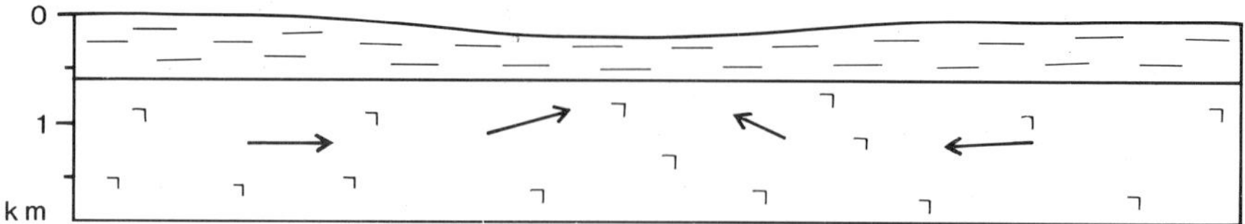

Figure 3. Schematic section. Morphology as initial disturbance for salt migration.

For salt pillows to further develop to salt stocks it is necessary that instability increases enough so that the overburden, normally also thickening, can be penetrated. In north-west Germany the majority of structures have developed to diapirs. Some of these required considerable time and broke through only in the Tertiary. On the other hand there are salt pillows which possess insufficient potential energy and will never be able to penetrate the overburden on their own.

IV. DATING THE DEVELOPMENT OF STRUCTURES

Trusheim (1957, 1960) and Sannemann (1963, 1968) have described the regularity of the development of salt structures in north-west Germany. In accordance with this description the development stages of the structure leave behind distinct indications in the surrounding rock and overburden. These indications permit dating of the structural development.

The salt pillow stage is marked by the fact that the surface layers above the areas of salt migration subside more than normal. Sediments of above average thickness then form in these so-called primary rim synclines. However, initially the differences between normal thickness and greater-than-normal thickness are very small in such primary rim synclines. More apparent are the less-than-normal thicknesses above the top of the salt pillow. This is because the structurally high parts of the salt pillow have a smaller area than the primary rim synclines so that a specific salt volume would cause greater thickness differences over the top of the pillow than in the primary rim synclines. After a salt stock has been formed parts of the thickness minima over the top of the former salt pillows remain intact and, moreover, permit the stage of formation of the salt pillow to be determined.

The salt, which has become abnormally thick as a result of the formation of the salt pillow, migrates from surrounding areas into the center of the structure and rises there. Consequently considerable layer thicknesses arise directly next to the salt stock and characterize the stage of salt stock formation. These increases in thickness are known as secondary rim synclines.

After a salt stock has broken through the overlying layers it often still has large amounts of salt at its base. This basal salt moves with time towards the salt stock and then upwards. Yet more rim synclines are formed over the salt migration areas, these are the "subsequent rim synclines" (the perhaps obvious term of "tertiary rim synclines" should be avoided as it would lead to confusion with rim synclines formed during the Tertiary). If suberosion has not taken place at the top of the salt stock then the less than normal thicknesses of the overlying layers give a measure of the subsequent movements.

Two examples should elucidate the development history of salt stocks; the first illustrates a rapid development from horizontal salt beds to a salt stock (Fig. 4), the second a very slow development up to the point of breakthrough (Fig. 5).

The Strackholt salt stock located approximately 20 km east of the Ems estuary is shown in Figure 4 with its various stages of development. A distinct salt pillow had already developed in the Bunter (Figure 4.1). The breakthrough took place in the Middle Keuper; Figure 4.2 illustrates the situation immediately prior to the beginning of the breakthrough. At the onset of Lias (Figure 4.3) a well formed salt stock was already present, nevertheless there was still a lot of salt outside of the

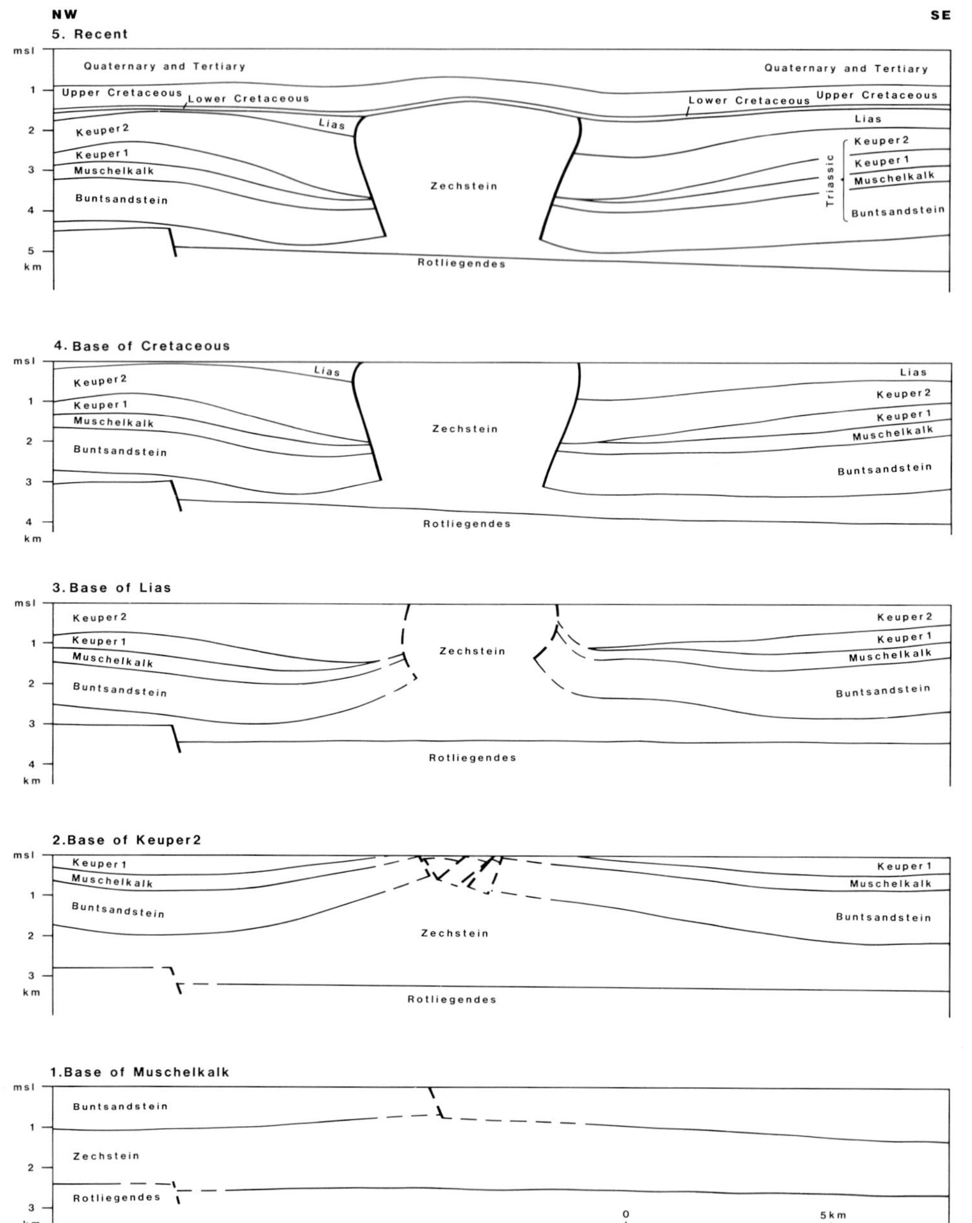

Figure 4. Development history of the Strackholt salt stock (after Jaritz, 1975).

formation. In figure 4.4 the continued development is shown during the Jurassic up to the base of the Cretaceous, the transgressive Hauterivian. At this stage the salt stock had essentially reached the form it has today. In the present day geological section (Figure 4.5) the doming of the layers above the salt stock and the slightly above-normal thicknesses around the salt stock indicate that small subsequent movements took place during the Cretaceous and Tertiary.

The development of the Dethlingen salt stock (Fig. 5) also started in the Bunter with the formation of a salt pillow (Figure 5.1). However, the salt migration occurred much slower, so that a well-formed salt pillow first became evident from the doming of the layers below the transgressive Lower Cretaceous. During the Cretaceous the salt pillow further developed and, at the beginning of the Tertiary, reached the stage just before break-through (Figure 5.3). The present day section (Figure 5.4) shows the young salt stock having broken through during the Tertiary. It is still surrounded by thick masses of salt, which indicate that considerable salt flow into the formation can be expected for several tens of millions of years.

The average speed at which salt rises in salt stocks is between 0.1 and 0.5 mm/year (Jaritz, 1980). During the stage of subsequent movement this drops rapidly to just a few hundredths of a millimeter per year. In some salt stocks the subsequent movements are even smaller, but nevertheless no salt stock ever comes to a complete standstill.

V. DEVELOPMENT SEQUENCE OF STRUCTURES IN NORTH-WEST GERMANY

During the Bunter and Muschelkalk (Lower and Middle Triassic) a large number of salt pillows had already formed from the former horizontal Permian salt. Even by the end of the Muschelkalk isolated salt stocks were probably already in existence. The Keuper (Upper Triassic) is the most

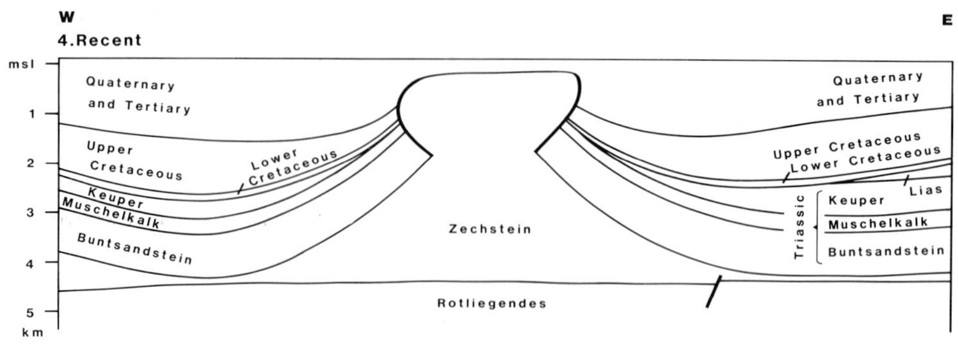

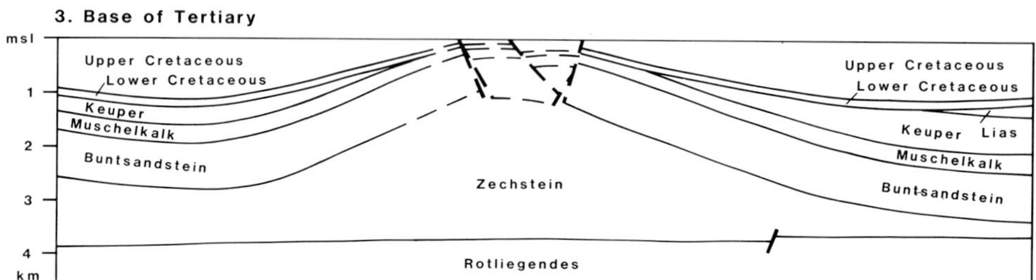

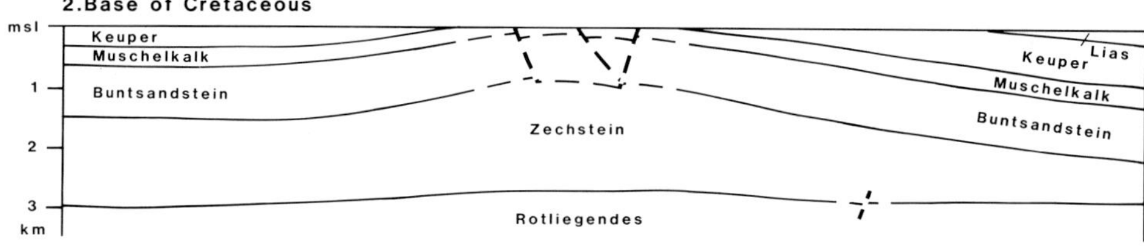

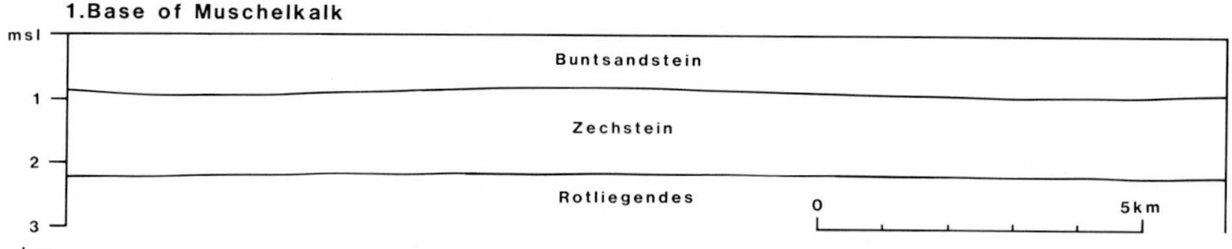

Figure 5. Development history of the Dethlingen salt stock (after Jaritz, 1977).

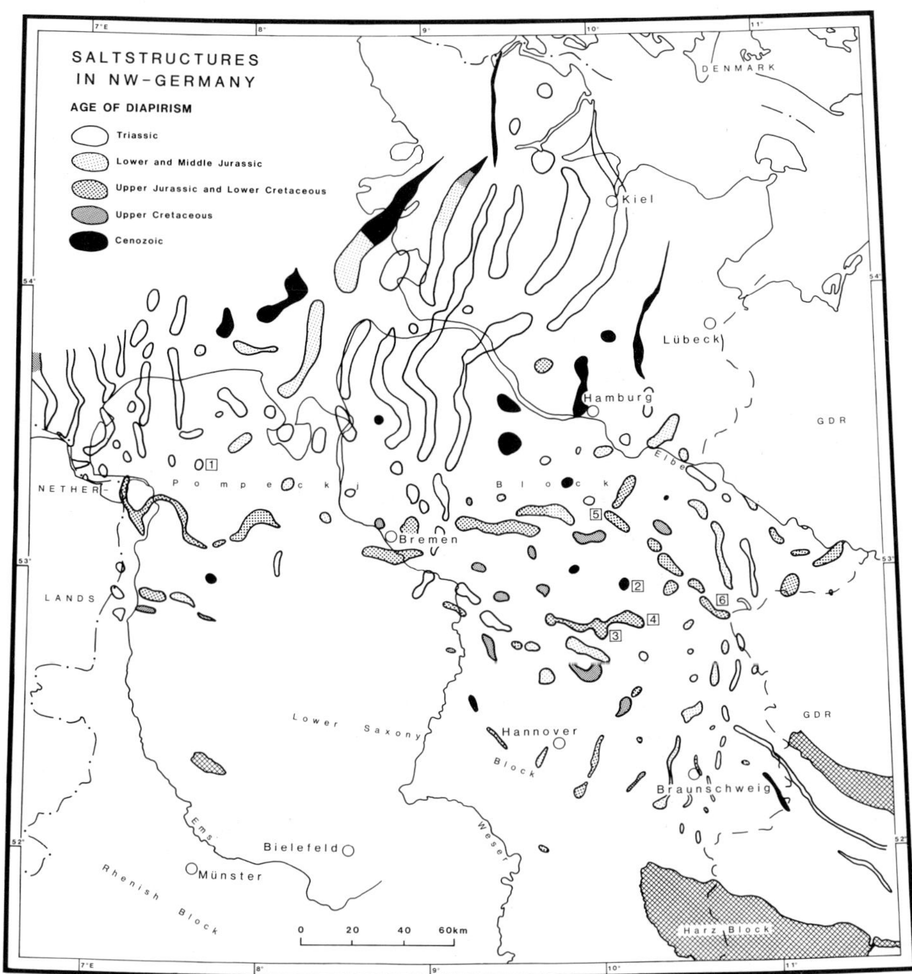

Figure 6. Age of salt stock breakthrough (after Jaritz, 1973 and Kockel et al., 1985). Salt stocks: 1 Strackholt, 2 Dethlingen, 3 Sulze, 4 Wessen-Lutterloh, 5 Egestorf-Soderstorf, 6 Bodenteich.

halokinetically active epoch; up to the end of the Triassic about half of the structures had either passed through or were in the diapir stage (Fig. 6). In some areas, such as Schleswig-Holstein and East Friesland, nearly all the salt walls and salt stocks were in existence.

In the Lias and Dogger numerous other salt pillows originated, and in Schleswig-Holstein, East Friesland and around Braunschweig (Brunswick) further salt walls and salt stocks were formed. These are mainly diapirs which belong to the later generations of the salt stock families in the sense of Sannemann (1963, 1968). The halokinetic activity was considerably less than in the Keuper.

The Upper Jurassic and Lower Cretaceous have to be considered together. The Lower Saxony Block was a basin from the Upper Jurassic onwards, bounded in most places by faulting. The basin was subdivided into different troughs and semi-troughs, and the majority of the faults had a strike direction of about 110°. On the Lower Saxony Block during this period the last pillows originated from Permian salt. Moreover, in the lowermost Lower Cretaceous salt pillows were formed from the Upper Jurassic salt. Salt stocks broke through mainly in the southern part of the Pompeckj Block.

In the middle of the Upper Cretaceous the Lower Saxony Block was lifted and overthrust at the marginal faults onto the surrounding blocks, the Pompeckj Block and Rhenish Block. At the same time deep troughs were formed at the margins of these adjacent blocks and have been referred to as "marginal troughs off block margins" by Voigt (1963). New salt stocks originated during this time especially on the Lower Saxony Block and in the southern marginal areas of the Pompeckj Block. Furthermore, salt was squeezed out along the mentioned overthrusts (Fig. 1). Additionally some salt pillows of Keuper salt were formed in Schleswig-Holstein and in the vicinity of the Lower Elbe.

Several more salt stocks originated in the Cenozoic, in particular in the marginal areas of the Schleswig-Holstein salt stock complex. Even today some very large salt pillows exist here. They will develop to salt stocks in the geological future.

VI. INTERACTION OF HALOKINESIS AND TECTONICS

As a result of the progress made in the investigation of the deeper subsurface down to the base of the Zechstein salt, it is now generally known where and when the formation of salt structures has been tectonically influenced. In this respect the development of salt pillows and diapirs must be differentiated from one another.

A. Salt Pillow Formation

As mentioned in section III it has been established at several places that the occurrence of troughs and semi-troughs has brought about salt migration and consequently the formation of salt pillows. This refers to some of the extended approximately N-S striking salt pillows, from which salt walls have later developed (Kockel et al., 1985). When parallel salt pillows exist only the oldest needs tectonic encouragement. The younger structures can then begin formation as subsequent generations without further impetus (Sannemann, 1963, 1968). Salt pillow formation modified by tectonics took place once more in the Upper Jurassic on the Lower Saxony Block under the influence of the formation of 110° striking troughs and semi-troughs. A pure halokinetic formation of salt pillows probably exists for some of the relatively small roundish salt stocks which broke through as early as the Triassic.

B. Diapirism

Generally as a salt pillow grows in size the density instability continuously increases so that the salt ascent is itself intensified until finally the salt breaks through the overlying layers and creates a salt stock. No additional impetus is required for these processes. Consequently it can be established that the majority of salt stocks have broken through independently. In particular this refers to salt stocks which have broken through in the Triassic as well as in the Lower and Middle Jurassic.

In the Upper Jurassic and in the Lower Cretaceous normal faulting was active on the Lower Saxony Block and to some extent spread over onto the southern margins of the Pompeckj Block. Some of the salt stock break-throughs in this period occurred closely associated with faulting and are therefore probably influenced by the faults, for example the salt stocks Sulze (3) and Weesen-Lutterloh (4) marked in Figure 6 as well as the salt stock chain from Egestorf-Soderstorf (5) to Bodenteich (6).

Tectonically influenced diapirism took place in the Upper Cretaceous at the marginal thrust faults of the Lower Saxony Block and at other paralleling thrust faults (Fig. 1). As the amount of salt was relatively small in this marginal area of salt structure formation, the salt was generally only squeezed out (Fig. 1) at the normal and thrust faults, and only a few real salt stocks could be formed.

REFERENCES

Best, G. (1986). Die Grenze Buntsandstein/Zechstein nach Bohrlochmessungen (Gamma-Ray und Sonic-Log) im Nordwestdeutschen Becken. - Unpubl. Report, 17 p., 4 fig., 2 tabl., 58 encl., Hannover.

Best, G., Kockel, F. and Schoneich, H. (1983). Geological History of the Southern Horn Graben. - Geologie en Mijnbouw 62, 25-33.

Brink, H.-J. (1984). Die Salzstockverteilung in Nordwestdeutschland. Geowiss. in uns. Zeit 2, 160-166.

Brink, H.-J. (1986). Salzwirbel im Untergrund Nordwestdeutschlands. Geowiss. in uns. Zeit 4, 81-86.

Hunsche, U. (1976). Modellrechnungen zur Entstehung von Salzstockfamilien. Diss. Tech. Univ. Braunschweig, 103 p., 28 fig. 8 tab., Braunschweig.

Jaritz, W. (1973). Zur Entstehung der Salzstrukturen Nordwestdeutschlands. Geol. Jb., v. 10, 77 p., Hannover.

Jaritz, W. (1975). Section in: Geologische Ubersichtskarte 1:100,000, CC 3110 Bremerhaven, Hannover.

Jaritz, W. (1977). Section in: Geologische Ubersichtskarte 1:200,000, CC 3126 Hamburg-Ost, Hannover.

Jaritz, W. (1980). Einige Aspekte der Entwicklungsgeschichte der nordwestdeutschen Salzstocke. Z. dt. geol. Ges. 131, 387-408.

Kading, K.-Chr. (1977). Salinarformation des Zechstein. - In Leper J.: Erl. Bl. Uslar Nr. 4323, 13-15.

Kockel, F., Best G., Jurgens, U., Baldschuhn, R., Frisch, U., Schmitz, J., Stancu-Kristoff, G., Deneke, E., Jaritz, W. and Sattler-Kosinowski, S. (1985). Geotektonischer Atlas von NW-Deutschland, Forschungsvorhaben 03 E 3014 A. Unpubl. Report, 68, p., Hannover.

Matthesius, G. (1974). Vertikale Dichte-, Porenanteil- und Druckdifferenzprofile an Sedimentgesteinen des Nordwestrandes des Gifhorner Troges. - Diss. Techn. Univ. Braunschweig, 149 p.

Plein, E. (1978). Rotliegend-Ablagerungen im Norddeutschen Becken. Z. dt. Geol. Ges. 129, 71-97.

Reichenbach, W. (1970). Die lithologische Gliederung der rezessiven Folgen von Zechstein 2 - 5 in ihrer Beckenausbildung. Probleme der Grenzziehung und Parallelisierung. Ber. dt. Ges. geol. Wiss., A, Geol. Palaont. 15, 555-563.

Richter-Bernburg, G. (1955). Stratigraphische Gliederung des Zechsteins. - Z. dt. geol. Ges. 105, 843-854.

Sannemann, D. (1963). Uber Salzstock-Familien in NW-Deutschland. Erdol.-Z. 79, 499-506.

Sannemann, D. (1968). Salt-stock Families in Northwestern Germany. In: Diapirism and Diapirs, Mem. 8, Americ. Ass. Petrol. Geol., p. 261-270.

Trusheim, F. (1957). Uber Halokinese und ihre Bedeutung fur die strukturelle Entwicklung Norddeutschland. Z. dt. geol. Ges. 109, 111-151.

Trusheim, F. (1960). Mechanism of Salt Migration in Northern Germany. Bull. Americ. Ass. Petrol. Geol. 44, 1519-1540.

Trusheim, F. (1971). Zur Bildung der Salzlager im Torliegenden und Mesozoikum Mitteleuropas. Beih. geol. Jb. 112, 51.

Voigt, E. (1963). Uber Randtroge vor Schollenrandern und ihre Bedeutung im Gebiet der Mitteleuropaischen Senke und angrenzender Gebiete. Z. dt. geol. Ges. 114, 378-418.

SECTION C

CAPROCK

TEXTURAL AND PALEOMAGNETIC EVIDENCE FOR THE MECHANISM AND TIMING OF ANHYDRITE CAP ROCK FORMATION, WINNFIELD SALT DOME, LOUISIANA

J. Richard Kyle

Department of Geological Sciences
University of Texas at Austin
Austin, TX 78713

Mark R. Ulrich

Mobil Exploration and Production Services, Inc.
P. O. Box 900
Dallas, TX 75221

Wulf A. Gose

Department of Geological Sciences and
Institute for Geophysics
University of Texas at Austin
Austin, TX 78713

I. INTRODUCTION

The nature and origin of the cap rocks for the Gulf Coast salt diapirs have been the subjects of many previous studies. It is now generally accepted that these deposits are the result of complex evolutionary processes associated with dissolution of the rising halite diapir, the accumulation of the relatively insoluble components, and the alteration of these materials, largely by bacterially controlled diagenetic processes (Murray, 1966; Martinez, 1980; Feely and Kulp, 1957). However, the detailed mechanisms, relations, and timing of the stages of these processes have not been addressed and are the subject of this investigation.

The Winnfield dome in the North Louisiana Basin provides an unequalled opportunity in the Gulf Coast to study a salt dome cap rock. Active quarrying operations provide access to three-dimensional exposures of a thick anhydrite and calcite cap rock

(Fig. 1). Concentrations of metallic sulfide minerals are present in the calcite and anhydrite zones (Ulrich et al, 1984). Within the anhydrite portion of the cap rock, sulfides occur as laminar zones that are parallel to the tightly intergrown,

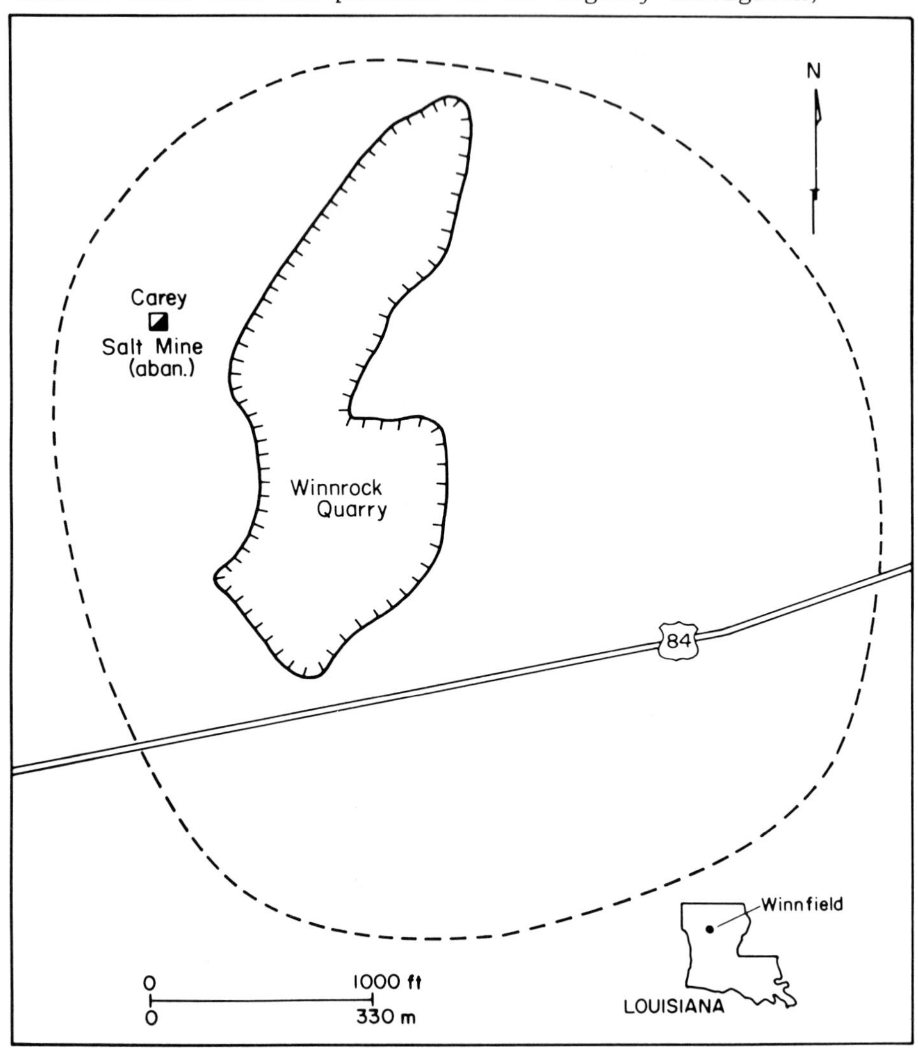

Figure 1. Outline of the Winnfield salt diapir showing the Winnrock quarry. Dashed line is the salt stock boundary at -75 m elevation; hachures outlines the quarry as defined by mining exposures of the anhydrite cap rock. Modified after Belchic (1960) and Ulrich et al. (1984).

layered anhydrite either as cement between relatively undeformed anhydrite grains or as large euhedral crystals. Pyrrhotite is the dominant sulfide, but marcasite, sphalerite, galena, and barite may be locally abundant.

It has been demonstrated that sulfide mineralization associated with many Gulf Coast salt domes is intimately related to cap rock formation (Kyle and Price, 1986). The sulfide concentrations within the Winnfield cap rock are believed to have originated from metalliferous formation water that rose along the salt dome flanks to the cap rock. Mixing of metalliferous fluids with reduced sulfur within the cap rock may have been the cause of sulfide precipitation. Textural relationships and chemical compositions of the sulfide layers in the anhydrite cap rock suggest that distinct episodic pulses of metalliferous brines entered the cap rock to form sulfide layers (Ulrich et al, 1984). Anhydrite grains that are completely surrounded by sulfides are euhedral and undeformed, similar to the anhydrite disseminated throughout the salt diapir. Anhydrite grains outside the mineralized areas are deformed and tightly intergrown. These textures suggest that mineralizing fluids entered the cap rock along the salt-anhydrite interface at the zone of salt dissolution, before that portion of the anhydrite zone was compressed and accreted to the overlying anhydrite cap rock (Fig. 2). As salt dissolution and accompanying anhydrite accumulation and compaction continued, sulfide layers were precipitated from episodic pulses of metalliferous fluids. Therefore, the oldest sulfides originating by this mechanism occur at the top of the present anhydrite cap rock zone while the youngest are found at the base.

This systematic cap rock stratigraphy can be studied to provide valuable information on the mechanism and timing of salt dome cap rock formation. The purpose of this paper is to present the results of field mapping, petrologic, petrographic, and paleomagnetic investigations of the Winnfield dome, particularly as they pertain to the origin of the anhydrite cap rock zone. Preliminary aspects of this research project were reported by Ulrich et al (1984) and Gose et al (1985).

II. GEOLOGY OF THE WINNFIELD DOME

Winnfield Dome is one of nineteen interior piercement salt domes of the nôrthwest-southeast trending North Louisiana Basin, a rift basin on the northern margin of the Gulf of Mexico. Seismic and drill hole data to approximately 1500 m indicate that sediments of Tertiary and Cretaceous age terminate up dip against the steeply dipping sides of the salt dome. The stratigraphy of the North Louisiana Basin is described in detail by Berryhill et al (1968).

The top of the salt stock is within 104 m of the surface and has a roughly circular outline which extends about 1350 m from north to south and 1200 m from east to west (Fig. 1). The stratigraphic position of the underlying "mother" mid-Jurassic

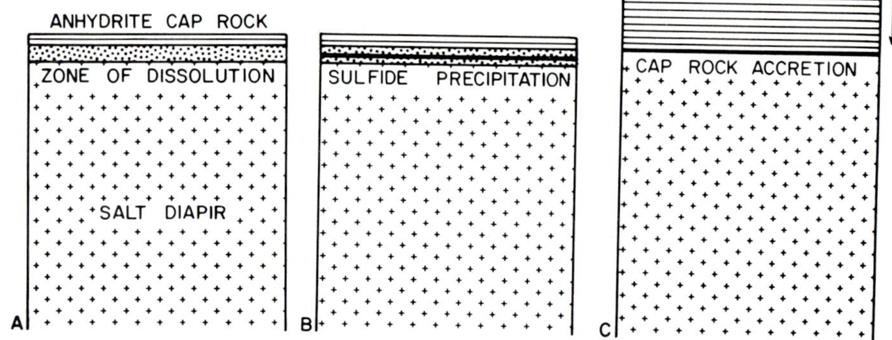

Figure 2. Generalized model for the development of the anhydrite cap rock. A. Salt stock is dissolved by halite-undersaturated pore waters leaving a residue of relatively unsoluble components, principally anhydrite. B. Sulfides precipitate in intercrystal pores to form local cement. C. Anhydrite residue zone is accreted to the base of the previously formed cap rock, accompanying diapiric uplift. Arrow indicates direction of increasingly younger anhydrite cap rock formed from multiple dissolution-accretion episodes.

Louann Salt is at a depth of over 5 km in this region (Eversull, 1984). The Winnfield salt stock is mantled by a typical cap rock sequence of calcite, gypsum, and anhydrite (Figs. 3,4). Erosion and mining have exposed the calcite portion of the cap rock. Because of the dip of the cap rock layers away from the center of the dome, the quarry provides access to the upper 100 m of the anhydrite zone stratigraphy. The extensive quarrying operations permit the study of the three-dimensional nature of cap rock features. Rotary cuttings provide information on the cap rock below the present quarry floor, as well as the material that has already been removed by mining. Although the workings of the Carey salt mine are no longer accessible, limited samples of the Winnfield salt mass are available for study.

A. Salt-Stock Lithology

Halite from the Winnfield salt diapir is massive, coarsely crystalline, white to dark gray, and foliated. Salt produced from the Carey mine averaged 97% NaCl with approximately 3% fine to medium grained, disseminated, idioblastic to subidioblastic anhydrite (Fig. 5). Trace quantities of dolomite, quartz, barite, pyrite, and calcite are also found in the salt. Steeply dipping anhydrite bands, generally 2 to 10 cm in width, parallel the foliation and contain 15% to 90% anhydrite. High pressure brine and gas inclusions are common throughout the salt, particularly within anhydrite bands (Belchic, 1960).

The top of the salt is 76 to 80 m below sea level, is horizontal, and is immediately overlain by anhydrite cap rock. In some drill holes, unconsolidated residual anhydrite sand was present at the contact. Other drill holes encountered large solution cavities filled with brine under high pressures. Locally, carbon dioxide or flammable gas, smelling of H_2S, were associated with these fluids (Belchic, 1960).

CENTER
PERIMETER
200
50
100
ELEVATION (meters)
ELEVATION (feet)
0
0
EXPLANATION
Calcite
Gypsum
Anhydrite
Halite
-100
-50
-200
-300

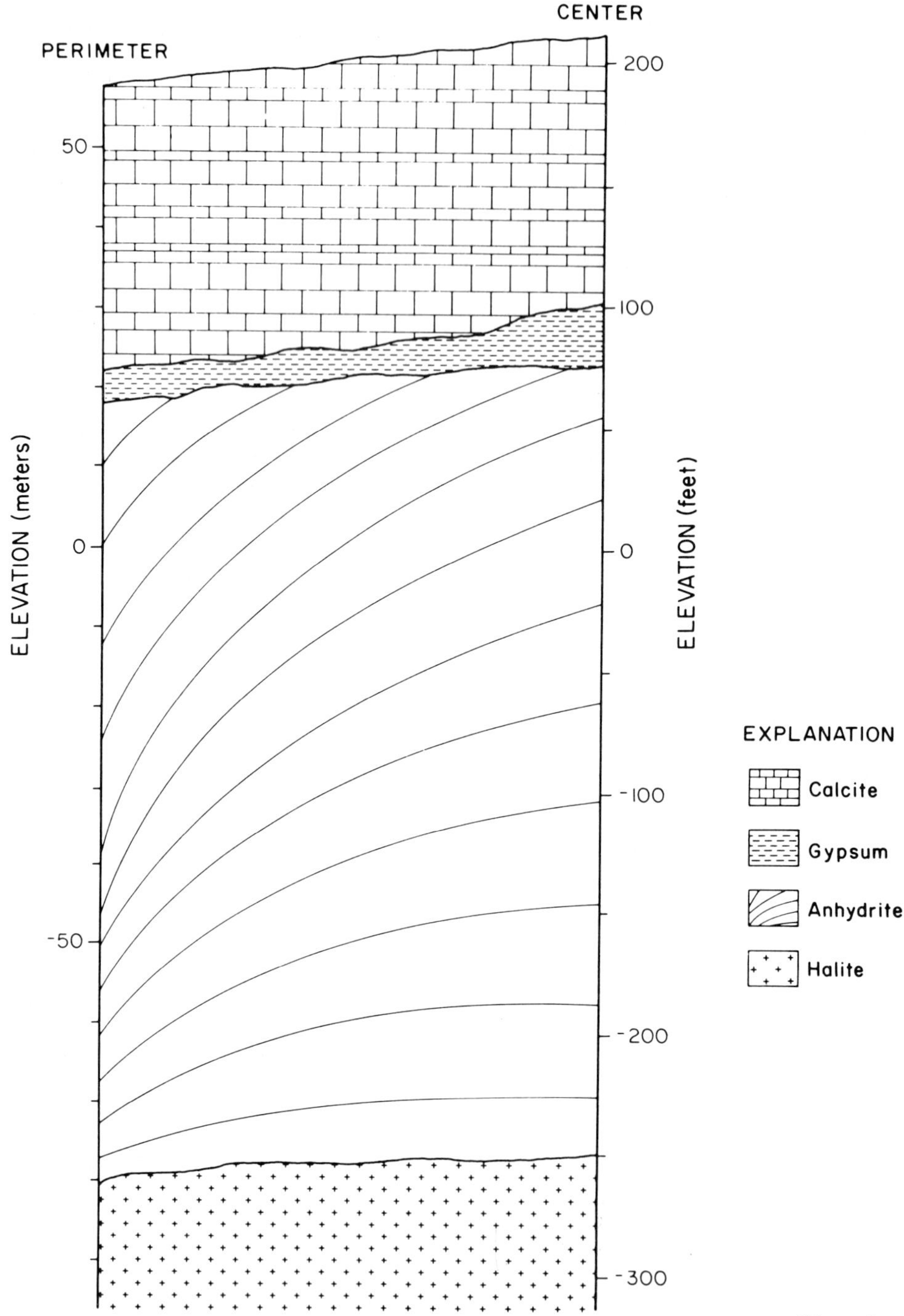

Fig. 3

Figure 4. View of northwest wall of the Winnrock quarry. Karsted calcite cap rock (CA) separated from the banded anhydrite zone (AN) by a gypsum zone of variable thickness. Anhydrite zone appears to dip from left to right, but true dip is into the plane of the figure toward the margin of the dome.

B. Anhydrite Zone Lithology

The anhydrite zone immediately overlies the salt and reaches a maximum thickness of 94 m over the center of the dome. Because of the dip of the anhydrite zone away from the crest of the dome, the total amount of anhydrite stratigraphy present along the cap rock margins is perhaps twice this amount (Fig. 3,4). An anhydrite zone is present along the flanks of the dome to a depth greater than 1500 m where it is still approximately 27 m thick (Belchic, 1960). The anhydrite cap rock is light to medium gray

Figure 3. Schematic lithologic section of the Winnfield dome cap rock. Modified from Gose et al. (1985)

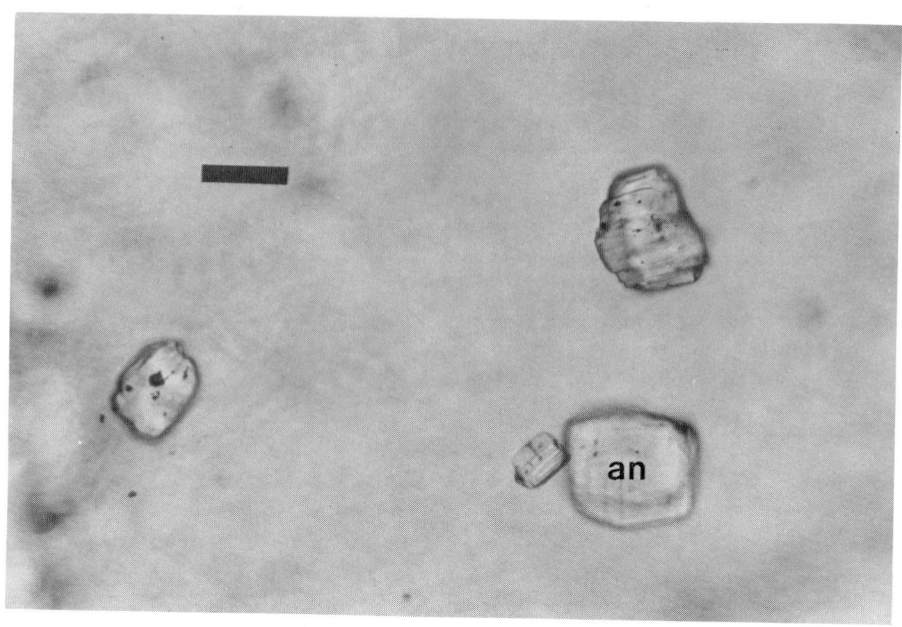

Figure 5. Transmitted light photomicrograph of anhydrite crystals (an) in halite. Dark spots on the anhydrite crystal surfaces are brine- and gas-filled inclusions. Scale = 0.5 mm

and generally layered (Fig. 4). The anhydrite consists of tightly interlocking, predominantly xenoblastic crystals with no visible preferred orientation (Fig. 6). Many of the grains have undergone strain and recovery as indicated by undulose extinction, glide twins and subgrain boundaries. Porosity and permeability are very low. Deformed, banded anhydrite clasts are common within the layers (Fig. 7). These clasts are derived from primary anhydrite beds within the Louann Salt which have been deformed during diapiric emplacement. Within the anhydrite portion of the cap rock, sulfides occur as laminar zones parallel to the tightly intergrown, layered anhydrite grains or as coarse euhedral crystals growing in open space (Fig. 8). Sulfide layers generally are less than one millimeter to several centimeters thick and can be traced laterally for tens of meters. These

Figure 6. Transmitted light photomicrograph of the anhydrite cap rock. Note sutured grain boundaries and included euhedral quartz crystal (qz). Crossed nicols. Scale = 0.5 mm.

layers locally thicken into laminated flat-bottomed, convex upward sulfide mounds. Although these mounds vary in size and shape, they are usually less than 2 m long and 1 m thick. The principal sulfide is pyrrhotite, but sphalerite, galena, and barite are abundant locally. Marcasite generally is the dominant mineral in oxidized samples and appears to be an alteration product of pyrrhotite.

The anhydrite/calcite interface is highly irregular with as much as 18 m of relief along the contact. A gypsum zone as much as 12 m thick locally occurs between the anhydrite and calcite zones. Gypsum occurs as light to medium gray, intergrown masses or as poikilotopic cement surrounding corroded anhydrite grains. Elsewhere at the anhydrite/calcite interface, fine-crystalline calcite surrounds corroded anhydrite grains.

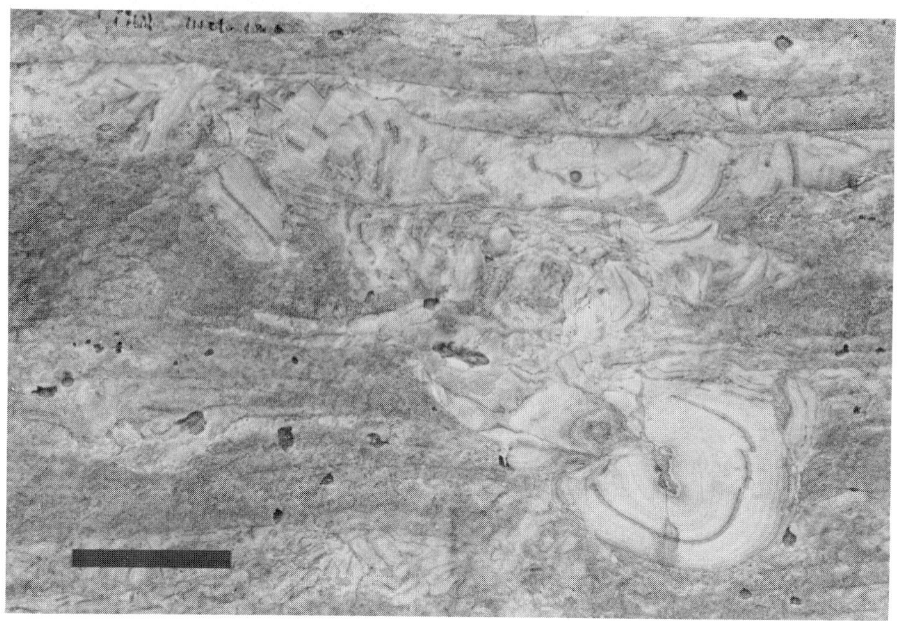

Figure 7. Deformed clasts of primary anhydrite beds within anhydrite cap rock. Scale = 5 cm.

C. Calcite Zone Lithology

The calcite zone varies in areal extent and in thickness from about a meter to over 90 m (Figs. 3,4). The nature and origin of this zone are discussed in detail by Posey et al (this volume). The upper portion of the calcite cap rock is a slightly fossiliferous, glauconitic, pelleted marine limestone (Fig. 9), the "marine false cap rock" of Posey et al. This unit also contains abundant euhedral quartz and dolomite crystals that are believed to have been derived from the salt mass, probably during dissolution on the sea floor (Fig. 10). The underlying "true" calcite cap rock is distinctly banded with layers of dark gray, fine-crystalline calcite separated by layers of white, coarse-crystalline calcite (Fig. 11); these bands vary in thickness from a few millimeters to several centimeters. The coarse-crystalline calcite appears to have formed in open spaces and to have formed

Figure 8. Polished slab of the anhydrite cap rock with a 1-cm thick sulfide-rich zone (s) and several thinner laminae.

later than the fine-crystalline calcite. Iron sulfides and barite commonly line solution cavities and fractures.

A roughly laminated massive sulfide lens is exposed at the calcite/anhydrite transition zone. The lens has a maximum thickness of about 5 m and extends for more than 30 m. The edges of the lens taper to a layer approximately 30 cm thick which separates the overlying calcite from anhydrite. The massive sulfide lens consists of marcasite and pyrite which are complexly intergrown with calcite, gypsum, and anhydrite. Common textures include sulfide overgrowths on calcite rhombs, sulfides cementing corroded anhydrite grains and sulfides rimming calcite pseudomorphs after anhydrite. The entire sulfide mass is fractured and brecciated. Colloform sphalerite, galena, iron disulfides, and barite line vugs and fractures. Minor amounts of

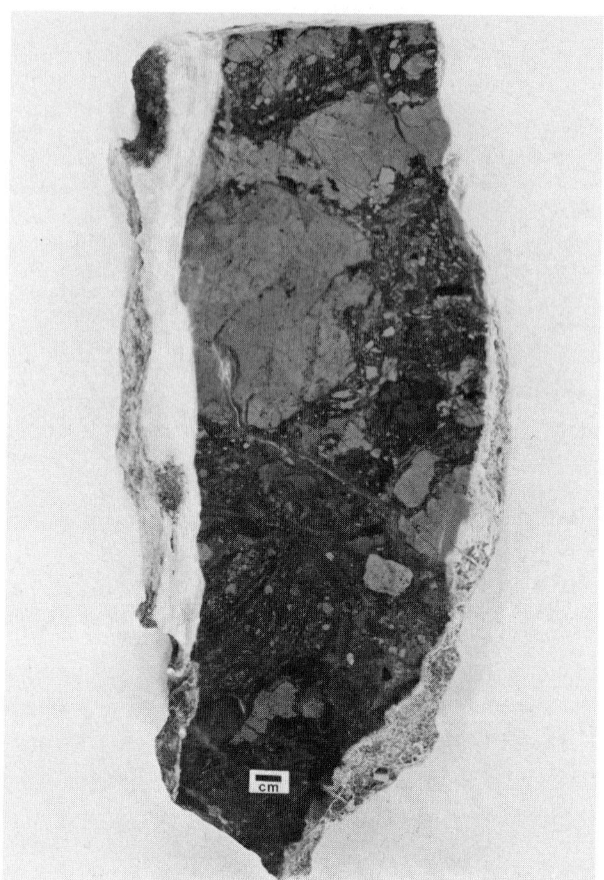

Figure 9. Polished slab of the marine limestone false cap rock cut by late calcite vein.

pyrrhotite occur within the sulfide lens. Examination of rotary drill samples suggests that massive sulfide concentrations similar to the exposed example are common along the calcite anhydrite transition zone.

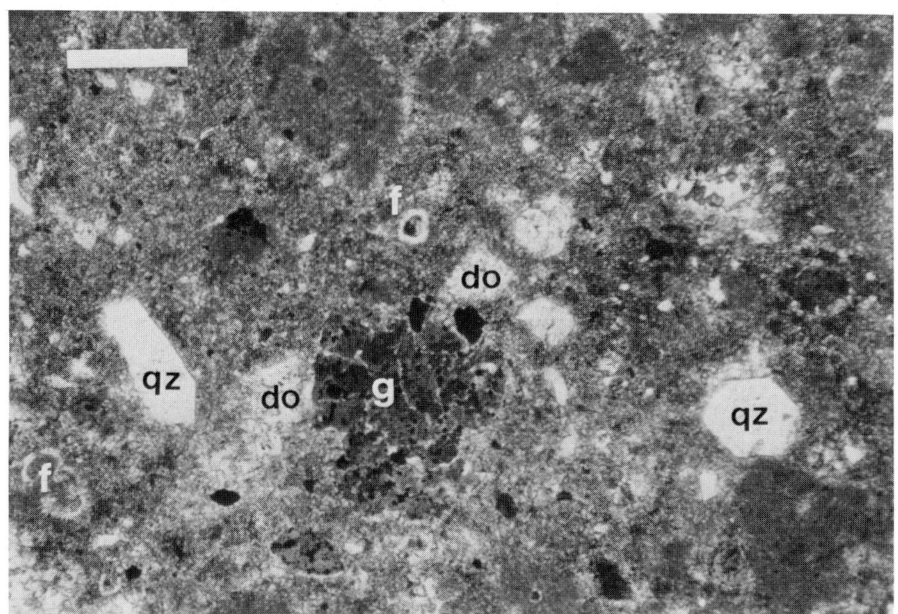

Figure 10. Transmitted light photomicrograph of the marine limestone false cap rock. Note abundant quartz (qz) and recrystallized dolomite (do) crystals derived from dissolution of the salt stock in addition to glauconite (g) and fossils (f) in the pelleted matrix. Scale is 0.5 mm.

III. INTERNAL STRUCTURES WITHIN THE ANHYDRITE ZONE

A. Morphology of "Mound" Structures

"Mound" structures are common within the anhydrite cap rock. The relative dimensions of these mounds vary considerably. Some mounds are broad and squat (Fig. 12); others are tall and spine-like (Fig. 13). The mounds generally are flat-bottomed and range between 0.3 and 2 m high. These mounds are completely encased by "typical", laminated anhydrite cap rock. The laminations commonly sweep upward and terminate along the flanks of the mound. The laminations immediately overlying the mound are anticlinal, mimicking the shape of the mound. The amplitude of the anticline abruptly diminishes upward from the mound. Mound-like structures may be concentrated along a particular horizon within

Figure 11. Polished slab of a true calcite cap rock consisting of alternating dark and light calcite bands.

the anhydrite zone, and in some areas appear to be stacked vertically, separated by "typical" laminated anhydrite (Fig. 14).

Internally, the mounds vary both compositionally and texturally. Most mounds are composed almost entirely of anhydrite. Mounds composed of sulfides (commonly >50%) with anhydrite or gypsum are less common. Mounds composed of coarse-crystalline salt with disseminated anhydrite (Figs. 14,15) are also uncommon (the salt quickly dissolves in the humid Louisiana climate once exposed by quarrying). Mound-shaped cavities,

Figure 12. Polished slab of a multiply laminated anhydrite mound.

which presumably once contained salt are present at several places within the quarry. Some mound-shaped cavities are partly filled with selenite and minor calcite.

ANHYDRITE MOUNDS. The mounds composed of anhydrite have three distinct internal textures. Some mounds consist of a chaotic, cemented breccia (Fig. 16). These breccia fragments are laminated, commonly contorted, and generally range from a few millimeters to several centimeters in length. Some anhydrite mounds are doubly plunging anticlines or domes defined by the laminated anhydrite (Fig. 12). Other anhydrite mounds display a fine grained, homogenous texture (Fig. 13). In all three types of anhydrite mounds, pressure solution bands, roughly conformable to the top of the mound, extend across the width of the mound.

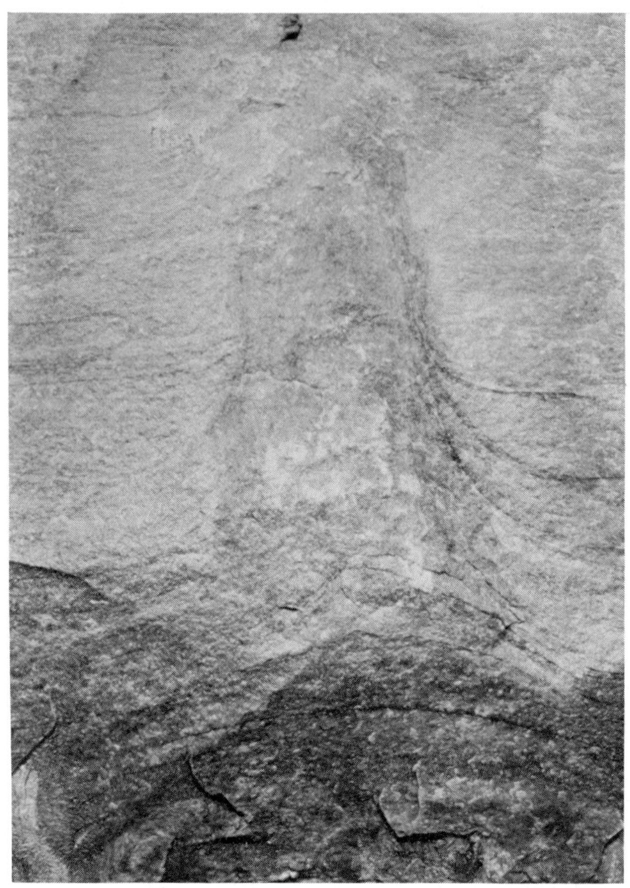

Figure. 13. Anhydrite mound with an homogeneous core. See Fig. 25 for a line drawing of this mound; horizontal dimension is about 1 m.

SULFIDE MOUNDS. In some horizons of the anhydrite cap rock, abundant sulfides occur within mounds, in the dipping laminae flanking mounds and in anticlinal laminae above mounds (Fig. 17). Unfortunately, the exposed sulfide-bearing mounds are all extremely oxidized, obscuring the details of the original texture. However, the sulfides within these mounds occur in distinct laminae. Some laminae are composed entirely of sulfides

Figure 14. Salt mound in anhydrite. Note laminated anhydrite mound near the scale separated from the salt mound by horizontally laminated anhydrite. Scale = 18 cm.

which are interlayered between bands of anhydrite (Fig. 8). The most abundant sulfide is monoclinic pyrrhotite. Where most of the anhydrite has been converted to gypsum, the dominant sulfide is marcasite which appears to be an alteration product of pyrrhotite. As anhydrite was converted to gypsum, iron appears to have been remobilized and precipitated as fine-grained marcasite with the gypsum.

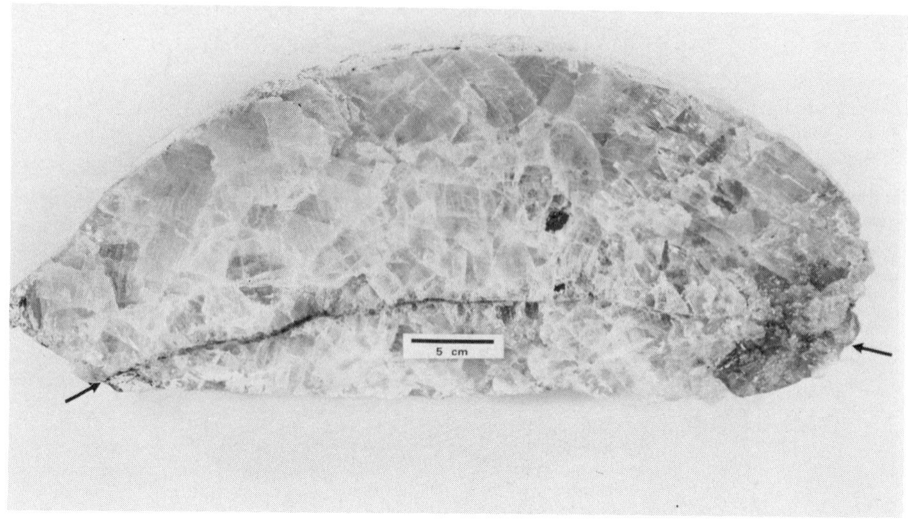

Figure 15. Polished slab of a salt mound. Note anhydrite crystal laminae (arrows) with pyrrhotite conforming to the overall shape of the salt mound.

SALT MOUNDS. The mounds composed of salt have the same overall morphology as the anhydrite mounds (Figs. 14,15), but internally, the salt mounds are composed of coarse-crystalline halite with disseminated euhedral anhydrite and numerous liquid and gaseous fluid inclusions. When samples of this salt are placed in water, a crackling, popping sound is heard as the gas-filled inclusions are released. The gas has a slight hydrocarbon smell but does not readily ignite. About 3 to 5 weight-percent of the "salt" from these mounds is insoluble in water. This residue is almost entirely anhydrite with trace amounts of euhedral dolomite, quartz, and iron sulfides. These residue crystals are similar in size and appearance to those present in the salt from the main salt stock, i.e. from the Carey salt mine.

Figure 16. Polished slab of stacked anhydrite breccia mounds.

Some salt mounds contain internal anhydrite crystal residue zones with local pyrrhotite cement which mimic the overall mound geometry (Fig. 15).

B. Origin of Mound Structures

The morphologic similarities between the various mound structures suggest that they are the products of similar processes. Several authors (Goldman, 1952; Dix and Jackson, 1982) have suggested that each layer within the anhydrite zone

Figure 17. Sulfide-rich mound in anhydrite.

represents a distinct cycle of salt dissolution, residual anhydrite accumulation, and residue accretion to the base of the overlying anhydrite cap rock. It may be further proposed that the undulose to "wavy" form of the laminae represents a mold of the top of the salt stock during that cycle of cap rock formation. The undulose laminae imply a gentle, hummocky surface across the top of the salt. During diapiric uplift, residual anhydrite sand from salt dissolution is compacted between the base of the cap rock and the top of the salt stock to form a new layer. The top of this new layer conforms to the previous base of the cap rock while the base of this layer retains the shape of the top of the salt stock. Pressure solution occurs along the boundaries between layers as indicated by partially dissolved breccia clasts at the tops of individual layers (Fig. 7). Dix and Jackson (1982) proposed that most of the compaction, intergrowth and pressure solution takes place at, or immediately above, the salt contact at Oakwood dome, a relation that can be demonstrated for other domes as well.

LAMINATED ANHYDRITE MOUNDS. Small bumps on the top of the salt stock are believed to be responsible for the simple anticlinal mounds observed in the anhydrite cap rock (Fig. 18). Impressions of these bumps are cast into the base of the cap rock as the residual anhydrite sand undergoes compaction and intergrowth during diapiric uplift.

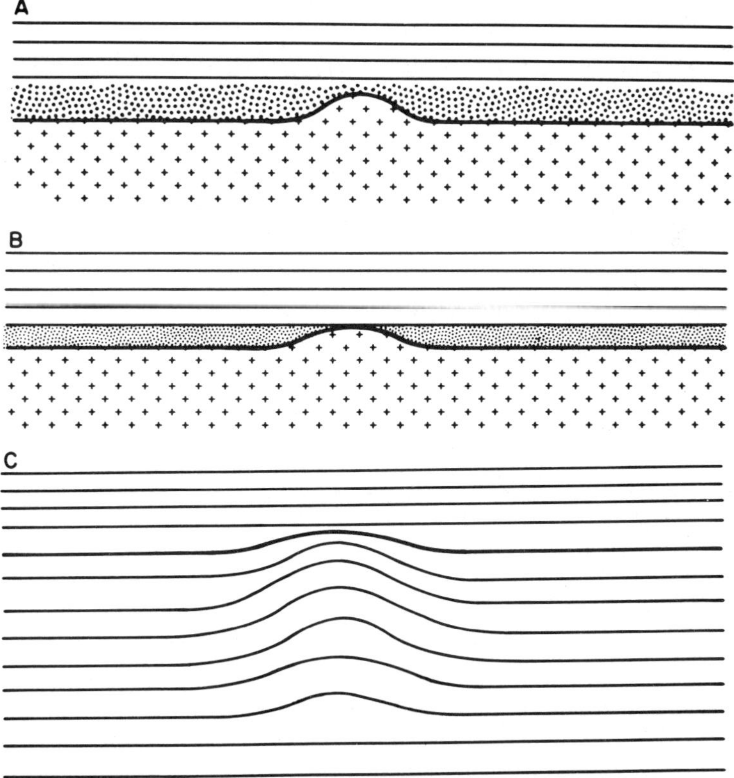

Figure 18. Model for the origin of the laminated anhydrite mounds. A. Dissolution produces irregular salt surface and residual anhydrite crystal sand in the brine-filled cavity. B. During diapiric uplift, anhydrite sand is accreted to base of cap rock. The morphology of the irregular salt surface is preserved in new layer. C. Final structure after several cycles of cap rock formation records growth and dissolution of the salt mound.

Sufficient anhydrite sand must be present during compaction to cushion the salt from the base of the cap rock; otherwise, the salt will flow to conform to the shape of the overlying cap rock. As the cycle is repeated, other layers are added to the base of the cap rock. Each new layer records the shape of the top of the salt stock at that time. If the salt bump becomes more pronounced over time due to selective dissolution, then the amplitude of the resultant anhydrite mound increases. If the salt bump diminishes with time, then the amplitude of the resulting mound decreases.

The position of the salt bumps may be controlled in part by the presence of steeply plunging, contorted thin anhydrite beds within the salt stock. These types of anhydrite beds are locally abundant in the salt stock as exposed in the Carey mine (Cameron, 1949; Hoy et al., 1962). The former presence of deformed anhydrite beds is demonstrated by the accumulations of anhydrite clasts in the cores of, and in association with, the cap rock mound structures (Fig. 16).

SALT MOUNDS. It appears that bumps along the top of the salt stock locally persisted through several cycles of cap rock formation (Fig. 19). Underplating of anhydrite sand is prevented where these salt bumps contact the overlying cap rock. Mound-like or even spine-like impressions in the cap rock develop depending upon how long the salt bumps existed.

Although it might be anticipated that salt dissolution would be greatest across the tops of salt bumps due to pressure from diapiric uplift, the relief that these bumps create in the base of the overlying cap rock also provides a possible mechanism to minimize salt dissolution. If gases are present along the zone of dissolution, they become trapped in the molds formed by the salt bumps in the overlying cap rock. The gases most likely to be found in this environment are CH_4, CO_2, or H_2S. Salt above the gas-water contact is protected from dissolution and therefore

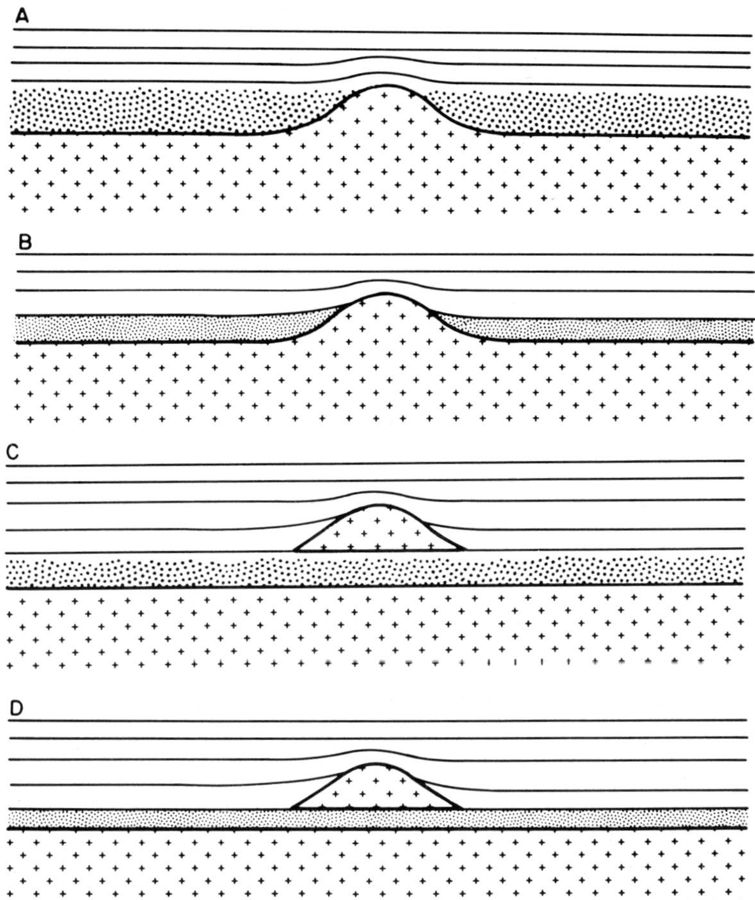

Figure 19. Model for the formation of halite mounds in laminated anhydrite. A. Salt mound and residual anhydrite sand form during salt dissolution as described in Figure 18. B. During diapiric uplift, the salt mound contacts overlying cap rock. At the salt mound/anhydrite contact, anhydrite accretion is prevented, and salt is removed by solution. C. Salt mound is separated from stock (see text for discussion); residual anhydrite sand accumulates in the brine-filled cavity. D. During the next cycle of uplift, the salt mound is sealed as an inclusion in the cap rock by accretion of the underlying anhydrite layer.

may persist over several cycles of cap rock formation. High-pressure gas inclusions within the salt from the mounds support this hypothesis; these small gas inclusions were probably trapped at the zone of dissolution. Similar inclusions were observed in salt-filled fractures in the anhydrite cap rock of the Oakwood dome (Dix and Jackson, 1982).

Salt solution probably does occur along the tops of the salt bumps from pressure solution in order to distribute pressure and space variations across the top of the salt during diapiric uplift. However, once the salt bump contacts the overlying cap rock, it then prevents significant accretion of residual anhydrite sand to the base of the cap rock in the area of contact (Fig. 19B). Salt flow and recrystallization probably also takes place along the salt-anhydrite contact.

Salt mounds within the anhydrite cap rock are believed to be simply salt bumps and spines that separated by horizontal fractures from the stock and remained in their impressions in the overlying cap rock (Fig. 19C). The 3 to 5 percent euhedral disseminated anhydrite within this salt clearly ties its origin to the salt stock. It is not clear how the salt bump or spine is separated from the stock. If the salt stock "sinks" with respect to the cap rock, then the salt mounds and spines protruding into the cap rock could be separated from the salt stock by extensional fractures. Alternatively or subsequently, the salt spines and mounds may be separated from the stock by dissolution; the resulting salt mound might be preserved from further dissolution by the presence of gas along the base of the overlying cap rock. Whatever the reason for separation, it generally occurred at the same time for all the mounds in a particular location in the cap rock as indicated by several mounds having their bases along a common anhydrite layer. During the next cycle of diapiric uplift and anhydrite accretion, the salt is sealed within the cap rock by the underplating of the new anhydrite layer.

HOMOGENEOUS ANHYDRITE MOUNDS. If the salt bump or spine which has now been cast in the overlying cap rock is removed by dissolution, then the impression in the overlying cap rock will be filled by injecting anhydrite sand into the cavity during the next period of uplift and compression (Fig. 20). Passage of low salinity water across the top of the stock would promote extensive dissolution.

ANHYDRITE BRECCIA MOUNDS. The origin of anhydrite breccia mounds is relatively straight forward (Fig. 21). Anhydrite layers within the salt were deformed into isoclinal folds during diapirism and eventually reached the zone of dissolution. As the salt dissolved, the anhydrite fragments accumulated on the solution table. The abundance of anhydrite within that portion of the salt may have made it locally more resistent to dissolution as well (Fig. 21B). During uplift, these breccia fragments along with anhydrite sand were incorporated into the cap rock as a breccia mound.

Vertical stacking of all types of mound-like features observed in quarry faces suggests that the character of the salt stock beneath that portion of the cap rock promoted mound formation (e.g. Fig. 14).

SULFIDE MOUNDS. The sulfide mounds are believed to form in the same fashion as the anhydrite mounds, but sulfide cements have filled open space associated with the mound prior to compaction. It is likely that the sulfides were introduced to the mound structures in a fashion similar to sulfide mineralization in "typical" laminated anhydrite. The salt impression provided a trap for reduced sulfur (H_2S gas). As the metalliferous brine passed through the zone of dissolution and mixed with the reduced sulfur, sulfides were precipitated in pore spaces and solution cavities associated with the salt bump. Each

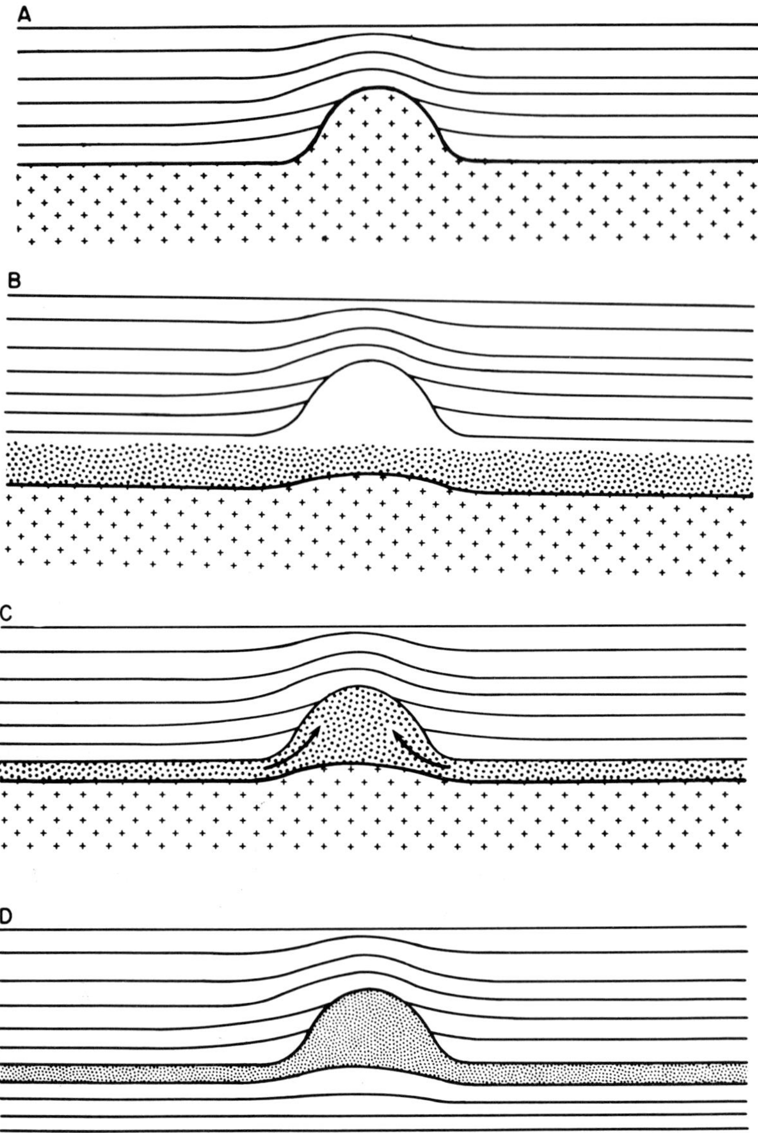

Figure 20. Model for the formation of the homogeneous anhydrite mounds. A. Salt mound forms as described in Figures 18 and 19 and persists over several cycles of cap rock formation. B. Salt mound dissolves leaving an impression in the cap rock overlying the residual anhydrite sand zone. C. During uplift, anhydrite sand is injected into the impression. D. Final structure after cap rock accretion.

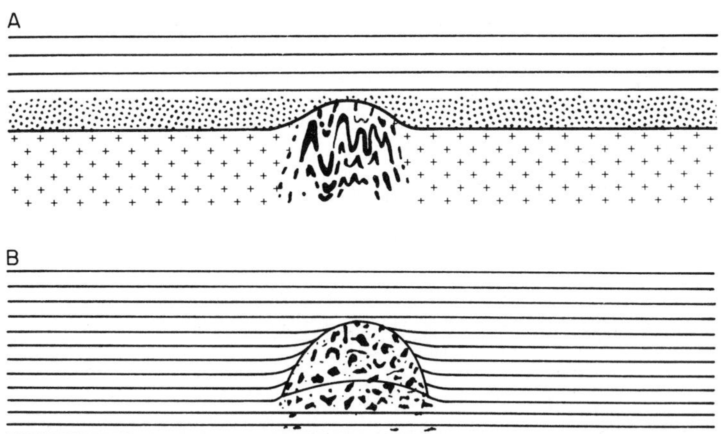

Figure 21. Model for the formation of the anhydrite breccia mounds. A. The presence of deformed primary anhydrite layers within the salt diapir promotes mound formation on salt surface. Salt-cemented breccia mound prevents accretion of anhydrite sand where mound contacts overlying cap rock. Down-building continues around the mound over several cycles of cap rock formation, forming an impression in the overlying cap rock. B. Eventually primary anhydrite is freed from salt stock by dissolution. During uplift, anhydrite breccia clasts and injected anhydrite sand fill impression and become intergrown with the cap rock.

layer represents one cycle of cap rock formation; the massive sulfide layers fill open space and the disseminated sulfides fill pore space between anhydrite sand grains.

IV. RELATION OF SULFIDE MINERALIZATION TO CAP ROCK FORMATION

Textural relationships and varying mineral composition of sulfide layers suggest that distinct pulses of metalliferous fluids entered the cap rock at the salt/anhydrite contact during anhydrite cap rock accumulation. The sample shown in Figure 22 provides an example of the preserved morphology of this contact. The upper portion of this sample is composed of tightly interlocking, xenoblastic anhydrite. No sulfides are contained within this layer. Coarse, subhedral pyrrhotite, with local small cavities between grains, occurs beneath this anhydrite layer. The coarse pyrrhotite is immediately underlain by xenoblastic anhydrite and disseminated, finer grained pyrrhotite. Some of these pyrrhotite grains cement inclusions of anhydrite (Fig. 23). The anhydrite inclusions are euhedral and undeformed, similar to the anhydrite disseminated throughout the salt stock (Fig. 5). The anhydrite grains surrounding the pyrrhotite are deformed and tightly intergrown with sutured grain contacts. Anhydrite grains that are partially cemented by pyrrhotite retain euhedral crystal faces only within the pyrrhotite. Pyrrhotite becomes finer grained and decreases in abundance downward through the sample.

In Figure 22B, the sample is shown in the setting of the salt/anhydrite contact during the time of its incorporation into the cap rock (the effects of compaction have not been removed from this figure). The uppermost anhydrite layer of the sample represents the bottom of the anhydrite cap rock. This layer was accreted to the base of the cap rock during the

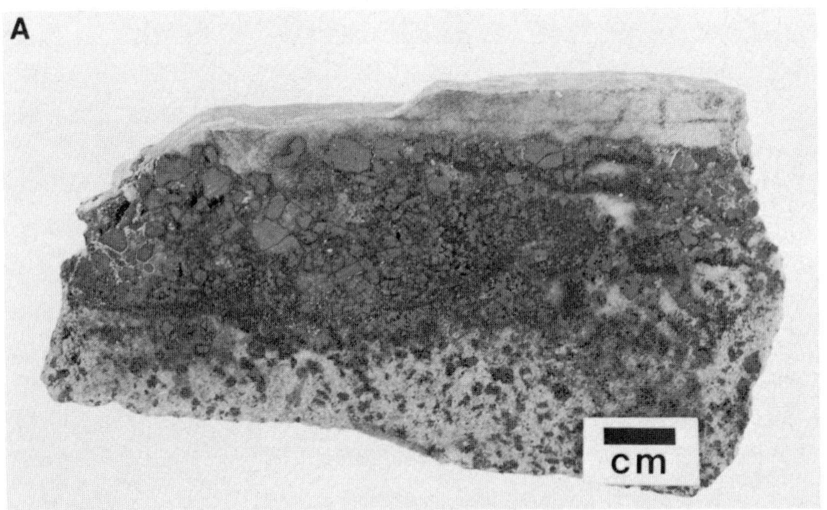

Figure 22. Pyrrhotite-rich anhydrite cap rock. A. Polished slab. B. Line drawing showing interpretation of the significance of the textures relative to processes of formation at the salt/anhydrite interface. See text for explanation.

B
CAP ROCK
COMPACTED ANHYDRITE
OPEN SPACE
METALLIFEROUS BRINE
ANHYDRITE SAND
INCREASINGLY CEMENTED BY SALT
ZONE OF SALT DISSOLUTION
SALT STOCK
SALT WITH DISSEMINATED ANHYDRITE

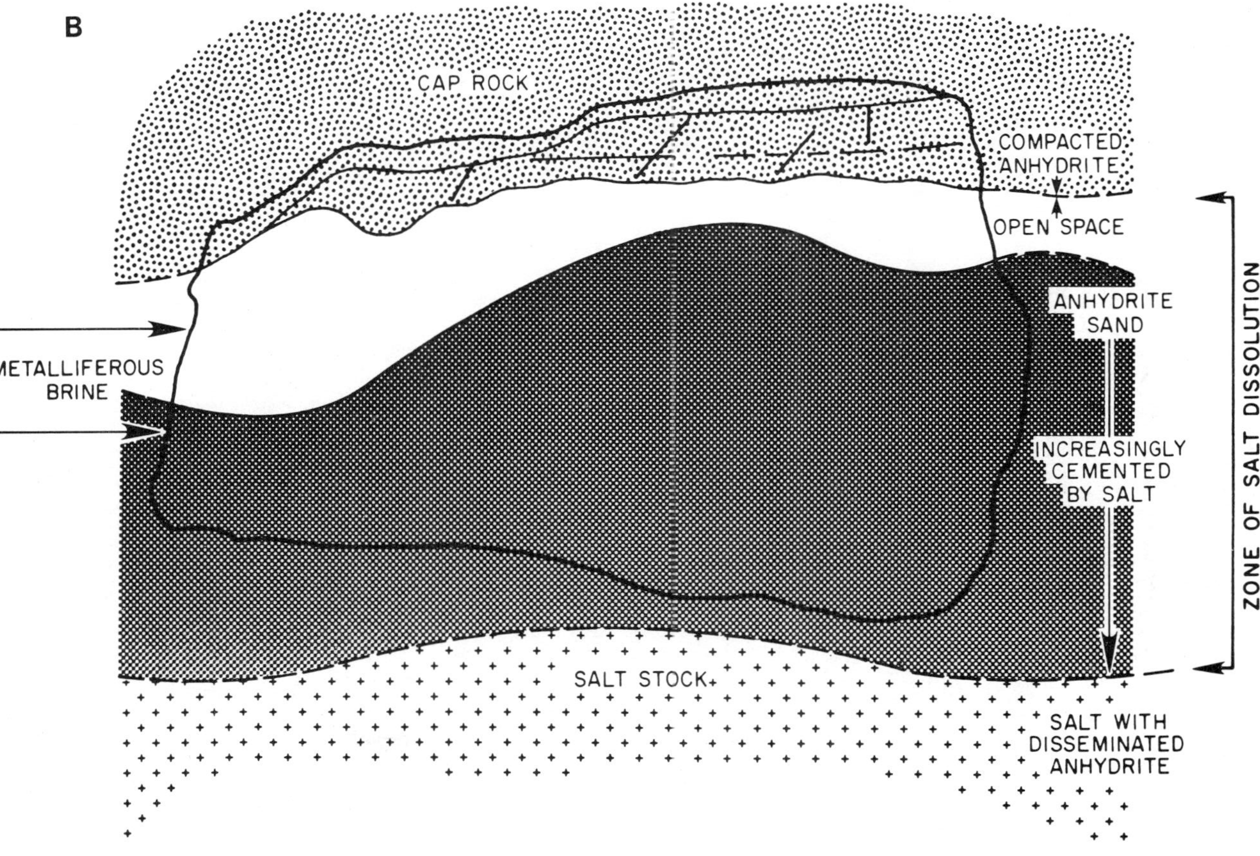

Figure 22B.

Figure 23. Transmitted light photomicrograph of euhderal undeformed anhydrite grains within pyrrhotite (po). Anhydrite (an) outside of the pyrrhotite grain has the interlocking xenoblastic texture typical of the anhydrite cap rock. Scale = 0.5 mm.

previous cycle of dissolution and uplift. The coarse pyrrhotite in the sample fills a solution cavity through which fluids responsible for salt dissolution and sulfide mineralization passed. An accumulation of the less soluble anhydrite sand covered the bottom of the solution cavity. Metalliferous fluids percolating through this solution cavity precipitated sulfides, filling open spaces and cementing anhydrite grains. The anhydrite sand may have been increasingly cemented by salt towards the top of the salt stock thus preventing complete cementation by sulfides.

With continued uplift and compaction, this anhydrite-sulfide package was accreted to the bottom of the overlying cap rock.

The sulfide-cemented anhydrite grains were armored from deformation during this and succeeding periods of compression.

This mechanism for sulfide emplacement implies that each sulfide layer within the anhydrite cap rock represents a distinct pulse of metalliferous fluid. Intervals between pulses, where only salt dissolution occurred, are represented by unmineralized laminated anhydrite. The earliest formed sulfides occur near the top of the anhydrite cap rock while the more recent sulfides occur near the base. Differing Fe-Zn-Pb ratios for sulfide layers suggest that metalliferous fluids of varying composition were introduced periodically along the salt/anhydrite contact.

V. PALEOMAGNETIC DATA

The sequential accretion of the cap rock and the presence of the magnetic mineral pyrrhotite suggested the intriguing possibility that a paleomagnetic study might provide considerable insight into the timing and rate of anhydrite cap rock formation. Encouraged by preliminary results (Gose et al., 1985), we collected additional samples for a total of 308 samples. Importantly, the new samples increased the stratigraphic coverage from 15 m to 40 m.

All samples were measured with a cryogenic magnetometer and subjected to stepwise alternating field (AF) demagnetization up to 30 mT and, in some cases, up to 50 mT. The intensities of magnetization varied from 1×10^{-5} to 5 A m^{-1} with most values falling into the 10^{-3} to 10^{-2} A m^{-1} range. The data were analyzed using orthogonal vector projections, examples of which are shown in Fig. 24.

Before interpreting the results, it is essential to assess whether corrections have to be applied to account for small-scale structures such as mounds as well as the overall doming of the cap rock. Figure 25 shows a line drawing of the sulfide layers around one mound (Fig. 13) and the locations of paleomagnetic

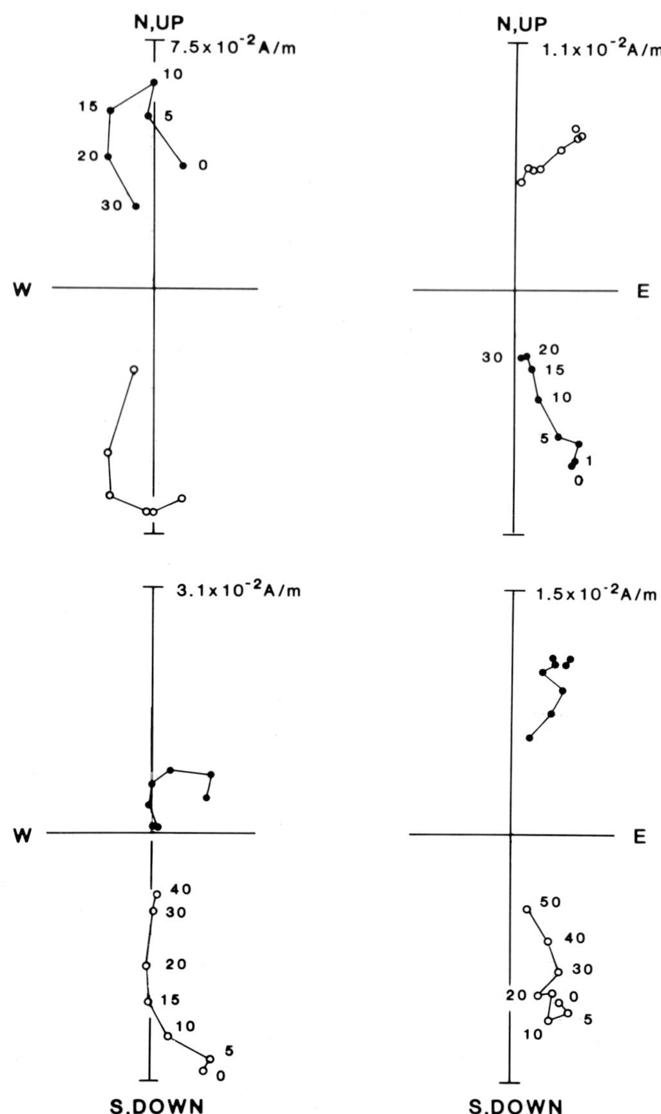

Figure 24. Orthogonal projections of the magnetic vector of the anhydrite cap rock samples. Open circles are in the N-E-S-W plane, and solid circles in the Up-E-Down-W plane. Numbers indicate the demagnetizing field in mT.

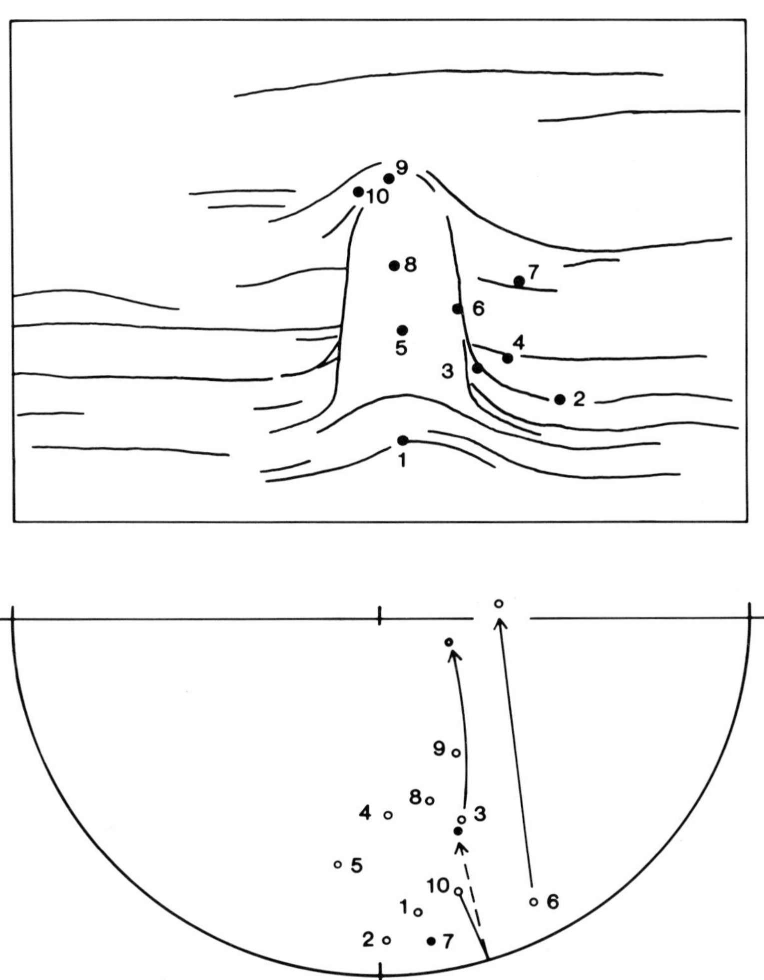

Figure 25. Line drawing of an homogeneous anhydrite mound showing paleomagnetic sample locations. See Fig. 13 for photograph of the mound. Bottom figure shows an equal area stereographic projection of the directions of magnetization. Open circles are in the upper hemisphere, solid circles are in the lower.

samples. In the stereographic projection, the directions of magnetization are displayed with out tilt corrections. For the three samples taken from steeply dipping sulfide layers, arrows point to the directions after correction for tilt. Clearly, the grouping is better for the uncorrected data implying that the mound is a primary structure and that the sulfide layers were not rotated after their formation. Furthermore, the interior part of the mound, consisting of deformed anhydrite fragments in anhydrite sand matrix, has the same direction of magnetization as the undisturbed surrounding area.

About two thirds of the samples were collected along the eastern face of the quarry where the strata have an easterly dip of 5° to 15°. The other samples were taken along the western exposure where the bedding dips about 20° to 30° to the west. The different bedding attitudes make it possible to apply the fold test which indicates whether magnetization was acquired before or after deformation. If the directions cluster better after unfolding, expressed statistically by an increase in the precision parameter k (a measure of the tightness of the cluster), then the magnetization was acquired before folding. For the Winnfield data, k is significantly larger if the structural correction is not applied. This is true for the total data population as well as for the normal and reversed subsets. This relationship implies that the anhydrite was magnetized before folding and that the domal structure of the anhydrite represents the original salt-cap rock interface.

The statistics in Table 1 are based on 107 samples rather that the 308 samples that were collected. There are three reasons for this discrepancy:

1. Samples from the old part of the quarry did not yield usable data due to pervasive weathering.
2. In order to increase the stratigraphic coverage, some samples were collected while standing in the shovel of a front end loader which affected the magnetic orientation of the samples. The magnetic polarity of these samples, however, could still be assessed.
3. While the data analysis clearly revealed the magnetic polarity of most samples, many samples did not yield a well-defined characteristic direction of magnetization. Figure 26 depicts the directions of the samples included in the statistics.

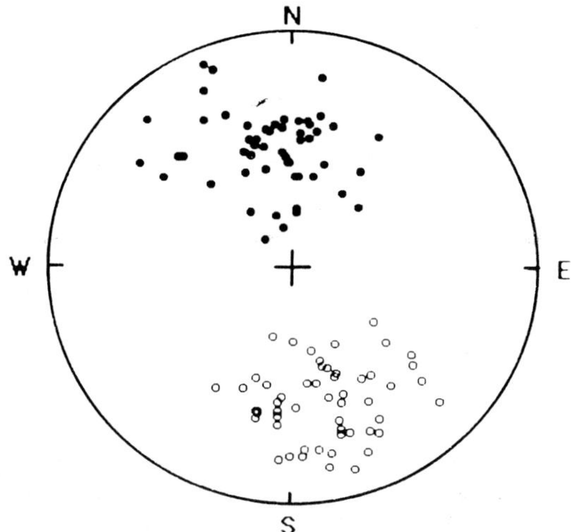

Figure 26. Schmidt equal area projection of the directions of the stable component of magnetization. Filled (open) circles are in the lower (upper) hemisphere. The directions are plotted without structural correction.

V. TIMING OF CAP ROCK FORMATION AT WINNFIELD DOME

A. Stratigraphic Relations Attributed to Halokinesis

Movement of bedded salt into halokinetic structures influences the thickness and facies of strata deposited immediately overlying and surrounding the growing salt structure (Seni and Jackson, 1983a,b). Thus, structural development can be dated by lithostratigraphic effects produced in the sediments that accumulated in the withdrawal basins or rim synclines formed by salt migration. However, it should be noted that the exact age of sedimentary units, particularly for fossil-poor siliciclastic strata, is difficult to define precisely. Furthermore, the units are generally time transgressive, as has been amply demonstrated by Coleman and Coleman (1981) for the relevant units in the North Louisiana Basin. Therefore, the timing of halokinetic events using these types of sedimentation events may not be exact relative to absolute geologic time. Nevertheless, this information provides a means of testing the results of the paleomagnetic investigations of the Winnfield dome

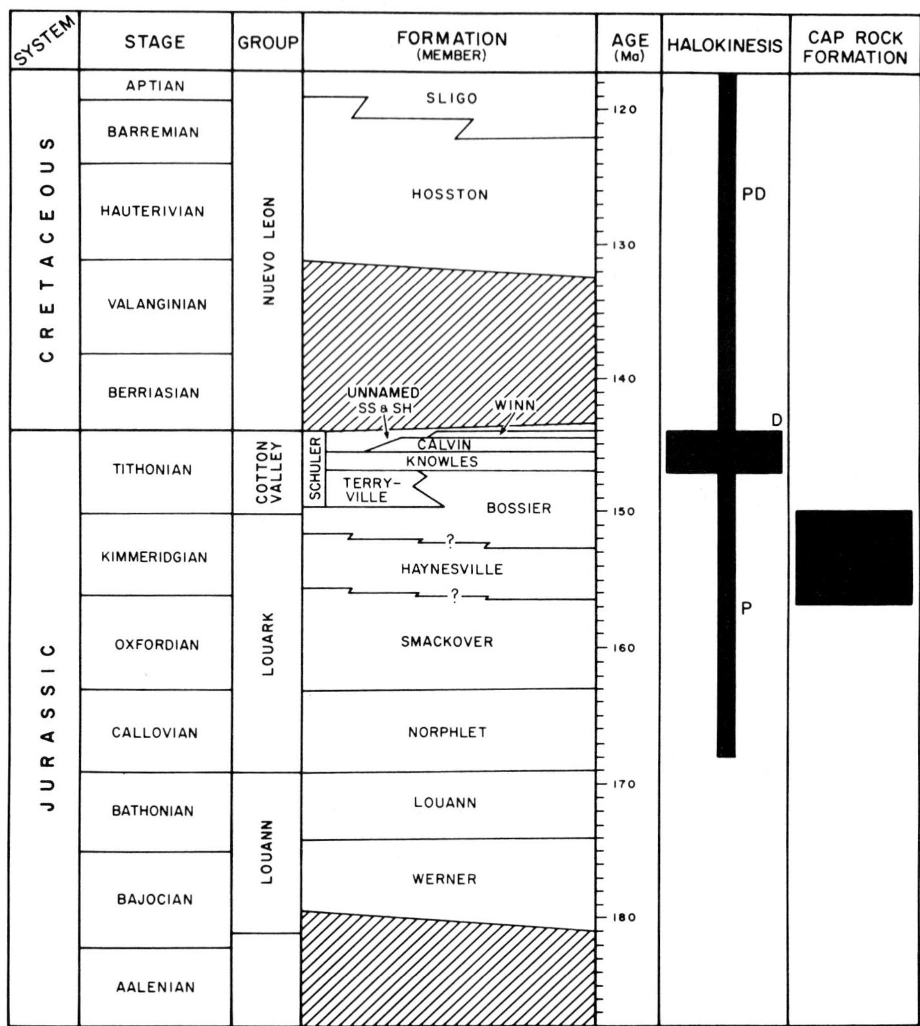

Figure 27. Comparison of the age of anhydrite cap rock formation as determined by paleomagnetic analysis with the age of halokinesis as determined from stratigraphic relations for Winnfield dome. Stratigraphic age of halokinesis after Lobao and Pilger (1985) modified as to absolute geologic age according to Harland et al. (1982), Palmer (1983), and Imlay (1980). P = Pillow Stage; D = Diapir Stage; PD = Postdiapir Stage.

cap rock. A comparison of the timing information derived by these methods for the North Louisiana Basin is provided in Fig. 27.

Hessenbruch (1975), using seismic and "scanty" electric log drill hole data, was the first to attempt to develop a growth history for Winnfield dome. He concluded that sediment response to withdrawal of salt from the southern portions of Milams sub-basin suggests that dome growth reached a maximum during the Late Jurassic and diminished gradually throughout the Early Cretaceous.

Lobao and Pilger (1985) used regional seismic and drill hole data to develop a detailed history of halokinesis in the North Louisiana Basin. Their analysis suggests that the main phase of diapir growth in the southern part of the North Louisiana Salt Basin occurred in the Late Jurassic during the deposition of the Calvin siliciclastics. Using 141 Ma for the age of the Jurassic/Cretaceous boundary, Lobao and Pilger (1985) assigned an age range from 141.5 to 142.5 Ma to the Calvin Sandstone. The implied accuracy is rather misleading because the age of the Calvin is poorly constrained (Coleman and Coleman, 1981). However, it is reasonable to conclude that the deposition of the Calvin sandstone represents a brief time span in the latest Jurassic. Before the deposition of the Calvin, the Winnfield salt pillow was breached by erosion as evidenced by a seismic isochore map of the Late Jurassic Smackover through Bossier Formations (Lobao and Pilger, 1985). This period of salt dissolution on the sea floor may correspond to the formation of the "marine false cap rock" that constitutes the upper part of the Winnfield calcite cap rock and contains abundant detritus probably derived from the salt mass (Figs. 9,10). Varying amounts of post-diapir halokinesis is indicated for the North

Louisiana Basin domes with salt movement extending into the early Tertiary at a continually decreasing rate (Lobao and Pilger, 1985).

The East Texas Basin appears to have evolved in a similar fashion and time to the North Louisiana Basin. Seni and Jackson (1983a,b) developed a detailed growth history for salt structures in the East Texas Basin and determined that salt pillows began to form during deposition of the Gilmer Limestone (Oxfordian). Diapirism was initiated by and reached a maximum during progradation of the Schuler and Hosston clastics during Tithonian through Aptian time. However, the age (115-137 Ma) assigned to this period by Seni and Jackson (1983 a,b) is considerably longer and younger than that assigned by Lobao and Pilger (1985) to the maximum diapiric event in North Louisiana, even though similar depositional units are involved. This discrepancy alludes to the problems with assigning exact ages to these siliciclastic formations. Salt diapirism in the East Texas Basin continued throughout the Cretaceous and Tertiary at a continually decreasing rate.

Anhydrite cap rock development presumably begins when the salt stock enters a zone with water undersaturated with respect to halite. However, anhydrite solubility data (Blount and Dickson, 1973) suggest a variety of scenarios that could also result in dissolution of anhydrite as well as halite in this geologic environment. Nevertheless, if it is assumed that periods of movement of large volumes of salt correspond to periods of extensive salt dissolution and anhydrite accumulation, then the timing of most intense cap rock formation at Winnfield dome was during latest Jurassic. During this time, deposition of the Calvin Sandstone and other siliciclastic units produced rapid dome growth. It is likely that the top of the salt stock remained relatively near the sediment-water interface where circulating pore waters allowed salt dissolution to proceed rapidly. Anhydrite cap rock accumulation may have continued throughout the Cretaceous and into the Tertiary but at a continually decreasing rate.

VI. PALEOMAGNETIC EVIDENCE

The age of the cap rock formation can be estimated from the paleomagnetic data in two different ways, by comparing the pole position with the North American apparent polar wander path (APWP) and by matching the reversal sequence with the sea floor magnetic anomaly pattern.

Figure 28 shows the APWP based on the compilation of Irving and Irving (1982) and the pole positions calculated from the normally and reversely magnetized samples with their 63% error limits (= one standard deviation). It is quite apparent that the pole position from the normal samples is shifted towards the geographic pole relative to the reversed pole position. This shift results because the normal direction is very close to the present field direction and it is not always clear whether the

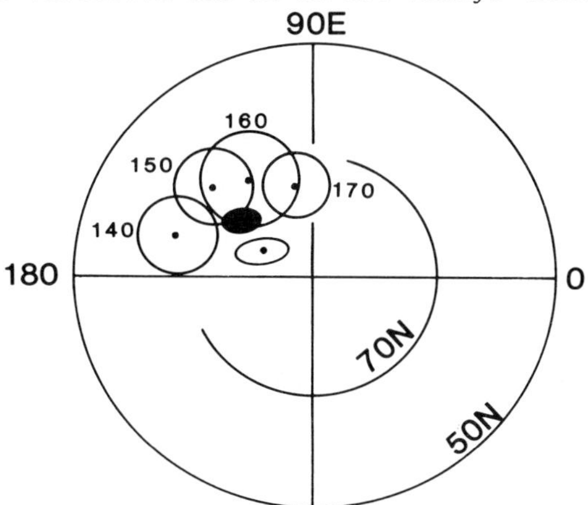

Figure 28. Comparison of the Winnfield pole positions with the polar wander path for stable North America (after Irving and Irving, 1982). The reference poles from 140 to 170 Ma are depicted with their 63% circles of confidence. The black (white) ellipse represents the pole position based on the reversed (normal) samples at the 63% confidence level. The reversed Winnfield pole position implies that these rocks were magnetized between 160 and 150 Ma.

TABLE 1: Statistical parameters of paleomagnetic analysis

	N	R	D	I	k	α_{95}	LAT	LONG
NORMAL + REVERSE								
NSC	107	99.72	348.2	42.5	14.6	3.7	77.3	145.3
SC	107	97.19	346.8	38.1	10.8	4.4	74.3	138.8
NORMAL								
NSC	53	49.26	350.1	45.9	13.9	5.4	80.3	151.9
SC	53	47.70	345.7	42.4	9.8	6.6	75.4	150.7
REVERSE								
NSC	54	50.67	166.6	-39.2	15.9	5.0	74.6	141.4
SC	54	49.76	167.8	-34.0	12.5	5.7	72.8	129.9

N = number of samples; R = resultant vector; D = declination; I = inclination;

k = precision parameter; α_{95} = radius of 95% circle of confidence; LAT = latitude; LONG = east longitude;

NSC = no structural correction; SC = structurally corrected.

primary magnetization has been isolated during demagnetization or whether the sample has been remagnetized. The error ellipse of the reversed pole position overlaps the error circles of the pole positions for 150 and 160 Ma.

When plotted stratigraphically, the paleomagnetic data define a distinct reversal sequence (Fig. 29). The unique correlation between the reversal patterns of the two quarry faces (columns 1 and 2) demonstrates the power of magnetostratigraphy, even when visual correlation is ambiguous as is the case at Winnfield. Column 3 in Fig. 29 depicts the combined Winnfield profile which has been inverted to facilitate correlation with the seafloor magnetic anomalies (column 4). The reversal pattern correlates rather well with the seafloor magnetic anomalies M20 through M22. Using the time scale of Harland et al (1982), an age range of 157 to 150 Ma is indicated for the formation of the sampled cap rock sequence, which is in excellent agreement with the age obtained from the pole position. Also it should be noted that Ogg and Steiner (1984) assigned ages to these reversals which are about 4 Ma younger.

The identification of the reversal pattern also makes it possible to calculate an average accretion rate for the anhydrite cap rock of 6 m/Ma. This value is considerably less than estimated growth rates for salt diapirs. Seni and Jackson (1984) calculated maximum growth rates in the range of 100 to 500 m/Ma for the East Texas Basin salt domes. Anhydrite concentrations in salt stocks vary between 2 and 10 volume percent (Taylor, 1937; Murray, 1966), and the Winnfield salt contains an average of 3 percent (Belchic,1960). If there is negligible loss of anhydrite during cap rock formation, the anhydrite accumulation rates should be 10 to 50 times less, i.e. in the range of 2 to 50 m/Ma. Thus, the value obtained from the paleomagnetic data is most compatible with previous independent estimates.

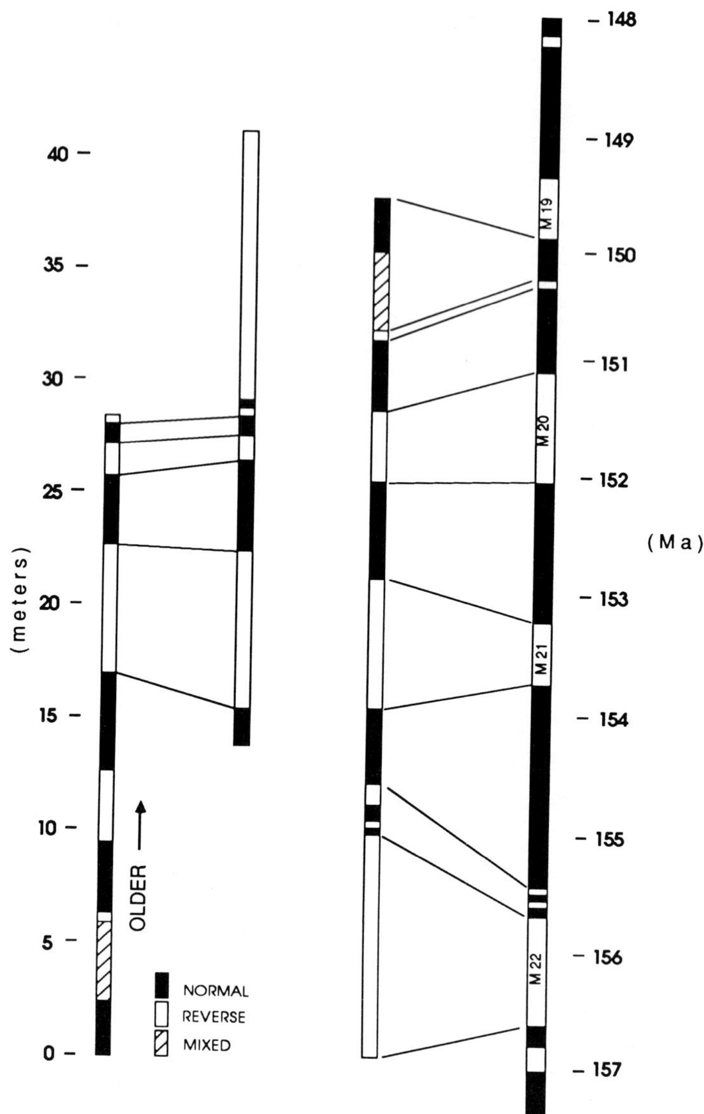

Figure 29. Reversal stratigraphy observed in the anhydrite. Left: correlation between the east face (column 1) and west face (column 2) of the quarry. Right: proposed correlation of the composite and inverted Winnfield section (column 3) with the sea floor magnetic anomaly time scale (column 4) of Harland et al. (1982).

The sampled anhydrite zone occurs roughly in the middle of the cap rock sequence (Fig. 3). At least 60 m of older cap-rock strata overlies the sampled sequence, without accounting for anhydrite that might have been lost via dissolution, including conversion to calcite cap rock. Therefore, if the 6 m/Ma growth rate is assumed for the total anhydrite cap rock, then the time of initiation and growth of the cap rock is extended considerably older than 157 Ma (Fig. 27). It might be speculated that the early stages of salt diapirism and anhydrite cap rock formation may proceed at more rapid rates, although the contemporaneous sedimentation, i.e. loading, rate should be the primary control. Similarly, the younger anhydrite strata that occur between the sampled zone and the present salt table may include strata that accumulated at progressively slower rates and at considerably younger geologic ages.

VII. CONCLUSIONS

The textures within the anhydrite cap rock of Winnfield dome are the result of cycles of halite dissolution and residual anhydrite accumulation at the top of the salt stock. Various types of domal structures within the anhydrite provide information on the contemporaneous local salt table topography. Periodic introduction of metalliferous fluids along the salt/anhydrite contact resulted in the formation of stratiform sulfide laminae. This systematic record of anhydrite accretion serves as the basis of this study to provide unique information on the mechanism and timing of anhydrite cap rock development. The paleomagnetic results support the petrographic/geochemical model and, in addition, provide critical age constraints on the accumulation of the anhydrite and thus the growth history of the salt dome. The correlation of the reversal pattern with the sea floor magnetic anomalies M20 through M22 implies that the sampled portion of the anhydrite formed between 157 and 150 Ma. The same age is obtained by comparing the paleomagnetic pole position with the apparent polar wander path for stable North America. If the time of anhydrite accretion corresponds to periods of major salt movement, then this age is also the time of diapir formation. In addition, the paleomagnetic data establish an average accretion rate of 6 m/Ma for the anhydrite.

VIII. ACKNOWLEDGMENTS

Research at Winnfield dome has been supported by National Science Foundation grant EAR-8407736 to J. R. Kyle. We are grateful to Marathon Oil Company for support of the initial geologic investigations and to Chevron Oil Field Research Company for support of the reconnaissance paleomagnetic investigations of the Winnfield cap rock. We acknowledge many discussions with our colleagues, P. E. Price and H. H. Posey, that contributed to the overall understanding of the origin of salt dome cap rocks. M. P. A. Jackson generously reviewed the manuscript and made several valuable comments. We also greatly appreciate the access to the quarry provided by Winn Rock, Inc., and many courtesies provided by Ray Daughtry during our trips to Winnfield. This article is the University of Texas Institute for Geophysics contribution number 668.

REFERENCES

Belchic, H. (1960). The Winnfield salt dome, Winn Parish, Louisiana, in Guidebook 1960 Spring Field Trip, Shreveport Geol. Soc., p. 29-48.

Berryhill, R. A., Champion, W. L., Meyerhoff, A. A., Sigler, G. C., and other Shreveport Geological Society Members, (1968), Stratigraphy and selected gas-field studies of North Louisiana, in Beebe, B. W., ed., Natural gases of North America: Am. Assoc. Petrol. Geol., Mem. 9, 1099-1137.

Blount, C. W., and Dickson, F. W. (1973). Gypsum-anhydrite equilibria in systems $CaSO_4$-H_2O and $CaSO_4$-NaCl-H_2O: Am. Mineral. 58, 323-331.

Cameron, H. V. (1949). The Winnfield, Louisiana, salt dome: unpubl. Masters thesis, Louisiana State Univ., Baton Rouge, 87 p.

Coleman, J. L. Jr., and Coleman C. J. (1981). Stratigraphic, sedimentologic and diagenetic framework for the Jurassic Cotton Valley Terryville massive sandstone complex, northern Louisiana, in Maggio, C. M., ed., Trans. Gulf Coast Assoc. Geol. Socs. 31, 71-79.

Dix, O. R. and Jackson, M. P. A. (1982). Lithology, microstructures, fluid inclusions and geochemistry of rock salt and of the cap rock contact in Oakwood Dome, East Texas: significance for nuclear waste storage: Univ. of Texas at Austin, Bur. Econ. Geol., Rept. Inv. 120, 59 p.

Eversull, L. G. (1984). Regional cross sections, North Louisiana: Louisiana Geol. Surv., Baton Rouge, Folio 7, 10 p.

Feely, H. W. and Kulp, J. L. (1957). Origin of the Gulf Coast salt dome sulphur deposits: Am. Assoc. Petrol. Geol. Bull. 41, 1802-1853.

Goldman, M. I. (1952). Deformation, metamorphism, and mineralzation in gypsum-anhydrite cap rock, Sulphur salt dome, Louisiana: Geol. Soc. Amer. Memoir 50, 169 p.

Gose, W. A., Kyle, J. R. and Ulrich, M .R. (1985). Preliminary paleomagnetic investigation of the Winnfield salt dome, Louisiana, in Ewing, T.E., ed., Trans. Gulf Coast Assoc. Geol. Socs. 35, 97-106.

Imlay, R. W. (1980). Jurassic paleobiogeography of the conterminous United States in its continental setting: U. S. Geol. Surv., Prof. Paper 1062, 134 p.

Irving, E., and Irving, G. A. (1982). Apparent polar wander paths Carboniferous through Cenozoic and the assembly of Gondwana: Geophys. Surveys 5, 141-188.

Harland, W. B., Cox, A., Llewellyn, P. G., Pickton, C. A. G., Smith, A. G. and Walters, R. (1982). A geologic time scale: Cambridge Univ. Press, 131 p.

Hessenbruch, J.M. (1975). Salt tectonics of Winn Parish, Louisiana: unpubl. Masters thesis, Louisiana State Univ., Baton Rouge, 176 p.

Hoy, R.B., Foose, R.M., and O'Neill, J.B., Jr. (1962). Structure of Winnfield salt dome, Winn Parish Louisiana. Am. Assoc. Petrol. Geol. Bull. 46, 1444-1459.

Kyle, J. R., and Price, P. E. (1986). Metallic sulphide mineralization in salt-dome cap rocks, Gulf Coast, U.S.A.: Trans. Instn. Min. Metall. 95, Sect. B, p. B6-B16.

Lobao, J. J., and Pilger, R. H., Jr. (1985). Early evolution of salt structures in the North Louisiana salt basin, in Ewing, T.E., ed., Trans. Gulf Coast Assoc. Geol. Soc. 35, 189-198.

Martinez, J. D. (1980). The nature and evolution of salt domes and their caprock: Louisiana Geol. Surv., Baton Rouge, Geol. Pamp. 6, 16 p.

Murray, G. E., Jr. (1966). Salt structures of the Gulf of Mexico Basin - a review: Am. Assoc. Petrol. Geol. Bull. 50, 439-478.

Ogg, J. G., and Steiner, M. B. (1984). Jurassic magnetic polarity time scale: current status and compilation, in Michelsen, O., and Zeiss, A., eds., International Symposium on Jurassic Stratigraphy: Geol. Surv. Denmark, Copenhagen 3, 777-794.

Palmer, A.R., director, (1983). Geologic time scale: Decade of North American Geology. Geol. Soc. Amer., 1 p.

Posey, H. H., Price, P. E., and Kyle, J. R. (1987) (this volume). Mixed carbon sources in calcite cap rocks of Gulf Coast salt domes, in Lerche, I., and O'Brien, J., eds., Dynamical Geology of Salt and Related Structures: Academic Press, Orlando.

Seni, S. J., and Jackson, M. P. A. (1983a). Evolution of salt structures, east Texas diapir province, part 1, sedimentary record of halokinesis: Am. Assoc. Petrol. Geol. Bull. 67, 1219-1244.

Seni, S. J., and Jackson, M. P. A. (1983b). Evolution of salt structures, east Texas diapir province, part 2, patterns and rates of halokinesis: Am. Assoc. Petrol. Geol. Bull. 67, 1245-1274.

Seni, S. J. and Jackson, M. P. A. (1984). Sedimentary record of Cretaceous and Tertiary salt movement, East Texas Basin: Univ. of Texas, Austin, Bureau of Economic Geology, Rept. Inv. No. 139, 89 p.

Taylor, R. E. (1937). Water-insoluble residues in rock salt of Louisiana salt plugs: Am. Assoc. Petrol. Geol. Bull. 21, 1268-1310.

Ulrich, M. R., Kyle, J. R. and Price, P. E. (1984). Metallic sulfide deposits in the Winnfield salt dome, Louisiana: evidence for episodic introduction of metalliferous brines during cap rock formation, in White, B. R., ed., Trans. Gulf Coast Assoc. Geol. Soc. 34, 435-442.

EVOLUTION OF BOLING DOME CAP ROCK WITH EMPHASIS ON INCLUDED TERRIGENOUS CLASTICS, FORT BEND AND WHARTON COUNTIES, TEXAS

Steven J. Seni

Bureau of Economic Geology
The University of Texas at Austin
University Station Box X
Austin, TX 78713

I. INTRODUCTION

The crest and upper flanks of shallow salt domes are usually overlain by cap rocks which are residual accumula-tions of salt stock impurities remaining after halite dissolution as well as products of complex diagenetic reactions. Stratified sediments and sulfide deposits occur within some cap rocks. Cap rocks occupy a range of subsurface environments from deep basinal to near surface in the vadose zone. A cap rock perched on the crest of a rising diapir is likely to have been transported through this range of subsurface environments during accumulation. A single cap rock may extend vertically through hundreds to thousands of meters and thus simultaneously be affected by a wide range of environmental conditions. Evolution of cap rock is influenced by a complex range of physical-chemical processes which vary in both time and space. The three-dimensional internal geometry of cap rock lithofacies provides clues for deciphering some of these complex diagenetic and geodynamic processes. Boling Dome cap rock is appropriate for such a study because of the availability of core data and because an understanding of these processes is critical for determining the suitability of using Boling Dome for disposal of toxic waste (Seni et al., 1984; Seni et al. 1985).

Boling Dome cap rock contains unusually large inclusions of terrigenous clastic strata. Dimensions include thicknesses exceeding 30 m and lengths of hundreds of meters. Understanding the mechanism by which included terrigenous clastics are incorporated in cap rocks would provide additional evidence pertaining to the evolution of cap rock and intrusion of Boling Dome salt stock.

Boling Dome underlies approximately 100 mi^2 (250 km^2) of eastern Wharton County and a small part of southwest Fort Bend County on the western margin of the Houston diapir province (Figure 1). Boling Dome hosts the only currently active Frasch sulfur operation from domal cap rock in Texas. Stratigraphic data from approximately 200 oil wells and cores from approximately 100 closely spaced sulfur-production holes (Figure 2), loaned by Texasgulf Chemicals Co., form the basis for this

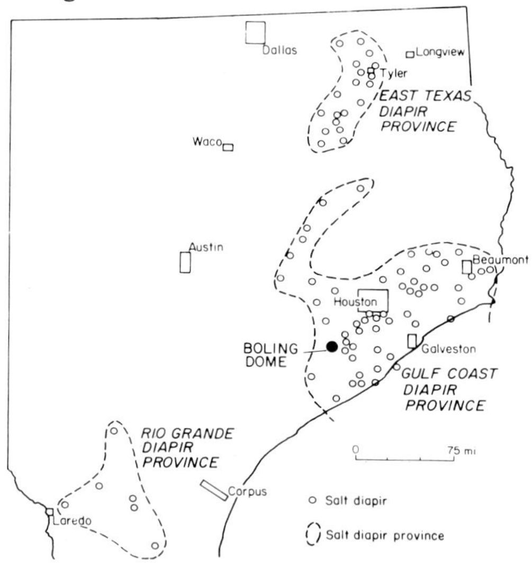

Figure 1. Location map of Gulf Coast diapir province, Gulf Interior Basins, and Frasch sulfur domes.

study. Cores from Boling Dome are from the area of active sulfur mining, and were collected from 1984 to 1986. Texasgulf Chemicals Co. is one of several companies with an extensive history of sulfur mining at Boling Dome. Texasgulf has drilled more than 8,500 core holes at Boling for production of sulfur by the Frasch method (F. Samuelson, Texasgulf Chemicals Co., personal communication, 1986). The cores are from a relatively narrow band oriented east-west along the strike of dome structure on the southern dome flank (Figure 2). The cores intersect the cap rock "shoulder," between the steeply dipping deep dome flank and the nearly horizontal dome crest. Additional cap-rock lithofacies data are also presented for Moss Bluff Dome (Figure 1). Moss Bluff Dome in Liberty County was another exceptional producer of sulfur within the Texas Gulf Coast. Sulfur production was from 1948 to 1982 at Moss Bluff Dome and from 1929 to present at Boling Dome. This long history of resource recovery has resulted in collection of an extensive suite of data (Railroad Commission of Texas Docket No. 03-72,099).

By concentrating on reconstruction of the three-dimensional geometry of internal cap-rock lithofacies this study attempts to decipher the evolution of Boling Dome cap rock and included sedimentary strata. Core from the cap rock and surrounding strata was studied 1) to determine the stratigraphy and facies of cap rock by binocular examination of core slabs, petrographic analysis of thin sections, SEM analysis, and geochemical analysis; 2) to integrate these findings to better understand cap rock evolution in general and the evolution of Boling Dome cap rock in particular.

Mineral identifications were made with standard thin sections, SEM grain mounts of acid-insoluble residue, and X-ray diffraction of $< 62\ \mu$ fraction on oriented clay slides. Thin sections were stained with alizarn red S for calcite, dolomite, and ferroan-calcites. All carbonate minerals were calcite.

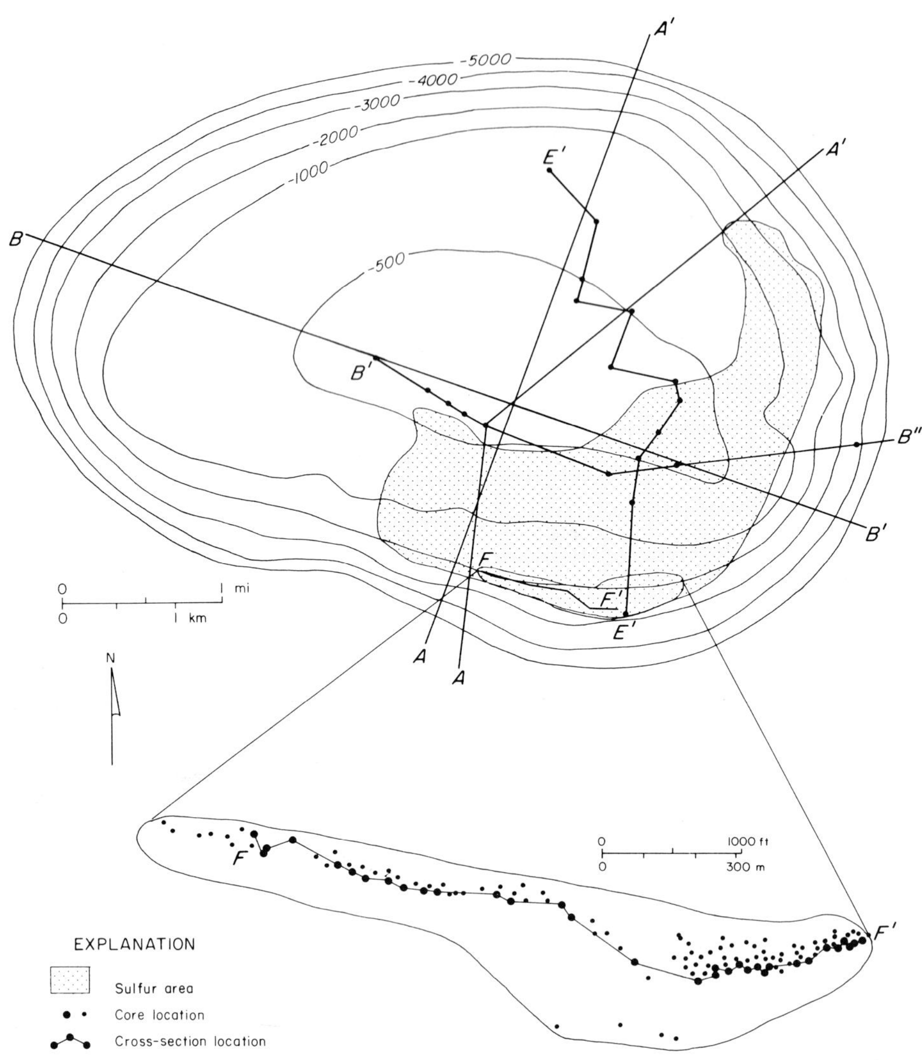

Figure 2. Location map of Boling Dome, cross sections, and area of sulfur production in relation to salt structure contours.

Included cap rock samples were stained with Na-cobaltinitrite to identify potassium feldspars and distinguish them from albite and quartz. One-hundred points were accumulated on framework detrital grains for most thin sections; total counts for individual sections ranged from 211 to 1033. All data on grain size, cement percentages, fabrics, and textures are based on thin section point-counts. Microprobe point count techniques are probably necessary to quantitatively and reliably distinguish quartz from albite. Although such techniques were not employed, SEM energy dispersive semi-quantitative analysis of random grain mounts revealed very low albite percentage. SEM studies employed an energy-dispersive semi-quantitative analyzer on acid-insoluble residues from included cap rock. X-ray diffraction analysis was performed on twenty-five oriented samples of clay-sized fraction from Boling cap rock and sediments surrounding Boling and Damon Mound Domes. Samples were air dried, run, and reanalyzed for expandable clays after ethylene glycolation.

A. General Model of Cap Rock Evolution

Historically significant salt dome and cap rock studies (Taylor, 1938; Goldman, 1933, 1952; Feely and Kulp, 1957; Murray, 1966; Bodenlos, 1970; Kreitler and Dutton, 1983; Price et al., 1983) have developed a model of cap rock evolution which accounts for the commonly observed cap rock lithologic sequence of anhydrite, gypsum and calcite. Two major mechanisms are inferred, 1) dissolution of the crest of the salt stock and accumulation of an anhydrite residuum and 2) reaction of the anhydrite with hydrocarbons, bacteria, and subsequent reaction products to precipitate calcite, gypsum, sulfur, and relatively minor sulfide phases (Murray, 1966; Price et al., 1983). The evolutionary pathway first requires that the salt stock enter a zone of halite-undersaturated, rapidly circulating, ground water. Halite at the crest of the salt stock is dissolved and removed. Cap-rock accumulation is concentrated in a planar zone at the diapir crest, thus shallow, circulating ground waters are thought responsible for most halite dissolution. Deep-basin fluids expulsed up the flanks of the salt stock may aid salt dissolution and cap-rock accretion if sufficient volumes of low-salinity fluids are available.

1. Cap-Rock Reactions

Relatively insoluble phases within the salt stock, principally anhydrite, accumulate as dissolution truncates the crest of the salt stock. Successively younger layers of anhydrite are added to the base of the cap rock. The earliest American workers on cap rocks (Goldman, 1925; Teas, 1931) reasoned, on the evidence of clearly sedimentary terrigenous clastics within cap rocks, that cap rocks were plugs of sedimentary rocks which were uplifted by the salt stock. The recognition that the cap rock rests on a flat-topped solution table that decapitates folds at the top of the salt stock solidified the current interpretation of the origin of anhydrite cap rock as a dissolution residuum (Goldman, 1933, 1952; Taylor, 1938; Feely and Kulp, 1957). Many domes contain a zone of loose anhydrite sand at the base of the cap rock where active salt dissolution is removing halite. Continued halite dissolution releases additional anhydrite sand as older anhydrite sand compacts and is cemented to the base of the cap rock. Domes with a tight contact between anhydrite and halite are inferred to have a quiescent interface between the cap rock and the salt stock.

The compositional similarity of residues in the salt stock and residues in the anhydrite cap rock is additional evidence for the dissolution model (Martinez, 1980). Anhydrite-cap rock has the same sulfur isotopic compositions as anhydrite inclusions within the salt stock (Feely and Kulp, 1957).

Reactions considered to be important in the evolution of cap rocks are shown below. These reactions are principally involved with alteration of anhydrite and calcite. Calcite and other transient phases are produced as a result of a geochemically complex set of reactions. In reaction 1, anhydrite is reduced by

CAP-ROCK REACTIONS

$CaSO_4 + 2CH_2O + \text{bacteria} \dashrightarrow CaCO_3 + H_2S + H_2O + CO_2$ (1)

Inferred from Sassen (1980) and Feely and Kulp (1957)

$CaSO_4 + CH_4 + \text{bacteria} \dashrightarrow CaCO_3 + H_2S + H_2O$ (2)

Ruckmick et al. (1979) and Davis and Kirkland (1979)

$3H_2S + SO_4^{--} \dashrightarrow 4S^o + 2H_2O + 2OH^-$ (3)

Ruckmick et al. (1979)

$2H_2S + O_2 \dashrightarrow 2S^o + 2H_2O$ (4)

Davis and Kirkland (1979)

$CO_2 + H_2O \dashrightarrow H_2CO_3$ (5)

Krauskopf (1967)

$CaCO_3 + H_2CO_3 \dashrightarrow Ca^{++} + 2HCO_3^-$ (6)

Krauskopf (1967)

sulfate-reducing and hydrocarbon (CH_2O)-oxidizing bacteria (Feely and Kulp, 1957, Milner et al., 1977; Sassen, 1980; Davis and Kirkland, 1979; Ruckmick et al., 1979) generating calcite, H_2S, H_2O, and CO_2. Some workers (e.g., Ruckmick et al., 1979) use CH_4 in reaction 2 as the hydrocarbon feedstock and do not generate CO_2 until later unspecified reactions, presumably oxidization of organics. CO_2 is also formed during generation of CH_4 and other hydrocarbon gases. The H_2S is oxidized to elemental sulfur by sulfate in reaction 3 (Feely and Kulp, 1957, Ruckmick et al., 1979) or by O_2 from oxygenated meteoric waters in reaction 4 (Davis and Kirkland, 1979). CO_2 and H_2O generated during sulfate reduction in reaction 1 will lower pH by yielding H_2CO_3 and HCO_3^-. However, the pH is likely to fluctuate as oxidization of H_2S by SO_4 produces excess OH^- and elevated pH. Precipitation of calcite, sulfur, and Fe-, Zn-sulfides indicates a complex interaction of hydrocarbon-bearing deep-basin fluids and shallow ground waters (Price et al., 1983). The hydrocarbon-bearing deep basin fluids provide metals and hydrocarbon fuel for the sulfate reducing bacteria, whereas shallow ground waters oxidize H_2S, hydrate anhydrite to gypsum, and dissolve carbonate. The complex karstic porosity and indications of solution-reprecipitation of calcite in some calcite-cap rocks results from widely variable pH and partial pressure of CO_2 in the shallow subsurface.

2. Included Cap Rocks

Included sedimentary material has been described both in salt stocks (Kupfer, 1974, 1980; Thoms and Martinez, 1980; Heald, 1924; Wilson, 1977; Kent, 1979) and in cap rock (Teas, 1931; Walker, 1974; Murray, 1966). Previous studies have not examined in detail the distribution or significance of terrigenous clastics within cap rocks, although Murray (1966) noted that some cap rocks are characterized by abundant incorporated terrigenous clastic material. Kupfer (1976, 1980) described sedimentary material in Louisiana salt stocks that was thought to represent both primary clastics deposited with the Jurassic Louann Salt and secondary sediments, such as Miocene shale, within boundary shear zones. Eocene shales also have been described at a depth of 1,040 to 1,050 m (3,413 to 3,446 ft) in a deep well (Dow Chemical Co. Cavern Well #1) near the center of Bryan Mound Salt Dome (Railroad Commission of Texas Docket No. 03-16282). In southern Iran, salt domes exposed at the surface contain igneous inclusions up to 3 km (1.9 mi) long, thought to have been rafted up from the base of the original salt stratum (Kent, 1979; Harrison, 1930, 1931).

The Boling Dome cap rock contains approximately 33 volume percent included terrigenous clastic rock within the calcite cap rock over a 2,000 m (6,000 ft) arc of the southern flank of the dome. These sedimentary rocks have at least three possible sources: 1) clastics deposited with the salt, 2) clastics originally deposited around the periphery of the growing salt stock and incorporated by boundary shear zones within salt, and 3) sediments deposited over the crest of the cap rock and partially replaced by calcite which is genetically related to the formation of calcite cap rock by sulfate reduction and oxidization of organics. Origin of false cap rocks present around many salt domes is related in some instances to depositional carbonate strata and in others to replacement or calcite cementation of strata over the dome (Collins, 1985; Sassen, 1980). False cap rock strata are external to the cap rock and are not emphasized here.

II. CAP-ROCK FACIES

Cap rocks are lithologically diverse, commonly being composed of an upper calcite zone, a central transitional zone of calcite, gypsum, sulfur, and anhydrite, and a basal zone of anhydrite (Goldman, 1925, 1933; Taylor, 1938; Martinez, 1980). Although a single well or core can give valuable information about the vertical distribution of lithology for that one hole (e.g. Goldman, 1952; Kreitler and Dutton, 1983), with a greater lateral distribution of downhole information, the geometry of cap rock lithofacies becomes more complex (Walker, 1974). The following section describes the vertical and lateral relationships of facies or zones within the cap rock. Lithofacies data are shown for Boling and Moss Bluff Domes. The lithofacies data at Moss Bluff Dome are more generalized than those presented for Boling Dome. However, the data from Moss Bluff Dome extend along a radial section from the crest of the cap rock to the deep flank and generally include the entire cap-rock thickness.

A. General Facies Relationships

A generalized cross section of Moss Bluff Dome (Figure 3) illustrates typical cap-rock facies, their distribution and variability. As a result of sulfur mining, lithofacies distributions are well known at Moss Bluff Dome, thus helping us to understand the genesis of cap rocks in general. The cap rock at Moss Bluff includes calcite and anhydrite lithofacies. A transitional zone between the calcite and anhydrite cap rock is characterized by abundant sulfur and some gypsum. Figure 3 extends from near the center of the cap rock over Moss Bluff Dome to the western flank. The line of section crosses the zone where sulfur was thickest and was mined. The calcite cap rock thins toward the center of the dome and thickens to a maximum over the shoulder of the dome. Thickness of anhydrite is relatively constant over the dome crest. Note that the base of anhydrite is remarkably planar. Anhydrite thins dramatically in a narrow band around the shoulder of the dome (Figure 3). Anhydrite thinning is a result of replacement by calcite. Despite the thinness of the anhydrite, the base of the anhydrite is still planar near the dome shoulder. Therefore, replacement of anhydrite by calcite was concentrated at the top of the anhydrite zone. The sulfur-bearing zone extends in a broad band from the area where the anhydrite is thin on the dome shoulder toward the cap rock center. The area of maximum thickness of the sulfur-bearing zone is offset towards the dome center from the area where the anhydrite is thinnest and the calcite is thickest.

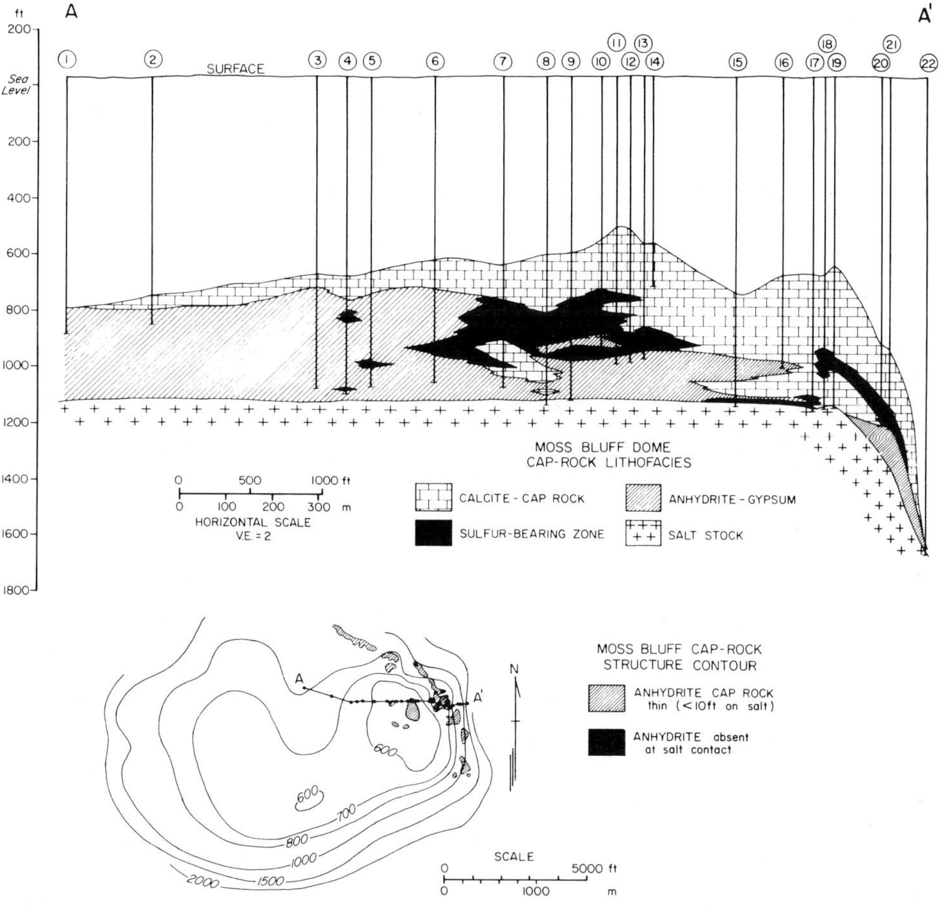

Figure 3. Cross section cap-rock facies, Moss Bluff Dome. Dip-oriented section shows interrelations of various cap-rock facies. Contact between salt stock and cap rock is strikingly planar and horizontal. Data synthesized and modified from Railroad Commission of Texas Hearing Files Docket # 3-72,099.

Distribution of these cap rock facies indicates that the zone of the most intense anhydrite alteration to calcite was on the dome shoulder. The alteration began at the top of the anhydrite on the dome shoulder and probably migrated down through the thickness of anhydrite. Fluids that altered the anhydrite to calcite concentrated their activity at the top of the anhydrite-cap rock. Presumably hydrocarbon-bearing fluids discharged up the flanks of the dome and moved across the top of the anhydrite cap rock mixing with meteoric recharge over the crest of the dome. The occurrence of sulfur concentrated in the area between the center part of the cap rock and the anhydrite minimum zone indicates that sulfur reaction products from the alteration of anhydrite penetrated from their source in the area of maximum anhydrite alteration toward the interior of the dome. The hummocky upper contact of the calcite zone does not mimic the more regular upper surface of the anhydrite zone. It is likely that some of the sediments over the dome were replaced or cemented by calcite, and that the top of the cap rock represents some irregularly distributed calcite-cemented sandstones.

1. Boling Dome Cap-Rock Facies

Spatial relationships and relative thicknesses of cap rock and strata surrounding the dome are shown in Figure 4. Plio-Pleistocene sand-rich fluvial facies pinch out over the crest of the dome (Figures 4A, 4B) where a 30 to 60 m (100 to 200 ft) thick mud-dominated lagoonal facies, bearing Crassostrea oyster shells, directly overlies the cap rock. The low permeability of the mud unit has aided trapping of hydrocarbons and may have influenced the preservation of abundant elemental sulfur by limiting contact of oxidizing meteoric waters with the cap rock environment. Older Miocene and Oligocene sandstone facies are uplifted toward the dome and most sand bodies thin and pinch out

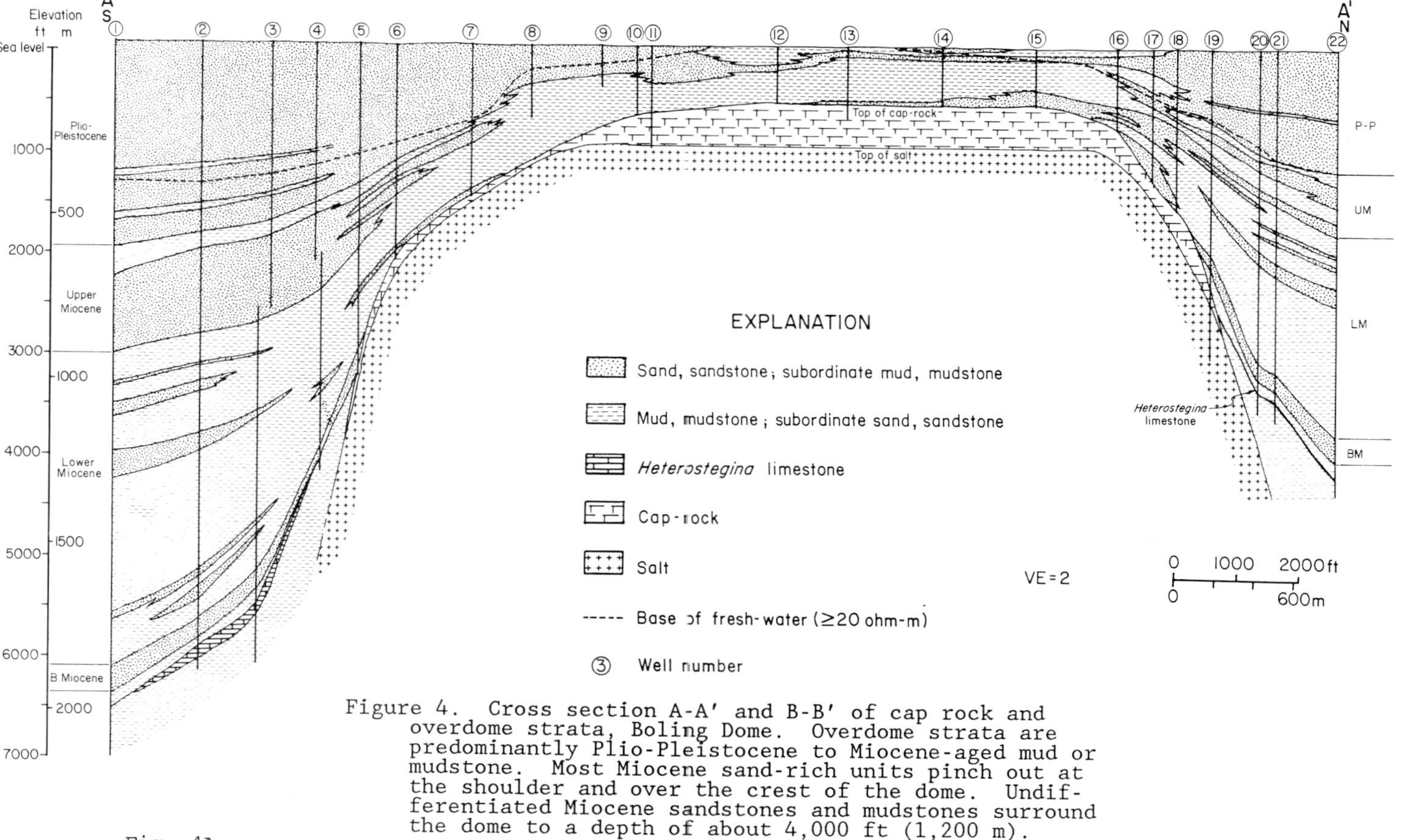

Figure 4. Cross section A-A′ and B-B′ of cap rock and overdome strata, Boling Dome. Overdome strata are predominantly Plio-Pleistocene to Miocene-aged mud or mudstone. Most Miocene sand-rich units pinch out at the shoulder and over the crest of the dome. Undifferentiated Miocene sandstones and mudstones surround the dome to a depth of about 4,000 ft (1,200 m).

Fig. 4A

B
W
B'
N 70° W | S56° E
N56° W | S70° E
N70° W | S84° W
B''
E
Elevation
ft m
Sea level
1000
500
2000
3000
1000
4000
5000
1500
6000
Plio-Pleistocene
Upper Miocene
Lower Miocene
B. Miocene
P-P
U.M.
L.M.
B.M.
Top of cap-rock
Top of salt
EXPLANATION
Sand, sandstone; subordinate mud, mudstone
Mud, mudstone; subordinate sand, sandstone
Cap-rock
Salt
Base of fresh water (≥ 20 ohm·m)
Well number
VE=2
0 2000ft
0 600m

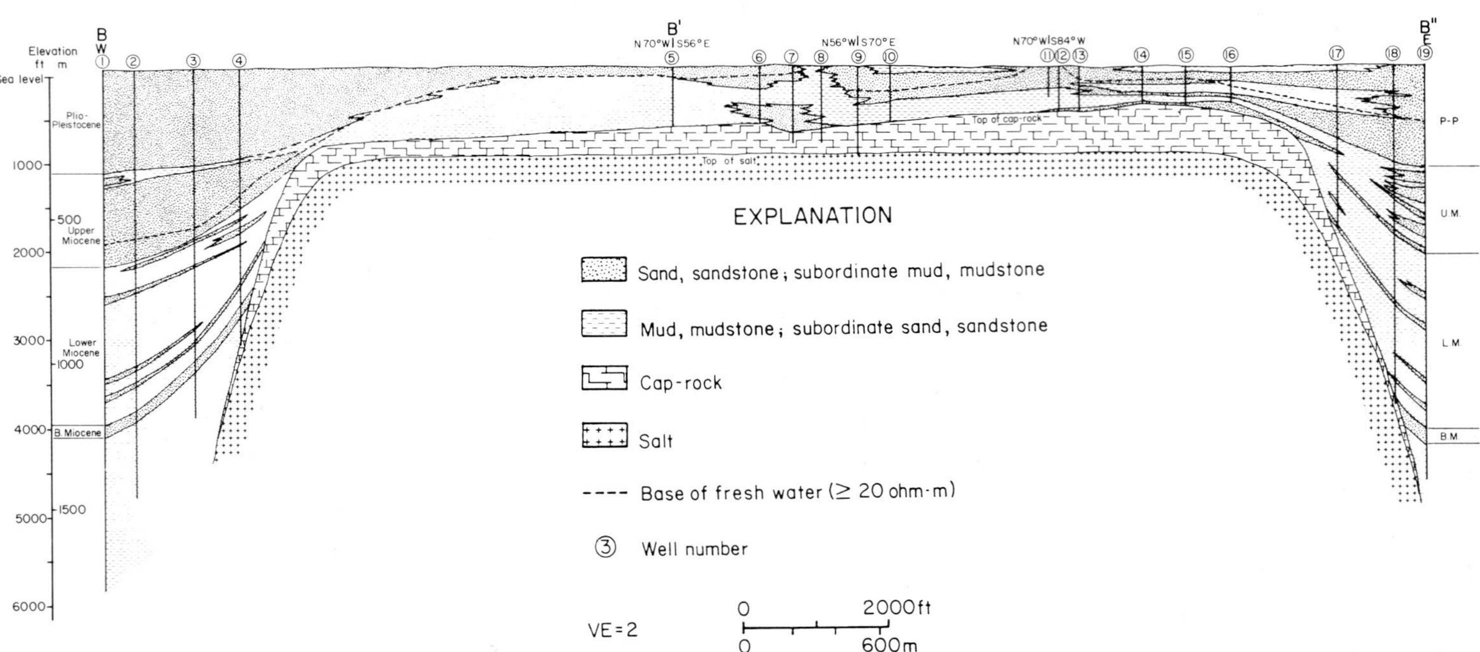

Fig. 4B

within 150 m (500 ft) of the dome flank. Hydrocarbons at Boling Pliocene strata over the dome crest, and in deep Oligocene (Frio Formation) flank sand reservoirs.

Cap-rock facies relationships at Boling Dome are similar to the generalized model and to Moss Bluff Dome, but with additional complexities documented with dense core control. General cap-rock features illustrated in Figure 5 include the thinning of cap rock strata down the flank of the salt stock, thickening of calcite cap rock to a maximum at the shoulder and thinning toward the crest, and relatively planar interface between the salt stock and the base of the cap rock. Particular attention was focused on the distribution and significance of the included terrigenous clastics within the area of dense core control. Many of the cores completely penetrate the calcite facies, but on the average only penetrate the upper 4.6 m (15 ft) of the anhydrite facies. A cross section of 33 cored wells spaced on the average 51 m (168 ft) apart illustrates facies variations along a 1830 m (6000 ft) arc of the southern flank of the dome (Figure 6).

Correlatable lithologic facies within the cap rock at Boling Dome include 1) included terrigenous clastics, 2) calcite, and 3) anhydrite. The transitional zone between the calcite cap rock and the anhydrite cap rock at Boling is poorly developed. The contact between the calcite and the anhydrite is commonly sharp for a distance of millimeters to decimeters. There is very little gypsum to mark the transition from calcite to anhydrite. Incorporated terrigenous clastics mark the contact between the calcite and anhydrite zones in 6 of the 33 wells (Figure 6). Sulfur may be present in any facies, but is most abundant in the calcite facies. Brecciation is also common. The calcite facies is more commonly brecciated than the anhydrite facies. Over much of the dome the contact between anhydrite and the salt stock is

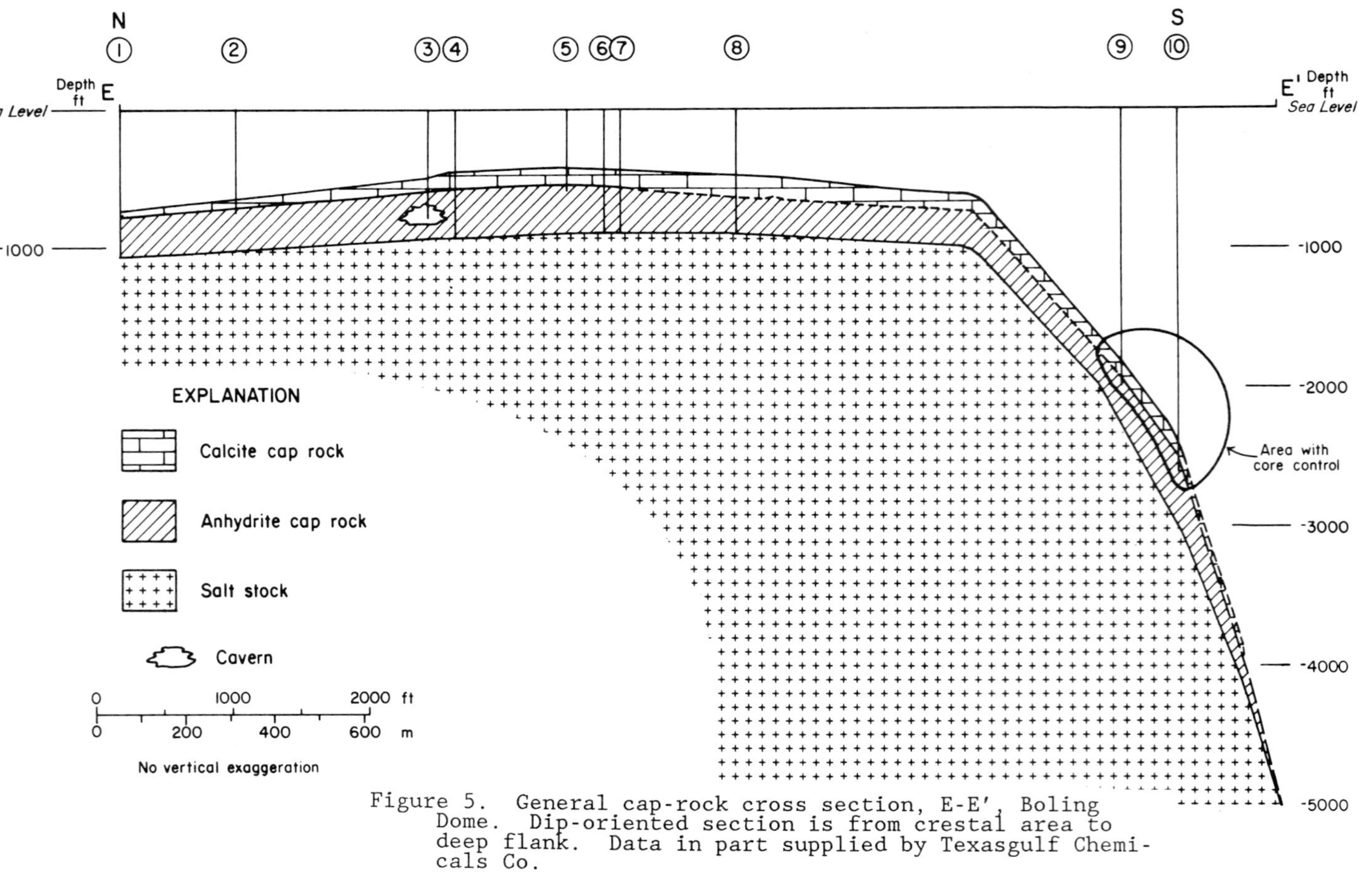

Figure 5. General cap-rock cross section, E-E', Boling Dome. Dip-oriented section is from crestal area to deep flank. Data in part supplied by Texasgulf Chemicals Co.

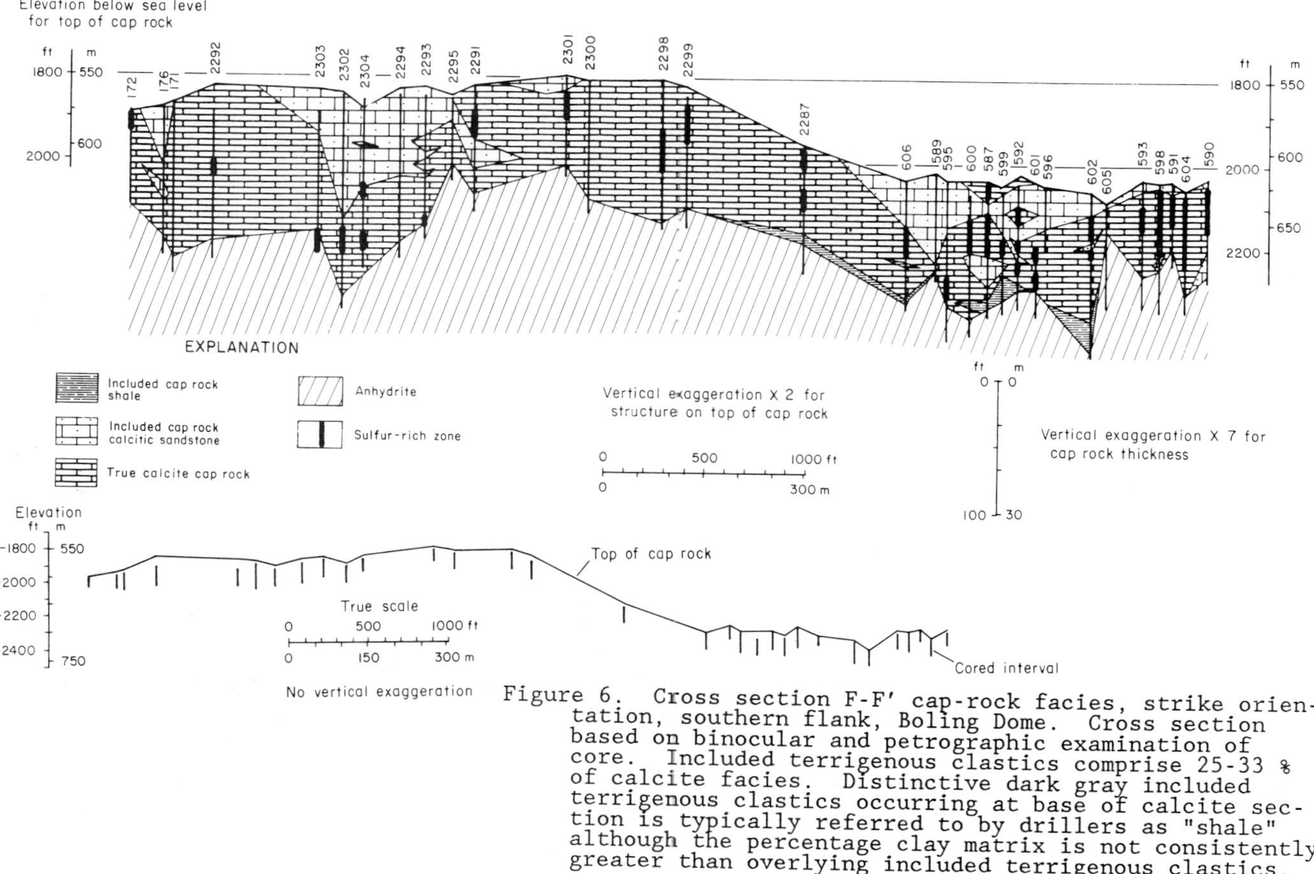

Figure 6. Cross section F-F' cap-rock facies, strike orientation, southern flank, Boling Dome. Cross section based on binocular and petrographic examination of core. Included terrigenous clastics comprise 25-33 % of calcite facies. Distinctive dark gray included terrigenous clastics occurring at base of calcite section is typically referred to by drillers as "shale" although the percentage clay matrix is not consistently greater than overlying included terrigenous clastics.

sharp and tight (F. Samuelson, personal communication, 1985). Three cap-rock facies will be discussed: 1) included cap rock, 2) calcite cap rock, and 3) anhydrite cap rock.

a. Included Terrigenous Clastics

Included cap rock comprises zones up to 30 m (100 ft) thick which extend for 300 m (1000 ft) along strike as discrete pods within the cap rock. The thickest bodies of included cap rock occur near the top of the calcite cap rock and are characteristically light gray calcitic very fine sandstone quartzarenite (Figure 7). A smaller volume of dark gray included cap rock occurs in thin beds from 2 cm to 3 m (0.1 to 10 ft) thick at the contact between the calcite and anhydrite cap rock or within the anhydrite cap rock. Although typically described by drillers as shale, the percentage of matrix in the dark gray included cap rocks is not systematically higher than in other included light gray calcitic sandstones. Included cap rock interfingers with calcite cap rock, but the distribution of included cap rock in a dip sense is poorly defined. Sulfur operators report that the distribution of included cap rock is commonly restricted to the margins of the cap rock. Included bodies are apparently pancake-shaped wedges that thin towards the center of the cap rock. Contacts between included cap rock and the calcite facies are sharp on a thin-section scale, but the facies interfinger over a vertical distance of a few meters.

The included cap rock is typically a very coarse, sandy, calcitic, bimodal, very fine sandstone quartzarenite. Terrigenous clastics within the included cap rock are dark to light-gray quartzose sandstone cemented by a variable but high percentage of finely crystalline calcite (Figure 8A, 8B). The

Figure 7. Cross section of cap rock, Boling Dome showing interrelation between included cap rock, calcite cap rock, and anhydrite cap rock. Brecciated zones in calcite cap rock are correlated between well spaced approximately 250 ft (75 m) apart (facing page).

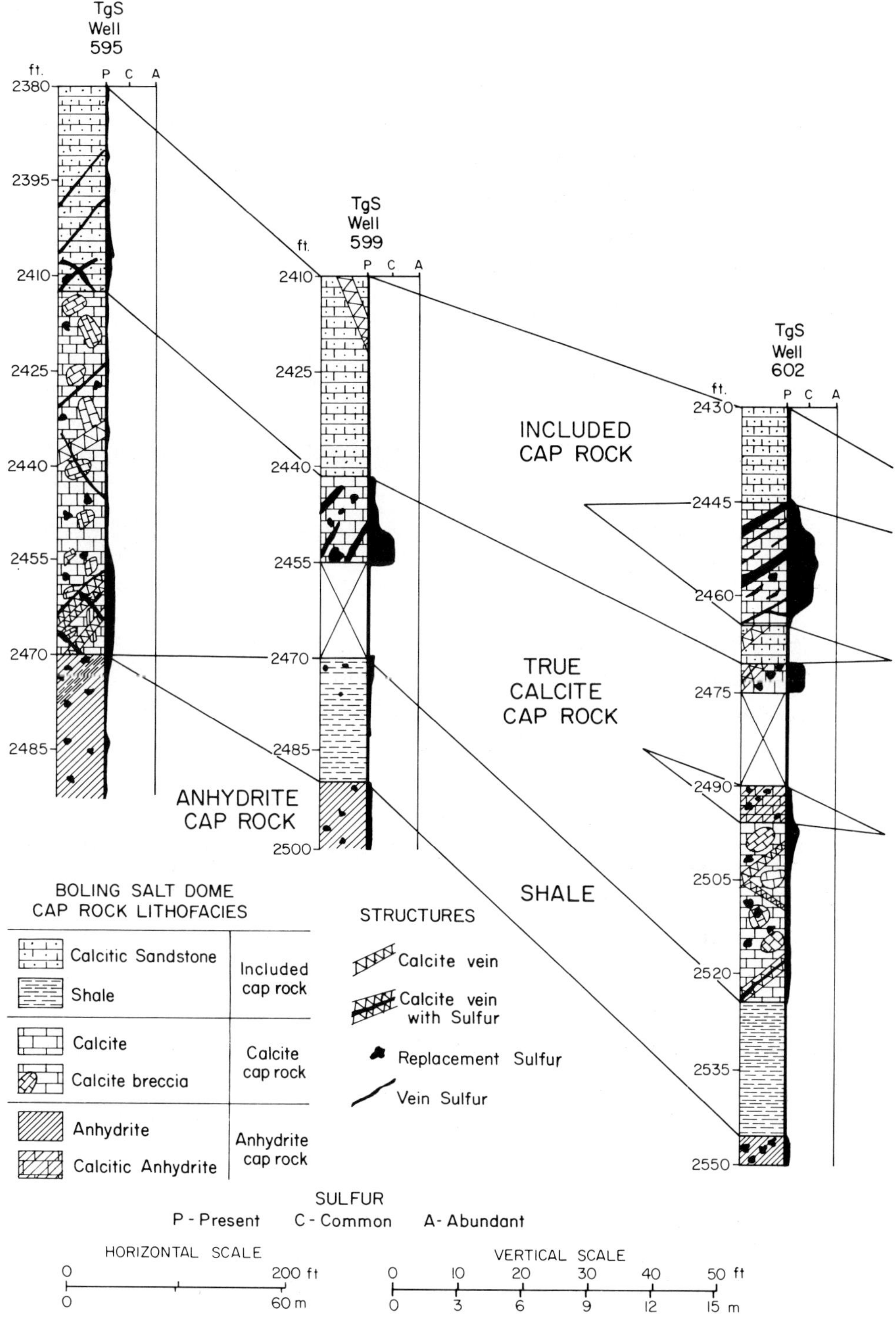
TgS Well 595
TgS Well 599
TgS Well 602
ft.
P C A
2380
2395
2410
2425
2440
2455
2470
2485
2500
2430
2445
2460
2475
2490
2505
2520
2535
2550
INCLUDED CAP ROCK
TRUE CALCITE CAP ROCK
ANHYDRITE CAP ROCK
SHALE
BOLING SALT DOME CAP ROCK LITHOFACIES
Calcitic Sandstone
Shale
Included cap rock
Calcite
Calcite breccia
Calcite cap rock
Anhydrite
Calcitic Anhydrite
Anhydrite cap rock
STRUCTURES
Calcite vein
Calcite vein with Sulfur
Replacement Sulfur
Vein Sulfur
SULFUR
P - Present
C - Common
A- Abundant
HORIZONTAL SCALE
0
200 ft
0
60 m
VERTICAL SCALE
0 10 20 30 40 50 ft
0 3 6 9 12 15 m

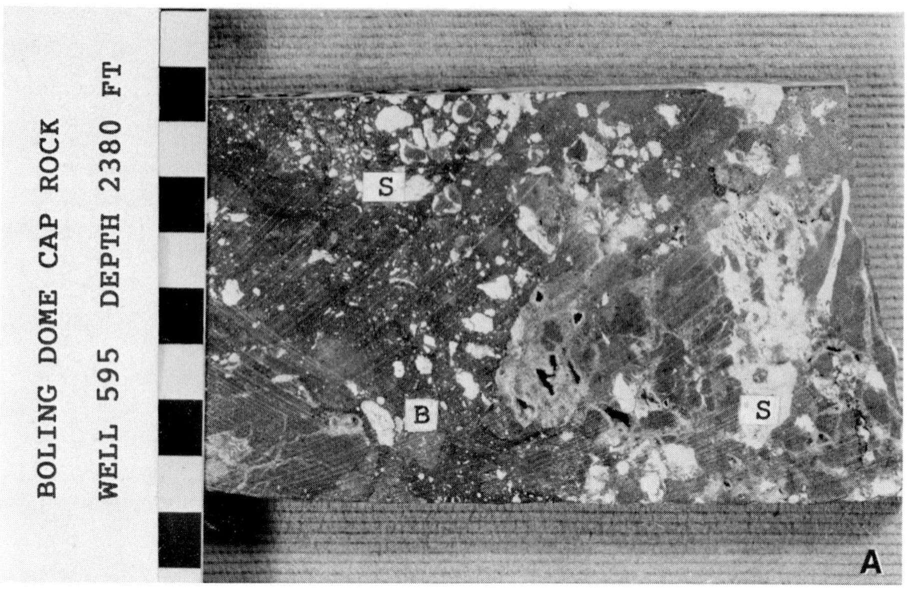

Figure 8. Core slab photograph (A) and photomicrograph (B) of included terrigenous clastics in calcite cap rock, Boling Dome, Well 595, 2380 ft (725 m) below surface (elevation 80 ft (24.4 m)). (A) Light areas are finely crystalline sulfur (S). Unusual, light gray circular patches have different percentages of terrigenous clastics and calcite. Patches appear to be burrow fills (B). Original matrix of terrigenous clay has been replaced by microspar calcite (centimeter scale). (B) Microspar calcite (mspar) content approximately 80 % except within circular burrowed areas where percentage cement is 10-30 %. Detrital quartz (qtz) and orthoclase feldspar (feld) are more abundant in burrows (B) (1 mm scale bar).

rock fabric is unusual in the common occurrence of elongate to circular, mottled patches covering areas up to 1 cm in diameter which strongly resemble tubular burrows. Coarse sand-sized detrital material is concentrated within the "burrowed" patches (Figure 8B).

In hand specimens, the included cap rock facies with veins of spar and sulfur can closely resemble the true calcite cap rock. Sulfur, although present, is not common in the calcitic sandstone. Averaged percentage sulfur over 3 m (10 ft) distance (vertical) is 3 %, whereas the percentage calcite in the included

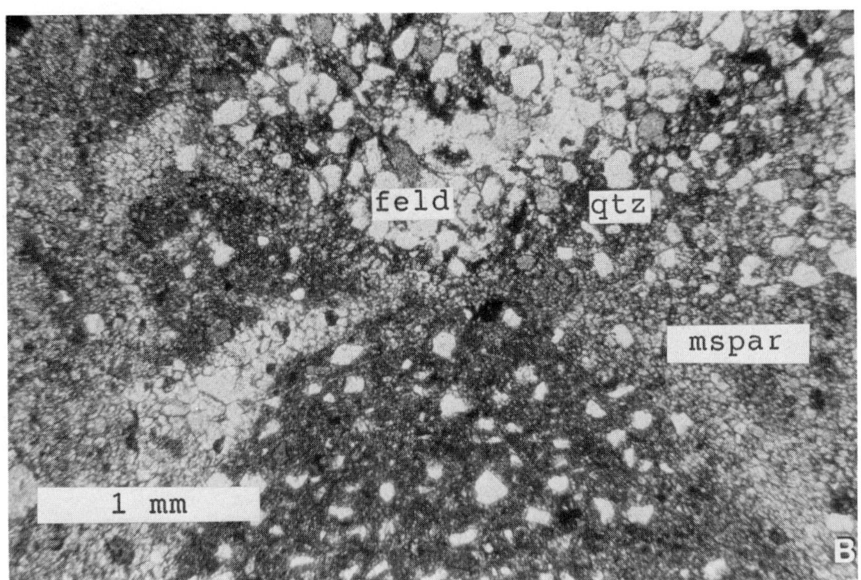

Figure 8B.

clastic rocks commonly ranges from 40 to 90 percent. Very fine quartz sand is the dominant detrital component, rock fragments and feldspars are rare. Well rounded coarse to very coarse sand-sized quartz and metamorphic rock fragments are a subordinate component comprising 1 to 10 percent of the detrital fraction. All the calcite is apparently authigenic. The texture and fabric of the calcite phases are especially diverse and complex suggesting the operation of both grain and matrix replacement processes in addition to simple pore filling cementation. The abundance of the authigenic carbonate suggests that the carbonate replaced the original matrix material. The original matrix may have been a micrite carbonate or clay minerals or some mixture of the two.

The detrital mineralogy of included sand-sized clastics within the cap rock has been reduced to a quartz-rich residuum. The mean percentage ratio of quartz to all framework grains for included cap rocks is 95 %. There is a slight positive correlation between the ratio of quartz to total framework grains and the percentage cement to total framework grains. The ratio of quartz to total framework grains shows a slight increase as the percentage cement increases (Figure 9). However, there was a slight negative correlation between the grain size of detrital grains and the percentage cement.

The mean percentage ratio of quartz to framework grains for sandstones surrounding Boling Dome is 81 %. The sandstones surrounding Boling Dome comprise a separate field distinct from those sandstones included within the cap rock in a graph of mean grain size versus percentage quartz to total framework grains (Figure 9). Sandstones surrounding the dome are subarkosic containing 3 to 33 % detrital orthoclase and microcline. Only 2 of 16 sandstones from within the cap rock contain orthoclase and microcline feldspars. The two included cap rock sandstones with significant feldspars plot in a field similar to that of the sandstones surrounding the dome. The only sedimentary rock fragments present within included cap rock samples are microcrystalline quartz (chert) and quartz-rich metamorphic rock fragments. Heavy minerals, other than authigenic pyrite, are very rare. Quartz is the only detrital component present in abundance.

The textural relationships of the detrital components within the included cap rocks are unusual. The calcitic sandstones are bimodal, coarse sandy, very fine sandstones. The coarse

Figure 9. Graphs of A) percentage quartz of total framework grains versus percentage cement, B) mean grain size of framework grains versus percent cement, and C) mean grain size of framework grains versus percentage quartz of total framework grains. The percentage quartz of total framework grains (A) illustrates a slight trend of increasing quartz percentage as percentage cement increases indicating greater consumption of non-quartz framework grains by calcite cement. However, the mean grain size (B) is not related to the percentage cement. The mean grain size (C) versus percentage quartz of framework grains discriminates population clusters dependent on feldspar content and grain size (facing page).

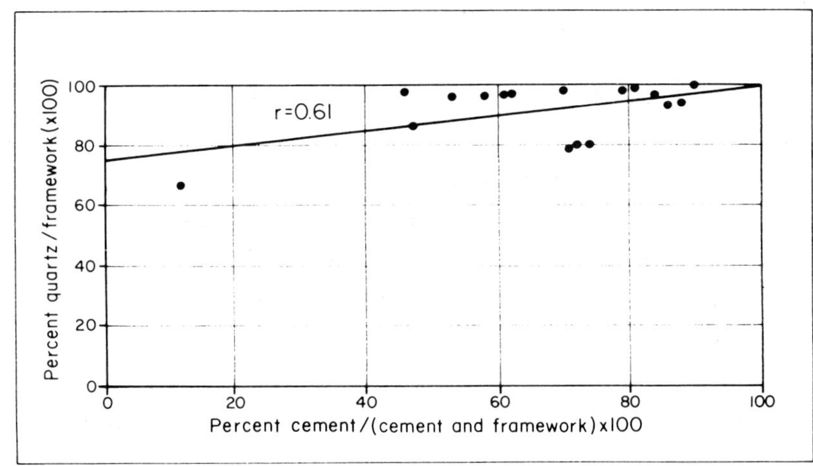
Percent quartz / framework (x100)
100
80
60
40
20
0
r=0.61
0
20
40
60
80
100
Percent cement/(cement and framework) x100
(a)

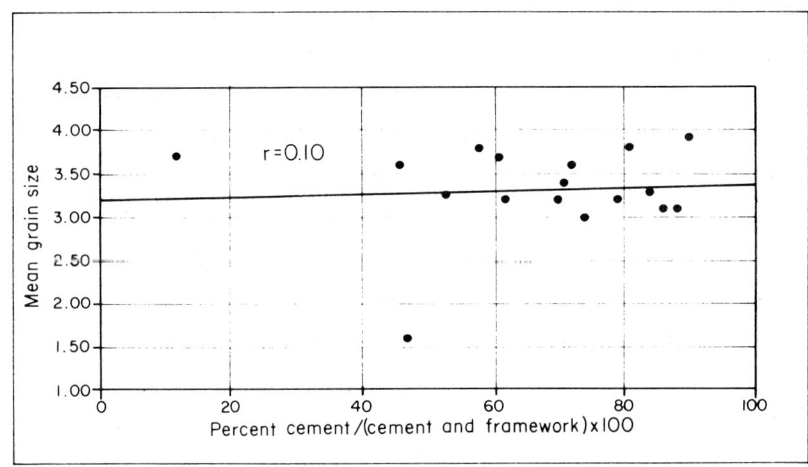
Mean grain size
4.50
4.00
3.50
3.00
2.50
2.00
1.50
1.00
r=0.10
0
20
40
60
80
100
Percent cement/(cement and framework) x100
(b)

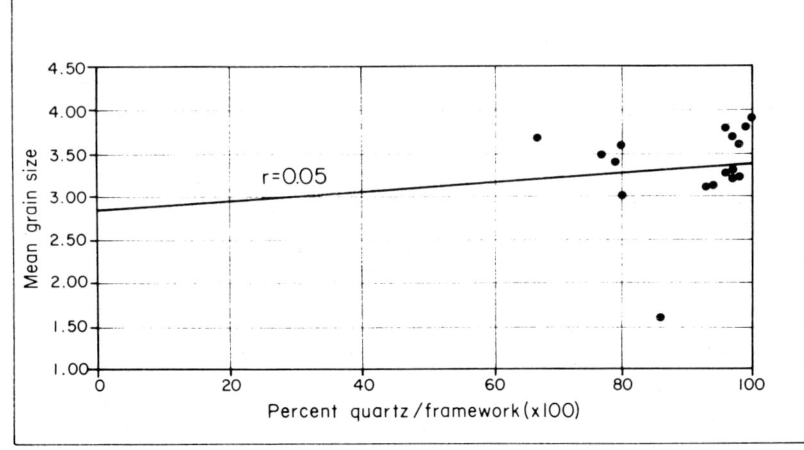
Mean grain size
4.50
4.00
3.50
3.00
2.50
2.00
1.50
1.00
r=0.05
0
20
40
60
80
100
Percent quartz/framework (x100)
(c)

component is very well rounded, medium to very coarse quartz, microcrystalline quartz, or quartz rich metamorphic rock fragments. The mean and modal size of the quartz detritus is a very fine sand (mode 3.6 phi; $mean_{50}$ 3.5 phi; $mean_G$ 3.4 phi).

The nonframework material of included cap rock includes calcite, primarily as cement and clay-sized matrix is composed of both illite clay minerals and calcite. Distinct calcite textures include microspar, coarse microspar, pseudospar, grain replacements, porphyroblasts, and vein spar. Grain replacements by fine-grained calcite is also locally common. The calcite is typically microspar to pseudospar (0.01-0.1 mm) (Folk, 1965). Coarser calcite occurs at vein filling spar and euhedral porphyroblasts. Most of the calcite matrix is clayey microspar (0.01-0.03 mm). Crystals are of quite uniform size and are anhedral masses to somewhat dirty rhombs. Clay minerals occur with the calcite, commonly as coatings around individual crystals of calcite. Where the microspar has neomorphosed to pseudospar, clay minerals and other impurities are segregated into discrete areas and layers commonly surrounding individual crystal. The uniformity of grain sizes during neomorphism indicates a coalescive crystal growth process (Folk, 1965).

Illite is the predominant clay mineral associated with included terrigenous clastics within the cap rock (Figure 10). The illite is well crystallized as indicated by the sharp 10A spike on the X-ray diffraction pattern. Smectite was not detected in the cap rock with standard X-ray diffraction patterns. Interestingly, sediments just outside the cap rock and at a similar depth surrounding Boling Dome and Damon Mound Dome are composed primarily of mixed layer smectite-illite.

Figure 10. X-ray diffraction patterns for clay size fraction of acid-digestion residue of included cap rock and samples from sediments surrounding the cap rock. Note the dominance of interlayered smectite-illite clay minerals in the sediments surrounding the cap rock of Boling Dome. In contrast, smectite is effectively absent from cap rock samples and well crystallized illite is the dominant clay mineral (facing page).

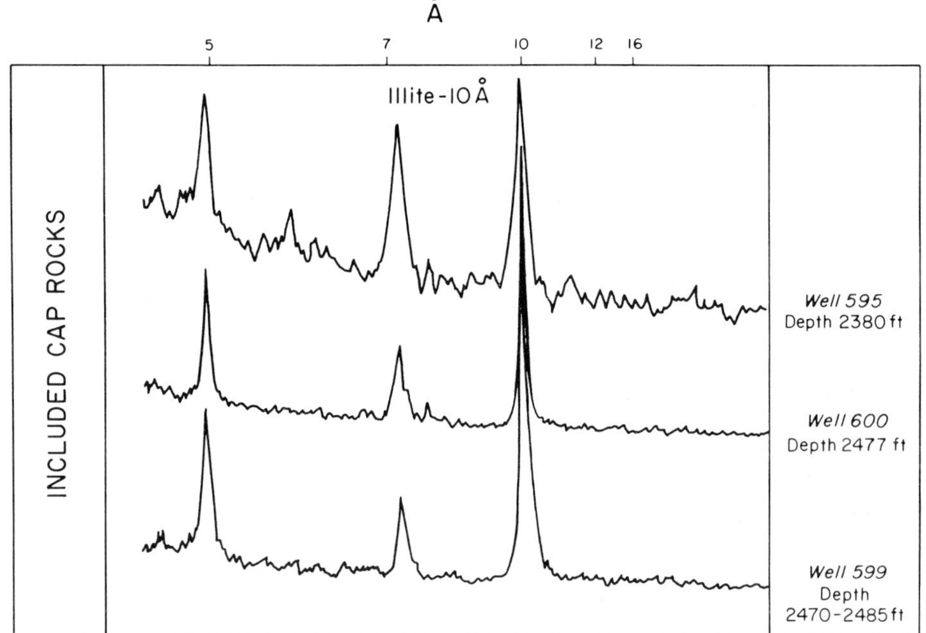
Å
5
7
10
12
16
INCLUDED CAP ROCKS
Illite-10 Å
Well 595
Depth 2380 ft
Well 600
Depth 2477 ft
Well 599
Depth
2470-2485 ft
NEAR DOME SEDIMENTS
Smectite-Illite Interlayer
12.5-14 Å
Well 2302
Depth
1960-1975 ft
Well 601
Depth 2432 ft
Well 607
Depth
2737-2739 ft
Banker Jr. #22
Depth 1407 ft
20°
10°
2°
Degrees 2θ

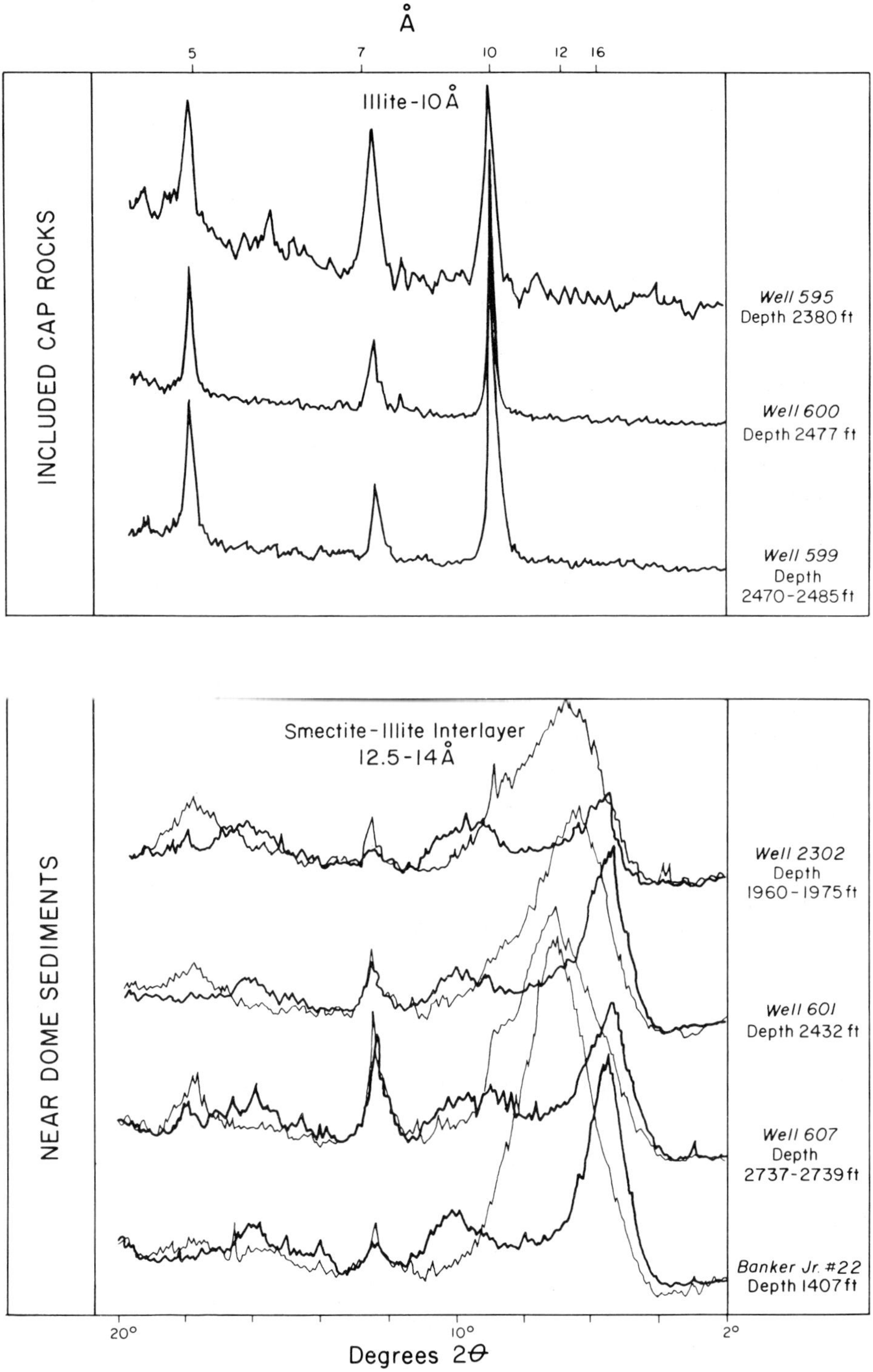

Figure 11. Core slab photograph (A) and photomicrograph (B) of included cap rock near contact with normal calcite cap rock, Boling Dome, Well 599, depth 2440 ft (744 m). (A) Rhombohedral porphyroblasts of calcite (calc rhom) are dark crystals ranging from 0.2 to 2.0 mm (centimeter scale). (B) Isolated calcite porphyroblasts (stained red with alizarn red S) have excluded or replaced fine-grained microspar calcite and detrital grains. Banded luminescence within calcite crystals suggests that they may have originally been dolomite (J. R. Kyle, personal communication). Remnant microcrystalline quartz grain with calcite-filled pits is visible near center of euhedral calcite crystal (calc rhom). Dark rim is original outline of quartz crystal (1 mm scale bar).

b. Included Cap Rock - Normal Cap Rock Contact

A sharp boundary commonly separates the included cap rock from the normal calcite cap rock on thin section scale, but the included sandstone and calcite cap rock facies interfinger over a scale of meters. As the boundary is traversed from included cap rock into true calcite cap rock, the grain size of the authigenic calcite increases through neomorphism of microspar to pseudospar and the detrital quartz disappears in the interval that is clearly normal calcite cap rock (Figure 11). Calcite of the true

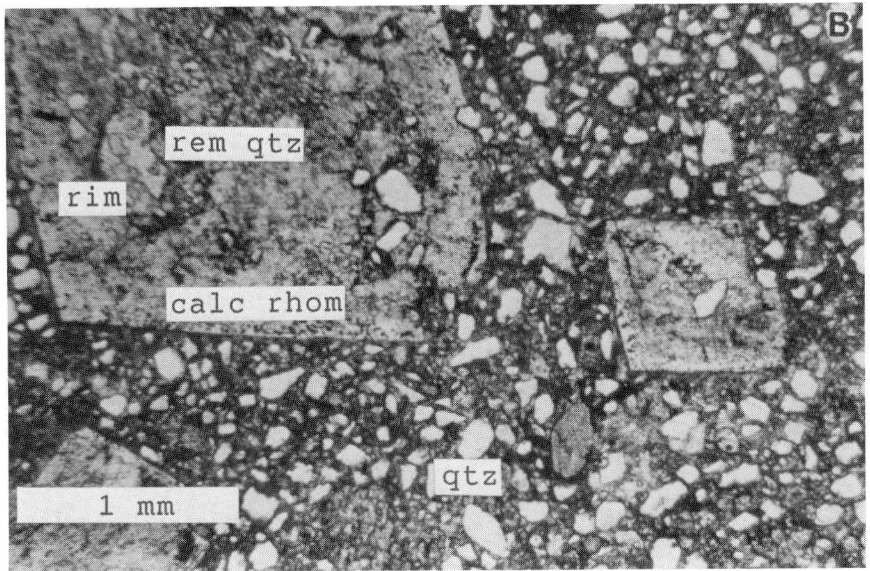

Figure 11B.

cap rock clearly first replaces the matrix of the sandstones, but whether this matrix was originally clay-sized calcite micrite, microspar, or terrigenous clays or a mixture is not known. The uniform grain size of the pseudospar crystals indicates that coalescive crystal growth caused the grain size of the calcite to increase through neomorphism. Impurities are segregated into discrete layers and zones (Figure 11). Contacts between true calcite and included cap rock are commonly marked by large euhedral porphyroblasts of rhombohedral calcite (1.0-2.0 mm) isolated in a matrix of pseudospar (0.1-0.3 mm). This porphyroid neomorphism has been described by Folk (1965) as typical of many partially dolomitized limestones. Many of these large calcite crystals have central ghosts about the same size and shape as clastic grains in the surrounding rock (Figure 12). A few of the porphyroblasts contain small remnants of quartz. SEM grain mounts were used to study the surface morphology of grains from

Figure 12. Photomicrograph of pseudospar calcite (pspar) crystals in included cap rock near contact with calcite cap rock, Boling Dome, Well 591, depth 2365 ft (721 m). Many irregular polyhedra to euhedral rhombohedra of calcite pseudospar crystals (pspar) have central ghosts and dark inclusion-rich areas (I) resulting from possible incorporation of original matrix (M) through coalescent crystal growth of individual pseudospar crystals (1 mm scale bar).

the area of the contact between true cap rock and proto cap rock. Actual replacement of quartz or other grains by calcite was not documented unequivocally. Quartz grains include well developed quartz overgrowths (Figure 13A) and authigenic bipyramids (Figure 13B). Although inconclusive, the surface morphology of the rare detrital feldspar grains (Figure 13C) have an increased surface roughness suggestive of active dissolution.

Significant but uncommon mineral species present along the contacts include authigenic quartz and sulfide minerals, principally pyrite. The authigenic quartz fills pores flanking the side of large veins. The quartz is clearly authigenic, as it occurs in single euhedral bipyramids (0.05-0.2 mm) and in 2-3 mm

Figure 13. Scanning electron photomicrograph of insoluble residue from contact between included cap rock and normal calcite cap rock, Boling Dome, Samples A, C, D Well 176, depth 2073 ft (632 m). Sample B Well 600, depth 2477 ft (755 m). (A) Quartz sand grain with well developed overgrowth, (B) Authigenic bipyramidal quartz, (C) Plagioclase feldspar with dissolution surface features, (D) Sulfur crystal with euhedral pits that formerly contained anhydrite (50 μ scale bar).

rosettes with sharp crystal boundaries. If large-scale dissolution of quartz is postulated to transform included cap rock composed of calcitic sandstone into true calcite cap rock, then large scale dissolution and transport of silica is required. However, occurrence of the authigenic quartz only indicates that, at present, silica is precipitating in the cap rock environment. No evidence from SEM observations indicates large scale dissolution of sand-sized quartz in the cap rock environment.

The sulfide minerals present in the cap rock include pyrite, chalcopyrite, and possibly sphalerite. Sulfide seams are up to 1 cm thick, but are rare. Pyrite occurs in cubic masses up to 2.0

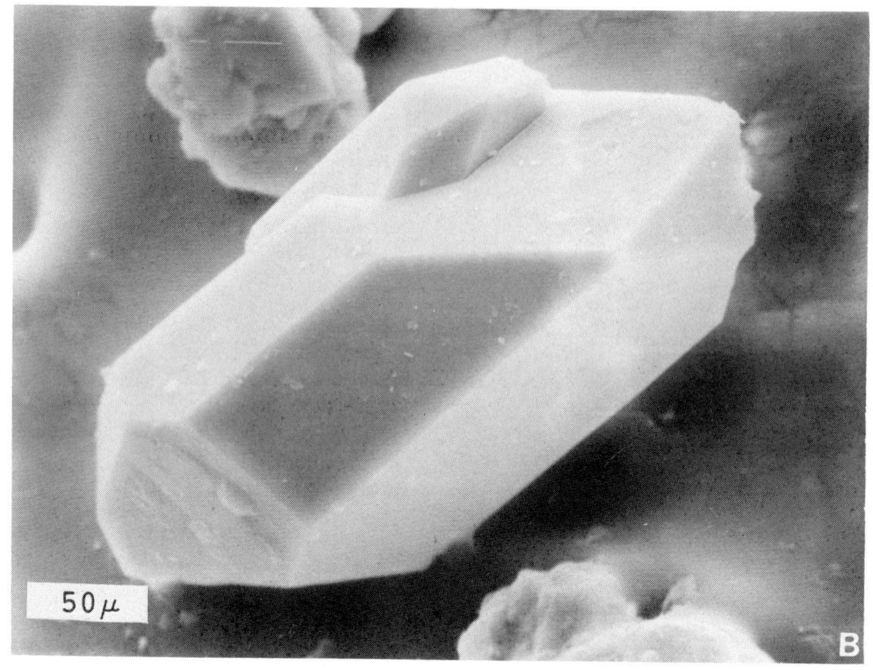

Figure 13B.

cm across. Spherules of an opaque mineral presumed to be pyrite are 0.01-0.02 mm in diameter within dark calcite grains and segregated in dark laminae 1-2 mm thick.

c. Calcite Cap Rock

The calcite cap rock at Boling Dome includes the commonly described dark and light normal calcite cap rock (Figure 14). The alternating dark and light banding of calcite so characteristic of some other cap rocks, for example Oakwood Dome (Kreitler and Dutton, 1983), is rare at Boling except near the base of the calcite section. In addition to zones of incorporated terrigenous clastics the upper calcite cap rock contains abundant veins of calcite spar and sulfur, brecciated zones of resedimented calcite, and void spaces ranging in size from 0.1 mm intergranular pores to caverns of unknown dimensions.

Figure 13C.

On the average there are 0.7 veins and faults per foot (2.2 veins and faults/m) of core in the calcite cap rock. Any void space whose dimension approaches the size of the core (5 in) will not be recovered in its original condition. The percentage recovery of the calcite section was about 50 percent, the lowest of any cap-rock facies at Boling Dome.

The calcite cap rock at Boling Dome is a light to dark-gray, fine to medium calcite. Most of the calcite is termed pseudospar (Folk, 1965). Calcite grain size may commonly be quite uniform (0.05-0.2 mm) (Figure 14). Crystal shape varies from anhedral to rhombic. The centers of the larger calcite grains (0.1-0.2 mm) often have internal ghosts at the center of the crystal,

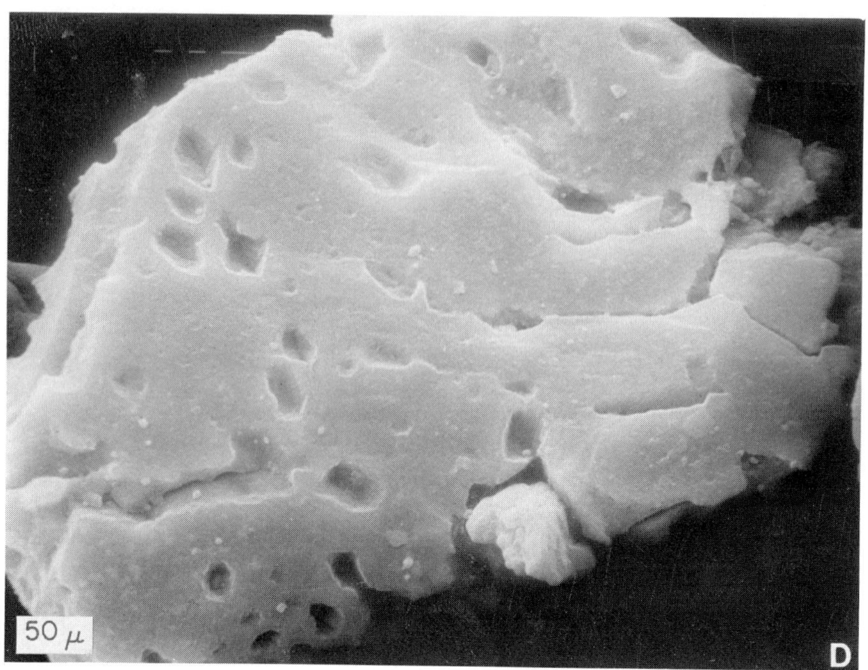

Figure 13D.

suggesting the calcite replaced some precursor mineral (Figure 12). A dark, opaque, spherical mineral, presumed to be pyrite framboids, is sprinkled in the central parts of the crystals.

The many veins of calcite and sulfur associated with the calcite section of the cap rock (Figures 14, 15) are typically white calcite spar, commonly with late-stage euhedral crystals of sulfur precipitated as the last mineral phase. Monominerallic veins of either pure calcite or sulfur are also common (Figure 14B). The percentage of sulfur within a vein increases as the vein decreases in size. The veins of calcite spar usually contain white, coarse crystalline, cm-sized anhedral crystals of spar. Crystal terminations are usually scalenohedrons or may be blocky and trigonal. Frequency of veins in the terrigenous clastics within the calcite cap rock is less than the frequency

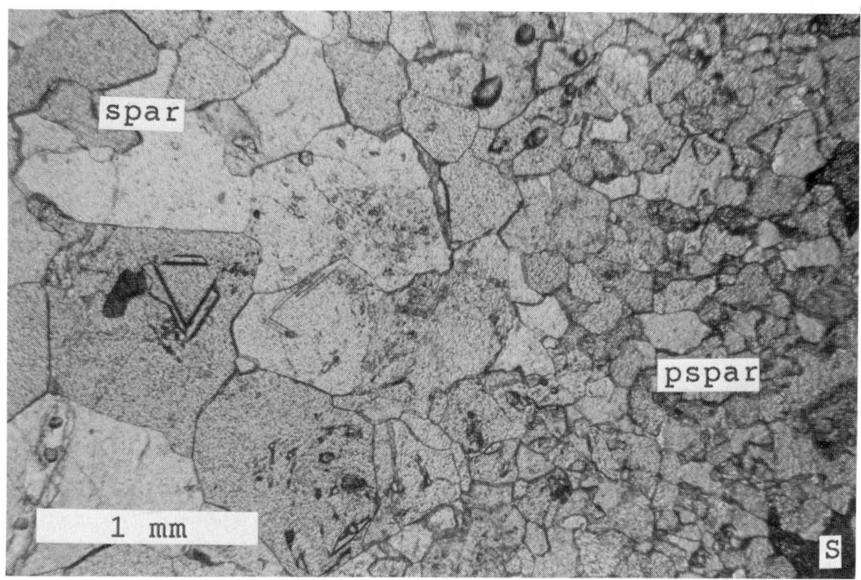

Figure 14. Photomicrograph of normal calcite cap rock, Boling Dome, Well 600, depth 2427 ft (740 m). A large vein of white calcite (spar) veins cuts gray pseudospar (pspar) calcite bearing small individual crystals of sulfur (S). Boundary between pseudospar calcite (right side) and coarse vein spar (left side) is marked by an abrupt increase in grain size from 0.1-0.3 mm to 0.5-1.0 mm. Dark crystals are sulfur (1 mm scale bar).

of veins within the pure calcite cap rock. Many veins have open interiors, suggesting the vein-filling minerals precipitated in open voids. In contrast, most of the veins in the anhydrite facies are diffuse planes where sulfur relaces anhydrite.

d. Sulfur Mineralization

The sulfur paragenesis is complex and will only be touched upon here. There are at least three distinct sulfur textures present at Boling Dome. An "early" sulfur texture is characterized by euhedral crystals of sulfur (0.5 to 10mm) which replace anhydrite (Figure 16). These euhedral sulfur crystals frequently have preserved within them fresh and embayed crystals of anhydrite as well as euhedral anhydrite-shaped pores. Such

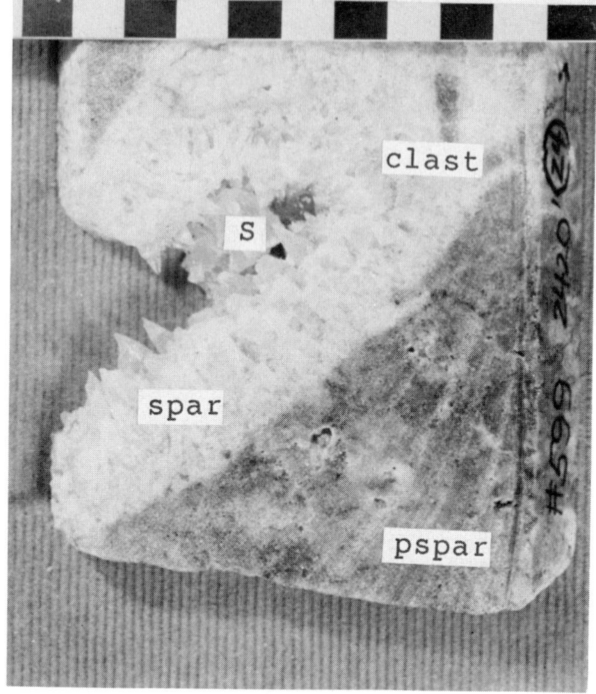

Figure 15. Core slab photograph of large open spar-sulfur vein in normal calcite cap rock, Boling Dome, Well 599, depth 2420 ft (738 m). Large vein (7 cm width) of brown calcite (spar) with interior of clear spar and coarse crystalline sulfur (S) (centimeter scale). Central area of vein is open. Angular clast of gray pseudospar calcite is apparently suspended within vein indicating support for clast at some other location along vein.

anhydrite-bearing sulfur crystals are common in the anhydrite cap rock but are also present within the calcite-cap rock immediately adjacent to the anhydrite and in very deep calcite-cap rock (greater than 700 m (2300 ft)). Preservation of anhydrite within the sulfur crystals present within the calcite cap rock indicates that the earliest sulfur mineralization must have replaced anhydrite prior to formation of the calcite-cap rock. The large euhedral sulfur crystals may be the product of slow oxidization of H_2S by SO_4^{--} as shown in reaction 3.

Relatively "late" and more extensive phases of sulfur mineralization (Figures 14, 15) are associated with the

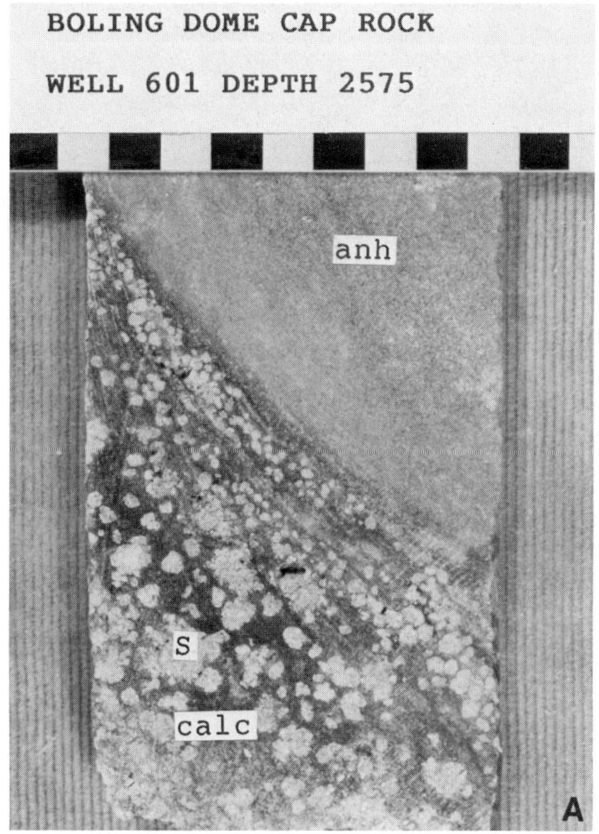

Figure 16. Core slab photograph and photomicrograph, sulfur texture, Boling Dome, Well 601, depth 2575 ft (785 m). (A) Slab photograph of light gray calcitic anhydrite (anh) with sparse sulfur and dark gray anhydritic calcite (calc) with euhedral to colliform sulfur (S) crystals 0.5 to 10 mm (centimeter scale). (B) Photomicrograph of dark gray anhydritic calcitic and euhedral sulfur. Paragenetic timing of sulfur mineralization is fixed by preservation of anhydrite within sulfur crystals that are surrounded by calcite. Thus sulfur mineralization predated replacement of anhydrite by calcite (1 mm scale bar).

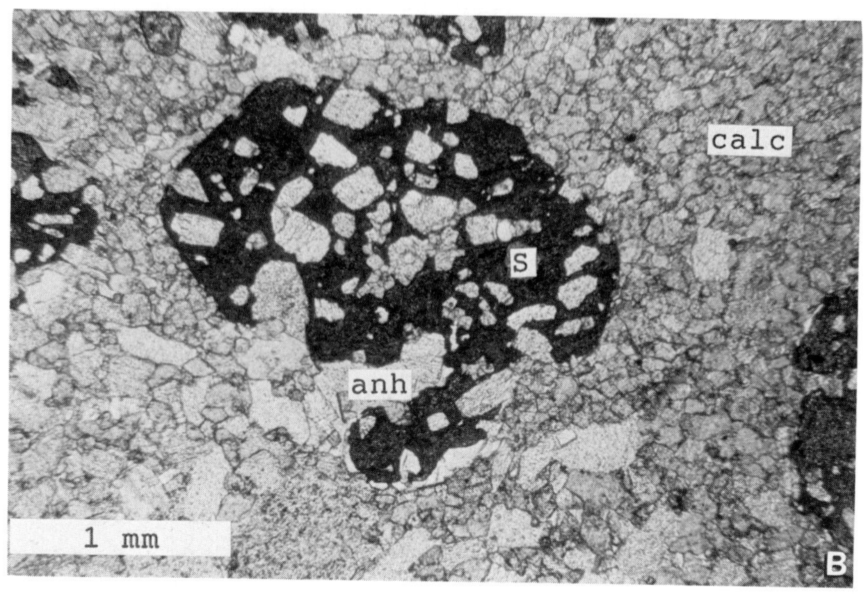

Figue 16B.

replacement of calcite cap rock by 1) fine-grained sulfur and by 2) coarse-grained, vein filling sulfur. These sulfur-bearing veins cross cut the euhedral sulfur. The "early" euhedral sulfur crystals were never observed to replace a calcite spar-sulfur vein. It is unknown if the initial generation of euhedral sulfur in anhydrite is the source for the late stage vein filling sulfur. The fine-grained sulfur that replaces calcite may result from rapid oxidization of H_2S by oxygenated meteoric ground "early" sulfur phase is presently precipitating and replacing anhydrite in deeper sections of the cap rock simultaneously with "late" stage sulfur precipitating in veins in shallow areas of the calcite cap rock.

e. Anhydrite Cap Rock

The anhydrite cap rock is a dense, dark-gray rock with idiotopic fabric composed predominantly of anhydrite (Figure 17). Individual crystals of anhydrite are euhedral and occur as

Figure 17. Core slab photograph and photomicrograph, anhydrite cap rock, Boling Dome, Well 600, depth 2483 ft (757 m). (A) Slab photograph of medium gray anhydrite with sulfur vein (S vein) oriented 45o to vertical (centimeter scale). Horizontal light fracture (frac) oriented horizontally is a coring induced fracture. (B) Photomicrograph of anhydrite (anh) with dark sulfur (S) vein. Remnant anhydrite (rem anh) crystal is located near terminus of vein (1 mm scale bar).

interlocking grains cemented by anhydrite. Near the contact with the calcite facies, anhydrite may be cemented by calcite. Anhydrite crystal size varies from 0.2-0.5 mm. Whereas the calcite cap rock is extremely heterogeneous with respect to mineralogy and structure, the anhydrite cap rock is much simpler. This simplicity, however, may be a function of the under-representation of the anhydrite cap rock within the core suite studied. Usually less than 15 ft (5 m) of the upper-anhydrite

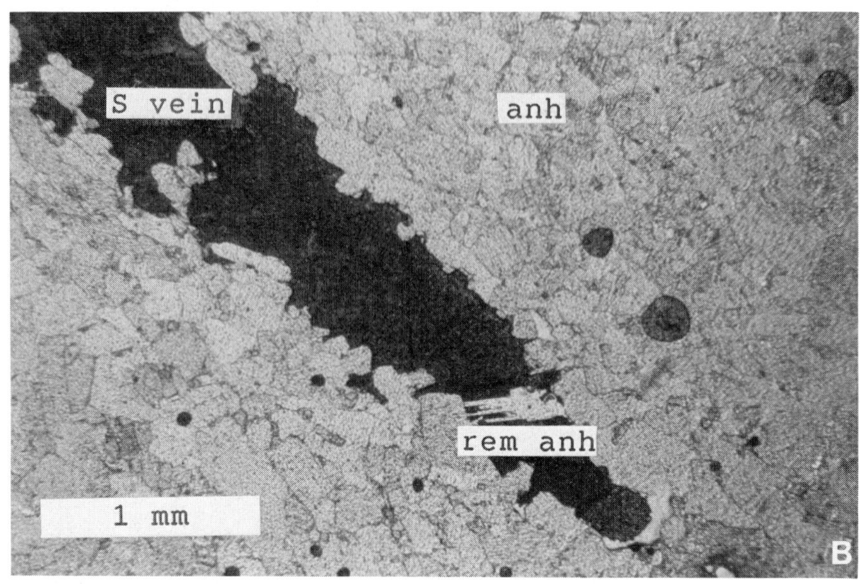

Figure 17B.

cap rock is drilled. However, recovery of the anhydrite averaged 90 %. Veins and thin layers of included terrigenous clastics within the upper part of the anhydrite cap rock dip uniformly at 45° to the vertical. The cap rock-salt stock interface and the cap rock-overburden interface also dip at about 45° in the area from where the cores were recovered. Distribution of mineralogical and structural zones within the main body of the anhydrite is unknown. Local geologists report that the contact between the anhydrite and salt stock is commonly tight and relatively impermeable (F. Samuelson, personal communication, 1985). The anhydrite section at Hockley Dome, 35 km (58 mi) northwest of Houston, contains several million tons of economic grade Zn-, Pb-, Au-, Ba-sulfides (Price and others, 1983; Kyle and Price, 1985).

At Boling Dome, all minerals found in the calcite cap rock are also present in the anhydrite cap rock. Anhydrite, however, comprises 90 % of the rock volume. Terrigenous clastics occur within the anhydrite cap rock and at the contact between the anhydrite and calcite cap rocks. Terrigenous clastics that are in anhydrite are dark gray, slightly calcitic, shaly, very fine sandstones. Calcite is present as an authigenic cement concentrated near the contact with the calcite cap rock. Calcite veins are not common and very large veins, common in the calcite cap rock, are absent. Veins of sulfur are common in anhydrite but are very different from those present in the calcite cap rock. The sulfur veins uniformly replace anhydrite and are not void fillings (Figure 17).

2. Discussion

Terrigenous clastics within the cap rock at Boling Dome are included within both the calcite and the anhydrite facies. The thickest and most laterally persistent included clastics occur at the top of the calcite facies. Significant features characteristic of the included terrigenous clastics are: 1) the dominance of quartz in the coarse detrital fraction; 2) subordinate bimodal, well rounded coarse and very coarse quartz sand; 3) the dominance of illite and the absence of smectite in the fine detrital fraction; and 4) dominantly authigenic fine calcite microspar matrix supporting 40 to 90 percent of the rock mass. Inclusion of sedimentary clastics within the anhydrite facies of the cap rock indicates that the sedimentary clastics associated with the anhydrite facies must have previously been surrounded by halite. Subsequent dissolution of the halite left the residual anhydrite and clastics. Most of the terrigenous clastics are inferred to have been incorporated within the calcite cap rock as the anhydrite cap rock was altered to calcite. The release of Ca^{2+} and CO_3^{--} during reaction of anhydrite with hydrocarbons and oxidization of organics also favored precipitation of calcite cement in sediments immediately overlying the salt dome.

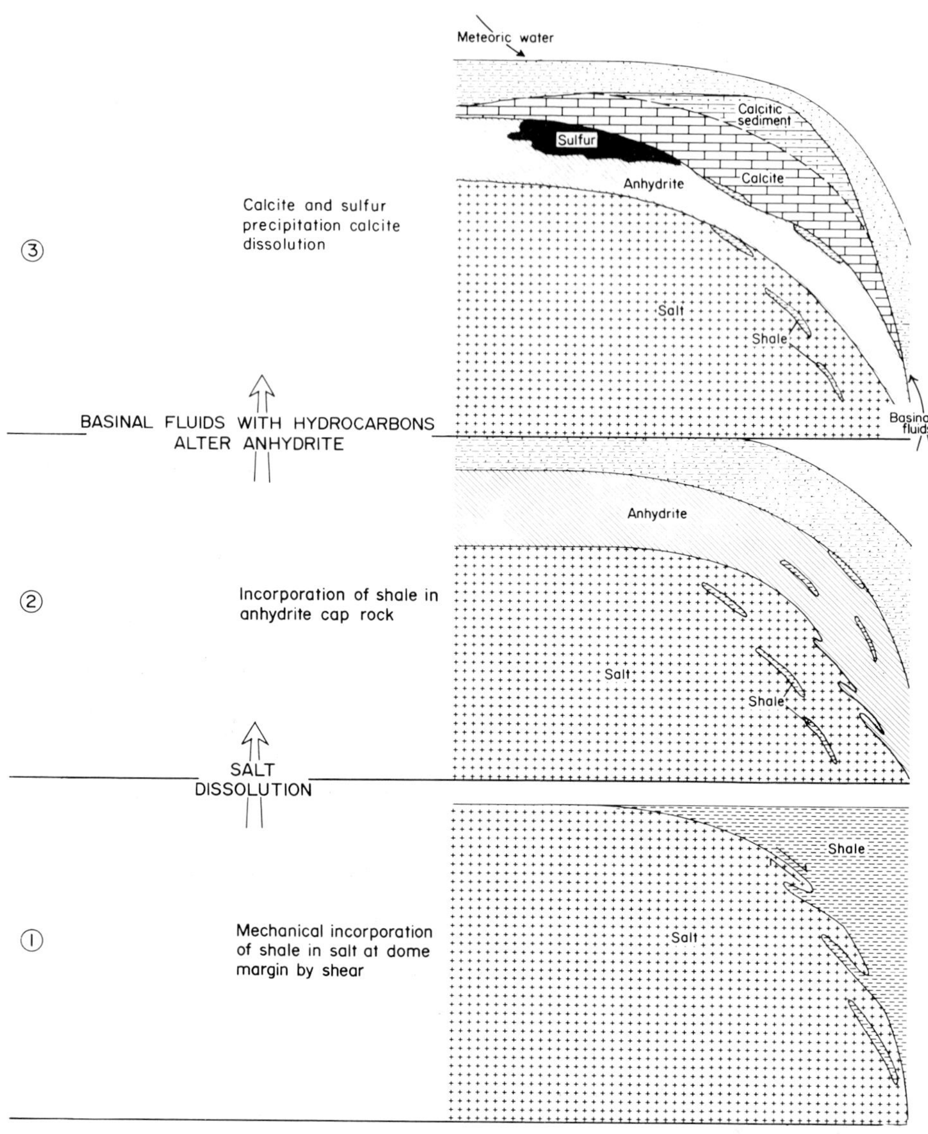

Figure 18. Schematic illustration of evolution of cap rock and mechanisms incorporating included terrigenous clastics within cap rock. Primary incorporation of Jurassic-aged clastics within salt is not illustrated.

The terrigenous sediments originally could have been incorporated within the salt stock through three mechanisms (fig. 18):

1) aeolian sediments deposited with the salt during Louann salt deposition (primary incorporation); 2) secondary incorporation within the salt at boundary shear zones; and 3) replacement of overlying sediment by calcite. In mechanisms 1 and 2, after the clastics are incorporated within the salt stock, they are underplated to the cap rock during salt dissolution.

Kupfer (1980) described quartz sand grains present in a thin (.1-.3 m) persistent zone within the salt of Cote Blanche Dome, Louisiana, as very well rounded, well sorted, and frosted. He inferred that the sands originally were aeolian deposits interstratified with the salt. Other domes in the Gulf Coast (Avery Island and Bryan Mound Domes) contain sedimentary material within boundary shear zones in salt (Kupfer, 1976) that has been dated on the basis of micropaleontology as Eocene to Miocene. No body fossils were recognized at Boling Dome, however the mottled fabric may be the result of burrowing. Such organic activity typifies the muddy sandstone and mudstones in Tertiary shelf and slope sediments of the Gulf Coast, and would have been anomalous in the arid environment prevailing during Louann deposition. The detrital mineralogy of included sediments within Boling Dome is a simplified suite in which much provenance information is lost. Cap rock sands are dominantly quartz with rare metamorphic rock fragments with stretched quartz and microcrystalline quartz. Only two cap-rock sandstone samples contained orthoclase or microcline feldspar as determined by feldspar staining. Twelve cap rock samples contained no orthoclase feldspar or twinned plagioclase. Tertiary sandstones from the Upper Texas coast typically contain 50-75% quartz with subequal feldspar and rock fragments in the sand fraction (Galloway, 1977; McCraken, 1967; Bebout et al., 1978). Sandstones currently surrounding Boling Dome are orthoclase subarkoses.

The bimodal fabric of the detrital mineralogical suite is also significantly different from that of sandstones surrounding the dome. From 1 to 10 % of the quartz sand fraction is coarse to very coarse and very well rounded. Such a mineralogical suite is not known from the northwestern margin of the Tertiary Gulf of Mexico. The bimodal character of the sandstones within the cap rock classically suggests aeolian origin, perhaps as sandstones blown into the salt basin during deposition of the Louann salt or Norphlet-type arkoses deposited immediately over the salt. The extreme simplicity of the detrital suite could also result from burial diagenesis and removal of potassic feldspar or an originally impoverished depositional suite. Massive alteration of both the coarse and fine detrital components of Gulf Coast Tertiary sandstones has been systematically documented in the northwestern Gulf of Mexico. Boles (1982), Land and Milliken (1981), Milliken et al., (1981), and Milliken (1985) have shown that the coarse detrital fraction of all the major Tertiary sequences has reacted extensively with pore fluids. The overall feldspar composition has shifted from a mixed potassium feldspar and plagioclase suite toward an albite-dominated suite at a depth mineral assemblages of intervening mudstones (Hower et al., 1976; Boles and Franks, 1979) have similarly reacted extensively with pore fluids and shifted from smectite-dominated suites to an illite-rich suite. Rb-Sr dating of the timing of the alteration of smectite to illite (Morton, 1985) holds promise as a possible technique to constrain the timing of incorporation of clastic strata within the salt.

Although the illite composition of clay minerals in included cap rock sediments is not diagnostic as to origin, the occurrence of well crystallized illite in shallow subsurface sediments and the absence of smectite is anomalous. Mixed layer smectite-illite dominates the clay mineral suite in shallow (less than 5,000 ft) Tertiary sediments in the northwestern Gulf of Mexico clastic depositional province. Such sediments surround the upper parts of Boling Dome and surrounding domes. The dominance of illite in the shallow subsurface cap rock strongly suggests that these rocks were subjected in the past to conditions necessary for conversion of smectite to illite. Upper Texas Gulf Coast shales, specifically Brazos-Colorado River sediments, commonly undergo a transition from smectite to illite in the temperature range of 105° to 138°C (220° to 285°F) over a depth range of 2,164 to 3,322 m (7,100 to 10,900 ft)(Bruce, 1984; Perry and Hower, 1970; Burst, 1969). According to Dunoyer de Segonzac (1970) there is a gradual reorganization of mixed-layer smectite-illite assemblage with increasing depth of burial. Tissot and Welte (1984) report that pure illite had not been formed at 4,000 m (13,000 ft) depth and 160°C. Thus the included sedimentary rocks in the cap rock at Boling Dome must have been originally buried to a similar or greater depth.

The absence of an illitic clay mineral suite in shallow sediments (surface to 800 m [2,600 ft] subsurface) immediately surrounding Boling Dome and nearby salt structures, indicates that the included sedimentary rocks within the cap rock were not incorporated by passive calcite precipitation of overlying sediments. Otherwise the clay mineral suite of the included rocks would more closely resemble that of the surrounding strata. The absence of diagenetic alteration of clay minerals in strata surrounding the salt dome also tends to indicate that deep, hot fluids expulsed up the flanks of Boling Dome were not responsible for the diagenetic alteration of smectite to illite within the cap rock buried at a shallow depth of 300 to 900 m (1,000 to 3,000 ft).

The strata around salt domes have slightly higher geothermal gradients than unaffected sediments. This heating is inferred to result from expulsion of hot basinal fluids up the salt flanks (Keen, 1983), and/or greater thermal conduction of the salt mass (Jensen, 1983; Geertsma, 1971; Selig and Wallick, 1966). The higher geothermal gradient and subsequent hotter subsurface temperature at a given depth, may decrease the depth required for conversion of smectite to illite near salt domes. Given the geothermal gradient of 36°C/km (2°F/100ft) (Seni and others, 1985) around Boling Dome, smectite conversion would occur over the depth range of 2,290 to 3,275 m (7,500 to 10,750 ft).

Included sedimentary strata within the cap rock of Boling Dome have clearly been incorporated with the salt stock and transported vertically from depths exceeding 1,500 to 4,500 m (5,000 to 15,000 ft). The original source of the detrital material is less clearly known owing to extensive diagensis and a simplified mineralogical suite. Two sources are inferred; 1) Tertiary sandstones deposited around the flank of the stock and 2) Jurassic sandstones deposited penecontemporaneously or immediately overlying the Louann Salt. Tertiary sandstones and shales are inferred to have been incorporated by shear along the margins of the salt stock at a depth exceeding that required for conversion of smectite to illite 2,290 to 3,275 m (7,500 to 10,750 ft) and alteration of feldspars 3,000 to 4,500 m (10,000 to 15,000 ft). Jurassic aeolian sandstones may have been originally deposited with the salt or immediately overlying the salt. By this hypothesis, sandstones would have been rafted up by the salt stock, survived subsequent salt dissolution, anhydrite dissolution and replacement by calcite, and arrived at the top of the cap rock as relatively undigested pods of quartz sand within a calcite matrix. The loss of orthoclase and plagioclase feldspars from Tertiary or Jurassic sandstones indicates that the sandstones were buried to great depths before incorporation within the salt stock and that subsequent dome growth brought these sediments to the near surface environment. The two cap rock sandstones containing potassium feldspars may have been incorporated within the salt stock at relatively shallow depths.

The cap-rock shoulder represents a mixing zone where hot, deep basinal fluids discharge up the diapir flanks and react with cool, meteoric waters recharging over the dome crest (Price et al., 1983). The upward migration of basinal fluids and hydrocarbons is indicated by the distribution of oil in cap-rock fractures, in deep Oligocene strata around Moss Bluff, Boling and many other domes, and in Miocene and Pliocene strata over the dome crest. The distribution of radial and concentric ring faults are thought to control access by the mineralizing fluids to the cap rock (Price et al., 1983; Samuelson, personal communication, 1985). The zone of mixing probably migrated back and forth across the margin of the cap rock in response to variations in meteoric recharge and discharge from basinal sources and rates of domal uplift. This, together with variations in pH and the partial pressure of CO_2 produced the characteristically complex diagenesis associated with cap rocks especially at the shoulder of the salt stock.

III. SUMMARY OF MAJOR FINDINGS

Major conclusions of this study are: 1) both calcite and anhydrite cap rock of Boling Dome, in the area studied, incorporate previously deposited terrigenous clastics; 2) terrigenous clastics present within the anhydrite section must have originally been incorporated within the salt stock; 3) terrigenous clastics were incorporated within the salt stock a) by shearing of deeply buried Tertiary strata during dome growth and/or b) by aeolian processes during deposition of the salt; 4) the impoverished quartz-rich coarse grained mineralogical suite and the illitic clay mineral suite indicate that incorporation of sedimentary material within the salt stock occurred at depths exceeding 1,500 m (5,000 ft) and possibly exceeding 4,000 m (13,000 ft); 5) the main mechanism for incorporating Tertiary sediments within the cap rock includes salt dissolution at the diapir crest and underplating of insoluble material and sediments to the base of the cap rock; 6) to a lesser extent, sandstones deposited over the crest of the cap rock also may have been included within the cap rock by replacement of matrix material with calcite from cap rock reactions. It is probable that some cap rock materials have been vertically uplifted thousands of meters. The attendent burial diagenesis and mineralogical simplification of formerly deeply buried materials requires caution when using mineralogical information for provenance information.

The presence of terrigenous clastics within the cap rock indicates that the salt stock of Boling Dome, and by inference other salt domes, may include large blocks of sediment especially around the margin of the dome. These exotic blocks can provide evidence of both the relatively insoluble stratified sediments present within the underlying salt stock, timing and diagenetic alteration of the included strata, and structural history during intrusion of the stock. Sedimentary strata within the cap rock may present the opportunity of sampling rocks not normally accessible owing to their transport by dome growth from deeply buried horizons.

IV. ACKNOWLEDGMENTS

Scant data exists outside the files of cap-rock sulfur mining companies documenting detailed lateral facies in cap rocks. This study only highlights our ignorance of the three dimensional structure and stratigraphy within cap rocks. Texasgulf Sulfur Chemical Co., through Fairis Samuelson, geologist, aided this study by their generous donation of core from Boling Dome and was the original source of data on file at the Railroad Commission of Texas on facies distribution at Moss Bluff Dome (Docket No. 3-72,099). Illustrations were drafted by Annie Kubert, Joel L. Lardon, Kerza A. Prewitt, and T. B. Samsel III under the supervision of Richard L. Dillon. The manuscript was reviewed by Drs. Robert J. Finley, Edward C. Bingler, and

Jules R. DuBar; their timely comments are acknowledged. Editing was by Mary Ellen Johansen and Amanda Masterson. Funding was provided by Texas Water Commission under interagency contract no. IAC(84-85)-2203.

REFERENCES

Bebout, D. G., Loucks, R. G., and Gregory, A. R., 1978, Frio sandstone reservoirs in the deep subsurface along the Texas Gulf Coast--Their potential for production of geopressured geothermal energy: The University of Texas at Austin, Bureau of Economic Geology Report of Investigations No. 91, 92 p.

Bodenlos, A. J., 1970, Cap-rock development and salt-stock movement, in Geology and technology of Gulf Coast salt--A symposium: Baton Rouge, School of Geoscience, Louisiana State Univ., p. 73-86c.

Boles, J. R., 1982, Active albitization of plagioclase, Gulf Coast Tertiary: American Journal of Science 282, 165-180.

Boles, J. R. and Franks, S. G., 1979, Clay diagenesis in Wilcox sandstones of southwest Texas: implications of smectite diagenesis on sandstone cementation: Journal of Sedimentary Petrology 49, 55-70.

Bruce, C. H., 1984, Smectite dehydration--its relationship to structural development and hydrocarbon accumulation in Northern Gulf of Mexico basin: American Association of Petroleum Geologists Bulletin 68, 673-683.

Burst, J. F., 1969, Diagenesis of Gulf Coast clayey sediments and its possible relation to petroleum migration: American Association of Petroleum Geologists Bulletin 53, 73-93.

Collins, E. W., 1985, Review of the geology and Plio-Pleistocene deformation at Damon Mound salt dome, Texas, in Seni, S. J., Collins, E. W., Hamlin, H. S., Mullican, W. F., III, and Smith, D. A., Phase III: Examination of Texas salt domes as potential sites for permanent storage of toxic chemical waste: The University of Texas, Bureau of Economic Geology, report prepared for the Texas Water Commission under Interagency Contract No. IAC(84-85)-2203, 310p.

Davis, J. B., and Kirkland D. W., 1979, Bioepigenetic sulfur deposits: Economic Geology, v. 74, pp. 462-468.

Dunoyer de Segonzac, G, 1970, The transformation of clay minerals during diagensis and low grade metamorphism, a review: Sedimentology 15, 281-286.

Feely, H. W., and Kulp, J. L., 1957, Origin of Gulf Coast salt-dome sulfur deposits: American Association of Petroleum Geologists Bulletin 41, 1802-1853.

Folk, R. L., 1965, Some aspects of recrystallization in ancient limestones: in Pray, L. C. and Murray, R. C., eds., Dolomitization and limestone diagenesis, a symposium: Society of Economic Paleontologists and Mineralogists Special Publication, no. 13, p. 14-48.

Galloway, W. E., 1977, Catahoula Formation of the Texas Coastal Plain: Depositional systems, composition, structural development, ground-water flow history, and uranium distribution: The University of Texas at Austin, Bureau of Economic Geology Report of Investigations No. 87, 59 p.

Geertsma, J., 1971, Finite-element analysis of shallow temperature anomalies: Geophysical Prospecting 19, 662-681.

Goldman, M. I., 1925, Petrography of salt dome cap rock: American Association of Petroleum Geologists Bulletin 7, 42-78.
Goldman, M. I., 1933, Origin of the anhydrite cap rock of American salt domes: U.S. Geological Survey Professional Paper 175, p. 83-114.
Goldman, M. I., 1952, Deformation, metamorphism, and mineralization in gypsum-anhydrite caprock, Sulfur Salt Dome, Louisiana: Geological Society of America Memoir 50, 169 p.
Harrison, J. V., 1930, The geology of some salt plugs in Laristan (Southern Persia): Geological Society of London Quarterly Journal 86, 463-522.
Harrison, J. V., 1931, Salt domes in Persia: Journal Institute of Petroleum Technology 17, 300-320.
Heald, K. C., 1924, Sandstone inclusions in salt in mine on Avery's Island: American Association of Petroleum Geologists Bulletin 8, 674-676.
Hower, J., Eslinger, E. V., Hower, M. E., and Perry, E. A., 1976, Mechanism of burial metamorphism of argillaceous sediment: I. Mineralogical and chemical evidence: Geological Society of America Bulletin 87, 725-737.
Jensen, P. K., 1983, Calculations on thermal conditions around a salt diapir: Geophysical Prospecting 31, 481-489.
Keen, C. E., 1983, Salt diapirs and thermal maturity; Scotian Basin: Canadian Petroleum Geology Bulletin 31, 101-108.
Kent, P. E. 1979, The emergent Hormuz salt plugs of southern Iran: Journal of Petroleum Geology 2, 117-144.
Kreitler, C. W., and Dutton, S. P., 1983, Origin and diagenesis of cap rock, Gyp Hill and Oakwood Salt Domes, Texas: The University of Texas at Austin, Bureau of Economic Geology Report of Investigations No. 131, 58 p.
Kupfer, D. H., 1974, Boundary shear zones in salt stocks: in Coogan, A. H., ed., Northern Ohio Geological Society, Fourth Symposium on Salt, Cleveland, Ohio, v. 1, p. 215-225.
Kupfer, D. H., 1976, Shear zones inside Gulf Coast salt stocks help to delineate spines of movement: American Association of Petroleum Geologists Bulletin 60, 1434-1447.
Kupfer, D. H., 1980, Problems associated with anomalous zones in Louisiana salt stocks, USA: in Coogan A. H. and Hauber, L., eds., Northern Ohio Geological Society, Fifth Symposium on Salt, Cleveland, Ohio, v. 1, p. 119-134.
Kyle, J. R., and Price, P. E., 1985, Mineralogical investigation of sulfide concentrations in the Hockley salt dome cap rock, Texas, in Park, W. C., Hausen, D. M., and Hagni, R. D., eds., Applied Mineralogy: Metallurgical Society of AIME, p. 1065-1082.
Land, L. S. and Milliken, K. L., 1981, Feldspar diagenesis in the Frio Formation, Brazoria County, Texas Gulf Coast: Geology 9, 314-318.
Martinez, J. D., 1980, Salt dome caprock--a record of geologic processes: in Coogan, A. H. and Hauber, L., eds., Northern Ohio Geological Society, Fifth Symposium on Salt, Cleveland, Ohio, v. 1, p. 143-151.
McCracken, W. A., 1967, Petrology of the Catahoula Formation (Miocene) in northeastern Gonzales County, northwestern Lavaca County, and southwestern Fayette County, Texas: University of Houston, Master's Thesis, 127 p.
Milliken, K.L., 1985. Petrology and burial diagenesis of Plio-Pleistocene sediments, northern Gulf of Mexico: Ph.D. dissertation. The University of Texas at Austin.
Milliken, K. L., Land, L. S., and Loucks, R. G., 1981, History of burial diagenesis determined from isotopic geochemistry, Frio Formation, Brazoria County, Texas: American Association of Petroleum Geologists Bulletin 65, 1397-1413.

Milner, C. W. D., Rogers, M. A., and Evans, C. R., 1977, Petroleum transformations in reservoirs: Journal of Geochemical Exploration 7, 101-153.
Morton, J. P., 1985, Rb-Sr evidence for punctuated illite-oirs: Journal of Geochemical Exploration 7, 101-153.
Morton, J. P., 1985, Rb-Sr evidence for punctuated illite-smectite diagenesis in the Oligocene Frio Formation, Texas Gulf Coast: Geological Society of America Bulletin 96 114-122.
Murray, G. E., 1966, Salt structures of Gulf of Mexico Basin--a review: American Association of Petroleum Geologists Bulletin 50, 439-478.
Perry, E. A., Jr. and Hower, J., 1970, Burial diagenesis in Gulf Coast pelitic sediments: Clays and Clay Minerals 18, 165-177.
Price, P. E., Kyle, J. R., and Wessel, G. R., 1983, Salt-dome related zinc-lead deposits: in Kisvarsanyi, G., and others, eds., Proceedings, International Conference on Mississippi Valley-Type Lead-Zinc Deposits: University of Missouri, Rolla, p. 558-571.
Ruckmick, J. C., Wimberly, B. H., and Edwards, A. F., 1979, Classification and genesis of biogenic sulfur deposits: Economic Geology 74, 469-474.
Sassen, R. 1980, Biodegradation of crude oil and mineral deposition in a shallow Gulf Coast salt dome: Organic Geochemistry 2, 153-166.
Selig, F. and Wallick, C. G., 1966, Temperature distribution in salt domes and surrounding sediments: Geophysics 31, 346-361.
Seni, S. J., Mullican, W. F., III, and Hamlin, H. S., 1984, Texas salt domes--aspects affecting disposal of toxic-chemical waste in solution-mined caverns: The University of Texas at Austin, Bureau of Economic Geology, Report prepared for Texas Department of Water Resources under interagency contract no. IAC(84-85)-1019, 94p.
Seni, S. J., Kreitler, C. W., Mullican, W. F., III, and Hamlin, H. S., 1985, Utilization of salt domes for chemical waste disposal: in Smerdon, E. T., and Jordan, W. R., Issues in Groundwater Management, Water Resources Symposium Number 12: Center for Research in Water Resources, The University of Texas at Austin, p. 419-460.
Taylor, R. E., 1938, Origin of the cap rock of Louisiana salt domes: Louisiana Geological Survey Bulletin 11, 191 p.
Teas, L. P., 1931, Hockley salt shaft, Harris County, Texas: American Association of Petroleum Geologists Bulletin 15, 465-469.
Thoms, R. L., and Martinez, J. D., 1980, Blowouts in domal salt: in Coogan, A. H., and Hauber, L., eds., Northern Ohio Geological Society, Fifth Symposium on Salt, Cleveland, Ohio, v. 1, p. 405-411.
Tissot, B. P., and Welte, D. H., 1984, Petroleum formation and occurrence: Springer Verlag, New York, 699 p.
Walker, C. W., 1974, The nature and origin of cap rock overlying Gulf Coast salt domes: Northern Ohio Geological Society, Fourth Symposium on Salt, Cleveland, Ohio, v. 1, p. 169-195.
Wilson, H. H. 1977, Shear zones inside Gulf Coast salt stocks help to delinate spines of movement (discussion): American Association of Petroleum Geologists Bulletin 61, 1090-1093.

MIXED CARBON SOURCES FOR CALCITE CAP ROCKS OF GULF COAST SALT DOMES

Harry H. Posey

The University of Texas at Austin
Bureau of Economic Geology
University Station Box X
Austin, TX 78713

Peter E. Price

Marathon Oil Company
P.O. Box 269
Littleton, CO 80160

J. Richard Kyle

Department of Geology
The University of Texas at Austin
Austin, TX 78713

I. INTRODUCTION

Salt dome calcite cap rocks have been the focus of more isotopic than petrographic or other geochemical examination. Early studies of carbon and sulfur isotope characteristics in cap rocks concluded that methane was the principal carbon source and that anhydrite was the sole sulfur source. More recent studies have reported oxygen as well as carbon and sulfur isotope characteristics, thus providing more information about the fluid sources and temperatures of mineral formation. In this section the principal papers on carbon and oxygen isotope geochemistry of Gulf Coast salt domes are reviewed. The salt domes to be discussed are shown in Figure 1 along with sample locations of major data sources. Most of the data are from Hockley dome (Fig. 1b) and Winnfield dome (Fig. 1c).

Thode et al. (1954) first recognized the extremely light carbon isotope values in calcite cap rocks and attributed them to a methane source. This established a link between cap rocks and hydrocarbon maturation, a subject that has been studied numerous times since. Feely and Kulp (1957), in a comprehensive study of salt dome sulfur formation, also reported light carbon isotope values, likewise attributing these light carbon isotope ratios to a methane source. Feely and Kulp (1957) concluded that calcium in calcite cap rocks is derived from the dissolution of anhydrite and that the reduction of sulfate by bacteria produces calcite (i.e., cap rock) and native sulfur as major by-products.

Sassen (1980) and Woo (1980) both studied cap rocks from the Damon Mound, Texas, but analyzed different materials that produced dramatically different results. Sassen analyzed "secondary" calcites from calcite cap rocks and from the late Oligocene Heterostegina limestone, a patch reef that occurs above the salt dome cap rock separated by several tens of feet of clastic rock. On the other hand, Woo analyzed porites corals and carbonate mudstones from the reef. Sassen reported that carbon isotope values from his samples were light, ranging from 22.1 to -29.2 ^{o}/oo (PDB), typical of other Gulf Coast salt dome values. However, Woo found that reef rocks have $\delta^{13}C$ typical of marine limestones, ranging from -1.7 to +1.7 ^{o}/oo (PDB). According to Sassen, microbial sulfate reduction and the consequent precipitation of calcite can proceed at temperatures no greater than 66°C. Recent studies (Brock, 1985) suggest that bacteria may survive to 200°C, but sulfate reduction is unlikely. $\delta^{18}O$ values reported by Woo are remarkably heavy, ranging from +1.1 to +7.2 ^{o}/oo (PDB). His explanation for the ^{18}O enrichment is that isotopically enriched anhydrite in the salt was added to the ground water by anhydrite dissolution and pumped to the

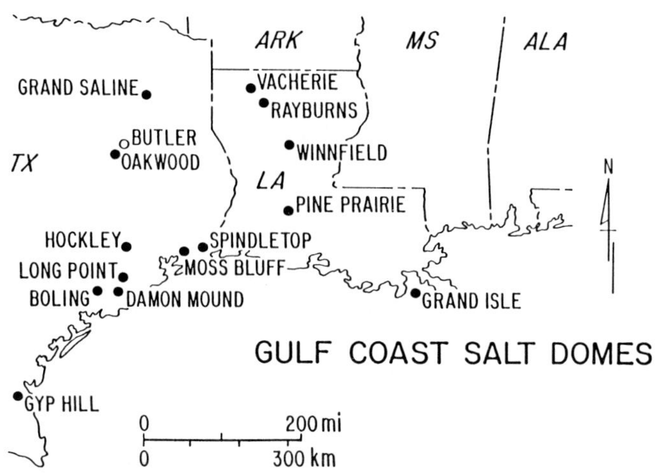

Figure 1A. Location of salt domes of this study and of sample locations for Hockley and Winnfield domes. Grand Saline, Butler, and Oakwood domes are located in the East Texas Basin. Vacherie, Rayburns, and Winnfield are located in the Northern Louisiana Basin. All other domes are located in the Gulf Coastal Basin.

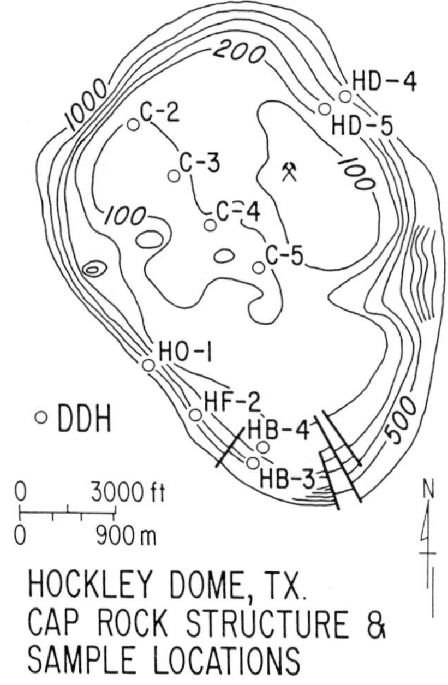

Figure 1B.

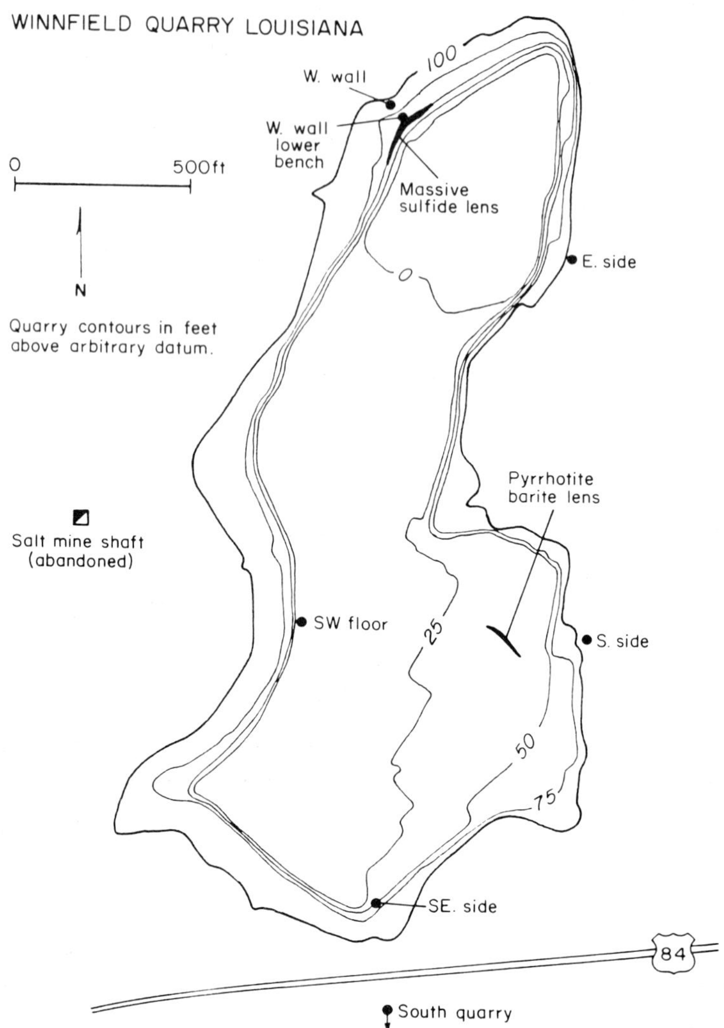

Figure 1C.

seafloor, where it mixed with seawater and produced calcite that is ^{18}O enriched. This is the only report of heavy $\delta^{18}O$ values for salt-dome-associated reefs.

Smith and Kolb (1981) studied cap rocks of Rayburns and Vacherie domes in Louisiana and reported the first oxygen isotope data for true calcite cap rocks. According to them, Vacherie cap rocks are typical of salt domes and have CH_4 carbon sources, whereas Rayburns calcite has recrystallized in the presence of ground water, thus incorporating both heavier carbon and oxygen values from the ground water.

Price and others (1983) reported $\delta^{13}C$ and $\delta^{18}O$ for cap rocks at Hockley dome in Texas. According to them, methanogenic carbon accompanied deep-basin, metal-rich brines to the cap rock, where they formed Fe-Zn-Pb sulfide concentrations.

Kreitler and Dutton (1983) presented a geochemical model for calcite at the Oakwood dome using combined petrographic, trace element, $\delta^{13}C$, and $\delta^{18}O$ analyses. They reported that $\delta^{13}C$ and $\delta^{18}O$ values in the Oakwood dome decrease with depth and that stage II calcites are isotopically heavier than stage I calcites. Kreitler and Dutton (1983) also showed that stage I calcites contain iron sulfides, quartz, and clay whereas stage II calcites are comparatively pure. According to them, stage I calcites have higher concentrations of Na, Cl, Fe, and Al than stage II calcites, which have comparatively high concentrations of Mg, Sr, and Mn. Their model states that two fluids participated in the formation of calcite cap rocks: an early, high-Na-Cl-Fe-Al fluid formed all of the stage I calcite during a single event whereas a late, high-Mg-Sr-Mn fluid dissolved part of the stage I calcite, then reprecipitated mineral-free calcite in the voids created.

Kreitler et al. (1985) reported $\delta^{13}C$ and $\delta^{18}O$ values for "false cap rocks" at the Butler dome, Texas. These consist of calcite-cemented sandstones that have light $\delta^{13}C$. According to them, the light values were derived from deeply-sourced methane that was focused into supracap sediments along a diapir-related fault.

Posey (1986) reported detailed $\delta^{13}C$ and $\delta^{18}O$ analyses for Winnfield, Hockley, and Boling domes and a few values for Long Point and Damon Mound domes, concluding that the lighter-with-depth $\delta^{13}C$ profile found by Kreitler and Dutton (1983) for the Oakwood dome is probably a general characteristic of calcite cap rocks but that stage II calcites are not characteristically heavier than stage I. Posey's model for calcite generation differs from previous ones and is the subject of the present paper. A general cap rock model is described by Light et al. (1986) elsewhere in this volume.

II. GEOLOGY AND GEOCHEMISTRY

A. Cap Rock Descriptions

New nomenclature was developed to explain cap rock variations seen in several salt domes of the Gulf Coast (Posey, 1986) and is shown schematically in Figure 2. Although it is possible that a given salt dome may contain all of the major lithotypes, none are known to contain all types. Major

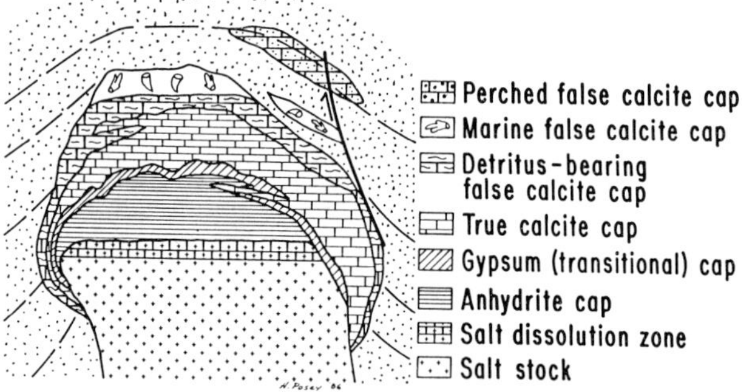

Figure 2. Schematic cross-section showing common cap rock lithotypes, generalized from all cap rocks of the Gulf Coast. Most domes are found to have only detritus-bearing false calcite cap rock, true calcite cap rock, gypsum and anhydrite, and no single dome is known to have all of the lithotypes shown.

lithologic units of cap rock include (1) perched calcite cap rock, (2) marine calcite cap rock, (3) detritus-bearing false calcite cap, (4) true calcite cap, (5) gypsum, and (6) anhydrite. Except as they relate to salt dome calcite cap rock formation, gypsum and anhydrite will not be discussed in this paper.

Perched calcite cap rock consists of calcite-cemented sediments above or adjacent to salt domes that have geochemical compositions indicating affiliation with diapirism. An example is calcite-cemented sandstone at Butler dome, Texas (Kreitler et al., 1985), which has very light $\delta^{13}C$ indicating a methane source (see Thode et al., 1957). Presumably the diapir focused the upward-moving fluids along the outer margins of the diapir where, under the proper conditions, calcite with light $\delta^{13}C$ precipitated as a cement in the adjacent or suprajacent sediments. Others have called this material "false cap rock" (Kreitler et al., 1985).

Marine false cap rock consists of marine carbonates and associated detritus that form on sea floor bulges created by salt diapirs. Examples include the Late Oligocene Heterostegina reef above Damon Mound (Baker, 1979) and the Recent East and West Flower Garden Banks in offshore Louisiana (Rezak, 1985). Similar materials have also been identified at Boling dome and Hockley dome (Posey, 1986) and Winnfield dome (Price, unpublished; and Kyle et al., this volume).

Detritus-bearing false calcite cap rocks are the second most common type of calcite cap rock. They consist of calcite-bearing clastics that grade downward into clastic-bearing calcite then into true calcite cap rock. There are at least two conceivable mechanisms by which detritus-bearing false calcite cap rock may form.

Posey (1986) proposed that detritus-bearing calcite cap rocks form at the anhydrite/sediment interface as the first components of calcite cap rock when anhydrite is dissolved and replaced by calcite. If calcite-calcium is derived from anhydrite as proposed by Feely and Kulp (1957) and Murray (1966), and if all calcium is conserved during the production of calcite, there should be a molar volume loss of 19.6 cm^3 per 100 cm^3 of anhydrite. This would create a void at the anhydrite/sediment interface into which sediment could fall. Assuming part of the anhydrite-Ca is lost to the fluid during the conversion of anhydrite to calcite, the volume loss will be greater. The void created by this process would induce brecciation or slumping of the overlying or adjacent clastics and previously formed calcite (Plate 1). As the front of anhydrite dissolution and replacement proceeds downward and inward, less of the disaggregated sediment would be able to reach the calcite and the facies would change from deritus-bearing calcite cap rocks to true calcite cap rock.

The sub-seafloor model (discussed above) for generating detritus-bearing false calcite cap rock seems applicable to most

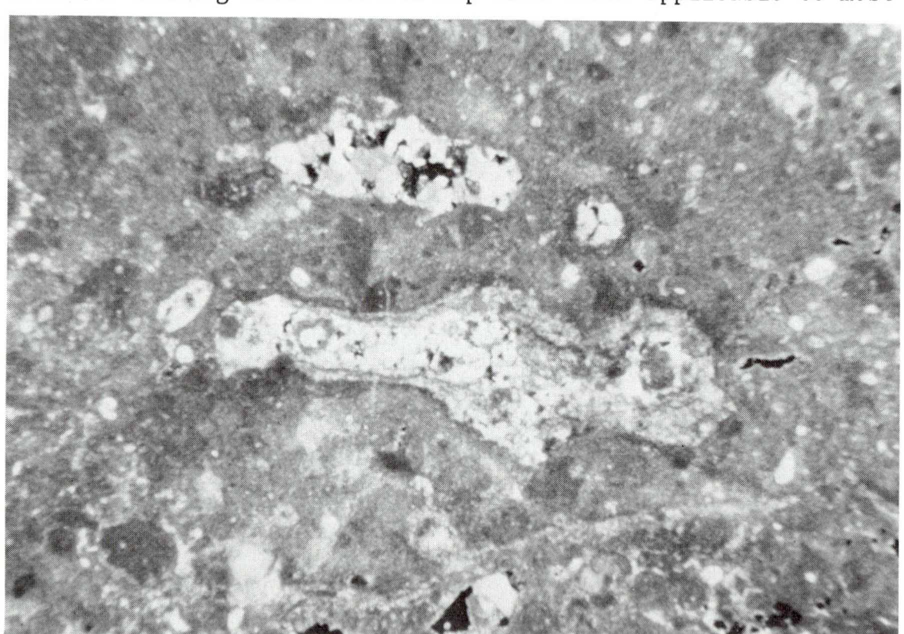

Plate 1. Detritus-bearing false calcite cap rock. HF-2 953; Hockley dome. Plain light. Long dimension of photograph = 6.9 mm.

cases we have studied. However, in most cases it is equally plausible that this type of cap rock may actually form at the sediment/water interface. Brecciation mechanisms would be similar but detritus would not necessarily be lithified or even compacted at the time of deposition.

True calcite cap rock consists of layers and irregular segregations of dark-colored (stage I) and light-colored (stage II) calcite that occur in widely varying proportions. For instance, Hockley calcite cap rock consists dominantly of dark calcite, whereas Winnfield is dominated by pale calcite, especially in the lower sections. True calcite cap rocks are generally the most abundant calcite cap rock type. The dark-colored segregations have phanerocrystalline granular texture, range from various shades of gray to black, and contain clay, quartz, dolomite, and/or pyrite. Stage I calcites provide substrates for stage II calcites, which are typically banded, sparry crystals with crystal terminations in an open cavity (Plate 2). Stage II calcites may be blocky or scalenohedral and range from shades of beige, yellow, and amber to white or colorless. The blocky and scalenohedral forms imply low Mg and high Mg, respectively (Folk, 1974). Both stage I and stage II calcites are commonly brecciated, and healed with stage III calcite. Stage III calcite is virtually identical in form to stage II calcite but occurs in veins that crosscut both stage I and stage II calcite. Brecciation is characteristic of both detritus-bearing false cap and true calcite cap.

Stage I and stage II calcite probably form successively in layers that parallel an advancing front of anhydrite dissolution and calcite precipitation (Posey, 1986). Stage I calcite probably crystallizes quickly from supersaturated solutions, thus providing a granular texture and incorporating clay, quartz, and dolomite residues left by the digestion of anhydrite (Posey, 1986). Pyrite generally occurs within stage I calcite; it

Plate 2. True calcite cap rock. PWD-5; Winnfield Dome. Crossed-nicols. Long dimension of photograph = 6.9 mm.

probably forms from iron in the carbon bearing brine and sulfur produced by local sulfate reduction (Feely and Kulp, 1957) or, as suggested by Price et al. (1983), from the brine. Stage II calcite probably precipitates slowly from fluids remaining after stage I calcite precipitation or from other fluid sources. Stage II calcite lines walls of stage I calcite, growing into the open spaces created by anhydrite dissolution.

Anhydrite cap rock appears to form before all other cap rock lithologies when salt is dissolved leaving a residue of anhydrite. However, a cover of marine false calcite cap rock or other marine carbonate may be a stabilizing feature that allows anhydrite to accumulate rather than dissolve. Originally, anhydrite and salt were derivatives of sea water, but anhydrite probably recrystallizes in the presence of other fluids (Posey,

1986). The phenomenon of anhydrite cap rock accumulation is well described in various reports, notably Murray (1966) and Martinez (1974).

Gypsum cap rock forms predominantly after all other cap rock lithologies when interstitial fluids cool and/or grow less saline. Gypsum is most prominent between anhydrite and calcite cap rocks and often fills open spaces between crystals and breccia fragments of both these lithologies. Hence, this cap rock type is often called transitional cap rock. According to Holland and Malinin (1979), gypsum is not stable above about 75°C, so deposition is a moderately low temperature phenomenon, probably related to the incursion of late meteoric water.

B. Carbon Isotope Data

$\delta^{13}C$ values for detritus-bearing false cap and true calcite caps are shown with $\delta^{18}O$ in Figure 3 and Table I (see Hudson, 1977 for comparison). In constructing this diagram, all samples from the Rayburns boulder zone and the uppermost samples of the Vacherie dome (Smith and Kolb, 1981) were considered to be detritus-bearing false cap rock because lithologic descriptions indicate considerable detritus and a general absence of dark-light banding (Nance and Wilcox, 1979; Nance et al., 1979). For similar reasons the uppermost samples at Oakwood dome (Kreitler and Dutton, 1983) were also interpreted to be detritus-bearing false calcite cap rocks.

Normal limestone $\delta^{13}C$ values generally cluster around 0±5 $^{o}/oo$ (PDB) (Hoefs, 1973; Hudson, 1977; James and Choquette, 1984; and numerous others). However salt dome cap rocks are quite ^{13}C-depleted compared with such limestone, ranging from -5 to -51 $^{o}/oo$ (Figs. 3 and 4). False calcite cap rocks are generally ^{13}C-enriched compared with true caps; false calcite cap ranges from about -5 to -25 $^{o}/oo$ whereas true cap ranges from about -12 to -51 $^{o}/oo$ for all cap rocks inclusive.

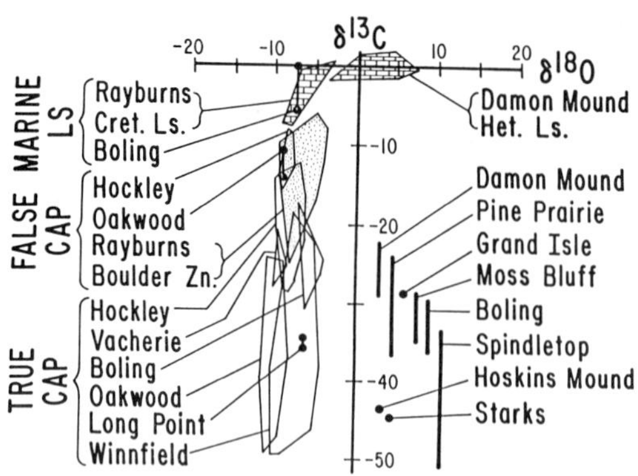

Figure 3. $\delta^{13}C$ vs $\delta^{18}O$ of calcite for all domes. Oxygen isotopes of salt dome cap rocks are approximately the same as the range of bulk rock analyses of marine limestones. However, carbon isotopes are extremely light compared with marine limestones. Detritus-bearing false calcite cap rocks are enriched in ^{13}C compared with true calcite cap rocks.

The lighter-with-depth $\delta^{13}C$ profile at Oakwood reported by Kreitler and Dutton (1983) was also observed at Hockley and Winnfield (Posey, 1986; and Price, unpublished). A profile for Winnfield is shown in Figure 5 and Table I. $\delta^{13}C$ values at Winnfield grow generally lighter with depth, ranging from an average of about -25 at the top to about -40 near the base. The lightest value, -53.3 o/oo is the lightest $\delta^{13}C$ value recorded for any calcite cap of the Gulf Coast.

Kreitler and Dutton (1983) reported that stage II calcites at Oakwood dome have heavier $\delta^{13}C$ than adjacent stage I calcites. This is also the case for most of the Winnfield dome samples (Posey, 1986; Price, unpublished). However, this is not necessarily so for Hockley dome calcite. A cross section of Hockley calcite (Fig. 6) shows that (1) $\delta^{13}C$ decreases with depth as well as toward the edge of the dome and (2) stage II calcite is typically lighter than adjacent stage I. The average difference in $\delta^{13}C$ between stage I and stage II calcites is similar for all domes. At Oakwood (Kreitler and Dutton, 1983) this difference ranges from +8.9 to -3.9 and averages +2.3. At

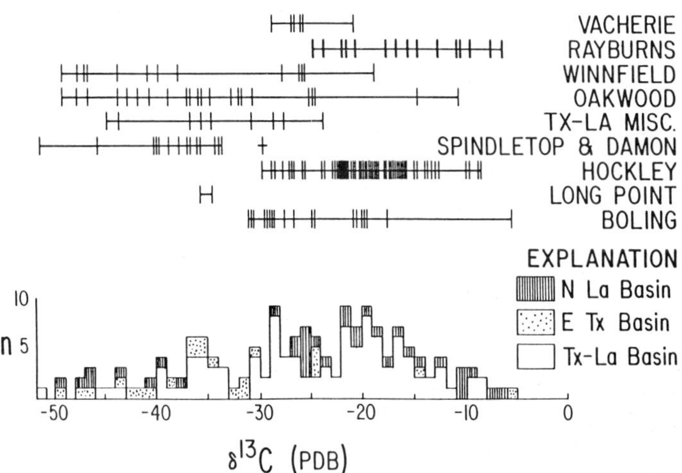

Figure 4. Histogram showing δ13C values for all reported calcite cap rocks from the Gulf Coast. Data from Thode et al., (1954), Feely and Kulp (1957), Sassen (1980), Woo (1980), Smith and Kolb (1981), Price et al. (1983), Kreitler and Dutton (1983), and Posey (1986). Data from Price (this report) is not shown. The largest percentage of heavy δ13C is from domes with a high percentage of detritus-bearing false calcite cap rocks. Data lines on right-hand side of diagram are for δ13C only. Oxygen isotope analyses are not available for these data.

Winnfield the difference ranges from +14.1 to -22.2 and averages -1.14 (Table 1). At Hockley the $\delta^{13}C$ difference between early- and late-stage calcite ranges from +3.5 to -8.4 and averages -1.6. Most (seven out of nine) of the stage II calcites at Oakwood are heavier than adjacent stage I calcites; at Winnfield, however, only about half (six out of eleven) are heavier; and at Hockley, only a few (three out of sixteen) are heavier. Thus, the process that leads to $\delta^{13}C$ differences between early- and late-stage calcites is not the same at each dome, because each shows different general isotopic characteristics between early- and late-stage calcites.

Winnfield, Oakwood, Spindletop, and Moss Bluff domes have lighter, average $\delta^{13}C$ than Vacherie, Rayburns, Hockley, Boling,

Table 1. Carbon and oxygen isotope analyses of calcite cap rock of Gulf Coast salt domes.

Sample No.	$\delta^{13}C$ (PDB)	$\delta^{18}O$ (PDB)	Calcite Type
HOCKLEY[1]			
C2 88.3	-18.4	-7.6	CA
C2 88.3	-19.5	-7.7	CB
C2 93.5	-19.6	-7.5	CA
C2 102.2	-19.0	-7.6	CA
C2 102.2	-20.2	-7.9	CB
C2 114.7	-18.9	-7.9	CA
C2 114.7	-19.3	-8.1	CB
C2 126.	-19.9	-8.0	CA
C3 84.3	-17.1	-7.4	CA
C3 84.3	-17.4	-8.1	CB
C3 93.8	-12.2	-7.0	CA
C3 93.8	-18.8	-7.6	CB
C3 101.7	-15.6	-7.5	CA
C3 101.7	-16.8	-6.4	CB
C4 106.3	-16.1	-6.5	CA
C4 115.	-16.9	-7.4	CA
C4 115.	-17.1	-7.4	CB
C4 121.	-16.1	-7.6	CA
C4 121.	-16.3	-7.7	CB
C5 131.4	-15.2	-7.7	CA
C5 131.4	-14.0	-8.1	CB
C5 142.6	-15.7	-7.4	CA
C5 153.	-15.8	-7.9	CA
HB4 304.2	-12.8	-7.6	CA
HB4 314.	-20.8	-7.9	CB
HD4 719.4	-15.9	-9.3	CA
HD4 719.4	-21.8	-6.5	CB
HD4 799.3	-26.4	-8.8	CA
HD4 799.3	-27.5	-7.6	CB
HD4 835.9	-23.4	-8.7	CA
HD4 835.9	-28.3	-9.0	CC
HD4 835.9	-22.9	-8.6	CD
HD4 863.6	-28.6	-8.2	CA
HD4 863.6	-26.5	-8.5	CB
HD4 836.6	-25.1	-9.0	CC
HO HD5 354	-9.2	-8.0	CA
HO HD5 354	-12.2	-8.2	CA
HO HD5 384	-9.0	-8.0	CA
HO HD5 384	-8.1	-8.5	CB
HO HF2 953	-8.9	-8.0	CA
HO HF2 1043	-13.7	-9.5	CA
HO HF2 1043	-14.5	-9.8	CB
HO HF2 1043	-20.1	-9.8	CC
HO HF2 1043	-22.1	-7.3	Vein
HO HF2 1092	-18.4	-8.2	CA
HO HO1 714.6	-21.4	-8.6	CB
HO HO1 714.6	-21.0	-8.6	CC
WINNFIELD[2]			
WD-2-T-C-RK	-27.23 ±0.010	-7.56 ±0.024	False Cap
WD-2-T-A	-28.47 ±0.028	-7.70 ±0.052	False Cap
WD-2-T-C-V	-28.02 ±0.037	-5.05 ±0.034	Vein
WD-2-T-10A-FBNDD	-28.61 ±0.028	-7.64 ±0.030	False Cap
WD-2-T-10A-RK	-36.12 ±0.028	-9.00 ±0.008	False Cap

TABLE 1. (cont.)

WD-2-T-20A1	-22.12 ±0.024	-5.09 ±0.018	False Cap
WD-2-T-20A2	-30.33 ±0.039	-7.61 ±0.027	False Cap
(Duplicate)	-30.24 ±0.029	-7.70 ±0.033	False Cap
WD-2-T-20A3	-36.02 ±0.014	-9.62 ±0.028	Vein
WD-2-T-20B	-30.54 ±0.027	-7.51 ±0.020	False Cap
WD-2-T-30A1	-38.27 ±0.026	-10.29 ±0.007	Vein
WD-2-T-30A2	-38.77 ±0.022	-10.15 ±0.020	Vein
WD-2-T-30A3	-38.85 ±0.024	-8.60 ±0.025	Light
WD-2-T-30A4	-38.42 ±0.028	-8.43 ±0.024	Dark
WD-2-T-30C	-46.02 ±0.022	-9.28 ±0.019	Vein
WD-2-T-30E	-38.61 ±0.033	-10.40 ±0.040	Vein
WD-2-T-40B1	-40.32 ±0.016	-9.11 ±0.043	Light
WD-2-T-40B2	-44.40 ±0.029	-9.63 ±0.030	Dark
WD-2-T-50A1	-38.08 ±0.037	-8.88 ±0.048	Air Leak
(Duplicate)	-38.29 ±0.014	-8.39 ±0.026	Dark
WD-2-T-50A2	-46.70 ±0.029	-9.43 ±0.019	Light
WD-2-T-50B1	-40.96 ±0.014	-8.51 ±0.020	Dark
WD-2-T-50B2	-49.64 ±0.028	-9.06 ±0.026	Light
WD-2-T-60A-L1	-41.51 ±0.037	-8.95 ±0.011	Light
WD-2-T-60A-L2	-40.56 ±0.020	-8.93 ±0.027	Light
WD-2-T-60A-D1	-45.73 ±0.027	-8.73 ±0.037	Dark
WD-SQ-B2-A	-52.46 ±0.034	-9.45 ±0.040	?
WD-SQ-B2-B	-52.75 ±0.027	-9.28 ±0.031	?
WD-SQ-B1	-53.29 ±0.021	-8.79 ±0.016	?
WD-SQ-Xls	-41.81 ±0.027	-7.38 ±0.024	Vein
WINNFIELD[1]			
WF PWD2 10	-18.4	-10.2	CA
WF PWD2 10	-40.6	-9.6	CD
WF PW51 10	-37.7	-9.4	CB
WF PWES 20	-49.5	-10.0	CA
WF PWES 20	-47.0	-10.2	CB
WF E 50	-47.2	-10.1	CA
WF E 50	-46.5	-9.6	CB
WF W 50	-19.6	-7.3	CB
WF W 50	-18.2	-7.7	CA
WF S 75	-39.2	-9.4	CA
WF S 75	-25.1	-6.4	CB
WF SQ 10	-47.0	-9.1	CA
WF SQ 10	-43.9	-8.5	CB
WF MS 100	-25.2	-6.5	CB
LONG POINT[1]			
LP A 848	-34.6	-6.2	CA
LP B 898	-36.6	-6.2	CA
BOLING[1]			
BO CH52 1924	-20.3	-5.9	CA
BO CH52 1940	-19.4	-6.3	CA
BO CH52 1940	-30.2	-6.2	CB
BO CH52 2345	-27.5	-5.8	CC
BO CH52 2345	-26.4	-4.4	CB
BO CH53 2251	-19.2	-6.9	CA
BO CH53 2251	-17.4	-7.1	CB
BO CH53 2251	-5.4	-7.2	CC
BO CH53 2251	-0.1	-7.4	CD
BP CH53 2277	24.7	-4.0	CA
BO CH53 2277	-24.4	-5.0	CB
BO CH53 2286	-20.9	-5.6	CA
BO CH53 2286	-19.7	-6.6	CB

TABLE 1. (cont.)

BO TA64 2091	-29.0	-6.5	CA
BO TA64 2115	-30.8	-6.1	CA
BO TA64 2137	-28.6	-5.9	CA
BO TA64 2165	-28.3	-7.6	CA
BO TA64 2165	-28.3	-5.6	CB
DAMON MOUND[1]			
DM CAP	-29.2	-5.7	CA
DM REEF	-2.0	-3.5	CA

[1]Data from Posey (1986). Precision for both carbon and oxygen isotope analyses is ±0.2 ^{o}/oo or better. Samples with lower in-run precision were analyzed a second or more times until precision was assured. Separate CO2 gases prepared from the same rock powder were repeatable within 2.2 percent or less. Accuracy varies with the relative value of the sample. For instance, isotopically light samples have higher calculated error (±2.2% for δ13C = -47.0 ^{o}/oo) than isotopically heavy samples (about ±0.1% for δ13C = -1.9 ^{o}/oo).

[2]Data from Price (unpublished)

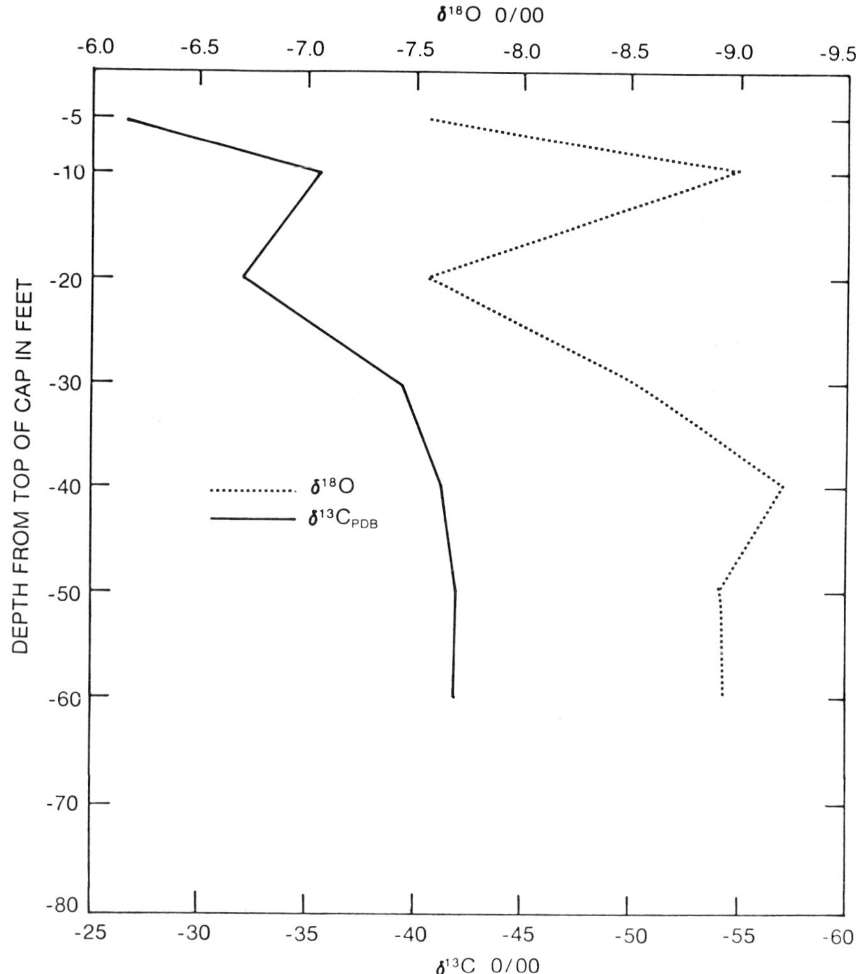

Figure 5. Profile of Winnfield Dome showing δ13C and δ180 of calcite cap rocks. The transition from detritus-bearing false cap rocks and true calcite cap rocks occurs between 20 and 30 ft depth. The top of the Winnfield Quarry is assumed to be 0 ft depth. Both δ13C and δ180 become lighter with depth.

and Damon Mound (Figs. 3 and 4). The cause of these variations may be variations in the source materials, differences in the temperature of formation, or both. Calcite precipitated at high temperatures in equilibrium with a fluid of fixed $\delta^{13}C$

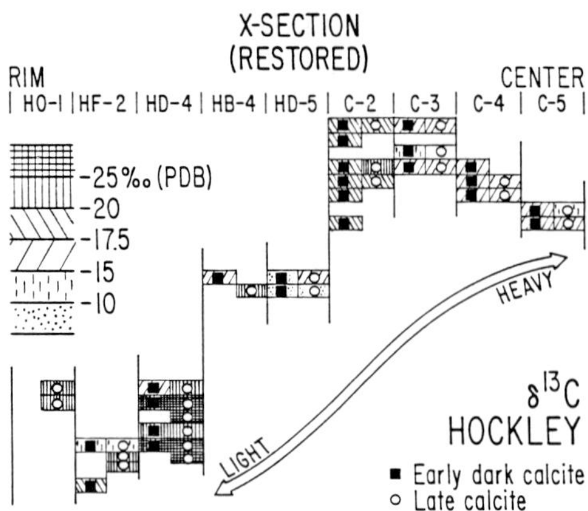

Figure 6. Circular cross section showing $\delta^{13}C$ values of stage I dark calcite and stage II pale calcite at Hockley dome (see Fig. 1 for location of samples). Values become generally lighter toward the margin of the dome and toward the base of each drill hole.

composition would have lighter $\delta^{13}C$ than calcite precipitated at lower temperatures from the same fluid, provided the fluid $\delta^{13}C$ and the carbon species were the same.

C. Oxygen Isotope Data

Calcite cap rock $\delta^{18}O$ values in Gulf Coast salt domes range from -4 to -11.2 $^{o}/oo$ (PDB) (Fig. 3). Like $\delta^{13}C$, $\delta^{18}O$ values also decrease, generally, with depth. Stage II calcite is typically lighter than adjacent stage I at Oakwood (Kreitler and Dutton, 1983) and Hockley (Posey, 1986), but Winnfield calcites have mixed trends. For nineteen dark-light calcite pairs at Hockley, the difference between stage I and stage II calcite $\delta^{18}O$ ranges from +2.8 to -0.8, but fifteen of those pairs show nil to slightly negative shifts. Oakwood ranges from +0.7 to -0.6 for nine dark-light pairs, seven of which have nil to negative shifts (Kreitler and Dutton, 1983). Winnfield is more mixed; differences range from +3.0 to -1.0 for eleven dark-light pairs, seven of which have negative shifts. Thus, if there is a general trend in difference between stage I and adjacent stage II calcites it is not clear from the data at hand.

Calcite cap rock $\delta^{18}O$ values lie within the same range as limestone of all ages as reported by Hudson (1977), though within the isotopically heavy portion of that range. Rayburns, Hockley, Boling, Long Point, and Damon Mound are typically heavier than -8 whereas Vacherie, Winnfield, Butler, and Oakwood are typically lighter than -8 (Fig. 7). If the Butler dome data, which represent perched calcite cap rocks, and the Rayburns data, all of which are false cap rock samples, are omitted from consideration, the East Texas Basin and Northern Louisiana Basin data segregate from the Gulf Coast Basin group which are, by comparison, isotopically heavy. As with carbon, heavy isotope enrichment indicates either heavier source materials or lower temperatures of crystallization.

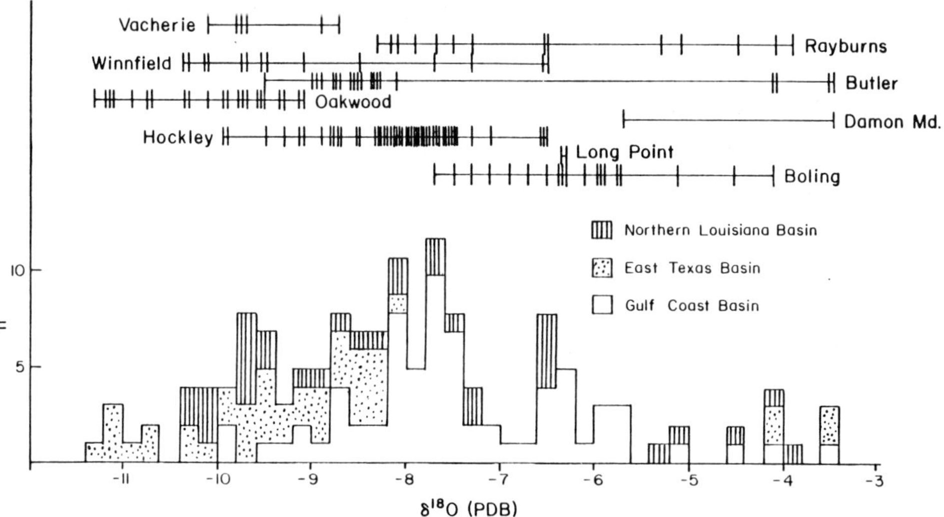

Figure 7. Histogram showing $\delta^{18}O$ values for all salt domes in the Gulf Coast. Data from Woo (1980), Smith and Kolb (1981), Price et al. (1983), Kreitler and Dutton (1983), and Posey (1986). New data for Winnfield (this report) are not shown.

III. DISCUSSION

For Gulf Coast salt dome cap rocks there are several characteristic relationships bearing evidence of cap rock origins, in general, and other characteristics bearing evidence about origins of individual salt domes or domes within a specific province or family. The following characteristics express general relationships about calcite cap rocks of the Gulf Coast:

1. The majority of all calcite cap rock consists of detritus-bearing false calcite cap and true calcite cap. False cap occurs generally above true cap and contains more siliciclastic material. Contacts between true and false cap rocks are sharp to gradational. True calcite contains more late-stage open-space-fill calcite. Both are crosscut by calcite veins.

2. Within true calcite cap rocks $\delta^{13}C$ values of stage I calcite are generally within 5 $^{o}/oo$ of adjacent, stage II calcite, and over half are within 2 $^{o}/oo$. The sense of change between stage I and II (whether positive or negative) is different for individual domes and may vary widely within a single dome.

3. $\delta^{13}C$ and $\delta^{18}O$ values grow generally lighter with depth, suggesting either that the calcite cap rock formed progressively from top to base from a finite carbon reservoir, preferentially incorporating ^{13}C in the early, upper zones, or that the carbon reservoir composition changed through time.

4. Most cap rocks in the Houston diapir province in the Gulf Coast basin have heavier $\delta^{13}C$ values than domes elsewhere in the Gulf Coast, East Texas, and Northern Louisiana Basins, suggesting either different carbon sources or different source- or precipitation-temperatures.

5. Most cap rocks in the Gulf Coast Basin have heavier $\delta^{18}O$ than the interior basins.

6. Stage II calcite has generally lighter $\delta^{18}O$ than adjacent stage I calcite, suggesting formation at higher temperature, formation from the remains of earlier formed calcite that preferentially incorporated heavy carbon, or mixing with a different fluid.

The wide $\delta^{13}C$ spread may be explained, in part, by temperature-controlled carbon isotope fractionation. However, if salt dome cap rocks formed over a narrow temperature range, say <50°C, temperature variation alone cannot explain all of the variation (see Kreitler and Dutton, 1983). CH_4-CO_2 fractionation decreases with increasing temperature so that, at high temperatures, methane in equilibrium with carbon dioxide is isotopically heavier than at low temperatures. Although this fractionation difference may explain part of the calcite $\delta^{13}C$ range (Bottinga, 1969), a temperature difference of almost 500°C would be necessary to explain the -5 to -50 ‰ $\delta^{13}C$ range observed in Gulf Coast calcite cap rocks. Virtually all of the $\delta^{13}C$ values require a methane-carbon source but, clearly, a second component is required to explain the very wide range of values under reasonable temperature conditions.

Studies of methane in Gulf Coast fluids compiled by Deines (1980) show that shallow methane $\delta^{13}C$ ranges from about -80 to -20 ‰ (PDB) and grows proportionally heavier with depth (Fig. 8). Rice (1980) also found this relationship for Gulf Coast brines and noted that the rate of change in methane $\delta^{13}C$ is constant over a given area but is related to the geothermal gradient, higher gradients having greater rates of change. Thus, it is likely that differences in cap rock $\delta^{13}C$ indicate CH_4 sources from different depths or methane sources of similar depths but with different geothermal gradients.

Under ideal conditions, carbon isotope fractionation may be used to calculate the temperature of formation of calcite and the temperature of the carbon source. However, where calcite precipitates from a parent mixture of methane and carbon dioxide, as is probably the case in cap rock environments, this temperature calculation cannot be made without knowing or assuming the proportions of the two carbon sources. Some simple modeling considerations demonstrate this point. Beginning with a methane carbon source, carbon isotope fractionation occurs between CH_4(g) and CO_2(g), between CO_2(g) and HCO_3^-, and between HCO_3^- and solid carbonate. At 25°C the fractionation between CO_2 and CH_4 is huge, almost 70 ^{o}/oo, and even quite large at higher temperatures; 49 ^{o}/oo at 100°C, for example (Bottinga, 1969). Assuming, for the sake of simplicity, that CH_4 is completely oxidized to CO_2 at 25°C so that all of the CH_4 carbon is incorporated in the CO_2 and no fractionation occurs, the resulting CO_2 must be further transformed to HCO_3^- (fluid) then combined with Ca to form calcite. According to Bottinga (1968) (corrected according to the conventions of Friedman and O'Neil, 1977) the fractionation between CO_2(g) and calcite is about 10.2 ^{o}/oo at 25°C or about 4.1 ^{o}/oo at 100°C. Thus, if methane carbon, which is completely oxidized to CO_2 at 100°C, produces calcite at that temperature the resulting calcite will be 4.1 ^{o}/oo heavier than the parent methane. Of course, incomplete oxidation of methane and lower temperatures will produce even greater differences between the methane source $\delta^{13}C$ and the product calcite $\delta^{13}C$.

Considering the above, the temperature range over which methane was generated can be approximated using the tabulation of Deines (1980) (Fig. 8). To produce calcite with $\delta^{13}C$ = -53.3 ^{o}/oo (the lightest value on record for Gulf Coast cap rocks), methane $\delta^{13}C$ would need to be a minimum of 11 ^{o}/oo lighter or -64.3 ^{o}/oo for precipitation at 25°C and a minimum of 4.2 ^{o}/oo lighter or -57.5 ^{o}/oo for precipitation at 100°C. Methane with these values is found only at temperatures below about 90 to 100°C (Figure 8). This assumes that carbon dioxide is not

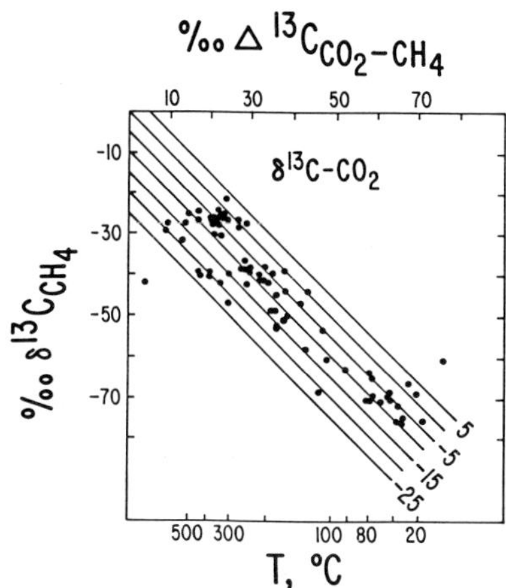

Figure 8. Chart showing the relationship between δ13C of CH_4 and the carbon isotope fractionation between CH_4 and CO_2 in natural gases (from Deines, 1980). Points represent methane carbon isotope and temperature measurements from natural gases. Canted lines represent carbon-dioxide δ13C equilibrium lines from Bottinga (1969). Although δ13C of methane shows considerable temperature related variation, CO_2 is fairly constant.

present, which is unlikely. The presence of carbon dioxide would increase the average $\delta^{13}C$, thus requiring even lower temperatures. According to Light et al. (1984) deep geopressured fluids in the Gulf Coast contain up to 10% carbon dioxide and are saturated with methane, meaning the carbon dioxide contribution will be small but measureable. Hence, the most likely source of extremely light carbon isotopes in calcite found at Winnfield, Spindletop, Oakwood, and Moss Bluff was shallow, low-temperature methane.

Calcite with heavy $\delta^{13}C$ could have been derived from several sources: (1) deeper, higher temperature fluids; (2) isotopically-enriched residual products of earlier calcite precipitation; (3) dissolved limestone or; (4) seawater. The first mechanism implies either an "infinite" or, at least, a very large carbon reservoir, whereas the latter mechanism implies a finite reservoir. The third and fourth alternatives imply simple mixing.

If, as proposed by Feely and Kulp (1957), bacteria use carbon or hydrogen as an energy source to metabolize SO_4^{--} and, in the process, produce calcite, the rate of sulfate reduction would be controlled by temperature (high temperatures would inhibit metabolism) and by the amount of carbon in the system. The production rate also depends on ionic strength of the solutions. For a given calcite pair (stage I and stage II calcite), it appears that the rate of calcite production waned in the later stages. Stage I calcites have granular texture indicative either of rapid precipitation from highly oversaturated solutions or nucleation around numerous nucleation sites provided by the abundant detritus. On the other hand, stage II and stage III calcites have larger crystals as well as banded, open-space-fill textures indicative either of slower crystal growth and, hence, lower ionic strength solutions or higher temperature, or both. The textures of stage I-stage II calcite pairs are similar throughout the calcite cap rock, top to bottom, although the ratio of stage I to stage II decreases with depth (Price, unpublished).

The infinite carbon reservoir mechanism seems unnecessarily complex. This mechanism requires literally hundreds of pulses of hot, deep-basinal fluid to explain the heavy $\delta^{13}C$ values found in some stage II calcites. However, not all stage II calcites are enriched compared with adjacent stage I calcite, so this mechanism cannot explain the light, stage II samples. On the other hand, the finite carbon reservoir seems equally implausible when considering that stage II calcites may be either enriched or depleted compared with stage I. Overall, the inconsistent isotopic characteristics of stage I and adjacent stage II calcites suggest origins by mixing rather than by any progressive precipitation phenomenon. On the weight of textural evidence, calcite cap rocks seem to have formed from two carbon reservoirs that periodically introduced light carbon (CH_4) and heavy carbon (dissolved limestone or seawater) to the calcite/anhydrite boundary. The boundary, in turn, probably migrated progressively downward.

$\delta^{18}O$ offers mixed interpretations. According to Hudson (1977), James and Choquette (1984), and Given and Lohmann (1985), $\delta^{18}O$ values lighter than about 0 $^{o}/oo$ ±1 are characteristic of meteoric water diagenesis. This implies fairly low temperatures indicative of shallow burial. However, Kreitler and Dutton (1983) noted a resemblance between calcite $\delta^{18}O$ values in deeply buried Gulf Coast sediments (as reported by Milliken et al., 1981) and the cap rock calcite $\delta^{18}O$ at Oakwood, and deduced that calcite cap rocks formed from these fluids. Implicit in this deduction is that calcite cap rock must have precipitated at the fluid's original reservoir temperature. Considering the loss of heat en route upwards and the likelihood that sulfate reduction proceeded at low temperatures (i.e., <70°C: see Sassen, 1980; or <100°C: see Trudinger et al., 1985), this latter hypothesis is unlikely, as noted by Kreitler and Dutton (1983).

It seems instead that the $\delta^{18}O$ values of calcite cap rocks represent a mixture between two or perhaps three fluids. Figure 9 shows the probable stability field for $\delta^{18}O$ in calcite and water for Gulf Coast salt domes. This field is assumed to lie between 30°C and 70°C. The upper limit, 70°C, is assumed to be the bacterial limit for sulfate metabolism, and the lower temperature, 30°C, is typical of very shallow burial conditions. If all the calcite precipitated at high temperature, then it could be explained as a mixture of hot, ^{18}O-enriched formation fluids and marine water. On the other hand, if the calcite precipitated at low temperature, the components would consist of marine water and isotopically light, meteoric water.

These parameters assume that (1) anhydrite is the principal source of Ca in calcite, (2) sulfate reduction is concomitant with calcite precipitation, and (3) sulfate reduction is

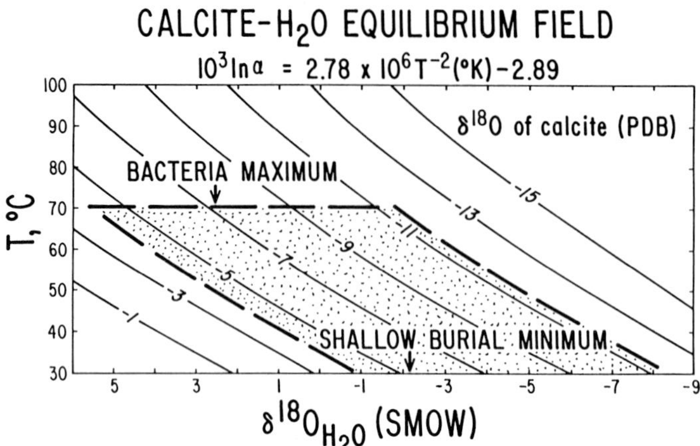

Figure 9. Chart showing the δ18O stability field for calcite in equilibrium with water. Equilibrium constant (α) is from Friedman and O'Neil (1977). Water values are shown relative to SMOW (standard mean ocean water) and calcite equilibrium lines are shown relative to PDB (Peedee belemnite). The stippled field outlines the range of conditions under which calcite may have formed. The upper and lower temperature limits are examples only, but represent the probably limit fixed by bacterial reduction of sulfate and very shallow burial. However, if bacteria can, in fact, reduce sulfate at higher temperatures, the field would shift to heavier δ18O-water values. Also, if calcite formed at or near the seafloor, the field would shift toward lighter values because of lower temperatures on the ocean bottom. From this diagram it appears that the range of calcite δ18O values in salt dome cap rocks (-4.0 to -11.2 $^o/oo$) can be explained either by precipitation over a broad temperature range or precipitation from two different fluids. (From Posey, 1986).

ineffective above 70°C. However, there is growing evidence that sulfate reduction may proceed biogenically above 70°C. For instance, Brock (1985) noted that bacteria that use sulfur are known to exist at temperatures as high as 110°C. However, they are known to reduce sulfate. If so, and if Ca is supplied by the C-bearing brine, then it is conceivable that all of the waters are formation fluids.

Considering that very light carbon isotope ratios could only have come from shallow burial depths (depths equivalent to <100°C) and considering that cap rocks form during late diapirism, probably when the surrounding sediments are bathed in seawater rather than meteoric water, it seems most plausible that calcite cap rocks formed at elevated temperatures (at most, 100°C and probably closer to 70°C) from a mixture of deep-basin formation fluids and seawater. The deep-basin formation fluids, themselves, have meteoric water signatures.

The comparative enrichment of heavy carbon and oxygen isotopes found in several of the domes, notably Hockley, Boling, Rayburns, and Vacherie, may be due to several factors. Lower precipitation temperatures, deeper carbon and oxygen fluid sourcing, greater abundances of dissolved limestone or dolomite, and, in the case of carbon, a higher CO_2:CH_4 ratio would all favor heavy isotope enrichment. The correlation between low, marine-type $^{87}Sr/^{86}Sr$ and heavy carbon and oxygen at Hockley dome (Posey, 1986) suggests that limestone dissolution added heavy carbon and oxygen to the uppermost calcite cap rock samples at Hockley. However, the variation in $^{87}Sr/^{86}Sr$ is low in other domes and not systematic (Posey, 1986). Thus, it is not certain that limestone dissolution can explain the carbon and oxygen isotope variation in all domes. Until more is known about the organic geochemistry of these carbonates, it may not be possible to determine the causes of carbon and oxygen isotope variation among domes.

Calcite with comparatively heavy carbon and oxygen isotopes occurs in detritus-bearing false calcite cap rocks (Fig. 10). The clastic detritus that occurs in these rocks is from one of two sources: (1) contemporaneous detritus deposited on the sea floor at the time of false calcite cap rock formation, or (2) older detritus that is intruded by the diapir and incorporated in the upper calcite units during early calcite cap rock formation. It is not clear whether the heavy isotopes in these rocks are a product of low formation temperatures, seawater carbon, or both.

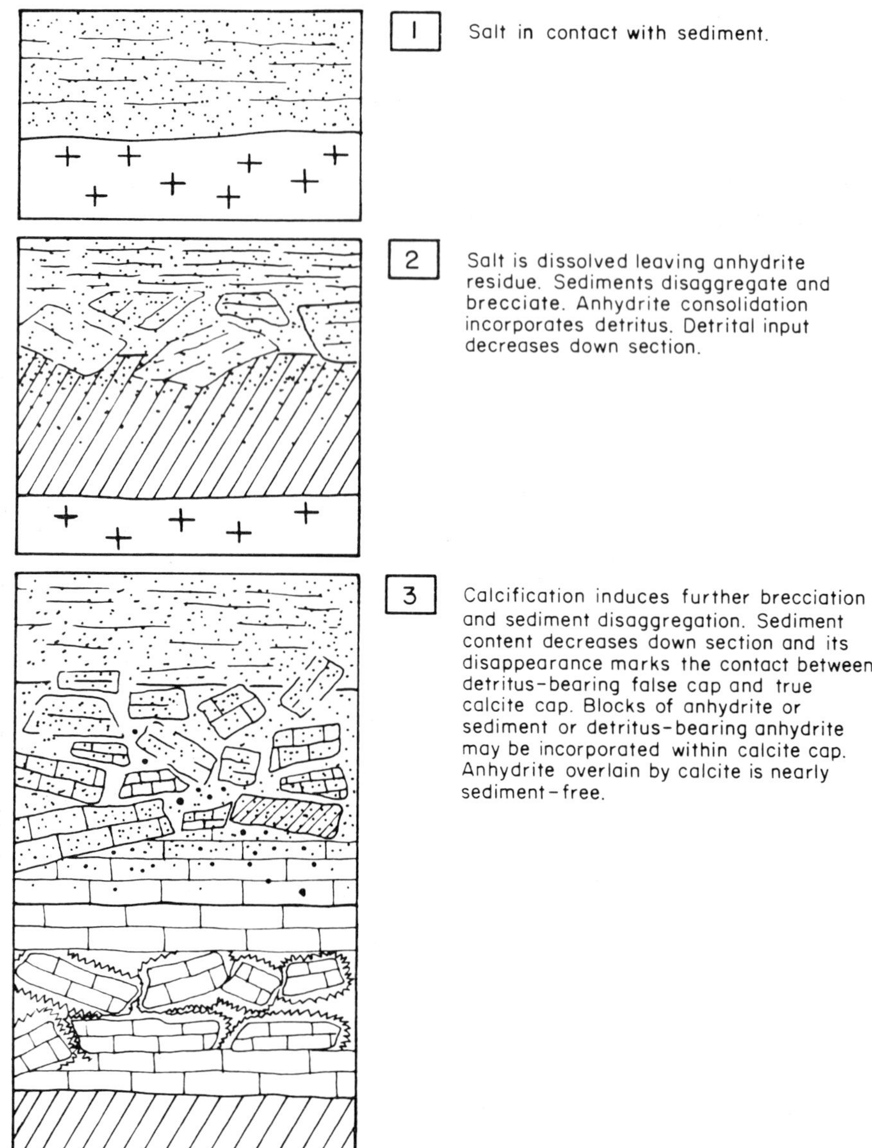

Figure 10. Model showing conceptual model for the generation of detritus-bearing false calcite cap rock.

A model showing the formation of interbanded stage I and stage II calcite is shown in Figure 11. Alternating bands of dark and light calcite probably formed sequentially from top to base along the calcite/anhydrite interface. According to the model, brines introduced at the calcite/anhydrite interface dissolve anhydrite, leaving a high-calcium solution (Fig. 11A). Because anhydrite solubility decreases with increasing temperature, it is likely that the fluid responsible for anhydrite dissolution was relatively cool, although the precise temperatures are not well understood. Dissolution of anhydrite creates a void at the calcite/anhydrite interface (Fig. 11B) into which overlying materials may fall. This explains at least part of the pervasive brecciation seen in calcite cap rocks.

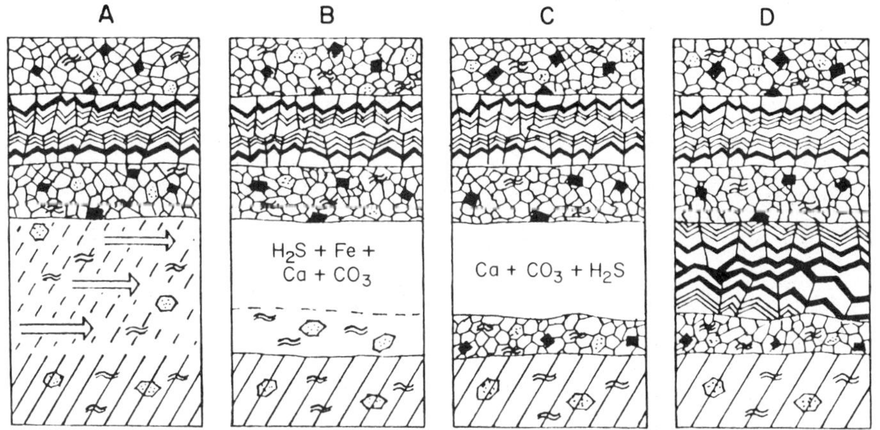

Figure 11. Model showing a conceptual model to explain banding in true calcite cap rock. Assuming calcite forms at the expense of anhydrite, brines or fresh water introduced to anhydrite would dissolve anhydrite (A) providing Ca and SO4 rich fluids. Insoluble materials in the anhydrite, mostly doubly-terminated quartz crystals and clay, would be incorporated in the calcite (B). Carbon rich fluids introduced into the cap rock area would promote bacterial reduction of sulfate (B) which, in turn, would produce H2S and CO3. H2S, when combined with iron in solution, would produce pyrite, and Ca when combined with CO3 would produce calcite. The absense of pyrite in late stage calcite indicates that Fe in solution was limited (C) and used up during precipitation of pyrite during the formation of early stage calcite. Residual solutions precipitated calcite slowly in an open void created during the early phases.

Bacterially-mediated reduction of dissolved sulfate causes calcite and pyrite to precipitate (Fig. 11C). Because the abundance of each depends on the activities of calcium, iron, bicarbonate, and sulfur in solution, the concentration of pyrite is probably controlled by the amount of Fe in brine because the system is saturated in sulfur. Both calcite and pyrite occur in the same samples (Fig. 11C) and have textures indicative of mutual growth. Quartz and clay, which also appear in stage I calcite, are insoluble residues left from the dissolution of anhydrite.

As calcite and pyrite precipitate, the fluids become less saturated with respect to those minerals and eventually become totally depleted of Fe. Afterwards, coarse-crystalline stage II calcite precipitates in the remaining open space, sometimes filling the void completely but, more commonly, leaving vug-lined cavities (Fig. 11D). Stage II calcites in Oakwood (Kreitler and Dutton, 1983; Posey, 1986) are typically heavier than adjacent stage I calcite, suggesting either that stage II calcite formed at lower temperatures or that they mixed with an isotopically heavy fluid prior to stage II calcite formation. However, many stage II calcites at Hockley, Winnfield, and Spindletop are isotopically lighter (Feely and Kulp, 1957; Posey, 1986), or opposite the trend found at Oakwood. This relation suggests that carbon components of stage I calcite fluid are present in the stage II fluid but that a second carbon source may be involved. Provided all domes have similar thermal histories, the variations between early and late calcite cannot be explained by variations in temperature of deposition. Rather, mixing must be called upon to explain the differences.

Isotopic similarities between stage I and adjacent stage II calcites suggest that both formed from the same fluid. However, other data suggest that fluid mixing was involved. Kreitler and Dutton (1983) proposed that stage II calcites formed by a process of dissolution of stage I calcite followed by reprecipitation as isotopically enriched, stage II calcite. However, the process of dissolution followed by reprecipitation seems unlikely because, although the calcite might dissolve up to solution saturation limits, it is hard to imagine how it could ever reach oversaturated conditions to induce new calcite precipitation. Also, for the system under study, the most likely solvent for calcite is CO_2. Excess CO_2 will dissolve calcite whereas CO_2 degassing will promote calcite precipitation. Although calcite might precipitate during periods of low CO_2 and dissolve during periods of high CO_2, part of the calcite carbon would be provided by the CO_2 itself. At Hockley and Spindletop where stage II calcites are isotopically lighter than adjacent stage I calcites, this would not be possible, because $\delta^{13}C$ of CO_2 is not known to be lighter than about -30 $^{o}/oo$ (Deines, 1980). Thus, it appears that stage II calcites represent a mixture of residual, stage I calcite fluids and some other source. At Winnfield and Oakwood, that source was isotopically heavier than the original whereas at Hockley and Spindletop it was isotopically lighter and almost certainly methanogenic.

Stage I calcite actually consists of several generations of calcite whose relative timing is not well understood. Typically, stage I calcite consists of diffuse bands of light gray to dark gray material and thin bands of very dark gray to black calcite. Both calcites contain residues of clay and doubly terminated quartz that were originally present in the anhydrite, but the concentrations of these residues is extremely high in the black bands compared with the gray bands. The fact that these residues are dispersed throughout gray, stage I calcite and concentrated along thin black bands suggests that residue concentration may have occurred through several mechanisms. Dispersal of residues within the gray bands suggests that anhydrite dissolution was concomitant with stage I calcite precipitation. However, high residue concentrations could have formed either by total dissolution of anhydrite with no concomitant calcite precipitation or by dissolution of gray stage I calcite, which itself contained dispersed residues. The black bands are found adjacent to both gray stage I calcite and stage II calcite, suggesting that they could have formed at more than one time in the sequence.

IV. MODEL

We discuss two models to explain the petrologic and geochemical features observed in salt dome calcite cap rocks. Two conditions required by both models are (1) that carbon is supplied by CH_4, and (2) CH_4 is drawn from source rocks that are cooler than about 100°C.

The first model (Fig. 12) assumes that most of the carbon is derived from hydrocarbon source rocks during the early stages of liquid petroleum maturation. As seen in Figure 12, the abundance of CO_2 at 100°C is about five times greater than CH_4 in humic source rocks and about two times less than in sapropelic source rocks. Because the CO_2 abundance is so great in humic source rocks and because it is inherently heavier than CH_4, if CH_4 is from a humic source it is probably biogenic.

As hydrocarbons mature, CO_2 and CH_4 are evolved during the early stages of maturation. As these gases migrate upward (Fig. 12B) or as overlying rocks adjacent to the diapir are buried (Fig. 12C), CO_2 dissolves whatever limestone it encounters. The dissolution of limestone thus increases the $\delta^{13}C$ by adding heavy carbon from the limestone. Introduction of this relatively heavy carbon along the anhydrite-sediment contact induces anhydrite dissolution and promotes bacterially-mediated sulfate reduction followed by calcite precipitation. This early calcite incorporates detritus from the overlying formations. Later solutions traversing the same strata encounter less limestone, so

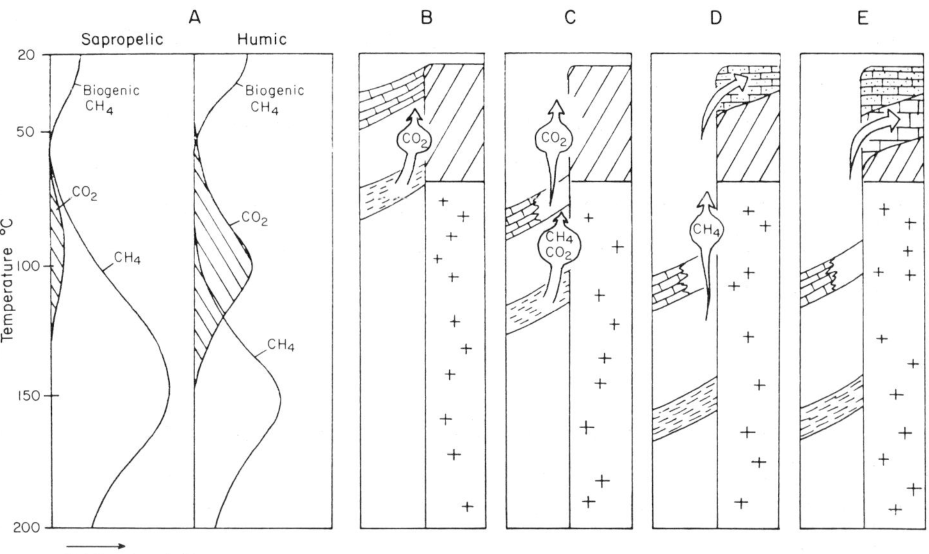

Figure 12. Model showing the possible relationship between CO2 and CH4 evolution, limestone dissolution, and calcite cap rock formation. Early hydrocarbon maturation first produced CO2, which, when introduced to overlying carbonate, dissolved the carbonate and lowered the δ13C of the fluid. This fluid thus produced relatively heavy calcite in the upper portions of the calcite cap rocks, and that calcite incorporated detritus from the overlying units. Later fluids not only contained more isotopically light methane, but traversed less carbonate due to earlier carbonate dissolution. Thus, these later fluids retained isotopically light methane-type signatures which produced isotopically light true calcite cap rocks beneath detritue-bearing calcite cap rocks. However, methane probably came from depths where temperature was no greater than about 100°C. Alternatively, the heavy carbon end-member could have been derived from seawater, and the isotopically light end-member could have been shallow, biogenic methane. Data currently available do not favor one hypothesis over another.

they retain their isotopically light character. Calcite that forms from later parts of the fluid train is thus depleted relative to the calcite above it, and contains less sediment because of the armoring effects of the overlying calcite. Oxygen in these formations, having equilibrated with rocks at higher temperatures, has $\delta^{18}O$ heavier than at surface temperatures and, upon limestone dissolution, this value grows even heavier.

The second model suggests that isotopically heavy carbon and oxygen are supplied by seawater, that salt dissolution takes place at the sediment/seawater interface, and that the light carbon source is biogenic methane. This model differs little from the first but carries important requirements. First, the carbon source is supplied from sources cooler than 50°C. Second, the sediment is supplied from newly deposited terrigenous clastics rather than pre-existing sediments or sedimentary rocks. This has important research implications because if the age of the sediments in the uppermost cap rocks can be dated, the age of cap rock formation can be determined. Third, the only HCO_3^- in the system would be supplied by seawater or, as with the former model, the oxidation of CH_4. It is possible that different salt domes contain components of both models, but neither model adequately explains the variable relationships between stage I and adjacent stage II calcites. As more studies focus on the organic geochemistry of the calcite cap rocks, it may be possible to distinguish between the two major carbon sources, and to determine whether oil and gas accumulations are related to calcite cap rock formation and, if so, whether their occurrence may be predicted from the compositions of salt dome cap rocks.

Provincial variations exist in $\delta^{13}C$ and $\delta^{18}O$ because of differences in carbon source and temperature of formation fluids. Sapropelic materials mature early and have higher $CH_4:CO_2$ ratios and thus lighter overall $\delta^{13}C$ than do humic sources, which mature later, at higher temperatures, and hence have heavier $\delta^{13}C$. From the evidence currently available, it appears that cap rocks above diapirs around the Hockley dome formed from higher temperature (deeper) fluids and possibly from humic source materials whereas all other domes formed from lower temperature fluids and possibly sapropelic sources. Corroborative evidence for this model may be found in Price and Kyle (1986).

V. CONCLUSIONS

Calcite cap rocks above salt diapirs grow in sequential layers, top to base, above anhydrite cap rocks. They may form on the seafloor at the sediment/water interface or beneath the seafloor below previously-deposited sediments. Carbon for calcite cap rocks is probably supplied by methane, which is isotopically light, and a heavier carbon source. The heavy carbon source may be from seawater, catagenic CO_2, or marine carbonate that is dissolved by CO_2. If carbon is supplied from CO_2 and CH_4 that are associated with maturing hydrocarbons, they represent an early fraction, one that formed at temperatures of 100°C or less. As these fluids reach anhydrite in salt dome cap rocks, the high salinities induce anhydrite dissolution, and carbon in the fluids promotes sulfate reduction by bacteria that utilize carbon as an energy source. As this process progresses calcite builds in layers in reverse stratigraphic sequence, early layers forming above later layers. Each layer forms in two stages: an early granular stage that precipitates quickly from supersaturated solutions and incorporates residues from the dissolved anhydrite and a late, coarsely crystalline stage that precipitates more slowly from the less saturated parts of the solution as open space fill. Upon introduction of new carbon-bearing fluid, excess CO_2 may promote calcite dissolution and, together with anhydrite dissolution, induce brecciation. Waning stages of calcite precipitation (during stage II) may incorporate carbon and oxygen from seawater but this does not always occur.

The entire calcite-forming process probably proceeds at temperatures below about 70°C and involves bacterial reduction of sulfate. The solutions that form early-stage calcite contain iron, which, in turn, combines with sulfide from the reduction of sulfate to form pyrite that is found in the early stage calcite. Later stage calcites fill open spaces left by the dissolution and reduction of sulfate and contain too little iron to precipitate pyrite. Throughout the course of calcite cap rock development, collapse-type brecciation opens previously formed calcite layers that are subsequently healed with very late stage calcite. The growth of calcite cap rock probably ceases when the supply of upward-migrating hydrocarbons is cut off, either by physical processes or by depletion of the source.

Anomalously heavy carbon and oxygen isotope compositions at the Hockley dome suggest either that it formed at a lower temperature, perhaps from hydrocarbon sources tapped from high temperature regimes, or that it contains a greater abundance of CO_2 or marine carbonate-derived carbon.

VI. ACKNOWLEDGEMENTS

The authors are grateful to Arnold Taylor, Steve Hurst, and Neil Sherrod of Conoco Oil Company, Ponca City, Oklahoma, for assistance in preparing the analyses for this paper. Special thanks are extended to Conoco and Marathon Oil for running the analyses. Samples were provided through the courtesy of Marathon Oil Company, Texasgulf Incorporated, and the University of Texas Bureau of Economic Geology. Rick Platt and Jim Morgan assisted with figure preparation; Diane Hall proofed the manuscript; and Steve Fisher kindly transferred the text to computer disc.

REFERENCES

Baker, Harold W. (1979). General geology of Damon Mound, in Eter, Evelyn M., ed., Damon Mound Field Trip Guidebook: Houston Geological Society, p. 10-25.

Bottinga, Y. (1968). Calculation of fractionation factors for carbon and oxygen exchange in the system calcite-carbon dioxide-water: Jour. Phys. Chemistry 72, 800-808.

Bottinga, Y. (1969). Calculated fractionation factors for carbon and hydrogen isotope exchange in the system calcite-CO_2-graphite-methane-hydrogen and water vapor: Geochim. et. Cosmochim. Acta, p. 49-64.

Brock, T. D. (1985). Life at high temperatures: Science 230, 132-138.

Deines, P., The isotopic composition of reduced orgenic carbon, in Fritz, P. and Fontes, J. Ch., 1980, Handbook of Environmental Isotope Geochemistry: Elsevier Scientific Publishing Co., New York, p. 329-406.

Feely, Herbert W., and Kulp, J. Laurence (1957). Origin of Gulf Coast salt-dome sulphur deposits: AAPG Bull. 41, 1802-1853.
Folk, R.L. (1974). The natural history of crystalline calcium carbonate: effect of magnesium content and salinity: Jour. Sed. Pet. 44, 40-53.
Friedman, Irving, and O'Neil, James R. (1977). Compilation of stable isotope fractionation factors of geochemical interest, in Fleischer, Michael, ed., Data of Geochemistry, 6th ed., Chapter KK, 12 p., 49 figures.
Given, R. K. and Lohmann, K. C. (1985). Derivation of the original isotopic composition of Permian marine cements: Jour. Sed. Pet. 55, 430-439.
Hoefs, J. D. (1973). Stable Isotope Geochemistry, in von Engelhardt, W., Hahn, T., Roy, R., and Wyllie, P. J., eds., Minerals, Rocks and Inorganic Materials, Monograph Series of Theoretical and Experimental Studies: Springer-Verlag, New York, 142 p.
Holland, Heinrich D., and Malinin, Sergey D. (1979). The solubility and occurrence of non-ore minerals, in Barnes, Hubert Lloyd, ed., Geochemistry of Hydrothermal Ore Deposits: John Wiley and Sons, New York, p. 461-508.
Hudson, J. D. (1977). Stable isotopes and limestone lithification: Journal of the Geological Society of London 133, 637-660.
Jackson, M.P.A., and Talbot, C.J. (1986). External shapes, strain rates, and dynamics of salt structures, GSA Bull. 97, 305-323.
James, Noel P., and Choquette, Phillip W. (1984). Diagenesis 9. Limestones--the meteoric diagenetic environment: Geoscience Canada 11, 161-194.
Kreitler, C.W., and Dutton, Shirley P. (1983). Origin and diagenesis of cap rock, Gyp Hill and Oakwood salt domes, Texas: Univ. Tx. Bur. Econ. Geol. Rep. Inv. No. 131, 58 p.
Kreitler, C.W., Collins, E.W., Fogg, G.E., Jackson, M.P.A., and Seni, S. J. (1985). Hydrologic characterization of the saline aquifers, Texas salt domes: The University of Texas Bureau of Economic Geology Report to U.S. Department of Energy under contract No. DE-AC97-80ET46617, 163 p.
Kyle, J. Richard, Ulrich, Mark R., and Gose, W. A., (this volume), Textural and paleomagnetic evidence for the mechanism and timing of anhydrite cap rock formation, Winnfield salt dome, Louisiana.
Light, M. P. R., Posey, Harry H., Kyle, J. R., and Price, P. E., (this volume), Integrated hydrothermal model for the Texas Gulf Coast Basin; origins of geopressured brines and lead-zinc, uranium, hydrocarbon and cap-rock deposits, in Lerche, Ian, and O'Brien, John, Dynamical Geology of Salt and Related Structures, Academic Press.
Light, M. P. R., 1974, Ewing, T.E., and Tyler, N. (1984). Report prepared for the U.S. Dept. of Energy, contract no. DE-AC08-79ET27111, The University of Texas at Austin, Bureau of Economic Geology, 97 p.
Martinez, Joseph D. (1974). Tectonic behavior of evaporites, in Coogan, Alan H., ed., Fourth Symposium on Salt, v. 1, p. 155-167: Cleveland, Northern Ohio Geological Society and Kent State University.
Milliken, K. L., Land, L. S., and Loucks, R. G. (1981). History of burial diagenesis determined from isotopic geochemistry, Frio Formation, Brazoria County, Texas,: AAPG Bull. 65, 1397-1413.
Murray, G.E., Jr. (1966). Salt structures of Gulf of Mexico Basin--a review. AAPG Bull. 50, 439-478.
Nance, D., and Wilcox, R. E. (1979). Lithology of the Rayburn's Dome salt core: Inst. Env. Studies, Louisiana State Univ., Topical Report ES11-02500-3, 307 p.

Nance, D., Rovic, J. and Wilcox, R. E. (1979). Lithology of the Vacherie salt dome core: Inst. Env. Studies, Louisiana State Univ., Topical Report ES11-02500-5, 343 p.

Posey, Harry H. (1986). Regional characteristics of strontium, carbon and oxygen isotopes in salt dome cap rocks of the Western Gulf Coast: Unpub. Ph.D. Dissert., Univ. North Carolina, Chapel Hill, 248 p.

Price, Peter E., Kyle, J. Richard, and Wessel, Gregory R. (1983). Salt dome related lead-zinc deposits, in Kisvarsanyi, Geza, Grant, Sheldon K., Pratt, Walden P., and Koenig, John W., eds., International Conference on Mississippi Valley Type Lead-Zinc Deposits, Proceedings Volume, p. 558-577.

Price, Peter E. and Kyle, J. Richard (in press). Genesis of salt dome hosted metallic sulfide deposits: the role of hydrocarbons and related fluids: Denver Region Exploration Geologists Society (DREGS) Proceedings.

Rezak, R. (1985). Local carbonate production on a terrigenous shelf: Gulf Coast Assoc. Geol. Soc. Trans. 35, 477-484.

Rice, J. M. (1980). Chemical and isotopic evidence of the origins of natural gases in offshore Gulf of Mexico: Transactions of the Gulf Coast Association of Geological Societies 30, 203-213.

Sassen, Roger (1980). Biodegradation of crude oil and mineral deposition in a shallow Gulf Coast salt dome: Org. Geochem. 2, 153-166.

Smith, Glenn, and Kolb, Charles R. (1981). Isotopic evidence for the origin of the Rayburns boulder zone of North Louisiana: Inst. Env. Studies, Louisiana State University, Baton Rouge, Louisiana, File Report QR.4.2, 23 p.

Thode, H.G., Wanless, R. K., and Wallouch, R. (1954). The origin of native sulfur deposits from isotope fractionation studies: Geochim. et Cos. Acta 5, 286-298.

Trudinger, P. A., Chambers, and Smith, J. W. (1985). Low-temperature sulphate reduction: biological versus abiological: Can. J. Earth Sci. 22, 1910-1918.

Woo, Kyung Sik (1980). Carbonate diagenesis and biostratigraphy of the Anahuac Formation at Damon Mound, Texas: M.S. Thesis, Texas A & M Univ., 69 p.

ORGANIC GEOCHEMISTRY OF SALT DOME CAP ROCKS, GULF COAST SALT BASIN

Roger Sassen

Basin Research Institute
Louisiana State University
Baton Rouge, LA 70803

I. INTRODUCTION

The biodegradation of crude oil in shallow reservoirs is a common occurrence that has resulted in the formation of billions of tons of heavy oil and tar worldwide (Philippi, 1977). Aerobic bacteria are the main agents of crude oil biodegradation, but sulfate reduction by anaerobic bacteria also frequently takes place in oil reservoirs (Jobson et al., 1979). CO_2 and H_2S are geochemically significant products resulting from microbial attack on crude oil. The ultimate fate in the geologic environment of these products appears to be in the form of characteristic mineral assemblages.

Calcite and other carbonate minerals frequently occur in association with biodegraded crude oil. For example, isotopically light calcite has been observed overlying some shallow oil fields in Oklahoma by Henry and Donovan (1978). In addition, similar occurrences of isotopically light calcite are found with biodegraded crude oil in subsurface reservoirs at Barrow Island, Western Australia (Gould and Smith, 1978).

Petroleum biodegradation and bacterial sulfate reduction have been linked to deposition of elemental sulfur in subsurface environments characterized by bedded anhydrite and gypsum of evaporite origin (Kirkland and Evans, 1976; Pawlowski et al., 1979; Barker et al., 1979). Moreover, microbial oxidation of crude oil and sulfate reduction have been suggested as factors in the formation of some lead-zinc ore deposits (Connan, 1979). Microbial generation of H_2S provides a mechanism for the localized precipitation of metallic sulfide minerals in Mississippi Valley type deposits.

II. GULF COAST SALT DOME CAP ROCKS

The cap rocks above shallow Gulf Coast salt domes in Texas and Louisiana contain massive accumulations of sulfate in the form of anhydrite and gypsum. Furthermore, the cap rocks and other traps that flank salt domes often serve as reservoirs for crude oil. The cap rocks of these salt domes also contain enormous volumes of isotopically light carbonate along with elemental sulfur and sulfide minerals.

A number of studies have implicated crude oil biodegradation and microbial sulfate reduction in the origin of elemental sulfur and isotopically light carbonate in the cap rocks of shallow salt domes onshore in Texas and Louisiana. Early studies relied on measurements of isotopic fractionation of carbon and sulfur to explain the origin of these minerals (Thode et al., 1954; Feely and Kulp, 1957). Additional research has included isotopic data but has also documented the effects of biodegradation on crude oils extracted from cap rocks (Milner et al., 1977; Sassen, 1980; Kreitler and Dutton, 1983). Moreover, there is evidence for the occurrence of microbial calcite and sulfur in the cap rock of Challenger Knoll, a shallow salt dome in the Gulf of Mexico at a water depth of 3600 m (Davis and Kirkland, 1979). A similar origin related to microbial activity appears likely for iron, zinc, lead, and silver sulfides in cap rocks of shallow Gulf Coast salt domes (Price and Kyle, 1983; Ulrich et al., 1984). A diagram summarizing the relationship of microbial activity to mineral deposition in salt dome cap rocks is shown in Figure 1.

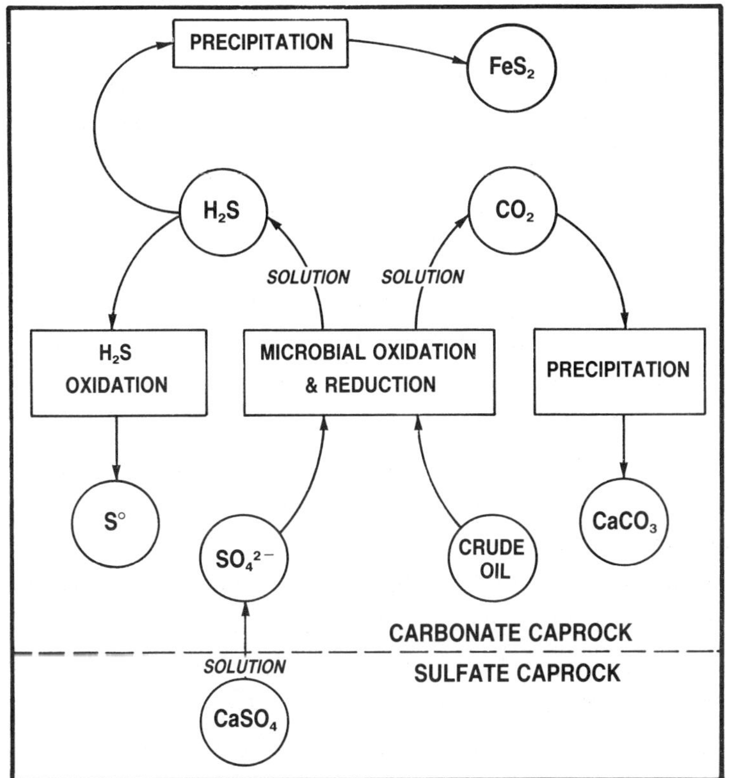

Figure 1. Summary diagram illustrating the relationship of crude oil biodegradation and microbial sulfate reduction to deposition of carbonate minerals, elemental sulfur, and metallic sulfides in cap rocks of shallow Gulf Coast salt domes. After Sassen (1980).

A. Microbial Versus Thermal Origin of Cap Rock Minerals

It is possible to argue that a suite of calcite, elemental sulfur, and sulfides could be the consequence of purely abiogenic reactions involving petroleum hydrocarbons and sulfate. There is little doubt, for example, that aqueous sulfate can rapidly oxidize some organic compounds in the laboratory at temperatures approximating 300°C (Toland, 1960). Temperatures for this reaction in the geologic environment where time is not a limiting factor have been suggested to be as low as 80° to 100°C (Macqueen and Powell, 1983). In addition, Orr (1974) provides compelling evidence that this reaction occurs in some deep, hot geologic environments where it is associated with the thermal cracking of crude oil.

The temperatures for formation of cap rock minerals in Gulf Coast salt domes, therefore, become important in evaluating the validity of the microbial hypothesis. One of the most direct geologic indicators of a low temperature of origin for cap rock minerals is shallow depth of occurrence. For example, Davis and Kirkland (1979) noted that most elemental sulfur had been produced from Gulf Coast salt dome cap rocks no deeper than about 550 m. At the present time, the deepest sulfur production is from cap rock at a depth of approximately 1000 m offshore Grand Isle, Louisiana. Although temperature anomalies occur overlying salt domes (O'Brien and Lerche, 1984), this factor does not appear to be geochemically significant in the shallow sulfur-productive cap rocks of the Gulf Coast.

The association of biodegraded crude oil with cap rock mineralization also places limits on the maximum temperatures of mineralization that exclude thermal reactions of hydrocarbons and sulfate. A number of studies have provided data on the maximum temperatures at which biodegradation occurs in the geologic environment (Bailey et al., 1973; Philippi, 1977; Winters and Williams, 1969). Maximum temperatures at which biodegraded oils often occur roughly correspond to temperature limits for thermophilic bacteria. Philippi concluded that the average maximum temperature for crude oil biodegradation is approximately 66°C. It is not unreasonable to assume that this maximum temperature also applies to mineralization in shallow Gulf Coast cap rocks.

Another line of evidence that tends to exclude abiogenic reactions is that the present is the key to the past. For example, crude oil biodegradation and mineral deposition are taking place at the present time at surface temperatures in outcropping cap rock of the Damon Mound salt dome in Texas, contributing additional evidence that similar processes have occurred in the geologic past (Sassen, 1980). Scanning electron micrographs of bacteria, calcite and sulfides found at Damon Mound in close association with biodegraded crude oil are shown in Figure 2.

Oil and gas seeps are widely distributed in sea bottom muds overlying shallow salt domes along the Gulf of Mexico slope that provide insight to minimum temperatures for deposition of isotopically light carbonate. Piston coring in the Green Canyon area of the slope has shown that crude oil biodegradation and deposition of calcite and aragonite are taking place at the present time at low sea bottom temperatures (<8°C).

Evidence for hydrocarbon migration to the sea bottom includes thermogenic gas hydrates that were recovered from Green Canyon cores (Brooks et al., 1984). Yellowish nodules of gas hydrate were found dispersed in the carbonate rubble of some oil-stained cores. Additional analytical work on these samples by Davidson et al. (1986) also indicates the presence of CO_2 and H_2S encaged within the gas hydrate structure. The high concentrations of gas hydrate and crude oil in many cores from the Green Canyon area reflect ongoing seepage. As in the case of salt domes, compositional variation in crude oil extracted from Green Canyon cores is largely the result of microbial attack (Anderson et al., 1983; Brooks et al., 1984).

The ongoing microbial oxidation of petroleum hydrocarbons is clearly associated with abundant deposition of isotopically light carbonate in the oil seeps (Sassen et al., 1985). The carbonate occurs as large masses, as discrete nodules, and as thin crusts. A scanning electron micrograph of isotopically light aragonite is shown in Figure 3.

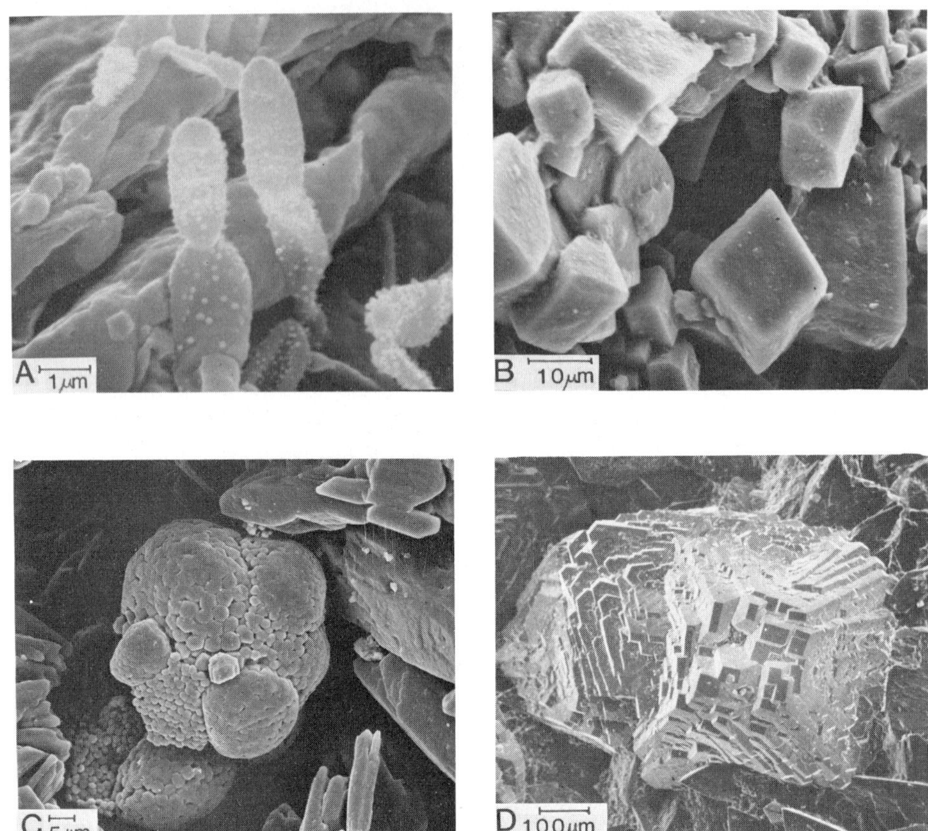

Figure 2. Scanning electron micrographs of microbes and minerals found in association with biodegraded crude oil in outcropping cap rock, Damon Mound salt dome. A. Bacteria on solid crude oil. B. Isotopically light rhombic calcite crystals on crude oil. Calcite $\delta^{13}C$ values were measured in the -22.1 to -29.2 ^{o}/oo PDB range at Damon Mound. C. Framboidal pyrite surrounded by gypsum crystals on crude oil. D. Pyrrhotite on calcite in an oil-bearing cavity.

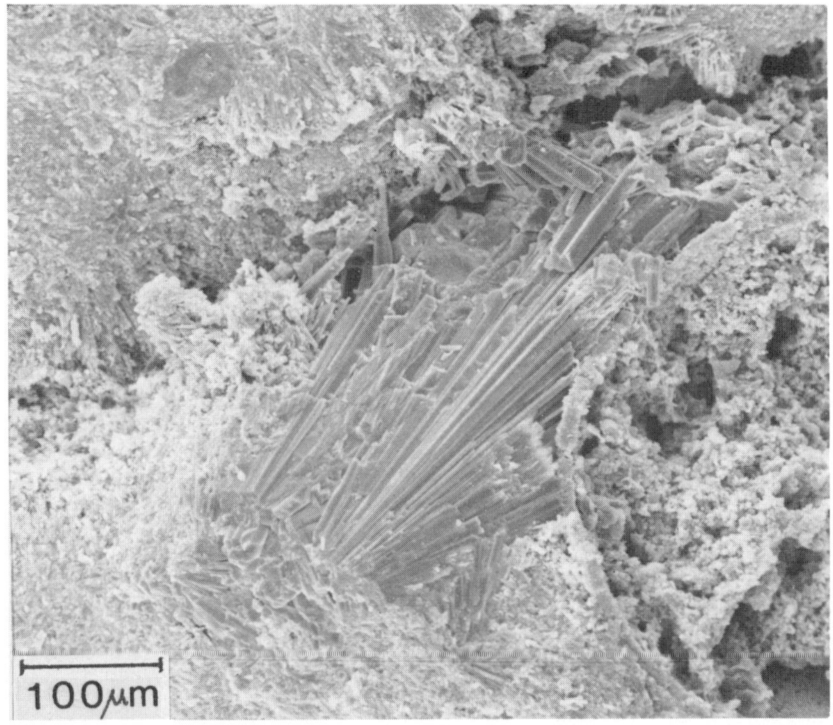

Figure 3. Sprays of aragonite crystals in a carbonate nodule recovered in association with biodegraded crude oil from a seep in the Green Canyon area of the Gulf of Mexico slope. The δ13C value of the nodule was -18.5 ^{o}/oo PDB.

III. ORIGIN OF GULF COAST OIL AND GAS

The generation of crude oil and thermogenic gas takes place at great depth in sediments of the Gulf Coast Basin. Enormous volumes of hydrocarbons have accumulated in shallow reservoirs of salt dome cap rocks and flanking sands as a consequence of vertical migration from deeply buried source rocks. The origin of petroleum in the Gulf Coast should be discussed because of its role in the deposition of cap rock minerals.

A. Hydrocarbon Generation and Destruction

The effects of time and temperature during burial result in generation of hydrocarbons from kerogen, the insoluble organic matter dispersed in sediments. The type of kerogen in sediments is a significant geochemical parameter that determines whether oil or gas will be generated. In general, marine organic matter is rich in hydrogen and prone to generation of crude oil, whereas terrestrial organic matter is more prone to generation of gas.

Meaningful generation of hydrocarbons will not occur until some minimum level of thermal maturity is attained during burial. Thermal maturity is the result of both time and temperature, and, within limits, one can substitute for the other. Thermal maturity can be measured by various techniques including vitrinite reflectance (Dow, 1977). Sediments capable of generating hydrocarbons but which are thermally immature are called potential source rocks. When thermally mature, such sediments become effective source rocks and migration commences.

The generalized relationship between thermal maturity and the generation and destruction of hydrocarbons is shown in Figure 4. The thermal maturity boundaries shown in the figure are approximate but useful nonetheless.

Significant generation of crude oil from hydrogen-rich kerogen occurs in some settings at a level of thermal maturity equivalent to a vitrinite reflectance of about 0.55%. Somewhat higher levels of maturity, however, are often necessary to initiate the main phase of hydrocarbon generation. Intense generation of medium-gravity crude oil appears to commence at a vitrinite reflectance of roughly 0.7%. The process of oil generation and migration approaches completion at a vitrinite reflectance of approximately 1%. Additional time-temperature exposure causes destruction of hydrocarbons.

The eventual destruction of liquid hydrocarbons after migration to reservoirs can also be controlled by the effects of

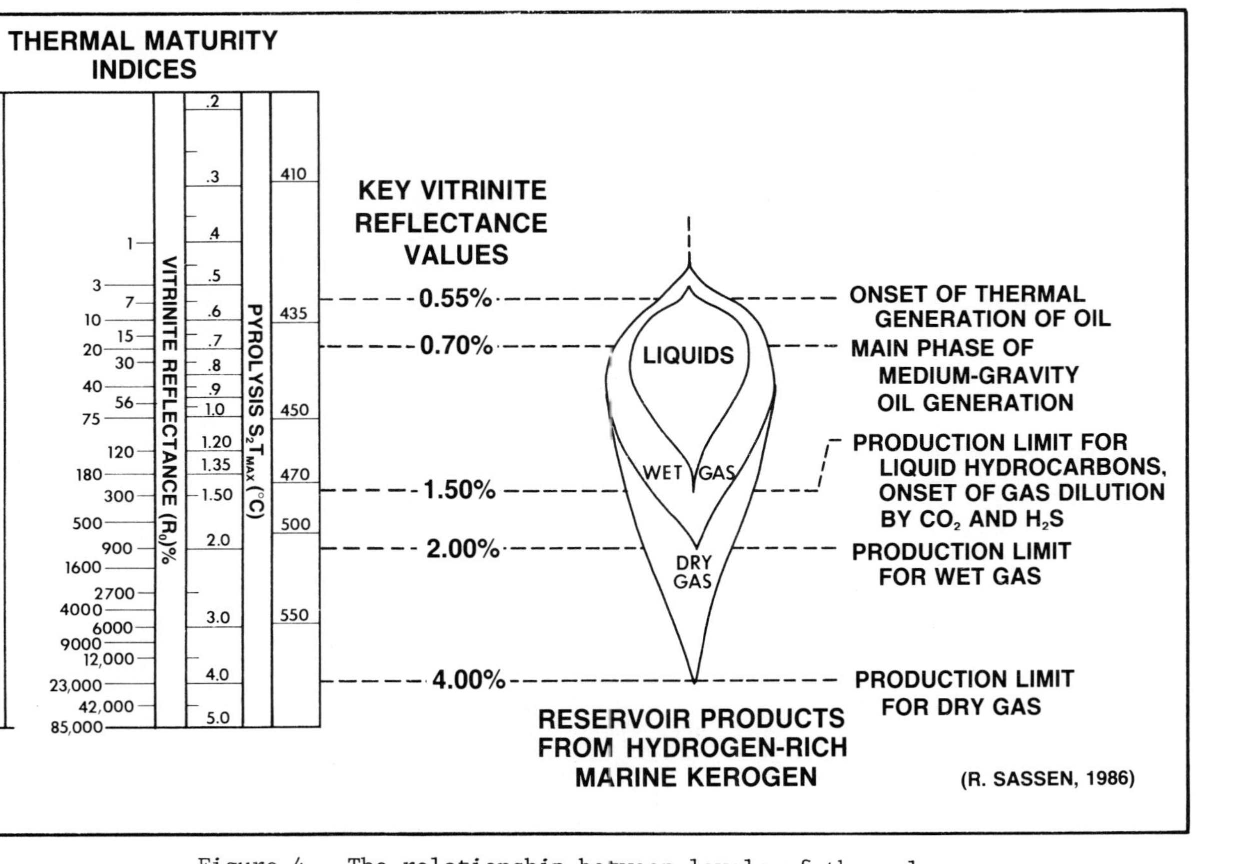

Figure 4. The relationship between levels of thermal maturity and the zones of hydrocarbon generation and destruction.

time and temperature. Once time-temperature effects in reservoirs reach the equivalent of a vitrinite reflectance of about 1.5%, the last liquid hydrocarbons are cracked to wet gas. In turn, wet gas and then dry gas are eventually destroyed at high levels of thermal maturity.

It is important to note that CO_2, H_2S, and other molecules such as N_2 become important reservoir constituents during the final stages of methane destruction. The origin of these molecules in deep, hot reservoirs is a consequence of inorganic oxidation of methane and reduction of sulfate. As discussed previously, temperatures of formation of cap rock minerals are too low to suggest that such essentially inorganic reactions are significant in shallow Gulf Coast salt domes.

B. Hydrocarbon Source Rocks of The Gulf Coast Salt Basin

The crude oil and thermogenic gas of the Jurassic salt basins of the Gulf Rim in east Texas, north Louisiana, south Arkansas, and central Mississippi are largely derived from hydrocarbon source rocks of Mexizoic age. Much of the petroleum in Jurassic reservoirs has been generated by carbonate source facies of the Smackover Formation (Oehler, 1984). Cretaceous source rocks for oil and gas also appear to be present (Koons et al., 1974).

The framework of hydrocarbon generation and migration in the Gulf Coast Salt Basin of Texas and Louisiana is less well understood because of a thick section of Pleistocene and Tertiary sediments. Most organic geochemists, however, accept the concept that crude oil and thermogenic gas in reservoirs of the Gulf Coast Salt Basin are the result of vertical migration from deep source rocks that are not usually encountered by drilling. Given the knowledge that source rock thermal maturity is the result of time and temperature, it should not be surprising that the levels of thermal maturity needed for intense generation of hydrocarbons do not occur at shallow depth in sediments of the Gulf Coast Salt Basin. The bulk of sediments are not only geologically young but they are also characterized by low geothermal gradients.

Some insight to source rocks is provided by the composition of hydrocarbons in reservoirs. There is enough variation in hydrocarbons of the Gulf Coast Salt Basin to suggest that multiple source facies are involved. For example, Walters and Cassa (1985) concluded that several families of crude oil exist offshore Louisiana which are the result of differences in environment of deposition, kerogen types, and thermal maturity history. The saturated hydrocarbon fractions of these Gulf of Mexico oils displayed $\delta^{13}C$ values in the -26.4 to -27.4 $^{o}/oo$ PDB range. In addition, Rice (1980) describes significant differences in carbon isotopic composition of methane in reservoirs across the Gulf of Mexico. The isotopically light methane of shallow Pleistocene reservoirs is of biogenic origin. Biogenic methane exhibited $\delta^{13}C$ values lighter than about -55 $^{o}/oo$ PDB. However, other methane found in Pleistocene, Pliocene, and Miocene reservoirs is isotopically heavier and derived from thermogenic processes associated with the generation and destruction of crude oil. Such methane is characterized by $\delta^{13}C$ values in the -55 to -35 $^{o}/oo$ PDB range.

It is possible that Cretaceous and Jurassic source rocks have contributed hydrocarbons to reservoirs of the Gulf of Mexico Salt Basin, but overmaturity is an objection to this hypothesis. Tertiary sediments are more frequently suggested as source rocks. However, Miocene and younger sediments are usually characterized by low contents of organic carbon, by dominance of gas-prone kerogen of terrestrial origin, and by low levels of thermal maturity. Given this widespread lack of source potential, Dow (1984) has proposed localized preservation of oil-prone kerogen during the lower Tertiary in anoxic intraslope basins. Even so, this single hypothesis is unlikely to explain all occurrences of thermogenic hydrocarbons in the Gulf Coast Salt Basin.

IV. EVIDENCE FOR CRUDE OIL BIODEGRADATION

The geochemical characteristics of crude oil vary from basin to basin, from field to field, and within individual oil fields. A portion of the compositional differences among Gulf Coast crude oils can be attributed to variations in kerogen types and thermal maturity histories of multiple source facies. Other transformations probably occur during migration from deep, hot source facies to generally shallower and cooler reservoir facies. However, the most obvious differences in composition among crude oils are attributable to biodegradation after emplacement in reservoirs. Wide variations in crude oil composition associated with Gulf Coast Salt Domes in Texas (Sassen, 1980) and in Louisiana (Milner et al., 1977) are the result of microbial activity.

A. Preferential Destruction of Saturated and Aromatic Hydrocarbons

Field studies of crude oils in reservoirs and in laboratory simulations have shown that microbes metabolize different components of petroleum at different rates. The progressive compositional changes observed during microbial attack are characteristic of the process and contribute evidence of its occurrence.

The saturated hydrocarbon fractions of crude oils are most rapidly degraded by microbes. The lighter saturated hydrocarbons are attacked first, followed by alteration of C_{15+} saturated hydrocarbons. An example of the progressive biodegradation of C_{15+} saturated hydrocarbons in the salt dome environment is shown in Figure 5.

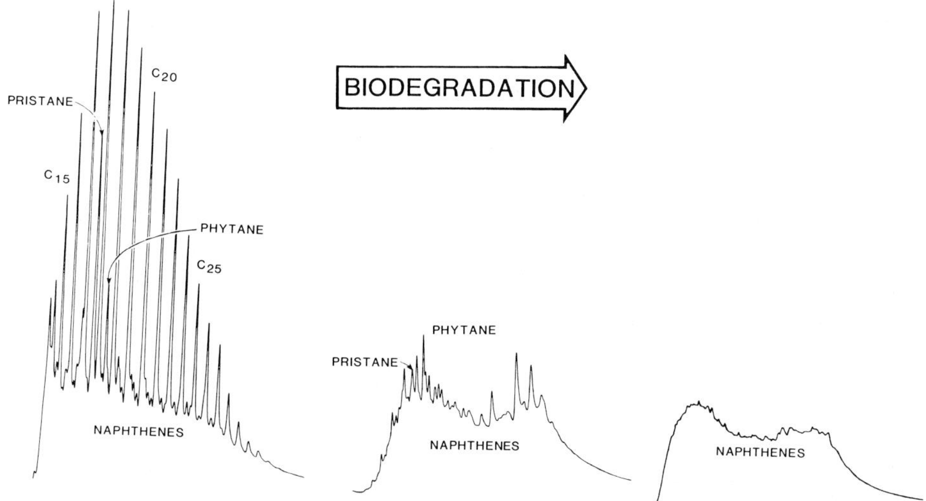

Figure 5. Chromatograms of C15+ saturated hydrocarbons illustrating the progressive effects of biodegradation. The unaltered crude oil at left first undergoes preferential destruction of n-paraffins followed by loss of the isoprenoids pristane and phytane. This progression is characteristic of microbial attack. Example from the vicinity of the Damon Mound salt dome.

There is abundant evidence that the n-paraffins of crude oils are degraded more rapidly than other hydrocarbons of similar molecular weight (Bailey et al., 1973; Evans et al., 1971; Philippi, 1977; Winters and Williams, 1969). When the biodegradation of the n-paraffins is advanced, the isoprenoids pristane and phytane often appear as prominent components in C_{15+} chromatograms. Studies of biodegradation indicate that once the n-paraffins are largely depleted, preferential biodegradation of pristane and phytane occurs (Bailey et al., 1973). As in the example of the Alberta tar sands described by Deroo et al. (1977), the naphthenes of crude oils are relatively resistant to microbial attack. It is for this reason that the chromatograms of heavily biodegraded crude oils frequently display naphthenic signatures.

TABLE 1. Typical compositions of biodegraded crude oils associated with cap rock mineralization in salt domes compared to non-biodegraded crude oils from various Gulf Coast salt basins. Experimental procedures have been described previously (Sassen, 1980).

NOTE: I = % SATURATED HYDROCARBONS
II = % AROMATIC HYDROCARBONS
III = % NSO COMPOUNDS
IV = % ASPHALTENES

Reservoir		$\%C_{15+}$ Fraction	C_{15+} Composition			
			I	II	III	IV
		Biodegraded Crude Oils				
Damon Mound	(Brazoria Co., TX)	*	3.4	10.9	25.7	40.0
Damon Mound	(Brazoria Co., TX)	*	5.4	14.6	21.5	58.5
Damon Mound	(Brazoria Co., TX)	*	26.9	23.9	26.1	24.1
Damon Mound	(Brazoria Co., TX)	*	19.5	12.1	29.0	39.4
Damon Mound	(Brazoria Co., TX)	*	10.3	13.5	15.3	47.9
Boling Dome	(Wharton Co., TX)	*	13.1	33.1	27.3	26.5
Boling Dome	(Wharton Co., TX)	*	15.3	3.0	28.9	52.8
Oakwood Dome	(Freestone Co., TX)	*	7.5	12.9	16.1	63.5
Oakwood Dome	(Freestone Co., TX)	*	1.0	6.7	5.5	88.4
		Nonbiodegraded Crude Oils				
Frio	(Brazoria Co., TX)	82.8	56.1	20.4	12.3	11.2
Frio	(Chambers Co., TX)	73.2	74.2	17.9	6.4	1.5
Wilcox	(Colorado Co., TX)	55.0	77.0	10.7	11.0	1.3
Wilcox	(Colorado Co., TX)	21.8	57.5	7.6	33.9	1.0
Haynesville	(Jones Co., MS)	59.0	79.2	15.4	4.4	1.0
Smackover	(Clarke Co., MS)	71.0	55.4	36.6	5.7	3.3
Smackover	(Columbia, Co., AR)	55.8	82.5	13.1	3.7	0.7
Lower Tuscaloosa	(Jefferson Co., MS)	82.2	74.2	17.0	6.2	2.6

*The biodegraded crude oils are tar-like solids almost totally composed of C_{15+} components

Studies by Seifert and Moldowan (1979) suggest that the steranes and terpanes are degraded at a slower rate than isoprenoids, but that they do undergo alteration. Nevertheless, Seifert et al. (1984) believe that sterane and terpane components can be used for source correlation purposes even after biodegradation.

Typical compositions of biodegraded crude oils associated with salt dome mineralization are compared to compositions of nonbiodegraded crude oils in Table 1. The biodegraded crude oils from the salt dome environment are frequently solid and lack light hydrocarbons. The nonbiodegraded crude oils are characterized by varying percentages of C_{15+} components, but substantial percentages of lighter components are always present. As can be seen, the preferential destruction of C_{15+} saturated hydrocarbons is reflected in lower percentages of these components when compared to crude oils that have not been altered by microbial attack.

Compositional components of crude oils that are more resistant to biodegradation tend to be concentrated during biodegradation. For example, the C_{15+} aromatic hydrocarbons are degraded more slowly than the saturated hydrocarbons (Evans et al., 1971). As shown by Volkman et al. (1984), the rate of biodegradation of aromatic components decreases with increasing numbers of aromatic rings. The decreased saturate/aromatic ratios of biodegraded crude oils, when compared to unaltered crude oils, reflect the relative stability of aromatic hydrocarbons in comparison to saturated hydrocarbons.

B. Concentration of Nonhydrocarbon Components

The NSO compounds contain nitrogen, sulfur and oxygen as well as carbon and hydrogen. These nonhydrocarbon components of crude oil frequently increase in relative abundance during biodegradation. There is some controversy as to whether the NSO compounds are enriched by depletion of other less stable oil components, or are actually introduced to crude oil from microbial biomass. Increases in organic nitrogen compounds, accompanied by increases in optical rotation, have been observed as a consequence of oil biodegradation (Winters and Williams, 1969). It has been suggested that this represents a contribution from microbes to crude oils. The detection of unsaturated even-carbon fatty acids in some biodegraded crude oils by Mackenzie et al. (1983) provides clear evidence that these microbial components can be introduced to crude oils.

Increased concentration of organic sulfur compounds are related to petroleum biodegradation (Milner et al., 1977). However, this does not necessarily imply that organic sulfur compounds have been added to the crude oil. Orr (1974) and Claret et al. (1977) suggest that the increased content of organic sulfur compounds is attributable to concentration in that fraction of oil not metabolized by microbes.

The asphaltenes, as noted for example by Bailey et al. (1973), are exceedingly resistant to biodegradation and, therefore, tend to dominate the compositions of highly biodegraded crude oils. Work by Rubenstein et al. (1979) further supports the contention that biodegradation does not significantly alter the composition of asphaltenes. The data of Table 1 show that concentrations of NSO compounds and especially asphaltenes are enhanced by the effects of biodegradation.

V. CONCLUSIONS

There is abundant evidence that the isotopically light carbonate minerals, elemental sulfur, and sulfides that characterize the cap rocks of shallow Gulf Coast salt domes are the result of microbial oxidation of petroleum and reduction of sulfate. This is shown by studies of isotopic fractionation of carbon and sulfur in cap rock hydrocarbons and minerals, as well as by studies of the effects of biodegradation on crude oils extracted from cap rocks. In addition, similar low-temperature processes are presently occurring in outcropping cap rocks and in deep-water oil seeps. Abiogenic reactions involving hydrocarbons and sulfate are unlikely to have occurred in the shallow and cool cap rocks of Gulf Coast salt domes.

REFERENCES

Anderson, R.K., Scalan, R.S., Parker, P.L. and Behrens, E.W. (1983). Seep Oil and Gas in Gulf of Mexico Sediment. Science 222, 619-621.

Bailey, N.J.L., Jobson, A.M., and Rogers, M.A. (1973). Bacterial Degradation of Crude Oil: Comparison of Field and Experimental Data. Chemical Geology 11, 203-221.

Barker, J.M., Cochran, D.E. and Semrad, R. (1979). Economic Geology of the Mishraq Native Sulfur Deposit, Northern Iraq. Economic Geology 74, 848-495.

Brooks, J.M., Kennicutt, M.C., II, Fay, R.R., McDonald, T.J., and Sassen, R. (1984). Thermogenic Gas Hydrates in the Gulf of Mexico. Science 225, 409-411.

Claret, J. Tchikaya, J.B., Tissot, B., Deroo, G., and van Dorsselaer, A. (1977). Un Example d'Huile Biodegradee a Basse Teneur en Soufre: Le Gisement d'Emeraude (Congo). Proceedings of the Ninth International Meeting Organic Chemistry, Madrid, 509-522.

Connan, J. (1979). Genetic Relation Between Oil and Ore in Some Pb-Zn-Ba Ore Deposits. Special Publications of the Geological Survey of South Africa 5, 263-274.

Davidson, D.W., Garg, S.K., Gough, S.R., Handa, Y.P., Ratcliffe, C.I., Ripmeester, J.A., Tse, J.S., and Lawson, W.F. (1986). Laboratory Analysis of a Naturally Occurring Gas Hydrate from Sediment of the Gulf of Mexico. Geochemica et Cosmochimica Acta 50, 619-623.

Davis, J.B., and Kirkland, D.W. (1979). Bioepigenetic Sulfur Deposits. Economic Geology 74, 462-468.

Deroo, G., Powell, T.G., Tissot, B., McCrossan, R.G., and Hacquebard, P.A. (1977). The Origin and Migration of Petroleum in the Western Canadian Basin, Alberta: A Geochemical and Thermal Maturation Study. Bulletin of the Geological Survey of Canada 262, 136 pp.

Dow, W.G. (1984). Oil Source Beds and Oil Prospect Definition in the Upper Tertiary of the Gulf Coast. Transactions of the Gulf Coast Association of Geological Societies 34, 329-339.

Dow, W.G. (1977). Kerogen Studies and Geological Interpretations. Journal of Geochemical Exploration 7, 79-99.

Evans, C.R., Rogers, M.A., and Bailey, N.J.L. (1971). Evolution and Alteration of Petroleum in Western Canada. Chemical Geology 8, 147-179.

Feeley, H.W., and Kulp, J.L. (1957). Origin of Gulf Coast Salt Dome Sulfur Deposits. Bulletin American Association of Petroleum Geologists 41, 1802-1853.

Gould, K.W., and Smith, J.W. (1978). Isotopic Evidence for Biologic Role in Genesis of Crude Oil from Barrow Island, Western Australia. Bulletin American Association of Petroleum Geologists 62, 455-462.

Henry, M.E., and Donovan, T.J. (1978). Isotopic and Chemical Data from Carbonate Cements in Surface Rocks Over and Near Four Oklahoma Oil Fields. United States Geological Survey Open File Report 78-927, 1-17.

Jobson, A.M., Cook, F.D., Westlake, D.W.S. (1979). Interaction of Aerobic and Anaerobic Bacteria in Petroleum Biodegradation. Chemical Geology 24, 355-365.

Kirkland, D.W., and Evans, R. (1976). Origin of Limestone Buttes, Gypsum Plain, Culberson County, Texas. Bulletin American Association of Petroleum Geologists 60, 2005-2018.

Koons, C.B., Bond, J.G. and Peirce, F.L. (1974). Effects of Depositional Environment and Postdepositional History on Chemical Composition of Lower Tuscaloosa Oils. Bulletin American Association of Petroleum Geologists 58, 1272-1280.

Kreitler, C.W. and Dutton, S.P. (1983). Origin and Diagenesis of Cap Rock, Gyp Hill and Oakwood Salt Domes, Texas. The University of Texas at Austin, Bureau of Economic Geology Report of Investigation 131, 58 pp.

MacKenzie, A.S., Wolff, G.A., and Maxwell, J. R. (1983). Fatty Acids in some biodegraded Petroleums, Possible Origins and Significance. in "Advances in Organic Chemistry 1981" (M. Bjoroy, ed.), 637-649.

Macqueen, R.W., Powell, T.G. (1983). Organic Chemistry of the Pine Point Lead-Zinc Ore Field and Region, Northwest Territories, Canada. Economic Geology 78, 1-25.

Milner, C.W.D., Rogers, M.A. and Evans, C.R. (1977). Petroleum Transformation in Reservoirs. Journal of Geochemical Exploration 7, 101-153.

O'Brien, J.J. and Lerche, I. (1984). The Influence of Salt Domes on Paleotemperature Distributions. Geophysics 49, 2032-2043.

Oehler, J.H. (1984). Carbonate Source Rocks in the Jurassic Smackover Trend of Mississippi, Alabama, and Florida: Petroleum Geochemistry and Source Rock Potential of Carbonate Rocks. American Association of Petroleum Geologists Studies in Geology 18 (J. G. Palacas, ed.), 63-69.

Orr, W.L. (1974). Changes in Sulfur Content and Isotopic Ratios of Sulfur During Petroleum Maturation -- Study of Big Horn Basin Paleozoic Oils. Bulletin American Association of Petroleum Geologists 50, 2295-3218.

Pawlowski, S., Pawlowska, K., and Kubica, B. (1979). Geology and Genesis of Polish Sulfur Deposits. Economic Geology 74, 475-483.

Philippi, Gt. (1977). On the Depth, Time and Mechanism of Origin of the Heavy to Medium-Gravity Naphthenic Crude Oils. Geochimica et Cosmochimica Acta 41, 33-52.

Price, P.E., and Kyle, J.R. (1983). Metallic Sulfide Deposits in Gulf Coast Salt Dome Cap Rocks. Transactions of the Gulf Coast Association of Geological Societies 33, 189-193.

Rice, D.D. (1980). Chemical and Isotopic Evidence of the Origins of Natural Gases in offshore Gulf of Mexico. Transactions of the Gulf Coast Association of Geological Societies 30, 203-213.

Rubenstein, I., Spyckerelle, C., and Strausz, O.P. (1979). Pyrolysis of Asphaltenes: A Source of Geochemical Information. Geochimica et Cosmochimica Acta 43, 1-6.

Sassen, R. (1980). Biodegradation of Crude Oils and Mineral Deposition in a Shallow Gulf Coast Salt Dome. Organic Geochemistry 2, 153-166.

Sassen, R. Kennicutt, M.C., II, and Brooks, J.M. (1985). Ongoing Deposition of Isotopically Light Carbonate in Oil Seeps, Gulf of Mexico Continental Slope. Sixth Annual Research Conference, Gulf Coast Section, Society of Economic Paleontologists and Mineralogists Foundation, Program and Abstracts, p. 27.

Seifert, W.K., and Moldowan, J.M. (1979). The Effect of Biodegradation on Steranes and Terpanes in Crude Oils. Geochimica et Cosmochimica Acta 43, 111-126.
Seifert, W.K., Moldowan, J.M. and Demaison, G.J. (1984). Source Correlation of Biodegraded Oils. Organic Geochemistry 6, 633-643.
Thode, H.G., Wanless, R.K., and Wallouch, R. (1954). The Origin of Native Sulfur Deposits from Isotope Fractionation Studies. Geochemica et Cosmochimica Acta 5, 286-298.
Toland, W.G. (1960). Oxidation of Organic Compounds with Aqueous Sulfate. Journal American Chemical Society 82, 1911-1916.
Ulrich, M.R., Kyle, J.R., and Price, P.E. (1984). Metallic Sulfide Deposits in the Winnfield Salt Dome, Louisiana: Evidence for Episodic Introduction of Metalliferous Brines During Cap Rock Formation. Transactions of the Gulf Coast Association of Geological Societies 34, 435-442.
Volkman, J.K., Alexander, R., Kagi, R.I., Rowland, S.J. and Sheppard, P.N. (1984). Biodegradation of Aromatic Hydrocarbons in Crude Oils from the Barrow Sub-Basin of Western Australia. Organic Geochemistry 6, 619-632.
Walters, C.C. and Cassa, M.R. (1985). Regional Organic Geochemistry of Offshore Louisiana. Transactions of the Gulf Coast Association of Geological Societies 35, 277-286.
Winters, J.C. and Williams, J.A. (1969). Microbiological Alteration of Petroleum in the Reservoir. American Chemical Society Division of Petroleum Chemistry, New York City Meeting, Preprints, E22-E31.

SECTION D

FLUID FLOW, SALT DISSOLUTION, HEAT FLOW, AND HYDROCARBON MIGRATION

DYNAMICS OF SUBSURFACE SALT DISSOLUTION AT THE WELSH DOME, LOUISIANA GULF COAST

Stephen S. Bennett

Sun Oil Co.
Lafayette, Louisiana 70501

Jeffrey S. Hanor

Department of Geology
Louisiana State University
Baton Rouge, Louisiana 70803

I. INTRODUCTION

Seni and Jackson (1984) estimate that nearly half of the total volume of salt which migrated upward in the East Texas Salt Basin during Early Cretaceous to Eocene has been lost by erosion and dissolution. While the presence of caprock on many South Louisiana salt domes and the pervasive occurrence of brines in the Louisiana subsurface attest to the existence of on-going dissolution of halite (Hanor and Bailey, 1983), there has been relatively little documentation of the mechanics and rates of this process in this region of the Gulf Coast. There are many important implications to subsurface dissolution of salt, however, including its role in driving fluid flow, in producing subsurface brines, and in compromising the integrity of salt domes as storage repositories.

Hanor and Bailey (1983), in a regional study of the hydraulic and salinity regime of southwestern Louisiana, first described the existence of an area of elevated pore water salinity spatially associated with the Welsh Dome in Jefferson Davis Parish. Of particular interest were the observations that the subsurface salinity high extended upward over 1.5 km above the top of salt and over 10 km laterally both to the north and to the south of the dome. The broad areal extent of the dispersion of dissolved salt at this site must reflect a dynamic, large-scale system of dissolution and mass transport. The purpose of this paper is to present the results of the first phase of a detailed investigation of the subsurface hydrology, geochemistry, and geology at Welsh undertaken to identify probable pathways, mechanisms, and rates of solute transport.

II. STUDY AREA AND TECHNIQUES

The Welsh Dome is located in Jefferson Davis Parish in southwestern Louisiana (Fig. 1) and is one of several similar domes situated along an east-west trend of salt structures in this portion of the South Louisiana salt basin. The dome was ultimately derived from depth by upward migration of the Louann Salt of Jurassic age (Martin, 1978). The Welsh oil field, which is spatially associated with the dome, is one of the oldest in the entire Gulf Coast. It was discovered in 1902 as a result of drilling a broad, 3-meter, topographic high in the ground surface above the subsequently-discovered dome (Reed, 1927).

The Welsh field has been extensively drilled, and numerous wire-line logs exist for boreholes in the area. The primary data base for this study consists of information derived from over fifty spontaneous potential (SP)/resistivity logs. Figure 2 shows the spatial distribution of borehole control. The logs were used in this study to correlate subsurface units at Welsh with units identified in the regional studies of Jones and Turcan (1954), Paine (1963), and Bailey (1978) and to prepare structural, stratigraphic, and lithologic cross sections and maps. Pore water salinities were calculated from SP response using conventional methods (Bateman and Konen, 1977).
Down-hole temperatures and fluid pressures, critical in assessing probable mechanisms of solute transport, were also calculated from log information. In-situ temperatures were calculated by

Figure 1. Map of Louisiana showing the location of the Welsh Dome.

applying a depth-related correction of up to 15°C to recorded bottom hole temperatures (BHT) to compensate for heat loss due to drilling fluid circulation (Kehle, 1971) and assuming that a linear temperature gradient exists from the shallowest corrected-BHT in a well to the ground surface. Pore fluid pressures were calculated using the shale resistivity method of George (1965). Extended discussion and documentation of all of the temperature, pressure, and salinity techniques employed in this study is given by Bennett (1985). The salinity, temperature, and pressure information thus derived was used to calculate hydraulic head and in-situ fluid densities, as will be described.

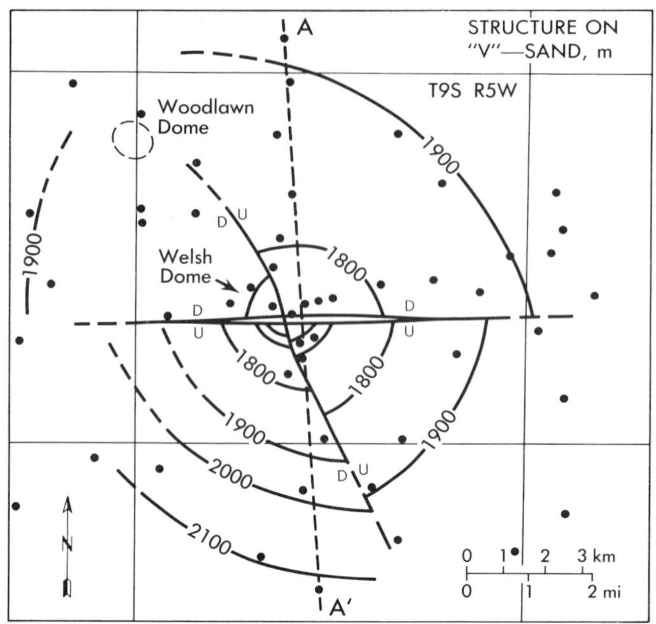

Figure 2. Map of the Welsh study area. Solid dots show the location of boreholes used as control. Contours show the depth from ground elevation, which averages 10 m above msl, to the top of "V"-sand (Fig. 3). Line A-A' marks the position of cross section A-A'.

III. PHYSICAL SETTING AND SALINITY REGIME

A. Stratigraphy and Structural Geology

The Welsh Dome is an intermediate depth, piercement diapir (Paine, 1963). The top of salt occurs at a depth of approximately 2050 m. There is no salt overhang and no reported occurrence of caprock. The dome is apparently circular in outline and roughly 1.6 km in diameter near its apex. The exact configuration of the dome surface at depth is not well known, however, and the shape shown on the cross-sections in this paper is strictly schematic.

The ages of the sediments represented by log information in the study area range from Recent to Oligocene (Fig. 3). The uppermost 300 m of the section are composed of Recent clays and sands and coarse-grained Pleistocene sands and gravels (Reed, 1927; Jones and Turcan, 1954). There is an erosional unconformity between the Pleistocene and underlying, 250 m thick, sequence of Pliocene shales and minor sands. This sequence is underlain in turn by a nearly 2 km thickness of Miocene sands, silts, and shales. Deposition of a 400 m thickness of shales during Upper Oligocene, Anahuac time, marks the only major marine transgression in the study area. Below the Anahuac there exists at least a kilometer thickness of alternating sands and shales of the Upper Frio Formation, also Oligocene in age. The variation in stratigraphic abundance of sands and shales, as deduced in this study from electric log response is shown in Figure 4. The dominantly sandy nature of the Miocene sequence, the massive shales of the Anahuac Formation, and the increase in shale content of the Frio with depth are evident.

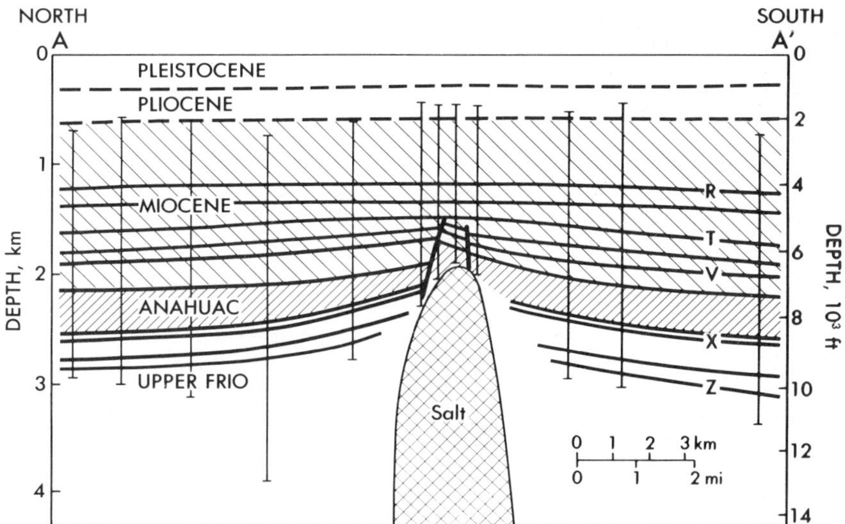

Figure 3. Cross section A-A', showing age and structure of sediments in the vicinity of the Welsh dome. Vertical lines show electric log control. Letters R through Z refer to structural markers. The shape of the salt dome as shown is schematic. Vertical exaggeration = 3X.

A number of sand beds can be readily correlated laterally throughout the study area and used to delineate structural deformation in the area (Figs. 2 and 3). Compared with many other salt domes in the region, deformation at Welsh has not been intense. There has been broad uplift of sediments around the margins of the dome with the development of faults over the dome proper. Dips of beds below depths of 1200 m approach 17 degrees near the margin of the dome. Maximum dips above depths of 1200 m are less than 1 degree. Faulting at Welsh is dominated by an east-west trending normal fault downthrown approximately 180 m to the north. Several faults of lesser displacement are transected in a west-east cross section described by Bennett (1985). Miocene beds cut by faults have equal thicknesses on both sides, implying that deformation was post-depositional (Gussow, 1968). There are no known growth faults in the immediate area (Paine, 1963). The faults shown in Figures 2 and 4 extend upward at least into the Miocene section. The existence of a 3-meter

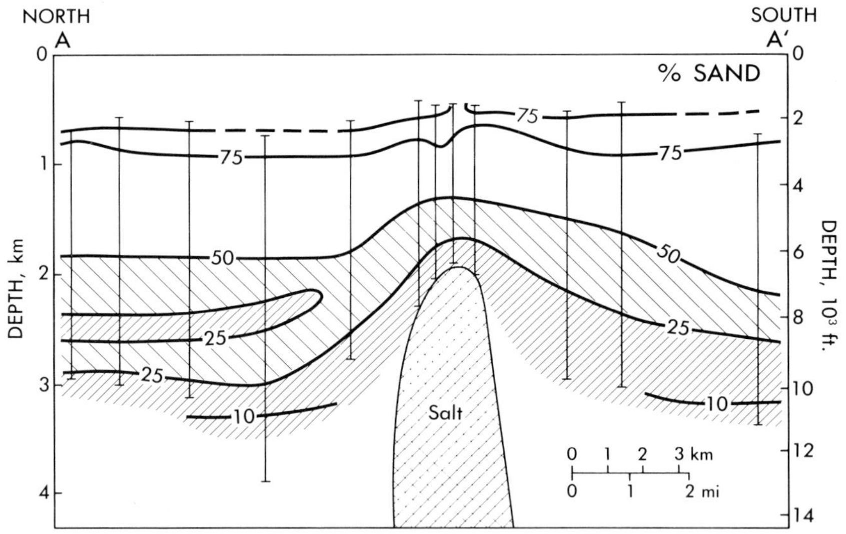

Figure 4. Variations in the proportions of sand and shale with depth as deduced from electric log response. Based on figures calculated for each 150-meter depth interval.

topographic anomaly above the dome at ground surface today is evidence for very recent and continued differential movement of sediment around the dome.

A second, deep-seated salt dome, the Woodlawn Dome, is located approximately 6 km northwest of Welsh (Fig. 2). The top of salt at Woodlawn occurs at a depth of approximately 3500 m. Its intrusion has had apparently little or no effect on the structural geology or salinity regime at Welsh.

B. Salinity Regime

The Welsh Dome is of particular interest in the general study of the dynamics of the dissolution of salt domes because of the associated pronounced pore water salinity anomaly (Hanor and Bailey, 1983). The existence of this salinity high has been verified in this study and mapped out in detail (Figs. 5 and 6; and Bennett, 1985). The sediments in the study area were deposited in fluvial-deltaic to normal marine environments (Rainwater, 1964) and thus originally contained pore water having salinities of 0 to 35 g/l. At the present time, however, most of the pore waters are much more saline. The north-south cross-section (Fig. 5) documents the existence of a plume of water characterized by elevated salinities in excess of 160 g/l which extends upward from the top of salt a distance of approximately 1 km. In map plan, this plume extends several kilometers laterally away from the dome to the northwest (Fig. 6) and is approximately coincident with the main area of hydrocarbon production at Welsh (Paine, 1963).

The overall region of elevated pore water salinity is not confined to the plume described above. Pore waters in the whole sedimentary sequence between depths of approximately 1 to 3 km over the entire study area have salinities in excess of 100 g/l. Pore waters freshen upward toward the base of the overlying fresh water zone and downward into the Frio Formation. The variation in salinity correlates in part with variations in lithology.

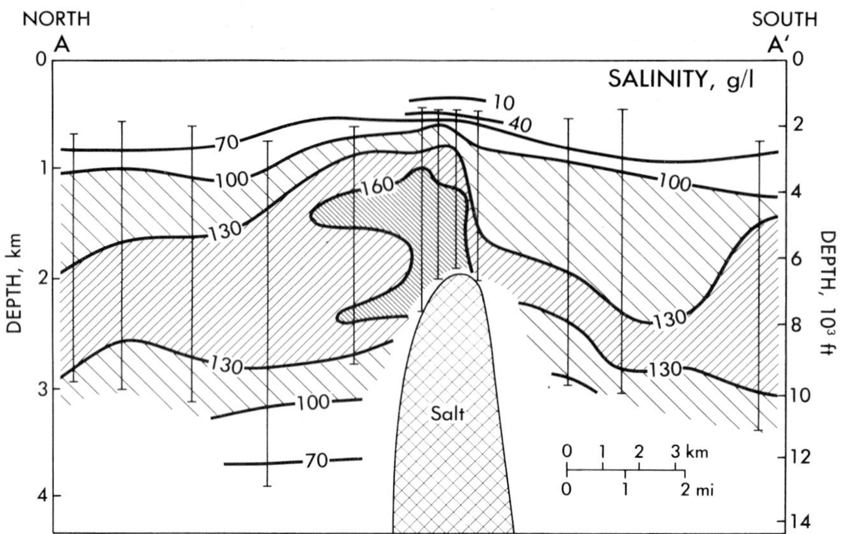

Figure 5. Variations in pore water salinity as deduced from spontaneous potential measurements.

The saltiest waters occur in the basal portion of the sand-dominated section. Salinities decrease in the underlying, shale-dominated part of the sequence. The apparent vertical bifurcation of the salinity plume to the north of the dome is spatially associated with the shales of the Anahuac Formation (Fig. 3). The preferential extension of the plume of highly saline brine to the northwest of the dome may similarly correlate with variations in sediment lithology, but our mapping is not yet detailed enough to verify this.

Three mechanisms are currently in vogue to explain the origin of brines in sedimentary basins (Hanor, 1983): subsurface dissolution of salt; burial of subareally-produced, so-called "connate" brines; and membrane filtration. We believe that the systematic variation in pore water salinities away from the Welsh dome provides convincing evidence for the dissolution of halite and the subsequent dispersion of dissolved salt at this particular site. The present brines cannot be connate in origin

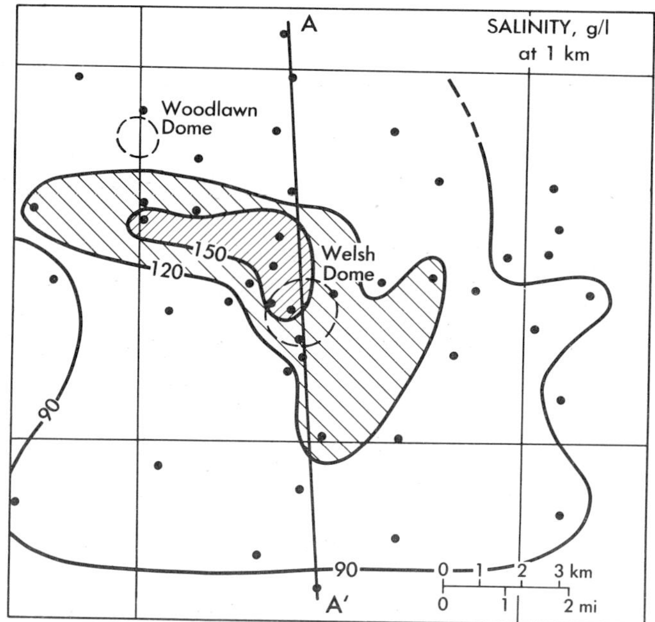

Figure 6. Map showing variations in pore water salinity at a depth of 1 km. The position of the 150 g/l plume corresponds approximately to the principal area of hydrocarbon production in the Welsh field (Paine, 1963).

because their host sediments were deposited in fresh water to marine environments, not under hypersaline conditions. The brines cannot have been generated by membrane filtration because, as we shall see, hydraulic drive is operating in the wrong direction to force reverse osmotic effects.

IV. MASS TRANSPORT REGIME

A. Introduction

What are the mechanisms of mass transport of dissolved salt at Welsh and how fast do they operate? In this paper, we will address the first of these questions and evaluate the potential fields that exist in the vicinity of the Welsh dome which have the capacity for driving fluid flow and/or solute transport. The strategy will be first to define the temperature and pore fluid pressure fields in the study area. Temperature and pressure information, in conjunction with salinity and elevation, can then be used to calculate important fluid properties such as hydraulic head and density which control or influence mass transport.

B. Temperature

Figure 7 shows the areal variation in estimated in-situ temperature in the study area at a depth of approximately 2000 m, just slightly above the elevation of the top of salt. There is a broad thermal high centered on the Welsh dome. Temperatures at the center of this high are approximately 10 to 15 degrees warmer than temperatures at comparable depths 6 to 7 km laterally away from the dome. An area of elevated temperature extends to the northwest and is approximately coincident with the salinity plume pictured in Figure 6.

Isotherms in cross section are shown in Figure 8. While control along the immediate margin of the dome is inadequate to

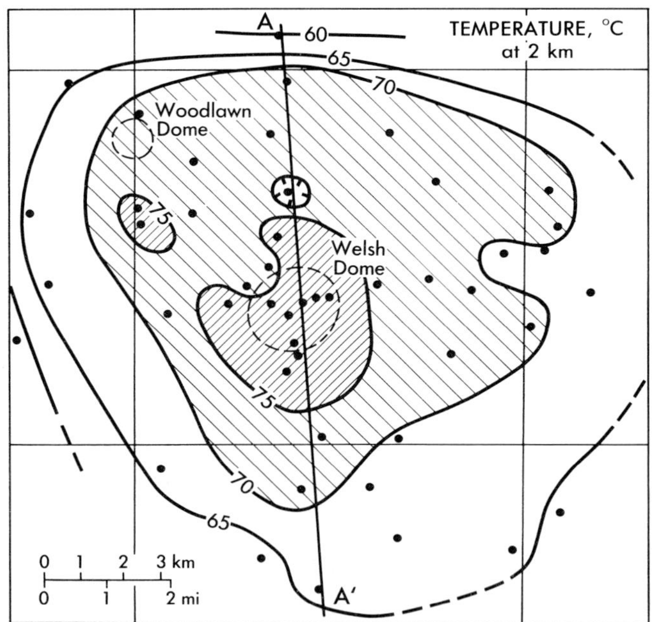

Figure 7. Areal variation in temperature at a depth of 2 km, as deduced from corrected bottom hole temperatures.

define the thermal regime within a kilometer of the salt stock below depths of 2 km, the regional displacement upward of isotherms is evident. The vertical compression of the 90 to 120°C isotherms to the north of the dome reflects the lower thermal conductivity of the undercompacted, overpressured shales of the Upper Frio. Of interest is the apparent finer-scale structure to the isotherms around the salt dome. The elevations of the 70° and 80° isotherms, for example, are slightly depressed at distances of from 3 to 4 km away from the dome. We believe these perturbations may reflect downwelling of fluid, as will be discussed.

C. Pore Fluid Pressure and Hydraulic Head

Results of the pore fluid pressure calculations in the study area (Bennett, 1985; not shown here) indicate that the Recent through Miocene portion of the section is hydrostatically pressured. There are, however, apparent areas of over-pressured

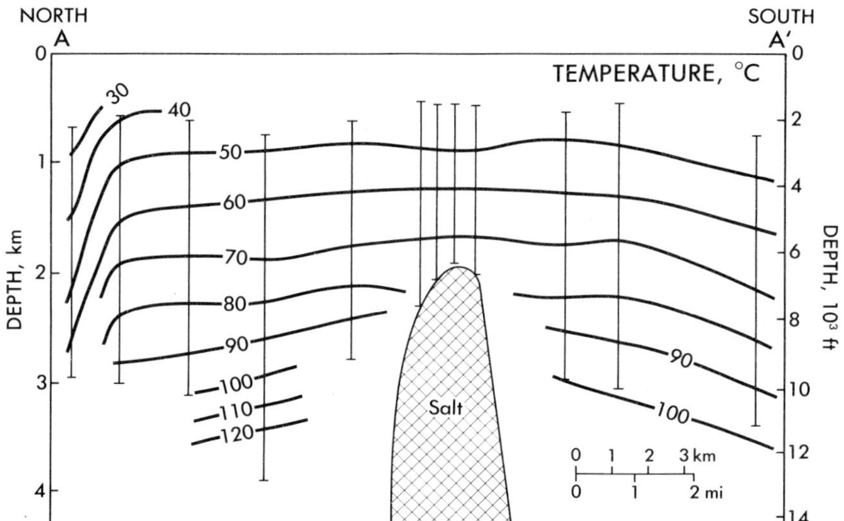

Figure 8. Cross section showing estimated position of isotherms in the vicinity of the Welsh dome. Further documentation is needed to confirm the apparent sharp depression of the isotherms in the northernmost control well.

fluid, where pressures exceed hydrostatic levels, within the shales of the Anahuac Formation and locally near the top of salt. The top of the regional overpressured section (Hanor and Bailey, 1983) is encountered at depths of approximately 3.0 to 3.5 km. The top of the overpressured section, as defined by the occurrence of fluids having geostatic or pressure/depth ratios greater than 0.70 psi/ft (1.6 bars/m), has been broadly uplifted in the vicinity of the Welsh dome. There is a suggestion of a northwest-southeast axis of elongation to this uplift (Fig. 9).

The principal interest in fluid pressure in this study is in the contribution of pressure to hydraulic head. Hydraulically-induced flow of fluid occurs in response to differences in hydraulic head, h (Hubbert, 1940). In isodensity systems, values of head can be calculated from the relation, $h = z + P/\rho g$, where z is elevation relative to an arbitrary datum, P is pore fluid pressure, ρ is fluid density, and g is the gravitational

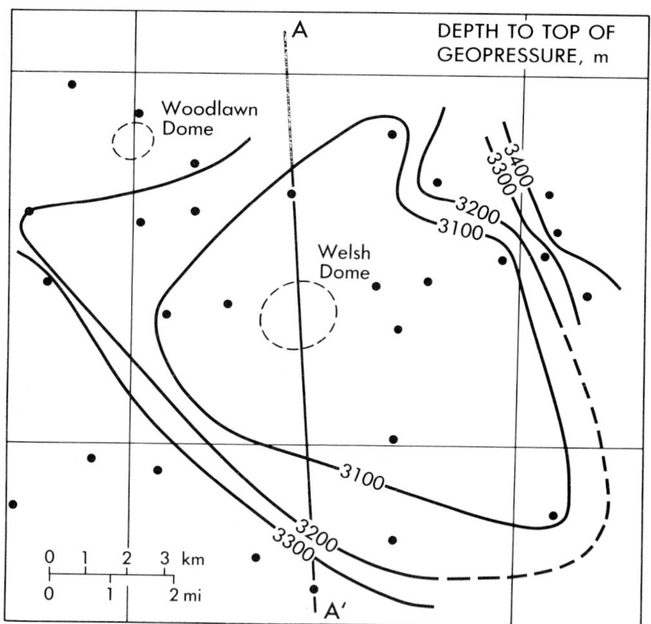

Figure 9. Depth to the top of the geopressured section, as defined by the first occurrence of pore fluids having a geostatic ratio of 1.6 bars/m (0.70 psi/ft).

constant. The shale resistivity techniques used to estimate fluid pressures in this study are not sensitive enough to delineate subtle deviations in fluid pressure in the hydrostatically- pressured zone. These techniques, however, are more than adequate for defining the large deviations in fluid pressure and hydraulic head which define the transition from hydropressured sediments into the underlying geopressured zone (Hanor and Bailey, 1983). Heads calculated in this study are normalized to a constant pore fluid density of 1.07 g/cm^3 and are expressed as elevations relative to mean sea level (msl). Variations in head resulting from differences in density (Garven and Freeze, 1984) are small relative to the variations resulting from differences in pressure in the overpressured sequences studied here, but must be incorporated in any quantitative modelling of hydraulic flow in hydropressured systems.

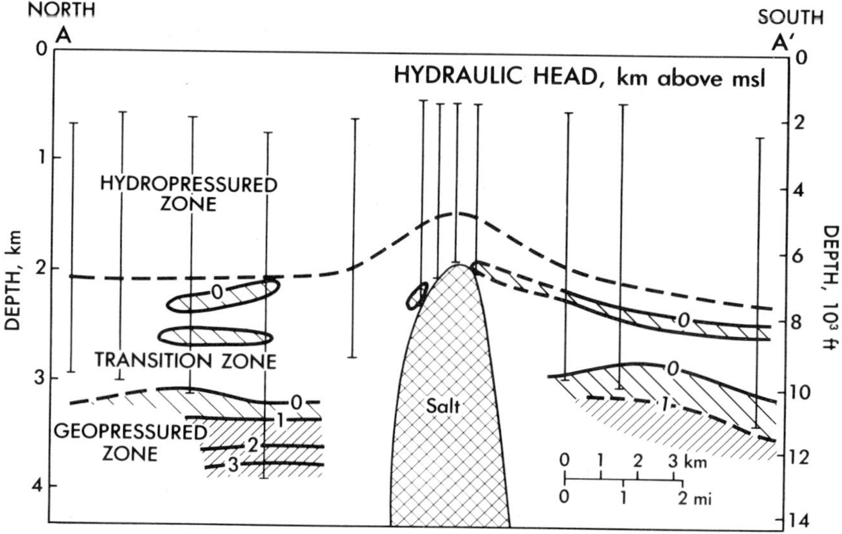

Figure 10. Cross-sectional variation in hydraulic head as calculated from log-derived pore fluid pressures. Heads are normalized to mean sea level and a constant fluid density of 1.07 g/cm3. The techniques employed here are not sensitive enough to define the small variations in head which undoubtedly exist in the hydropressured zone.

Figure 10 shows the variation in hydraulic head, as calculated above, for the north-south cross-section. Hydraulic heads are less than 300 m above msl down to the Anahac Formation at depths of approximately 2 to 2.5 km. Within this formation are one or more shaly units that are apparently overpressured and have heads in excess of 300 m. Zones of elevated head alternate with zones of lower head, a feature common to what Hanor and Bailey (1983) have termed the hydraulic transition zone in the South Louisiana Gulf Coast. Below depths of approximately 3 to 3.5 km, heads progressively increase to values in excess of 3000 m above sea level in the deepest boreholes for which information is available.

If the hydraulic conductivity of the sediments is known or can be estimated, it is possible from head variations to determine both the direction and rate of fluid flow (Garven and Freeze, 1984). While such calculations have not yet been completed for the study area, it is possible to make some qualitative interpretations. In the absence of other effects, fluids flow from regions of high to low head. At Welsh, it is reasonable to assume that overpressured fluids in the Upper Frio are migrating upward into the overlying sandy, Miocene section. The hydraulic conductivity of the Upper Frio section of shales and sands must be strongly anisotropic as a consequence of the difference in permeability across and along bedding. Where such sediments have been tilted, as along the margins of the dome, this differential hydraulic conductivity will result in a pronounced updip component to the flow direction. We believe it likely that overpressured fluids in the immediate vicinity of the dome are being focused upward and toward salt. Such a flow pattern has been documented at the Iberia Dome 60 km east of Welsh (Workman and Hanor, 1985) on the basis of spatial variations in the geochemistry of formation waters. Our hypothesis at Welsh could be tested if more subsurface information were available close to the salt stock. We would expect, for example, to find a depression in the head contours in the immediate vicinity of the dome, reflecting preferential migration of fluid toward and up the margins of the dome.

It is unlikely that membrane filtration or reverse osmosis (Hanor, 1983) is in any way contributing to the high salinities of the pore fluids at Welsh. If it were, the salinity of pore waters would progressively increase with depth in the geopressured section instead of freshening with depth, as observed.

D. Pore Fluid Densities

While the techniques of fluid pressure calculation employed here provide useful information on fluid flow and mass transport up from the deep geopressured zone, alone they are inadequate to permit an evaluation of mass transport within the overlying hydropressured zone. The present data base, however, does permit calculation of values for in-situ pore fluid density, variations in which provide a qualitative insight into some of the driving forces for fluid flow which exist in the hydropressured section and which are presently being used by us to generate density-corrected heads.

Pore fluid density is dependent on temperature, pressure, and composition. We have calculated in-situ pore fluid densities corrected for all three of these factors from the algorithm of Phillips et al. (1981). In making these calculations we have assumed that the formation waters are NaCl-dominated solutions (Hanor, 1984).

The variations in pore fluid density around the Welsh Dome are complex but systematic (Fig. 11). The densest waters occur in the warm, salty region just above the dome and in less saline but cooler areas approximately 10 km horizontally from the dome. The least dense waters in the hydropressured zone at depths of 0.5 to 2.5 km occur at distances of from 5 to 6 km from the dome.

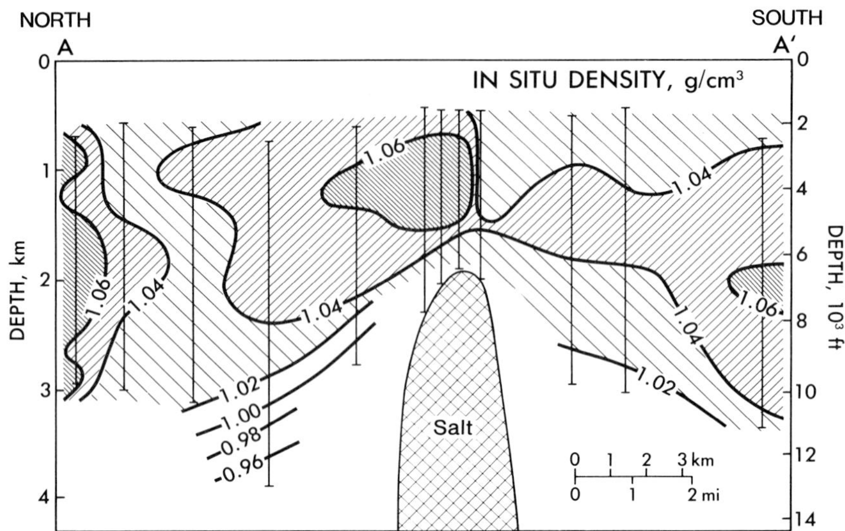

Figure 11. Calculated values of in-situ pore fluid density, corrected for temperature, pressure and salinity. Note pronounced density inversion and consequent buoyant instability in the pore fluid column below depths of from 1 to 2 km.

Densities decrease upward over the entire area toward the fresh water zone and decrease downward into the hot and less saline regime of the overpressured zone.

Most of the pore fluid column at Welsh below a depth of approximately 1 km is gravitationally unstable, i.e., pore fluid densities decrease with depth. The potential thus exist for large-scale overturn of fluid. Results of preliminary calculations suggest that pressure-dominated hydraulic gradients in the Upper Frio are sufficient to counter the increase in density upward and to force flow of fluid up the section. Pressure differences in the Miocene section, however, are insufficient to counter the instabilities induced by density differences. At a given depth within the hydropressured sections, areas of high fluid density should represent zones of downwelling, areas of lower fluid density zones of upwelling. If

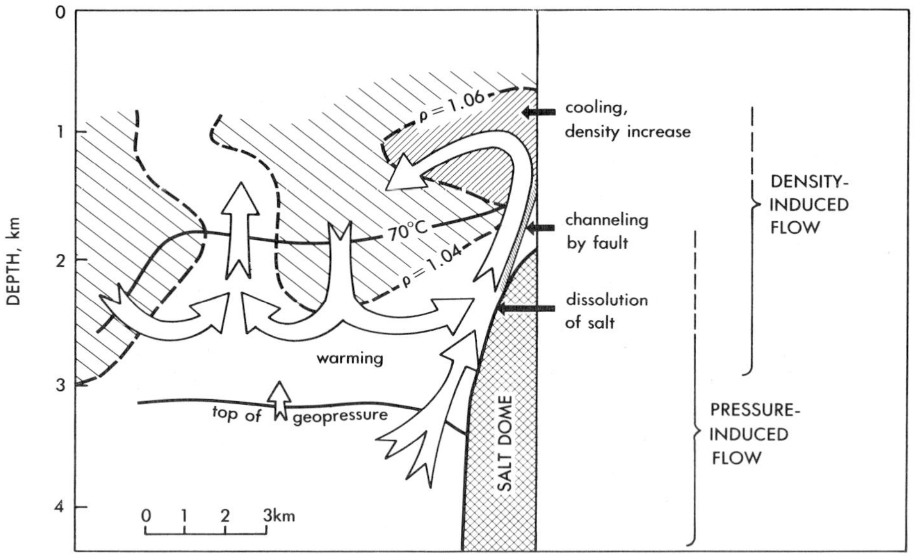

Figure 12. Possible pattern of large-scale solute transport on the north flank of the Welsh dome along section A-A′ as inferred from variations in hydraulic head, density, salinity, and temperature. Arrows indicate the probable net directions of subsurface transport. Actual flow paths are undoubtedly more complex, with a large component of transport occurring laterally across the plane of section A-A′ (c.f. Fig. 6).

such circulation is sufficiently rapid, there should be a depression of isotherms in downwelling zones and an elevation of isotherms in zones of upwelling.

The variations in density (Fig. 11) and temperature (Fig. 8) at Welsh are consistent with the existence of such circulation patterns. These we have illustrated schematically and in two dimensions in Figure 12. Actual mass transport pathways are undoubtedly more complex. The preferential skewing of the high-salinity plume to the northwest implies significant transport across the plane of section A-A′.

E. Intraformational Fluid Convection

The degree to which solute transport at Welsh involves cross-formational flow as pictured in Figure 12 or flow confined to individual sand beds with minor transport across shales is not yet known. Considerable attention has been given in the literature in the last few years to the possible existence of intraformational fluid convection in sedimentary sequences (Wood and Hewitt, 1982, 1984; Blanchard and Sharp, 1985), a process sometimes referred to as Bernard-Rayleigh convection. If the temperature at the base in a porous and permeable sedimentary bed filled with an isosalinity fluid is higher than the temperature at the top of the bed, fluid density differences induced by this difference may be sufficient to cause convective overturn of the pore fluid (Cambarnous and Bories, 1975). The criterion for convective overturn in a horizontal layer can be expressed by the dimensionless parameter known as the Rayleigh filtration number, Ra, which is the ratio of the driving forces of buoyancy to the dissipative mechanisms of viscous drag and heat conduction. In theory, sufficient force exists to induce overturn of fluid in a horizontal bed if Ra exceeds 40. For fluids in an inclined layer, however, there is no stability criterion, and the fluid will always be in motion if a temperature difference exists across the layer (Bories and Combarnous, 1973). The lateral velocity of pore fluid in an inclined bed can be estimated if the dip of the bed and other parameters including the coefficient of thermal expansion of the fluid and sediment, intrinsic permeability, thermal conductivity, and the temperature gradient are known (Wood and Hewitt, 1982).

In an attempt to assess the possible magnitude of the importance of Bernard-Rayleigh convection as a mass transport process at Welsh, values for Ra and for lateral convective fluid velocities were calculated by Bennett (1985) for individual sand beds within the section studied. Bennett found that the principal control on Ra and on calculated lateral convective velocity within the individual sand beds he studied is most dependent on the dip of the beds. Nearly horizontal beds within the study area have Ra values of less than 20 and calculated fluid velocities of less than 0.2 m/y. In the immediate vicinity of the salt dome, where the beds have been uplifted, Ra values exceed 20 and lateral fluid velocities exceed 0.5 m/y. Bennett has found the conditions for intraformational convection as defined above are apparently most favorable on the south side of the dome in section A-A', where structural dips are greatest. As has been demonstrated, however, the position of the salt plume is preferentially skewed to the north and northwest, suggesting that larger-scale transport may be more important.

F. Other Mass Transport Mechanisms

Mass transport processes in addition to pressure-induced and density-induced hydraulic flow may be operating in the Welsh area. These include molecular diffusion, thermal diffusion, and osmotic transport. Of these, calculations by Bennett (1985) based on estimated in-situ sediment and fluid properties suggest that molecular diffusion is likely to be the most important. Although molecular diffusion is often conceived of as being a negligibly slow process, the increase in magnitude of diffusion coefficients with increasing temperature and decreasing fluid viscosities can increase the efficiency of transport with depth (Ranganathan and Hanor, 1986). In addition, diffusion can operate in sedimentary sequences of very low hydraulic conductivity. Some of the areas in the Welsh section where molecular diffusion may be an important mass transport process include: 1) the geopressured zone, where salinity gradients may be inducing the downward migration of dissolved salt counter to the upward migration of pore fluid; 2) in the uppermost kilometer of the study area, where pore fluid densities and salinities both decrease upward toward the fresh water zone, and 3) across shales of low permeability in the Miocene and Anahuac.

G. Total Mass of Salt Dissolved

It is possible to make a reasonable estimate of the minimum total mass of halite which has been dissolved at Welsh. The total study area is 16 by 16 km (Fig. 2). Between a depth of 0.5 to 3.0 km, the average pore water salinity is approximately 130 g/l (Fig. 5). It is reasonable to assume that the fluvio-deltaic and marine sediments deposited at Welsh originally contained pore fluids having salinities of 0 to 35 g/l. Thus, an additional 100 g/l has been added by salt dissolution. Taking 0.20 as the average porosity of the sedimentary section between depths of 0.5 to 3.0 km (Bebout and Gutierrez, 1982), the total volume of fluid between this depth interval over the entire study area is 1.3×10^{14} l. The total mass of salt which dissolved to yield observed pore water salinities is 1.3×10^{13} kg, a value which is equivalent to 6.0×10^{9} m^{3} NaCl.

Assuming that the Welsh Dome has a radius of approximately 1 km at depth, then the volume of salt which dissolved to account for observed pore water salinity levels is equivalent to the loss of a 2000-m vertical extent of the salt diapir. This mass is approximately equivalent to the mass of salt shown in the cross sections in this paper. The figures cited above represent minimum values for the amount of halite which has dissolved at Welsh. Presumably there has been continuing loss of dissolved salt up section into the overlying fresh water zone and downward into less saline geopressured sediments. The intrusive history of the Welsh salt diapir has not yet been documented. If salt dissolution has been occurring since the beginning of Miocene, i.e., 25 My, then the minimum average rate of dissolution at Welsh has been 5.2×10^{5} kg NaCl/year. It is possible that the current rate of dissolution of salt at Welsh is balanced by the rate of upward intrusion of halite.

V. DISCUSSION

It is clear from the spatial variations in pore water salinity at the Welsh dome that active subsurface dissolution of salt is occurring at this site and that there exist processes of solute dispersion capable of transporting dissolved salt great distances laterally and vertically from the dome. We believe the following to be reasonable first hypotheses regarding the nature of the mass transport regime at Welsh.

There appear to be two major hydraulic drives operating in the vicinity of the Welsh Dome: first, large differences in fluid pressure, and second, significant differences in pore fluid density. It is reasonable to conclude on the basis of our calculated hydraulic heads that pore fluid is being driven up and out of the geopressured zone and toward the Welsh Dome. This water has salinities of less than 50 g/l and thus has the capacity for dissolving salt.

We do not know yet whether the fluids responsible for this dissolution are primarily upwelling waters from the geopressured zone, recycled waters from the hydropressured zone, or some blend of both. On the basis of the spatial variations in pore water salinities above the dome, we believe it likely that the major east-west fault pictured in Figures 2 and 3 is serving as a preferred conduit for fluid flow. The presence of the fault helps to explain how dissolved salt can be efficiently transported upward across a kilometer thickness of sands and shales. The differences in hydraulic head which are driving this upward transport are likely due in part to fluid density differences. Differences in pore fluid pressure might also be a factor if waters from the geopressured zone are being focused up the margins of the dome.

If the waters responsible for dissolution were saturated with respect to halite, they would contain in excess of 350 g/l dissolved salt. Instead, the maximum recorded salinities as deduced from the electric logs are about half this, or approximately 180 g/l. One explanation is that dissolved NaCl is diffusing through a thin shale sheath known to surround some domes (Gilreath, 1968) and that observed concentrations reflect the relative rates of diffusion and fluid flow. Another explanation is that there is a small core of 350 g/l water that we simply have not transected with our logs before it is diluted by dispersion.

As hot salty fluids move up and away from the fault above the Welsh Dome, they cool and become more dense. The density inversion produced by cooling results in a sinking downward of fluid away from the dome. The degree of cross-formation flow is not known, but may be significant. There is a dilution of the dense brines by dispersive mixing with less saline waters. This and warming at depth decreases fluid densities and produces conditions favorable for upwelling. The spatial variations in salinity, density, and temperature documented in this paper support the hypothesis that large-scale vertical transport exists in the hydropressured section around Welsh. The rates of vertical transport of warm fluid up and cooler fluid down appear, from our preliminary work, to be sufficient to cause discernable deflections in the thermal field around the dome (Figs. 8 and 12). The base of the region of apparent overturn coincides with the top of the section containing approximately 25% or less sand beds.

The processes by which dissolved salt is transported in the uppermost kilometer of the study area are not well understood by us. There is significant increase in pore fluid density down from the fresh water zone into the region of saline formation waters, which should preclude large-scale vertical instability. The systematic decrease in salinity above the dome, however, is evidence for the existence of some type of mixing process resulting in the net transport of dissolved salt up and out of the system. Transport by dispersion accompanying upward advection, intraformational convection, and molecular diffusion are three possibilities.

VI. CONCLUSIONS

The pattern of dissolution and dispersion of salt at Welsh reflects the existence of a dynamic pore water flow regime in the upper several kilometers of the sedimentary section in the region of the dome. Fluid transport here appears to be largely driven by spatial variations in the temperature and the pore water salinity fields. The configuration of temperature field is largely controlled by differences in thermal conductivity between salt and surrounding sediment. Superimposed on this conductive-dominated field are perturbations induced by fluid flow. The salinity field is maintained by two quite different, but equally important, processes. The first is continuing introduction of dissolved salt into the system by dissolution. The second is loss of dissolved salt by transport upward into the overlying fresh water zone and by dilution resulting from diffusive or dispersive mixing with less salty fluids of the underlying geopressured sediments. In the deeper part of the section, the pressure fluid becomes significant. Large hydraulic gradients induced by significant increases in pore fluid pressure and enhanced by decreases in density with depth are driving deep, less saline fluids upward into the hydropressured zone.

The details of the subsurface geology at Welsh play a critical role in determining the directions and rates of mass transport. Structural dip, the presence of faults, and the spatial variations in the relative abundance of sand and shale all appear to be important. Worthy of further study is the possible relation between dissolved salt and hydrocarbon transport.

Many of the conclusions presented in this paper can and are being tested by means of additional field study and the application of numerical modeling of heat and solute transport. The modeling techniques, in particular, will permit an estimate to be made of the absolute rates at which mass and thermal transport are occurring. While the observations and results obtained in the present paper are specific to Welsh, the general principles should be applicable to a large number of other salt domes.

VII. ACKNOWLEDGEMENTS

This study was funded in part by Grant No. 83-02-31 to J.S.H. from the LSU Center for Energy Studies. We would like to thank Paul Aharon, Charles Groat, Michael Simms. and Vishnu Ranganathan for useful comments and discussion.

REFERENCES

Bailey, J.E. (1978). Occurrence of dissolved methane resources in the formation water of the hydropressure zone in southern Louisiana. M.S. Thesis, Louisiana State Univ., 306 p.

Bateman, R.M., and Konen, C.E. (1977). Wellsite log analysis and the programmable pocket calculator. Society Professional Well Log Analysts Trans. 18, B1-B35.

Bebout, D., and Guteirrez, D.M. (1982). Regional cross-sections, Louisiana Gulf Coast. Louisiana Geol. Surv. Folio Series 2, 54 p.

Bennett, S.C. (1985). Mechanisms of mass transport of brines and hydrocarbons in the Welsh field, Jefferson Davis Parish, Louisiana. M.S. Thesis, Louisiana State Univ., 307 p.

Blanchard, P.E. and Sharp, J.M. (1985). Possible free convection in thick Gulf Coast sandstone sequences. Trans. Southwest Section Amer. Assoc. Petrol. Geol. 1985, 6-12.

Bories, S.A. and Combarnous, M.A. (1973). Natural convection in a sloping layer. J. Fluid Mechanics 57, 63-79.

Combarnous, M.A. and Bories, S.A. (1975). Hydrothermal convection in saturated porous media. Advances in Hydroscience 10, 231-306.

Garven G., and Freeze, R.A. (1984). Theoretical analysis of the role of groundwater flow in the genesis of stratabound ore deposits. Amer. Jour. Sci. 284, 1085-1174.

George, R. (1965). Graph for estimating formation pressures from shale resistivity data. Schlumberger Well Surveying Corp., New Orleans, 1 p.

Gilreath, J.A. (1968). Electric-log characteristics of diapiric shales. Amer. Assoc. Petrol. Geol. Memoir 8, 243-256.

Gussow, W.C. (1968). Salt diapirism: importance of temperature, and energy source of emplacement. Amer. Assoc. Petrol. Geol. Memoir 8, 123-142.
Hanor, J.S. (1983). Fifty years of thought on the origin and evolution of subsurface brines. In "Revolution in the Earth Science" (S.J. Boardsman, ed.), pp 99-111. Kendall/Hunt, Dubuque.
Hanor, J.S, and Bailey, J.E. (1983). Use of hydraulic head and hydraulic gradient to characterize geopressured sediments and the direction of fluid migration in the Louisiana Gulf Coast. Trans. Gulf. Coast Assoc. Geol. Soc. 33, 115-122.
Hanor, J.S. (1984). Salinity and geochemistry of subsurface brines in South Louisiana and their potential for reaction with injected hydrothermal waste waters. U.S. Dept. Energy Rept. DOE/NV/10174-3, 277-346.
Hubbert, M.K. (1940). The theory of groundwater motion. Jour. Geol. 48, 785-944.
Jones, P.H. and Turcan, A.N. (1954). Geology and groundwater resources of southwestern Louisiana. Louisiana Geol. Surv. Bull. 30, 285 p.
Kelhe, R.O. (1971). Geothermal survey of North America, 1971 annual progress report. Amer. Assoc. Petrol. Geol., Tulsa, 31 p.
Martin, R.G. (1978). Northern and eastern Gulf of Mexico continental margin. Amer. Assoc. Petrol. Geol. Studies in Geology 7, 21-42.
Paine, W.R. (1963). Geology of Acadia and Jefferson Davis parishes. Louisiana Geol. Surv. Bull 36, 277 p.
Phillips, S.L., Igbene, A., Fair, J.A., Ozbeck, H., Tavena, M. (1981). A technical databook for geothermal energy utilization. Lawrence Berkeley Lab. Rept. LBL-12810, 46 p.
Ranganathan, V. and Hanor, J.S. (1986). Diffusion of dissolved NaCl from evaporites in sedimentary basins. EOS, Trans. Amer. Geophys. Union 67, 274.
Reed, L.C. (1927). The Welsh, Louisiana oil field. Amer. Assoc. Petrol. Geol. Bull. 11, 464-477.
Seni, S.J., and Jackson, M.L. (1984). Sedimentary record of Cretaceous and Tertiary salt movement, East Texas basin. Texas Bureau Econ. Geol., Rept. Inv. 139, 89 p.
Wood, J.A. and Hewitt, T.A. (1982). Fluid convection and mass transfer in porous sandstones - a theoretical model. Geochim. Cosmochim. Acta 46, 1707-1713.
Wood, J.R. and Hewitt, T.A. (1984). Reservoir diagenesis and convective fluid flow. Amer. Assoc. Petrol. Geol. Memoir 37, 99-109.
Workman, A.L., and Hanor, J.S. (1985). Evidence for large-scale vertical migration of dissolved fatty acid in Louisiana oil field brines: Iberia field, south-central Louisiana. Gulf Coast Assoc. Geol. Soc. Trans. 35, 293-300.

THE DEVELOPMENT BY SALT DIAPIRS OF SUPERFICIAL OVERHANG FEATURES, AND EFFECTS ON ASSOCIATED SEDIMENTS

by

Malcolm K. Jenyon
Seismograph Service (England) Limited

I. INTRODUCTION

The emplacement of a salt diapir frequently involves in its intermediate or later stages a morphological development wherein a part or the whole of the shallower portion of the salt body extends laterally beyond the horizontal cross-sectional area of the diapir root. Such a development is known as an overhang. It may result from a change in the direction of upward emplacement of the rising salt, or from an upward increase in the horizontal cross-section of the salt mass. Some types of overhang which develop are illustrated schematically in the vertical sections through diapirs shown in fig. 1. Real earth examples of such shapes are provided by 1(a) Cote Blanche Island Dome, Louisiana (see, e.g., Woodbury et al., 1980); 1(b) Hainesville Dome, Texas (see, e.g., Seni and Jackson, 1984); 1(c) Butler Dome, Texas (ibid.); 1(d) Gorleben salt diapir, Germany (see, e.g., Trusheim, 1960); and 1(e) Wienhausen salt diapir, Germany (see e.g., Schott, 1956).

Definitive explanations for these varied shapes cannot be given, but many suggestions have been made. The most likely of these are probably as follows. Type 1(a), it is suggested, is due to a change in the direction of upward emplacement of the salt resulting from changes in the pattern of overburden deposition, or the influence of tectonic stress, or faulting. The massive increase in horizontal cross-section upwards in Types 1(b) and 1(c) may be related to the passage of the salt from a zone of high confining pressure upwards into a zone of lower pressure related, in many cases, to the occurrence of less dense, less competent lithologic types in the shallower section. The morphology of Type 1(d) may be due in part to this same factor, and in part to the causes of development of the overhang of Type 1(e), the superficial spreading overhang which evolves at the crest of some diapirs and which will be the main subject for discussion here.

II. PRINCIPAL FORCES INVOLVED IN DIAPIR EMPLACEMENT

It is not intended in the present context to discuss the rheological behavior of halite or the ab initio causes of salt structures, about which there is still no general agreement (see, e.g. Jenyon, 1986). However, some of the questions on which agreement is still lacking carry over into the piercement stage of salt structure evolution, and a word must be said on this.

Some geologists favour the idea of active upward movement of salt and, in the piercement stage, forceful injection of salt through overburden materials. Others (see, e.g. Woodbury et al., 1980) believe that all the effects seen may be explained by the application of the "downbuilding" hypothesis first advanced by Barton (1933), whereby the crest of a salt structure retains its position relative to the geoid, while the superincumbent (and later, the surrounding) overburden sediments subside around the pillow/diapir with continuing sedimentation at the depositional surface and subsidence on the regional scale. The writer, while acknowledging that the downbuilding hypothesis may explain the evolution of some diapirs over either a part or the whole of their history, does not believe that it is universally applicable. Basin-edge diapiric features have been described elsewhere (Jenyon, 1985a), the evolution of which is difficult to reconcile with the downbuilding process; if this is one exception, there may be others.

Notwithstanding the differences of view mentioned, it would probably be agreed by their proponents that reduced to the simplest terms, the main forces involved in the emplacement of a salt diapir are as follows.

A. Forces promoting uplift

1. The upward force produced by salt flowing into the diapir from the source layer.
2. The upward force by buoyancy of the salt, which depends on the density differential between salt and overburden sediments.

B. Forces inhibiting uplift

1. Gravity, where the salt enters overburden sediments, the bulk density of which is less than the salt density.
2 Overburden competency - the static resistance of the overburden material to piercement by the rising salt.
3. Frictional resistance to the upward passage of the salt of sediments truncated against the diapir flank.
4. Overburden loading.

It is noted that in A.(1), the upward force of salt flow may be entirely a gravitational effect (the pressure engendered by subsidence or a denser overburden), or to lateral, updip or downdip, flow within the salt interval beneath an overburden which is less dense than the salt (see, e.g., Humphris, 1979, for an offshore Gulf Coast example), or to both effects.

Also, factor A.(2) will, under normal circumstances, become decreasingly important as salt rises through the overburden, becoming zero at some level where the salt and average overburden densities equalize. Above this level, additional "work" must be done by the upward force of salt flow to lift the salt mass against gravity. It is a commonplace that salt flow rates at late stages in diapir development show very marked reduction in comparison to flow rates at earlier stages (see, e.g., Seni and Jackson, 1984). This effect is also implicit in factor B.(1).

Factor B.(2), the overburden competency, depends on the density and other elastic properties inherent in the fabric of the various lithologic units overlying the salt. It has been defined in terms of physical models by Ramberg (1981), who referred to its model equivalent as the plunging strength of model overburden material in centrifuged model studies of diapiric piercement.

The frictional resistance factor, B.(3), appears to have been very little studied, and it is not possible to say with any certainty how it rates in importance with the other forces mentioned. Selig and Wallick (1966) carried out a mathematical modelling investigation into the isogeothermal patterns in and around a salt intrusion, and one of their conclusions was that the flank zone of a salt diapir would show sensible effects in temperature distribution resulting from the frictional resistance to movement of the salt flanks through overburden material. It may be that this effect, combined with other forces, could be of importance in the piercement process. However, insufficient is known of the effect for it to be included as a significant factor in the present discussion.

III. PRINCIPAL FACTORS GOVERNING DIAPIR MORPHOLOGY

Apart from the basic importance of the amount of mobile salt available in the source layer for diapiric development, other factors influence the form taken by a piercement structure, including (not in any order of importance),

A. Thickness of overburden
B. Pattern of deposition in the overburden
C. Regional subsidence rates
D. Influence of tectonism
E. Competency of overburden rocks, and variations therein
F. Any faulting developed as a cause or effect of the piercement structure
G. Differential between average overburden density and salt density

H. Any tendency for spine development from the main salt mass

As a generalization it is probably true to say that factors D, F, and H are the most instrumental in determining the plan-view form of a diapir as either an elongate or a subcircular feature. Any or all of the factors listed may influence the size and vertical cross-sectional shape of a diapir. Their relative importance in the formation of the overhang features, which are the subject of this discussion, will be considered later.

IV. SUPERFICIAL SPREADING OVERHANG FEATURES

The specific type of overhang feature to be considered here is that sketched in fig. 1(e), where salt has spread laterally at the crest of a diapir.

A well-known example of this type is shown by the Wienhausen diapir in northern Germany, which is illustrated in the sketch profile of fig. 2. This structure, produced by the Upper Permian Zechstein salt, is a funnel-shaped salt mass, increasing in

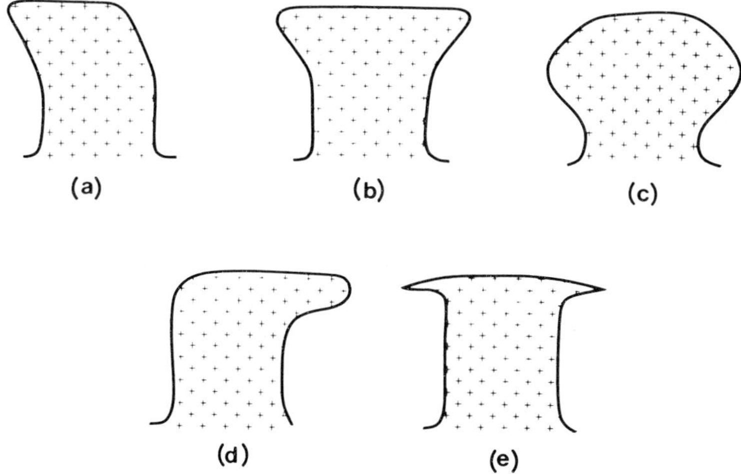

Figure 1. Schematic vertical sections through salt diapirs showing some of the most frequently-encountered types of overhang feature.

diameter upwards and terminating superficially in a thin spreading overhang beneath Upper Cretaceous and later sediments. The crest of the diapir proper is a flat dissolution/deflation surface overlain by a thin veneer of caprock which laterally thickens around the overhang; material of caprock type may also be present as a sheath on the diapir flanks.

Several suggestions have been put forward to explain spreading overhangs of the type seen in this example; the three most plausible of these will now be described.

A. Surface Extension of Salt

A number of workers, including Trusheim (1960), have suggested that these features may be due to the extrusion of salt at a depositional surface.

The so-called "salt glaciers" of Iran are perhaps the most well-known example of present day surface extrusion, being interpreted as the emergent crests of salt diapirs. Elsewhere in this volume they are discussed by Kent, who previously (Kent, 1979) has given a striking description of one of these features, Kuh-e-Namak, as a "..mound of shining salt rising to 1300 m above the plains.."

The writer has examined emergent features related to those in Iran (the source salt being the same stratigraphic unit) in the deserts of Central Oman, where a group exists including two named features, Qarn Alam and Q'arat Kibrit. The group consists of low mounds of detrital material together with euhedra of gypsum and calcite rhombs, accompanied by a strongly sulfurous smell (hence the name "Kibrit", which is the Arabic name for the old sulfur-based match). No salt is present, which suggests that only disintegrated caprock material has reached the surface in this locality.

Clearly, masses of salt extruded at the surface can, unless continuously fed from below, survive for any appreciable length of time only if the contemporary climate is highly arid, as in

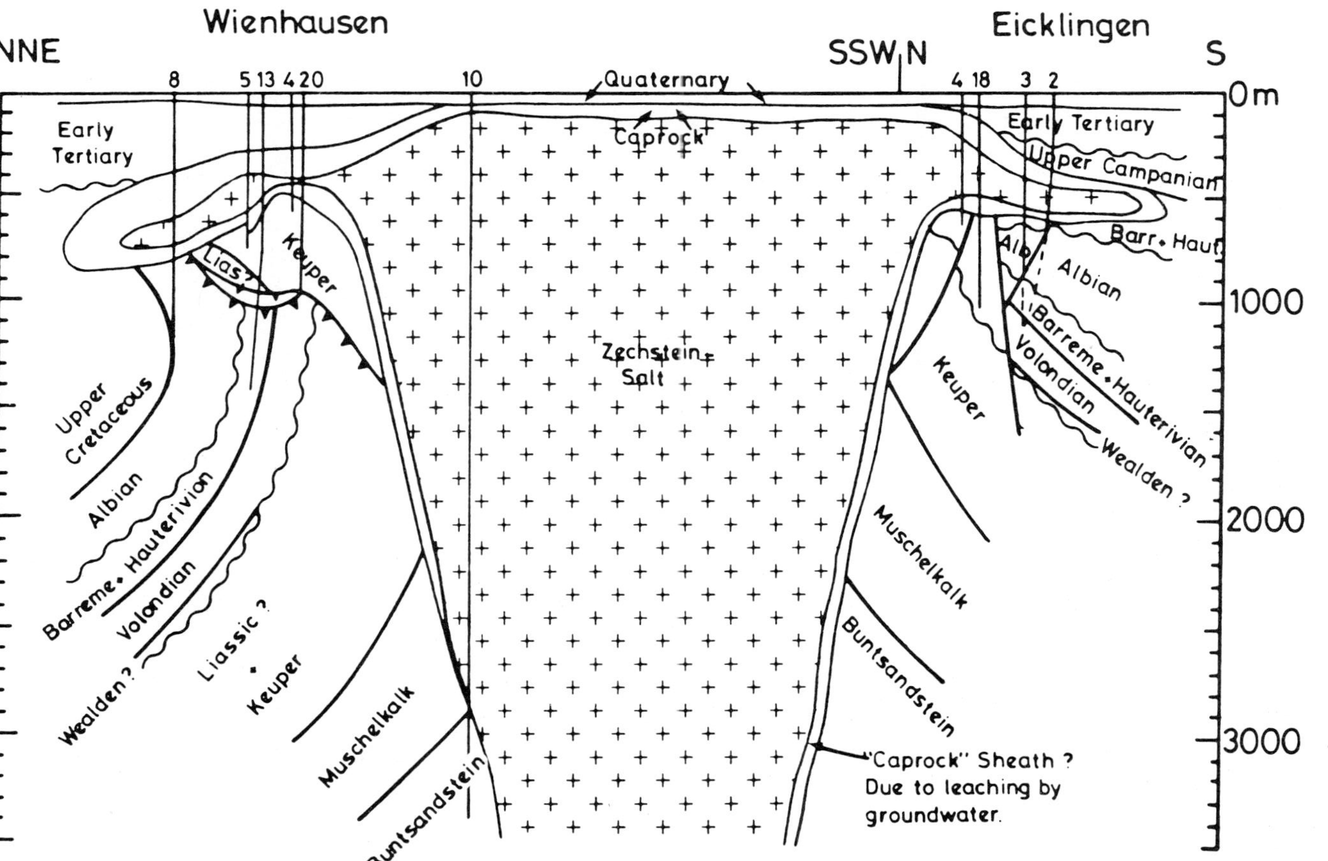

Figure 2. A superficial spreading overhang feature at the crest of a salt diapir. Vertical section through the Wienhausen salt structure, Germany. After Gussow (1968), courtesy of the American Association of Petroleum Geologists.

present day Iran and Oman. Aridity is not, of course, confined to low latitudes, and in this connection it is of some interest to note that Tolmachoff (1926) has described an emergent diapir of (probably) Silurian salt at Solenaya Sopca, Northern Siberia, which was probably extruded in Tertiary times, but has survived in the intensely cold and arid climate of the region.

If salt were to be extruded from a "feeder" channel of restricted, subcircular plan, it could be expected to form a dome-shaped feature, with the central part immediately above the feeder having the thickest salt where the latter is first extruded above the surface. It is presumed that, as in the Iranian examples, the salt would then spread laterally under gravity. The dome shape would be maintained only as long as the salt supply was in excess of the rate of lateral spreading; if the supply ceased, then over a period of time the extruded salt would presumably flow to form, eventually, a flat sheet.

If, on the other hand, the salt were to be extruded from an elongate "feeder", such as a fault zone, then the situation would be similar, except that a linear surface anticline of salt rather than a dome would be produced.

The question of whether surface salt bodies of this type can survive for a sufficiently long period of time to be preserved by later deposition must remain open. On the whole, it seems to be unlikely. It would depend on the continuing supply of salt from below, the absence of any major dissolution effects, the supply of aeolian sediments, etc. Halite is not a particularly hard mineral (2.5), and it is not easy to imagine it being preserved for geologically long periods in an environment of normal terrigenous deflationary processes, even were dissolution effects to be absent. It is even more difficult to imagine such a salt body surviving an inundation of marine or other waters for any length of time, unless the salt supply was continuous and sufficient to counterbalance the rapid removal of salt which would occur at the depositional surface. This applies also to the possibility of submarine extrusion at a surface already covered by water. The results of such a subaqueous extrusion may be imagined by reference to present day examples of collapse and subsidence at the seabed observed in various parts of the world related to the emergence at, or close approach to, the seabed of evaporites (see, e.g. Belderson et al. (1978) for features at the seabed in the Eastern Mediterranean related to Messinian evaporites, and Trabant and Presley (1977) for the Orca Basin subsidence feature in the Gulf of Mexico). Again, it is possible that sustained outflow of salt from a very thick and extensive source layer could maintain a subaqeous salt structure long enough for an impermeable "barrier" layer of sediment to be deposited over the salt, but the writer is not aware of any unequivocal examples of this.

In some areas, there is considerable evidence, of an inevitably circumstantial nature, that large volumes of salt have been extruded through a depositional surface at some geological periods. The Southern Zechstein Basin of the North Sea is one such area, where large volumes of Zechstein salt are believed to have been extruded at the surface during the Late Cimmerian (Upper Jurassic/Lower Cretaceous) erosional phase. Relatively thin Triassic salt intervals - particularly the lowest of these, the Rot Salt - in the section overlying the Permian are believed by many, including the writer, to owe their provenance at least in part to re-solution and re-precipitation of Zechstein salt which had been extruded in this area. In other areas, evidence for surface extrusion of salt is less apparent. Of 15 salt diapirs described by Seni and Jackson (1984) in the East Texas salt basin, only one - the Hainesville Dome - shows any evidence of extrusion and erosion of the salt.

B. Lateral Spread of Salt Beneath a Competent Overburden Unit

The suggestion that a rising salt mass may, on encountering a particularly competent unit in the overburden, be prevented from further upward movement and, as a result, spread laterally by injection of bedding planes, arose primarily out of some physical model studies.

Parker and MacDowell (1955) noted the occurrence of this phenomenon in their series of model diapiric experiments when a rising mass of the asphalt which formed the mobile material encountered a more than usually competent layer in the model overburden. The asphalt tended to spread laterally beneath the competent layer, forming an overhang. Ramberg (1981) in his centrifuged model studies of diapir formation made much the same observations. He also noted that the model diapiric material, on approaching the surface of the overburden, tended to spread laterally _within_ any low density layer in the overburden material, and also that any spreading before extrusion at the model surface was dependent upon the overburden material being relatively ductile. In real earth terms, this would seem to suggest either the injection of salt laterally into bedding planes, or to the lateral displacement ahead of the spreading overhang of lower density, relatively ductile or incompetent sediments by deformation - crumpling and shearing - or perhaps both these processes occurring at the same time. There is a fairly general belief amongst salt geologists that lateral injection of salt into adjacent sedimentary bedding planes can take place only under a relatively light overburden load, and this is probably true. However, from what evidence exists, it would seem that superficial spreading overhangs are formed at very shallow depths, when the salt has almost reached the surface, if not having been actually extruded, and this will be discussed further in the next section.

There is probably a considerable amount of overlap between the process considered here, and that to be dealt with next. The difference lies in the suggestion that where salt rising through the overburden meets a very competent unit, it will spread laterally not just very close to the surface, but at greater depths. The next process to be considered implies a similar mechanism, but operating only at very shallow depth in the overburden.

C. Lateral Spread of Salt Beneath a Superficial Sediment Layer

The concept of salt at the crest of a rising diapir spreading laterally as a "salt glacier", but beneath a veneer of consolidated sediment was advanced by Gripp (1958), and quoted by Trusheim (1960), who also mentioned the observation of outflow breccias of broken roof and caprock fragments associated with these features; it is assumed that these have been entrained by the salt paste being extruded, remaining as "saline morraines" after dissolution and removal of the salt. This is of interest in the light of later discussions in this paper.

Leaving aside the frictional drag factor mentioned earlier, it can be said that a rising mass of salt which is approaching the depositional surface has to overcome three sources of resistance to further upward movement. These are the overburden loading, gravity, and the intrinsic competency of the overburden. As noted earlier, in the later stages of piercement, the energy of emplacement of the salt is much reduced. With the salt approaching the surface, the bulk loading of the overburden may be diminished to the extent that the competency of the overlying beds determines the course of events. Lateral injection, with lifting of the overburden, may represent the line of least resistance to the salt rather than trans-formational piercement of the bedding.

The resistance to trans-formational piercement can be very high - as demonstrated by the frequency of occurrence of large salt pillows, sometimes rising 1000 meters or more above the adjacent salt source surface, which have not developed into diapirs. This resistance may exceed the effects of both overburden loading and gravity, whereupon lateral injection of salt at the diapir crest may occur. Above the level of zero density differential with the surrounding sediments, the salt in the upper part of the diapir has become a dead weight to be lifted by any further salt flow from beneath. Having lost the effect of buoyancy, this salt is, as it were, caught in a vice between any continuing upward salt flow, and the resistance of the overlying sediments to piercement. It will thus tend to spread laterally.

V. A SEISMIC EXAMPLE OF LATERAL SPREADING OF SALT

Fig. 3 shows an example of what is interpreted as the lateral injection of salt beneath a superficial layer of sediments. The illustration is of a portion of a (migrated) seismic section across a zone of basin-edge diapirism in the Southern Zechstein Sub-basin of the British North Sea. The processes involved in this zone have been discussed in some detail elsewhere (Jenyon, 1985a), and so only a brief general description will be given here.

The base and top of the Zechstein interval containing the salt are marked BZ, TZ in the figure, with the basin center off section to the south-west. There is a competent band - the Plattendolomit - at 'd' within the total salt interval, and the major "Late Cimmerian" Unconformity of Late Jurassic/Early Cretaceous age is at CU, whilst T indicates the Base Tertiary. A prism of subsided Mesozoic overburden is seen at P. To the south-west of P, the Zechstein is in basinal facies, whilst to the north-east of P, there is a change to marginal facies with

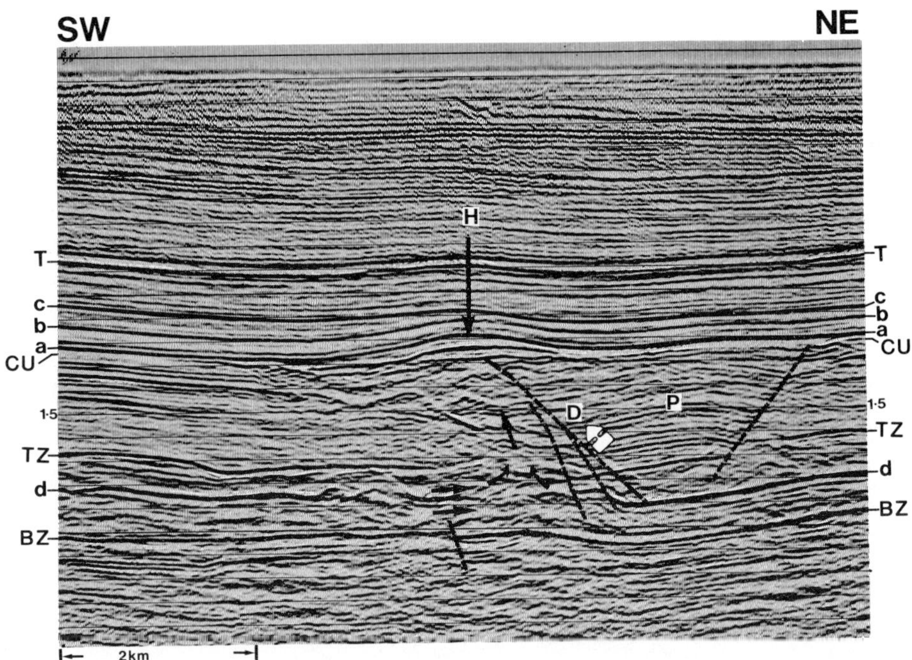

Figure 3. A salt mound, H, resulting from lateral injection of salt at a surface of unconformity CU, beneath a thin overburden layer. See text for key and discussion. Courtesy of Seismograph Service (England) Limited.

some interbedded halite which has remained effectively non-mobile. Lateral (updip) movement of mobile basinal salt towards the north-east in pre-CU times was dammed by the facies change, and a pillow formed which later became a diapiric piercement, the salt having been injected upwards through the overburden approximately in the zone marked D. Withdrawal of salt from beneath the overburden P caused the latter to <u>tend</u> to subside rotationally (counter-clockwise when looked at in this sense). The stresses set up in the overburden by this tendency caused the normal fault to the north-east of P to develop, and since the salt had pierced the overburden up the zone D and probably extruded at the depositional surface, the prism P was eventually unsupported, and slid down the fault plane to come to rest on the lower part of the Zechstein. The major indications of the "primary" diapir which was fed with salt up the feeder zone D were later removed by the CU erosional phase.

Having set the scene with regard to the primary features present, later events can now be considered. During the immediate post-CU period, deposition recommenced undisturbed, as can be seen from the uniform time-thickness of the interval 'a - b' which was deposited on the unconformity surface. Subsequently, or possibly during the deposition of the uppermost part of interval 'a - b', a secondary salt piercement commenced, with the probable route of the salt flow being indicated by the black arrows to the south-west of D. Salt formed a mound, H, doming-up the thin overburden and spreading laterally along the surface of unconformity CU syndepositionally. Evidence for this is seen in the marked drape-thinning over the mound H of the interval 'b - c'. Salt flow continued into the Tertiary, and almost to the period of the shallowest seismic events on the section. It should be borne in mind that the primary features seen - P, and D - and the mound H, are elongate parallel to the marginal shelf-edge, and perpendicular to the section.

The depositional features above the mound H are somewhat asymmetric, since most of the post-CU salt flow - and hence, more subsidence - occurred to the south-west of D, as there was little mobile salt remaining to be withdrawn to the north-east of D where the Zechstein passes into marginal facies.

The feature illustrated in Fig. 3 is of considerable interest in this context, since it shows an early stage of a process to be discussed, and the drape-thinning above the uniform depositional interval a - b is of particular significance in the light of what follows. It is suggested that the evolution of the mound of salt, H, may be the process by which a diapir develops a spreading overhang at its crest.

VI. THE EFFECTS OF DISSOLUTION

A simplified conceptual model of the situation described so far related to fig. 3 is embodied in the block diagram of fig. 4(a), with the diapiric piercement at D imagined to have injected laterally along a bedding plane 9 (or unconformity surface) below unit A. The latter is a uniform, pre-movement depositional unit analogous to the interval 'a - b' in fig. 3. A caprock development is shown over the crest of the salt mound, and units B, C, D, are indicated as showing drape-thinning caused by the syndepositional rise of the salt in post-A time.

If it is imagined that, at a later stage, dissolution removes the salt of the spreading mound and part of the salt in the diapiric "feeder" D, the resulting subsidence effects would most likely result in the (simplified) situation of the block diagram in Fig. 4(b). The maximum amount of subsidence will take place at the location of the crest of the mound, and the effects of this will be exaggerated by the drape-thinning of units B, C, and D, even if unit A undergoes very little subsidence into the "feeder" channel. More likely, however, but not illustrated here, will be that the zone along the central axis of the subsidence will disintegrate. This seems likely to occur because this axial zone in the overburden will undergo the maximum of flexing and is thus likely to develop fractures along and parallel to the axis of subsidence (the central dashed line in fig. 4(b)). In this case, debris from the disintegrated overburden sediments and any cap rock development will accumulate in the upper part of the feeder, D, as the salt retreats down the latter with continuing dissolution.

It should be stressed that it is the fracturing and disintegration along the axial zone of the thinner, antiform sediments above the salt which is an important factor in this process. This fracturing allows penetration of what was previously a barrier of impermeable sediments by undersaturated

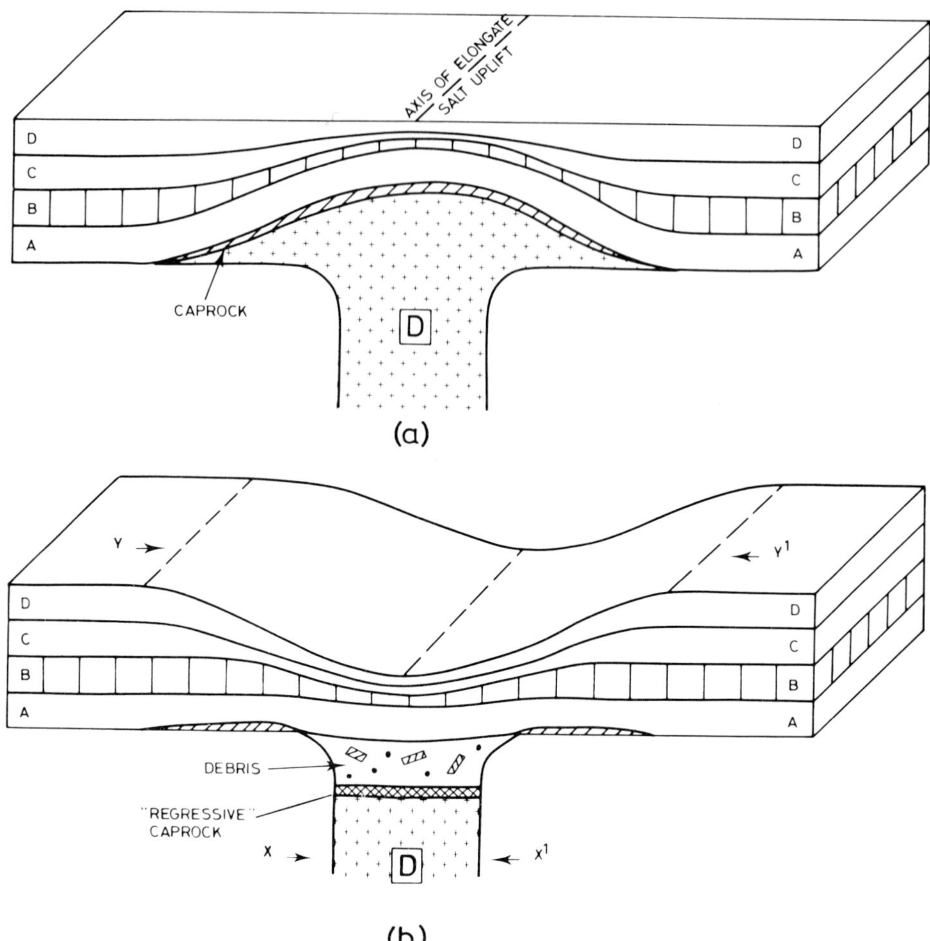

Figure 4. Simplified block-diagram models of (a) the process of formation of a spreading overhang at the crest of a salt piercement by lateral injection of a bedding plane or unconformity surface, and (b) the results of dissolution removal of the salt from the upper part of the piercement, and (b) the results of dissolution removal of the salt from the upper part of the piercement, and collapse of the overburden units.

water (brine) from the depositional surface above, leading to dissolution and removal of the salt when a circulatory system has been established.

The results of this process will be that two distinct collapse zones will develop: a narrower zone delimited by X-X′

in the upper part of the diapiric feeder, and much a wider zone Y-Y′ in the overburden. The shape of the latter will approximate in cross-section to an inversion, or mirror-image, of the shape of the salt mound before it was removed by dissolution.

Can the end-product of such a complex, two-stage collapse process be seen in nature? The writer believes that it can. The migrated seismic section in fig. 5 is thought to provide an example of the process suggested, and comes from another location in the Southern Zechstein sub-basin over 100 km from the example in fig. 3. The seismic line here runs across the axis of an elongate salt-wall type diapir, the crest of which has locally undergone dissolution collapse. The base and top of the salt

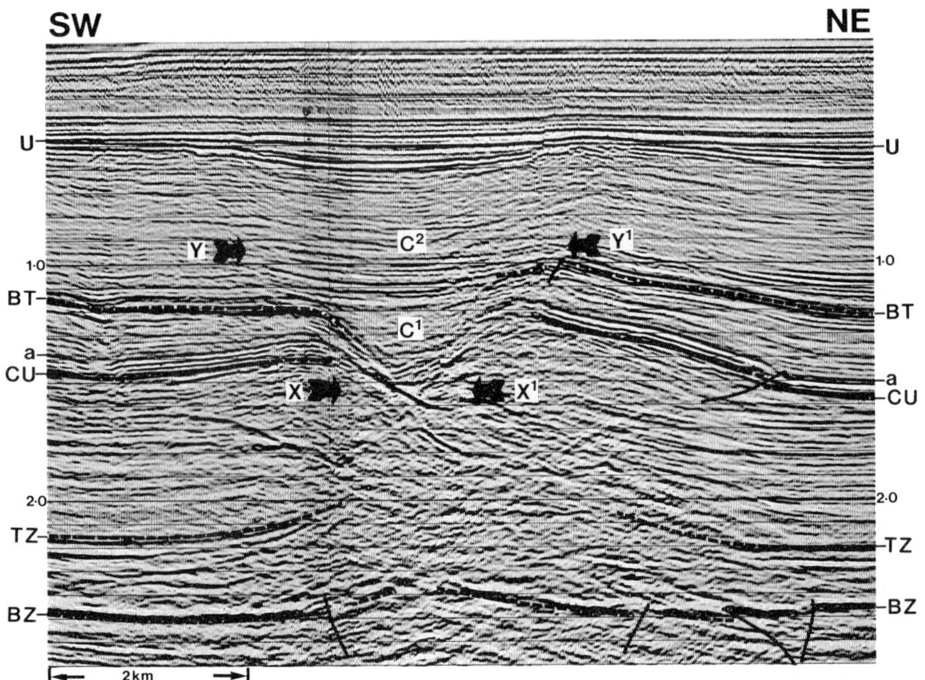

Figure 5. Migrated seismic section showing the results of the process modeled in fig. 4. Courtesy of Seismograph Service (England) Limited.

interval are marked BZ, TZ, the Late Cimmerian Unconformity is CU, the Base Tertiary BT, and there is a shallow mid-Tertiary unconformity at U.

It is clear that there are two distinct zones of collapse, C^1 and C^2, the latter being very much wider than the former and extending considerably beyond the confines of the C^1 collapse on both sides. The maximum extent of the C^1 collapse is delimited approximately by X-X', and that of the C^2 collapse by Y-Y'. Above the Late Cimmerian Unconformity CU, there is an interval of uniform time-thickness CU-a, but the following interval a-BT exhibits considerable thinning in towards the collapse feature from both directions.

It is submitted that the observed facts in this example fit well with the simplified model suggested in fig. 4(b). This would imply that a spreading overhang of salt, injected laterally at the CU unconformity surface below a superficial depositional interval CU-a, existed at one period, the injection taking place syndepositionally with the drape-thinned units above 'a'. After the mounding-up of the overlying sediments, dissolution and collapse occurred, removing the salt mound and also some salt in the upper part of the feeder zone of the diapir which became filled with debris at C^1, above which the wider, concave-upwards collapse of the overburden, C^2, exists as evidence of the removal of the spreading overhang mound. Some subsidence was still taking place as late as the post-shallow unconformity U period.

This situation can be compared with a more usual type of collapse graben as seen in Fig. 6, where a salt pillow has developed from the Zechstein interval, the top of which is at T. The pillow has never developed into a piercement, but extensional stress zone fracturing of the overburden due to the salt uplift has provided migratory paths for undersaturated formation water, and dissolution collapse has occurred, with the development of a normal collapse graben in the overburden and the crest of the

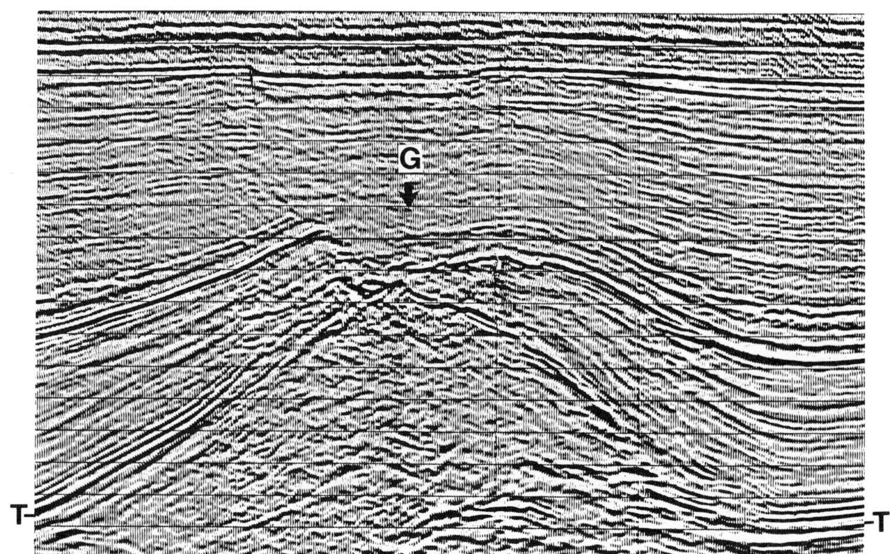

Figure 6. Seisic section showing a normal type of collapse feature resulting from fracturing of the overburden above a salt pillow, followed by dissolution removal of injected salt from the fracture zone, and from the upper part of the pillow, followed by collapse graben formation. Courtesy of Seismograph Service (England) Limited.

pillow, at G. The bounding graben faults in the overburden are effectively within the limits of development of the crestal collapse of the pillow, and there is no sign of the evolution of any two-stage process as previously discussed.

In this general area, several examples of features similar, if not identical, to the example in fig. 5 have been identified, and are now taken to be locations at which crestal spreading overhangs once existed. The exact morphology varies, but certain basic characteristics are always present - and in particular the relatively restricted deep collapse, the much wider shallow collapse features of the two-stage process, and the uniform depositional interval which is succeeded by an interval thinning

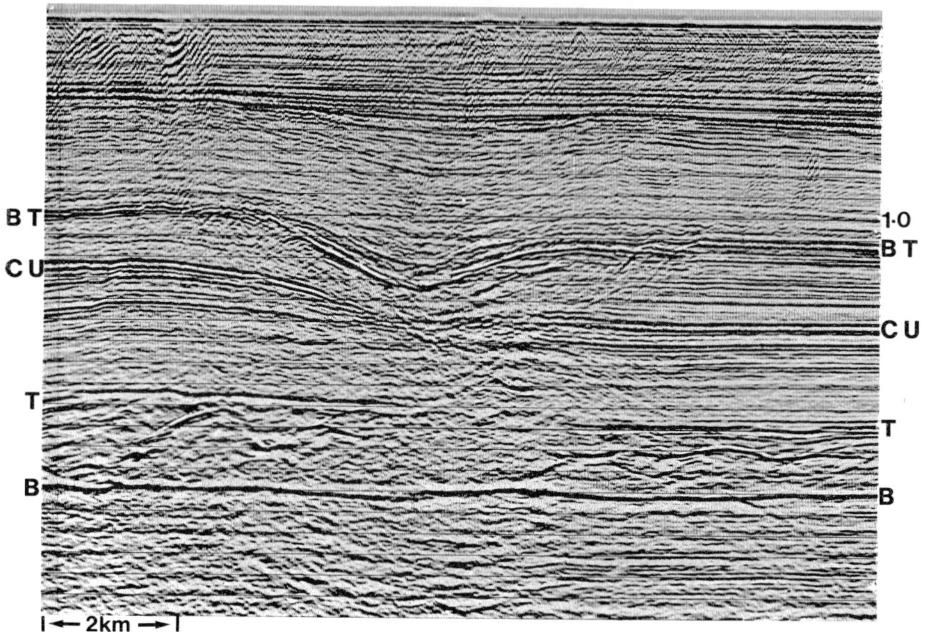

Figure 7. Another example of the more complex two-stage collapse feature resulting from the dissolution removal of a lateral salt injection. Courtesy of Seismograph Service (England) Limited.

markedly inwards towards the collapse location. One such similar type of feature is illustrated in Fig. 7, with the base and top of the salt interval marked B and T, the Late Cimmerian Unconformity CU, and the Base Tertiary, BT. In this example, the configuration of the unconformity CU related to the strong event about 200 ms below it suggests that this collapse mechanism may have occurred twice, with the earlier episode being very asymmetric, the spreading overhang injection having taken place mainly to the left of the collapse feature.

In an area of hydrocarbon exploration, if the suggested interpretation of these features is correct, some interest must be attached to possible linear zones of fracture porosity existing, where substantial flexing of elongate zones in the overburden, and collapse brecciation and megabrecciation may have occurred. Structures associated with dissolution collapse are seldom regarded as prospective; rather, they are normally considered more in the way of breaches through which underlying hydrocarbons may migrate into the upper section. However, depending on the geometry of the particular feature, the age of development in relation to the maturation and migration of hydrocarbons in the area, the presence of any intact seal formations in the immediate overburden sequence, and the effects of any later cementation and diagenetic effects either adverse or favorable, it is suggested that these situations could be of interest in some areas.

VII. OVERBURDEN THICKNESS DURING LATERAL INJECTION

It is of interest to obtain some idea, if possible, of the order of thickness of overburden present where lateral injection of salt into adjacent bedding planes can take place, since there is a general belief that the process can happen under only very limited loading. For this purpose, fig. 8 shows a migrated seismic section across a small salt diapir of elongate plan, the long axis running perpendicular to the section. The diapir, D, which is located a few tens of kilometers from the example of fig. 3, illustrates a somewhat similar situation to that in the latter figure, except that the salt "mound" of fig. 3 has developed here into a larger diapiric feature. The stratigraphic situation is almost identical, with the salt source layer between BZ and TZ; a competent band, the Plattendolomit, P, is present within the total salt interval. Above the BZ-TZ interval, there are three major unconformities: the Late Cimmerian, CU; the Base Tertiary, BT, and a shallow mid-Tertiary unconformity, U.

The diapir, which is of narrow-waisted cross-section similar to the schematic in fig. 1(c), has developed a spreading overhang as seen by the evidence of "splay" of the stratal reflections at

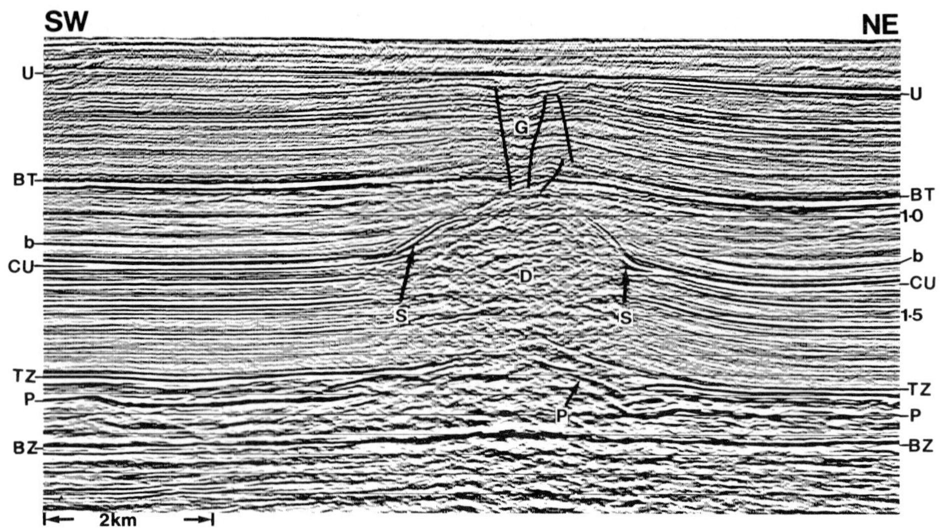

Figure 8. Seismic section across a small diapir, D, which is showing evidence of lateral salt injection at the unconformity CU. Compare with fig. 3. Courtesy of Seismograph Service (England) Limited.

S - S. These lateral injection effects have occurred at the Late Cimmerian Unconformity surface, as was the case in fig. 3.

Above CU, there is a uniform interval with a two-way time of about 100 ms, the top of which is at 'b', indicating that upward salt movement into the diapir did not commence until time level 'b' or slightly later. Above 'b', thinning of the stratal reflections from both sides in towards the diapir is evident, marking the period of salt uplift. From seismic interval velocities, the overburden thickness CU-b is estimated to be about 500-600 feet. Since the salt rise did not begin until level 'b' time or later, it is estimated that at the time of occurrence of the lateral salt injection at the CU level, overburden thickness was probably within the range of 1,000 to 1,500 feet.

It is reasonable to assume from the evidence of figs. 3 and 8 that the CU unconformity surface is a preferred location for

lateral salt injection in this a
low cohesion between the surface
immediately overlying sedimentary u

It should be borne in mind that
of overburden loading is the most impo
the presence or absence of lateral salt
thickness of overburden will vary from on
depending on the bulk density of the sedime

A point of incidental interest regardin
8 is that it has developed as a result of very
flow within the salt interval at the margin of
withdrawal basin seen to the NE of the diapir. T
effects in this withdrawal basin are believed to ha
fault affecting only the top of the salt interval, l
stress-relief injection of the overburden and formatio
diapir. Evidence that salt flowed only from the NE int
piercement is presented by (i) the much thinner BZ-TZ in
on the NE side, (ii) the overburden subsidence effects on
side; similar effects are absent from the SW side, and (iii)
deformation and disintegration of the competent band P on the
side, with no similar effects on the SW side. Further, consideration of the BZ-P and P-TZ intervals on both sides of the diapir indicate that most or all of the salt which moved into the diapir came from the BZ-P interval (the Zechstein Z2 - Stassfurt - salt) on the NE side. Splay effects in the BZ-TZ interval on the SW side, together with lack of overburden subsidence above, show that the Z2 salt from the NE side, apart from moving up into the diapir, also exerted lateral pressure on the salt in the BZ-TZ interval to the SW of the diapir. This suggests that the salt on the SW side of the structure was much more viscous (and hence less mobile) than the salt on the NE side, although at roughly the same burial depth. A more detailed discussion of this interesting structure will be found in Jenyon (1985b).

...jor dissolution
... has occurred
... events in

...ty U.
...ed above
...n be
...e
...ould
...e of
...uences of
...e

...elopments associated with
... this volume, and it is not
...ode of origin of this type of
...ply, however, the general consensus
... caprock results from the accumulation at
...d in some cases down the flanks of - a rising
... of (relatively) insoluble components of salt rock
...issolution of halite by undersaturated water in the
...ations through which the salt passes in its uprise, and (ii) the main primary constituent of caprock is the mineral anhydrite, and that an apparent layered structure may evolve in the caprock by various diagenetic processes, including bacterial action, resulting typically in upward passage from unaltered anhydrite through a transition zone which is a mixture of anhydrite and other minerals, to an upper zone of calcite.

Thus there is the circumstance of the formation of an increasingly thick, dense and indurated "crust" over the crest of the salt piercement as it rises towards the surface. Halbouty

(1967) noted that caprocks are poorly-developed or absent in diapirs at present depths greater than 10,000 feet in the Gulf Coast area, but are increasingly common, and large, in diapirs at progressively shallower depths; thicknesses vary in general from a few hundred feet to over 1,000 feet. These facts fit well the model of the progressive dissolution of halite, and accumulation of insolubles, with rise through the overburden.

Such massive, capping crust of dense, compact material is in sharp contrast to the less dense mobile salt beneath, and the clastic sediments above (anhydrite density is about 2.98 gm/cm^3), and could be expected to produce a strong acoustic impedance contrast. If the caprock is sufficiently thick, having regard to the dominant seismic wavelength at any specific level in the section, this will result in a strong reflection event. Where a spreading overhang develops which includes caprock (as in fig. 2, which was constructed from drilling results rather than seismic data), the latter should assist materially in determining the form of the overhang in the seismic section.

An example of a small salt diapir, D, in the Northern North Sea Zechstein Basin is shown in fig. 9, in a migrated seismic section, with the base and top of the source salt interval marked at B and T. In spite of severe obscuration of the diapir by near-horizontal multiples, a very strong antiform event, C, is clearly evident, which is interpreted as the response to a caprock development over the dome-shaped crest of the diapir. The lateral extent of this event suggests the presence of a spreading overhang at the crest.

Another example in this general area is seen in the diapir D in fig. 10. Here again, the strength of the event marked "overhang" leads to marked stand-out on a very noisy section, and its lateral extent is strong evidence for the presence of a major spreading overhang. Salt interval base and top are again marked B and T.

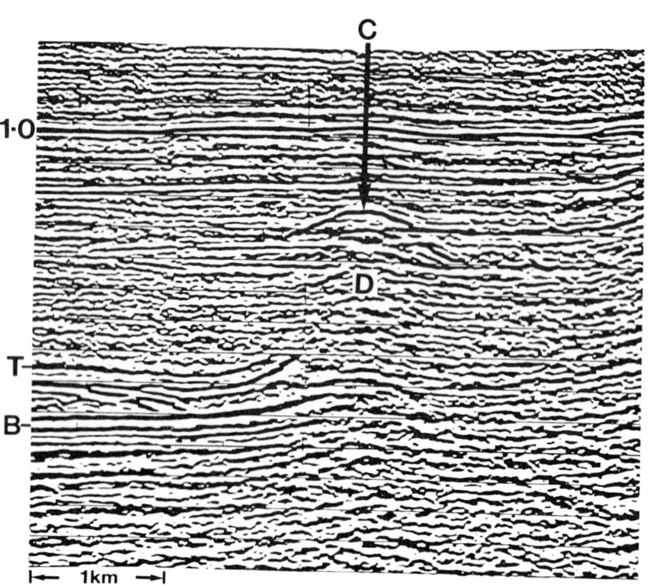

Figure 9. Probable caprock event, C, related to a spreading overhang at the crest of a small salt diapir in the Northern North Sea. Courtesy of Seismograph Service (England) Limited.

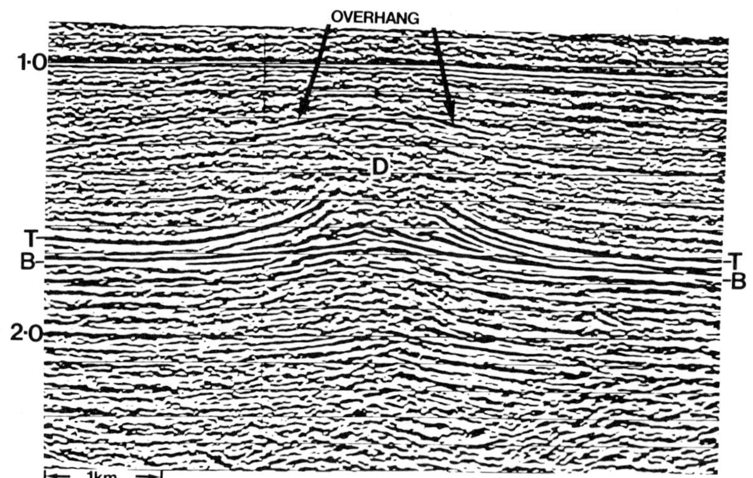

Figure 10. Spreading overhang at crest of a salt diapir, D, in the Northern North Sea. Courtesy of Seismograph Service (England) Limited.

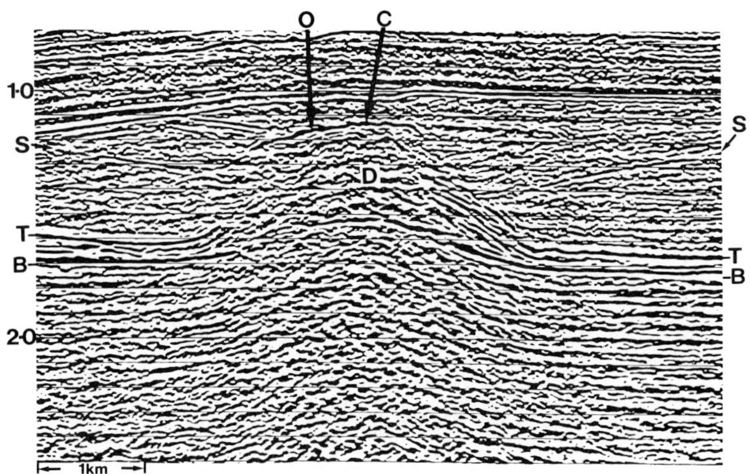

Figure 11. Spreading overhang, O, intact on left side of diapir D, but disintegrated on right side due to greater subsidence into the secondary rim syncline whose basal surface is marked at S. Courtesy of Seismograph Service (England) Limited.

In many areas, caprock developments are not seen in association with diapirs. This does not mean that they were never present, however. They are probably the norm rather than the exception, but - particularly if they are associated with a superficial spreading overhang - they are very susceptible to late-stage disintegration at the diapir crest due to faulting, graben formation, dissolution collapse, and subsidence effects. This is well illustrated in fig. 11, where the caprock event C over the crest of diapir D continues out as a spreading overhang feature O to the left of the diapir, but not to the right. The reason for this may be deduced from the events S on each side of the structure, which mark the basal surface of a secondary rim syncline associated with salt flow into the diapir. The subsidence into the rim syncline is clearly very asymmetric, being much greater to the right of the diapir. This greater subsidence is likely to have led to disintegration of the overhang and caprock on the right, with the caprock fragments probably being incorporated as a melange of anhydrite, calcite, and other mineral fragments within the clastics of the rim syncline.

IX. "REGRESSIVE" CAPROCK

In the diagram of fig. 4(b), a layer at the top of the retreating salt D in the upper part of the diapiric "feeder" is marked as "regressive" caprock. This term has been coined to suggest that a caprock development may form not only during the <u>rise</u> of a salt diapir through the overburden, but also during the <u>retreat</u> downwards of the salt by dissolution at a later stage. If the basic suggested mode of formation of caprock related to a rising diapir is accepted, it is not necessary to suggest reasons why a "regressive" caprock should form - in fact it is more apposite to demand a reason why it should <u>not</u> form.

In numerous examples of dissolution collapse related to major salt bodies in the North Sea, the frequent occurrence has been noted of a strong seismic event near the current upper part of the salt body. This event is often of somewhat irregular form, showing strong dip and sometimes an undulatory shape. A good example of this is seen in fig. 12, where a migrated seismic section crosses a major salt-wall diapir, D, which has undergone, over part of its length, considerable dissolution. In this example, a large collapse graben has formed above the diapir, with a series of step-faults of progressively later age from E to L having occurred. The top of the salt interval is marked, as is the Late Cimmerian Unconformity CU, and the Base Tertiary, A.

A strong, somewhat irregular event is seen dipping to the right at T. Although there is no supporting evidence from drilling, it is believed that this event could mark the top of a "regressive" caprock sequence, and that the shorter strong event some 50 ms beneath it could mark the base of the sequence, and the top of the salt rock proper. If this interpretation is

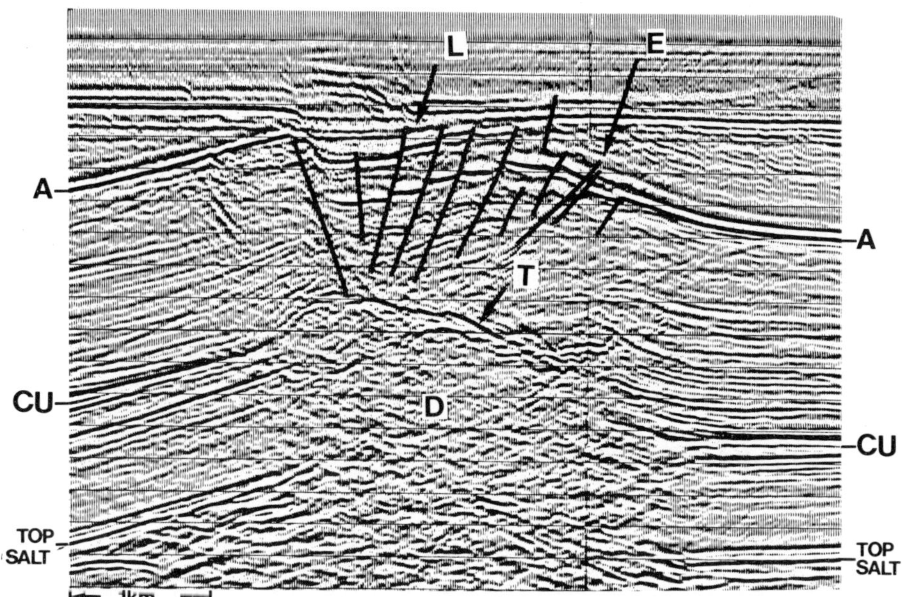

Figure 12. Event T is interpreted as a reflection from the top of a "regressive" caprock sequence associated with the salt-wall diapir, D, seen in cross-section. The salt has retreated downwards as a result of dissolution removal, with a major overburden collapse graben above the crest including progressive step faulting of early (E) to late (L) age. Courtesy of Seismograph Service (England) Limited.

correct, it indicates the presence of a substantial thickness - perhaps 500 feet or more - of caprock here. Such events are not uncommon in seismic data in this area. They could be of interest in hydrocarbon exploration in that, in rare instances, selective leaching of minerals in caprock sequences has resulted in high porosities and good permeabilities.

X. CONCLUSIONS

Evidence in seismic data of the North Sea Zechstein Basin suggests that superficial spreading overhangs at the crests of salt diapirs have been formed rather frequently in this area.

They probably owe their origin to lateral injection of salt beneath relatively thin near-surface sediments in the later stages of diapiric emplacement. Even where the salt has later been removed by dissolution, the effects of the overhang injection are preserved in the form of depositional patterns in the overburden sediments, and specifically in distinctive two-stage collapse graben.

Caprock developments can be of interpretative assistance in providing strong reflection events which delineate well the form of existing spreading overhangs. The occurrence of "regressive" caprock sequences related to dissolution retreat of salt rock is postulated.

Certain aspects of the dissolution collapse processes which are associated with some spreading overhangs, and the leaching of caprock sequences, may be of significance in hydrocarbon exploration.

XI. ACKNOWLEDGEMENTS

Thanks are due to the Directors of Seismograph Service (England) Limited for permission to use the excellent seismic illustrations, and to submit this paper for publication.

REFERENCES

Barton, D.C. (1933). Mechanics of formation of salt domes with special reference to Gulf Coast salt domes of Texas and Louisiana, AAPG Bulletin, 17, 1025-1083.

Belderson, R.H., Kenyon, N.H. and Stride, A.H. (1978). Local submarine salt-karst formation on the Hellenic Outer Ridge, eastern Mediterranean, Geology, 6, 716-720.

Gripp, K. (1958). Salzspiegel and Salzhut auf dem Lande und unter dem Meere, Abh. Naturw. Ver. Bremen, 35, 249-258.

Gussow, W.C. (1968). Salt diapirism: importance of temperature, and energy source of emplacement, AAPG Memoir 8, "Diapirism and Diapirs", 16-52.

Humphris, C.C. (1979). Salt movement on continental slope, Northern Gulf of Mexico, AAPG Bulletin, 63, 782-798

Jenyon, M.K. (1985a). Basin-edge diapirism and updip salt flow in Zechstein of Southern North Sea, AAPG Bulletin, 69, 53-64.

Jenyon, M.K. (1985b). Differential movement in salt rock. Oil and Gas Journal, 30th Spet. 1985, 73-75.

Jenyon, M.K. (1986). Salt Tectonics (in press), Elsevier Applied Science Publishers, Barking, Essex, U.K.
Kent, P.E. (1979). The emergent Hormuz salt plugs of Southern Iran, Jour. Petr. Geol., 2, 117-144.
Parker, T.J., and McDowell, A.N. (1955). Model studies of salt-dome tectonics, AAPG Bulletin, 39, 2384-2470.
Ramberg, H. (1981). Gravity, Deformation, and the Earth's Crust (2nd ed.), Academic Press, London, 452 pp.
Schott, W. (1956). Das niedersachsische Becken ostlich der Weser, In: 20th Int. Geol. Cong., Mexico, vol. 5, 59-64.
Sellig, R. and Wallick, G.C. (1966). Temperature distribution in salt domes and surrounding sediments, Geophysics, 31, 346-361.
Seni, S.J. and Jackson, M.P.A. (1984). Sedimentary record of Cretaceous and Tertiary salt movement, East Texas Basin, Bureau of Economic Geology, Univ. of Texas at Austin.
Tolmachoff, I.P. (1926). A salt dome, Solenaya Sopca, in northern Siberia, Econ. Geol, 31, 774-781.
Trabant, P.K. and Presley, B.J. (1977). Orca Basin, an anoxic depression on the continental slope, Northwest Gulf of Mexico, AAPG Studies in Geology, 7, 303-311.
Trusheim, F. (1960). Mechanism of salt migration in northern Germany, AAPG Bulletin, 44, 1519-1540.
Woodbury, H.O., Murray, I.B. and Osborne, R.E. (1980). Diapirs and their relation to hydrocarbon accumulations. In: Facts and Principles of World Petroleum Occurrence, Can. Soc. Petr. Geol. Memoir 6, (Ed. Miall), 119-142.

HEAT FLOW AND THERMAL MATURATION NEAR SALT DIAPIRS

J. J. O'Brien[1]
I. Lerche[2]

[1]Standard Petroleum of Alaska
900 East Benson Blvd.
Anchorage, Alaska 99502

[2]Department of Geology
University of South Carolina
Columbia, South Carolina 29208

I. INTRODUCTION

The temperature distribution in the subsurface has a marked impact on the occurrence of oil and gas, especially through the influence of temperature on hydrocarbon maturation. Thermal maturation can be modeled as a first-order chemical reaction in which the reaction rate doubles for every 10°C rise in formation temperature (Lopatin, 1971). Thus any effect which causes a significant variation in temperature distribution from the regional trend may have a substantial influence on hydrocarbon accumulations in the subsurface.

A common cause of such localized temperature variations is the presence of salt domes in the subsurface. Rock salt has a much higher thermal conductivity than typical sedimentary rocks. The salt bodies occur in massive diapirs having large vertical relief, and hence provide a path of low thermal resistance for the conduction of heat to the surface from depth. In this paper, we present a simple analytical model of this heat flow problem. Our goal is to emphasize the physics underlying this situation, which can be quite subtle in places. The results of this analytical model then are compared with observations from a salt dome in the Gulf of Mexico.

The problem which we wish to model is that of heat flux in the vicinity of a salt dome embedded in a semi-infinite uniform medium. We assume that a steady-state heat flow has been established and that heat is transported only by conduction. Thus we wish to solve the heat flow equation in an appropriate coordinate frame, i.e.,

$$\nabla \cdot \kappa \nabla T(\mathbf{r}) = -S(\mathbf{r}) \tag{1}$$

where κ denotes thermal conductivity, $T(\mathbf{r})$ denotes temperature, and $S(\mathbf{r})$ describes the spatial distribution of heat sources.

Equation (1) is the general heat flow equation which is valid for an arbitrary thermal conductivity and heat source distribution. For our model studies we introduce some simplifying assumptions. First, we assume constant thermal conductivity both within the salt and in the surrounding sediments. In doing this, we ignore any lithologic variations in the geologic column, any thermal conductivity variations within a given lithology due to compaction, and any temperature dependence of thermal conductivity. This approximation is made so that we can underscore the physics of heat flow in the vicinity of an idealized salt dome. In a practical case where such variations in thermal conductivity must be considered, a numerical approach may be preferred to the present analytical approach. Numerical models were described by Selig and Wallick (1966), Geertsma (1971), and Jensen (1983).

For the second assumption, we refer to the manner in which we model heat sources. The only source we consider is a uniform conductive heat flux through the basement. Thus, in the absence of a salt dome, we would expect a constant vertical heat flux throughout the basin. Our model excludes the effects of hot intrusives and of radioactive decay within the sediments. We assume the simplest form of crustal heating: uniform vertical heat flux across the basement in the absence of any salt accumulations in the overlying basin.

With these assumptions in mind, our model reduces to the problem of heat flow in the vicinity of a cylindrical salt dome embedded in a source-free semi-infinite medium of uniform thermal conductivity under appropriate boundary conditions (Figure 1).

(1) The sediment surface may be considered as a constant temperature surface, especially if it lies under water because the water will act as a constant temperature heat bath.

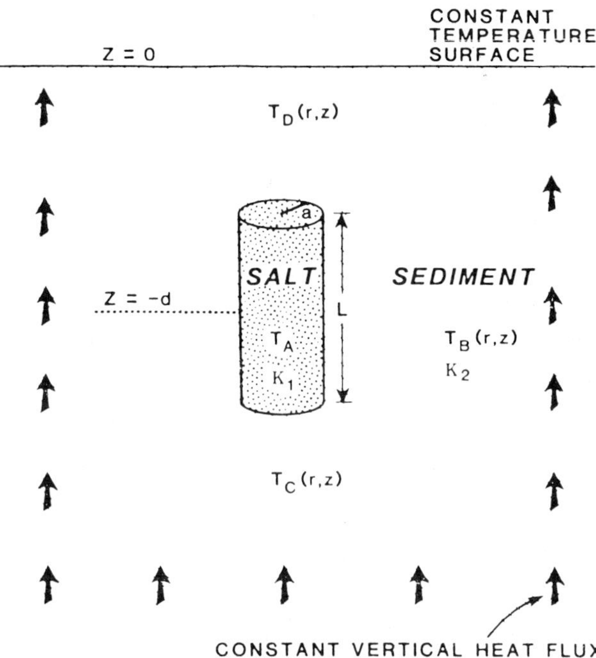

Figure 1. Geometry of the problem of heat flow in the vicinity of a cylindrical salt dome in a semiinfinite medium. T is temperature and κ is thermal conductivity.

(2) Since the influence of the salt dome is localized, we expect a constant vertical heat flux at large lateral distances from the salt dome ($r \to \infty$).

(3) For the same reason, we expect a constant vertical heat flux far beneath the salt dome ($z \to -\infty$).

(4) The temperature must be continuous across the sediment-salt interface since the thermal conductivity is finite.

(5) The component of heat flux normal to the sediment-salt interface must be continuous across this interface since we have no heat generation or loss at this interface.
The solution of the heat flux problem is completely specified by these boundary conditions.

II. SALT DOME EMBEDDED IN AN INFINITE MEDIUM

The problem as formulated in the previous section presents some difficulties if we wish to derive an analytic solution. Boundary conditions have been stated along two unconnected surfaces, the sediment surface and the salt-sediment interface. Satisfying both sets of boundary conditions simultaneously requires that we solve for the solution throughout all regions of space simultaneously. This involves solving a set of simultaneous integral equations. It is more convenient to solve the problem sequentially in different regions of space. To this end, we consider a related problem, i.e., heat flux in the vicinity of a salt dome embedded in an *infinite* uniform medium. Exact analytical solutions can be obtained for this problem. These can then be combined to yield an approximate solution for the case of a semi-infinite medium.

To formulate this problem, we must modify the first boundary condition specified in the previous section. Rather than having an isotherm at the sediment surface, we specify a constant heat flux at large vertical distances above the salt dome ($z \to +$

∞, all r). The horizontal isotherm is now provided by the plane through the middle of the salt dome rather than by the sediment surface, as in the previous section. This change in boundary conditions simplifies the problem in that it permits us first to determine solutions within the salt and within the sediments on the flanks of the dome (i.e., at the same vertical position as the salt). We then extend the solution to sediments lying both shallower and deeper than the salt and so solve the problem for the temperature distribution in a sequential rather than in a simultaneous manner.

Consider a cylindrical salt dome of height L and radius a embedded in an infinite uniform medium such that the plane through the center of the dome is located at z = -d and the z-axis coincides with the axis of symmetry (Figure 2). The thermal conductivity within the dome is denoted by κ_1 while that in the

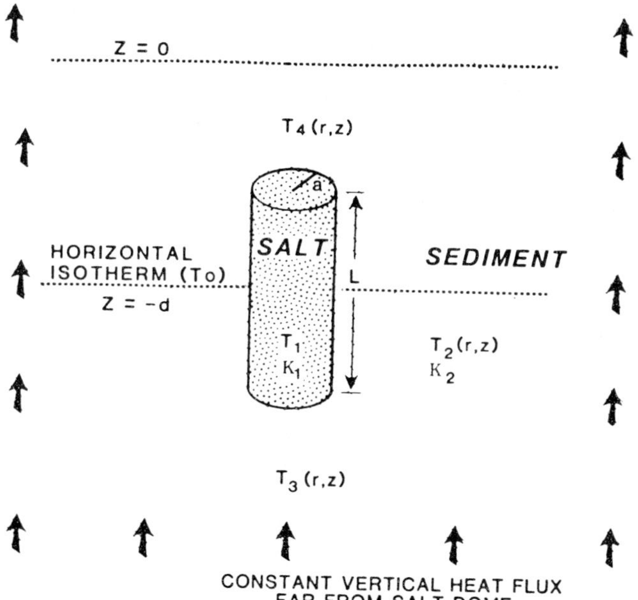

Figure 2. Geometry of the problem of heat flow in the vicinity of a cylindrical salt dome embedded in an infinite medium.

surrounding sediments is κ_2. Four different temperature regions are identified as follows: $T_1(r,z)$ is the temperature within the salt; $T_2(r,z)$ is the temperature in the sediments at the same depth as the salt ($-d - L/2 \leqslant z \leqslant -d + L/2$, $r \geqslant a$); $T_3(r,z)$ is the temperature distribution deeper than the salt ($z \leqslant -d - L/2$, all r); and $T_4(r,z)$ is the temperature distribution shallower than the salt ($z \geqslant -d + L/2$, all r).

In a source-free region of uniform thermal conductivity (for example, within the salt dome or within the surrounding sediment), the general heat flow equation (1) reduces to

$$\nabla^2 T = 0. \tag{2}$$

Assuming azimuthal symmetry, the general solution to equation (2) in cylindrical coordinates can be expressed in a Fourier- Bessel series as

$$T(r,z) = T_o - \frac{Q_o}{\kappa}(z + d) + \sum_{n=0}^{\infty} \left[\alpha_n \sin\left(2n\pi\frac{z + d}{L}\right) + \beta_n \cos\left(2n\pi\frac{z + d}{L}\right)\right]\left[\gamma_n I_o\left(2n\pi\frac{r}{L}\right) + \partial_n K_o\left(2n\pi\frac{r}{L}\right)\right] \tag{3}$$

where I_o and K_o are modified Bessel functions of the first and second kind of order zero, and Q_o is the basement heatflux in the absence of the salt dome. The quantities α_n, β_n, γ_n and ∂_n are arbitrary constants to be determined by the boundary conditions. The term linear in z is included explicitly since, in the absence of any effects introduced by the contrast in conductivities, we expect the solution to have this form. The series expansion in equation (3) reflects the fact that we do not have a uniform medium, but rather that there is a spatial variation in thermal conductivity. Equation (3) is valid both within the salt dome and in the surrounding sediments with the appropriate choice of thermal conductivity κ.

Now consider the temperature distribution inside the salt dome $T_1(r,z)$. Since T_1 is finite at $r = 0$ while $K_o[2n\pi(r/L)]$ diverges for small values of its argument, it follows that ∂_n must be zero in the expansion of $T_1(r,z)$ for all values of the index n. Furthermore, since the plane $z = -d$ is an isothermal surface ($T = T_o$ at all radial distances), then β_n, the coefficient of $\cos\{2n\pi[(z+d)/L]\}$, must be zero for all values of n. Thus $T_1(r,z)$ may be written

$$T_1(r,z)=T_o - \frac{Q_o}{\kappa_1}(z + d)+\sum_{n=1}^{\infty} A_n \sin\left(2n\pi \frac{z+d}{L}\right)I_o\left(2n\pi \frac{r}{L}\right) \tag{4}$$

for $-d - L/2 \leqslant z \leqslant -d + L/2$; $r \leqslant a$, where we have written the product of coefficients $\alpha_n\gamma_n$ as A_n.

Now consider the temperature distribution $T_2(r,z)$ within the sediment at the same depth as the salt dome. Since $T_2(r,z)$ remains finite as $r \rightarrow \infty$ while $I_o[2n\pi(r/L)]$ diverges for large values of its argument, it follows that the coefficient of $I_o[2n\pi(r/L)]$ must be identically zero in the expansion of $T_2(r,z)$ for all values of the index $n > 0$. Moreover, since the temperature $T_2(r,z)$ is a constant, T_o, along the plane $z = -d$, the coefficient of the term $\cos\{2n\pi[(z+d)/L]\}$ must be zero for all values of n. Hence, the expansion of $T_2(r,z)$ reduces to

$$T_2(r,z) = T_o - \frac{Q_o}{\kappa_2}(z + d) + \sum_{n=1}^{\infty} B_n \sin\left(2n\pi\frac{z+d}{L}\right)K_o\left(2n\pi\frac{r}{L}\right) \tag{5}$$

for $-d - L/2 \leqslant z \leqslant -d + L/2$; $r \geqslant a$, where we have written the product of coefficients $\alpha_n\partial_n$ as B_n.

Equations (4) and (5) are expansions for $T_1(r,z)$ and $T_2(r,z)$ which have the proper asymptotic behavior ($r = 0$ and $r \to \infty$). Far from the salt-sediment interface, the temperature profiles reduce to a constant vertical gradient, both within the salt and within the sediment. However, within the salt the temperature gradient is that appropriate to salt, while within the sediment it is the gradient appropriate to the sediment.

Two sets of coefficients remain to be determined, A_n and B_n. Since there are two independent boundary conditions which must be satisfied simultaneously at the interface (i.e., continuity of temperature and continuity of normal heat flux), both sets of coefficients may be determined uniquely.

First consider continuity of the normal component of heat flux across the salt-sediment interface:

$$\kappa_1 \left.\frac{\partial T_1}{\partial r}\right|_{r=a} = \kappa_2 \left.\frac{\partial T_2}{\partial r}\right|_{r=a} \tag{6}$$

for $-d - L/2 \leqslant z \leqslant -d + L/2$. Substituting expressions (4) and (5) for T_1 and T_2 yields the following relationship between A_n and B_n:

$$\kappa_1 \sum_{n=1}^{\infty} nA_n \sin\left(2n\pi \frac{z+d}{L}\right) I_1\left(2n\pi \frac{a}{L}\right)$$

$$= -\kappa_2 \sum_{n=1}^{\infty} nB_n \sin\left(2n\pi \frac{z+d}{L}\right) K_1\left(2n\pi \frac{a}{L}\right) \tag{7}$$

for $-d - L/2 \leqslant z \leqslant -d + L/2$, where we have used the identities

$$\frac{\partial I_o(x)}{\partial x} = I_1(x), \quad \text{and} \quad \frac{\partial K_o}{\partial x} = -K_1(x).$$

Since the functions $\sin\{2n\pi[(z + d)/L]\}$ forms a set of linearly independent functions over the height of the salt dome for nonnegative values of the integer n, we may equate the coefficients of sin $\{2n\pi[(z + d)/L]\}$ on either side of equation (7) as

$$\kappa_1 A_n I_1 \left(2n\pi \frac{a}{L} \right) = -\kappa_2 B_n K_1 \left(2n\pi \frac{a}{L} \right) . \tag{8}$$

Now consider temperature continuity along the salt-sediment interface, i.e.,

$$T_1(r = a, z) = T_2(r = a, z)$$

for $-d - L/2 \leqslant z \leqslant -d + L/2$. This condition implies the following relationship between the expansion coefficients A_n and B_n:

$$\sum_{n=1}^{\infty} \left[A_n I_o \left(2n\pi \frac{a}{L} \right) - B_n K_o \left(2n\pi \frac{a}{L} \right) \right] \sin\left(2n\pi \frac{z+d}{L} \right)$$

$$= \left(\frac{Q_o}{\kappa_1} - \frac{Q_o}{\kappa_2} \right) (z + d) \tag{9}$$

for $-d - L/2 \leqslant z \leqslant -d + L/2$. Let us now expand the term $(z + d)$ on the right-hand side of equation (9) in a Fourier series over the range $-d - L/2 \leqslant z \leqslant -d + L/2$. Because the function $(z + d)$ is odd over this range, the Fourier series reduces to the sine series

$$z + d = \sum_{n=1}^{\infty} C_n \sin \left(2n\pi \frac{z + d}{L} \right) \tag{10a}$$

for $-d - L/2 \leqslant z \leqslant -d + L/2$, where

$$C_n = \frac{2}{L} \int_{-d-L/2}^{-d+L/2} (z + d) \sin \left(2n\pi \frac{z + d}{L} \right) dz$$

$$= \frac{L}{n\pi} (-1)^n \tag{10b}$$

Substituting this expansion for (z + d) into equation (9) and again equating coefficients of sin $\{2n\pi[(z + d)/L]\}$, we find the following relationship between A_n and B_n:

$$A_n I_o\left(2n\pi \frac{a}{L}\right) - B_n K_o\left(2n\pi \frac{a}{L}\right) = C_n\left(\frac{Q_o}{\kappa_1} - \frac{Q_o}{\kappa_2}\right), \tag{11}$$

where C_n is given by equation (10b). Combining equations (8) and (11), we find

$$A_n = \frac{-L}{n\pi}(-1)^n\left(\frac{Q_o}{\kappa_1} - \frac{Q_o}{\kappa_2}\right) K_1\left(2n\pi \frac{a}{L}\right) \times \left[I_o\left(2n\pi\frac{a}{L}\right)K_1\left(2n\pi\frac{a}{L}\right) + \frac{\kappa_1}{\kappa_2} I_1\left(2n\pi\frac{a}{L}\right)K_o\left(2n\pi\frac{a}{L}\right)\right]^{-1} \tag{12a}$$

and

$$B_n = \frac{\kappa_1}{\kappa_2}\frac{L}{n\pi}(-1)^n\left(\frac{Q_o}{\kappa_1} - \frac{Q_o}{\kappa_2}\right) I_1\left(2n\pi \frac{a}{L}\right) \times \left[I_o\left(2n\pi\frac{a}{L}\right)K_1\left(2n\pi\frac{a}{L}\right) + \frac{\kappa_1}{\kappa_2} I_1\left(2n\pi\frac{a}{L}\right)K_o\left(2n\pi\frac{a}{L}\right)\right]^{-1} \tag{12b}$$

Knowing the coefficients A_n and B_n we have fully determined the temperature distribution within the salt dome and on its flanks.

We note that both sets of coefficients A_n and B_n are oscillatory due to the nature of the Fourier expansion coefficients for the function (z + d). The physical parameters which enter equations (12) are (1) the ratio of the salt dome radius to its length, and (2) the contrast in thermal gradients in uniform infinite salt and sedimentary media $(Q_o/\kappa_1 - Q_o/\kappa_2)$. Thus in the case of a salt dome embedded in an infinite medium, the coupling between heat flow inside and outside the salt dome depends only on these two parameters.

Next, consider the temperature distribution at positions shallower than or deeper than the salt dome. If the temperature anomaly introduced by the salt dome is confined to a volume of radius R centered about the dome[1], then the temperature profile below the dome $T_3(r,z)$ can be described over the range $0 \leqslant r \leqslant R$ by a Fourier-Bessel series of the form

$$T_3(r,z)=T_o - \frac{Q_o}{\kappa_2}(z+d) + \sum_{n=1}^{\infty} D_n J_o(\beta_n r) e^{\beta_n (z+d+L/2)}, \tag{13}$$

for $z \leqslant -d - L/2$, where J_o is the Bessel function of the first kind of order zero and we again assume cylindrical symmetry. In equation (13) we introduced the quantity β_n as

$$\beta_n = x_n/R, \tag{14}$$

where x_n is the nth root of the equation $J_o(x) = 0$[2]. In writing equation (13), we used the fact that the solution $T_3(r,z)$ must reduce to a constant vertical temperature gradient far from the salt dome (i.e., $z \rightarrow -\infty$ or $r \rightarrow \infty$). From continuity of temperature across the surface $z = -d - L/2$, obtained from the solutions T_1 and T_2, the set of coefficients can be determined as

$$D_n = \frac{2}{R^2 J_1^2(x_n)} \int_o^R r \left[T(r,z=-d-L/2) - T_o - \frac{Q_o L}{2\kappa_2} \right] J_o(\beta_n r)\, dr. \tag{15}$$

----

[1] In later numerical examples we use a value R of ten times the salt dome radius.

[2] The following series is suitable for calculating all (except the first) roots of $J_o(x)$ correctly to at least five digits:

$$x_n = \frac{\pi}{4}(4n-1) + \frac{1}{2\pi(4n-1)} - \frac{31}{6\pi^3(4n-1)^3} + \frac{3779}{15\,{}^5(4n-1)^5} - \ldots$$

The lowest root x_1 is approximately 2.405.

Using similar logic, the temperature distribution at depths above the salt dome $T_4(r,z)$ can also be determined. Alternatively from symmetry arguments we note that the influence of the salt dome on temperature is an odd function about the plane z = -d - L/2. From inspection of $T_3(r,z)$ we may immediately write the solution $T_4(r,z)$ as

$$T_4(r,z) = T_o - \frac{Q_o}{\kappa_2}(z + d) - \sum_{n=1}^{\infty} D_n J_o(\beta_n r) e^{-\beta_n (z+d-L/2)}, \tag{16}$$

where β and D_n are as before.

We have now determined the temperature distribution in all regions of space for the case of a cylindrical salt dome embedded

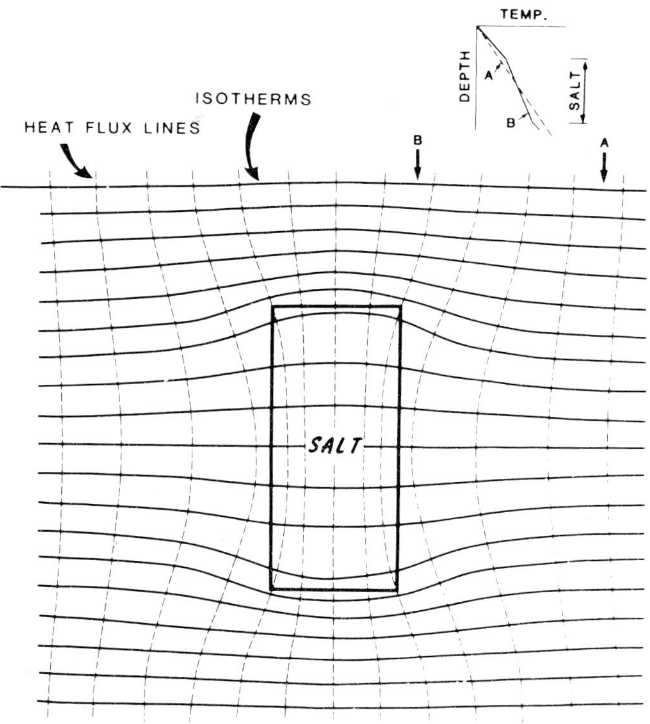

Figure 3. Schematic diagram of the heat flow and temperature distribution in the neighborhood of a salt dome. Insert shows the temperature profiles at locations A and B.

in an infinite uniform medium. These results are sketched schematically in Figure 3, where one may see the influence of the salt dome on the temperature distribution and on the heat flux. Because of the contrast in thermal conductivities, heat flux lines are distorted such that heat flux/area is increased in the vicinity of the salt. Isotherms are also distorted since isothermal surfaces are orthogonal to heat flux lines. The influence of the salt dome decreases, of course, as one moves away from the salt, being limited to a distance beyond the salt-sediment interface which is on the order of the salt dome radius. A more complete and more quantitative discussion of the thermal influence of salt domes is deferred until we first develop a model of heat flux in the vicinity of a cylindrical dome buried in a semi-infinite uniform medium.

III. SALT DOME EMBEDDED IN A SEMI-INFINITE MEDIUM

In this section we consider a cylindrical salt dome of radius a and height L positioned such that the horizontal plane $z = -d$ bisects the cylinder while the free surface $z = 0$ is an isotherm (Figure 1). To obtain the solution to this problem, we use the method of images. Consider two identical coaxial cylinders of radius a and height L positioned symmetrically above and below the plane $z = 0$ in an infinite medium. In the previous section we derived the temperature distribution in all regions of space due to the lower salt dome alone. By a simple translation of axes we can also write the temperature distribution due to the upper salt dome alone. Thus $T'_3(r,z)$, the temperature profile at depths below the upper salt dome due to the upper salt dome alone, is given by

$$T'_3(r,z) = T'_o - \frac{Q_o}{\kappa_2}(z - d) + \sum_{n=1}^{\infty} D_n J_o(\beta_n r) e^{\beta_n (z-d+L/2)}, \tag{17}$$

where β and D_n are again given by equations (14) and (15).

Now let us superpose the solutions due to the two salt domes in isolation. Since the surface $z = 0$ is now a plane of symmetry, it must be a horizontal isotherm. Thus this superposition satisfies the surface boundary condition at $z = 0$ exactly; it must also satisfy the heat flow equation since each component of the superposition satisfies the heat flow equation individually. If the result of the superposition also satisfies the boundary conditions along the salt-sediment interface, then it will provide an exact solution to the problem under consideration. To the extent that the influence of the upper salt dome is not felt by the lower dome, these boundary conditions are also satisfied. (This breaks down if the two salt domes are too close together vertically because the solution due to the upper dome alone was derived by assuming that the volume occupied by the lower salt dome was in fact occupied by sediment). Thus if the buried salt dome and its image have a sufficient vertical separation, the boundary conditions on the salt-sediment interface will be approximately satisfied, and our solution will be a good approximation to the exact one. Since the influence of a salt dome extends a distance approximately equal to its radius, this condition can be stated as

$$2(d - L/2) \gg a. \tag{18}$$

Under these conditions, we can write the temperature distribution in the vicinity of a salt dome buried in a semi-infinite medium. The temperature inside the salt dome $T_a(r,z)$ can be expressed as

$$T_a(r,z) = \frac{1}{2}\left(T_1(r,z) + T'_3(r,z)\right)$$

$$= T_s - \frac{1}{2}\frac{Q_o}{\kappa_1}(z+d) - \frac{1}{2}\frac{Q_o}{\kappa_2}(z+d)$$

$$+ \frac{1}{2}\sum_{n=1}^{\infty} A_n \sin\left(2n\pi\frac{z+d}{L}\right) I_o\left(2n\pi\frac{r}{L}\right)$$

$$+ \frac{1}{2}\sum_{n=1}^{\infty} D_n J_o(\beta_n r) e^{\beta_n (z-d+L/2)}, \qquad (19)$$

for $-d - L/2 < z < -d+L/2$; $0 < r < a$, where $T_s[=1/2(T_o+T'_o)]$ is the surface temperature.

The temperature distribution in the sediment at the same depth as the salt dome $T_b(r,z)$ is given by

$$T_b(r,z) = \frac{1}{2}\left[T_2(r,z) + T'_3(r,z)\right]$$

$$= T_s - \frac{Q_o}{\kappa_2} z$$

$$+ \frac{1}{2}\sum_{n=1}^{\infty} B_n \sin\left(2n\pi\frac{z+d}{L}\right) K_o\left(2n\pi\frac{r}{L}\right)$$

$$+ \frac{1}{2}\sum_{n=1}^{\infty} D_n J_o(\beta_n r) e^{\beta_n (z-d+L/2)}, \qquad (20)$$

for $-d - L/2 < z < -d + L/2$; $r > a$.

The temperature distribution in the sediment beneath the salt dome $T_c(r,z)$ is given by

$$T_c(r,z) = \frac{1}{2}\left[T_3(r,z) + T'_3(r,z)\right]$$

$$= T_s - \frac{Q_o}{\kappa_2} z + \frac{1}{2} \sum_{n=1}^{\infty} D_n J_o(\beta_n r) \times$$

$$[e^{\beta_n (z+d+L/2)} + e^{\beta_n (z-d+L/2)}], \quad (21)$$

for $z < -d - L/2$; $r > 0$.

The temperature distribution in the sediment above the salt dome $T_d(r,z)$ is given by

$$T_d(r,z) = \frac{1}{2} [T_4(r,z) + T'_3(r,z)]$$

$$= T_s - \frac{Q_o}{\kappa_2} z + \frac{1}{2} \sum_{n=1}^{\infty} D_n J_o(\beta_n r) \times$$

$$[-e^{-\beta_n (z+d-L/2)} + e^{\beta_n (z-d+L/2)}] \quad (22)$$

for $0 < z < -d + L/2$; $r > 0$.

In equations (19)-(22) the coefficients A_n, B_n, β_n, and D_n are as given in equations (12), (14) and (15).

The heat flux $Q(r,z)$ can be determined by evaluating the temperature gradient at the position (r,z) as

$$Q(r,z) = -\kappa \nabla T(r,z) \quad (23)$$

where κ is the thermal conductivity at (r,z). Since explicit expressions are available for the temperature distribution $T(r,z)$, analytic expressions for the heat flux $Q(r,z)$ can be obtained by differentiating equations (19)-(22). As an example, the radial and tangential components of heat flow within the sediments on the flanks of the salt dome, Q_r and Q_z, are found by differentiating equation (20):

$$Q_r(r,z) = -\kappa_2 \frac{\partial T_b(r,z)}{\partial r}$$

$$= \frac{\kappa_2}{2} \sum_{n=1}^{\infty} (2n\pi/L) B_n \sin\left(2n\pi \frac{z+d}{L}\right) K_1(2n\pi r/L)$$

$$+ \frac{\kappa_2}{2} \sum_{n=1}^{\infty} \beta_n D_n J_1(\beta_n r) e^{\beta_n (z-d+L/2)} , \qquad (24a)$$

and

$$Q_z(r,z) = -\kappa_2 \frac{\partial T_b(r,z)}{\partial z}$$

$$= Q_o - \frac{\kappa_2}{2} \sum_{n=1}^{\infty} (2n\pi/L) B_n \cos\left(2n\pi \frac{z+d}{L}\right) K_o\left(2n\pi \frac{r}{L}\right)$$

$$- \frac{\kappa_2}{2} \sum_{n=1}^{\infty} \beta_n D_n J_o(\beta_n r) e^{\beta_n (z-d+L/2)} , \qquad (24b)$$

for $-d - L/2 < z < -d + L/2$; $r > a$. Similar expressions can be derived for the heat flux in other regions of space.

IV. ANALYTICAL MODEL RESULTS

The expressions derived in the previous section have been evaluated numerically for several sample cases. To generate these models, we require as input the regional geothermal gradient as well as the thermal conductivity contrast between the salt dome and the surrounding sedimentary formations (κ_1/κ_2). The thermal conductivity of clastic rocks generally lies in the range 1.5-2.5 W/m/°C (Clark, 1966), depending primarily on lithology and degree of compaction. The thermal conductivity of halite is about 6 W/m/°C at room temperature and decreases with increasing temperature (Birch and Clark, 1940). In the temperature range 50°C-100°C, the thermal conductivity of salt is about 4.5 W/m/°C. The analytical model presented here considers the thermal conductivity of both rock salt and surrounding sediments to be constant. In the following examples, we assume the thermal conductivities to be in the ratio 3 : 1 and the regional geothermal gradient to be 3.6×10^{-2}°C/m. Modifying the contrast in thermal conductivities will change the magnitude of the thermal anomaly introduced by the salt, while changing the regional thermal gradient will alter the overall scale of the temperature distribution.

The first model consists of a salt dome of radius 600 m and height 3,000 m which is buried under 1,500 m of sediment. In this and all subsequent models, the salt is buried sufficiently deeply that all boundary conditions are well satisfied. The temperature profile corresponding to this model is illustrated in Figure 4. At large radial distances from the salt dome, the constant vertical regional thermal gradient is regained. However, the gradient in the interior of the salt dome is considerably reduced below the regional trend. In the vicinity of the salt-sediment interface, the two temperature distributions join smoothly; thus, the thermal influence of the salt dome is not restricted to the interior of the dome, but is also felt in the sediments immediately adjacent to the salt.

While the vertical separation of isotherms expands within the salt, the spacing of isotherms correspondingly compresses in the overlying and underlying sediments. This implies an enhancement of heat flux relative to the regional trend. This effect persists even to the sediment surface, although the magnitude of the effect observed at the surface obviously decreases with increasing depth of burial of the salt. Thus, while this model predicts no surface temperature anomaly over a salt dome (since we postulated the surface as being an isotherm), it does predict

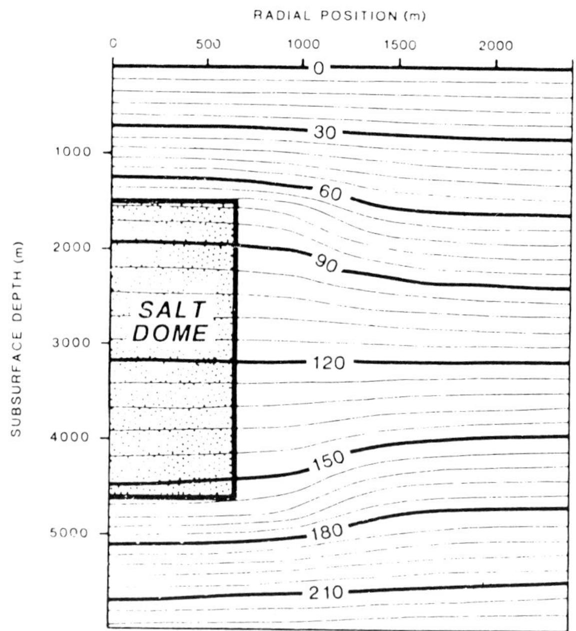

Figure 4. Temperature distribution (°C) in the vicinity of a salt dome of radius 600 m, height 3,000 m, buried at a depth of 1,500 m. Horizontal exaggeration is 2:1.

a surface heat flux anomaly. Such effects are well known from field observations (Epp et al., 1970; Poley and Van Steveninck, 1970; Von Hertzen et al., 1972; Jensen, 1983), and also from previous modeling studies (Selig and Wallick, 1966; Geertsma, 1971; Jensen, 1983).

Figure 5 presents the surface heat flux above a salt dome as a function of radial position for various depths of burial. This shows an enhancement of the surface heat flux extending beyond the radius of the salt dome. The width of this anomaly increases with increasing depth of burial, while at the same time the magnitude of the anomaly at r = 0 decreases. Thus for a given salt dome geometry, a shallow depth of burial gives a surface heat flux anomaly of larger magnitude and of sharper

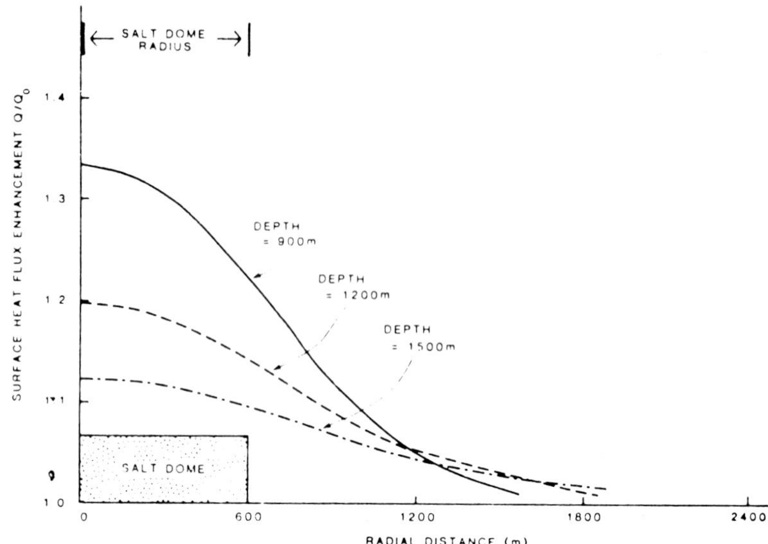

Figure 5. Enhancement of surface heat flux above a salt dome as a function of radial position for salt domes buried under 900, 1,200 or 1,500 m of overlying sediment. In each case the radius of the salt dome is 600 m, and the height of the dome is 3,000 m.

profile; increasing depth of burial gives a less distinct anomaly until eventually it merges with measurement noise. A depth of burial less than the shallowest calculated depth would presumably yield an even larger surface anomaly than that seen in Figure 5; however, our solution becomes less accurate at these shallow depths of burial.

In addition to depth of burial, surface heat flux anomalies are also sensitive to the physical dimensions of the salt dome. Figure 6 shows the influence of salt dome height on the enhancement of surface heat flux. Changing the height of the salt dome alters the magnitude of the surface anomaly but does not significantly influence its shape. Thus, we see that the surface heat flux anomaly explicitly reflects the geometry of the buried salt dome; its shape reflects the radius and depth of

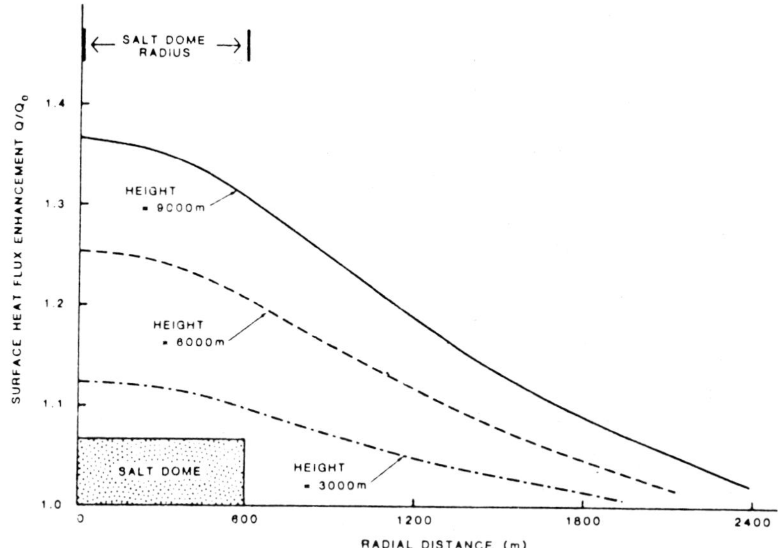

Figure 6. Enhancement of surface heat flux as a function of radial position above salt domes of height 3,000, 6,000 and 9,000 m. In each case the dome radius is 600 m, and the top of the salt is 1,500 m below the surface.

burial of the salt, while its magnitude is a function of the height of the salt dome and the thermal conductivity contrast.

Figure 4 shows the subsurface temperature profile for the case of a salt dome buried at 1,500 m. Figure 7 contours the temperature anomaly associated with this model, defined as the difference between the temperature at a point and the regional trend at that subsurface depth. Figure 7 shows the largest positive temperature anomaly immediately above the salt dome, the largest amplitude negative anomaly immediately beneath the dome, while the anomaly goes to zero along a surface which passes through the middle of the dome. This is easily understood since the salt dome provides a path of high thermal conductivity enabling heat to flow away easily from the base of the salt dome, thus cooling sediments at the base. This enhanced heat flux flowing toward the top of the salt increases the temperature of sediments at the top of the salt.

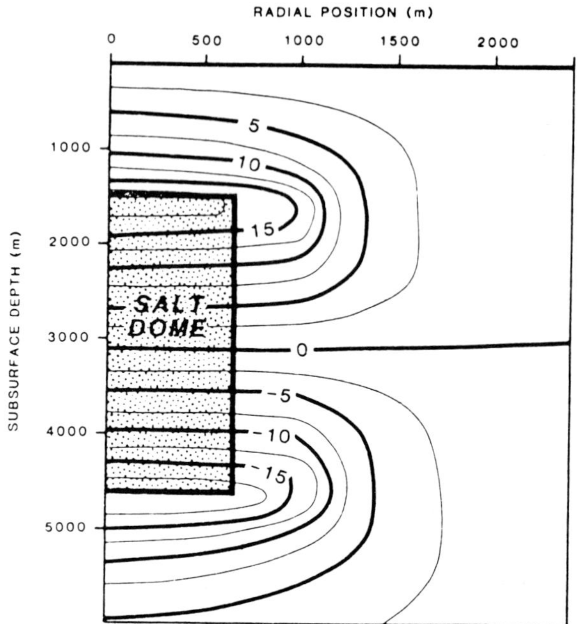

Figure 7. Contour plot of subsurface temperature anomaly (°C) for a salt dome of height 3,000 m and radius 600 m. Horizontal exaggeration is 2:1.

This anomalous temperature distribution is not restricted to regions immediately above and below the salt, but is also felt by sediments along the flanks of the salt dome. This is illustrated in Figure 8, which plots temperature anomaly as a function of radial distance from the axis of the dome at various subsurface depths. Along the upper flanks of the salt dome a positive anomaly is observed (temperature increase), while along the lower flanks a negative anomaly is observed (temperature decrease). This figure shows, furthermore, that the magnitude of the anomaly increases in proceeding from the midplane of the dome toward either end and also in proceeding radially inward toward the salt-sediment interface. The temperature anomaly extends a radial distance beyond the edge of the salt dome which is comparable to the salt dome radius and can have an appreciable

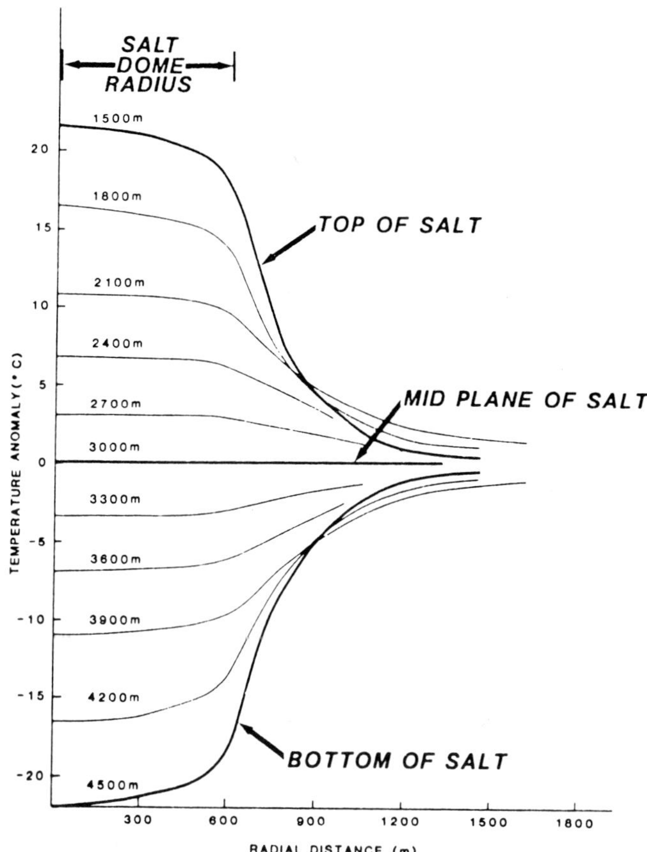

Figure 8. Temperature anomaly (°C) as a function of radius at various subsurface depths for a salt dome of radius 600 m, height 3,000 m, covered by 1,500 m of sediment.

magnitude. At a distance of 150 m beyond the edge of the salt dome, the model illustrated in Figure 8 predicts a temperature anomaly of ~7°C near the top of the salt dome, while an anomaly of the same magnitude but of opposite sign is predicted near the base of the dome.

Figure 9 plots the temperature anomaly at a radial distance of 150 m beyond the edge of the salt for salts domes of height 3,000, 6,000 and 9,000 m. This shows a positive anomaly over the

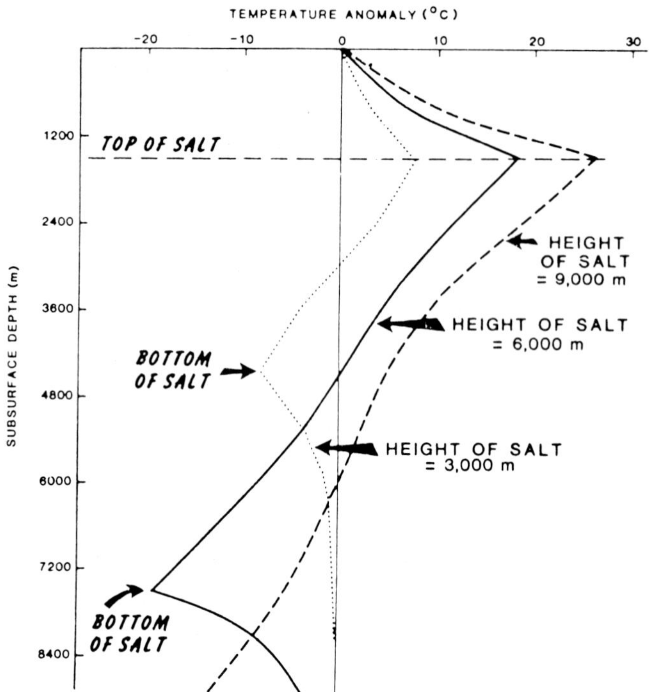

Figure 9. Temperature anomaly profile at a radial location 150 m beyond edge of salt dome for domes of height 3,000, 6,000 and 9,000 m. In each case the dome has a radius of 600 m and is buried under 1,500 m of sediment.

upper flanks, a negative anomaly on the lower flanks, while the magnitude of the anomaly increases as one approaches the upper and lower extremities of the salt dome. It is also seen that the magnitude of the anomaly increases markedly with increasing salt dome height; Figure 9 shows anomalies as large as 25°C at a radial distance of 150 m from a salt dome of height 9,000 m. Closer to the salt dome the temperature anomaly is even larger. Thus, we see that the impact of a salt dome on the temperature of the surrounding sediments can be quite considerable.

Based on the model calculations illustrated in Figure 9, one would hope to observe an anomalous profile when downhole temperature measurements are made in the vicinity of a salt dome. Heroy (1962) reported such an effect in a hole drilled through the Pescadito salt dome. Anomalous profiles should also be observed in holes drilled on the flanks of domes. In a later section, we will present such data for wells drilled on the flanks of a West Bay, Louisiana salt dome.

The model which we have considered here is an extremely simple one. In practice salt domes are not cylindrical in shape; thermal conductivity is not uniform within a basin but varies depending upon lithology, porosity, and temperature; additionally, the thermal conductivity of halite is also temperature-dependent. However, these caveats do not place serious restrictions on this analysis of the basic physical processes which are operating. Our results are based on the existence of a significant contrast in thermal conductivities over an extended region of space. While fine-scale details of the results will depend upon the geometry and on the exact form of the thermal conductivities used, the effects reported here are sufficiently large that we believe the gross effects should be as described in this paper.

Salt domes are often not found in isolation, as has been modeled here, but instead may be found in association with a basal salt mass of greater radial extent. Our present results may be generalized to this situation by considering two salt cylinders of different dimensions located on-axis one above the other. Qualitatively we see that heat flux lines tend to converge toward the lower salt body and to diverge again above it. The upper salt then tends to force convergence of the heat flux lines once more. Having passed through the upper salt body, the flux lines diverge and are incident normally on the sediment surface. Thus this geometry also yields surface heat flux and subsurface temperature anomalies, but their values now depend upon the dimensions of the two salt bodies and also on the separation between the two.

V. INFLUENCE ON HYDROCARBON MATURATION

Since chemical reaction rates are temperature-dependent, reaction rates on the flanks of a salt dome will differ from those which would be predicted based on regional trends. Reaction rates on the upper flanks will be increased while rates on the lower flanks will be decreased. These findings are particularly relevant in connection with hydrocarbon thermal maturation because hydrocarbon accumulations are commonly found in association with salt domes. The positive temperature anomaly on a salt dome's upper flanks implies an enhancement of thermal maturation of any organically rich source rocks near the top of the salt. The degree of enhancement is expected to depend upon the height of the salt dome, the position of the source rock relative to the salt dome, and the dependence of the chemical reaction rates on temperature. Likewise thermal maturation rates on the lower flanks of a salt dome will be suppressed.

To assess quantitatively the impact of a salt dome on hydrocarbon maturation, one must consider not only the present day temperature anomaly, but also past values of the anomaly and so the development history of the salt dome. The accuracy of the results will then depend on how well salt dome emplacement is modelled, which in turn depends on the quantity and quality of the available geologic data and the veracity of the model of salt dome development. For the present, let us assume that salt dome emplacement is adequately described by the quasi-equilibrium model discussed by Lerche and O'Brien elsewhere in this volume. In this model diapirism is assumed to result solely from the progressive buoyancy of salt upon burial and the compaction of the overlying sedimentary formations. A critical thickness of overburden above the salt is required before diapirism is initiated. Neglecting the mechanical strength of the overburden which would inhibit the initiation of diapirism, we estimate the critical thickness of overburden to be approximately 3000 m. As the mother salt is buried progressively deeper a salt dome develops, the height of the salt dome being estimated from considerations of eustacy. This then provides us with a model of the salt dome geometry since the initiation of diapirism; by specifying the sedimentation rate we can calculate the salt dome height through geologic time. At each stage of development we calculate the temperature distribution in the subsurface, using the formalism provided here. In this way the thermal history of each sedimentary unit is traced. This information can be used to gauge the thermal maturation of organic material in these sediments through the use of a maturation indicator such as Lopatin's (1971) Time Temperature Index (TTI),

$$\mathrm{TTI} = \int_{\tau_o}^{\tau_{MAX}} \left[\exp\left\{ \frac{T-105}{10} \right\} \ln 2\right] d\tau, \qquad (25)$$

In this equation T represents the formation temperature in °C, while the time, in Ma, between τ_o and τ_{MAX} represents the interval over which thermal maturation occurred.

Figure 10 illustrates the results of a calculation along the lines described above. In this figure, we plot a depth profile of the TTI of a sedimentary unit located 80 m beyond the edge of a salt dome of radius 800 m at the time at which the top of the salt is 600 m below the sediment surface and the base of the dome is 5000 m below the surface. In this model we assume a constant sedimentation rate of 60 m/Ma, a geothermal gradient of 2.3×10^{-2} °C/m in formations unaffected by the influence of the

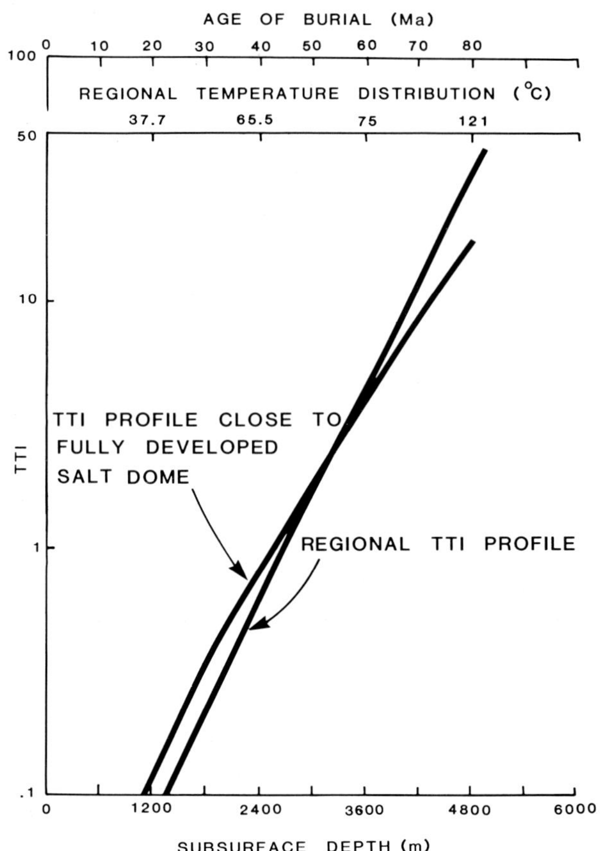

Figure 10. Sketch of the enhanced TTI profile in the vicinity of a rising salt dome. Deposition is 60 m/Ma, and regional geothermal gradient is 2.3×10^{-2} °C/m.

salt dome, and a thermal conductivity contrast of 3:1 between salt and sediment. For comparison Figure 10 also shows the TTI profile for a similar sedimentary unit in the absence of a salt column.

Figure 10 indicates an enhancement in thermal maturation associated with the positive temperature anomaly on the upper flanks of a salt dome, as well as a decrease in maturation along the lower flanks. In the present model the presence of the salt diapir can modify the TTI index by as much as a factor of two. An even greater enhancement can be expected for salt domes with a greater associated temperature anomaly. This could occur for taller domes or at points closer to the edge of the salt body.

The effect of this change in maturation on the generation of hydrocarbons is dependent on the overall level of maturation. In the case illustrated in Figure 10, organic material located near the top of this dome is buried at a shallow depth and has undergone little thermochemical alteration, even if we include the enhanced heat flow associated with the salt. In this instance the negative temperature anomaly on the lower flanks of the salt dome may be a more significant effect, serving to lower the level of maturation of organic material at those depths. In general the impact of salt-related enhancement of heat flow on thermal maturation depends on the location of the oil generation window with respect to the salt dome. If the oil generation window lies near the top of the salt dome then the degree of maturation is expected to be enhanced. This enhancement is greater for taller salt columns and so this effect is expected to be most significant in deep basins in which tall salt diapirs have developed, extending upwards into the oil generation window.

On the lower flanks of the salt dome, a negative thermal anomaly is predicted (i.e., temperatures less than the regional trend); this serves to reduce the thermochemical rate constants and so inhibit the hydrocarbon reactions. If the lower flanks of the salt dome are buried deeply below the subsurface, then any associated hydrocarbons are most likely mature; the temperature perturbation introduced by the salt dome serves to inhibit over-maturation of the hydrocarbons and the chemical decomposition of hydrocarbons (through sulfur reactions, for example). The effect of the deeper part of a salt dome will be to lower the lower limit of the hydrocarbon window. Hydrocarbons may be found at greater depths in the vicinity of a salt dome than would otherwise be expected.

Thus we see that a major impact of the thermal influence of a salt dome is to enlarge the hydrocarbon window. Given a salt dome of appropriate dimensions and depth of burial, the upper limit of the hydrocarbon window may be moved shallower, the lower limit may be moved deeper, and under favorable conditions both limits may be moved.

As we have previously remarked: To assess quantitatively the impact of a salt dome on hydrocarbon maturation, one must consider not only the present day temperature anomaly but also past values of the anomaly and the development history of the salt dome. The accuracy of the results will then depend upon how well salt dome emplacement is modeled, which in turn depends upon the quantity and quality of the available geologic data and the veracity of the model of salt dome development. Furthermore, the timing of trap formation and hydrocarbon migration must also be considered. The point we emphasize is that modification of the subsurface temperature distribution in the vicinity of a salt dome can have a significant influence on hydrocarbon maturation and any calculation of gas/oil windows must include these effects.

VI. COMPARISON OF THEORY WITH OBSERVATIONS

Vizgirda et al. (1985) provide a detailed analysis of downhole temperature measurements from several wells at different distances from a salt dome in the West Bay, Louisiana, region. That analysis makes use of a large body of ancillary geologic data which are available to assist in the interpretation of the downhole temperature data. A numerical approach was used in that work so that full advantage could be taken of all these data. In

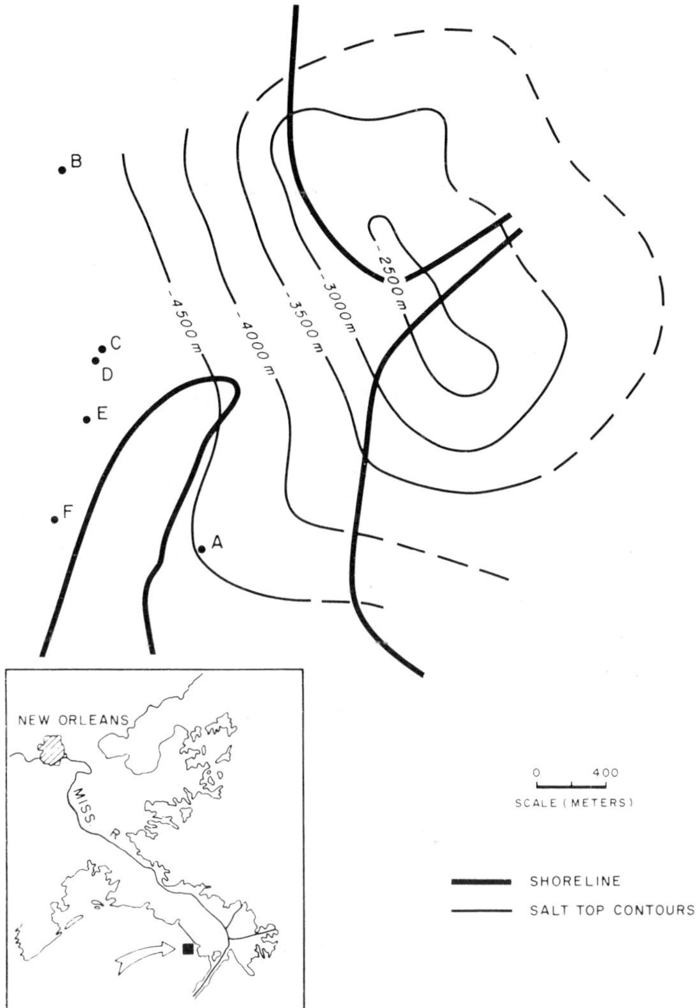

Figure 11. Map of the western portion of the West Bay salt dome showing the subsurface salt contours and well locations (A to F).

other surveys, however, such supporting information may not be available. In such situations, a meaningful interpretation of the downhole temperature data may still be achieved because conductive heat flow in the subsurface is such a robust physical process. Using a simple analytical model, we show here how downhole temperature data alone may be interpreted to gain a better understanding of heat flow in the vicinity of a salt dome.

The locations of the six wells relative to the salt dome are shown in Figure 11. While we model the salt dome as cylindrical body, Figure 11 shows that the upper surfaces of the salt, at least, do not possess the regular geometry of a cylinder. The uppermost reaches of the salt dome are not circular in cross-section, but instead are somewhat elongated, the north-south axis being approximately 50 percent greater than the east-west axis. Furthermore, the upper flanks of the salt dome are not vertical but are inclined at approximately 45-55 degrees to the vertical. However, these departures from cylindrical geometry are expected to have only a minor impact on the integrated heat flux anomaly for the following reasons. First, these features are relatively minor in size when compared with the overall dimensions of the salt dome which, as we shall show below, has a height in the range 9,000-15,000 m. Second, the integrated heat flow anomaly is sensitive primarily to .pa the mass of the salt producing the anomaly, i.e., this is a bulk effect. The actual distribution of the salt is only of secondary importance in this respect.

However, the shape of the salt dome does bear on the present work in one important way, namely, that the temperature and heat flow anomalies at any given point depend upon the location of that point with respect to the salt dome. Thus, since the radius of the West Bay salt dome increases over the depth range 3,000-5,000 m (see Figure 11), temperature anomalies recorded in boreholes drilled vertically on the flanks of the dome will be modified due to the fact that the radial separation of the hole from the salt dome decreases with depth. An approximate compensation for this effect can be obtained by honoring the radial

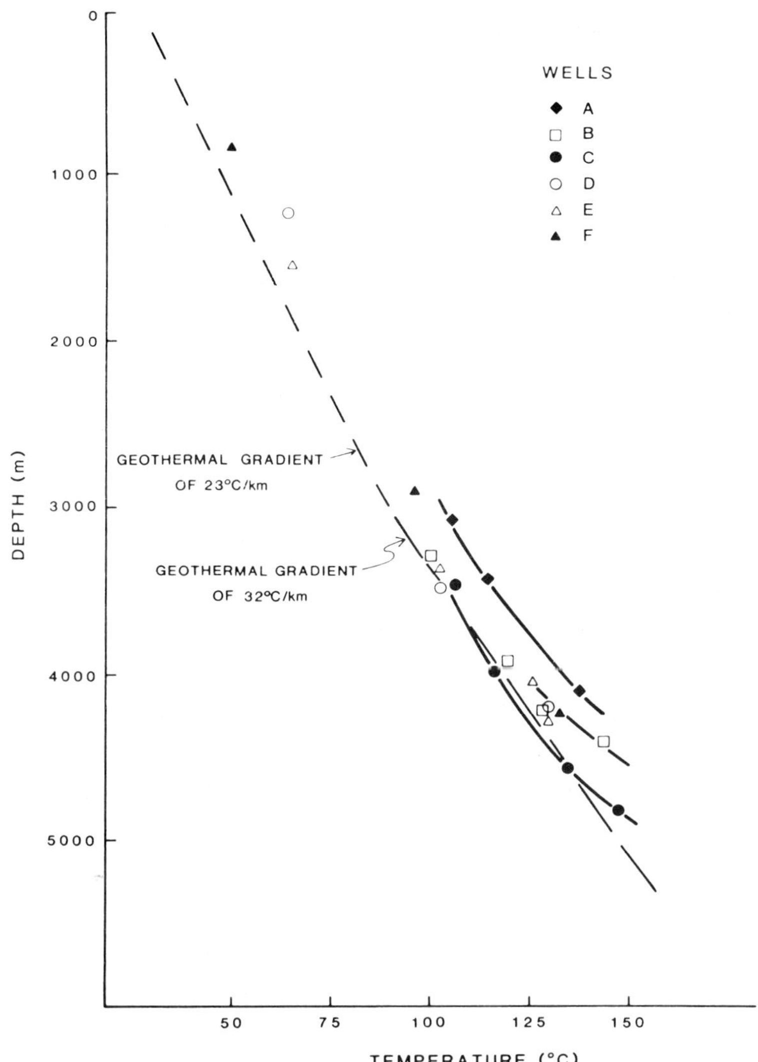

Figure 12. Estimates of true formation temperature on the flanks of the salt dome (from Vizgirda et al., 1985). Curves are only to guide the eye.

separation of the borehole from the salt dome at all depths. Thus, while measurements are recorded in vertical holes drilled on the flanks of a salt dome whose upper surface is inclined at an angle α, calculations are performed for the equivalent case of inclined holes approaching a vertical cylindrical salt dome at the same angle α.

True formation temperatures have been estimated by applying the empirical correction of Kehle (1971) for borehole effects to the measured data. Figure 12 plots the estimated true formation temperatures obtained for these six wells. The temperature anomalies of largest amplitude are predicted to occur at the top and bottom surfaces of the salt dome. With the exception of data from well D, the range of temperatures measured as a function of offset appears to increase with depth down to 4,200 m in the vicinity of the West Bay salt dome. To illustrate this effect, we point out the difference in temperature profiles in well E and well F, where well F is approximately 180 m farther from the salt dome in a radial direction than well E. These two wells yield very similar temperature measurements at a depth of 3,300 m. At 4,000 m, the temperature in well E, which is closer to the salt, is higher by 4.5°C than the temperature in well F. This temperature difference between the two wells becomes even more pronounced with increasing depth down to total depth. Thus, in modeling the thermal anomalies introduced by the salt dome, we consider the conductive heat flow through an equivalent cylinder of salt whose upper surface is located 4,200 m below the sediment surface.

In Figure 13 we present a plot of the temperature distributions predicted by the analytical model described above. In this calculation, the height of the salt column is set at 9,000 m, its radius at 2,400 m, the upper surface of the salt column is located at 4,200 m below the sediment surface, and the regional geothermal gradient is assumed to be 2.6×10^{-2}°C/m (Vizgirda et al., 1985). For clarity in presentation, only data points from wells closer to the salt column are plotted. In the wells far-

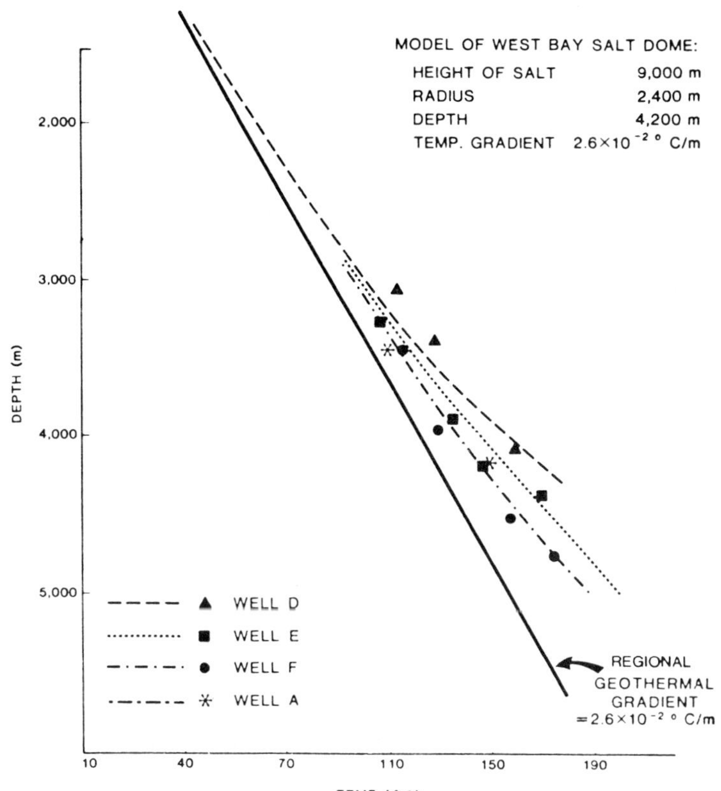

Figure 13. Subsurface temperature distribution on the upper flanks of the West Bay salt dome and the theoretical fit to these data (broken curves) assuming a regional geothermal gradient of 2.6×10^{-2} °C/m.

ther from the salt dome (wells B and C), the variation in downhole temperature is less than the uncertainty in recording these temperatures, which we estimate to be ±2°C. Thus, for reasons of clarity the data from wells B and C are omitted.

In Figure 13 the calculated curves in general provide an excellent fit to the data. The dependence of temperature with depth is clearly reproduced in each well, as are the differences in temperatures between wells at a given depth which we attribute to the increase in lateral distance from the flanks of the salt dome. Thus we find excellent agreement between the observed temperature distribution and the theoretical predictions which are based on a model of enhanced conductive heat flow in the vicinity of a salt dome. The only disagreement is provided by the two data points of 3,030 m and 3,360 m in well D, which we discuss later.

This analysis does not provide a unique solution. We have repeated these calculations assuming geothermal gradients ranging from 2.2×10^{-2}°C/m to 3.6×10^{-2}°C/m. In each of these calculations the height of the salt column was varied until an optimum fit to the data was found, as determined by visual inspection. All other parameters were held constant in these models, including the ratio of thermal conductivities in salt and surrounding sediments. These calculations show that assuming a lower geothermal gradient means that the observed temperature distribution implies a relatively larger thermal anomaly. To reproduce this effect, we must invoke a larger salt mass, i.e., we are forced to increase the height of the salt dome. We can obtain equally acceptable fits to the temperature data with any geothermal gradient which lies in the range 2.2×10^{-2}°C/m-3.6×10^{-2}°C/m, but with different salt dome heights. Thus these data, by themselves, do not define uniquely the regional geothermal gradient at depth.

Model calculations of the temperature distribution are sensitive to the height of the salt column and, in fact, the data can be used to obtain an estimate of the height of the salt dome. However, this estimate is dependent upon the geothermal gradient one assumes. In Table 1, we present estimates of the salt dome height for a range of geothermal gradients. These estimates were obtained from theoretical fits to the temperature distribution on the upper flanks of the salt dome, as described above. For geothermal gradients of (2.2-2.6) x 10^{-2}°C/m, we obtain estimates of the salt dome height of 9,000 to 15,000 m, while for more extreme geothermal gradients (3.3-3.6) x 10^{-2}°C/m, we find a salt height of 5,000 to 6,000 m. Since the geothermal gradient at depth is believed to lie most probably in the range (2.2-2.6) x 10^{-2}°C/m, this yields a total sediment thickness overlying the basal salt of 13,200-19,200 m, which is a reasonable estimate in this area of the Gulf Coast.

Finally, we note the anomalous temperatures recorded in well D at depths of 3,030 and 3,360 m. These data points are anomalously high and are not satisfactorily explained by any of the models we have considered. Referring to Figure 11, we see that seismic data indicate a limb of the salt dome extending toward the location of well D. Due to the finite lateral resolution of the seismic technique which tends to smooth out such convex features, it is probable that the lateral extent of this

Table 1. Estimates of the height of the West Bay salt dome, assuming various geothermal gradients at depth. These estimates were obtained by modeling the temperature distribution on the upper flanks of the salt dome, as described in the text.

Geothermal gradient (°C/m)	Salt dome height (m)
2.2 x 10^{-2}	15 000
2.6 x 10^{-2}	9 000
2.9 x 10^{-2}	7 000
3.3 x 10^{-2}	6 000
3.6 x 10^{-2}	5 000

limb is not accurately imaged. Thus, at subsurface depths of 3,000-3,300 m, well D may be nearer to the salt than estimated, resulting in higher downhole temperatures at these depths.

In summary, we find that the observed downhole temperature distribution in the West Bay area is well described by the enhanced conductive heat flow resulting from the high thermal conductivity of the salt dome. We also find that the available temperature data enable us to place constraints on the height of the salt column.

VII. CONCLUSIONS

In this paper we presented an analysis of the flow of heat and the temperature distribution in the vicinity of a cylindrical salt dome buried in a uniform semi-infinite medium. Because of the contrast in thermal conductivities of salt and sedimentary formations, heat flux is enhanced within a buried salt mass. This heat flux enhancement is not confined to the interior of the salt dome but is also experienced in the surrounding sediments. As a result, two measurable effects are predicted: an increase in surface heat flux above the salt dome; and a modification of the temperature distribution along the flanks of the dome. The surface heat flux anomaly has the following characteristics.

(1) It is largest in magnitude immediately above the salt dome, but extends beyond the radius of the salt mass.

(2) The shape (i.e., radial dependence) of the heat flux anomaly depends upon the radius and depth of burial of the salt dome.

(3) Increasing salt dome height results in an increased enhancement of surface heat flux at all radial locations.

The subsurface temperature anomaly, defined as the difference between the temperature at a point and the regional trend at that depth, has the following characteristics.

(1) The temperature anomaly is largest in magnitude immediately above and below the salt dome. It also occurs on the flanks of the salt, with the magnitude of the anomaly decreasing as one approaches the mid-plane of the dome.

(2) On the upper flanks of the dome a positive temperature anomaly is predicted (i.e., an increase in temperature relative to the regional trend) while temperatures below the regional trend are predicted on the lower flanks.

(3) The temperature anomaly decreases with increasing radial separation from the salt dome; it is essentially confined to within a distance of one salt dome radius from the salt-sediment interface.

(4) The maximum amplitude of the temperature anomaly increases with increasing height of the salt dome.

Depending upon (a) the contrast of thermal conductivities and (b) the dimensions of the salt mass, the temperature anomaly introduced by the salt dome can be quite substantial, acting to decrease the temperature differential between sediments on the upper and lower flanks of the salt. Thus, temperatures on the upper flanks are increased while temperatures on the lower flanks are decreased relative to the regional trend. This may have a significant impact on the thermochemical generation of hydrocarbons within any potential source rocks in close proximity to the salt dome. To illustrate this point, we note that if the vertical extend of the salt dome essentially coincides with the hydrocarbon generation window, then the positive temperature anomaly on the upper flanks of the salt dome will enhance maturation at that depth, while the negative anomaly on the lower flanks will impede overmaturation. In this case, the present day influence of the salt dome will be to enlarge the hydrocarbon window by moving both the upper and lower bounds of the window.

A complete calculation of the location of the hydrocarbon window would, of course, have to consider the subsurface temperature distribution as modified by the salt dome in the geologic past as well as in the present. This requires an analysis of all available geologic data concerning emplacement of the individual salt dome. The point we emphasize is that, at all times since its emplacement, a salt dome modifies the temperature in the surrounding sediments, thereby creating a localized enhanced environment for hydrocarbon maturation and preservation. Depending upon the development history of the salt dome, this effect can be quite large and surely should be taken into account when evaluating the hydrocarbon potential of a salt dome.

REFERENCES

Birch, F., and Clark, H. (1940). The thermal conductivity of rocks and its dependence upon temperature and composition: Am. J. Sci. 238, 529-558 and 613-655.

Clark, S.P. (1966). Thermal conductivity *in* Handbook of physical constants: Clark, S.P., Ed., Memoir 97, Geol. Soc. Am., 459-481.

Epp, D., Grim, P.J., and Langseth, M.G., Jr. (1970). Heat flow in the Caribbean and Gulf of Mexico: J. Geophys. Res. 75, 5655-5669.

Geertsma, J., (1971). Finite element analysis of shallow temperature anomalies: Geophys. Prosp. 19, 662-681.

Heroy, W. (1962). Thermicity of salt as a geologic function: Presented at the International Conference on Saline Deposits, Houston.

Jensen, P.K. (1983). Calculations on the thermal conditions around a salt diapir: Geophys. Prosp. 31, 481-489.

Kehle, R.O. (1971). Geothermal survey of North America: 1971 Annual Progress Report, unpublished, Research Committee, Am. Assoc. Petr. Geol., Tulsa.

Lopatin, N.V. (1971). Temperature and geologic time as factors in coalification: Akad. Nauk SSSR Izv. Ser. Geol., no. 3, 95-106.

Poley, J. Ph., and Van Steveninck, J. (1970). Geothermal Prospecting-- delineation of shallow salt domes and surface faults by temperature measurements at a depth of approximately two meters: Geophys. Prosp. 18, 666-700.

Selig, F., and Wallick, G.C. (1966). Temperature distribution in salt domes and surrounding sediments: Geophys. 31, 346-361.

Vizgirda, J., O'Brien, J.J., and Lerche, I. (1985). Thermal anomalies on the flanks of a West Bay, Louisiana, salt dome: Geothermics 14, 553-565.

Von Hertzen, R.P., Hoskins, H., and Van Andel, T. (1972). Geophysical Studies in the Angola diapir field: Bull., Geol. Soc. Am. 83, 1901-1910.

SALT DOMES, ORGANIC-RICH SOURCE BEDS AND RESERVOIRS IN INTRASLOPE BASINS OF THE GULF COAST REGION

Douglas F. Williams
Ian Lerche

Department of Geology
University of South Carolina
Columbia, SC 29208

I. INTRODUCTION

The Gulf Coast geosyncline has already proven to be one of the major hydrocarbon-producing areas of the world, yet new drilling efforts in deep offshore areas of Texas and Louisiana are confirming resource estimates which indicate large quantities of crude oil and natural gas remain to be discovered in the outer shelf-upper slope region (Rice, 1980; Foote et al., 1983; Miller et al., 1975). It has long been known that much of the oil and gas generation within the Gulf Coast geosyncline occurred in deep water muds and shales, with the overlying sands providing suitable reser voirs. A significant number of these reservoirs occur in sections as young as Pleistocene both onshore and in some deep water tracts (Sabate, 1968; Powell and Woodbury, 1971). Compared with Pliocene and Pleistocene reservoirs, however, the volume of oil and natural gas discovered in Miocene sandstone and shale magnafacies far outweighs the younger reservoirs (Rice, 1980).

It is also well known that the structural styles of salt structures and the surrounding depositional environments must be considered in exploration strategies of the Gulf Coast Tertiary (Halbouty, 1979). However, our purpose here is to describe a larger, and until now, underappreciated, role of salt structures and their relationship to the occurrence of commercial petroleum reservoirs. For example, salt structures and their associated tectonics have done more than produce traps for hydrocarbons. Salt dynamics has also had a major role in a) determining the origin of source beds in the Gulf Coast geosyncline throughout the Tertiary, b) controlling the source of the organic matter in those source beds, and c) moderating the maturation history of crude oil and/or natural gas from the organic matter. We believe that salt flowage and dissolution have periodically created special environments throughout the Tertiary for enhancing both the preservation and maturation of marine organic matter in a prograding shelf-slope margin setting. Exploration strategies in both the Gulf Coast region and other evaporitic basins would benefit greatly from the considerations presented in this contribution.

Here we discuss an empirical model in an attempt to integrate and better understand the known associations among salt domes, organic-rich sediments and oil/gas in the Gulf Coast region. For the purpose of this paper we shall consider the Gulf Coast region to include the areas of Texas, Louisiana and Mississippi south and southeast of the Ouachita System (McGookey, 1975). As a secondary aim, we show that it is possible for oil and/or gas to be generated from Pleistocene source beds in intra-slope regions of the Gulf of Mexico, i.e., that "young" oil/gas is possible when the regional time-temperature relationships are changed. Our ultimate goal is to use structural, geochemical and geophysical data from well-studied Gulf Coast salt domes in order to derive a predictive model. Such models will be necessary guides for future oil and gas exploration strategies in deeper waters of the continental slope of the northern Gulf of Mexico, and in other evaporitic basins analagous to the salt provinces of the Gulf Coast geosyncline.

II. SALT STRUCTURES AND EXPLORATION PLAYS IN THE GULF COAST REGION

Thick evaporitic salt deposits of mid- to late Jurassic (Louann) age exist under vast areas of the Gulf Coast region (Antoine and Ewing, 1963; McGookey, 1975; Halbouty, 1979) (Figure 1). Since the end of the Cretaceous, the stable but dynamic margin of the northern Gulf of Mexico has been characterized by rapid out- building and progradation on the continental shelf and slope. Sediment loading caused extensive movement of the underlying Louann salt and associated shales. Depotroughs or depocenters formed and reformed between the salt ridges, other salt structures and the shelf-slope break throughout the Tertiary. Achievement of isostatic balance between sediment infilling and coastal warping is accompanied by the seaward

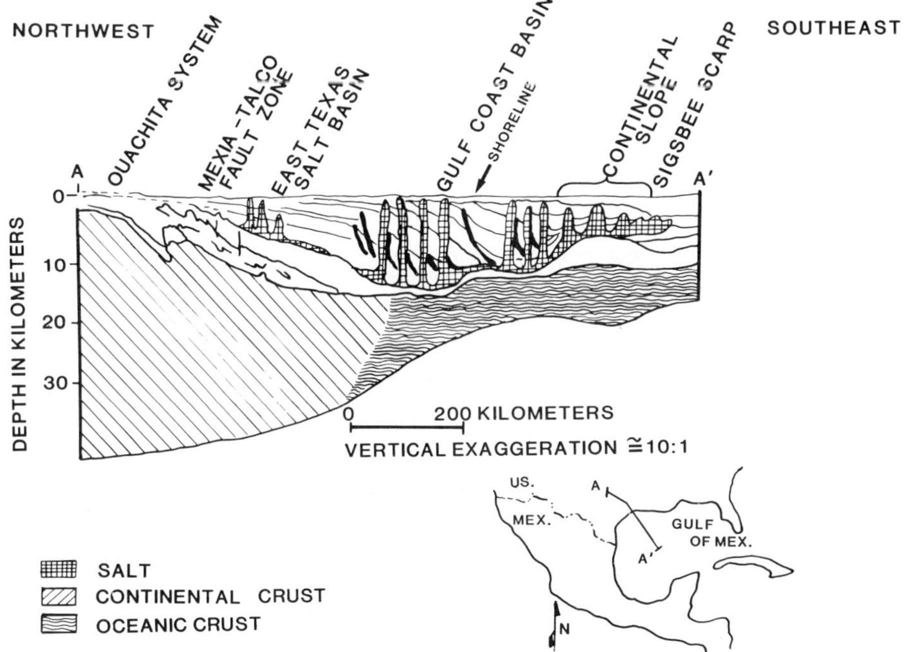

Figure 1. Schematic regional profile of the Gulf of Mexico Basin illustrating the lateral and vertical movement of salt due to sediment loading in a progradational margin setting. (Figure modified from McGookey, 1975).

migration of the shoreline and the shelf-slope break, and the progradation of predominantly fluvial sediments out over the previous slope (Bruce, 1973; Martin, 1973). The interplay between (a) rates of deposition and removal of sediments, on the one hand, and (b) upward intrusion and lateral flowage of underlying Louann salt and shale facies, on the other, thus becomes one of the major factors governing sediment deformation and the structural development of the Gulf Coast Region since the late Cretaceous (Hanna, 1934; Halbouty and Hardin, 1956; Hardin and Hardin, 1961; Lehner, 1969; Murray, 1966; Johnson and Bredeson, 1971; among others).

With increasing overburden, salt structures have a distinctive evolutionary development and become deformed into a wide variety of structures consisting of diapirs or domes, massifs and anticlinal ridges (Martin, 1973; Stude, 1978; Halbouty, 1979; Jackson and Seni, 1983). These salt structures produce basins, knolls, troughs and linear or irregular hills through uplift and faulting which, in turn, are responsible for much of the hummocky topography characterizing the present day continental slope of the northern Gulf of Mexico (Gealey, 1955; Ewing and Antoine, 1966; Antoine and Bryant, 1969; Emery and Uchupi, 1972; Martin and Bouma, 1978; Lehner, 1969; Halbouty, 1979). This interplay has been active throughout the Tertiary. As the North American continental platform underwent continued erosion after drainage of the Interior basin, thick depocenters of interbedded alternating deep shales and fluvial sands shifted southward from Eocene to Oligocene time, to the east during the Miocene and almost southwestward from Miocene to Pleistocene time (Figure 2). The Houston Salt Basin during Eocene-Oligocene time was thus analagous to the present shelf-slope region (Lehner, 1969; McGookey, 1975). During some of the periods of most rapid sedimentation, sediment thicknesses exceeding 50,000 feet occur locally (Murray, 1961). The underlying salt deposits

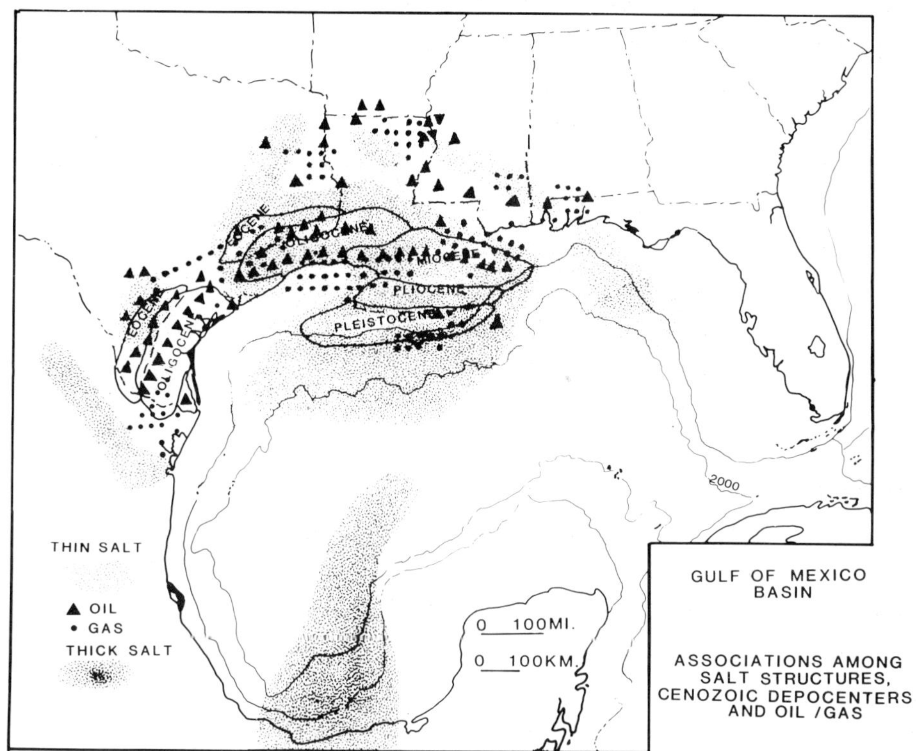

Figure 2. The distribution of Tertiary depocenters, salt basins, and oil/gas in the northern Gulf of Mexico Basin (modified from McGookey (1975) and Martin (1981)). Note how the depocenters through the Tertiary prograde first in a southeast-ward and then eastward from the Oligocene to the Miocene through Pliocene and Pleistocene.

subsequently responded to the overburden pressures of the depocenters, while in some cases, diapirism and burial were concurrent processes. Drilling of the depocenters and their sometimes buried, sometimes exposed, salt structures, revealed an unmistakable association with the widespread occurrence of oil and gas.

Salt domes, or diapirs, quickly became widely known as loci for oil and gas exploration plays in the Gulf Coast region (Figure 2). The most productive belts in the Tertiary onshore and offshore consist of massive sandstone-shale sequences (magnafacies) of marine shelf-slope facies deposited during eustatic sea level cycles (Sabate, 1968; Berg, 1981; Rice, 1980). The hydrocarbon-producing trends for oil and gas closely follow the path of the shifting depocenters and loci of salt structures (Woodbury et al., 1973; 1980; McGookey, 1975; among others). In the Miocene producing belt, a sequence of deep water shales which are devoid of sand and overlain by salt (Johnson and Bredeson, 1971; Rainwater, 1964) lie under a succession of shale and sand units which are themselves overlain by continental sandstone and shale. Optimum production occurs in the zone of interbedded transitional and shallow marine sands. In these units, Johnson and Bredeson (1971) suggested that salt uplift occurred penecontemporaneously and that the salt remained at or near the sediment-water interface during deposition of the flanking marine shales and sands.

The hydrocarbon-producing trend follows the southward shift of the thick Pliocene and Pleistocene depocenters onto thick underlying salt (Woodbury et al., 1973). Maximum sediment thickness and production in the Pliocene depocenter occurs in offshore Louisiana north of the present shelf edge and at the Pliocene shelf-slope location. The Pleistocene depocenter is located still further to the south, and new drilling into deeper tracts continues to define the limits of Pleistocene production (Woodbury et al., 1973; among others). The close association with salt structures and these depocenters is now undeniable.

Currently, salt domes and their tectonic displacement of Tertiary depocenters are thought to provide a major control for the structural framework within which the oil and gas producing belts are found (Levorsen, 1954; Halbouty, 1979). The hydrocarbon-trapping capacity of a particular salt diapir or dome is related to the depositional history of its overburden (i.e., sedimentation rate, progradation rate and lithology of the overlying overburden) (Bishop, 1978). No attempt is made here to review completely the mechanisms of salt diapirism and associated structures (The reader is referred to excellent reviews by Levorsen (1954), Halbouty (1979), among others). Salt structures may become deeply buried after maintaining a near surface position for long periods of time like those in the salt provinces of Louisiana, Texas and Mississippi or they may be classified as piercement structures. Piercement of overburden strata may occur contemporaneously or penecontemporaneously with its deposition or via catastrophic displacement of overburden (Smith and Reene, 1970; Bishop, 1978; Halbouty, 1979).

Any given salt structure may exhibit a variety of diapiric structural styles during its development (Jackson and Seni, 1983; Seni and Jackson, 1983a,b). Along any specific segment of the prograding margin, differential vertical uplift occurs as the result of its location relative to the operative depocenter, differences in sedimentation rate, bathymetric position, and the relationship of the salt structures to the mother salt. These complicated histories for both individual salt domes and localized portions of the margin contribute to the apparent selective nature of oil and gas occurrence around on-shore and off-shore salt structures. Difficulties in establishing predictable patterns to these occurrences hampered early attempts to find a comprehensive explanation and production strategy in the Gulf Coast region (Atwater and Forman, 1959; Halbouty, 1979).

Here we present a model such that, when salt structures are closest to the surface in outer shelf to intraslope regions and are behaving as piercement structures, they create special environments which account for the later association of salt structures with oil and/or gas. Briefly, salt flowage creates the depressions or basins on the margin which become ideal loci for deposition of marine organic matter (Dow, 1978). Dissolution of the near-surface salt elevates the salinity of the nearby or downslope waters of some basins. This chemical change leads to anoxic conditions in some basins which enhance the preservation of marine organic matter. We use as our modern analogues the anoxic Orca Basin in the intraslope region offshore Texas (Trabant and Presley, 1978) and the anoxic Tyro and Bacino Bannock basins recently discovered in the Eastern Mediterranean (Jongsma et al., 1983; de Lange and ten Haven, 1983; Cita et al., 1985). These hypersaline basins represent only one type of basin which are potential preservation sites of organic-rich sediments in a prograding margin setting underlain by thick salt. We begin the development of our model by reviewing the types of intraslope basins known to exist on the slope of the northern Gulf of Mexico.

III. INTRASLOPE BASINS AS SITES FOR SOURCE BEDS ON THE GULF COAST MARGIN

Hundreds of intraslope basins are known to exist on the Texas-Louisiana continental slope (Bouma et al., 1978a; Martin, 1973, 1981). The importance of salt structure movement to the development of these intraslope basins cannot be over-emphasized. Only a handful of these basins have been studied systematically with high resolution seismic and chemical oceanographic surveys. The Orca Basin is therefore the only basin in the Gulf of Mexico presently known to contain basin waters which are so highly saline that an anoxic brine is found (Figure 3) (Shokes et al., 1977; Trabant and Presley, 1978; Sackett et al., 1979). The basin was formed structurally by diapiric salt uplift. Subsequent dissolution of the surrounding and underlying salt produced a dense anoxic brine approximately 190 meters (590 feet) thick with salinities over 300 parts per thousand (Figure 3). Because of this high salinity, the density of the basin water is sufficient to prevent overturn/mixing with normal marine waters. Oxidation of organic matter consumes available oxygen and

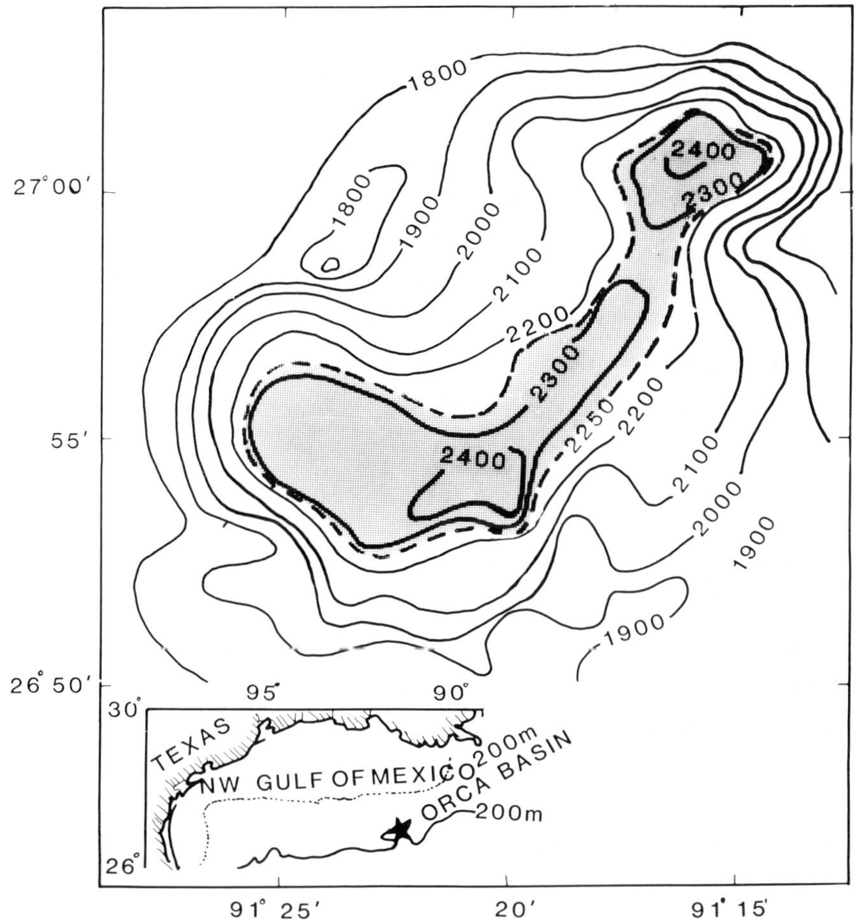

Figure 3. Location and bathymetric contours in meters of the anoxic Orca Basin in the northern Gulf of Mexico (from Sackett et al., 1979).

eliminates almost all macro- and micro-benthic life. The basin thus becomes an ideal environment for the enhanced preservation of marine organic matter. Studies of piston cores indicate that the Orca brine is only 7,900 to 8,600 years old, and that organic-rich sediments are accumulating beneath the anoxic brine at a rate exceeding 40 cm (1 foot) per thousand years (McKee et al., 1978; Addy and Behrens, 1980; Leventer et al., 1982). The sediment consists of a black mud with a) metastable metal sulfides (McKee et al., 1978), b) high water contents (Tompkins and Shephard, 1979), and c) organic carbon contents (largely of marine origin) up to and exceeding 1-4% by dry weight (Sackett et al., 1979; Northam et al., 1981). The anoxic Tyro and Bacino Bannock basins in the Eastern Mediterranean have characteristics similar to the Orca Basin (Jongsma et al., 1983; Cita et al., 1985).

Because the Orca Basin is young in age, small in extent and recently created, we believe that this intraslope basin, its brine and its depositional environment represent a modern analog for the enhanced preservation of organic carbon and the origin of potential source beds at continental slope water depths. The counter argument, that the Orca Basin is the only such basin presently known to exist, is weak when one considers the number of basins which must surely have existed during different parts of the Tertiary. It is also important to realize that hypersaline brines are not necessary in order to enhance the preservation of organic carbon in these potential basins. The average bottom water salinities and temperatures on the continental slope of the Gulf of Mexico are about 35 parts per thousand (ppt) and 4°C (Pequegnat, 1972). A salinity increase of only 2 to 5 ppt would produce an increase in the density stratification of a basin sufficient to prevent mixing. In as little as 100 years, such a stratification could lead to euxinic if not anoxic bottom conditions in the basin.

Therefore, to avoid controversy over the necessity of an hypersaline brine, we introduce the concept of <u>H</u>igh <u>S</u>alinity <u>B</u>asin <u>W</u>aters, defined as having salinity and temperature characteristics which make them more dense than the normal bottom waters of the margin. The hypersaline waters of the Orca and Mediterranean basins thus become extreme end-members of the basinal waters with density characteristics needed to enhance the preservation of marine organic carbon.

IV. HIGH SALINITY BASIN WATERS AND INTRASLOPE BASINS

In Figures 4-6, we present conceptual examples of how high saline basin waters in existing intraslope basins of the northern Gulf of Mexico could lead to the origin of organic-rich sediments. We discuss the three principal types of intraslope basins commonly found today in the northern Gulf of Mexico (Bouma, 1981; 1982). The three types are based on acoustic profiles and sediment cores. Bouma (1981) explained the variation in size and shape of these intraslope basins in terms of their mode of origin, rate of sediment in-filling, and relationship to underlying salt structures.

Type I (Fig. 5) is based on the interdomal configuration in which upward coalescing diapirs isolate a depressed, non-canyon region of the seafloor (Bouma, 1982). The present day Orca Basin is a good example of this basinal type. The Orca Basin has an approximate area of 400 sq. km and approximately 500 to 700 meters of relief from the basin rim to maximum depths (Trabant and Presley, 1978). The enclosed brine pool is constrained by the 2,230m bathymetric contour. Surficial sediments within the upper 2-6 meters consist of homogeneous black clays of hemipelagic origin with a mineralogy typical of Mississippi River sources (Trabant and Presley, 1978). Post-depositional slumping is a common occurrence in interdomal basins as seen by parabolic reflections in seismic surveys. The extent of this slumping was later confirmed by Leg 96 drilling of the Deep-Sea Drilling Project (Bouma et al., 1986). Bulk chemistry of the Orca Basin brine shows that its major constituents are consistent with leaching of either an underlying or exposed salt mass (Stokes et al., 1977). On the basis of high dissolved sulfate and low

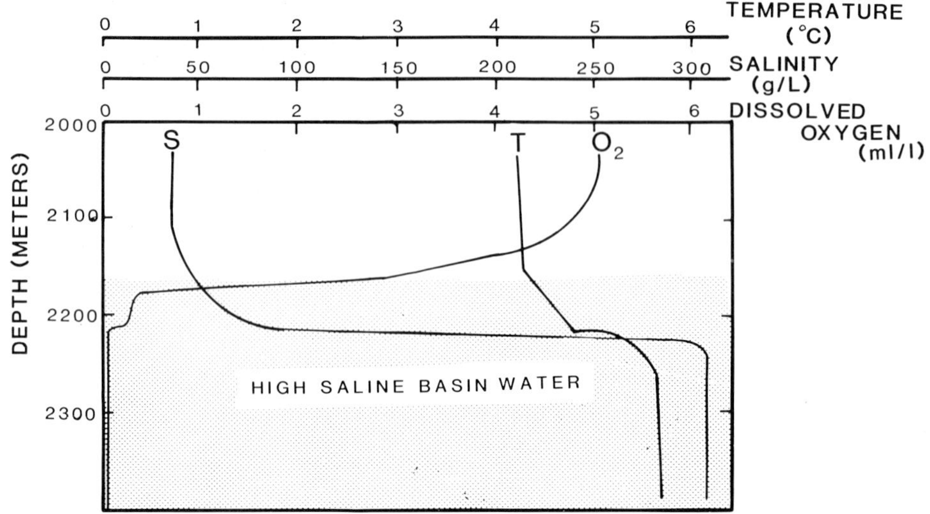

Figure 4. Temperature, salinity and dissolved oxygen profiles in the Orca Basin on the Texas-Louisiana continental slope illustrating the Orca Basin brine as an extreme end member of the high saline basin waters which we argue have periodically permeated the various types of intraslope basins in the Gulf of Mexico during the Tertiary. Water column chemical profiles modified from Trabant and Presley, 1978.

hydrogen sulfide concentrations, little sulfate reduction appears to be occurring within the Orca Basin brine (Stokes et al., 1977; Sackett, et al., 1979). Organic carbon contents in the upper 5 meters average 1.5% (range from 2.9 to 0.8%) on a salt and carbonate-free basis (Sackett et al., 1979). These organic carbon contents are considerably above the levels considered to be source sediments for oil and gas production in the Louisiana Gulf Coast region (Dow and Pearson, 1975; Dow, 1978). Beneath the black anoxic muds lie gray clays typical of oxygenated slope sediments found today in other intraslope environments (Bouma et al., 1978b).

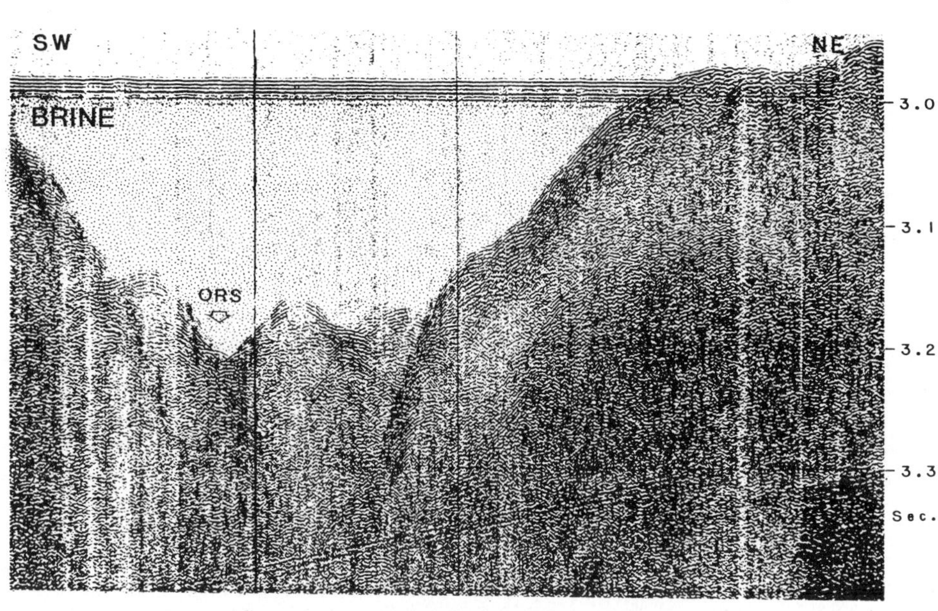

Figure 5. A seismic reflection profile across the Orca Basin to illustrate an interdomal basin as one type of intraslope basin capable of trapping High Saline Basin Waters during the Tertiary (from Bouma, 1982). The Orca Basin brine represents one end member of the high saline basin waters which we argue are needed to prevent mixing and overturning of the waters. Organic-rich sediments represent the enhanced preservation of marine organic carbon beneath the high saline basin waters as oxygen demand exceeds supply and the high saline basin waters become euxinic.

Carbon isotope values for Orca Basin organic matter are consistent with a marine origin (Sackett et al., 1979). Marine organic matter is the most ideal form for producing oil and/or gas depending on the conditions which control its preservation, its burial history and the geothermal gradient (Tissot et al., 1974; Dow, 1978). We believe that rather than being unique environments, basins with high saline basin waters have represented sites for enhanced preservation of marine organic carbon in the Gulf of Mexico throughout the Tertiary. Changes in

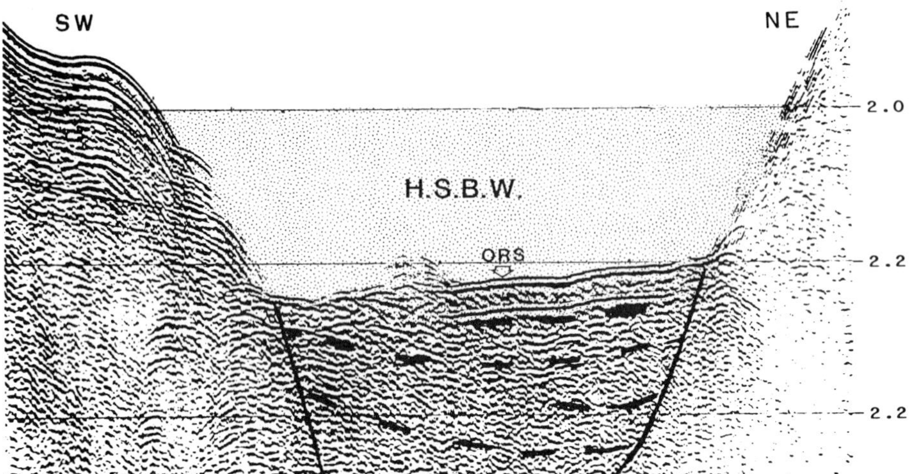

Figure 6. A seismic reflection profile across the intraslope Gyre Basin in the Gulf of Mexico (modified from Bouma, 1982). The Gyre Basin typifies basins in which the diapiric uplift of salt has blocked a former submarine canyon, thereby causing sedimentation to change from primarily turbiditic to hemipelagic. Schematically shown is the presence of an high saline basin waters beneath which organic-rich sediments begin to accumulate as the preservation of marine organic carbon becomes enhanced in the euxinic conditions of the high saline basin waters.

organic productivity of Gulf of Mexico surface waters through enhanced nutrient run-off (Leventer et al., 1983) or upwelling (Dow, 1978) would tend to reinforce our preservation mechanism but such productivity changes are largely unnecessary.

Type II basins are intraslope basins comprised of former submarine canyons blocked by salt diapirism and are represented by the Gyre Basin (27°15'N, 94°10'W) (Figure 6) (Bouma, 1982).

The Gyre Basin is located on the Texas-Louisiana slope west of the Orca Basin. Sedimentation with in Type II basins is often characterized by sequences with sandy deposits from debris flows and turbidity currents which onlap onto the upper flanks of the diapirs (Bouma, 1982). The Gyre basin has an approximate area of 260 sq. km., a maximum depth of 1,650 meters and a basinal rim which lies between 900 to 1,000 meters water depth (Bouma et al., 1978b). Seismic surveys and drilling indicate that canyon blockage by salt diapirism occurred sometime in the late Pleistocene (Lehner, 1969; Bouma et al., 1986). After blockage, the mode of sedimentation changed from debris and turbidity flows with chaotic or transparent acoustic signals to predominantly hemipelagic sedimentation with parallel reflectors on the basin floor and localized slumping with hyperbolic reflectors along the steeper parts of the basinal rim (Bouma et al., 1978b; Bouma, 1981). Although the Gyre Basin does not presently contain high saline basin waters, the basin represents an example of an hypothetical intraslope catchment basin for high saline basin waters and associated organic-rich sediments.

The third basin type with the potential of containing high saline basin waters in the intraslope region is a collapsed basin (Figure 7). This basin type is represented by the East Breaks Basin (27°45N, 94°50'W) (Bouma, 1982). Such collapsed basins are characterized by graben-like features and normal and/or growth faults at the crests of diapirs due to the dissolution of salt.

We believe that these basinal types, along with other possibilities such as trough-like features common along past continental slopes of the Gulf Coast region, all serve as potential sites for the catchment of high saline basin waters, the development of anoxic conditions appropriate for the accumulation of organic-rich sediments, and the origin of source beds through time. Due to local and regional differences in diapiric uplift and sedimentation, these basinal types would have

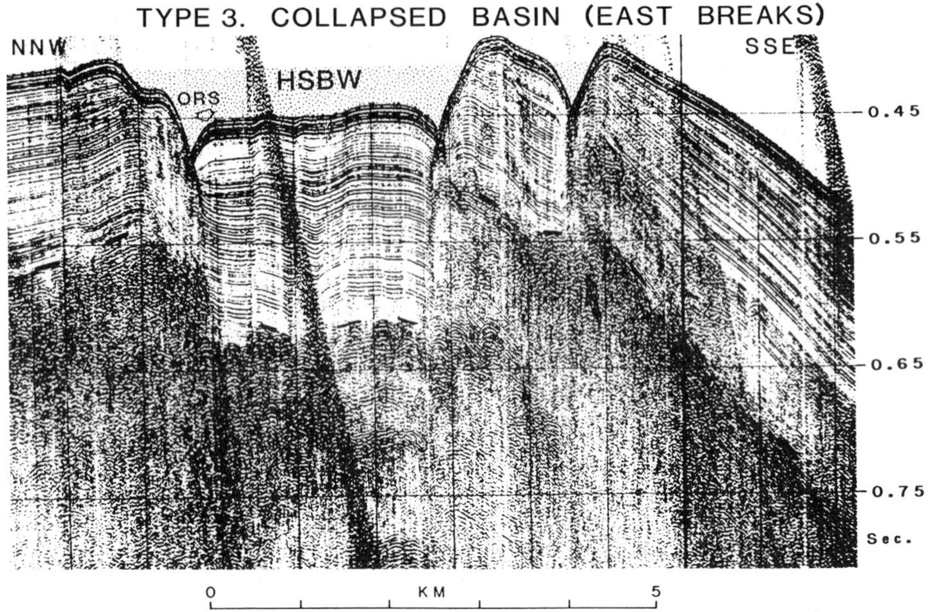

Figure 7. A seismic reflection profile across the third type of predominant intraslope basin formed by salt withdrawal in the intraslope region (from Bouma, 1982). Illustrated is the potential presence of high saline basin waters in the East Breaks collapsed basin and the subsequent accumulation of organic-rich sediments beneath the high saline basin waters (a blocked canyon basin as classified by Bouma (1981,1982)).

occurred along the paleo-slope of the Gulf Coast Region throughout its development in the Tertiary. Seismic reflection and deep drilling surveys have shown that some basins contain several thousands of meters of Pleistocene and late Tertiary sediments (Lehner, 1969; Antoine and Bryant, 1969; Garrison and Martin, 1973). The occurrence of organic-rich sediments in the vicinity of salt where the regional thermal gradient has been modified would change the timetable for maturation of the organic carbon to oil and/or gas as predicted by vitrinite reflectance (Dow, 1978).

V. THE "RADIATOR EFFECT" OF SALT DOMES: THERMAL ANOMALIES AND ENHANCED HEAT FLUX IN SURROUNDING SEDIMENTS

Relationships among time, temperature and burial depth in the conversion of organic matter to oil and gas have been established in several models (Lopatin, 1971; Tissot et al., 1974; Dow, 1978). According to these models and vitrinite reflectance data, most hydrocarbon production in the Gulf Coast Tertiary is from immature source rocks (Dow, 1978). We believe that salt domes and other structures can significantly modify these relationships by altering the local and/or regional geothermal gradient. Thus, salt diapirs would have an important impact on the maturation history of the organic-rich source beds. Previously, Rashid and McAlary (1977) found direct evidence for enhanced thermal heat fluxes above the Jurassic Argo Salt dome on the Scotian Shelf and for maturation of amorphous (marine) organic matter, in what are otherwise immature shales, into oil and gas source rocks.

The aspect of our model calling for modification of the maturation timetable for organic matter in the vicinity of salt structure is based on steady-state considerations by O'Brien and Lerche (1984). The thermal conductivity of salt is 3-5 times greater than that of sediments typical of clastic margins such as Gulf Coast Region. Salt maintains this conductivity difference even when considerations are made for variations in lithology, porosity, pore fluids, etc. Attachment of salt structures to the deeply buried Louann salt provides an efficient pathway for enhancing conductive heat flow from depth. O'Brien and Lerche (1984) have modelled the effects of salt domes on subsurface heat flow and changes in the geothermal gradient as a function of diameter and shape of the salt dome, height of the salt dome above the mother salt, and the thermal conductivity of the surrounding sediments (Figs. 8-10). This theoretical model has been confirmed by down-hole measurements made in the vicinity of a West Bay, Louisiana, salt dome (Vizgirda et al., 1985) and

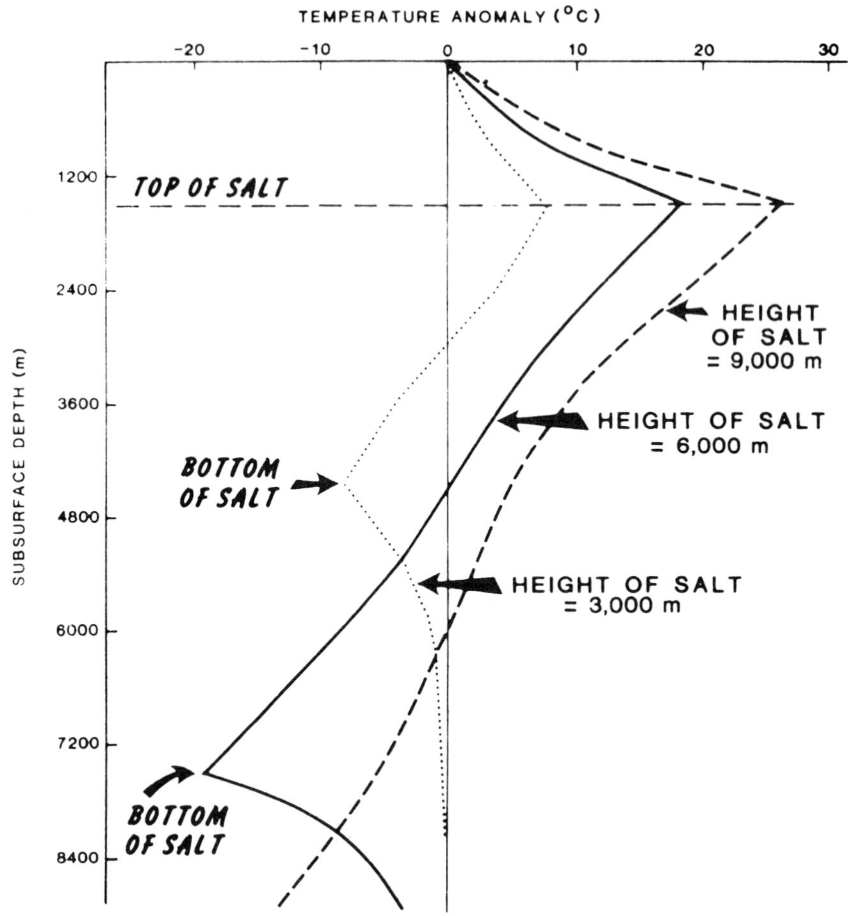

Figure 8. The predicted vertical distribution of temperature anomalies (in degrees centigrade) relative to the top of diapiric salt dome as a function of diapir height above the mother salt (3,000, 6,000 and 9,000 meters from the top of the salt dome to the underlying mother salt (from O'Brien and Lerche, 1984).

provides a functional theory for the present model. For example, changing the geothermal gradient in the vicinity of organic-rich sediments can alter the time/temperature maturation profile for a given system (Rashid and McAlary, 1977; Dow, 1978).

If one assumes steady state conservation of energy, then an enhanced heat flow relative to that determined from a regional

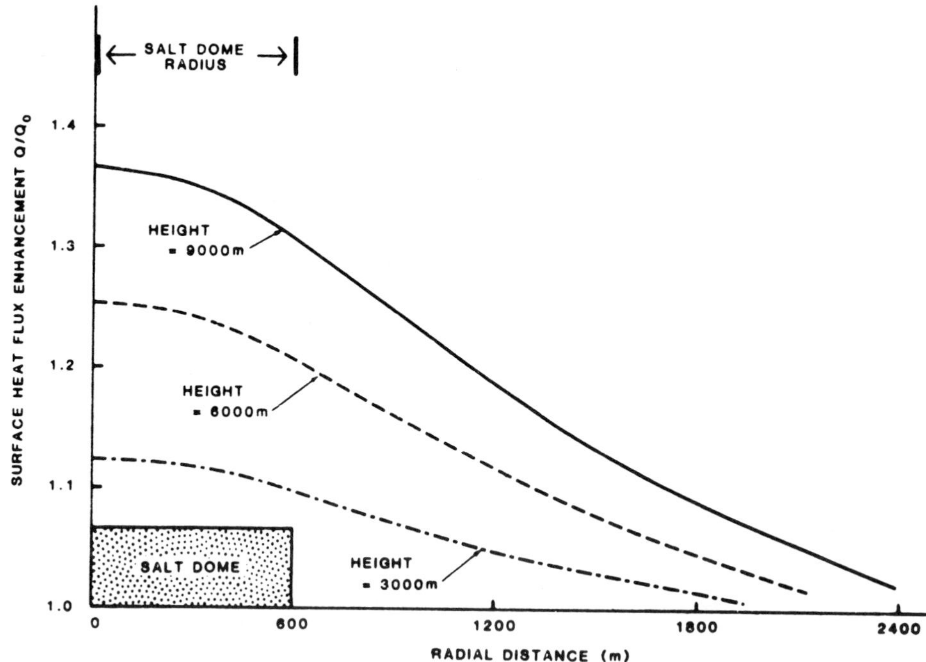

Figure 9. Enhancement of the surface heat flux (Q/Q_o) above salt domes of A) heights of 3,000, 6,000 and 9,000 meters above the mother salt for a salt dome with a radius of 600 m and a subsurface burial depth to the top of the salt at 1,600 m. B) Enhancement of surface heat flux above salt domes buried at subsurface depths of 900, 1,200 and 1,500 meters for a dome of a radius of 600 m and a height of 3,000 m from the mother salt to the top of the dome (taken from O'Brien and Lerche, 1984).

geothermal gradient will occur immediately above and below the salt dome (Figure 8). The enhanced heat flow results in positive temperature anomalies near the top of the salt dome and negative temperature anomalies along the domal flanks (Figure 8). The degree of enhancement of heat flux increases with increasing height of the salt dome structure (Figure 9A) and with decreasing subsurface burial (Figure 9B). The magnitude and exact profile of the surface heat flow anomaly depend on several factors:

1. the contrast in the thermal conductivity between salt and the surrounding sediments,
2. the height and radius of the salt dome, and
3. the thickness of the sediment overburden.

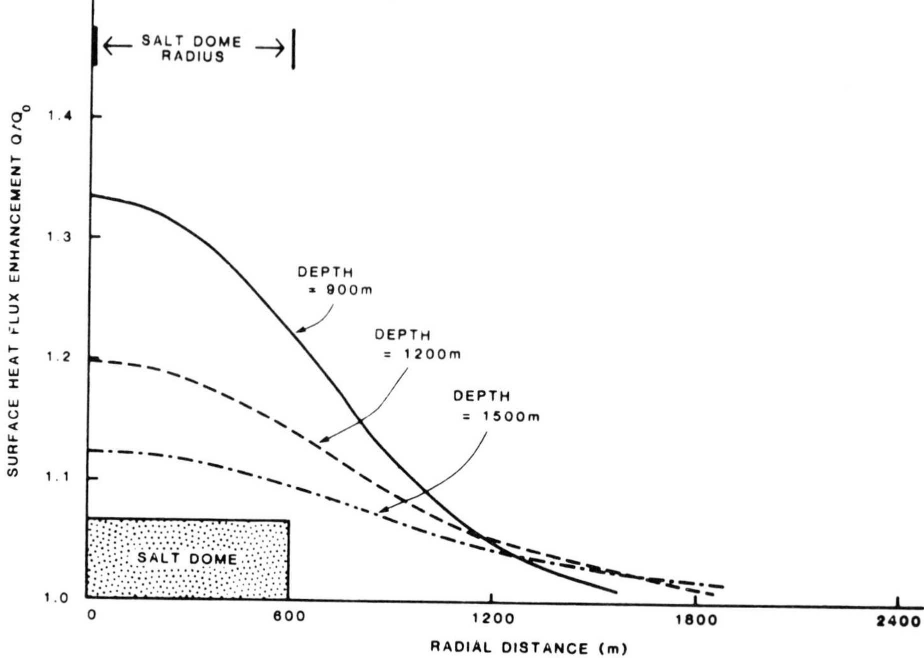

Figure 9B.

Increasing the height of the salt dome increases the magnitude of the temperature anomaly in the sediments near the top of the salt dome (Figures 8, 9A). Increasing the radius of the salt dome increases the lateral distance over which the temperature anomaly impacts the surrounding sediments (O'Brien and Lerche, 1984). In general, increased overburden thickness decreases the magnitude of the temperature anomaly at the top of the salt dome but increases the effective lateral distance over which the anomaly is felt.

The effectiveness with which salt domes conduct heat has another important effect on the regional geothermal profile. Just as enhanced heat flow increases the temperature of the sediments overlying the salt dome, the enhanced heat flow removes heat from the sediments along the lower flanks of the salt dome, causing the domal flanks to have temperatures which are lower (i.e., negative thermal anoma lies) than the regional trend (Figure 8). The consequence of this "radiator effect" of salt structures is to increase the regional temperature regime and thereby enhance the hydrocarbon maturation rate of organic-rich sediments near the top portions of the salt dome (Figure 10); the negative temperature anomalies and lowered temperatures on the lower flank of the salt dome result in a zone of inhibited hydrocarbon maturation relative to the regional behavior far from the salt dome (Figure 10). The exact temperature distribution and the magnitude of the temperature changes above or below the regional geothermal profile depend, of course, on the dimensions of the salt dome. As an example consider a salt dome with a height of 4,000 meters and a diameter of approximately 700 meters, under approximately 1,000 meters of overburden. In this case a zone of enhanced maturation with elevated temperatures 30°F above the regional profile would have a lateral and vertical extent of approximately 300 m from the flank and top of the salt dome (Figure 10). Dimensions of the zone of inhibited maturation and maximum temperature decrease on the lower flanks would be similar. Such temperature distributions have been confirmed through observations of down-hole temperatures on the upper flanks of a West Bay salt dome in Louisiana (Vizgirda et al., 1985) and the Argo salt dome on the Scotian Shelf (Rashid and McAlary, 1977).

One might ask: are the changes in geothermal gradient due to the "radiator effect" sufficiently above the regional trend to enhance significantly the rates or degree of hydrocarbon maturation? As O'Brien and Lerche (1984) point out, it is necessary to know the history of a particular salt structure's dynamics to adequately answer the question. However, the degree of maturation inhibition near the salt base opens up the question of deep reservoirs. Of course, other factors like localized

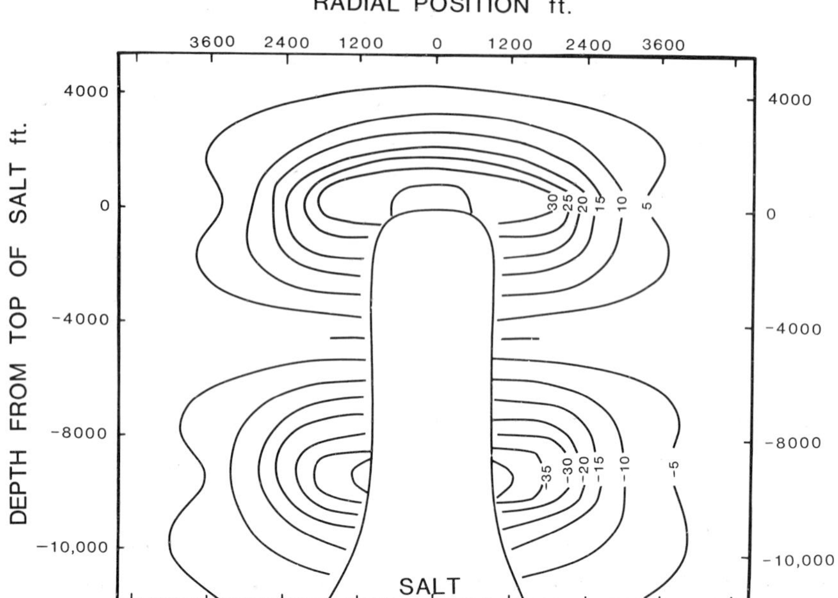

Figure 10. A contour plot of the predicted subsurface temperature anomalies (degrees centigrade above or below the regional geothermal gradient) for a salt dome of height 9,000 meters and radius 600 m buried under 1,600 m of sediment (modified from O'Brien and Lerche, 1984).

faulting, fluid flow, sediment heterogeneities, the presence of flanking diapiric shales (Atwater and Forman, 1959; Woodbury et al., 1973), the types of chemical reactions in the organic-rich sediments maturation process, the effects of elevated salinities in pore fluids on the rates of organic-rich sediments maturation are all important complications in need of more research.

VI. SCHEMATIC REPRESENTATION OF SALT MOVEMENT, BASINAL DEVELOPMENT AND ORGANIC-RICH SOURCE BEDS ON THE GULF COST MARGIN

In Figures 11-13, we illustrate schematically how the geographic distribution of salt structures and associated basins along a prograding margin would lead to the stratigraphic occurrence of alternating deep-water shale/sand sequences, High Saline Basin Waters and Organic-Rich Sediments. The schematic drawings illustrate hypothetical intraslope basins along a transect on the Texas-Louisiana upper continental slope during several Miocene through Pliocene time slices. For discussion purposes, the initial time period (t_o) is late Miocene, showing the upper slope as it might have existed in water depths of 1000 to 3000 meters (3,000 to 9,000 feet) (Figure 11). The three dimensional sketches are schematic for it is not possible to show to scale all the structural and tectonic processes related to the salt movement. Paleobathymetric and stratigraphic reconstructions of offshore and onshore Louisiana well data (Stude, 1978; Johnson and Bredeson, 1971; among others) show that the present-day Texas-Louisiana coastal basin and continental shelf were shallow (shallow (shelf) to deep-water (slope) marine environments) during the Miocene. The Miocene salt structures were primarily aligned in elongate, laterally extensive salt ridges, with occasional pillow and domal structures (Stude, 1978). Salt flowage and the beginning of diapirism (either shale or salt) created subsiding sedimentary basins of varying sizes. These basins may be in synclinal depressions, grabens, collapse depressions related to faulting or blocked submarine canyon systems (Figure 11). Such sedimentary basins were the catchment sites at various times for brines or high saline basin waters formed via dissolution of either subsurface or exposed salt structures on the margin. Numerous brine lakes formed in collapse features on the shelf as well (Rezak and Bright, 1981; Brooks et al., 1979). Brine lakes themselves may act as feeder systems for producing dense brines or high saline basin waters that spill over the rims of faulted depressions and flow down into the troughs and basins on the intraslope region. These intraslope brines collected in the basins, increased the salinity of the bottom waters contained therein and thus created high

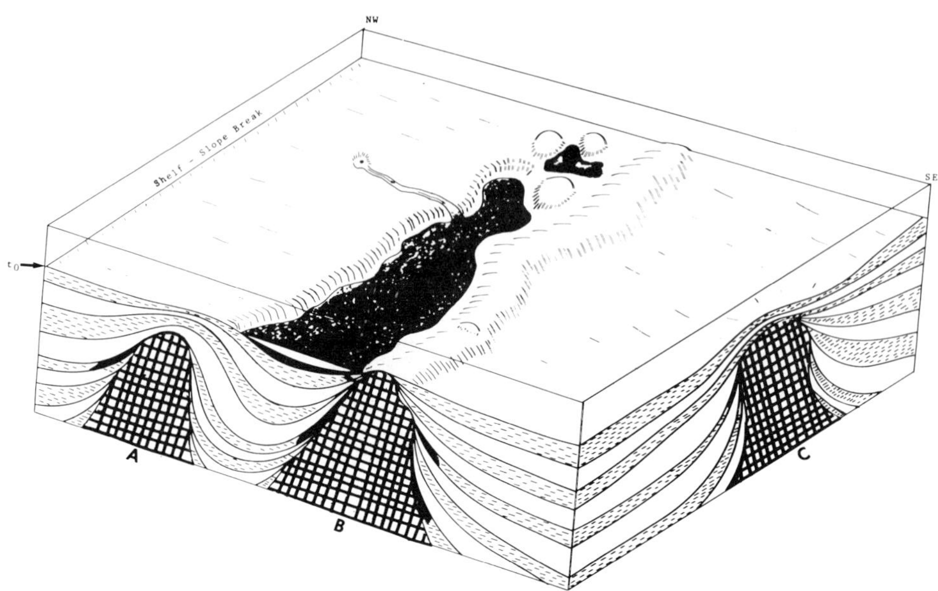

Figure 11. A schematic representation of the Texas-Louisiana continental slope during the late Tertiary illustrating the relationship between salt movement, intraslope basinal development and the origin of organic-rich source beds on the Gulf Coast margin. This figure represents a hypothetical Miocene time (t_o) and the occurrence of elongate linear salt ridges on the continental slope. Previously deposited alternating sand and shale magnafacies are shown vertically. Organic-rich sediments are shown in black; salt structures are hatchured.

saline basin waters. Thus, the basins would become ideal loci for the accumulation and enhanced preservation of marine organic carbon. Shaley deep-water organic-rich sediments thus accumulated for periods ranging from 10,000 to 100,000 years. The duration of each basin, its lateral extent and the maintenance of the correct preservational characteristics are all controlled by the depositional characteristics and tectonic settings along the prograding margin. At any given location, the proper oceanographic and depositional conditions may occur only once in a million years or as frequently as several times in a given 100,000 year period.

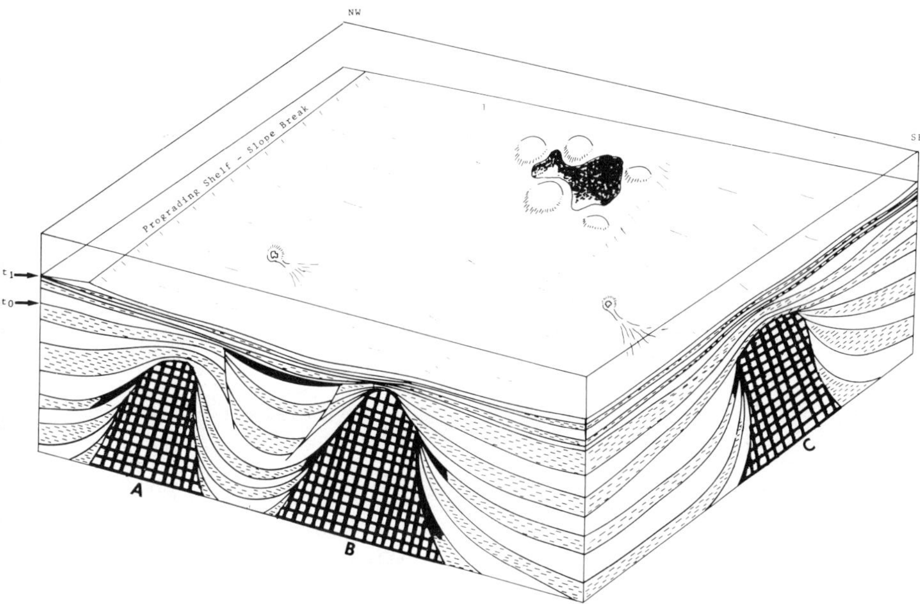

Figure 12. A schematic representation of the Texas-Louisiana continental slope during the early Pliocene to middle Pliocene (t_1 depositional surface). Progradation of the shelf slope break has occurred as the Pliocene depocenter begins to develop further out on the margin. The previous linear basins with high saline basin waters have become covered leaving occasional seeps and possibly brine lakes on the margin and only occasional intraslope basins like the intradomal basin shown here schematically. Salt structures A and C have remained relatively stationary from t_0 to t_1 while salt structure B has undergone some penecontemporaneous piercement of the overlying strata.

With progradation of the shelf-slope break in a southeastward direction during early Pliocene time (t_1 time surface) (Figure 12), the organic-rich sediments are buried by a shale-sand magnafacies. Salt movement can be temporarily halted, and areas of the margin will experience penecontemporaneous or contemporaneous uplift. Later stages of salt development give rise to thin elongate spines or salt domes, primarily dependent on the thickness and rate of deposition of the shale-sand magnafacies. Brine seeps and brine lakes continue to be active along various parts of the margin, as in regions of the

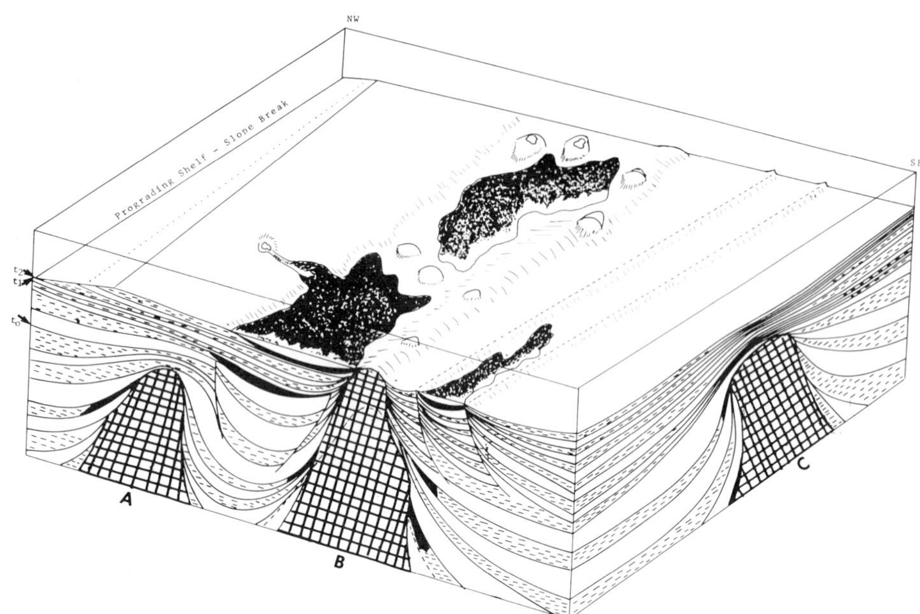

Figure 13. This schematic represents time (t_2) or approximately late Pliocene to early Pleistocene time when continued progradation of the shelf-slope break occurred as a result of rapid deposition on the continental margin. Salt structures A and C have remained relatively stationary while salt structure B has continued to pierce the overlying strata. In this scenario growth faulting further down the margin has produced another linear basin in which dissolved salt products either from underlying salt, exposed salt or brine seeps could infill the linear feature with an high saline basin waters. Between salt structure A and B large intraslope basins have developed once again and have been infilled with an high saline basin waters under which organic-rich sediments begin to accumulate.

continental shelf today (Rezak and Bright, 1981; Brooks et al., 1979). During some time periods, the intraslope region may have witnessed an apparent dearth of anoxic depressions with high saline basin waters.

With further development of the prograding margin to late Pliocene or early Pleistocene time (t_2 time slice) (Figure 13), the shelf-slope break continues to move southeastward. The Pleistocene depocenter begins to shift slightly southeastward relative to the entire Pliocene depocenter (Figure 2). The t_1

time surface is covered by upwards thinning sand-shale sequences related to depositional cycles caused by glaciation of North America. Large-scale fluctuations in glacio-eustatic sea levels during the Pleistocene provide an explanation for the thick Pleistocene sands (Beard et al., 1982; Shackleton and Opdyke, 1976; Williams, 1984; Williams and Trainor, 1986). New intraslope depressions develop, depending upon salt movement, and some provide anoxic depositional environments for the occurrence of organic-rich sediments. As depicted in Figure 13, new anoxic high saline basin waters may develop either from brine seeps or salt dissolution of either subsurface or exposed salt structures. To illustrate differential vertical uplift along a given segment of continental margin (Fig. 13), salt domes/ridges/pillows A and C have remained relatively stationary with further burial from t_o to t_2 time, but the salt ridge at B and the spines/domes to the north of B and part of ridge B have now pierced the t_1 to t_2 sediments. This piercement has the effects of bringing the earlier formed (t_o and pre-t_o) organic-rich sediments from the zones of high heat flow and enhanced thermal maturation at the tops of the salt structures, down to the flanks of the salt structure where the thermal gradients and thus maturation rates are dampened relative to the regional trend. These dampened rates may still be somewhat higher that maturation rates of sediments at shallower depths behaving under regional thermal conditions. The duration in which a particular source bed is positioned in a given maturation zone will determine the presence/absence and proportion of oil and natural gas in former intraslope regions which are now buried beneath continental fluvial clastic sediments of the Texas-louisiana coastal basins and continental shelf.

To illustrate the vertical (stratigraphic) sequence of events for the occurrence of high saline basin waters and organic-rich sediments in intraslope basins and maturation of the organic-rich sediments to hydrocarbons in a prograding margin, we use the schematic illustrations of Stude (1978, his Figure 26, p. 644) (Figure 14). Stude's (1978) paleobathymetric study of the late Miocene to Pleistocene age 20,000 feet of alternating sands and shales of the South Timbalier 54 field areas, and of other producing salt domes in the coastal region and continental shelf of Louisiana, led him to propose several general models for the relationship of salt dome evolution and depositional environments during sediment progradation into a deep water basin. During stage 1 (Figure 14), late Miocene (E) time, salt structures in lower slope-upper abyssal depths of 1,000 to 2,500 meters (3000-8000 feet) resemble the linear swells or ridges depicted in Figure 11. During this time, high saline basin waters and organic-rich source beds could have formed in intraslope depressions of varying sizes and along various portions of the margin. Mounding of the salt structure in stage II during early to middle Pliocene (D-B) time would bury the organic-rich sediments of E time. Uplift of the salt mound would bring the organic-rich sediments of E time into the thermal halo at the top of the salt structure. This zone of enhanced heat flux would begin the thermal maturation process at an earlier time and shallower depth than predicted using the accepted regional thermal gradient-time relationship (Dow, 1978). During the deposition and uplift of sediment units D-B, faulting and domal development would create new depositional opportunities (i.e., new intraslope depressions) for the occurrence of high saline basin waters and organic-rich sediments (Figure 14).

Continued progradation of the margin and deposition of thick shale-sand magnafacies, which become increasingly more sandy into the late Pliocene-early Pleistocene, lead to (1) evolution of the salt mound into a more spine-like structure (stage III) in upper slope water depths of 300 to 3000 feet (100 to 1000 meters), (2) the burial of organic-rich sediments from D, C and/or B times by

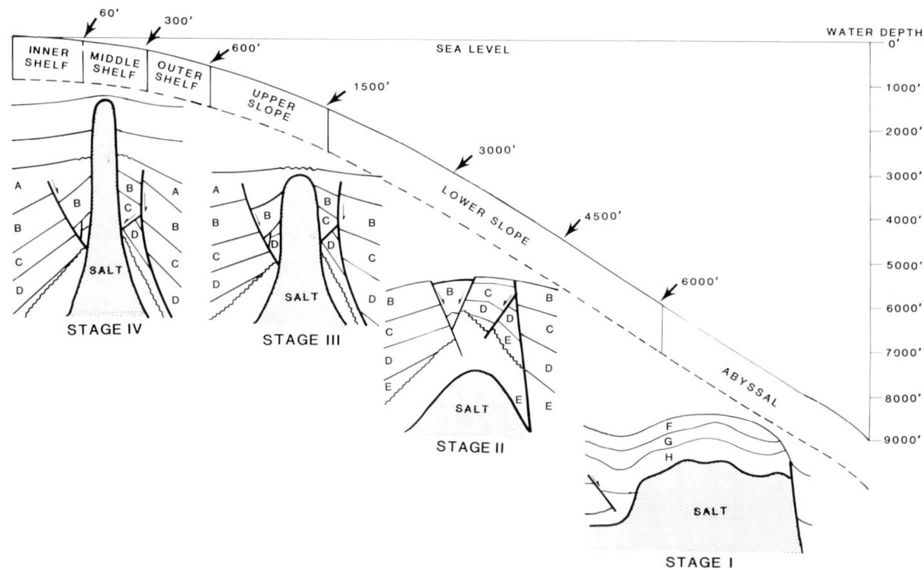

Figure 14. A hypothetical representation of the deposition of organic-rich sediments (blackened zones) in intraslope basins and the evolution of their maturation history as a function of salt movement and margin progradation. The structural reconstructions are directly from Stude (1978) for a salt dome from the South Timbalier Block 54 salt dome field, offshore, Louisiana.

strata of A time (B time shown in Figure 14 for simplicity), (3) movement of the now matured hydrocarbons of E time into the zone of dampened maturation along the dome flank (Figure 10) and (4) the implacement of organic-rich sediments from B time in the thermal halo at the top of the dome (Figure 14).

Evolution of the salt structure from stage III to an elongated dome-like structure (spine) in stage IV occurs as sands and fluvial-clastic deposits prograde onto the shelf (Figure 14) (Stude, 1978). In our model, we would predict that mature hydrocarbons and oil/gas would migrate from source beds originating during E and B time into producing reservoirs of sands (early Pliocene D time or early Pleistocene A time) depending on the time of organic-rich sediments formation and the maturation history of the organic-rich sediments relative to the uplift history of the related salt structure.

VII. CONCLUSIONS

For simplicity, we have presented basic stratigraphic, structural and depositional environments to illustrate our model. We recognize that different depositional environments existed during the Tertiary which are much more complex. We believe that the model's strengths rest on its simplicity, its ability to explain many known associations between oil/gas and salt in the Gulf Coast region, and its unifying theme. The model incorporates and integrates the periodic accumulation of Organic-Rich Sediments in anoxic intraslope basins with a mechanism for enhancing the maturation of hydrocarbons due to modification of the regional thermal gradient by salt conductivity. Our overriding working hypothesis is that salt domes and other salt structures provide more than the structural framework for hydrocarbon traps; they serve as the loci for source beds, the conduits for the heat needed for hydrocarbon maturation at earlier times and shallower depths than predicted using normal time-depth relationships (Dow, 1978). Our model is therefore consistent with (1) known production trends in the Gulf Coast region from "immature" reservoirs based on vitrinite reflectance considerations (Dow, 1978), (2) the occurrence of oil and gas production in this region and its structural evolution throughout the Tertiary, and (3) a mechanism by which marine, not terrestrial, organic carbon can be preserved in quantities sufficient to produce the hydrocarbons of commercial potential.

Our empirical model can be summarized as follows:

(1) Salt diapirism and salt flowage create intraslope depressions, grabens and subsiding sedimentary basins which are the ideal loci for the accumulation of marine organic carbon.

(2) In turn the salt structures are the source of the dissolved salts to produce High Saline Basin Waters which, in turn, are capable of providing density stratified depositional environments in intraslope basins. These environments enhance the preservation of marine organic carbon on continental slopes, at levels sufficient to produce organic-rich source beds in the range of 1 to 4 % amorphous marine organic carbon.

(3) The interdomal Orca Basin, the collapsed basin at East Breaks and the blocked canyon basin in the Gyre Basin, each constitute present day analogs to the types of intraslope basins which have served as loci for high saline basin waters and organic-rich sediments in the Tertiary. The hypersaline brine of the Orca Basin is one type of high saline basin waters whose water chemistry inhibits oxygen replenishment and thus produces an euxinic or anoxic environment ideal for the enhanced preservation of marine organic carbon.

(4) The higher thermal conductivity of salt enables salt structures to act as geological "radiators", producing thermal "halos" which enhance thermal anomalies near the top of the salt stock and diminish the thermal values near the salt base below the regional temperatures. These thermal halos thereby modify the time/temperature/depth relationships controlling the rate of hydrocarbon maturation in the Gulf Coast region.

(5) The maturation of organic carbon to oil, gas or methane in the salt provinces of the Gulf Coast region depends on the length of time in which the organic-rich sediments remain in the enhanced maturation window of the thermal halo at the top of a salt dome. The salinity of the high saline basin waters may also play a role in the maturation process by (a) chemistry, (b) porosity cementation, (c) "salting out" of hydrocarbons, and (d) changing fluid flow patterns because of density changes and thermal expansion changes. While we recognize these other parameters are important, they have not been considered in detail

in this paper. A more detailed discussion of their role will be forthcoming. Our ultimate goal is to determine a predictive model for future drilling efforts in the vicinity of salt structures of the Gulf Coast region.

VII. ACKNOWLEDGEMENTS

An outline of this model was presented at the 1984 annual meeting of the American Association of Petroleum Geologists in San Antonio, Texas. Dwight Trainor is thanked for assistance in assembling and presenting the poster. We also acknowledge discussions with Robert Ehrlich, Chris Kendall, and Howard Spero at the Department of Geology, USC, Richard Fillon of Texaco, and A. Bouma, J. O'Brien and other erstwhile colleagues at Gulf Research and Development Corporation (R. I. P.), Houston and Pittsburgh.

REFERENCES

Addy, S.K., and Behrens, E.W., (1980). Time of accumulation of hypersaline anoxic brine in Orca Basin (Gulf of Mexico): Marine Geology 37, 241-252.

Antoine, J.W. and Ewing, J., (1963). Seismic Refraction Measurements on the Margins of the Gulf of Mexico, 1: Geophys. Res. 68, 1975-1996.

Antoine, J.W. and Bryant, W. R., (1969). Distribution of salt and salt structures of the Gulf of Mexico: American Association of Petroleum Geologists Bulletin 53, 2543-2550.

Atwater, G.I. and Forman, M.J., (1959). Nature of growth of southern Louisiana salt domes and its effect on petroleum accumulation: American Association of Petroleum Geologists Bulletin 43, 2592-2622.

Beard, J.H., Sangree, J.B., and Smith, L.A., (1982). Quaternary chronology, paleoclimate, depositional sequences and eustatic cycles. American Association of Petroleum Geologists Bulletin 66, 158-169.

Berg, R. R., (1981). Deep-water reservoir sandstones of the Texas Gulf Coast. Transactions Gulf Coast Association of Geol. Societies 31, 31-39.

Bishop, R.S., (1978). Mechanism for emplacement of piercement diapirs, American Association Petroleum Geologists Bulletin 9, 1561-1583.

Bouma, A.H., (1981). Depositional sequences in clastic continental slope deposits, Gulf of Mexico: Geomarine Letters 1, 115-121.

Bouma, A.H., (1982). Intraslope basins in northwest Gulf of Mexico: key to ancient submarine canyons and fans, in J.S. Watkins and C.L. Drake, Geologic Evolution of Continental Margins, American Association Petroleum Geologists Memoir 34, 567-581.
Bouma, A.H., Moore, G.T., and Coleman, J.M., (1978a). Framework, Facies, and Oil-Trapping characteristics of the upper continental margin: American Associaton of Petroleum Geologists, Studies in Geology 7, 334 p.
Bouma, A.H., Smith, L.B., Sidner, B.R. and McKee, T.R., (1978b). Intraslope basin in northwest Gulf of Mexico, in A.H. Bouma, G.T. Moore, and J.M. Coleman, eds., Framework, facies, and oiltrapping characteristics of the Upper continental margin, American Assoc. Petro. Geol. Studies in Geology 7, 289-302.
Bouma, A.H., Stelting, C.E., and Sedimentologists, (1986). Seismic stratigraphy and sedimentary processes in Orca and Pigmy Basins, In: Initial Reports of the Deep-Sea Drilling Proj., ed. by A.H. Bouma and J. Coleman, V. 96, US Gov't. Print. Office, Washington, D.C., in press.
Brooks, J.M., Bright, T.J., Bernard, B.B., and Schwab, C.R., (1979). Chemical aspects of a brine pool at the East Flower Garden Bank, northwestern Gulf of Mexico: Limnology and Oceanography 24, 735-745.
Bruce, C.H., (1973). Pressured shale and related sediment deformation: Mechanism for development of regional contemporaneous faults: American Association of Petroleum Geologists Bulletin 57, 878-886.
Cita, M.B., Kastens, K.A., McCoy, F.W., Aghib, F., Cambi, A., Camerlenghi, A., Corselli, C., Erba, E., Giamastiani, M., Herbert, T., Leon, C., Malinverno, P., Nostto, A., and Parisi, E. (1985). Gypsum precipitation from cold brines in an anoxic basin in the eastern Mediterranean. Nature 314, 152-154.
de Lange, G.J. and ten Haven, H.L. (1983). Recent sapropel formation in the eastern Mediterranean. Nature 305, 797-798.
Dow, W.G. (1978). Petroleum source beds on continental slopes and rises: American Association of Petroleum Geologists Bulletin 62, 1584-1606.
Dow, W.G. and Pearson, D.B. (1975). Organic matter in Gulf Coast sediments. Preprints, Seventh Annual Offshore Technology Conf., paper no. Offshore Technology Conference 2343.
Emery, K.O. and Uchupi (1972). Western North Atlantic Ocean: topography, rocks, structure, water, life and sediments, American Association Petroleum Geologists Memoir 17, 532 p.
Ewing, M., and Antoine, J. (1966). New seismic data concerning sediments and diapiric structures in Sigsbee deep and upper continental slope, Gulf of Mexico: American Association of Petroleum Geologists Bulletin, 50, 479-504.
Foote, R.Q., Martin, R.G., and Powers, R.B. (1983). Oil and gas potential of the Maritime Boundary Region in the Central Gulf of Mexico, American Association Petroleum Geologists Bulletin 67, 1047-1065.
Garrison, L.E., and Martin, R.G. (1973). Geologic structures in the Gulf of Mexico Basin: U.S. Geological Survey Professional Paper 773, 85 p.
Gealy, B.L. (1955). Topography of the continental slope in northwest Gulf of Mexico: Geological Society of America Bulletin 66, 203-227.
Halbouty, M.T. (1979). Salt Domes, Gulf Region, United States and Mexico, second ed., Gulf Publishing Co., Houston, Texas.
Halbouty, M.T. and Hardin, G.C., Jr., (1956). Genesis of Salt Domes of the Gulf Coastal Plain, American Association Petroleum Geologists Bulletin 40, 737-746.

Hanna, M.A. (1934). Geology of the Gulf Coast salt domes, in, W. Wrather and F. Lahee, eds., Problems of Petroleum Geology: American Association of Petroleum Geologists Bulletin, 629-678.

Hardin, F.R., and Hardin, G.C., Jr. (1961). Contemporaneous normal faults of Gulf Coast and their relation to flexures: American Association of Petroleum Geologists Bulletin 45, 238-248.

Jackson, M.P.A. and Seni, S.J. (1983). Geometry and evolution of salt structures in a marginal rift basin of the Gulf of Mexico. Geology 11, 131-135.

Johnson, H.A., and Bredeson, D.H. (1971). Structural Development of Some Shallow Salt Domes in Louisiana Miocene Productive Belt, American Association Petroleum Geologists Bulletin55 204-226.

Jongsma, D., Fortuin, A.R., Huson, W., Troelstra, S.R., Klaver, G.T., Peters, J.M., van Harten, D., de Lange, G.J., and ten Haven, L. (1983). Discovery of an anoxic basin within the Strabo Trench, eastern Mediterranean. Nature 305, 795-797.

Lehner, P. (1969). Salt Tectonics and Pleistocene Stratigraphy on continental slope of Northern Gulf of Mexico, American Association Petroleum Geologists Bulletin 53, 2431-2479.

Leventer, A., Williams, D.F., and Kennett, J.P. (1982). Relationships between anoxia, glacial meltwater and microfossil preservation in the Orca Basin, Gulf of Mexico. Marine Geology 53, 23-40.

Leventer, A., Williams, D.F. and Kennett, J.P. (1982). Dynamics of the Laurentide Ice Sheet during the last deglaciation: Evidence from the Gulf of Mexico, Earth and Planetary Science Letters, 59, 11-17.

Levorsen, A. (1954). Geology of Petroleum, W.H. Freeman and Company, San Francisco, CA.

Lopatin, N.V. (1971). Temperature and geologic time as factors in coalification. Akad. Nauk SSSR Izv. Ser. Geol. 3, 95-106.

Martin, R.G. (1973). Salt structure and sediment thickness, Texas- Louisiana continental slope northwestern Gulf of Mexico: U.S. Geological Survey Open File Report.

Martin, R.G. (1981). Regional Geology of the Gulf of Mexico, in, R.B. Powers, ed., Geologic framework, Petroleum potential, Petroleum-Resource estimates, Mineral and geothermal resources, Geologic Hazards, and deep-water drilling technology of the Maritime Boundary region in the Gulf of Mexico: U.A. Geological Survey Open File Report, 81-265, 19-29.

Martin, R.G., and Bouma, A.H. (1978). Physiography of Gulf of Mexico, in, A.H. Bouma, G.T. Moore and J.M. Coleman, editors, Framework, Facies, and Oil Trapping Characteristics of the Upper Continental Margin, American Assoc. Petroleum Geologists Studies in Geology 7, 3-19.

McGookey, D.P. (1975). Gulf Coast Cenozoic Sediments and structure: An excellent example of extra-continental sedimentation. Transactions Gulf Coast Association of Geological Societies 28, 627-646.

McKee, T.R., Jeffrey, L.M., Presley, B.J., and Whitehouse, U. G. (1978). Holocene sediment geochemistry of continental slope and intraslope basins areas, Northwest Gulf of Mexico, in, A.H. Bouma, G.T. Moore, and J.M. Coleman, eds., Framework, Facies, and Oil Trapping Characteristics of the Upper Continental Margin, American Association Petroleum Geologists Studies in Geology 7, 313-326.

Miller et al. (1975). Geological estimates of undiscovered recoverable oil and gas resources in the United States: U.S. Geological Survey Circular 725, 78 p.

Murray, G.E. (1961). Geology of the Atlantic and Gulf coastal province of North America: New York, Harper and Row Bros.

Murray, G.E. (1966). Salt structures of Gulf of Mexico Basin - A review: American Association of Petroleum Geeologists Bulletin 50, 439-478.
Northam, M.A., Curry, D.J., Scanlan, R.S. and Parker, P.L. (1981). Stable carbon isotope ratio variations of organic matter in Orca Basin sediments. Geochimica et Cosmochima Acta 45, 257-260.
O'Brien, J.J. and Lerche, I. (1984). Influence of salt domes on paleotemperature distributions, Geophysics 49, 2032-2043.
Pequegnat, W.E. (1972). A deep bottom current on the Mississippi Cone. In: Contributions on the Physical Oceanography of the Gulf of Mexico, ed. by L.R.A. Capurro and J.L. Reid, Gulf Publishing Co., Houston, TX, 65-87.
Powell, L.C. and Woodbury, H.O. (1971). Possible future Petroleum potential of Pleistocene, western Gulf Basin, in, Future Petroleum Provinces of the United States - their Geology and Potential: American Association of Petroleum Geologists Memoir 15, 813-823.
Rainwater, E.H. (1964). Regional stratigraphy of the Gulf Coastal Miocene: Gulf Coast Association Geological Societies Transactions 14, 81-124.
Rashid, M.A., and McAlary, J.D. (1977). Early Maturation of organic matter in genesis of hydrocarbons as a result of heat from a shallow piercement salt dome. Journal of Geochemical Exploration 8, 549-569.
Rezak, R.N., and Bright, T.J. (1981). Seafloor instability at east flower Garden Bank, northwest Gulf of Mexico, Geology Marine Letters 1, 97-103.
Rice, D.D., (1980). Chemical and isotopic evidence of the origins of natural gases in offshore Gulf of Mexico, Transactions-Gulf Coast Association of Geological Societies 30, 203-213.
Sabate, R.W. (1968). Pleistocene Oil and Gas in Coastal Louisiana. Transactions-Gulf Coast Association of Geological Societies 18 373-386.
Sackett, W.M., Brooks, J.M., Bernard, B.B., Schwab, C.R., Chung, H. and Parker, R.A. (1979). A carbon inventory for Orca Basin Brines and sediments, Earth and Planetary Letters 44, 73-81.
Seni, S.J. and Jackson, M.P.A. (1983a). Evolution of salt structures, East Texas Diapir Province, part I: sedimentary record of halokinesis. American Association of Petroleum Geologists Bulletin 67, 1219-1244.
Seni, S.J. and Jackson, M.P.A. (1983b). Evolution of salt structures, East Texas Diapir Province, part 2: patterns and rates of halokinesis. American Association of Petroleum Geologists Bulletin 67, 1245-1274.
Shackleton, N.J. and Opdyke, N.D. (1976). Oxygen-isotope and paleomagnetic stratigraphy of Pacific core V28-239: late Pliocene-earliest Pleistocene. Geological Society of America Memoir 145, 449-464.
Shokes, R.F., Trabant, P.K., Presley, B.J. and Reid, D.F. (1977). Anoxic hypersaline basins in the Northern Gulf of Mexico, Science 196, 1443-1446.
Smith, D.A., and Reeves, F.A.E. (1970). Salt piercement in shallow Gulf Coast salt structures. American Association of Petroleum Geologists Bulletin 54, 1271-1289.
Stude, G.R. (1978). Depositional environments of the Gulf of Mexico South Timbalier Block 54 salt dome growth models, Transactions-Gulf Coast Association of Geological Societies 28, 627-646.
Tissot, D., et al. (1974). Influence of nature and diagenesis of organic matter and formation of petroleum: American Association of Petroleum Geologists Bulletin 58, 499-506.

Tompkins, R.E., and Shephard, L.E. (1979). Orca Basin: Depositional Processes, Geotechnical properties and clay mineralogy of Holocene sediments within an anoxic hypersaline basin, northwest Gulf of Mexico: Marine Geology 33, 221-238.

Trabant, P.K., and Presley, B.J. (1978). Orca Basin, Anoxic depression on the continental slope, Northwest Gulf of Mexico, in, A.H. Bouma, G.T. Moore, and J.M. Coleman, editors, Framework, facies and oil trapping characteristics of the upper continental margin, American Association Petroleum Geologists Studies in Geology 7, 303-311.

Vizgirda, J., O'Brien, J.J., and Lerche, I. (1985). Thermal anomalies on the flanks of a salt dome: Geothermics 14, 553-565.

Williams, D.F. (1984). Correlation of Pleistocene marine sediments of the Gulf of Mexico and other basins using oxygen isotope stratigraphy, In: Principles of Pleistocene Stratigraphy applied to the Gulf of Mexico, N. Healy-Williams, ed., IHRDC Press, Boston, MA, pp. 67-118.

Williams, D.F. and Trainor, D.M. (1986). Application of isotope chronostratigraphy in the northern Gulf of Mexico. Transactions Gulf Coast Association of Geological Societies, in press.

Woodbury, H.O., Murray, I.B., Jr., and Osborne, R.E. (1980). Diapirs and their relation to hydrocarbon accumulation, In: Facts and principles of world petroleum occurrence, A.D. Maill, ed., Calgary, Canadian Society of Petroleumj Geologists, Calgary, p. 119-142.

Woodbury, H.O., Murray, I.B., Pickford, P.J., and Akers, W.H. (1973). Pliocene and Pleistocene depocenters, outer continental shelf, Louisiana and Texas: American Association Petroleum Geologists Bulletin 57, 2428-2439.

MODEL FOR THE ORIGINS OF GEOPRESSURED BRINES, HYDROCARBONS, CAP ROCKS AND METALLIC MINERAL DEPOSITS: GULF COAST, U.S.A.

Malcolm P. R. Light
Harry H. Posey

Bureau of Economic Geology
The University of Texas at Austin
Austin, Texas 78713

J. Richard Kyle

Department of Geological Sciences
The University of Texas at Austin
Austin, Texas 78713

Peter E. Price

Marathon Oil Company
Denver Research Center
P.O. Box 269
Littleton, Colorado 80160

I. INTRODUCTION

A. General

Land (1984) has outlined evidence for vertical movement of diagenetic fluids at least several kilometers from Jurassic to Pleistocene formations in the Gulf Coast. This evidence includes the following: (1) discharge at the land surface of Mesozoic-derived brines as "bad water"; (2) emplacement of Mississippi Valley-type iron-zinc-lead mineralization by fluids derived from Mesozoic formations in salt dome cap rocks at or near the land surface; (3) emplacement of uranium in Tertiary aquifers as a result of reduction by ascending reduced sulfur, presumably of Mesozoic origin; (4) emplacement of calcite cement derived from

Mesozoic strata in Tertiary sandstones; and (5) presence of fluids in Plio-Pleistocene rocks with chemical signatures that could only have been derived from Mesozoic strata.

An integrated hydrothermal model has been developed that shows how the sequence of diapirism, cap rock formation, and cap rock mineralization may be related to the sequential generation of diagenetic minerals, hydrothermal fluids and hydrocarbons that form as a consequence of source rock and 'mother' salt burial. This model draws principally from studies of the Gulf Coast (Fig. 1) but should be applicable to other subsiding sedimentary basins undergoing salt deformation and hydrocarbon maturation. A

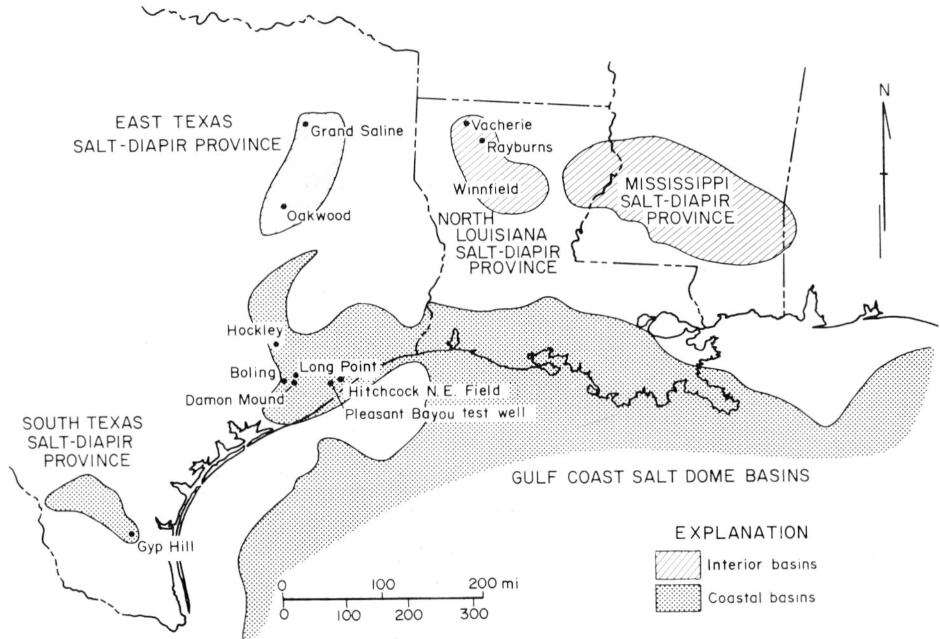

Figure 1. Location of Pleasant Bayou geopressured-geothermal test well, and Hitchcock N.E. field compared to the interior and exterior salt dome basins (modified from Kyle after Halbouty (1979)).

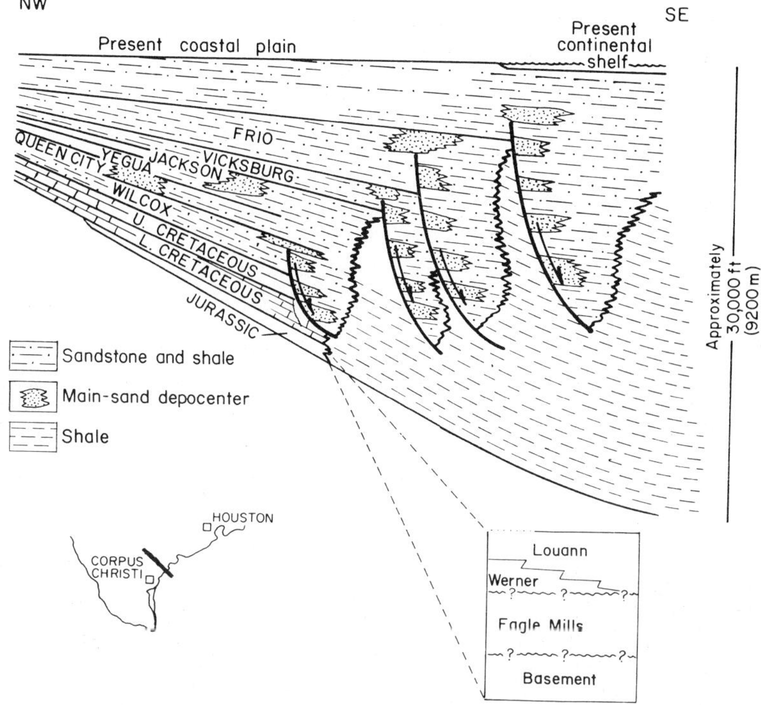

Figure 2. Depositional and structural style of the Tertiary along the Texas Gulf Coast (from Gregory et al. (1979)).

generalized stratigraphic column showing major depositional units and depositional style of the Gulf Coast is provided for reference (Fig. 2).

Components of the model consider some of the physical aspects of geopressuring/hydrofracturing, salt diapirism, and growth faulting as well as geochemical contributions made by shale de-watering (the transformation of smectite to illite), albitization, hydrocarbon maturation at various levels, and metal contributions made by fluids in different burial regimes.

B. Background

1. Anhydrite in Salt

Part of a basin's history may be deduced from anhydrite that is found in salt diapirs. Though generally considered with halite as a primary mineral phase, it incorporates trace element or isotopic compositions that tell of diagenetic fluids. Anhydrite accounts for up to several percent of the evaporites in Gulf Coast salt domes. Most of these anhydrites have $^{87}Sr/^{86}Sr$ ratios indicative of a mid-Jurassic seawater source. Others have slightly higher ratios indicative of contributions to the principal mid-Jurassic component by continental source materials (Posey, 1986). However, several domes in the Gulf Coast basin have anomalously high strontium isotope ratios (Posey, 1986) and require explanation.

Posey (1986) has proposed several possible scenarios for generating high strontium isotope ratio anhydrites present in the salt diapirs within the Gulf Coast Basin. The strontium isotope ratios present in these anhydrites could have been incorporated in the primary depositional environment from high $^{87}Sr/^{86}Sr$ continental clastics, meteoric waters that traversed continental materials, or hydrothermal waters extracted from distant shale or arkosic (red bed) successions. However, a continental origin for the radiogenic strontium is ruled out, in part because the strontium isotopic ratios of anhydrites in salt domes in the interior basins are lower than those of the Gulf Coast (exterior) Basin--even though they are closer to the continental source--and because the salts themselves are marine type rather than continental type (Posey, 1986). Water with high $^{87}Sr/^{86}Sr$ could also be extracted from arkosic (red-bed) successions underlying or adjacent to the Louann salt, but the waters would be required to pass through the salt to tranfer the radiogenic strontium to the anhydrites (Posey, 1986). A third method of generating high strontium isotopic anhydrites in the salt would require the recrystallization of the anhydrites in the presence of a radiogenic strontium-bearing fluid which migrated through the salt domes during halokinesis but prior to cap-rock formation (Posey, 1986). Such a mechanism would require fluid conduits to be present through the salt and the dissipation within it of the high $^{87}Sr/^{86}Sr$ waters which would result in the recrystallization of the anhydrites.

Evaporites that precipitated within the Louann salt during Mid-Jurassic incorporated $^{87}Sr/^{86}Sr$ ratios of Mid-Jurassic seawater (0.7068-0.7076, Posey, 1986). However the $^{87}Sr/^{86}Sr$ of anhydrites in the interior basins are currently much lower than the isotopic ratios of anhydrites in the Houston and the South Texas diapir provinces (Posey, 1986) (Fig.1).

In the model presented here, radiogenic strontium-rich fluids in the salt-hosted anhydrite are considered to be a direct consequence of the burial and geopressuring of the underlying Triassic to Jurassic arkosic Eagle Mills and Werner clastics. These clastics were deposited as the initial synrift deposits in the Gulf Coast Basin (Fig. 2) (Murray, 1966) and subsequently de-watered via geopressuring. Presumably radiogenic strontium contained within these connate fluids was transported upward, out of the clastic units, and into the salt where it exchanged with strontium in anhydrite.

2. Fluid Regimes

Knowledge of the various fluid regimes is important for the model as it helps account for the various sources of metals and their associated temperatures. Galloway (1982) compiled a description of many of the physical and chemical aspects of the three major formation fluid regimes in the Texas Gulf Coastal sedimentary prism: the meteoric regime which extends down to about 1,500 m; the Elysian (compactional) regime from about 1,500 to 3,700 m; and the deeply buried abyssal (thermobaric) regime at greater depths. In the abyssal regime the fluid pressures approach lithostatic pressure, gas generation and clay dehydration accelerate, and microfractures develop (Bogomolov et al., 1978).

a. Meteoric Regime

According to Galloway (1982), waters in the meteoric regime are commonly fresh to brackish in the Gulf Coast and consist of mixed calcium-sodium bicarbonate-chloride types. The Cu, Zn, Pb and Ni concentrations in the meteoric regime are very low (less than 0.1 to 0.01 mg/l); iron is less than 0.1 to 1 mg/l; and total dissolved solids range from 100 and 10,000 mg/l. Local pockets of methane have been encountered in the meteoric regime where fluids tend to be neutral to alkaline (pH 7 to 8), and oxidizing (containing less than 10 mg/l H_2S). Though the U_3O_8 content of the meteoric fluids is less than 0.1 mg/l it is 10,000 times greater than in the Abyssal regime fluids.

b. Elysian Regime

Fluids in the compactional, Elysian regime in the Gulf Coast differ from surface waters in that they are sodium chloride brines of the marine connate type, similar to the abyssal regime fluids with calculated pH values between 6 and 8 (White, 1965; Kharaka et al., 1977). The rocks here are dominantly reduced, contain iron sulfides, dispersed organic matter, or both, and the hydrogen sulfide content of the fluids is low (less than 10 mg/l, Galloway, 1982). The Cu, Zn, Pb and Ni content of fluids in the Elysian regime is generally less than 1 mg/l, and Fe is 0.1 to 10 duced gases (methane, heavier fractions, and distillate) commonly contain 0.1 to 1 mole percent of carbon dioxide (Galloway, 1982) much less than the carbon dioxide content of abyssal waters (± 10 percent in the Lower Frio at the Pleasant Bayou well, Morton, 1981).

c. Abyssal Regime

Hot thermobaric waters are generated in the abyssal regime, and are moderately to highly saline (10,000 to 200,000 mg/l, Gustavson and Kreitler, 1977) with the greatest values in those parts of the Gulf Coast characterized by salt diapirism (Galloway, 1982). Light hydrocarbons (principally methane), hydrogen sulfide, and carbon dioxide are generated from organic matter at temperatures characteristic of the abyssal zone (Galloway, 1982). For instance, methane and carbon dioxide are generated between 80°C and 200°C (Kharaka et al., 1985), and thermal gas predominates between 170°C and 180°C (Dozy, 1970; Karstev et al., 1971). Abyssal regime fluids contain 0.1 to 10 mg/l Cu, Zn, Pb, and Ni, and higher concentrations of Fe. These elements may reach 10 to 100 mg/l in carbonate rocks, but the uranium content is very low (10^{-5} mg/l) (Galloway, 1982).

3. Smectite - Illite Transformation

Shale dewatering is important in sedimentary diagenesis as it accompanies volume losses and associated tectonic adjustments, potentially carries dissolved constituents up-dip and up-section, and allows the bulk rock chemistry to adjust to changes induced by higher temperature and compaction states. The major contributer of water is smectite and its major product at sub-metamorphic temperatures is illite.

Smectite undergoes transformation to the ordered mixed-layer 20 percent smectite/80 percent illite as temperature and pressure increase owing to burial (Burst, 1969; Boles and Franks, 1979). This transformation is an important water producer as well as a potential source of iron, magnesium, calcium, and silica if aluminum is conserved (Burst, 1969; Hower et al., 1976). Shale de-watering has been considered a source of radiogenic strontium. However, the role of illitization in producing strontium is questionable since illite incorporates more strontium than smectite can produce (Perry and Turekian, 1974). Smectite may

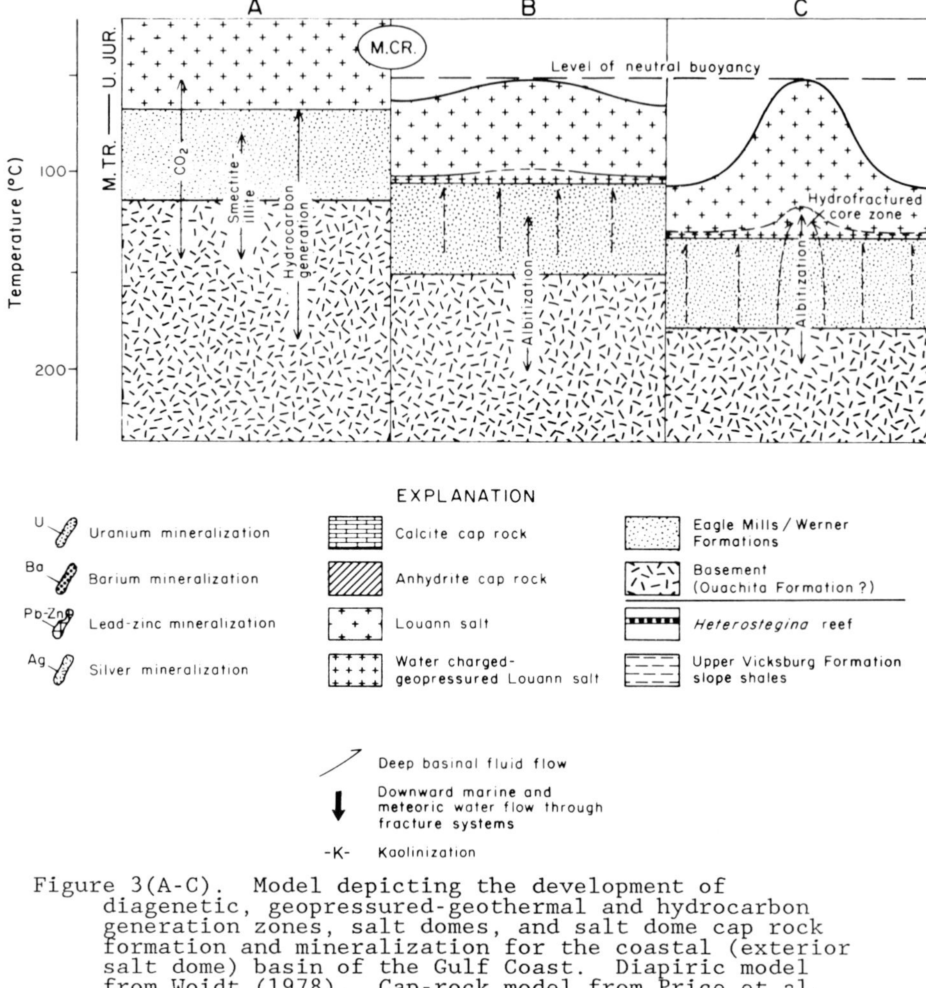

Figure 3(A-C). Model depicting the development of diagenetic, geopressured-geothermal and hydrocarbon generation zones, salt domes, and salt dome cap rock formation and mineralization for the coastal (exterior salt dome) basin of the Gulf Coast. Diapiric model from Woidt (1978). Cap-rock model from Price et al. (1983). Diagenetic temperatures from Loucks et al. (1981). Hydrocarbon generation temperatures from Hunt (1979) and Kharaka et al. (1985).

not be a contributer of radiogenic strontium unless that strontium is released and removed from the shales prior to transformation to illite. Although this transformation is partly controlled by temperature (Loucks et al., 1981), other factors (rock composition, activity, kinetics) must also play a role.

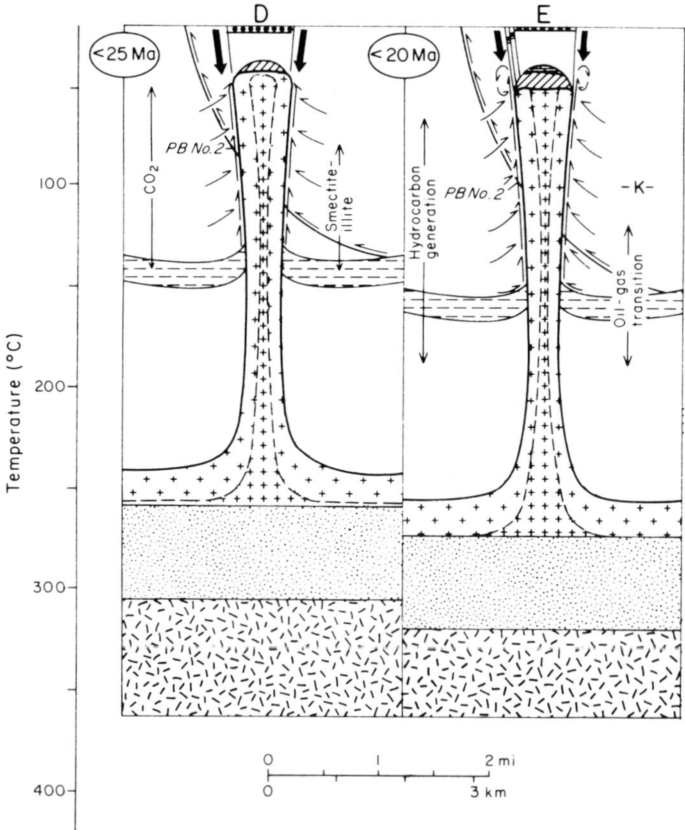

Figure 3(D,E). Model depicting the development of diagenetic, geopressured-geothermal and hydrocarbon generation zones, salt diapirs, and salt dome cap rock formation and mineralization for the coastal (exterior salt dome) basin of the Gulf Coast after major salt diapirism.

In the Pleasant Bayou geopressured-geothermal test well this transformation occurs in the upper Frio Formation (Fig. 5) (Freed, 1979) at about 3,500 m or 120°C. This is 25°C to 30°C higher than in Hidalgo County in South Texas (Loucks et al., 1981). The smectite-illite transformation occurs near the uppermost major marine sandstone unit in the Frio, the uppermost appearance of geopressure in the sandstones, and the upper limit of anomalously high vitrinite reflectance (Fig. 6).

The amount of water lost by the conversion of smectite to illite represents ten to fifteen percent of the compacted bulk volume of the argillaceous sediments (Burst, 1969; Mooney et al., 1952). The temperature of clay dehydration coincides with a maturity level of 0.5 percent Ro in the Gulf Coast (Foscolos et al., 1976; Powell et al., 1978).

Boles and Franks (1979) have shown that smectite-illite reactions with aluminum as an immobile component release significantly more cations (silica release increases more than five times) than do reactions in which aluminum is considered a mobile component (Loucks et al., 1981). The transformation reaction is outlined below:

$$\text{Smectite} + Al^{3+} + K^{+} = \text{illite} + H_4SiO_4{}^{o} + \text{various amounts of } Ca^{2+} + Mg^{2+},\ Fe^{2+} \text{ and interlayer } H_2O \quad (1)$$

(Hower et al., 1976)

The smectite-illite transformation is a major source of Si^{4+}, Ca^{++}, Mg^{++}, Fe^{++}, and a major sink for Al^{3+}, and K^{+}. Elements which substitute for any of these constituents and are metals, will also be produced or absorbed by the reaction. Of special concern to our model are the ore metals.

Rubidium, which substitutes for K^{+} in the clays and other minerals, is the source of ^{87}Sr. During the conversion of smectite to illite the potassium content of associated brines is
9
depleted (Collins, 1975) owing to potassium absorption by illite (Lyon and Buckman, 1960). Rubidium, which commonly substitutes for potassium, is more easily absorbed by clays than potassium and is removed more readily from solution than is potassium (Collins, 1975).

The Eagle Mills and Werner clastics entered the smectite-illite transition zone when temperatures exceeded about 80°C, and this resulted in the generation of relatively fresh (low

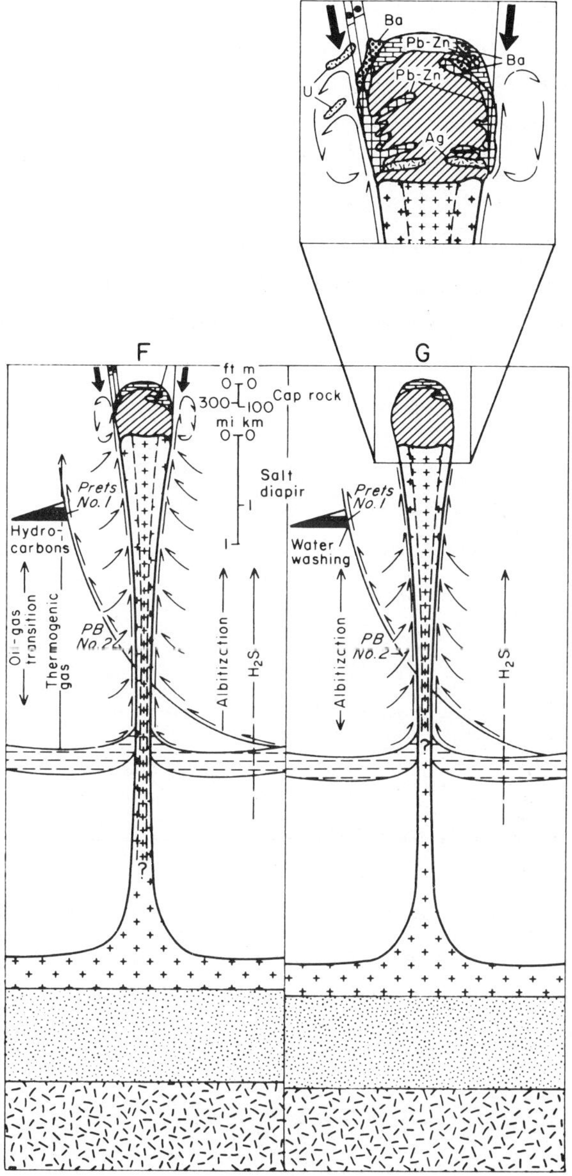

Figure 3(F,G). Model depicting the late sage development of diagenetic, geopressured-geothermal and hydrocarbon generation zones of salt dome and salt dome cap rock formation and mineralization for the coastal (exterior salt dome) basin of the Gulf Coast.

salinity) water (Burst, 1969; Powers, 1967; Loucks et al., 1981) below the Louann salt. This fluid was probably enriched in radiogenic strontium and, where blocked from migrating laterally, was probably trapped by, and collected below, the impermeable Louann salt where it probably initiated salt dissolution (Fig. 3b). This region probably became aquathermally pressured from the continuous temperature increase with progressive increase in depth of burial (Palciauskias and Domenico, 1980), an imposition that eventually led to hydrofracturing within the Eagle Mills.

4. Albitization

With regard to the model, albitization of Ca-plagioclase is an important source of calcium and associated non-radiogenic strontium. On the other hand, K-feldspar albitization is an important source of rubidium and attendant radiogenic ^{87}Sr, and an important source of metals. When the Eagle Mills and Werner clastics had been buried sufficiently (at temperatures in excess of 120°C; Loucks et al., 1981), they entered the zone of albitization. This reaction and the destruction of K-feldspar are probably capable of releasing considerable amounts of lead and barium as well as minor amounts of zinc and iron (see typical concentrations reported, for instance, in Deer et al., 1966). According to Loucks et al. (1981), albitization takes place concurrently with the peak of oil generation, the maximum being at 150°C. Insofar as this model is concerned, conduits created by hydrofracturing may, in some cases, remain open for some time because albitization, the most reasonable source of radiogenic strontium, may occur some distance below the zone of hydrofracturing, at a higher temperature (see section 6 on microfracturing). However, the two events may have a considerable range of overlapping temperatures.

5. Hydrocarbon Maturation

Burial of the Eagle Mills-Werner clastics below the 50°C to 110°C isotherms (Fig. 3b and 4) resulted in the generation of carbon dioxide and subsequently hydrocarbon gases in rocks with sufficient organic matter. According to Kharaka et al. (1985) methane and carbon dioxide can be generated by the decarboxylation of acetic acid between about 80°C to 200°C over this depth range. However, Lundegard (1985) discounts the role this reaction plays in generating CO_2, showing instead the importance of hydrous pyrolosis reactions between organics and water. Lundegard (1985) showed that, locally, profiles of CO_2 concentrations increase down-section in response to temperature increases. The net effect of CO_2 generation, regardless of its source, is to enhance overpressuring and hydrofracturing.

The peak of carbon dioxide generation during the maturation of sapropelic organic matter occurs at about 100°C, about 20°C cooler than the peak of oil generation in humic rocks (Hunt, 1979). Hence initial fluids will have high carbon dioxide contents prior to formation of liquid hydrocarbons (Fig. 4) (Hunt, 1979). The generation of carbon dioxide would acidify the fluids and result in the formation of secondary porosity by the dissolution of carbonates (Selley, 1979). Light and D'Attilio (in press) applied burial history methods and found that secondary porosity formation in the lower Frio Formation at the Pleasant Bayou geopressured-geothermal well began about 20 Ma and peaked between 14 and 7.5 Ma. This period of carbon dioxide generation post-dates the 23.6 Ma smectite-illite transition event found at this same level (Morton, 1981) and predates hydrocarbon gas and condensate migration into the sandstones in the Hitchcock N.E. field (Light and D'Attilio, in press). Upward migration of the carbon dioxide bearing fluids through shallower Cretaceous and Tertiary carbonate sequences probably

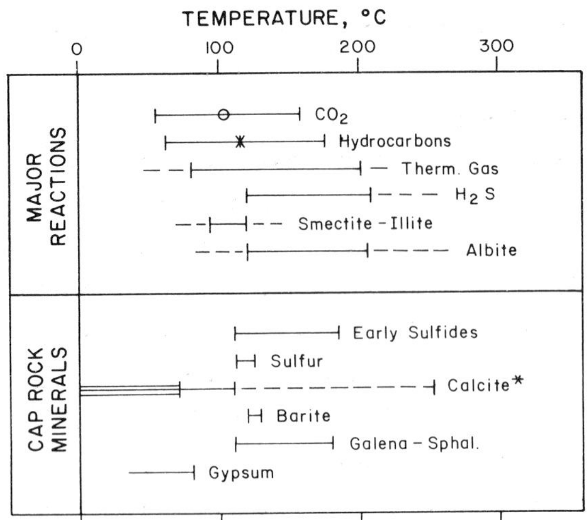

Figure 4. Temperature range of major hydrocarbon and mineral reactions, and measurd temperatures of diagenetic minerals in salt dome cap rocks. Fluid inclusion and mineral stability data from Price et al. (1983); Kreitler and Dutton (1983); Kaiser and Richmann (1981); Loucks et al. (1981); Posey (1986); Weaver (1980); and Knauth et al. (1980). Hydrocarbon generation temperatures from Kharaka et al. (1985) and Hunt (1979).

resulted in carbonate dissolution followed by precipitation at shallower levels in salt dome cap rocks owing to the drop in CO_2 pressure (Posey, 1986; Posey et al, this volume).

6. Microfracturing

Extensive microfractures develop in rocks when the cumulative fluid pressure exceeds the minimum principal stress. An expansion of pore volume in the rocks provides pressure relief to accomodate the volumetric expansion of the heated water, and the microfractures provide conduits for the escape of fluids (Palciauskas and Domenico, 1980). For sedimentary rocks containing progressively heated pore fluids, microfracturing and dilatancy have been observed where the ratio of the pore fluid pressure to the geostatic pressure exceeds 0.8 (Handin et al., 1963; Palciauskas and Domenico, 1980). This result is probably correct for sandstone, limestone, siltstone and shale as it is not controlled by the rheology of the rocks (Palciauskas and Domenico, 1980). However, it may be different for evaporites, which can behave more plastically.

The occurrence of dilatancy corresponds with the peak of oil generation (Hunt, 1979) and major albitization of the feldspars (Loucks et al., 1981). Furthermore, the fluid volume increase which results from the transformation of smectite to illite (Burst, 1969; Powers, 1967) acts much in the same way as the thermally expanded fluid and can contribute to the initiation of dilatancy (Palciauskas and Domenico, 1980). The initiation of dilatancy is independent of the rate of burial; for a given geothermal gradient and depth to the dilatancy zone there is a wide range of organic maturation possibilities prior to microfracture development (Palciauskas and Domenico, 1980), although as pointed by W. Macpherson (personal communication, 1986) this effect may be overstated since illite is more dense than smectite thus lowering the net rock volume during this transformation.

II. SALT DIAPIRISM

According to our model (Fig. 3) fluids beneath the Louann evaporites and alongside salt diapirs evolved in response to burial and participated in cap rock formation and metallization, in part as a consequence of diapiric focusing of heat (Figs. 3a-g). Following Triassic rifting and mid-Jurassic evaporite deposition, salt pillow formation in the Louann salt was initiated in Late Jurassic by the deposition of thick limestones and shales followed by the diapiric stage, occurring generally between Early Cretaceous and Early Tertiary (135-60 Ma), and post-diapir movement, some of which continues to the present (Seni and Jackson, 1983). In general, salt domes in the Gulf

Coastal basin experienced major diapiric movement after those in the interior basins, but studies of their timing are less definitive.

Fluids within the Eagle Mills-Werner sequence probably remained below the Louann salt during initial salt dome movement until fluid pressures rose to the point where widespread dilation of the Eagle Mills-Werner and the base of salt occurred. At this point, connate fluids escaped from the Eagle Mills-Werner into overlying formations. The formation of salt pillows or ridges may have confined geopressured fluids below the cores of salt-centered structures (Fig. 3c), and the base of the salt probably developed an erratic dissolution zone when shale waters collected below it (Fig. 3c).

During halokinesis, fluids were progressively drawn into the centers of salt ridges or domes as they evolved. Anhydrites with Jurassic $^{87}Sr/^{86}Sr$ then recrystallized in the presence of the geopressured fluids, which contained variable amounts of radiogenic ^{87}Sr derived from albitization reactions and, possibly, illitization reactions (Posey, 1986). Recrystallization of anhydrites probably occurred simultaneously with salt dome movement, incorporating radiogenic strontium according to the amount available in the fluids and the proportions of mixing (Posey, 1986). This resulted in the variable but high $^{87}Sr/^{86}Sr$ in the anhydrite, much higher in parts of the Gulf Coast basin than in the interior basins, owing to a greater lapse of time since re-equilibration and/or greater volumes of K-bearing materials.

Sulfur evolved along with organics. High concentrations of water soluble hydrogen sulfide are confined primarily to carbonate reservoirs where they are enriched in ^{34}S, as is typical of deep thermal sulfide (Goldhaber et al., 1979; Galloway, 1982). The temperature of initial hydrogen sulfide generation in organic matter is about 120°C (Hunt, 1979), but continues at higher temperatures. The temperature range of H_2S generation is about the same as the temperature of albitization reactions in feldspars (Loucks et al., 1981) which releases Pb, Ba, and minor Zn and Fe (Moller, 1983), suggesting that sulfur and metals, which are found in lead-zinc deposits in salt dome cap rocks, may be generated in the same initial fluid.

Several lines of evidence indicate that salt may have been buried to considerable depths before reaching its present position as diapirs. For instance, isotope data from mine water at Weeks Island, Jefferson Island, and Belle Isle salt domes, Louisiana, suggest that these waters are formation waters introduced at temperatures of 150°C to 180°C or depths of 3,000 to 4,000 m (Knauth et al., 1980; Kreitler and Dutton, 1983) (Fig. 4). In addition, inclusions in rock salt at Winnfield Dome consist largely of carbon dioxide, water, and smaller amounts of heavier hydrocarbons under pressures ranging from 500 to 1,000 atmospheres at 0°C, and have caused blowouts during salt mining (Hoy et al., 1962). These pressures correspond with depths of 4,300 m to 6,400 m using a normal hydrostatic and a geopressure gradient of 0.7 psi/ft for the Gulf Coast. These combined composition and pressure (depth) data from fluid inclusions indicate that salt recrystallized in the presence of geopressured fluids, probably sourced from beneath or adjacent to the salt, in the shale dewatering, carbon dioxide and hydrocarbon generation zones. The presence of graphite in salt at the Weeks Island dome indicates temperatures above 300°C and, hence, greater burial depths (Weaver, 1980). Similar inclusions occur in the Oakwood Dome and appear to have been trapped between 213 m and 765 m (Dix and Jackson, 1982). These graphite occurrences are extreme examples but emphasize that some salt must reach extreme temperatures prior to diapiric ascent to shallow levels where it is now found.

Price et al. (1983) have shown that doubly terminated quartz crystals within the cap rocks at Hockley Dome (Gulf Coastal Basin) contain fluid inclusions which crystallized at temperatures of about 240°C. This isolated example suggests that quartz crystallized within the salt at depths of 6,700 m. Geopressured fluids below the Louann probably became saturated in silica during salt burial, perhaps owing to the breakdown of clay or feldspar in the Eagle Mills, and were drawn into the core of the salt diapirs. These fluids were subsequently heated then precipitated as quartz, either because of a drop in temperature (to about 240°C, according to the fluid inclusion temperatures) or an increased activity coefficient of $H_4SiO_4^o$ in the presence of saline brines (Figs. 3d and 4).

Thus, there is clear evidence that thermobaric fluids from the abyssal regime have been introduced into salt domes during burial. Various mineralogical consequences of these fluids include anhydrite recrystallization, quartz precipitation, and, in extreme cases, graphitization.

III. FORMATION OF SALT DOME CAP ROCKS

Gose et al., (1985), and Kyle et al., (this volume) have dated the anhydrite cap rock of the Winnfield dome in the Northern Louisiana basin as late Jurassic (150 to 160 Ma) using paleomagnetic data. This age is coincident with the onset of diapirism during late Jurassic-Early Cretaceous in the East Texas Basin as inferred by thickness variation in sediments adjacent to the domes (Seni and Jackson, 1983). However, major widespread diapirism in the coastal basins (Fig. 1) was initiated later, during the Tertiary by the deposition of Wilcox and younger, thick, fluvio-deltaic sequences (Fig. 2). These sediments quickly loaded the basinal sequence, causing a large buoyancy differential, the external driving mechanism for salt dome diapirism (Fig. 3c). Whereas diapirism and fluid maturation are ongoing phenomena of a subsiding basin, it is not yet possible to equate the actual time of fluid maturation and cap rock formation in the Gulf Coast.

The presence of fluids in salt will greatly increase its mobility (Jackson and Talbot, 1986). Initially the core zones of salt domes would be more buoyant than the rims as they would contain hot, low density, geopressured brines. This buoyancy differential could result in the formation of a convection system within the salt which would further increase the rate of uplift of the core zone and mixing and dissipation of the geopressured fluids throughout the salt. The occurrence of radiogenic strontium in an otherwise nil-Rb environment is a critical component for recognizing the role of extraneous fluids in salt diapirism.

A. An Example of Fluid Migration during Oligocene

Clauer (1976) has shown that the clay fractions of shales (< 2 microns) should generate isochrons clearly indicative of strontium isotope homogenization. These will date either the time of deposition, diagenesis, or burial metamorphism (Clauer, 1976)--most likely diagenetic or burial metamorphic events (Cahen et al., 1984).

Using rubidium-strontium isotope data, Morton (1983) documented an homogenization event in Frio Formation clay in the Pleasant Bayou geopressured-geothermal test well (Fig. 5). This isochron is 23.6 ± 0.8 Ma (Oligocene-Miocene transition) with an initial ratio of 0.7088 ± 0.0004. Shales above 3,200 m give older ages (34 Ma) than the deeper shales coincident with their time of deposition (Morton, 1983). The age of 23.6 Ma coincides with deposition of the Middle Anahuac Formation some 150 to 600 m shallower, and the shallowest shale sample (3,200 m) of this age corresponds with a break in a linear trend of the oxygen isotope ratios for clay minerals in the area (Ewing et al., 1983; Morton, 1983) (Fig. 5).

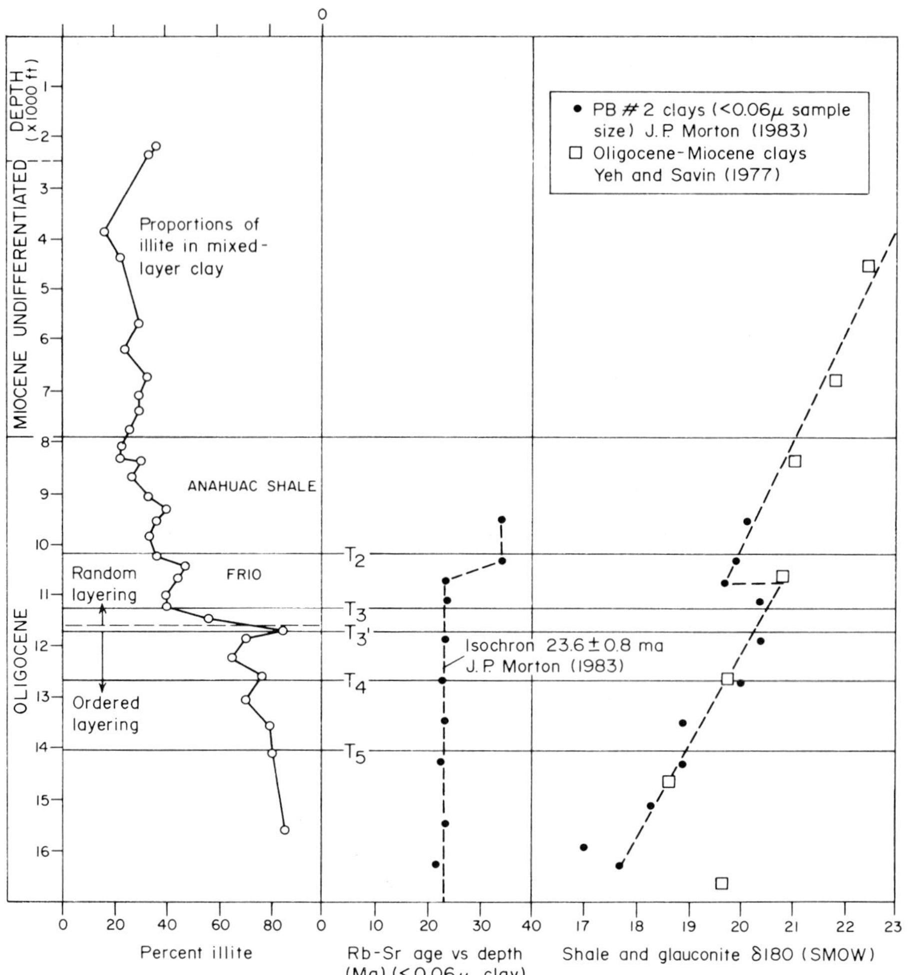

Figure 5. Proportion of illite and smectite in mixed-layer clays versus depth, Pleasant Bayou No.1 well from Freed (1979); apparent Rb-Sr age of clays <0.06 millimicron in diameter versus depth in the test well from Morton (1983); and oxygen isotope values for shale and glauconite versus depth in the test well from Yeh and Savin (1977) and Morton (1983). Modified from Ewing et al. (1983).

Morton (1983) interpreted the 23.6 Ma age to be a time of major smectite to illite transformation and suggested that at this time, the Frio, which had been rapidly depressed, was invaded by deeply sourced fluids containing potassium that was generated during feldspar dissolution. This clay conversion occurred between 65° and 75°C (estimated from oxygen isotope data, Morton, 1983). Evidently the sediments below the Frio (<3,200 m) in the Pleasant Bayou test well were flushed by hot fluids 23.6 Ma ago (Morton, 1983).

We speculate that this is not an isolated incident of fluid flushing but rather is one of numerous incidents in the Gulf Coast. On the weight of evidence in this paper, the process seems to be a normal consequence of basin maturation. Exactly how the 23.6 Ma event recorded by Morton (1983) relates to matters in this paper is not known, but we highlight this example for its potential to produce significant volumes of fluid and associated clay-derived elements and for its age similarity with other fluid flushing events.

There is evidence for widespread salt dome diapirism in the Gulf Coast basin prior to the Oligocene-Miocene transition (T. E. Ewing, personal communication, 1985). Seismic evidence indicates that major diapirism was initiated in the Hockley Dome - Katy field area about 60 Ma ago (lower Wilcox), and terminated about 45 Ma ago (Yegua Formation) (Ewing, 1983; Price, unpublished). Also, the age of the Heterostegina reef on Damon Mound (Mid-Anahuac, 22-24.8 Ma, T. E. Ewing, personal communication, 1984) corresponds closely to the 23.6 Ma age found for the smectite-illite transition (Morton, 1983). This suggests that widespread diapirism represented by the Heterostegina reefs and a significant diagenetic event in the Gulf Coast shales might be related.

Correspondence between diapirism in the Gulf Coast basin and the period of conversion of smectite to illite by waters at elevated temperatures (Morton, 1983) suggests that prior to the Oligocene-Miocene boundary there was a period of rapid uplift of the Houston Diapir province salt stocks and concomitant removal of the basal salt layer which culminated in the stocks puncturing the Anahuac shelf and being capped by Heterostegina reef facies at the end of the Oligocene (Fig. 3d). The removal of salt would have resulted in the subsidence and reactivation of the Frio growth fault system at this time and could also be related to major subsidence of the Gulf Coast basin which resulted in the late Oligocene transgression. Formation of fracture conduits would allow deeply sourced saline brines to migrate up around salt diapirs and away from them along growth faults through the Frio Formation sandstones and shales where they subsequently reset the Rb-Sr age of the shales to 23.6 Ma (Morton, 1983) (Fig. 3d). The driving mechanism for a period of major diapirism in the Oligocene in the Gulf Coast basin may have been the deposition of the thick fluvio-deltaic Frio Formation (Houston and Norias Delta systems, Galloway, 1982). Likewise, at least part of the driving mechanism may have been the introduction of fluid from below the Frio into the salt.

B. Cap-Rock Formation

Murray (1966) proposed a general model for salt dome cap-rock formation. He stated that when a salt diapir approaches the surface, a solution table forms as the salt undergoes dissolution by undersaturated fluids, and, as a result, residual anhydrite accumulates. At the Winnfield dome, this fluid was probably saturated in pyrrhotite, dolomite, and anhydrite as pyrrhotite rims, euhedral anhydrite (Ulrich et al., 1984) and anhydrite-hosted dolomite remained stable during pyrrhotite formation (Posey, 1986). However, this fluid was undersaturated with respect to halite as salt was dissolved. The iron, calcium, and magnesium contents of the fluid suggest that it may have been derived from the smectite-illite transition zone (Fig. 3d). Fracture conduits developed above and around the margins of the salt domes during diapiric movement probably allowed the moderately deep sourced fluids passage to the surface where they would mix with downward percolating meteoric or marine waters and form the initial salt dissolution table (Fig. 3d).

Most cap rock anhydrites appear to have lost strontium upon cap rock formation as they have lower strontium concentrations than most salt-hosted anhydrites (Posey, 1986). Most cap rock anhydrites are slightly radiogenic compared with mid-Jurassic, but those in the Gulf Coast basin are extremely so (Posey, 1986). Anhydrites in the salt from Hockley have unusually high $^{87}Sr/^{86}Sr$ ratios and appear to be the result of mixing of two components, one with a low $^{87}Sr/^{86}Sr$ ratio (0.7071 or less) probably mid-Jurassic evaporitic source material, and one with a high $^{87}Sr/^{86}Sr$ ratio (0.7094 or greater) probably radiogenic strontium sourced from K-feldspar dissolution, albitization processes and/or breakdown of clay minerals (Posey, 1986). The cap rock anhydrites are within the same $^{87}Sr/^{86}Sr$ range but have lower strontium concentrations (Posey, 1986). Combined, data from the two types of anhydrite suggest that the salt-hosted anhydrites recrystallized in the presence of radiogenic waters and that the cap rock anhydrites recrystallized in the presence of (a) fluids with a late Cretaceous marine carbonate component which migrated up the flanks of the salt dome, or (b) early Miocene marine waters which migrated down the flanks of the cap rock after the salt dome had punctured the shelf. Although this relationship may be fortuitous, the isotope data suggest that an early Miocene age for the marine fluids is most likely and the salt domes must therefore have been close to or have punctured the sea bed at that time to have produced peripheral open fracture systems down which marine waters could have leaked into the salt dissolution zone and anhydrite cap-rock area. These fluids would mix with deep basinal brines, exchange strontium during anhydrite recrystallization, and produce the variable $^{87}Sr/^{86}Sr$ ratios found in the cap-rock anhydrites.

During salt dome cap rock formation, hydrocarbon-bearing saline brines pass through the cap rock and dissolve anhydrite; hydrogen sulfide and carbon dioxide are produced by bacteria in the presence of hydrocarbons at a transition zone (Fig. 3f) (Murray, 1966) which moves progressively downwards (Posey, 1986). Calcites formed initially are dark, full of pyrite and residual materials, and cement early anhydrites or replace euhedral dolomite rhombohedrons (Kreitler and Dutton, 1983; Posey, 1986; Posey et al., this volume). These calcites are granular, sucrosic and probably crystallized quickly from concentrated saline solutions (Posey, 1986). Later void-filling calcites are larger, light-colored crystals that probably have crystallized from more dilute solutions (Posey, 1986).

Posey et al. (this volume) have described the carbon-isotope characteristics found in Gulf Coast salt domes, and appeal to several carbon and oxygen sources for calcite, namely formation fluids, seawater, and meteoric water. Posey (1986) indicated that, in general, detritus-bearing carbonate false cap rocks have depleted $\delta^{13}C$ (-5 to -25 ppt PDB) whereas the underlying true calcite cap rocks are more depleted, reaching -50 ppt in some salt domes.

The layered early-dark, late-light calcite pairs appear to have formed consecutively in a time-reversed stratigraphy (Posey, 1986; Posey et al., this volume). A void probably formed at the base of the calcite cap rock from the anhydrite-calcite volume decrease (Kreitler and Dutton, 1983) or block settling of the anhydrite. Fluids could enter the peripheries of this dissolution zone and alter the top of the anhydrite to calcite by biogenic activity in the presence of hydrocarbons. This would proceed until all the hydrocarbons were used up or channels were choked off or diverted away from the cap rock. Finally, calcite would precipitate from less saline residual fluids as a void fill (Posey, 1986).

The strontium isotopes in Hockley dome cap-rocks become more enriched in ^{87}Sr and the carbon isotopes grow lighter radially outwards and downwards, which suggests a deep peripheral source for the high $^{87}Sr/^{86}Sr$ and light carbon, assuming the calcite forms from top to base (Posey, 1986; Posey et al., this volume).

C. Lead-Zinc-Barite Mineralization

1. Introduction

Mississippi Valley-type lead-zinc deposits occur primarily in carbonates but also in some sandstone hosts (Anderson and Macqueen, 1982; Ohle, 1952; Beales and Onasick, 1970). They are associated with pinch-outs that surround domal structures, basin growth faults and nearshore reefs. Mississippi Valley-type lead and zinc mineralization is also present in salt dome cap rocks which lie at or near the land surface in Texas and Louisiana (Kyle and Price, 1986; Land, 1984; Price et al., 1983).

There appears to be a progressive temperature drop from the early high temperature sulfides (110°C to 180°C; temperatures estimated from sulfur isotope pairs from sulfides) to later sulfates (110°C to 140°C; temperatures estimated from fluid inclusions in barite) (Price et al., 1983). This temperature decrease could simply be a result of the mixing of increasing quantities of cool seawater with hot deeply sourced brines in the salt dissolution zone and anhydrite cap rock area.

In many Mississippi Valley-type ore deposits the elemental abundances generally decrease from iron to zinc to barium to lead. The major exception is southeast Missouri where iron and lead are dominant over zinc. Barite, if related at all to the lead deposits, forms separate concentrations and is more abundant than zinc. Although barium tends to be more abundant than zinc or lead in salt dome cap rocks, the paragenetic sequence is early iron sulfides followed by lead-zinc sulfides and finally barium sulfates (Price et al., 1983). The later sulfides and sulfates include marcasite, pyrite, sphalerite, galena, rare acanthite, and late stage barite (Price et al., 1983; Price and Kyle, 1986; Ulrich et al., 1984). The parageneses of salt dome cap rock sulfides vary and are repetitive. At Winnfield they begin with early pyrrhotite which is present in the anhydrite cap rocks and which preceded the main period of lead-zinc formation (Ulrich et al., 1984).

Price et al. (1983) suggested that calcite cap-rock formation was contemporaneous with the formation of Mississippi Valley-type lead-zinc and barite deposits along the margins of salt domes (Fig. 3g). Sulfur isotope and fluid inclusion geothermometry indicate a crystallization temperature of 110°C to 180°C for the sulfide minerals. Metalliferous brines appear to have leaked up the margins of the salt stocks (Price et al., 1983). These fluids would therefore need to be deeply sourced from either the smectite-illite transition zone or the zone of albitization or both (80°C to 200°C) in the Abyssal regime (Galloway, 1982; Price et al., 1983; Posey, 1986).

Hydrocarbons have been observed bleeding from freshly cut sections of massive sulfides at Hockley dome and may result from primary capture during the precipitation of the mineralization or later introduction into the porous mineralization zone (Price et al., 1983). The overall processes responsible for calcite cap rock formation and sulfide precipitation are roughly contemporaneous. However, on the local scale, sulfides are fractured or form breccia clasts in calcite cap rock; that is, sulfide mineralization pre-dated local cap rock formation. Conversely, sulfide and sulfate minerals may occur in vugs within the calcite cap rock and therfore post-dated local cap rock development. Hence hydrocarbon introduction may have preceded, been contemporaneous with, or post-dated sulfide mineralization.

2. Potential Sources of Zinc and Lead

Land (1984) presented evidence that Mississippi Valley-type lead-zinc mineralization in salt dome cap rocks of the Texas Gulf Coast was a result of fluids derived from the Mesozoic, particularly Jurassic carbonate source rocks. However, dissolution and albitization of feldspars (Milliken et al., 1981), and other water rock interactions (smectite-illite transition), could account for the release of metals to the formation fluids (Price et al., 1983; Moller, 1983; Thornton and Seyfried, 1985).

Zinc has a very similar ionic radius and geochemistry to Mg^{+2}, Co^{+2}, Fe^{+2}, and Mn^{+2} and tends to be concentrated in limestones (Collins, 1975). Deeply sourced acidic brines, with a high CO_2 content (Morton, 1981), could extract zinc from Jurassic and Cretaceous age carbonates (Land, 1984).

The large radius of the Pb^{+2} ion (1.33 A) results in its substitution for both Ca^{+2} (radius 1.06 A) in plagioclase and K^{+} (1.33 A) in potash feldspar (Ahrens, 1965b). The replacement of potassium by lead is greater than its replacement of calcium (Ahrens, 1965a), hence the breakdown or albitization of potassium feldspars should release abundant lead.

The zinc:lead ratios of fluids produced by experimental clay-rich marine shale reactions at 200°C and 300°C (8:1 to 3:1; Thornton and Seyfried, 1985) are within the zinc:lead ratios of salt dome cap rocks and Mississippi Valley lead-zinc deposits. The magnesium and iron content of formation fluids present in the Gulf Coast (Kharaka, et al., 1980) can be produced by marine shale reactions in the 200°C to 300°C range under oxidizing or reducing conditions (Thornton and Seyfried, 1985). This temperature range is high compared with normal Mississippi Valley ore fluid temperatures (Cathles and Smith, 1983). Though they used modern marine shales, Thornton and Seyfried (1985) deduced that fluid data and mineralogical assemblages of modern shales are similar to those of the Gulf Coast. They stated that the observed differences are due to a higher degree of reaction in the geologic environment and increased stability of illite and chlorite in geologic fluids. Their data suggest that copper-bearing solutions accompanied by lead and zinc may evolve in sedimentary basins characterized by oxidizing conditions whereas zinc and lead will predominate in more reducing environment (Thornton and Seyried, 1985). Reducing conditions probably exist in the Abyssal zone in the Gulf Coast (Galloway, 1982). The lack of copper mineralization in salt dome cap rocks implies either that shale reactions occurred under reducing conditions forming the mineralizing fluids, or that copper was not available in the primordial materials.

Sverjensky (1984) reviewed evidence suggesting that oil field brines do not have the capacity to become mineralizers unless they react with rocks (carbonates or siliciclastics) somewhere enroute to their ultimate ore-depositing site. A theoretical evaluation of the consequences of reacting brine of Pleasant Bayou composition with either dolomite or sandstone each with 3% K-feldspar, 0.2% muscovite, and 0.01% pyrite showed that the destruction of K-feldspar led to a decrease in the Na/K atomic ratio and an increase in the Zn/Pb atomic ratio in the fluid. Such solutions would lead to early stage sphalerite-rich solutions with Na/K ratios typical of Mississippi Valley type deposits. The early-stage, sphalerite-rich deposits are characteristic of the sulfide/barite deposits found in salt dome cap rocks.

3. Potential Source of Barium

Barite, which commonly forms a late stage mineralization event in salt dome cap rocks, (Figs. 3g and 4), has a high strontium isotope ratio suggesting that the fluid from which it crystallized contained radiogenic strontium derived from either albitization, feldspar dissolution, or shale breakdown or dewatering (Posey, 1986) . Fluid inclusions within the barite indicate temperatures of 120°C to 140°C (Kyle and Price, 1986; Price et al., 1983) suggesting that the fluids were derived from the albitization zone or deeper. The salinity of the fluid inclusions within the barite varies from 3-12 percent NaCl equivalent and within the range of deep geothermal brines in the Houston Delta System.

The barite, which contains isotopically heavy strontium, also contains extremely heavy sulphur (up to 80 ppt, Kyle and Price, 1986). This very heavy sulfur may result from partial sulfate reduction whereby light isotopes are fractionated into reduced phases (pyrite and other sulfides) such that the residual fluid becomes isotopically heavier. The sulfur is too heavy to have been simply sourced at high temperatures (Kyle and Price, 1986). At Moss Bluff dome, hydrogen sulfide with a $\delta^{34}S$ of -15ppt coexists with dissolved sulfate with a $\delta^{34}S$ of +40 (Feely and Kulp, 1957) which supports this contention (Kyle and Price, 1986).

The source of the barium is probably clay or K feldspars (Barnes, 1979). The solubility of barite increases with temperature up to 150°C, at salinities above 0.2 molar NaCl and with pressure (Barnes, 1979). Hence barite will precipitate from simple cooling (Holland and Malinin, 1979), a pressure decrease, or mixing with less saline waters. Barium-rich geothermal brines could also precipitate barite if brought into contact with sulfate-rich fluids from the anhydrite cap rock zone, as high concentrations of barium are only possible at low sulfate contents (Collins, 1975). Mixing is the most plausible mechanism for barite precipitation in salt-dome cap-rocks.

The most recent major mineral events in salt dome cap rocks are the formation of late stage gypsum after anhydrite (Posey,

1986) and sulfur as cross-cutting dikes and pore fillings. These probably formed from mixing of oxidizing groundwaters with hot H_2S-bearing brines.

4. Other mineralization

Galloway (1982) has described the sequence of sulfide mineralization associated with uranium deposits at Ray Point in the Wilcox fault zone in South Texas. At Ray Point early (pre-uranium ore) pyrite psuedomorphing iron titanium oxide is rimmed with marcasite which, in turn, is overgrown by smectite. Sulfur in pyrite in this zone is isotopically heavy (+10 and +28 o/oo, CDT) whereas in marcasite it is isotopically very light ($\delta^{34}S$ averages about -55 ppt). The very light $\delta^{34}S$ values are consistent with bacterial or inorganic reduction of groundwater sulfate and the heavy sulfur is indicative of very deeply-sourced hydrocarbons, near the upper limit of the oil window (Galloway, 1982; Goldhaber et al., 1979). Isotopically heavy sulfur isotopes imply that the sulfur bearing waters were generated near the base of the Gulf Coast sequence in this area of South Texas (Land, 1984).

Some of the sulfides associated with uranium in sediments adjacent to Hockley dome also have high $\delta^{34}S$ values (+26 to +30 o/oo) (Price and Kyle, 1986), indicating a deep source from temperatures similar to the Ray Point values demonstrated by Goldhaber et al. (1979). However, sulfides within Hockley cap rock have lower $\delta^{34}S$ compositions (<15 o/oo; Kyle and Price, 1986). The similarity between Hockley cap rock sulfide $\delta^{34}S$ compositions and H_2S compositions from South Texas gas fields (Busche et al., 1982) suggest that externally reduced sulfur might have migrated to the structural trap provided by the salt diapir (Kyle and Price, 1986). However, the coincidence of these values with other light sulfur isotopes leaves open the possibility that the sulfide sulfur is a product of bacterial reduction of cap rock anhydrite as described by Feely and Kulp (1957).

Roll-front uranium mineralization at the Felder Uranium deposit, Live Oak County, Texas, gives a well-defined $^{207}Pb/^{204}Pb$-$^{235}U/^{204}Pb$ isochron age of 5.07 ± 0.15 Ma (Ludwig et al., 1982). At this deposit, ore-stage marcasites have light δ^{34} sulfur values while post-ore pyrites have heavy δ^{34} sulfur indicating a late (less than 5 Ma old) resulfidization by deeply sourced H_2S. This age is probably significantly younger than cap rock mineralization at Hockley. However, it demonstrates that basin maturation and fluid escape are ongoing features in the Gulf Coast.

D. Gulf Coast Brines

Systematic vertical and lateral variations in composition and salinity that are found in Gulf Coast hydrocarbon reservoir fluids have implications for our model. Fluids presently in the Lower Frio Formation below 14,000 ft. (4270 m) at the Pleasant Bayou geopressured geothermal test well (Brazoria County, Texas) are highly saline and enriched in iron, manganese, lead and zinc (Kharaka et al., 1980). These formation waters also contain 29 standard cubic feet of gas per barrel of water of which 85.5 volume percent is methane and 10.5 volume percent is carbon dioxide (Morton, 1981), a $CH_4:CO_2$ ratio of about 8:1.

Morton et al., (1983) have studied depth related variations in salinity at the Chocolate Bayou field and Alta Loma fields close to the Pleasant Bayou test well. In the Chocolate Bayou field the salinities increase with depth below 3660 m, whereas the reverse is seen in data from the Alta Loma field (Morton et al., 1983). However the sodium:calcium ratio in both fields decreases with increasing depth below 3048 m, and becomes constant at greater depths (Morton et al., 1983). The increases in the fluid content of calcium, decrease in sodium:calcium and magnesium:calcium ratios and increase in the potassium:sodium ratio with depth can be related to albitization which occurs below about 90°C to 120°C (Loucks et al., 1981; Milliken et al., 1981) (Figs. 3f, 3g, and 4).

Salinities in the Pleasant Bayou formation waters are much higher (131,320 mg/l; Morton, 1981) than salinities of adjacent oil and gas fields and are also depleted in bromide compared with chloride (Kharaka et al., 1980). This is probably due to dissolution of evaporites (Kharaka et al., 1980). Iron, zinc, and lead ratios of these Gulf Coast oil-field brines are similar to cap-rock deposits (Kyle and Price, 1986). Reservoir fluids in these Gulf Coast oil and gas fields are enriched in calcium, strontium, barium, fluorine, and depleted in sodium, potassium, magnesium, and sulfate relative to seawater (Hanor, 1979).

The lower Frio sandstones are highly geopressured (>0.7 psi/ft, 15.83 kPa/m) so that updip flow of hot, basinal fluids is retarded and the thermal maturity has remained relatively unchanged (Light et al., 1984). These fluids also have sulfate contents of 5.4 to 21 mg/l, and the strontium varies from 867 to 1020 mg/l (Kharaka et al., 1980). Fluids developed in marine shales, reacted at 200°C and 300°C contain 146 to 2743 mg/l sulfate, and 0.8 to 9.3 mg/l strontium (Thornton and Seyfried, 1985), even though similar concentrations of strontium are present in these and Gulf Coast shales. Evidently high temperature clay reactions produce 100 times too much sulfate, and two orders too little strontium to account for higher strontium and lower sulfur concentrations in the geopressured brines. Strontium may have been absorbed by illite during conversion to smectite (Perry and Turekian, 1974). However, the fate of excess sulfate is not known. The calcium content is also too high in the lower Frio waters at Pleasant Bayou (Kharaka et al., 1980) to have been derived from high temperature clay reactions. Thus high Ca, Ba and Sr are probably a consequence of albitization of K feldspars and plagioclases at temperatures greater than 120°C (Loucks et al., 1981; Morton et al., 1983; Land and Prezbindowski, 1981).

E. Age of Mineralization

The lead isotope composition of galenas in the Hockley Dome cap rocks are non-radiogenic and lie within the field of Cenozoic pelagic sediments and Holocene metalliferous saline brines (Price et al., 1983; Doe and Zartman, 1979). Thus, lead isotope data support a model of very recently evolved metalliferous fluids and recent sulfide deposition sourced from Gulf Coast formation fluids (Price et al., 1983). However, as stated earlier, part of the mineralization formed before calcite cap rock, and calcite cap rocks at Hockley are potentially as old as Oligo-Miocene (the age of the Heterostegina reef above Damon Mound). Thus, we cannot yet offer precise age estimates for all of the Hockley dome sulfides.

The lead-zinc ratio of 1:3 also supports derivation of the metals from a shaly basin of the character of the Gulf Coast basin (Sangster, 1983). These data suggest that the fluids in the lower Frio Formation that were trapped by geopressure are the parent fluids or are similar to those causing the salt dome cap-rock mineralization.

The metallic sulfides appear to have been sourced from depths similar to the saline brines at the Pleasant Bayou well (Light et al., 1984). Formation of the cap rock and metal sulfide deposition resulted from the leakage of rela tively hot, saline metalliferous formation waters along the margins of the rising salt diapir when it was within 5,000 ft (1,500 m) of the surface (Price et al., 1983); subsequent work suggests that cap rock formation and mineralization may be much closer to the surface or sediment-water interface.

At Damon Mound, warped late Pleistocene sediments, correlated with the Beaumont Formation, are thought to have accumulated during a Wisconsin interstadial that occurred approximately 34,000 years before present (Collins, 1985; Jenkins, 1979). However, the Beaumont sediments have also been correlated with the Sangamon interstadial which ranged in age from 115,000 to 130,000 years before present (Dubar in Collins, 1985). The Pleistocene Beaumont strata at Damon Mound have been uplifted 24 m indicating an uplift rate of about 0.6 m per 1,000 years (Beaumont Fm., 34,000 years old) or 0.19 m per 1,000 years (Beaumont Fm., 130,000 years old) (Collins, 1985). These high recent uplift rates measured at Damon Mound are in contrast to the 0.08 m per 1,000 years since the Oligocene (Collins, 1985) and could result in the reactivation of faults peripheral to the diapirs and major growth faults in the Gulf Coast Basin by additional removal of the "mother" salt basal layer. Such activity would reactivate conduits for fluid flow and allow upward migration of lead-zinc bearing basinal brines generated in the abyssal zone which would precipitate as the late cross-cutting sulfide concentrations in the salt-dome cap rocks (Fig. 3g). An anomaly present in the geothermal profile in the upper Frio Formation at the Pleasant Bayou test well suggests that hot updip basinal brine migration still continues at this level (Fig.6, Light and D'Attilio, in press).

The period of sulfide mineralization at Hockley is contemporaneous with and post-dates the true calcite cap-rock formation but precedes late gypsum and cross cutting sulphur veinlet formation (Posey, 1986). However, there are no absolute dates for the cap rock sulfide mineralization other than the constraint of the apparent initiation of salt movement and cap rock accumulation (60 Ma; Lower Wilcox). It would appear from this and the Pleasant Bayou data that the potential for mineralization spans a considerable period that may include the present.

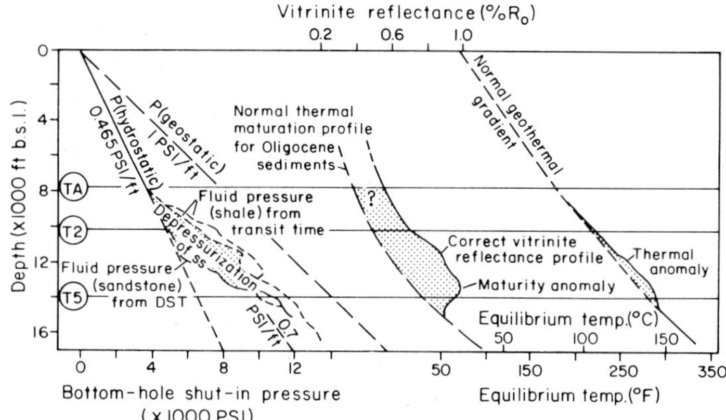

Figure 6. Fluid pressure, geothermal, and thermal maturity profiles for the Pleasant Bayou No. 2 test well from Light and D'Attilio (in press).

IV. GROWTH FAULT SYSTEMS

In addition to upward fluid migration along salt dome margins, fluids are also expelled along growth faults. Ewing (1984) indicated that the Frio style of growth faulting consists of several major, sinuous, moderately listric faults, possibly flattening into a regional detachment at 6,100 m to 7,600 m. There is substantial mobilized shale in diapirs and ridges though the influence of salt tectonics is surprisingly small (Ewing, 1984).

Hoskins Mound and Stratton Ridge lie southeast and south of the Pleasant Bayou-Chocolate Bayou fault block and intersect the major growth fault system forming the southeastern margin of the Chocolate Bayou oil and gas field (Light et al., 1984). The upper Frio sandstones appear to have formed a migration pathway up and out of the basin for hydrocarbons now trapped in the Chocolate Bayou field (Morton, 1983). This migration was partially driven by a horizontal pressure gradient which exists between the Pleasant Bayou-Chocolate Bayou fault block and more seaward fault blocks (Light et al., 1984).

Marked changes in salinity and chemical composition over short lateral distances suggest that fluids have migrated out of the basin along faults and discontinuities (Morton, 1983). As the major growth faults sole out into the lower Frio or Vicksburg slope shales between 6,100 m to 7,600 m (Ewing et al., 1983), both the marine source for the hydrocarbons (gas and oil) and the migration pathways (growth faults) are present at this level. In addition, abundant generation of gas and hydrocarbons in the slope shales probably contributed to the formation of high geo-pressures, rock failure and the "base Frio" growth fault detachment plane at these levels 6,100 m to 7,600 m. Furthermore, deep dissolution of the Stratton Ridge and Hoskins Mound salt domes by hot basinal fluids migrating along these growth faults could produce the high chloride to bromide ratios of the Pleasant Bayou brines (Kharaka et al., 1980). Evidently dewatering of the lower Frio or Vicksburg slope shales has occurred between 6,100 m and 7,600 m assisted by the generation of methane and carbon dioxide from the decarboxilation of organic acids (Kharaka et al., 1985). The increased salinity of these brines results from the dissolution of salt as the fluid migrated past deep salt zones (Light et al., 1984). Fast migration of these hot brines through the upper Frio sandstones raised the ambient temperature of these rocks (Fig.6). Ewing (1983) indicated that the Frio fault system intersected with the Hockley salt stock, further establishing the link between a fluid source and cap rock generating mechanisms.

V. CONCLUSIONS

Data from salt domes and oil and gas wells document several periods of fluid migration through the Gulf Coast Tertiary sequence. The actual timing of the fluid events and mineralizing events are not necessarily related, but they each follow a similar sequence. Thus, fluid flushing events have probably occurred at numerous times during basin burial and maturation.

There appear to have been at least three major periods of fluid migration which have affected Tertiary formations, and one of these may be related to the formation of salt dome cap rocks. An early period of fluid migration (23.6 Ma) was coincident with major salt diapirism and resulted in the transformation of smectite to illite in the Frio (Morton, 1981). This fluid may have introduced early anhydrite rimming sulfides and sulfur into salt dome cap rocks and isotopically heavy sulfur into pre-ore sulfides. This type of fluid event could be related to the type of sulfidizing event that resulted in the 5 Ma uranium deposits in south Texas (Ludwig et al., 1982). These fluids were at temperatures in excess of 120°C but cooled to about 70°C by mixing with late Oligocene marine waters near the surface in fracture zones surrounding and within salt dome cap rocks which had begun to form from salt dissolution. This fluid flush might also be the one responsible for Mississippi Valley-type mineralization.

One episode of fluid migration introduced warm (>100°C), saline, hydrocarbon, and metalliferous brines from depths of 6,900 to 7,600 m (the abyssal regime) within the lower Frio or Vicksburg marine slope shales which acted as a detachment plane for the Frio growth faulting (Ewing, 1984). Initial fluid migration was updip along the growth faults and then through the upper Frio sandstones which formed continuous conduits to surface formations.

A second flush of deep basinal fluids resulted in the introduction of hydrocarbons into the upper Frio Formation less than 20 Ma. This fluid contained the hydrocarbon constituents necessary for calcite cap rock formation as well as oil and gas. The fluid was derived from intermediate depths (the Elysian regime).

During the third fluid flush oils in the upper Frio Formation were waterwashed by the deeply sourced hot, aromatic-rich brines (Fig.3g). This fluid was also derived from the abyssal regime. Geothermal data at Pleasant Bayou No.2 suggest that updip migration of hot brines is still active in the Upper Frio (Light and D'Attilio, in press). Rapid uplift of Damon Mound since 130,000 to 34,000 years ago (Collins, 1985) suggest that faults may have been activated very recently in the Gulf Coast Basin, and formed conduits for the upward migration of fluids now still active in the upper Frio Formation (Light and D'Attilio, in press).

The sequence of fluid migration from early abyssal, diapir-related fluids to Elysian, hydrocarbon and metals-related fluids to late, abyssal, aromatic-rich fluids is consistent with the general sequence salt dome cap-rock and cap-rock mineralization (Posey, 1986).

VI. ACKNOWLEDGMENTS

Major funding for this research was provided by the Gas Research Institute under contract no. 5084-212-0924. Research on salt dome mineralization was supported by National Science Foundation grant EAR-8407736. We gratefully acknowledge assistance given by Steve Hurst and Arnold Taylor of Conoco Oil Company, Ponca City, Oklahoma, and by Marathon Oil Company, Littleton, Colorado. We would also like to thank L. L. Anderson, K. P. Peterson, and W. A. Parisi of Eaton Industries of Houston, Inc. We wish to thank E. C. Collins, W. E. Galloway, P. Mukhopadhyay and J. G. Price for valuable discussions. Various drafts of this manuscript were reviewed by R. J. Finley, Wendy Macpherson, and Randy Farr. Illustrations were drafted by J. L. Lardon, J. T. Ames, and T. Burke under the direction of Richard L. Dillon. Publication authorized by the Director, Bureau of Economic Geology, The University of Texas at Austin.

REFERENCES

Ahrens, L.H. (1965a). Distribution of the elements in our planet. McGraw-Hill, New York, 110 p.

Ahrens, L.H. (1965b). Introduction - the comparative geochemistry of potassium, rubidium, calcium, argon, strontium, uranium, thorium and lead. In: Applied Geochronology. (I.E.Hamilton, ed.) pp.vi-xiv. Academic Press, New York.

Anderson, G.M., and Macqueen, R.W. (1982). Ore deposits models-6.Mississippi Valley lead-zinc deposits. Geosci. Canada 9, 108-117.

Barnes, H.L. ed. (1979). Geochemistry of Hydrothermal Ore Deposits. John Wiley, New York, 798 p.

Beales, F.W., and Onasick, E.P. (1970). The stratigraphic habit of Mississippi Valley ore bodies. Inst. Mining and Metallurgy Trans. 79B, B145-B154.

Bogomolov, Y.G., Kudelsky, A.V., and Lapshin, N.N. (1978). Hydrogeology of large sedimentary basins. In: Hydrogeology of great sedimentary basins. Internat. Assoc. Sci. Hydro. Pub. 120, 117-122.

Boles J.R., and Franks, S.G. (1979). Clay diagenesis in Wilcox sandstones of southwest Texas, implications of smectite diagenesis on sandstone cementation. Jour. Sed. Pet. 49, 55-70.

Brock, T.D. (1985). Life at high temperatures. Science 230, 132-138.

Burst, J.F. (1969). Diagenesis of Gulf Coast clayey sediments and its possible relation to petroleum migration. AAPG Bull. 53, 73-93.

Busche F.D., Reynolds R.L., and Goldhaber M.B. (1982). Fault leaked H_2S and the origin of South Texas uranium deposits: implications of sulfur isotopic studies. Preprint, Soc. Mining Engineers AIME annual meeting, Dallas, 13 p.

Cahen, L., Snelling, N.J., Delhal, J., and Vail, J.R. (1984). The geochronology and evolution of Africa. Clarendon Press, Oxford, 512 p.

Cathles, L.M., and Smith, A.T. (1983). Thermal constraints on the formation of Mississippi Valley-Type lead-zinc deposits and their implications for episodic dewatering and deposit genesis. Econ. Geol. 78, 983-1002.

Clauer, N. (1976). Geochimie isotopique du strontium des milieux se dimentaires.Application a la geochronologie de la couverture de craton ouest-africain. Mem. Sci. Geol. no. 45, 256 p.

Collins, A.G. (1975). Geochemistry of Oil Field Waters. Elsevier, New York.

Collins, E.C. (1985). A review of the geology and Plio-Pleistocene deformations at Damon Mound salt dome, Texas. In: Phase III, Examination of Texas salt domes as potential sites for permanent storage of toxic chemical waste: Univ. Texas Bur. Econ. Geol. report prepared for the Texas Water Commission, contract no. IAC(84-85)-2203, p. 275-305.

Deer, W.A., Howie, R.A., and Zussman, J. (1969). An Introduction to the Rock-Forming Minerals. Longmans, London, 528 p.

Dix, R.O., and Jackson, M.P.A. (1982). Lithology, micro-structures, fluid inclusions, and geochemistry of rock salt and of the cap-rock contact in the Oakwood Dome, East Texas: significance for nuclear waste storage. Texas Univ. Austin Bur. Econ. Geology Rept. Inv. no. 120, 59 p.

Doe, B.R.and Zartman, R.E. (1979). Plumbotectonics, the Phanerozoic. In: Geochemistry of hydrothermal ore deposits (H.L.Barnes, ed.), pp. 22-70. John Wiley, New York.

Dozy, J.J. (1970). A geological model for the genesis of lead-zinc ores of the Mississippi Valley, U.S.A. Inst. Mining and Metallurgy Trans. 79, B163-B172.
Ewing, T.E. (1984). Section I.Regional fault compartment geometries of the Wilcox and Frio growth-fault trends, Texas Gulf Coast. In: Consolidation of geologic studies of geopressured geothermal resources of Texas. Texas Univ. Austin Bur. Econ. Geology report prepared for U.S. Dept. of Energy, contract no.DE-AC08-79ET27111, 97 p.
Ewing, T.E., Light, M.P.R., and Tyler, N. (1983). Section IV. Integrated geologic study of the Pleasant Bayou - Chocolate Bayou area, Brazoria County, Texas - first report. Texas Univ.Austin Bur.Econ.Geology report prepared for the U.S. Dept. of Energy, contract no.DE-AC08-79ET27111, p. 90-142.
Feely, H.W., and Kulp, J.L. (1957). Origin of Gulf Coast salt-dome sulphur deposits. Am. Assoc. Petroleum Geologists Bull. 41, 1802-1853.
Foscolos, A.E., Powell, T.G., and Gunter, P.R. (1976). The use of clay minerals, inorganic and organic geochemical indicators for evaluating the degree of diagenesis and oil generating potential of the shales. Geochim. et Cosmochim. Acta 40, 953-960.
Freed, R.L. (1979). Shale mineralogy of the No. 1 Pleasant Bayou geothermal test well: a progress report. In Proc. 4th Conf. Geopressured Geothermal Energy Austin 1, 153-167.
Galloway, W.E. (1982). Epigenetic zonation and fluid flow history of uranium-bearing fluvial aquifer systems, South Texas Uranium Province. Texas Univ. Austin Bur. Economic Geology Rept. Inv. no. 119, 31 p.
Galloway, W.E., Hobday, D.K., and Magara, K. (1982). Frio Formation of the Texas Gulf Coast Basin - depositional systems, structural framework, and hydrocarbon origin, migration, distribution and exploration potential. Texas Univ. Austin Bur. Economic Geology Rept. Inv. no. 122, 78 p.
Goldhaber, M.B., Reynolds, R.L., Rye, R.O., and Grauch, R.I. (1979). Petrology and isotope geochemistry of calcite in a South Texas roll-type uranium deposit. U.S Geol. Survey Open-File Rept. no. 79-828, 21 p.
Goldman, M.I. (1952). Deformation, metamorphism, and mineralization in gypsum-anhydrite cap rock, Sulphur salt Dome, Louisiana. Geol. Soc. America Mem. no. 50, 169 p.
Gose, W.A., Kyle, J.R., and Ulrich, M.R. (1985). Preliminary paleomagnetic investigation of the Winnfield salt dome cap rock, Louisiana. Gulf Coast Assoc. Geol. Socs. Trans. 35, 97-106.
Gregory A.R., Dodge, M.A., and Posey, J.S. (1979). Progress report on evaluation of entrained methane in deep reservoirs, Texas Gulf Coast. In: Proc. 4th Cong. Geo-pressured Geothermal Energy, Austin 1, (M.H.Dorfman and W. L. Fisher eds.), 1 414-510.
Gustavson, T.C., and Kreitler, C.W. (1977). Geothermal resources of the Texas Gulf Coast - environmental concerns arising from the production and disposal of geothermal waters. In: Geology of alternate energy resources in the south-central United States (M. D. Campbell ed.), Houston Geol. Soc. 297-335.
Handin, J., Hager, R.V., Jr., Friedman, M, Feather, J.N. (1963). Experimental deformation of sedimentary rocks under confining pressure: pore pressure tests. Am. Assoc. Petroleum Geologists Bull. 47, 717-755.
Hanor, J.S. (1979). The sedimentary genesis of hydrothermal fluids. In: Geochemistry of hydrothermal ore deposits (L.H.Barnes ed.), pp. 137-168. Wiley, New York.
Hay, R.L. (1966). Zeolites and zeolitic reactions in sedimentary rocks. Geol. Soc. America Spec. Paper 85, 130.

Holland, H.D., and Malinin, S.D. (1979). Solubility and occurrence of non-ore deposits. In: Geochemistry of Hydrothermal Ore Deposits. (H.L.Barnes, ed.), pp. 461-508. Wiley, New York.

Hower, J., Eslinger, E.V., Hower, M.E., and Perry, E.A. (1976). Mechanism of burial metamorphism of argillaceous sediment, I.: mineralogical and chemical evidence. Geol. Soc. America Bull. 87, 725-737.

Hoy, R.B., Foose, R.M., and O'Neill, B.J., Jr. (1962). Structure of Winnfield salt dome, Winn Parish, Louisiana. Am. Assoc. of Petroleum Geologists Bull. 46, 1444-1459.

Hunt, J.M. (1979). Petroleum Geochemistry and Geology. W.H. Freeman, San Francisco, 617 p.

Jackson, M.P.A., and Seni, S.J. (1983). Geometry and evolution of salt structures in a marginal rift basin of the Gulf of Mexico, East Texas. Geology 11, 131-135.

Jackson, M.P.A., and Talbot, C.J. (1986). External shapes, strain rates, and dynamics of salt structures: G.S.A. Bull. 97, 305-323.

Jenkins, J.T., Jr. (1979). Geology and paleontology of the upper clastic interval at Damon Mound. In: Damon Mound, Texas. Houston Geol. Soc. Guidebook, p. 45-62.

Kaiser, W.R., and Richmann, D.L. (1981). Predicting diagenetic history and reservoir quality in the Frio Formation of Brazoria County, Texas, and Pleasant Bayou test wells. Proc. 5th Conf. Geopressured Geothermal Energy, Austin, p. 67-74.

Karstev, A.A., Vassoevich, N.B., Geodekian, A.A., et al. (1971). The principal state in the formation of petroleum. Proc. 8th World Petroleum Congr., Moscow, 2, 3-10.

Kharaka, Y.K., Callender, E., and Carothers, W.W. (1977). Geochemistry of geopressured waters of the northern Gulf of Mexico basin, Part 1, Brazoria and Galveston Counties, Texas. In: Proc. 2nd Intern. Symp. Water-Rock Interaction. Sect.II., Strasbourg, (H.Paquet and Y.Tardy, eds.), p. 32-41.

Kharaka, Y.K., Lico, M.S., Wright, V.A., and Carothers, W.W. (1980). Geochemistry of formation waters from Pleasant Bayou No. 2 well and adjacent areas in coastal Texas. In: Proc. 4th U.S. Gulf Coast Geopressured Geothermal Energy Conf., Austin 1, (M.H.Dorfman and W.L.Fisher, eds.), 168-199.

Kharaka, Y.K., Carothers, W.W., and Law, L.M. (1985). Origin of gaseous hydrocarbons in geopressured geothermal waters. In Proceedings 6th U.S. Gulf Coast Geopressured Geothermal Energy Conference. (M.H.Dorfman and R.A.Morton eds.), p. 125, Pergamon, New York.

Knauth, L.P., Kumar, M.B., and Martinez, J.D. (1980). Isotope geochemistry of water in Gulf Coast salt domes. Journal Geophys. Research 85, 4863-4871.

Kreitler C.W., and Dutton, S.P. (1983). Origin and diagenesis of cap rock, Gyp Hill and Oakwood salt domes, Texas. Texas Univ. Austin Bur. Econ. Geology Rept. Inv. no. 131, p. 58.

Kyle, J.R.and Price, P.E. (1985). Mineralogical investigation of sulfide concentrations in Hockley salt dome cap rock, Texas. In Applied Mineralogy (W.C.Park, D.M.Hausen, and R.D., Hagni eds.) Metallurgical Soc.of AIME, 1065-1082.

Kyle, J.R.and Price, P.E. (1986). Metallic sulfide deposits in salt dome cap rocks, Gulf Coast, U.S.A.: Genetic Relationships among ore deposits, metalliferous formation waters, and hydrocarbon reservoirs. Inst.Mining and Metallurgy Trans. 95, B6-B16.

Kyle, J.R., Ulrich, M.R. and Gose, W.A., (this volume). Textural and paleomagnetic evidence for the mechanism and timing of anhydrite cap rock formation, Winnfield salt dome, Louisiana, in: Dynamical Geology of Salt and Related Structures (I. Lerche and J. J. O'Brien, eds.) Academic Press, Orlando.

Land, L.S. (1984). Evidence for vertical movement of diagenetic fluids, Texas Gulf Coast. Am. Assoc. Petroleum Geologists Annual Convt., San Antonio, Abs. Land, L.S., and Prezbindowski, D.R., 1981, The origin and evolution of saline foramion water, Lower Cretaceous carbonates, south-central Texas, U.S.A. in: W. Back and R. Letolle (Guest-editors), Symposium on Geochemistry of Groundwater--26th International Geological Congress; J. Hydrol. 54, 51-74.

Light, M.P.R., and D'Attilio, W. (in press). Section III. Integrated geologic study of the Hitchcock N,.E. field (Galveston County) and the Pleasant Bayou-Chocolate Bayou area, Brazoria County, Texas. Texas Univ.Austin Bur. Econ. Geology, report prepared for U.S. Dept. of Energy, Contract no. DE-AC08-79 ET27111-12.

Light, M.P.R., Ewing, T.E., and Tyler, N. (1984). Section II. Thermal history and hydrocarbon anomalies in the Frio Formation, Brazoria County, Texas - An indicator of fluid flow and geopressure history. Univ. Texas Austin Bur. Econ. Geology, report prepared for the U.S. Dept. of Energy, contract no. DE-AC08-79ET27111, 97.

Loucks, R.G., Richmann, D.L., and Milliken, K.L. (1981). Factors controlling reservoir quality in Tertiary sandstones and their significance to geopressured geothermal production. Texas Univ. Austin Bur. Econ. Geology. Rept.Inv. 111, 41p.

Ludwig, K.R., Goldhaber, M.B., Reynolds, R.L., Simmons, K.R. (1982). Uranium-lead isochron age and preliminary sulfur isotope systematics of the Felder uranium deposit, south Texas. Econ. Geology 77, 557-563.

Lundergard, P.D. (1985). Carbon dioxide and organic acids: origin and role in diagenesis (Texas Gulf Coast Tertiary): Unpub. Ph.D. Thesis Univ. Texas, Austin, 145 p.

Lyon, T.L.and Buckman, H.D. (1960). The nature and properties of soils. Macmillan, New York.

Martin, R.G. (1978). Northern and eastern Gulf of Mexico continental margin: stratigraphic and structural framework. Am. Assoc. Petroleum Geologists Studies in Geology 7, 21-42.

Milliken, K.L., Land, L.S., and Loucks, R.G. (1981). History of burial diagensis determined from isotopic geochemistry, Frio Formation, Brazoria County, Texas. Am. Assoc. Petroleum Geologists Bull. 65, 1397-1413.

Moller, P. (1983). Lanthanoids as a geochemical probe and problems in Lanthanoid geochemistry. Distribution and behavior of Lanthanoids in non-magmatic phases. In: Systematics and the properties of Lanthanides (S.P. Sinha, ed.), Reidel, Dordrecht.

Mooney, R.W., Keenan, A.G., and Wood, L.A. (1952). Adsorption of water vapour by montmorillonite. 1. Heat of desorption and application of BET theory. Jour. Am. Chem. Soc. 74, 1367-1371.

Morton, J.P. (1983). Age of clay diagenesis in the Oligocene Frio Formation, Texas Gulf Coast. pp. 33. Texas Univ. Austin Ph.D. dissertation.

Morton, R.A. (1983). Site reviews. In Mins.DOE/Industry Geopressured Geothermal Res. Dev. Prog. Working Group Meeting, Houston, 71-82.

Morton, R.A. (1981). Pleasant Bayou No.2-- a review of rationale, ongoing research and preliminary test results. Proc. 5th Conf. Geopressured Geothermal Energy, Austin, 55-57.

Morton, R.A., Ewing, T.E.,and Tyler, N. (1983). Continuity and internal properties of Gulf Coast sandstones and their implications for geopressured fluid production. Texas Univ. Austin Bur. Econ. Geology Rept. Inv. 132, 70.

Morton, R.A., Han, J.H., and Posey, J.S. (1983). Variations in chemical compositions of Tertiary formation waters, Texas Gulf Coast. Texas Univ. Austin Bur. Econ. Geology report prepared for the U.S., Dept.of Energy, contract no.DE-AC08-79ET27111, 83-36.

Murray, G.E. (1966). Salt structures of Gulf of Mexico Basin - a review. Am. Assoc. Petroleum Geologists Bull. 50, 439-478.

Ohle, E.L. (1951). The influence of permeability on ore distribution in limestone and dolomite. Econ. Geol. 46, 477-483.

Palciauskas, V.V., and Domenico, P.A. (1980). Microfracture development in compacting sediments, relation to hydrocarbon maturation kinetics. Am. Assoc. Petroleum Geologists Bull. 64, 927-937.

Perry and Turekian (1974). The effect of diagenesis on the redistribution of strontium isotopes in shales. Geochim et Cosmochim. Acta 38, 929-935.

Posey, H.H. (1986). Regional characteristics of strontium, carbon, and oxygen isotopes in salt dome cap rocks of the western Gulf Coast. pp. 239. Univ. North Carolina at Chapel Hill, Ph.D.dissertation.

Posey, H.H., Price, P.E. and Kyle, J.R. (this volume). Mixed carbon sources for calcite cap rocks of Gulf Coast salt domes. In: Dynamical Geology of Salt and Related Structures (I. Lerche and J. J. O'Brien, eds.) Academic Press, Orlando.

Powell, T.G., Foscolos, A.E., Gunther, P.R., and Snowdon, L.R. (1978). Diagenesis of organic matter and fine clay minerals: a comparative study. Geochim. et Cosmo-chim. Acta 42, 1181-1197.

Powers, M.C. (1967). Fluid release mechanisms in compacting marine mudrocks and their importance in oil exploration. Am. Assoc. Petroleum Geologists Bull. 51, 1240-1254.

Price, P.E., Kyle, J.R., and Wessel, G.R. (1983). Salt dome related lead zinc deposits. In Proc. Intern. Conf. Mississippi Valley type lead-zinc deposits (G. Kisvarsanyi, S.K. Grant, W.P. Pratt, J.W. Koenig, eds.,), 558-577. Rolla.

Price, P.E., and Kyle, J.R. (1986). Genesis of salt dome hosted metallic sulfide deposits: the role of hydrocarbons and related fluids. In Proceedings of the Symposium on Organics and Ore Deposits Denver Reg.Exploration Geol. Soc, (W.Dean ed.), p.171-184.

Sangster, D.F. (1983). Mississippi Valley type lead-zinc deposits: a geological melange. In Proc. Intern. Conf. Mississippi Valley type lead-zinc deposits (G. Kisvarsanyi, S.K. Grant, W.P. Pratt, J.W. Koenig, eds.), pp. 558-571. Rolla.

Selley, R.C. (1979). Concepts and methods of subsurface facies analysis. Am. Assoc. Petrol. Geol. Contin. Educ. Course Note Ser. No. 9, 82 p.

Seni, S.J., and Jackson, M.P.A. (1983). Evolution of salt structures, East Texas diapir province, Part 2: pattern and rates of halokinesis. Am. Assoc. Petroleum Geologists Bull.67, 1245-1274.

Sverjensky, D.A. (1984). Oil field brines as ore-forming solutions: Econ. Geol. 79, 23-37.

Thornton E.C., and Seyfried W.E., Jr. (1985). Sediment-sea-water interaction at 200 and 300°C, 500 bars pressure: The role of sediment composition in diagenesis and low-grade metamorphism of marine clay. Geol. Soc. America Bull. 96, 1287-1295.

Ulrich, M.R., Kyle, J.R., and Price, P.E. (1984). Metallic sulfide deposits in the Winnfield salt dome, Louisiana: evidence for episodic introduction of metalliferous brines during cap rock formation. In Gulf Coast Assoc. Geol. Socs. Trans., (B.R.White ed.), 435-442.
Weaver, C.E. (1980). Graphite in Gulf Coast salt. U.S. Dept. Energy, ONWI No.E512-0010, 3.
White D.E. (1965). Saline waters of sedimentary rocks. In Fluids in subsurface environments. Am. Assoc. Petroleum Geologists Mem. 4, (A. Young and J.E. Galley, eds.) 342-366.
Winkler, H.G.F. (1976). Petrogenesis of metamorphic rocks. Springer-Verlag, New York.
Woidt, W.-D. (1978). Finite element calculations applied to salt dome analysis. Tectonophysics 50, 369-386.
Yeh, H.W., and Savin, S.M. (1977). The mechanism of burial metamorphism of argillaceous sediments: 3. Oxygen isotopic evidence. Geol. Soc. America Bull. 88, 1321-1330.

Index